Excel

Get the Results You Want!

SUCCESS ONE® HSC TOPIC-BY-TOPIC

MATHEMATICS ADVANCED

Past HSC questions arranged into topics with worked answers **1992–2022**

BRAND NEW SERIES

PASCAL PRESS

ISBN 978 1 74125 730 4

Pascal Press
PO Box 250
Glebe NSW 2037
(02) 8585 4050
www.pascalpress.com.au

Publisher: Vivienne Joannou
Project editor: Rosemary Peers
Edited by Rosemary Peers
Answers checked by Peter Little
Cover and page design by Sonia Woo
Page layout and typesetting by lj Design (Julianne Billington)
Printed by Vivar Printing/Green Giant Press

Photo of windfarm on page 15 reproduced with the kind permission of Miguel Saavedra.

CONTENTS

About this book iv
Self-assessment summary vi

YEAR 11

Functions

- Algebraic techniques 1
  Worked answers 4
- Introduction to functions 7
  Worked answers 9
- Linear, quadratic and cubic functions 10
  Worked answers 18
- Further functions and relations 24
  Worked answers 29

Trigonometric functions

- Trigonometry 34
  Worked answers 43
- Radians 51
  Worked answers 56
- Trigonometric functions and identities 61
  Worked answers 64

Calculus

- Introduction to differentiation 67
  Worked answers 69
- The derivative function and its graph 71
  Worked answers 73
- Calculating with derivatives 75
  Worked answers 77

Exponential and logarithmic functions

- Introducing logarithms 80
  Worked answers 81
- Logarithmic laws and applications 82
  Worked answers 84
- Exponential function and natural logarithms 86
  Worked answers 87
- Graphs and applications of exponential and logarithmic functions 89
  Worked answers 92

Statistical analysis

- Probability and Venn diagrams 95
  Worked answers 104
- Discrete probability distribution 112
  Worked answers 113

YEAR 12

Functions

- Graphing techniques 114
  Worked answers 117

Trigonometric functions

- Trigonometric functions and graphs 120
  Worked answers 123

Calculus

- Differentiation of trigonometric, exponential and logarithmic functions 126
  Worked answers 127
- Rules of differentiation 128
  Worked answers 131
- The first and second derivative 135
  Worked answers 139
- Applications of the derivative 142
  Worked answers 156
- The antiderivative 182
  Worked answers 188
- Areas and the definite integral 195
  Worked answers 208

Financial mathematics

- Modelling investments and loans 222
  Worked answers 227
- Arithmetic sequences and series 230
  Worked answers 232
- Geometric sequences and series 234
  Worked answers 238
- Financial applications of sequences and series 243
  Worked answers 249

Statistical analysis

- Data and summary statistics 260
  Worked answers 268
- Bivariate data analysis 272
  Worked answers 277
- Continuous random variables 279
  Worked answers 281
- The normal distribution 284
  Worked answers 290

Formulae sheet 294

ABOUT THIS BOOK

Revise the smart way and use this guide during the year to complete HSC questions whenever you have completed a topic in class and need to study for a topic test, an assessment task or lastly for the HSC Exam.

This book contains over 1000 questions from 31 years of HSC past papers arranged into syllabus topics.

Each topic:

- has questions arranged in order with the most recent HSC papers first
- has questions marked with one of three levels of difficulty: Easy, Medium and Hard
- has questions relevant only to the current course (the Mathematics Advanced course replaced the Mathematics course and was first examined in 2020).

Extra information on the questions

- CQ This symbol is next to questions that are common to both the HSC Mathematics Standard 2 and the HSC Mathematics Advanced papers.
- **Bonus question** Some questions have been written to give extra practice on a topic. These are usually new syllabus topics where there are only past HSC questions from 2020 onwards.
- The exam from which each question is drawn is indicated at the end of each question. **HSC** indicates the exam is Mathematics Advanced or a version of the course that preceded it: Mathematics (2001–2019) or Mathematics 2 Unit (1992–2000). Due to the changes in the syllabus, questions from different courses are also included. **ME1 HSC** refers to Mathematics Extension 1 or the versions of the course that preceded it: Mathematics 3 Unit (1992–2000). **MS2 HSC** refers to Mathematics Standard 2 or the versions of the course that preceded it: Mathematics General 2 (2014–2018) or General Mathematics (2001–2013).
- Some questions do not have marks included because these were not indicated in the HSC Exam paper from which they came. The answer section will then include the note: (**No mark allocation in exam**).
- Some questions only have total marks included because marks for the individual parts were not indicated in the HSC Exam paper from which they came. The answer section will then include the note: (**Total mark allocation only in exam**).
- In some instances if there is a multipart question and one part is not on the chapter's topic or one part is no longer part of the syllabus for this course, it will be omitted. However, if the answer to the omitted part is needed to complete the remaining part it will be **provided in square brackets**.

Self-assessment summary

Students can complete the self-assessment summary on page vi to assess their progress.

Past HSC Examination papers

- The past HSC Examination questions contained in this publication have been reproduced under licence from the NSW Education Standards Authority (NESA) in whom copyright is vested.
- NESA takes no responsibility for errors in the reproduction of past HSC Examination questions contained in this publication.
- NESA was the first publisher of the HSC Examination papers from which the questions come in the years 1992–2022.
- HSC Examination papers in Mathematics 2 Unit 1992–2000 © NSW Education Standards Authority for and on behalf of the Crown in right of the State of New South Wales 1992–2000
- HSC Examination papers in Mathematics 2001–2019 © NSW Education Standards Authority for and on behalf of the Crown in right of the State of New South Wales 2001–2019
- HSC Examination papers in Mathematics Advanced 2020–2022 © NSW Education Standards Authority for and on behalf of the Crown in right of the State of New South Wales 2020–2022
- HSC Examination papers in Mathematics 3 Unit 1992–2000 © NSW Education Standards Authority for and on behalf of the Crown in right of the State of New South Wales 1992–2000
- HSC Examination papers in Mathematics Extension 1 2001–2021 © NSW Education Standards Authority for and on behalf of the Crown in right of the State of New South Wales 2001–2021
- HSC Examination papers in General Mathematics 2001–2013 © NSW Education Standards Authority for and on behalf of the Crown in right of the State of NSW 2001–2013
- HSC Examination papers in Mathematics General 2 2014–2018 © NSW Education Standards Authority for and on behalf of the Crown in right of the State of NSW 2014–2018
- HSC Examination papers in Mathematics Standard 2 2019–2021 © NSW Education Standards Authority for and on behalf of the Crown in right of the State of NSW 2019–2021

Worked answers

- The worked answers contained in this publication are examples of answers which the authors believe would score full marks.
- They are not necessarily ideal or model answers, nor are they the only answers which would score full marks.
- They are not endorsed by NESA.
- For Mathematics Advanced (and the previous courses Mathematics and Mathematics 2 Unit) worked answers for the 1992–2003 HSC Examination papers were written by Keith Saines, for the 2004–2005 HSC Examination papers by Lynne Knapman and Berra Mossemenear, for 2006–2009 by Berra Mossemenear, Graeme Downward, Bronwyn Opferkuch and Robert Russell, and for 2010–2021 by Allyn Jones.
- For Mathematics Standard 2 (and the previous courses Mathematics General 2 and General Mathematics) worked answers for the 2001–2009 HSC Examination papers were written by Barbara D'Angelo and for the 2010–2021 HSC Examination papers by Lyn Baker.
- For Mathematics Extension 1 (and the previous course Mathematics 3 Unit) worked answers for the 1992–2003 HSC Examination papers were written by José Carreno, for the 2004–2008 HSC Examination papers by Barbara D'Angelo and Megan McKeown, for 2009 by Barbara D'Angelo, for 2010–2020 by Lyn Baker and for 2021 by Mitchell Jones.
- Bonus questions and answers were written by Allyn Jones and Lyn Baker.
- The ticks in the solutions are provided as a guide only to the division of marks and are not endorsed by NESA.

SELF-ASSESSMENT SUMMARY

Note: For the purposes of the functionality of this summary, all parts of questions which have been assigned a level are counted as a whole question.

YEAR 11	EASY	MEDIUM	HARD	MASTERY
FUNCTIONS				
Algebraic techniques	/50	/11	/1	1 2 3 4 5
Introduction to functions	/7	/9	/1	1 2 3 4 5
Linear, quadratic and cubic functions	/56	/21		1 2 3 4 5
Further functions and relations	/54	/22	/1	1 2 3 4 5
TRIGONOMETRIC FUNCTIONS				
Trigonometry	/37	/45	/10	1 2 3 4 5
Radians	/24	/20	/8	1 2 3 4 5
Trigonometric functions and identities	/14	/14	/7	1 2 3 4 5
CALCULUS				
Introduction to differentiation	/14	/9		1 2 3 4 5
The derivative function and its graph	/6	/11		1 2 3 4 5
Calculating with derivatives	/21	/8	/4	1 2 3 4 5
EXPONENTIAL AND LOGARITHMIC FUNCTIONS				
Introducing logarithms	/6	/4		1 2 3 4 5
Logarithmic laws and applications	/7	/15	/1	1 2 3 4 5
Exponential function and natural logarithms	/9	/11	/4	1 2 3 4 5
Graphs and applications of exponential and logarithmic functions	/26	/13	/2	1 2 3 4 5
STATISTICAL ANALYSIS				
Probability and Venn diagrams	/60	/58	/8	1 2 3 4 5
Discrete probability distribution	/7	/4		1 2 3 4 5

YEAR 12	EASY	MEDIUM	HARD	MASTERY
FUNCTIONS				
Graphing techniques	/14	/13	/5	1 2 3 4 5
TRIGONOMETRIC FUNCTIONS				
Trigonometric functions and graphs	/11	/17	/7	1 2 3 4 5
CALCULUS				
Differentiation of trigonometric, exponential and logarithmic functions	/7	/11		1 2 3 4 5
Rules of differentiation	/52	/9		1 2 3 4 5
The first and second derivative	/6	/20	/4	1 2 3 4 5
Applications of the derivative	/81	/99	/62	1 2 3 4 5
The antiderivative	/65	/29	/17	1 2 3 4 5
Areas and the definite integral	/66	/81	/26	1 2 3 4 5
FINANCIAL MATHEMATICS				
Modelling investments and loans	/21	/16	/2	1 2 3 4 5
Arithmetic sequences and series	/19	/6	/1	1 2 3 4 5
Geometric sequences and series	/26	/19	/14	1 2 3 4 5
Financial applications of sequences and series	/21	/35	/24	1 2 3 4 5
STATISTICAL ANALYSIS				
Data and summary statistics	/42	/22		1 2 3 4 5
Bivariate data analysis	/31	/8		1 2 3 4 5
Continuous random variables	/6	/17	/8	1 2 3 4 5
The normal distribution	/32	/19	/7	1 2 3 4 5

1 Solve $x + \frac{x-1}{2} = 9$. *(2 marks)*

CQ (Q11, **2021 HSC**) Easy

2 Simplify $\frac{x+2}{x} \div \frac{x^2-4}{2x}$. *(2 marks)*

Bonus question (see page iv) Medium

3 What is the value of p so that $\frac{a^2a^{-3}}{\sqrt{a}} = a^p$?

A -3 **B** $-\frac{3}{2}$

C $-\frac{1}{2}$ **D** 12 *(1 mark)*

(Q3, **2019 HSC**) Easy

4 What is the value of $3x^0 + 1$?

A 1 **B** 2

C 3 **D** 4 *(1 mark)*

(Q2, **2018 MS2 HSC**) Easy

5 Which expression is equivalent to $\frac{k^2}{10} \div \frac{5}{k}$?

A $\frac{2}{k}$ **B** $\frac{k}{2}$

C $\frac{50}{k^3}$ **D** $\frac{k^3}{50}$ *(1 mark)*

(Q16, **2018 MS2 HSC**) Easy

6 Which of the following expresses v as the subject of $k = \frac{1}{2}mv^2$?

A $v = \pm\sqrt{\frac{2k}{m}}$ **B** $v = \pm\frac{\sqrt{k}}{2m}$

C $v = \pm\sqrt{\frac{k}{2m}}$ **D** $v = \pm\frac{2\sqrt{k}}{m}$ *(1 mark)*

(Q25, **2018 MS2 HSC**) Easy

7 Find the values of x and y which satisfy the following equations simultaneously.

$$y = x + 5$$
$$3y - x = 7$$

(3 marks)

(Q27b, **2018 MS2 HSC**) Easy

8 Rationalise the denominator of $\frac{3}{3+\sqrt{2}}$. *(2 marks)*

(Q11a, **2018 HSC**) Easy

9 Which expression is equal to $3x^2 - x - 2$?

A $(3x-1)(x+2)$ **B** $(3x+1)(x-2)$

C $(3x-2)(x+1)$ **D** $(3x+2)(x-1)$

(1 mark)

(Q2, **2017 HSC**) Easy

10 Rationalise the denominator of $\frac{2}{\sqrt{5}-1}$. *(2 marks)*

(Q11a, **2017 HSC**) Easy

11 By letting $m = t^{\frac{1}{3}}$, or otherwise, solve $t^{\frac{2}{3}} + t^{\frac{1}{3}} - 6 = 0$. *(2 marks)*

(Q13c, **2017 HSC**) Medium

12 Simplify $\frac{8x^4y}{24x^3y^5}$. *(2 marks)*

(Q26b, **2016 MS2 HSC**) Easy

13 Factorise fully $3x^2 - 27$. *(2 marks)*

(Q11b, **2015 HSC**) Easy

14 Express $\frac{8}{2+\sqrt{7}}$ with a rational denominator. *(2 marks)*

(Q11c, **2015 HSC**) Easy

15 Which of the following is $3x^0 + 5x$ in its simplest form?

A $6x$ **B** $8x$

C $1 + 5x$ **D** $3 + 5x$ *(1 mark)*

(Q11, **2015 MS2 HSC**) Easy

16 Rationalise the denominator of $\frac{1}{\sqrt{5}-2}$. *(2 marks)*

(Q11a, **2014 HSC**) Easy

17 Simplify $6w^4 \times \frac{1}{3}w^2$.

A $2w^6$ **B** $2w^8$

C $18w^6$ **D** $18w^8$ *(1 mark)*

(Q11, **2014 MS2 HSC**) Easy

18 Expand $4x(7x^4 - x^2)$. *(1 mark)*

(Q26a, **2014 MS2 HSC**) Easy

19 Solve the equation $\frac{5x+1}{3} - 4 = 5 - 7x$. *(3 marks)*

(Q26c, **2014 MS2 HSC**) Easy

20 Solve these simultaneous equations to find the values of x and y.

$$y = 2x + 1$$
$$x - 2y - 4 = 0$$

(3 marks)

(Q26d, **2014 MS2 HSC**) Easy

21 What are the solutions of $2x^2 - 5x - 1 = 0$?

A $x = \dfrac{-5 \pm \sqrt{17}}{4}$ **B** $x = \dfrac{5 \pm \sqrt{17}}{4}$

C $x = \dfrac{-5 \pm \sqrt{33}}{4}$ **D** $x = \dfrac{5 \pm \sqrt{33}}{4}$ *(1 mark)*

(Q1, **2013 HSC**) Easy

22 Which of the following is $\dfrac{12T^2}{3T \times 2W}$ in its simplest form?

A $\dfrac{2T}{W}$ **B** $2TW$

C $\dfrac{8T}{W}$ **D** $8TW$ *(1 mark)*

(Q5, **2013 MS2 HSC**) Easy

23 Leanne wants to build a rectangular vegetable garden in her backyard. She has 20 metres of fencing and will use a wall as one side of the garden. The plan for her garden is shown, where x metres is the width of her garden.

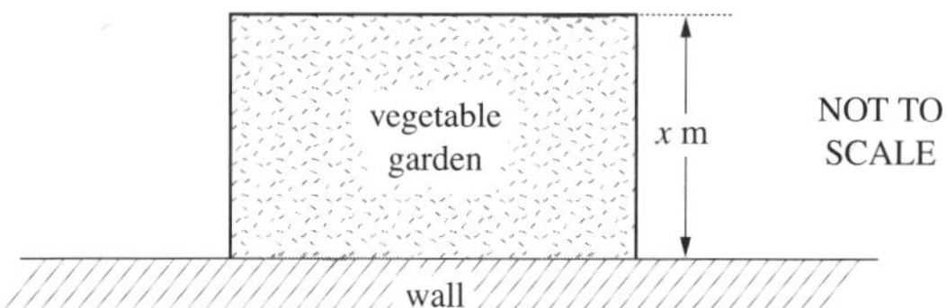

Which equation gives the area, A, of the vegetable garden?

A $A = 10x - x^2$ **B** $A = 10x - 2x^2$

C $A = 20x - x^2$ **D** $A = 20x - 2x^2$ *(1 mark)*

(Q22, **2013 MS2 HSC**) Easy

24 Which of the following is equal to $\dfrac{1}{2\sqrt{5} - \sqrt{3}}$?

A $\dfrac{2\sqrt{5} - \sqrt{3}}{7}$ **B** $\dfrac{2\sqrt{5} + \sqrt{3}}{7}$

C $\dfrac{2\sqrt{5} - \sqrt{3}}{17}$ **D** $\dfrac{2\sqrt{5} + \sqrt{3}}{17}$ *(1 mark)*

(Q2, **2012 HSC**) Easy

25 Which of the following expresses $2x^2(5 - x) - x(x - 2)$ in its simplest form?

A $-2x^3 + 9x^2 + 2x$ **B** $-2x^3 - 9x^2 - 2x$

C $9x^2 - x + 2$ **D** $9x^2 - x - 2$ *(1 mark)*

(Q14, **2012 MS2 HSC**) Easy

26 Simplify fully $\dfrac{18ab}{3a^2} \times \dfrac{c}{b}$. *(2 marks)*

(Q28b, **2012 MS2 HSC**) Easy

27 Simplify $\dfrac{n^2 - 25}{n - 5}$. *(1 mark)*

(Q1b, **2011 HSC**) Easy

28 Solve $2^{2x+1} = 32$. *(2 marks)*

(Q1c, **2011 HSC**) Easy

29 Rationalise the denominator of $\dfrac{4}{\sqrt{5} - \sqrt{3}}$.

Give your answer in the simplest form. *(2 marks)*

(Q1f, **2011 HSC**) Easy

30 **i** Rationalise the denominator in the expression

$$\frac{1}{\sqrt{n} + \sqrt{n+1}}$$

where n is an integer and $n \geq 1$. *(1 mark)* Hard

ii Using your result from part **i**, or otherwise, find the value of the sum

$$\frac{1}{\sqrt{1} + \sqrt{2}} + \frac{1}{\sqrt{2} + \sqrt{3}} + \frac{1}{\sqrt{3} + \sqrt{4}} + \cdots + \frac{1}{\sqrt{99} + \sqrt{100}}$$

(2 marks) Easy

(Q9d, **2011 HSC**)

31 Solve $x^2 = 4x$. *(2 marks)*

(Q1a, **2010 HSC**) Easy

32 Find integers a and b such that $\dfrac{1}{\sqrt{5} - 2} = a + b\sqrt{5}$. *(2 marks)*

(Q1b, **2010 HSC**) Medium

33 Solve $\dfrac{5x - 4}{x} = 2$. *(2 marks)*

(Q1b, **2009 HSC**) Easy

34 Simplify $\dfrac{2}{n} - \dfrac{1}{n+1}$. *(2 marks)*

(Q1c, **2008 HSC**) Medium

35 Expand and simplify $(\sqrt{3} - 1)(2\sqrt{3} + 5)$. *(2 marks)*

(Q1e, **2008 HSC**) Easy

36 Solve $2x - 5 > -3$ and graph the solution on a number line. *(2 marks)*

(Q1b, **2007 HSC**) Medium

37 Factorise $2x^2 + 5x - 12$. *(2 marks)*

(Q1e, **2007 HSC**) Medium

38 Rationalise the denominator of $\frac{1}{\sqrt{3}-1}$. *(2 marks)*

(Q1c, **2007 HSC**) Easy

39 Solve $3 - 5x \leq 2$. *(2 marks)*

(Q1e, **2006 HSC**) Easy

40 Express $\frac{(2x-3)}{2} - \frac{(x-1)}{5}$ as a single fraction in its simplest form. *(2 marks)*

(Q1d, **2005 HSC**) Easy

41 Solve $\frac{x-5}{3} - \frac{x+1}{4} = 5$. *(2 marks)*

(Q1c, **2004 HSC**) Medium

42 Find integers a and b such that

$(3-\sqrt{2})^2 = a - b\sqrt{2}$. *(2 marks)*

(Q1d, **2004 HSC**) Easy

43 Solve $x^2 = 5x$. *(2 marks)*

(Q1c, **2002 HSC**) Easy

44 Solve $3x - \frac{2x-5}{2} = 6$. *(3 marks)*

(Q1e, **2002 HSC**) Easy

45 Solve the pair of simultaneous equations

$x - 2y = 8$
$2x + y = 1$. *(2 marks)*

(Q1f, **2002 HSC**) Easy

46 Solve $x^2 - 2x - 8 = 0$. *(2 marks)*

(Q1c, **2001 HSC**) Medium

47 Simplify $\frac{x}{x^2-4} + \frac{2}{x-2}$. *(2 marks)*

(Q1e, **2001 HSC**) Medium

48 Solve the pair of simultaneous equations:

$x - y = 2$
$3x + 2y = 1$ *(2 marks)*

(Q1e, **2000 HSC**) Easy

49 Solve $3 - 2x \geq 7$. *(2 marks)*

(Q1c, **1999 HSC**) Easy

50 Solve the simultaneous equations

$x + y = 1$
$2x - y = 5$. *(2 marks)*

(Q1d, **1999 HSC**) Easy

51 Find integers a and b such that

$(5-\sqrt{2})^2 = a + b\sqrt{2}$. *(2 marks)*

(Q1e, **1999 HSC**) Easy

52 By rationalising the denominators, express

$\frac{1}{3-\sqrt{2}} + \frac{1}{3+\sqrt{2}}$ in simplest form. *(2 marks)*

(Q1f, **1998 HSC**) Medium

53 By rationalising the denominator, express

$\frac{8}{3-\sqrt{5}}$ in the form $a + b\sqrt{5}$. *(3 marks)*

(Q1g, **1997 HSC**) Medium

54 Rationalise the denominator of $\frac{4}{\sqrt{5}+2}$. *(2 marks)*

(Q1c, **1996 HSC**) Easy

55 Simplify $\frac{2}{3} - \frac{x-1}{4}$. *(2 marks)*

(Q1d, **1996 HSC**) Easy

56 Solve the equation $u^2 - u - 1 = 0$, correct to three decimal places. *(2 marks)*

(Q5b, **1996 HSC**) Easy

57 Simplify $\frac{x}{3} + \frac{3x-1}{2}$. *(2 marks)*

(Q1d, **1995 HSC**) Easy

58 Find the value of $4\pi\sqrt{\frac{a}{g}}$ if $a = 4.1$ and $g = 9.8$.

Give your answer to 2 significant figures.

(Q1a, **1993 HSC**) Easy

59 Solve the pair of simultaneous equations:

$2x + y = 7$
$x - 2y = 1$.

(Q1e, **1993 HSC**) Easy

60 Solve $4x + 7 = 3(x - 2)$.

(Q1c, **1992 HSC**) Easy

61 Solve the pair of simultaneous equations:

$x + y = 2$
$2x - y = 7$.

(Q1e, **1992 HSC**) Easy

Year 11 Algebraic techniques—Worked answers

1
$$x+\frac{x-1}{2}=9$$
$$2x+x-1=18 \quad ✓$$
$$3x=19$$
$$x=\frac{19}{3} \quad ✓$$
(2 marks)

2
$$\frac{x+2}{x}\div\frac{x^2-4}{2x}$$
$$=\frac{x+2}{x}\times\frac{2x}{x^2-4}$$
$$=\frac{x+2}{x}\times\frac{2x}{(x-2)(x+2)} \quad ✓$$
$$=\frac{2}{x-2} \quad ✓$$
(2 marks)

3
$$\frac{a^2a^{-3}}{\sqrt{a}}=a^2\times a^{-3}\div a^{\frac{1}{2}}$$
$$=a^{-1}\times a^{-\frac{1}{2}}$$
$$=a^{-\frac{3}{2}}$$
$$\therefore p=-\frac{3}{2}$$
Answer B

4 $3x^0+1=3\times1+1$
$=4$
Answer D

5
$$\frac{k^2}{10}\div\frac{5}{k}=\frac{k^2}{10}\times\frac{k}{5}$$
$$=\frac{k^3}{50}$$
Answer D

6
$$k=\frac{1}{2}mv^2$$
$$2k=mv^2$$
$$v^2=\frac{2k}{m}$$
$$v=\pm\sqrt{\frac{2k}{m}}$$
Answer A

7 $y=x+5$ (i)
$3y-x=7$ (ii)
Substitute (i) into (ii):
$3(x+5)-x=7$
$3x+15-x=7$ ✓
$2x+15=7$
$2x=-8$
$x=-4$ ✓
Substitute into (i):
$y=-4+5$
$y=1$
$\therefore\ x=-4$ and $y=1$ ✓
(3 marks)

8
$$\frac{3}{3+\sqrt{2}}=\frac{3}{3+\sqrt{2}}\times\frac{3-\sqrt{2}}{3-\sqrt{2}} \quad ✓$$
$$=\frac{3(3-\sqrt{2})}{9-2}$$
$$=\frac{3(3-\sqrt{2})}{7} \quad ✓$$
(2 marks)

9 $3x^2-x-2=(3x+2)(x-1)$
Answer D

10
$$\frac{2}{\sqrt{5}-1}=\frac{2}{\sqrt{5}-1}\times\frac{\sqrt{5}+1}{\sqrt{5}+1} \quad ✓$$
$$=\frac{2\sqrt{5}+2}{5-1}$$
$$=\frac{2(\sqrt{5}+1)}{4}$$
$$=\frac{\sqrt{5}+1}{2} \quad ✓$$
(2 marks)

11 Let $m=t^{\frac{1}{3}}$:
$m^2+m-6=0$
$(m+3)(m-2)=0$
$m=-3, 2$ ✓
$\therefore\ t^{\frac{1}{3}}=-3 \qquad t^{\frac{1}{3}}=2$
$t=-27 \qquad t=8$
$\therefore t=-27$ or 8 ✓
(2 marks)

12
$$\frac{8x^4y}{24x^3y^5}=\frac{x}{3y^4} \quad ✓✓$$
(2 marks)

13 $3x^2-27=3(x^2-9)$ ✓
$=3(x-3)(x+3)$ ✓
(2 marks)

14
$$\frac{8}{2+\sqrt{7}}=\frac{8}{2+\sqrt{7}}\times\frac{2-\sqrt{7}}{2-\sqrt{7}} \quad ✓$$
$$=\frac{8(2-\sqrt{7})}{4-7}$$
$$=\frac{8(2-\sqrt{7})}{-3}$$
$$=\frac{8(\sqrt{7}-2)}{3} \quad ✓$$
(2 marks)

15 $3x^0+5x=3\times1+5x$
$=3+5x$
Answer D

16
$$\frac{1}{\sqrt{5}-2}=\frac{1}{\sqrt{5}-2}\times\frac{\sqrt{5}+2}{\sqrt{5}+2} \quad ✓$$
$$=\frac{\sqrt{5}+2}{1}$$
$$=\sqrt{5}+2 \quad ✓$$
(2 marks)

17 $6w^4\times\frac{1}{3}w^2=2w^6$
Answer A

18 $4x(7x^4-x^2)=28x^5-4x^3$ ✓
(1 mark)

19
$$\frac{5x+1}{3}-4=5-7x$$
$5x+1-12=15-21x$ ✓
$26x=26$ ✓
$x=1$ ✓
(3 marks)

20 $y=2x+1$ (1)
$x-2y-4=0$ (2)
Substitute (1) into (2)
$x-2(2x+1)-4=0$ ✓
$x-4x-2-4=0$
$-3x=6$
$x=-2$ ✓
Substitute into (1)
$y=2\times-2+1$
$=-3$ ✓
$\therefore x=-2$ and $y=-3$
(3 marks)

21 $2x^2-5x-1=0$
$$x=\frac{5\pm\sqrt{(-5)^2-4(2)(-1)}}{2(2)}$$
$$=\frac{5\pm\sqrt{33}}{4}$$
Answer D

22
$$\frac{12T^2}{3T\times2W}=\frac{12T^2}{6TW}$$
$$=\frac{2T}{W}$$
Answer A

23 One side of the garden is a wall and does not need a fence.
Let the length of the garden be l m.

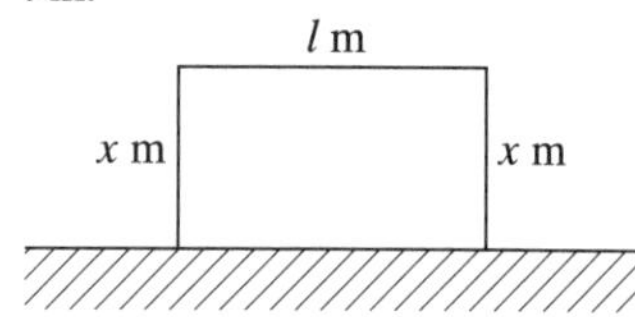

There is 20 m of fencing.
so $l+2x=20$
$l=20-2x$
Area = length × width
$A=(20-2x)\times x$
$=20x-2x^2$
Answer D

24
$$\frac{1}{2\sqrt{5}-\sqrt{3}}$$
$$=\frac{1}{2\sqrt{5}-\sqrt{3}}\times\frac{2\sqrt{5}+\sqrt{3}}{2\sqrt{5}+\sqrt{3}}$$

$= \dfrac{2\sqrt{5} + \sqrt{3}}{20 - 3}$

$= \dfrac{2\sqrt{5} + \sqrt{3}}{17}$

Answer D

25 $2x^2(5 - x) - x(x - 2)$

$= 10x^2 - 2x^3 - x^2 + 2x$

$= -2x^3 + 9x^2 + 2x$

Answer A

26 $\dfrac{18ab}{3a^2} \times \dfrac{c}{b} = \dfrac{18abc}{3a^2b}$ ✓

$= \dfrac{6c}{a}$ ✓

(2 marks)

27 $\dfrac{n^2 - 25}{n - 5} = \dfrac{(n - 5)(n + 5)}{n - 5}$

$= n + 5$ ✓

(1 mark)

28 $2^{2x+1} = 32$

$2^{2x+1} = 2^5$

$\therefore 2x + 1 = 5$ ✓

$2x = 4$

$x = 2$ ✓

(2 marks)

29 $\dfrac{4}{\sqrt{5} - \sqrt{3}}$

$= \dfrac{4}{\sqrt{5} - \sqrt{3}} \times \dfrac{\sqrt{5} + \sqrt{3}}{\sqrt{5} + \sqrt{3}}$ ✓

$= \dfrac{4(\sqrt{5} + \sqrt{3})}{5 - 3}$

$= \dfrac{4(\sqrt{5} + \sqrt{3})}{2}$

$= 2(\sqrt{5}+\sqrt{3})$ ✓

(2 marks)

30 **i** $\dfrac{1}{\sqrt{n} + \sqrt{n+1}} \times \dfrac{\sqrt{n} - \sqrt{n+1}}{\sqrt{n} - \sqrt{n+1}}$

$= \dfrac{\sqrt{n} - \sqrt{n+1}}{n - (n+1)}$

$= \dfrac{\sqrt{n} - \sqrt{n+1}}{-1}$

$= \sqrt{n+1} - \sqrt{n}$ ✓

(1 mark)

ii From **i**,

$\dfrac{1}{\sqrt{n} + \sqrt{n+1}} = \sqrt{n+1} - \sqrt{n}$

$\therefore$ As $n = 1$, $\dfrac{1}{\sqrt{1} + \sqrt{2}} = \sqrt{2} - \sqrt{1}$

$n = 2$, $\dfrac{1}{\sqrt{2} + \sqrt{3}} = \sqrt{3} - \sqrt{2}$

$n = 3$, $\dfrac{1}{\sqrt{3} + \sqrt{4}} = \sqrt{4} - \sqrt{3}$

$\therefore$ series is $(\sqrt{2} - \sqrt{1}) + (\sqrt{3} - \sqrt{2}) + (\sqrt{4} - \sqrt{3}) + \ldots + (\sqrt{100} - \sqrt{99})$ ✓

$= -\sqrt{1} + \sqrt{100}$

$= -1 + 10$

$= 9$ ✓

(2 marks)

31 $x^2 = 4x$

$x^2 - 4x = 0$

$x(x - 4) = 0$ ✓

$\therefore x = 0$ or 4 ✓

(2 marks)

32 $\dfrac{1}{\sqrt{5} - 2}$

$= \dfrac{1}{\sqrt{5} - 2} \times \dfrac{\sqrt{5} + 2}{\sqrt{5} + 2}$

$= \dfrac{\sqrt{5} + 2}{5 - 4}$ ✓

$= \sqrt{5} + 2$

$= 2 + \sqrt{5}$, which is in the form $a + b\sqrt{5}$, where $a = 2$ and $b = 1$. ✓

(2 marks)

33 $\dfrac{5x - 4}{x} = 2$

$\therefore x\left(\dfrac{5x - 4}{x}\right) = 2x$

$5x - 4 = 2x$ ✓

$3x = 4$

$\therefore x = \dfrac{4}{3}$ ✓

(2 marks)

34 $\dfrac{2}{n} - \dfrac{1}{n+1} = \dfrac{2(n+1) - n}{n(n+1)}$ ✓

$= \dfrac{2n + 2 - n}{n(n+1)}$

$= \dfrac{n + 2}{n(n+1)}$ ✓

(2 marks)

35 $(\sqrt{3} - 1)(2\sqrt{3} + 5)$

$= \sqrt{3} \times 2\sqrt{3} + \sqrt{3} \times 5 - 2\sqrt{3} - 5$ ✓

$= 6 + 5\sqrt{3} - 2\sqrt{3} - 5$

$= 1 + 3\sqrt{3}$ ✓

(2 marks)

36 $2x - 5 > -3$

$2x > 2$

$\therefore x > 1$. ✓

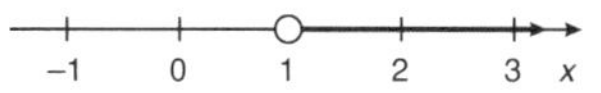

✓

(2 marks)

37 $2x^2 + 5x - 12$

$= 2x^2 + 8x - 3x - 12$

$= 2x(x + 4) - 3(x + 4)$

$= (2x - 3)(x + 4)$. ✓✓

(2 marks)

38 $\dfrac{1}{\sqrt{3} - 1} = \dfrac{1}{\sqrt{3} - 1} \times \dfrac{\sqrt{3} + 1}{\sqrt{3} + 1}$

$= \dfrac{\sqrt{3} + 1}{(\sqrt{3})^2 - (1)^2}$ ✓

$= \dfrac{\sqrt{3} + 1}{3 - 1}$

$= \dfrac{\sqrt{3} + 1}{2}$ ✓

(2 marks)

39 $3 - 5x \leqslant 2$

$-5x \leqslant -1$ ✓

$x \geqslant \dfrac{1}{5}$ ✓

(2 marks)

40 $\dfrac{(2x - 3)}{2} - \dfrac{(x - 1)}{5}$

$= \dfrac{5(2x - 3) - 2(x - 1)}{10}$ ✓

$= \dfrac{10x - 15 - 2x + 2}{10}$

$= \dfrac{8x - 13}{10}$ ✓

(2 marks)

41 $\dfrac{x - 5}{3} - \dfrac{x + 1}{4} = 5$

$\therefore 12\left(\dfrac{x - 5}{3}\right) - 12\left(\dfrac{x + 1}{4}\right) = 5 \times 12$

$4(x - 5) - 3(x + 1) = 60$ ✓

$4x - 20 - 3x - 3 = 60$

$x - 23 = 60$

$\therefore x = 83$ ✓

(2 marks)

42 $(3 - \sqrt{2})^2 = a - b\sqrt{2}$

Now $(3 - \sqrt{2})^2 = 9 - 6\sqrt{2} + 2$ ✓

$= 11 - 6\sqrt{2}$

$= a - b\sqrt{2}$

where $a = 11$ and $b = 6$. ✓

(2 marks)

43 $x^2 = 5x$

$\therefore x^2 - 5x = 0$

$\therefore x(x - 5) = 0$ ✓

$\therefore x = 0$ or 5 ✓ *(2 marks)*

44
$$3x - \frac{2x-5}{2} = 6$$
$\therefore 6x - (2x - 5) = 12$ ✓
$6x - 2x + 5 = 12$
$4x + 5 = 12$ ✓
$4x = 7$
$x = \frac{7}{4}$ ✓
(3 marks)

45
$x - 2y = 8$ . . . (1)
$2x + y = 1$ . . . (2)
(2) × 2 $4x + 2y = 2$ . . . (3)
(1) + (3) $5x = 10$
$\therefore x = 2$ ✓

Substituting $x = 2$ into (2):

$2(2) + y = 1$
$4 + y = 1$
$y = -3$ ✓
(2 marks)

46
$x^2 - 2x - 8 = 0$
$(x - 4)(x + 2) = 0$ ✓
$\therefore x = 4$ or -2 ✓
(2 marks)

47
$$\frac{x}{x^2-4} + \frac{2}{x-2}$$
$$= \frac{x}{(x-2)(x+2)} + \frac{2}{x-2}$$
$$= \frac{x+2(x+2)}{(x-2)(x+2)}$$ ✓
$$= \frac{3x+4}{(x-2)(x+2)}$$ ✓
(2 marks)

48
$x - y = 2$ . . . (1)
$3x + 2y = 1$ . . . (2)
(1) × (2) $2x - 2y = 4$ . . . (3)
(2) + (3) $5x = 5$
$\therefore x = 1$ ✓

Substituting $x = 1$ into (1).

$1 - y = 2$
$-y = 1$
$\therefore y = -1$ ✓
(2 marks)

49
$3 - 2x \geqslant 7$
$\therefore -2x \geqslant 4$ ✓
$x \leqslant -\frac{4}{2}$
$\therefore x \leqslant -2$ ✓
(2 marks)

50
$x + y = 1$ . . . (1)
$2x - y = 5$ . . . (2)

(1) + (2) $3x = 6$
$\therefore x = 2$ ✓

Substituting $x = 2$ into (1);

$2 + y = 1$
$\therefore y = -1$ ✓ *(2 marks)*

51
$(5 - \sqrt{2})^2 = 25 + 2 - 10\sqrt{2}$ ✓
$= 27 - 10\sqrt{2}$
$= a + b\sqrt{2}$
where $a = 27$ and $b = -10$. ✓ *(2 marks)*

52
$$\frac{1}{3-\sqrt{2}} + \frac{1}{3+\sqrt{2}}$$
$$= \frac{1}{3-\sqrt{2}} \times \frac{3+\sqrt{2}}{3+\sqrt{2}} + \frac{1}{3+\sqrt{2}} \times \frac{3-\sqrt{2}}{3-\sqrt{2}}$$
$$= \frac{3+\sqrt{2}}{9-2} + \frac{3-\sqrt{2}}{9-2}$$ ✓
$$= \frac{3+\sqrt{2}}{7} + \frac{3-\sqrt{2}}{7}$$
$$= \frac{3+\sqrt{2}+3-\sqrt{2}}{7}$$
$$= \frac{6}{7}$$ ✓
(2 marks)

53
$$\frac{8}{3-\sqrt{5}} = \frac{8}{3-\sqrt{5}} \times \frac{3+\sqrt{5}}{3+\sqrt{5}}$$ ✓
$$= \frac{8(3+\sqrt{5})}{9-5}$$
$$= \frac{8(3+\sqrt{5})}{4}$$
$= 2(3 + \sqrt{5})$ ✓
$= 6 + 2\sqrt{5}$
$= a + b\sqrt{5}$
where $a = 6$ and $b = 2$. ✓
(3 marks)

54
$$\frac{4}{\sqrt{5}+2} = \frac{4}{\sqrt{5}+2} \times \frac{\sqrt{5}-2}{\sqrt{5}-2}$$ ✓
$$= \frac{4\sqrt{5}-8}{5-4}$$
$= 4\sqrt{5} - 8$ ✓
(2 marks)

55
$$\frac{2}{3} - \frac{x-1}{4} = \frac{4(2)-3(x-1)}{4 \times 3}$$ ✓
$$= \frac{8-3x+3}{12}$$
$$= \frac{11-3x}{12}$$ ✓
(2 marks)

56
$$u = \frac{-b \pm \sqrt{b^2-4ac}}{2a}$$
$$= \frac{-(-1) \pm \sqrt{(-1)^2-4(1)(-1)}}{2(1)}$$
$$= \frac{1 \pm \sqrt{1+4}}{2}$$
$$= \frac{1 \pm \sqrt{5}}{2}$$ ✓
$$= \frac{1+\sqrt{5}}{2} \text{ or } \frac{1-\sqrt{5}}{2}$$
$= 1.6180 \ldots$ or $-0.6180 \ldots$
$= 1.618$ or -0.618 ✓
to 3 decimal places
(2 marks)

57
$$\frac{x}{3} + \frac{3x-1}{2}$$
$$= \frac{2(x)+3(3x-1)}{6}$$ ✓
$$= \frac{2x+9x-3}{6}$$
$$= \frac{11x-3}{6}$$ ✓
(2 marks)

58
$$4\pi\sqrt{\frac{a}{g}} = 4\pi\sqrt{\frac{4.1}{9.8}}$$
$= 4\pi\sqrt{0.4183\ldots}$
$= 4\pi(0.6468\ldots)$
$= 8.1280\ldots$
$= 8.1$ to 2 sig. figures.
(No mark allocation in exam)

59
$2x + y = 7$ . . . (1)
$x - 2y = 1$ . . . (2)
(1) × 2 $4x + 2y = 14$ . . . (3)
(2) + (3) $5x = 15$
$\therefore x = 3$
Substituting $x = 3$ into (1):
$2(3) + y = 7$
$6 + y = 7$
$\therefore y = 1$
(No mark allocation in exam)

60
$4x + 7 = 3(x - 2)$
$4x + 7 = 3x - 6$
$x = -13$
(No mark allocation in exam)

61
$x + y = 2$ . . . (1)
$2x - y = 7$ . . . (2)
(1) + (2) $3x = 9$
$\therefore x = 3$
Substituting $x = 3$ into (1):
$3 + y = 2$
$\therefore y = -1$
(No mark allocation in exam)

1 Which of the following is the range of the function $f(x) = x^2 - 1$?

A $[-1, \infty)$ **B** $(-\infty, 1]$
C $[-1, 1]$ **D** $(-\infty, \infty)$ *(1 mark)*

(Q4, **2022 HSC**) Easy

2 The graphs of $y = f(x)$ and $y = g(x)$ are shown.

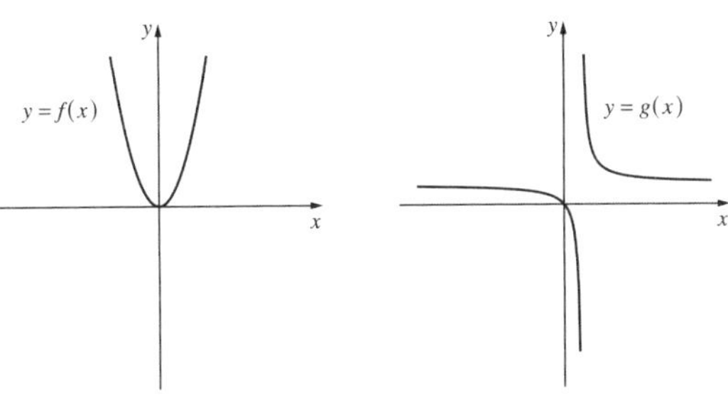

Which graph best represents $y = g(f(x))$?

A **B**

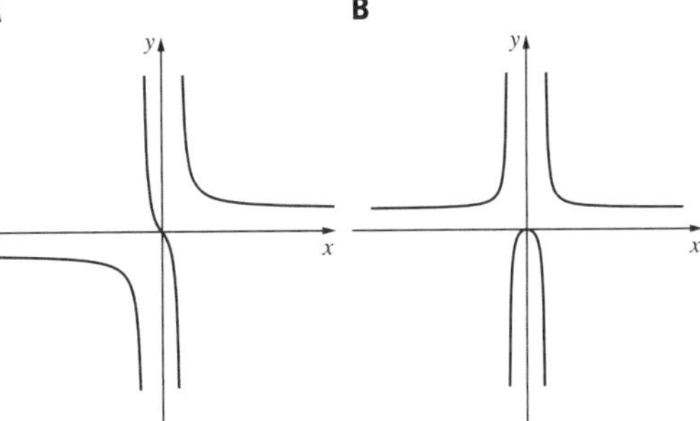

C **D**

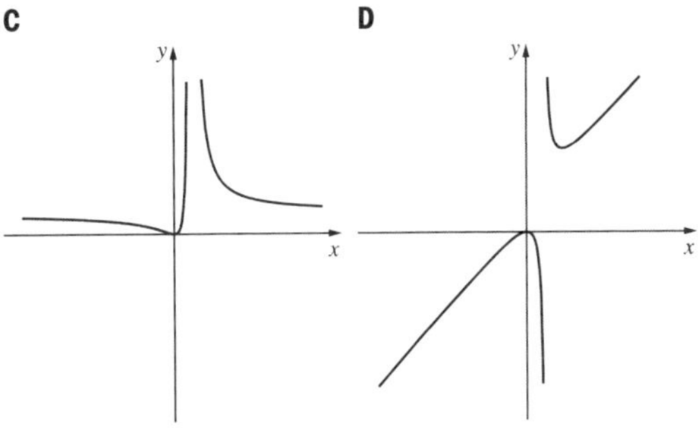

(1 mark)

(Q10, **2022 HSC**) Hard

3 Which inequality gives the domain of $y = \sqrt{2x - 3}$?

A $x < \frac{3}{2}$ **B** $x > \frac{3}{2}$

C $x \leq \frac{3}{2}$ **D** $x \geq \frac{3}{2}$ *(1 mark)*

(Q1, **2020 HSC**) Medium

4 Let f and g be two functions such that $f(3) = 2$, $f(5) = 3$, $g(2) = 3$, $g(4) = 5$.

What is the value of $f(g(4))$?

A 2 **B** 3
C 4 **D** 5 *(1 mark)*

Bonus question (see page iv) Medium

5 Which of these is not a one-to-one function?

A $f(x) = 2x^3$ **B** $f(x) = 2^x + 1$
C $f(x) = \sqrt{x - 2}$ **D** $f(x) = x^3 + 2x^2$

(1 mark)

Bonus question Easy

6 Given that $f(x) = x^2 - 3$ and $g(x) = \dfrac{4x}{x+2}$.

i Write an expression for $g(f(x))$.

(1 mark) Easy

ii Hence or otherwise find the value of $g(f(-2))$. *(1 mark)* Easy

Bonus question

7 What type of relation is shown?

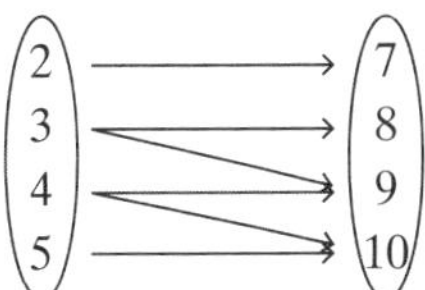

A one-to-one **B** one-to-many
C many-to-one **D** many-to-many

(1 mark)

Bonus question Easy

8 $f(x) = x^2 + 4x + 4$ and $g(x) = x^2 - 4$.

i Find $\dfrac{f(x)}{g(x)}$ in simplest form.

(2 marks) Medium

ii What is the domain of $\dfrac{f(x)}{g(x)}$?

(2 marks) Medium

Bonus question

9 If $f(x) = ax^3 + bx^2 + cx + d$ find $f(x) - f(-x)$.

(2 marks)

Bonus question Medium

10 Find the range of $f(x) = \dfrac{9 - 2x}{5}$ over the domain $[-3, 2)$. (Give the answer in interval form.) *(2 marks)*

Bonus question Medium

11 Which is **not** the graph of a function?

A

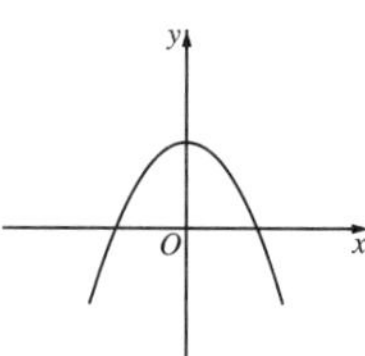

B

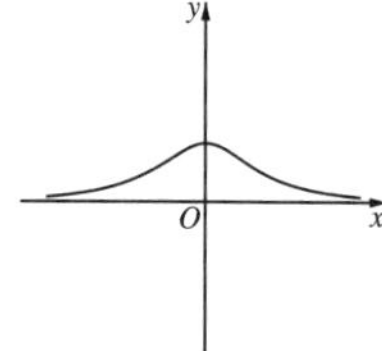

C

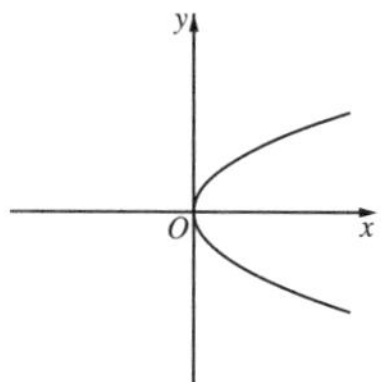

D

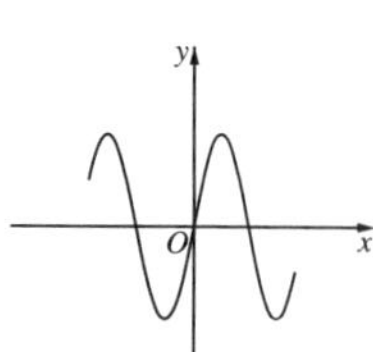

(1 mark)

Bonus question Easy

12 Find the domain of the function $f(x) = \sqrt{3 - x}$. *(2 marks)*

(Q11h, **2017 HSC**) Medium

13 Which diagram shows the graph of an odd function?

A

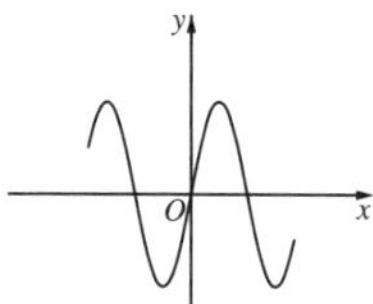

B

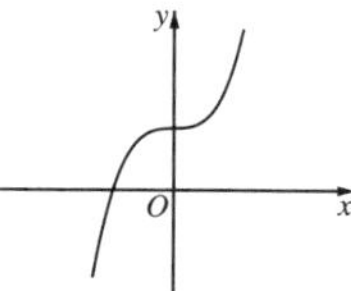

C

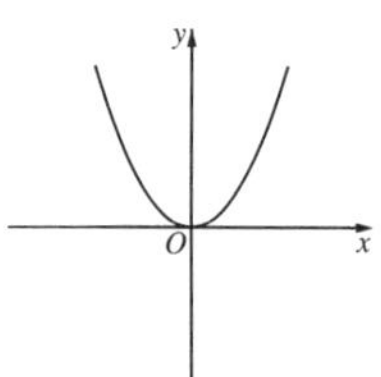

D

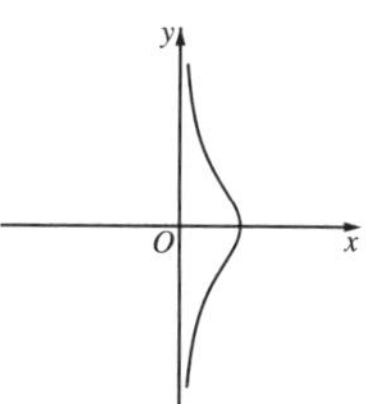

(1 mark)

(Q4, **2016 HSC**) Easy

14 Which inequality defines the domain of the function $f(x) = \dfrac{1}{\sqrt{x+3}}$?

A $x > -3$ **B** $x \geq -3$

C $x < -3$ **D** $x \leq -3$ *(1 mark)*

(Q3, **2013 HSC**) Medium

15 Let $f(x) = \sqrt{x - 8}$. What is the domain of $f(x)$?

(1 mark)

(Q1g, **2010 HSC**) Medium

Year 11 Introduction to functions—Worked answers

1

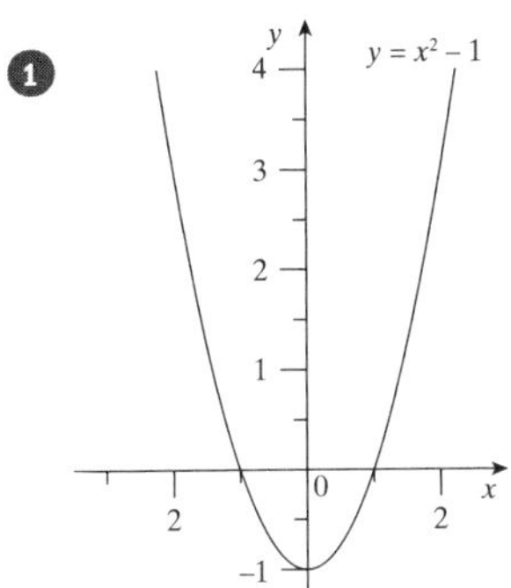

Range is $[-1, \infty)$

Answer A

2 Suppose $f(x) = x^2$ and $g(x) = \dfrac{1}{x-1} + 1$.

Hence $g(f(x)) = \dfrac{1}{x^2-1} + 1$.

As $y = \dfrac{1}{x^2-1} + 1$, then discontinuities occur at $x = 1, -1$, and hence possible options are A and B.

But $y = \dfrac{1}{x^2-1} + 1$ is an even function (reflects about y-axis), hence the correct graph is B.

Answer B

3

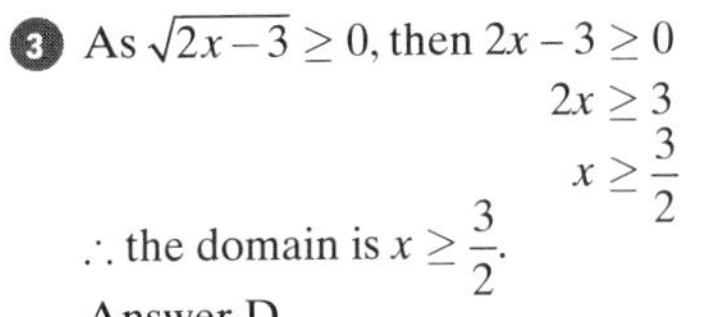

As $\sqrt{2x-3} \geq 0$, then $2x - 3 \geq 0$

$$2x \geq 3$$
$$x \geq \frac{3}{2}$$

$\therefore$ the domain is $x \geq \dfrac{3}{2}$.

Answer D

4

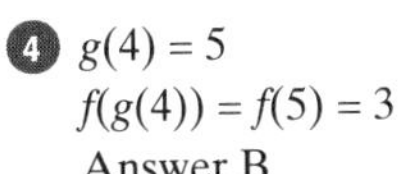

$g(4) = 5$

$f(g(4)) = f(5) = 3$

Answer B

5 Sketch each function and use the horizontal line test.

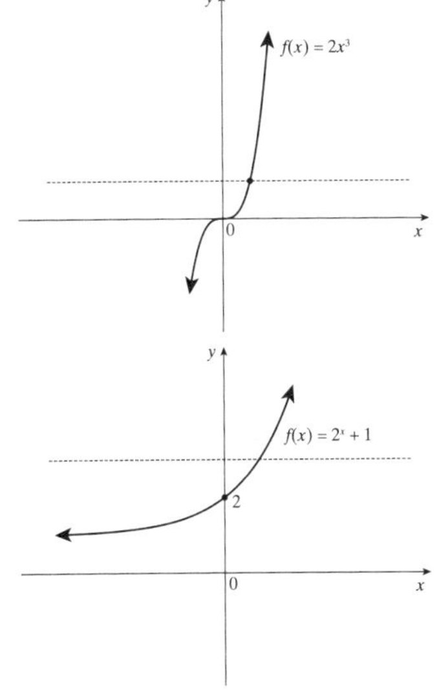

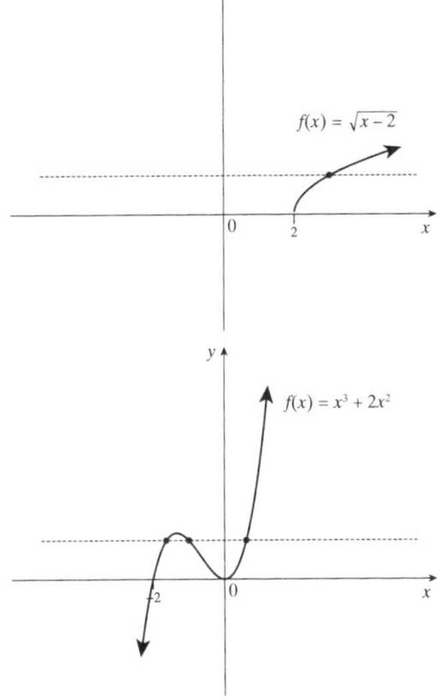

A horizontal line can be drawn on $f(x) = x^3 + 2x^2$ and it cuts at more than one point.

Answer D

6 **i** $g(x) = \dfrac{4x}{x+2}$

$$g(f(x)) = \frac{4(x^2-3)}{(x^2-3)+2}$$
$$= \frac{4x^2-12}{x^2-1} \quad ✓ \quad \textit{(1 mark)}$$

ii $g(f(-2)) = \dfrac{4(-2)^2-12}{(-2)^2-1}$

$$= \frac{4}{3} \quad ✓ \quad \textit{(1 mark)}$$

7 Some elements of the range correspond to more than one element of the domain and some elements of the domain correspond to more than one element of the range. So the relation is many-to-many.

Answer D

8 **i** $f(x) = x^2 + 4x + 4$,

$g(x) = x^2 - 4$

$$\frac{f(x)}{g(x)} = \frac{x^2+4x+4}{x^2-4}$$
$$= \frac{(x+2)^2}{(x+2)(x-2)} \quad ✓$$
$$= \frac{x+2}{x-2} \quad ✓ \quad \textit{(2 marks)}$$

ii $f(x) = \dfrac{x+2}{x-2}$

Now $x - 2 \neq 0$ ✓

$$x \neq 2$$

The domain is all real values of x except $x = 2$. ✓ *(2 marks)*

9 $f(x) = ax^3 + bx^2 + cx + d$

$f(-x) = a(-x)^3 + b(-x)^2 + c(-x) + d$

$= -ax^3 + bx^2 - cx + d$ ✓

$f(x) - f(-x)$

$= ax^3 + bx^2 + cx + d - (-ax^3 + bx^2 - cx + d)$

$= ax^3 + bx^2 + cx + d + ax^3 - bx^2 + cx - d$

$= 2ax^3 + 2cx$ ✓ *(2 marks)*

10 $f(x) = \dfrac{9-2x}{5}$, $[-3, 2)$

The graph of the function is a straight line.

The domain is $-3 \leq x < 2$.

Now $f(-3) = \dfrac{9-2\times(-3)}{5}$

$= 3$

$f(2) = \dfrac{9-2\times 2}{5}$

$= 1$ ✓

So the range is $1 < y \leq 3$.

The range is $(1, 3]$. ✓ *(2 marks)*

11 A vertical line can be drawn through the graph in C and intersects the graph at more than one point. So, using the vertical line test, the graph in C is not a function.

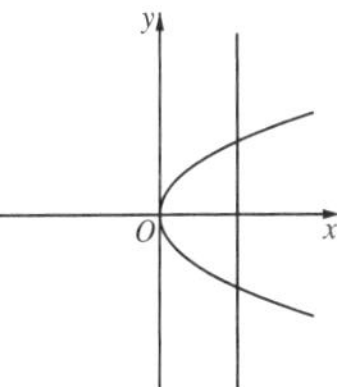

Answer C

12 As $3 - x \geq 0$, then ✓

$x \leq 3$ ✓ *(2 marks)*

13 The graph of an odd function has point symmetry about the origin.

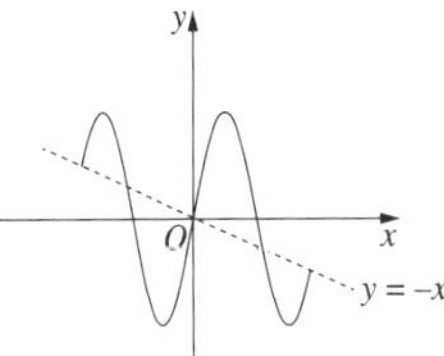

Answer A

14 $f(x) = \dfrac{1}{\sqrt{x+3}}$

Consider $x + 3 > 0$

$$x > -3$$

Answer A

15 $f(x) = \sqrt{x-8}$.

Domain is $x - 8 \geq 0$

$$x \geq 8$$

$\therefore$ The domain is $x \geq 8$. ✓ *(1 mark)*

1 Which of the following could be the graph of $y = -2x + 2$?

A

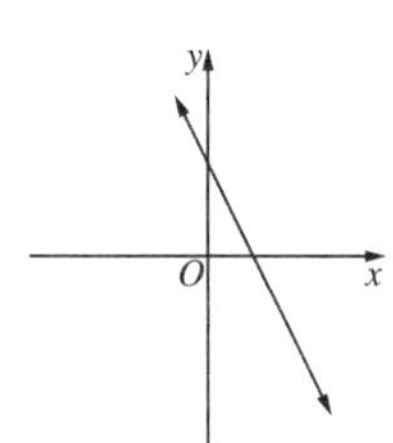

B

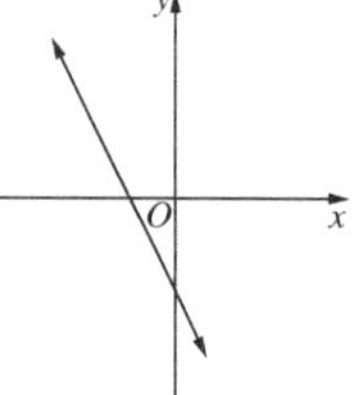

C

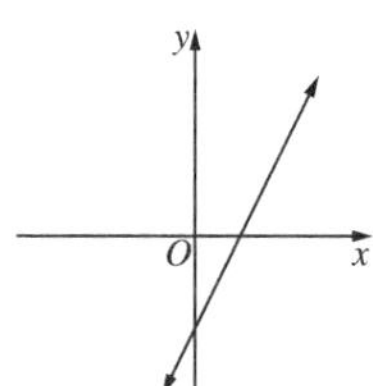

D

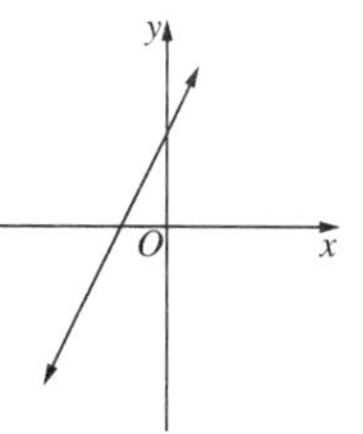

(1 mark)

 (Q1, **2022 HSC**)  Easy

2 Which of the following could represent the graph of $y = -x^2 + 1$?

A

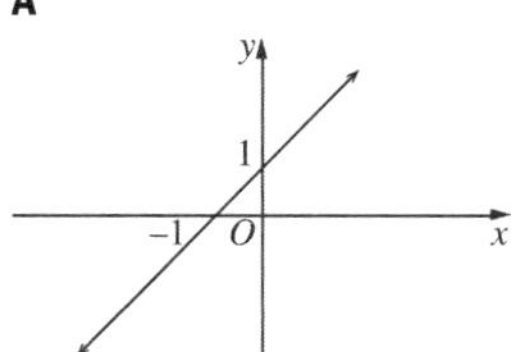

B

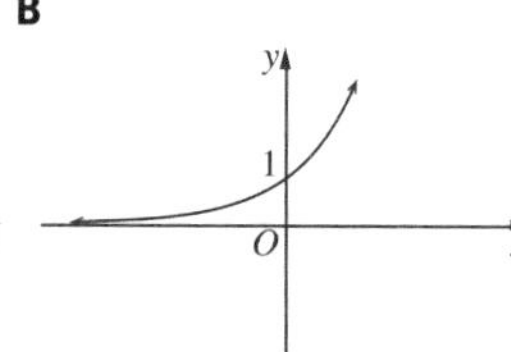

C

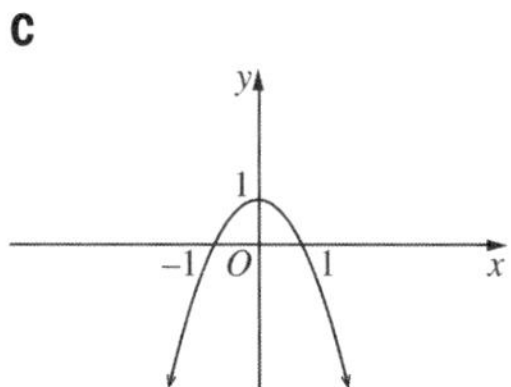

D

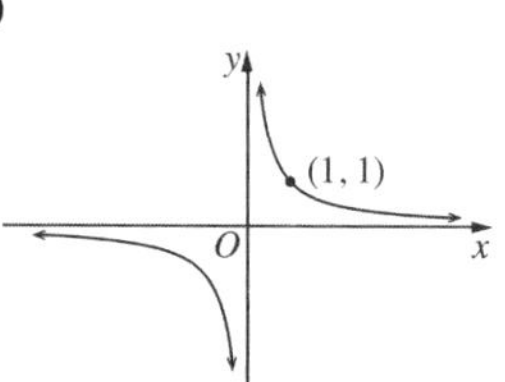

(1 mark)

(Q1, **2020 MS2 HSC**) Easy

3 Which of the following could represent the graph of $y = -x^2 + bx + 1$, where $b > 0$?

A

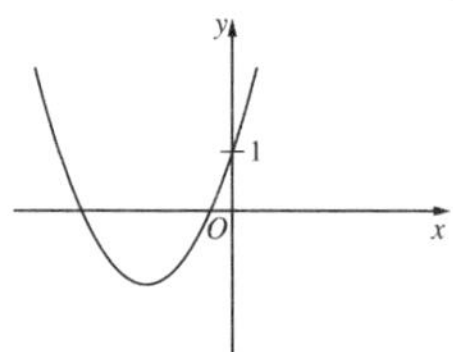

B

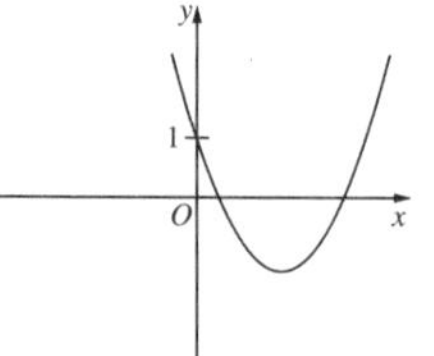

C **D**

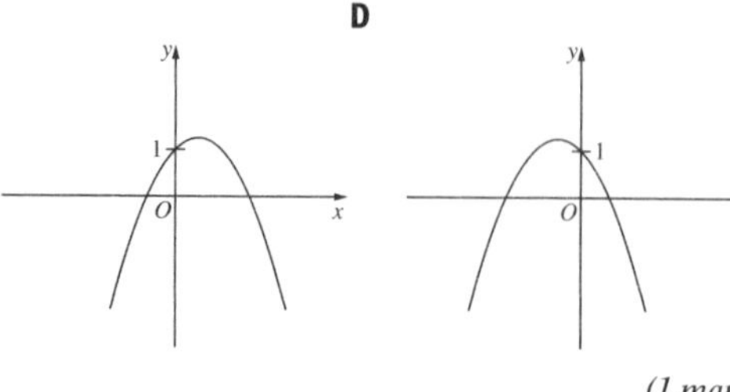

(1 mark)

(Q5, **2020 HSC**) Medium

4 There are two tanks on a property, Tank A and Tank B. Initially, Tank A holds 1000 litres of water and Tank B is empty.

a Tank A begins to lose water at a constant rate of 20 litres per minute. The volume of water in Tank A is modelled by $V = 1000 - 20t$ where V is the volume in litres and t is the time in minutes from when the tank begins to lose water. On the grid below, draw the graph of this model and label it as Tank A. *(1 mark)* Easy

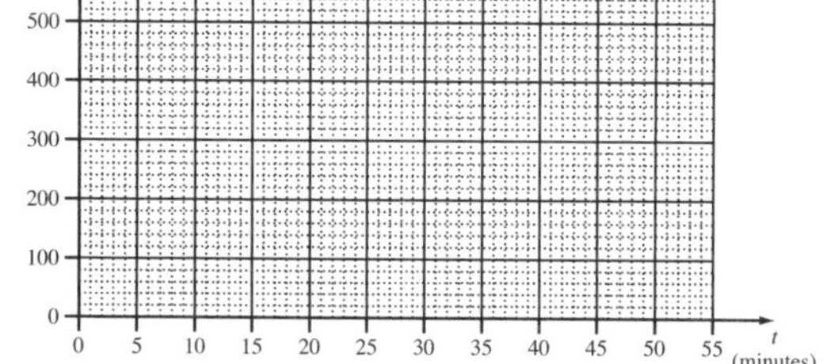

b Tank B remains empty until $t = 15$ when water is added to it at a constant rate of 30 litres per minute.
By drawing a line on the grid on the previous page [above], or otherwise, find the value of t when the two tanks contain the same volume of water. *(2 marks)* Easy

c Using the graphs drawn, or otherwise, find the value of t (where $t > 0$) when the total volume of water in the two tanks is 1000 litres. *(1 mark)* Medium

CQ (Q11, **2020 HSC**)

5 Swarthy runs a small business. She uses a straight-line depreciation model of 20% of the original price per annum to depreciate the cost of her office desk originally valued at \$1850.

Construct a function to represent the value y of the furniture after x years, and state the domain and range of the function. *(2 marks)*

Bonus question (see page iv) **Medium**

6 Find the range of values of k for which the expression $x^2 - 3x + (k + 2)$ is always positive. *(2 marks)*

Bonus question **Easy**

7 Find the maximum value of y given $y = 5 + 4x - 2x^2$. *(1 mark)*

Bonus question **Easy**

8 Find the coordinates of the point P on the line $x = -4$, such that the line joining P to $Q(1, -3)$ is perpendicular to the line $3x - 2y = 12$. *(3 marks)*

Bonus question **Medium**

9 The line $2x + 3y - 6 = 0$ cuts the x-axis at A. The line $x - 2y + 4 = 0$ cuts the x-axis at B.

The lines $2x + 3y - 6 = 0$ and $x - 2y + 4 = 0$ intersect at C.

Find the area of triangle ABC. *(3 marks)*

Bonus question **Medium**

10 The line ℓ, with equation $x - 2y + 4 = 0$, passes through the point $A(8, 6)$ and intersects the x-axis at the point B.

The line through A perpendicular to ℓ intersects the x-axis at the point C.

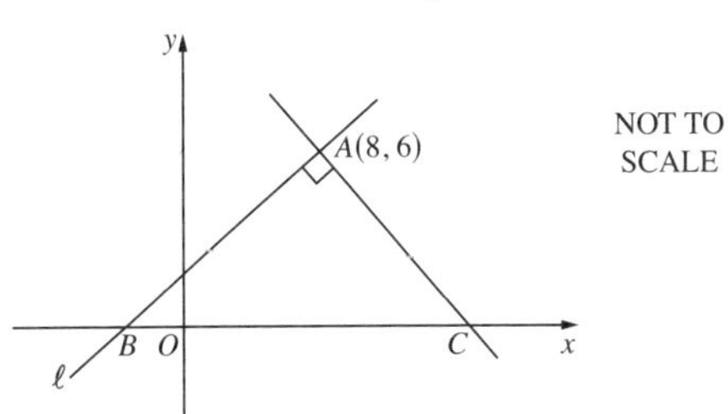

i Find the equation of the line AC. *(2 marks)* **Easy**

ii Find the area of triangle ABC. *(2 marks)* **Medium**

(Q12a, **2019 HSC**)

11 Last Saturday, Luke had 165 followers on social media. Rhys had 537 followers. On average, Luke gains another 3 followers per day and Rhys loses 2 followers per day.

If x represents the number of days since last Saturday and y represents the number of followers, which pair of equations model this situation?

A Luke: $y = 165x + 3$ Rhys: $y = 537x - 2$

B Luke: $y = 165 + 3x$ Rhys: $y = 537 - 2x$

C Luke: $y = 3x + 165$ Rhys: $y = 2x - 537$

D Luke: $y = 3 + 165x$ Rhys: $y = 2 - 537x$

(1 mark)

(Q14, **2019 MS2 HSC**) **Easy**

12 A rectangle has width w centimetres. The area of the rectangle, A, in square centimetres, is $A = 2w^2 + 5w$.

The graph of $A = 2w^2 + 5w$ is shown.

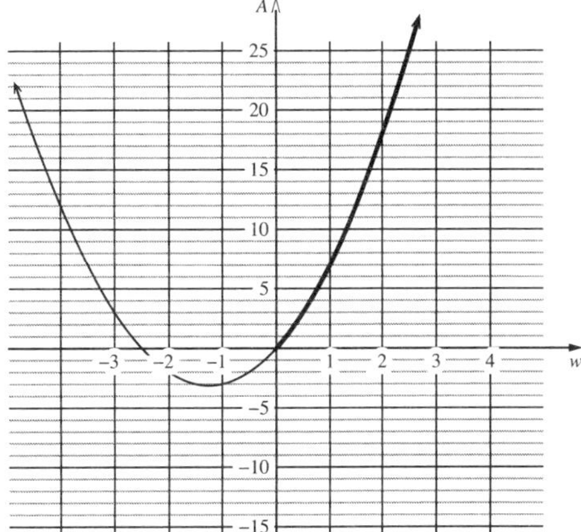

a Explain why, in this context, the model $A = 2w^2 + 5w$ only makes sense for the bold section of the graph. *(1 mark)* **Easy**

b The area of the rectangle is 18 cm^2. Calculate the perimeter of the rectangle. *(2 marks)* **Easy**

(Q31, **2019 MS2 HSC**)

13 The relationship between British pounds (p) and Australian dollars (d) on a particular day is shown in the graph.

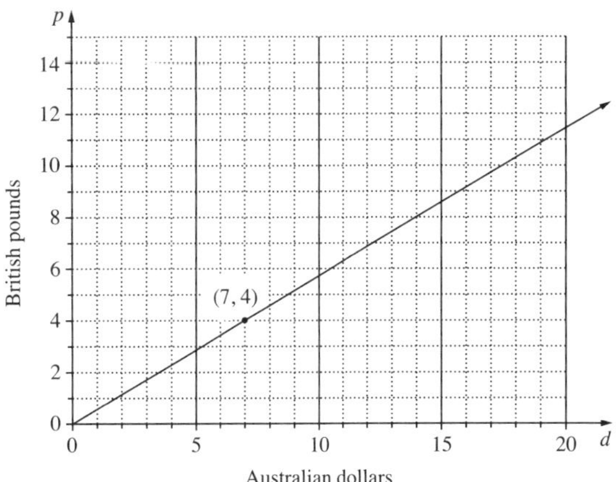

a Write the direct variation equation relating British pounds to Australian dollars in the form $p = md$. Leave m as a fraction. *(1 mark)* **Easy**

b The relationship between Japanese yen (y) and Australian dollars (d) on the same day is given by the equation $y = 76d$.
Convert 93 100 Japanese yen to British pounds. *(2 marks)* Medium

(Q34, **2019 MS2 HSC**)

14 A small business makes and sells bird houses. Technology was used to draw straight-line graphs to represent the cost of making bird houses (C) and the revenue from selling bird houses (R). The x-axis displays the number of bird houses and the y-axis displays the cost/revenue in dollars.

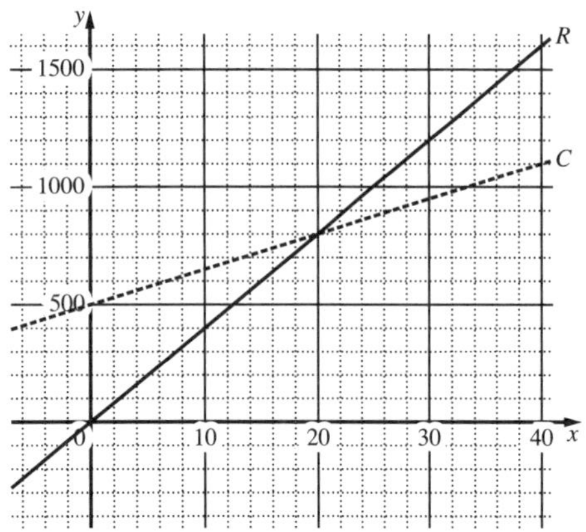

a How many bird houses need to be sold to break even? *(1 mark)* Easy

b By first forming equations for cost (C) and revenue (R), determine how many bird houses need to be sold to earn a profit of \$1900. *(3 marks)* Medium

(Q36, **2019 MS2 HSC**)

15 What is the x-intercept of the line $x + 3y + 6 = 0$?

A $(-6, 0)$ **B** $(6, 0)$
C $(0, -2)$ **D** $(0, 2)$ *(1 mark)*

(Q3, **2018 HSC**) Easy

16 Which graph best represents the equation $y = x^2 - 2$?

A

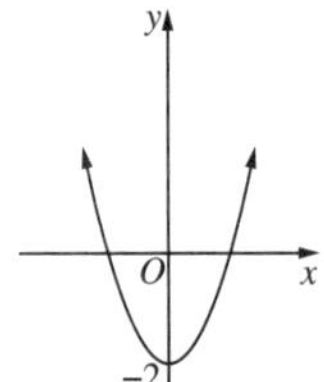

B

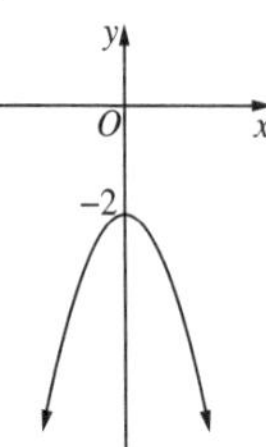

C

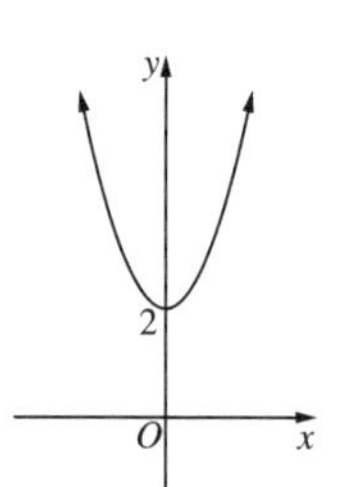

D

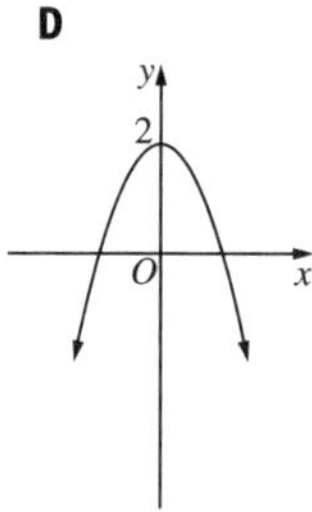

(1 mark)

(Q4, **2018 MS2 HSC**) Easy

17 The graph displays the cost (\$$c$) charged by two companies for the hire of a minibus for x hours.

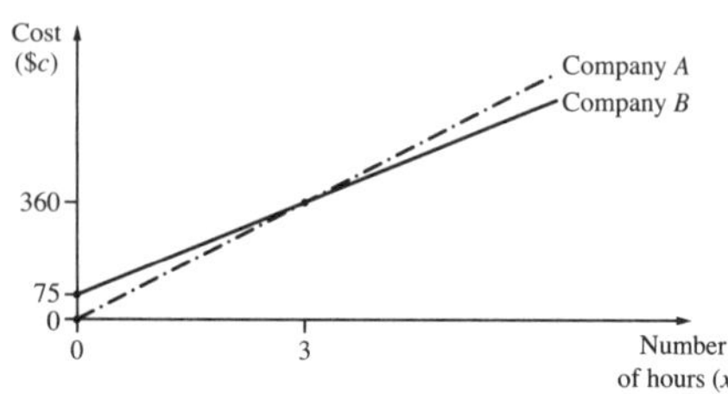

Both companies charge \$360 for the hire of a minibus for 3 hours.

i What is the hourly rate charged by Company A? *(1 mark)* Easy

ii Company B charges an initial booking fee of \$75.
Write a formula, in the form of $c = mx + b$, for the cost of hiring a minibus from Company B for x hours. *(2 marks)* Easy

iii A minibus is hired for 5 hours from Company B.
Calculate how much cheaper this is than hiring from Company A. *(2 marks)* Easy

(Q27d, **2018 MS2 HSC**)

18 What is the gradient of the line $2x + 3y + 4 = 0$?

A $-\frac{2}{3}$ **B** $\frac{2}{3}$
C $-\frac{3}{2}$ **D** $\frac{3}{2}$ *(1 mark)*

(Q1, **2017 HSC**) Easy

19 The graph of the line with equation $y = 6 - 2x$ is shown.

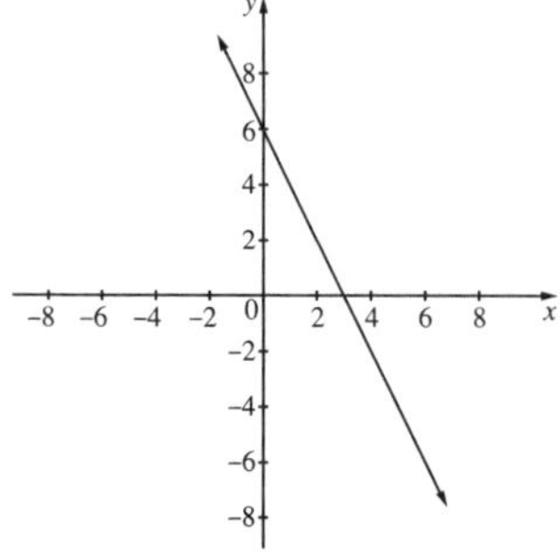

When the graph of the line with equation $y = x + 3$ is also drawn on this number plane, what will be the point of intersection of the two lines?

A $(0, 6)$ **B** $(1, 4)$
C $(2, 2)$ **D** $(3, 0)$ *(1 mark)*

(Q17, **2017 MS2 HSC**) Easy

---

20 Temperature can be measured in degrees Celsius (C) or degrees Fahrenheit (F).
The two temperature scales are related by the equation $F = \frac{9C}{5} + 32$.

i Calculate the temperature in degrees Fahrenheit when it is –20 degrees Celsius. *(1 mark)* Easy

ii Solve the following equations simultaneously, using either the substitution method or the elimination method.

$F = \frac{9C}{5} + 32$

$F = C$ *(2 marks)* 

iii The graphs of $F = \frac{9C}{5} + 32$ and $F = C$ are shown below.

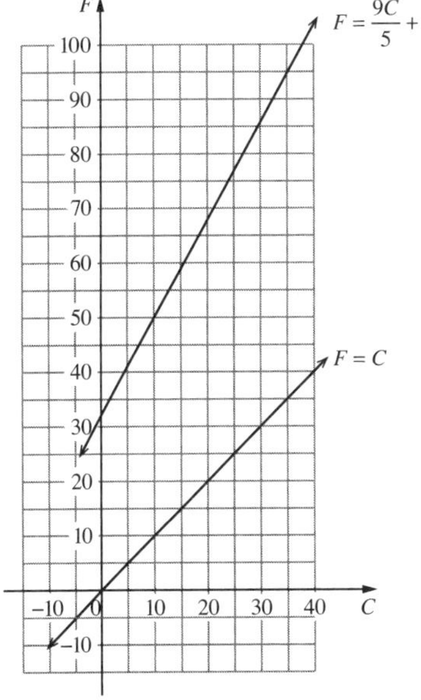

What does the result from part **ii** mean in the context of the graph? *(1 mark)* Easy

(Q28a, **2017 MS2 HSC**)

---

21 The cost of a jewellery box varies directly with the cube of its height.

A jewellery box with a height of 10 cm costs \$50.

Calculate the cost of a jewellery box with a height of 12 cm. *(2 marks)*

(Q30b, **2017 MS2 HSC**) Medium

---

22 Which diagram best shows the graph of the parabola $y = 3 - (x - 2)^2$?

A

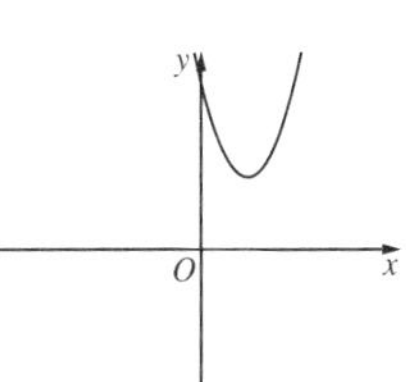

B

C

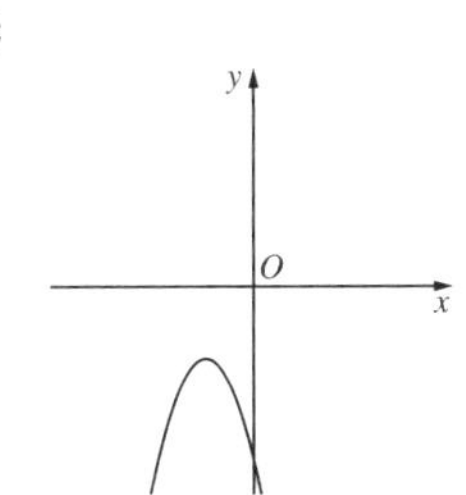

D

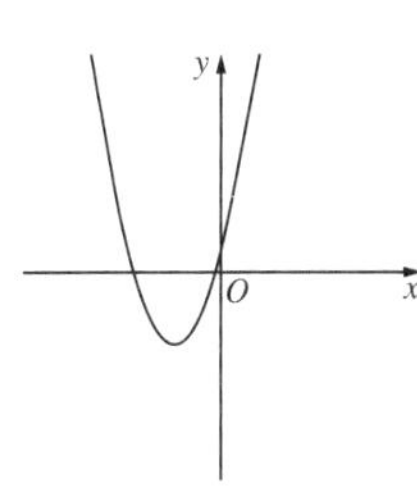

(1 mark)

(Q3, **2016 HSC**) Easy

---

23 Find the points of intersection of $y = -5 - 4x$ and $y = 3 - 2x - x^2$. *(3 marks)*

(Q11e, **2016 HSC**) Easy

---

24 A company manufactures phones. The company's income equation and cost equation are drawn on the same graph.

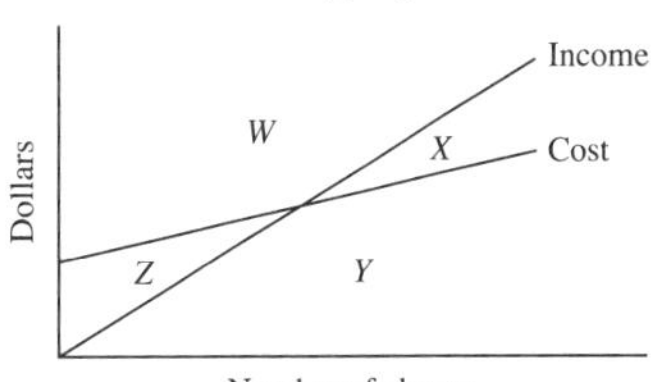

Which region of the graph is the profit zone?

A W **B** X
C Y **D** Z *(1 mark)*

(Q4, **2016 MS2 HSC**) Easy

---

25 The graph shows a line which has an equation in the form $y = mx + b$.

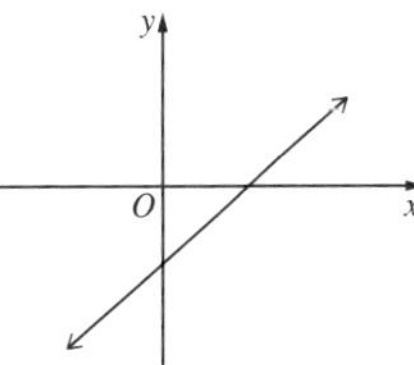

Which of the following statements is true?

A m is positive and b is negative
B m is negative and b is positive
C m and b are both positive
D m and b are both negative *(1 mark)*

(Q14, **2016 MS2 HSC**) Easy

26 The value of E varies directly with the square of S.

It is known that $E = 20$ when $S = 10$.

What is the value of E when $S = 40$?

A 40 **B** 80

C 320 **D** 400 *(1 mark)*

(Q18, **2016 MS2 HSC**) **Medium**

27 What is the slope of the line with equation $2x - 4y + 3 = 0$?

A -2 **B** $-\frac{1}{2}$

C $\frac{1}{2}$ **D** 2 *(1 mark)*

(Q2, **2015 HSC**) **Easy**

28 The diagram shows the rhombus $OABC$.

The diagonal from the point $A(7, 11)$ to the point C lies on the line ℓ_1.

The other diagonal, from the origin O to the point B, lies on the line ℓ_2 which has equation $y = -\frac{x}{3}$.

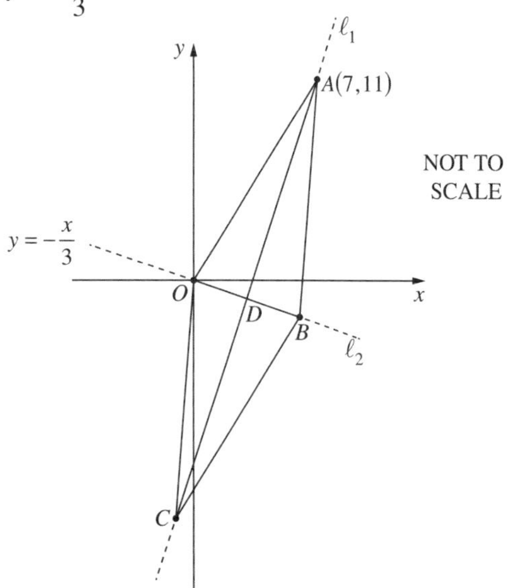

i Show that the equation of the line ℓ_1 is $y = 3x - 10$. *(2 marks)* **Easy**

ii The lines ℓ_1 and ℓ_2 intersect at the point D. Find the coordinates of D. *(2 marks)* **Easy**

(Q12b, **2015 HSC**)

29 For what values of k does the quadratic equation $x^2 - 8x + k = 0$ have real roots? *(2 marks)*

(Q12d, **2015 HSC**) **Medium**

30 Which of the following graphs best represents the equation $y = x^3 + 1$?

A

B

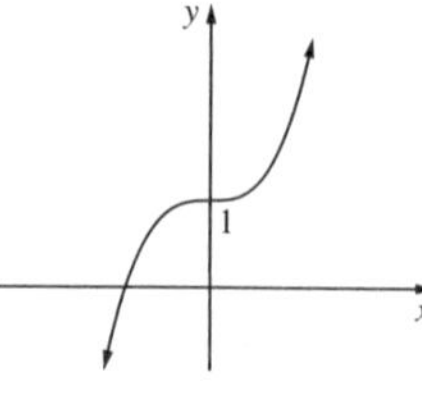

C

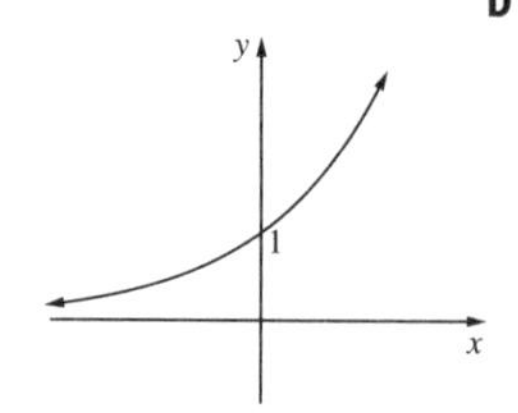

D

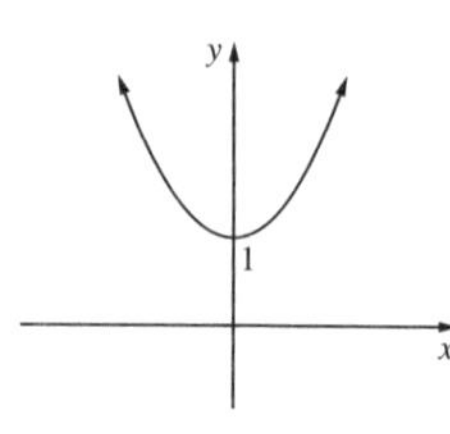

(1 mark)

(Q5, **2015 MS2 HSC**) **Easy**

31 What is the equation of the line ℓ?

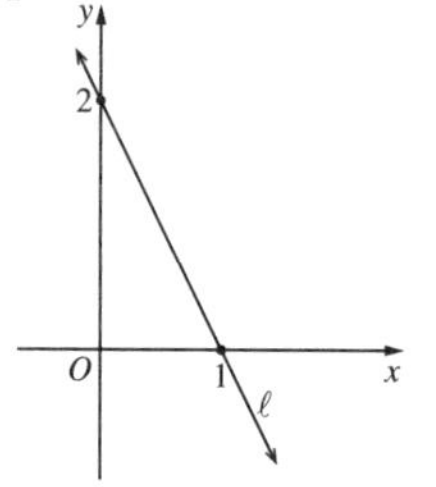

A $y = -2x + 2$ **B** $y = 2x + 2$

C $y = -\frac{x}{2} + 2$ **D** $y = \frac{x}{2} + 2$ *(1 mark)*

(Q13, **2015 MS2 HSC**) **Easy**

32 Ariana's parents have given her an interest-free loan of \$4800 to buy a car. She will pay them back by paying \$$x$ immediately and \$$y$ every month until she has repaid the loan in full.

After 18 months Ariana has paid back \$1510, and after 36 months she has paid back \$2770.

This information can be represented by the following equations.

$$x + 18y = 1510$$
$$x + 36y = 2770$$

i Solve these equations simultaneously to find the values of x and y. *(2 marks)* **Medium**

ii How many months will it take Ariana to repay the loan in full? *(2 marks)* **Easy**

(Q27c, **2015 MS2 HSC**)

33 A diver springs upwards from a diving board, then plunges into the water. The diver's height above the water as it varies with time is modelled by a quadratic function. Graphing software is used to produce the graph of this function.

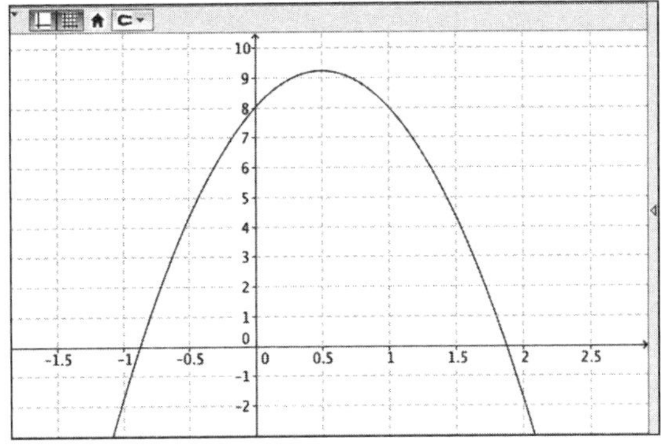

Explain how the graph could be used to determine how high above the height of the diving board the diver was when he reached the maximum height. *(2 marks)*

(Q29e, **2015 MS2 HSC**) Easy

34 Which graph best represents $y = (x - 1)^2$?

A

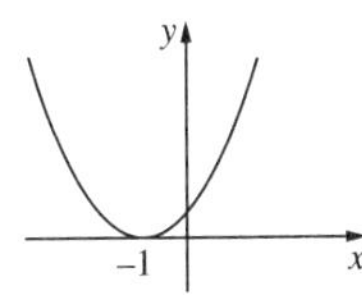

B

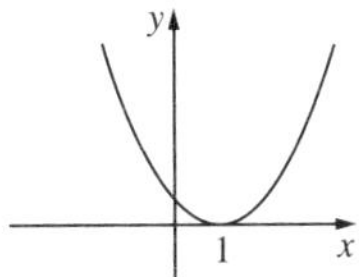

C

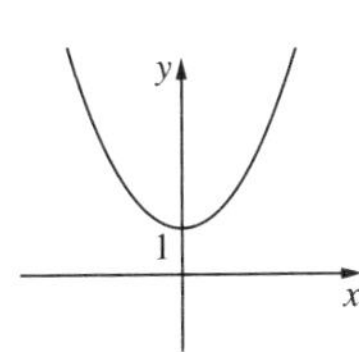

D

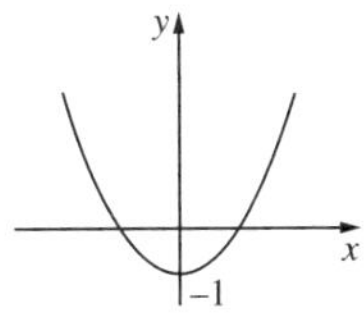

(1 mark)

(Q2, **2014 HSC**) Easy

35 Which equation represents the line perpendicular to $2x - 3y = 8$, passing through the point $(2, 0)$?

A $3x + 2y = 4$ **B** $3x + 2y = 6$

C $3x - 2y = -4$ **D** $3x - 2y = 6$ *(1 mark)*

(Q5, **2014 HSC**) Easy

36 The diagram shows the graph of an equation.

Which of the following equations does the graph best represent?

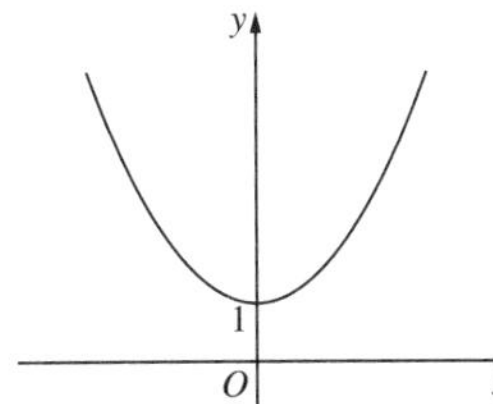

A $y = \frac{3}{x} + 1$ **B** $y = 3^x + 1$

C $y = 3x^2 + 1$ **D** $y = 3x^3 + 1$ *(1 mark)*

(Q3, **2014 MS2 HSC**) Easy

37 Which of the following is the graph of $y = 2x - 2$?

A

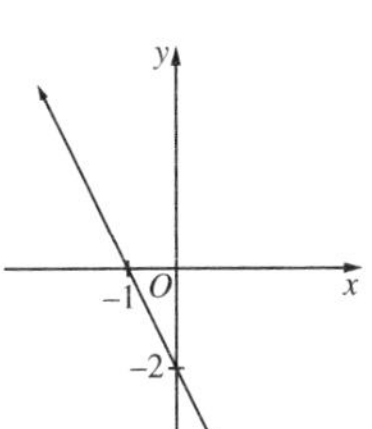

B

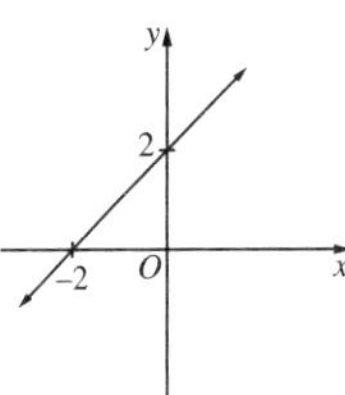

C

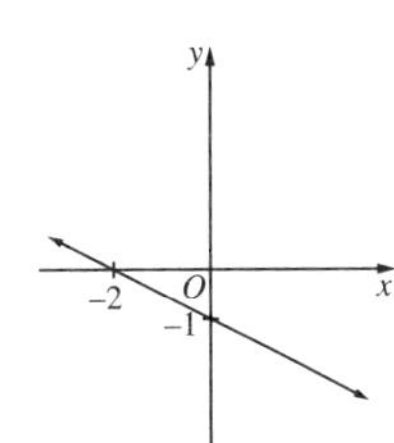

D

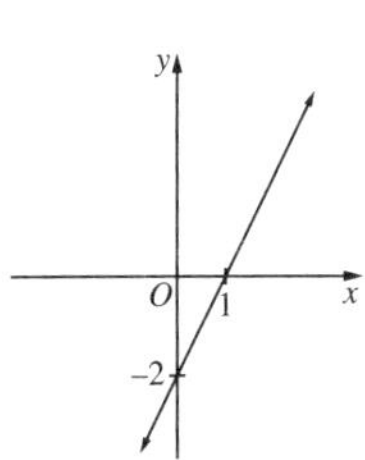

(1 mark)

(Q7, **2014 MS2 HSC**) Easy

38 The weight of an object on the moon varies directly with its weight on Earth. An astronaut who weighs 84 kg on Earth weighs only 14 kg on the moon.

A lunar landing craft weighs 2449 kg when on the moon. Calculate the weight of this landing craft when on Earth. *(2 marks)*

(Q26f, **2014 MS2 HSC**) Medium

39 The diagram shows the line ℓ.

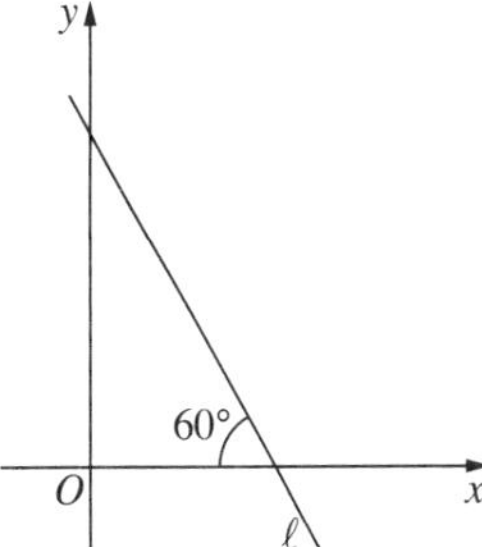

What is the slope of the line ℓ?

A $\sqrt{3}$ **B** $-\sqrt{3}$

C $\frac{1}{\sqrt{3}}$ **D** $-\frac{1}{\sqrt{3}}$ *(1 mark)*

(Q2, **2013 HSC**) Medium

40 Wind turbines, such as those shown, are used to generate power.

In theory, the power that could be generated by a wind turbine is modelled using the equation

$$T = 20000w^3$$

where
T is the theoretical power generated, in watts
w is the speed of the wind, in metres per second.

i Using this equation, what is the theoretical power generated by a wind turbine if the wind speed is 7.3 m/s? *(1 mark)* Easy

ii In practice, the actual power generated by a wind turbine is only 40% of the theoretical power.
If A is the actual power generated, in watts, write an equation for A in terms of w. *(1 mark)* Easy

The graph shows both the theoretical power generated and the actual power generated by a particular wind turbine.

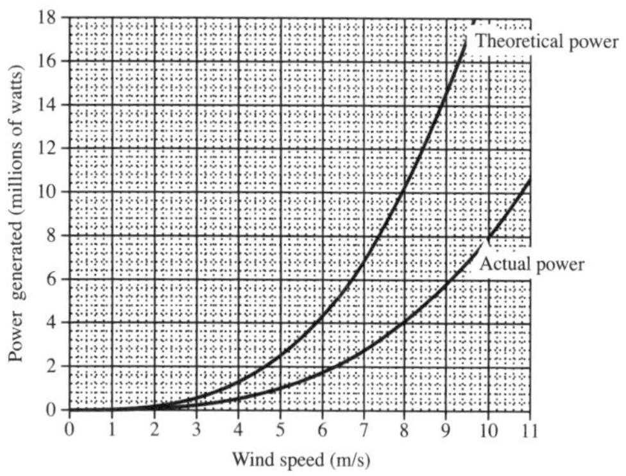

iii Using the graph, or otherwise, find the difference between the theoretical power and the actual power generated when the wind speed is 9 m/s. *(1 mark)* Easy

iv A particular farm requires at least 4.4 million watts of actual power in order to be self-sufficient.
What is the minimum wind speed required for the farm to be self-sufficient? *(1 mark)* Easy

v A more accurate formula to calculate the power (P) generated by a wind turbine is

$$P = 0.61 \times \pi \times r^2 \times w^3$$

where
r is the length of each blade, in metres
w is the speed of the wind, in metres per second.

Each blade of a particular wind turbine has a length of 43 metres. The turbine operates at a wind speed of 8 m/s.

Using the formula above, if the wind speed increased by 10%, what would be the percentage increase in the power generated by this wind turbine? *(3 marks)* Medium

(Q30a, **2013 MS2 HSC**)

---

41 The line ℓ has intercepts p and q, where p and q are positive integers.

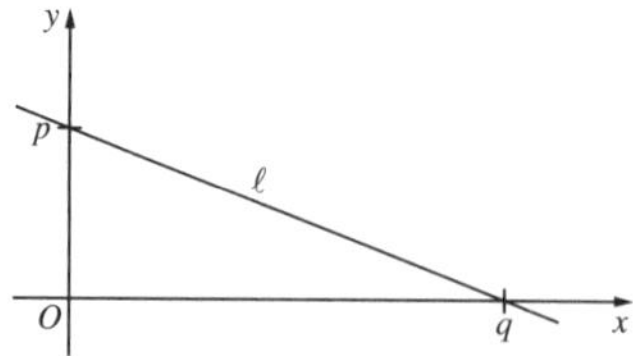

What is the gradient of line ℓ?

A $-\dfrac{p}{q}$ **B** $-\dfrac{q}{p}$

C $\dfrac{p}{q}$ **D** $\dfrac{q}{p}$ *(1 mark)*

(Q5, **2012 MS2 HSC**) Easy

---

42 Sketch the graph of $y - 2x = 3$, showing the intercepts on both axes. *(2 marks)*

(Q1a, **2009 HSC**) Easy

---

43 Find the values of k for which the quadratic equation $x^2 - (k + 4)x + (k + 7) = 0$ has equal roots. *(3 marks)*

(Q4b, **2009 HSC**) Medium

---

44 In the diagram, the points A and C lie on the y-axis and the point B lies on the x-axis. The line AB has equation $y = \sqrt{3}x - 3$. The line BC is perpendicular to AB.

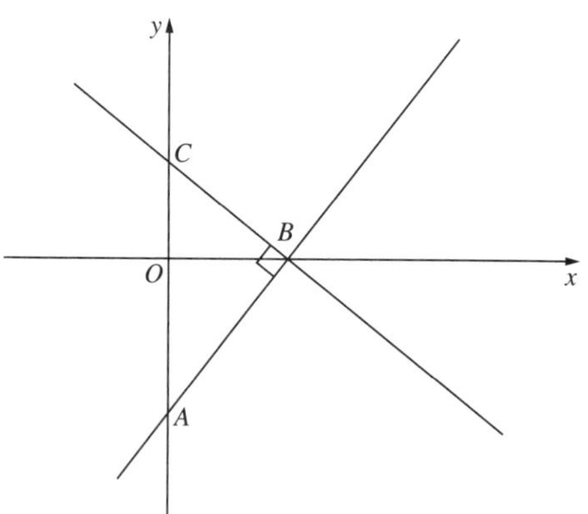

i Find the equation of the line BC. *(2 marks)* Easy

ii Find the area of the triangle ABC. *(2 marks)* Medium

(Q5a, **2009 HSC**)

---

45 Let M be the midpoint of $(-1, 4)$ and $(5, 8)$.

Find the equation of the line through M with gradient $-\dfrac{1}{2}$. *(2 marks)*

(Q2b, **2008 HSC**) Easy

---

46 In the diagram, $ABCD$ is a quadrilateral. The equation of the line AD is $2x - y - 1 = 0$.

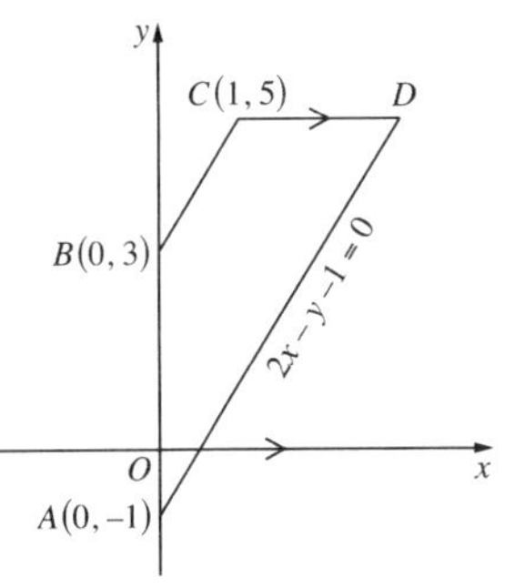

i Show that $ABCD$ is a trapezium by showing that BC is parallel to AD. *(2 marks)* **Medium**

ii The line CD is parallel to the x-axis. Find the coordinates of D. *(1 mark)* **Easy**

(Q3a, **2008 HSC**)

47 Find the equation of the line that passes through the point $(-1, 3)$ and is perpendicular to $2x + y + 4 = 0$. *(2 marks)*

(Q1f, **2007 HSC**) **Easy**

48 In the diagram, A, B and C are the points $(6, 0)$, $(9, 0)$ and $(12, 6)$ respectively. The equation of the line OC is $x - 2y = 0$. The point D on OC is chosen so that AD is parallel to BC. The point E on BC is chosen so that DE is parallel to the x-axis.

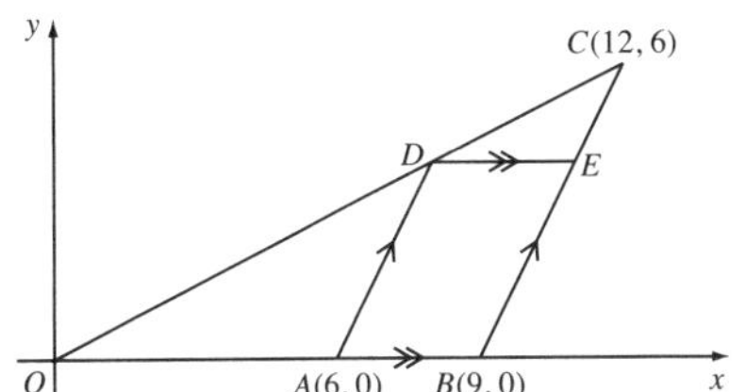

i Show that the equation of the line AD is $y = 2x - 12$. *(2 marks)* **Medium**

ii Find the coordinates of the point D. *(2 marks)* **Medium**

iii Find the coordinates of the point E. *(1 mark)* **Easy**

(Q3c, **2005 HSC**)

49 Sandra is on holiday in Mexico and she plans to buy some silver jewellery. Various silver pendants are on sale. The cost varies directly with the square of the length of the pendant. A pendant of length 30 mm costs 130 Mexican pesos.

How much does a pendant of length 40 mm cost? (Answer correct to the nearest Mexican peso.) *(2 marks)*

(Q28a, **2003 MS2 HSC**) **Medium**

50 Sketch the line $y = 2x + 3$ in the Cartesian plane. *(2 marks)*

(Q1g, **2000 HSC**) **Easy**

51 **i** Write down the discriminant of $3x^2 + 2x + k$. **Easy**

ii For what values of k does $3x^2 + 2x + k = 0$ have real roots? **Easy**

(2 marks) (Q7a, **1998 HSC**)

52 Let A and B be the points $(0, 1)$ and $(2, 3)$ respectively.

Find the slope of the line AB.

(Q3b ii, **1997 HSC**) **Easy**

53 The line ℓ is shown in the diagram. It has equation $x + 2y + 8 = 0$ and cuts the x axis at A.

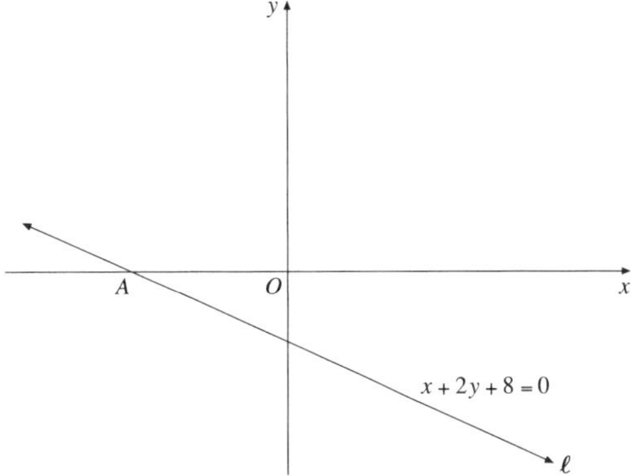

The line k has equation $y = -\frac{1}{2}x + 6$, and is not shown on the diagram.

Copy or trace the diagram into your Writing Booklet.

a Find the coordinates of A. *(1 mark)* **Easy**

b Explain why k is parallel to ℓ. *(1 mark)* **Easy**

c Draw the graph of k on your diagram, indicating where it cuts the axes. *(2 marks)* **Easy**

(Q2, **1996 HSC**)

54 The line $6x - ky = 4$ passes through the point $(3, 2)$. Find the value of k.

(Q1b, **1993 HSC**) **Easy**

1 $y = -2x + 2$
As y-intercept = 2, possibly A or D.
As gradient = $-2 < 0$, must be option A.
Answer A

2 $y = -x^2 + 1$ is a parabola.
It passes through $(-1, 0)$, $(0, 1)$ and $(1, 0)$.

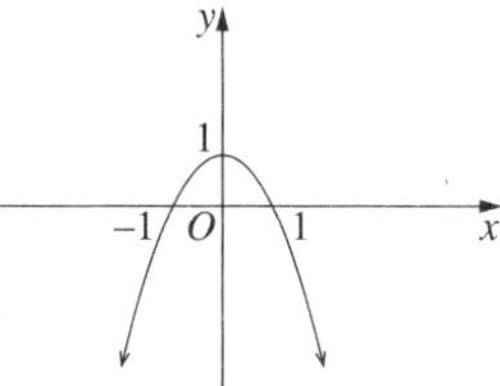

Answer C

3 $y = -x^2 + bx + 1$ is in the form $y = ax^2 + bx + c$.
As $a = -1 < 0$, the graph is concave down (could be C or D).
Axis of symmetry is

$$x = -\frac{b}{2a} = -\frac{b}{-2} = \frac{b}{2}, \text{ and}$$

$b > 0$. This means the axis of symmetry is to the right of the y-axis.
Answer C

4 **a**

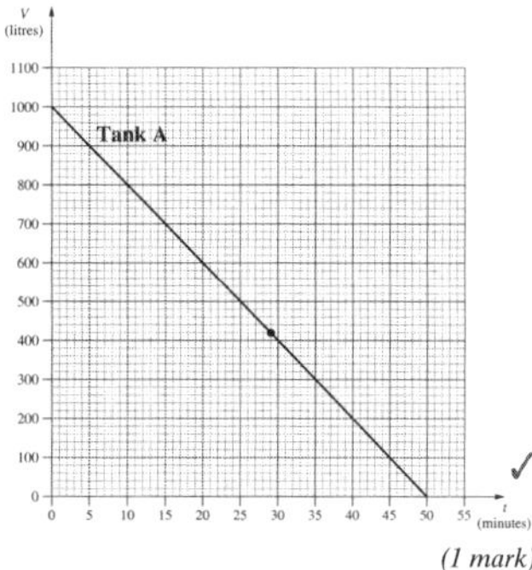

✓ *(1 mark)*

b

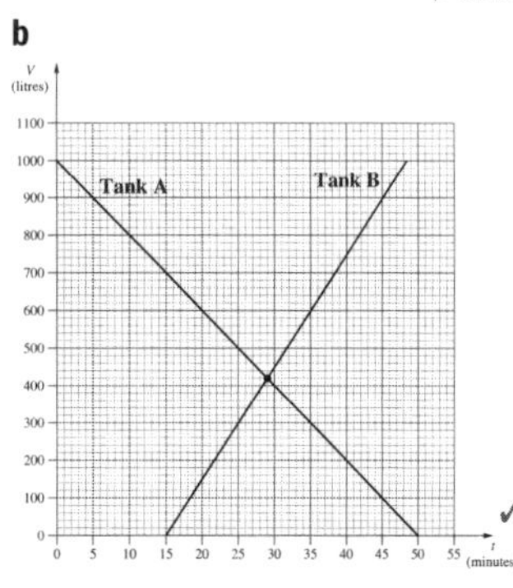

✓

Look for the intersection of both graphs. After 29 minutes the tanks contain the same amount.
Alternatively, solve simultaneously the equations $V = 1000 - 20t$ and $V = 30(t - 15)$.

$$1000 - 20t = 30(t - 15)$$
$$1000 - 20t = 30t - 450$$
$$1000 + 450 = 30t + 20t$$
$$50t = 1450$$
$$t = 29$$ ✓

(2 marks)

c By inspection, after 45 minutes, Tank A contains 100 L and Tank B contains 900 L, which is a total of 1000 L.
Alternatively, Tank A is modelled by the equation $V_A = 1000 - 20t$.
Tank B is modelled by the equation $V_B = 30(t - 15)$.
Now solve

$$1000 - 20t + 30(t - 15) = 1000$$
$$1000 - 20t + 30t - 450 = 1000$$
$$10t = 450$$
$$t = 45$$ ✓

(1 mark)

5 Depreciation $= 0.2 \times 1850 = 370$
$\therefore y = 1850 - 370x$ ✓
As $\frac{1850}{370} = 5$, then domain is $0 \le x \le 5$, and range is $0 \le y \le 1850$.
Function is $y = 1850 - 370x$, with domain $0 \le x \le 5$ and range $0 \le y \le 1850$. ✓

(2 marks)

6 $\Delta = b^2 - 4ac$

$$= (-3)^2 - 4(1)(k + 2)$$
$$= 9 - 4k - 8$$
$$= 1 - 4k$$ ✓

When $\Delta < 0$,

$$1 - 4k < 0$$
$$4k > 1$$
$$k > \frac{1}{4}$$ ✓

As $a > 0$ and $\Delta < 0$, then $k > \frac{1}{4}$.

(2 marks)

7 Using $x = -\frac{b}{2a}$:

$$x = -\frac{4}{2(-2)} = 1$$
$$y(1) = 5 + 4(1) - 2(1)^2 = 7$$

Maximum value of 7. ✓

(1 mark)

8 Gradient of $3x - 2y = 12$:

$$2y = 3x - 12$$
$$y = \frac{3x - 12}{2}$$

$\therefore$ gradient $= \frac{3}{2}$
$\therefore$ gradient of perpendicular $= -\frac{2}{3}$ ✓

Now, find equation of PQ, using $m = -\frac{2}{3}$ and passing through $(1, -3)$:

$$y + 3 = -\frac{2}{3}(x - 1)$$ ✓

Substitute $x = -4$:

$$y + 3 = -\frac{2}{3}(-4 - 1)$$
$$y + 3 = -\frac{2}{3}(-5)$$
$$3y + 9 = 10$$
$$3y = 1$$
$$y = \frac{1}{3}$$

The coordinates of P are $(-4, \frac{1}{3})$. ✓

(3 marks)

9

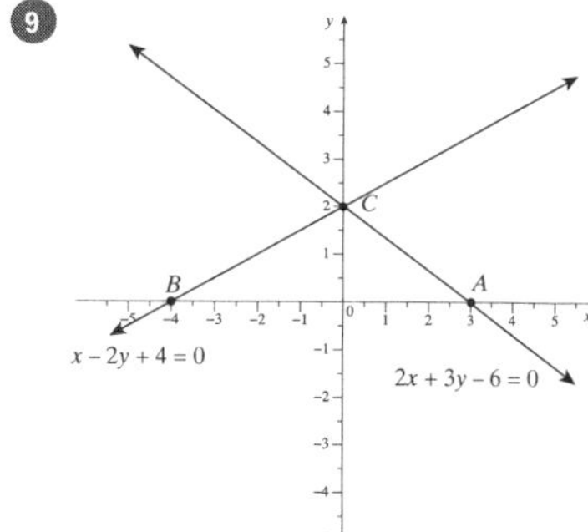

✓

Substitute $y = 0$ in $2x + 3y - 6 = 0$. $\therefore A(3, 0)$
Substitute $y = 0$ in $x - 2y + 4 = 0$. $\therefore B(-4, 0)$
Solving simultaneously, $C(0, 2)$. ✓

$$\text{Area} = \frac{1}{2} \times 7 \times 2 = 7$$

$\therefore$ area is 7 units^2. ✓

(3 marks)

10 **i** From $x - 2y + 4 = 0$,

$$y = \frac{x + 4}{2}$$

$\therefore$ gradient of AB is $\frac{1}{2}$.

Hence gradient of $AC = -2$. ✓
Equation of AC:

$$y - 6 = -2(x - 8)$$
$$y - 6 = -2x + 16$$
$$2x + y - 22 = 0$$ ✓

(2 marks)

ii Substitute $y = 0$ in
$x - 2y + 4 = 0$:
$x = -4$
$\therefore B(-4, 0)$
Substitute $y = 0$ in
$2x + y - 22 = 0$:
$2x = 22$
$x = 11$ ✓
$\therefore C(11, 0)$
$\therefore BC = 15$ units, and height is 6 units:
$\therefore$ Area $= \frac{1}{2} \times 15 \times 6$
$= 45$
$\therefore$ area is 45 units^2. ✓
(2 marks)

11 Luke has 165 followers plus 3 per day.
So for Luke $y = 165 + 3x$
Rhys has 537 followers less 2 per day.
So for Rhys $y = 537 - 2x$
Answer B

12 a w is the width of the rectangle. The width cannot be negative so the model only makes sense for non-negative values of w. ✓
(1 mark)

b From the graph, when $A = 18, w = 2$.
Now, Area = length × width
So $l \times 2 = 18$
$l = 9$ ✓
$P = 2(l + w)$
$= 2 \times (9 + 2)$
$= 22$
The perimeter of the rectangle is 22 cm. ✓
(2 marks)

13 a $m = \frac{4}{7}$
So $p = \frac{4}{7}d$ ✓
(1 mark)

b $y = 76d$
When $y = 93\,100$
$76d = 93\,100$
$d = 1225$ ✓
When $d = 1225$
$p = \frac{4}{7} \times 1225$
$= 700$
So 93 100 yen is 700 British pounds. ✓ *(2 marks)*

14 a 20 bird houses need to be sold to break even. ✓
(1 mark)

b Revenue:
y-intercept = 0
Gradient $= \frac{800}{20}$
$= 40$
R: $y = 40x$ ✓
Cost: y-intercept = 500
Gradient $= \frac{300}{20}$
$= 15$
C: $y = 15x + 500$ ✓
Now profit
= revenue – cost
$= 40x - (15x + 500)$
$= 40x - 15x - 500$
$= 25x - 500$
When the profit is \$1900,
$25x - 500 = 1900$
$25x = 2400$
$x = 96$
So 96 bird houses need to be sold to earn a profit of \$1900. ✓
(3 marks)

15 Let $y = 0$:
$x + 3(0) + 6 = 0$
$x = -6$
$\therefore (-6, 0)$
Answer A

16 $y = x^2 - 2$ is concave up and has y-intercept -2.

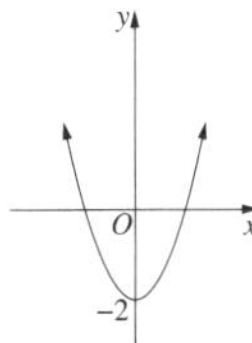

Answer A

17 i Rate for company A
= \$360 ÷ 3
= \$120 per hour ✓
(1 mark)

ii $m = \frac{\text{vertical change in position}}{\text{horizontal change in position}}$
$= \frac{360 - 75}{3}$
$= 95$ ✓
$b = 75$
$c = mx + b$
$c = 95x + 75$ ✓
(2 marks)

iii Cost from company B:
$c = 95x + 75$
When $x = 5$,
$c = 95 \times 5 + 75$
$= 550$
It costs \$550 to hire the minibus from company B. ✓
Cost from company A
$= 5 \times \$120$
$= \$600$
So it is \$50 cheaper to hire the minibus from company B than from company A. ✓
(2 marks)

18 $2x + 3y + 4 = 0$
$3y = -2x - 4$
$y = \frac{-2x - 4}{3}$
$\therefore m = -\frac{2}{3}$
Answer A

19 At the point of intersection:
$x + 3 = 6 - 2x$
$3x + 3 = 6$
$3x = 3$
$x = 1$
When $x = 1, y = 1 + 3 = 4$.
The point of intersection is $(1, 4)$.

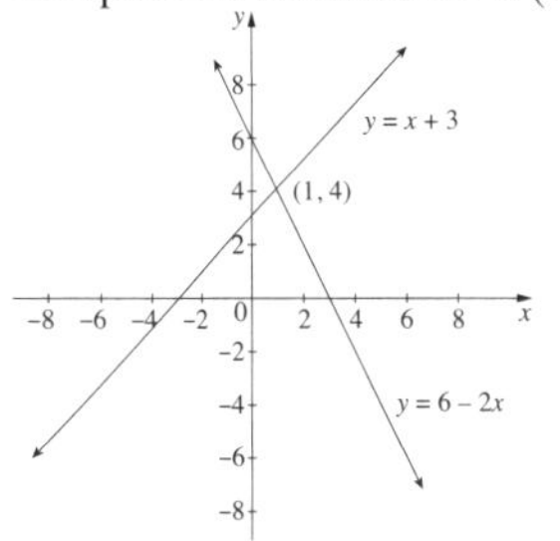

Answer B

20 i $F = \frac{9C}{5} + 32$
When $C = -20$
$F = \frac{9 \times -20}{5} + 32$
$= -4$
The temperature is –4 °F. ✓
(1 mark)

ii $F = \frac{9C}{5} + 32$ (I)
$F = C$ (II)
[Substitution method]
Substitute (II) into (I):
$C = \frac{9C}{5} + 32$
$5C = 9C + 160$
$-4C = 160$
$C = -40$ ✓
Now $F = C$ from (II)
So $F = C = -40$ ✓
[Or: Elimination method:
(I) – (II)
$0 = \frac{4C}{5} + 32$
$0 = 4C + 160$
$-4C = 160$
$C = -40$
Now $F = C$ from (II)
So $F = C = -40$]
(2 marks)

iii The point of intersection of the two lines on the graph will be at (–40, –40). ✓ *(1 mark)*

21 Let $C be the cost and h cm be the height.

Now $C \propto h^3$

i.e. $C = kh^3$ (k constant)

When $h = 10$, $C = 50$

$50 = k \times 10^3$

$k = 50 \div 10^3$

$k = 0.05$

$\therefore C = 0.05h^3$ ✓

When $h = 12$,

$C = 0.05 \times 12^3$

$= 86.4$

The cost of a jewellery box of height 12 cm is $86.40. ✓

(2 marks)

22 $y = (x-2)^2$ has an axis of symmetry $x = 2$, and $y = 3-(x-2)^2$ is concave down.

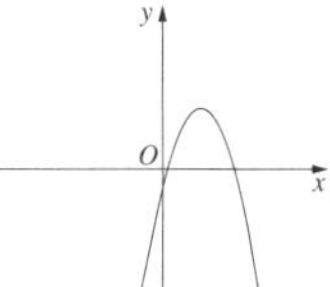

Answer B

23 $y = -5-4x$ ... ①

$y = 3-2x-x^2$ ... ②

Let ① = ②:

$-5-4x = 3-2x-x^2$

$x^2-2x-8 = 0$

$(x-4)(x+2) = 0$

$x = 4, -2$ ✓✓

Subs. in ①:

$y(4) = -5-4(4)$

$= -21$

$y(-2) = -5-4(-2)$

$= 3$

$\therefore (4, -21)$ and $(-2, 3)$ ✓

(3 marks)

24 The profit zone is X. [where the income is greater than the cost]

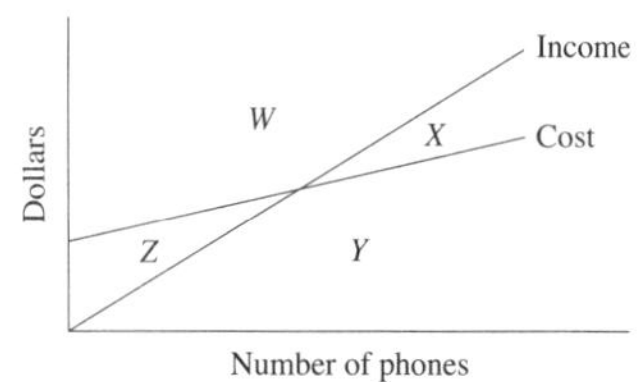

Answer B

25 The gradient, m, is positive.
The y-intercept, b, is negative.

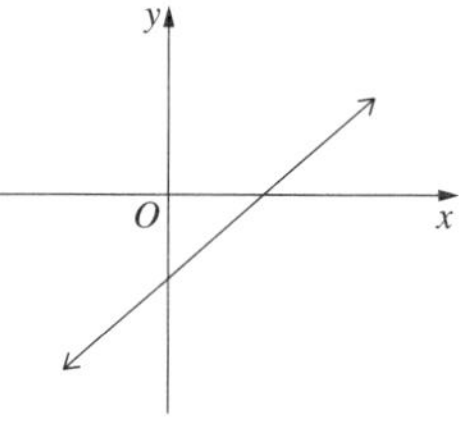

Answer A

26 $E = kS^2$

When $E = 20$, $S = 10$

$20 = k \times 10^2$

$= 100k$

$k = 20 \div 100$

$= 0.2$

$\therefore E = 0.2S^2$

When $S = 40$

$E = 0.2 \times 40^2$

$= 320$

Answer C

27 $2x-4y+3 = 0$

$4y = 2x+3$

$y = \frac{2x}{4} + \frac{3}{4}$

$y = \frac{x}{2} + \frac{3}{4}$

$\therefore$ slope is $\frac{1}{2}$.

Answer C

28 i For ℓ_2:

$y = -\frac{x}{3}$, gradient is $-\frac{1}{3}$.

As diagonals of rhombus are perpendicular, then gradient of ℓ_1 is 3. ✓

For line ℓ_1, using (7, 11) and $m = 3$:

$y-y_1 = m(x-x_1)$

$y-11 = 3(x-7)$

$y-11 = 3x-21$

$y = 3x-10$ ✓

(2 marks)

ii $y = 3x-10$ ..... ①

$y = -\frac{x}{3}$ .... ②

Let ① = ②: $3x-10 = -\frac{x}{3}$

$9x-30 = -x$

$10x = 30$

$x = 3$ ✓

Subs. in ②: $y = -\frac{3}{3}$

$= -1$

$\therefore D(3, -1)$ ✓

(2 marks)

29 If $x^2-8x+k = 0$,

$\Delta = b^2-4ac$

$= (-8)^2-4 \times 1 \times k$

$= 64-4k$ ✓

$\Delta \geq 0$ for real roots.

$64-4k \geq 0$

$4k \leq 64$

$k \leq 16$ ✓

(2 marks)

30 $y = x^3+1$ is cubic curve with y-intercept 1

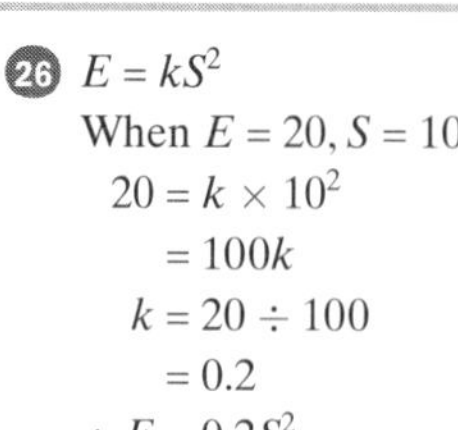

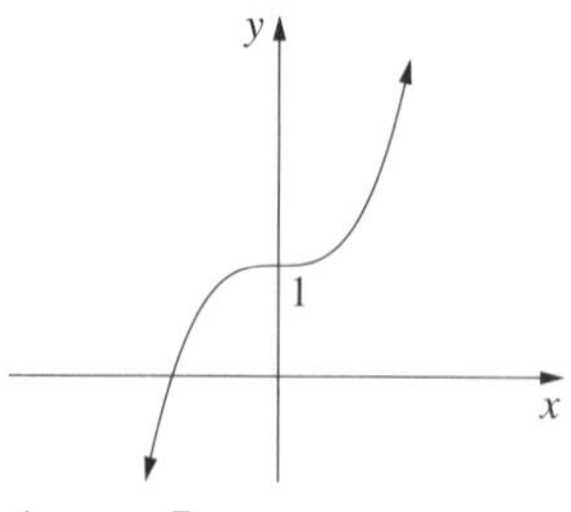

Answer B

31 gradient $= \frac{-2}{1} = -2$

y-intercept $= 2$

Equation is $y = -2x+2$

Answer A

32 i

$x+18y = 1510$ (1)

$x+36y = 2770$ (2)

(2) – (1) $18y = 1260$

$y = 70$ ✓

Substitute into (1):

$x+1260 = 1510$

$x = 250$

So $x = 250$ and $y = 70$ ✓

(2 marks)

ii Let m be the number of months to repay the loan.

Now $250+70m = 4800$ ✓

$70m = 4550$

$m = 65$

It will take Ariana 65 months to repay the loan in full. ✓

(2 marks)

33 The curve should cut the y-axis at the height of the diving board. ✓ The difference between the maximum height and the height at which the curve cuts the y-axis is how high the diver is above the height of the diving board when he reached the maximum height. ✓

(2 marks)

34 $y = (x-1)^2$

Axis: $x = 1$

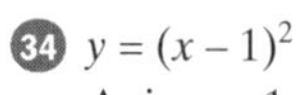

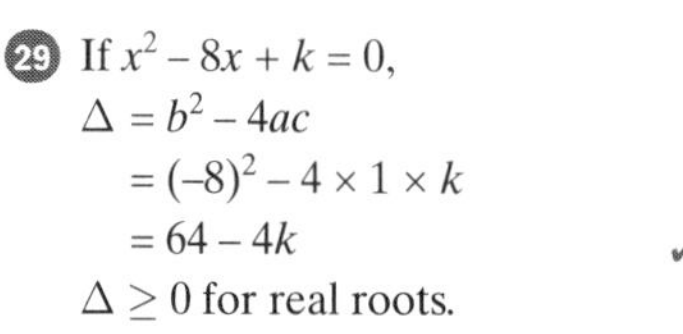

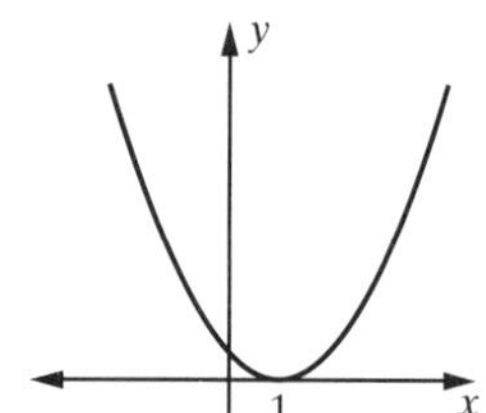

Answer B

35 $2x - 3y = 8$

$$3y = 2x - 8$$
$$y = \frac{2x - 8}{3}$$

$\therefore$ gradient of perpendicular is $\frac{-3}{2}$.

$$y - 0 = \frac{-3}{2}(x - 2)$$
$$2y = -3x + 6$$
$$3x + 2y = 6$$

Answer B

36 The graph is a parabola. The only equation of a parabola is $y = 3x^2 + 1$.

Answer C

37 $y = 2x - 2$ has y-intercept -2.

[So the graph could only be A or D.]

The gradient is 2.

So the graph can only be D.

[A has a negative gradient.]

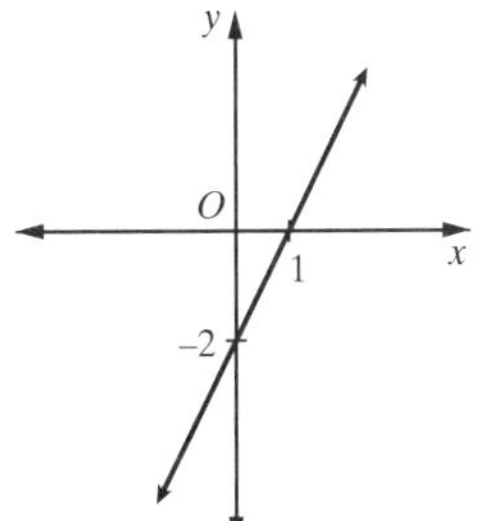

Answer D

38 Let m be the weight on the moon and e the weight on earth.

Now $m \propto e$

$$m = ke$$

When $e = 84, m = 14$

$$14 = k \times 84$$
$$k = \frac{14}{84}$$
$$= \frac{1}{6}$$

So $m = \frac{1}{6}e$ ✓

When $m = 2449$,

$$2449 = \frac{1}{6}e$$
$$e = 2449 \times 6$$
$$= 14\,694$$

On Earth, the weight of the landing craft would be 14 694 kg. ✓

(2 marks)

39

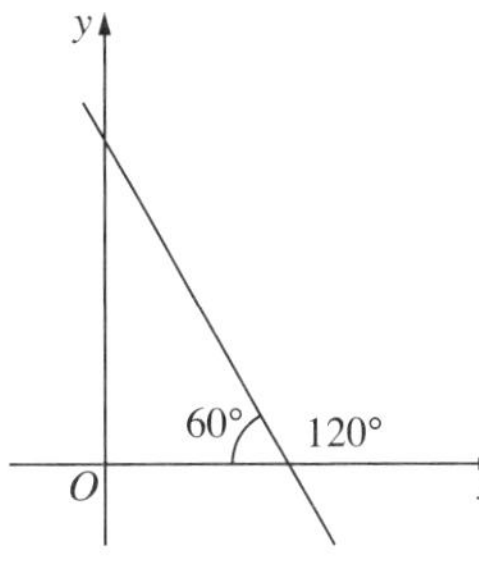

The line forms an angle of 120° with the positive direction of the x-axis:

$$\therefore m = \tan\theta$$
$$= \tan 120°$$
$$= -\tan 60°$$
$$= -\sqrt{3}$$

Answer B

40 **i** $T = 20\,000w^3$

When $w = 7.3$,

$$T = 20\,000 \times 7.3^3$$
$$= 7\,780\,340$$

The theoretical power generated is 7 780 340 watts. ✓

(1 mark)

ii $A = 40\%$ of T

$$= 0.4 \times 20\,000w^3$$
$$= 8000w^3$$ ✓

(1 mark)

iii From the graph, when the wind speed is 9 m/s the actual power is 5.8 million watts and the theoretical power is 14.6 million watts.

Now, $14.6 - 5.8 = 8.8$

So the difference is 8.8 million watts. ✓

(1 mark)

iv From the graph, to produce 4.4 million watts of actual power the wind speed needs to be 8.2 m/s.

So the minimum wind speed for the farm to be self-sufficient is 8.2 m/s. ✓

(1 mark)

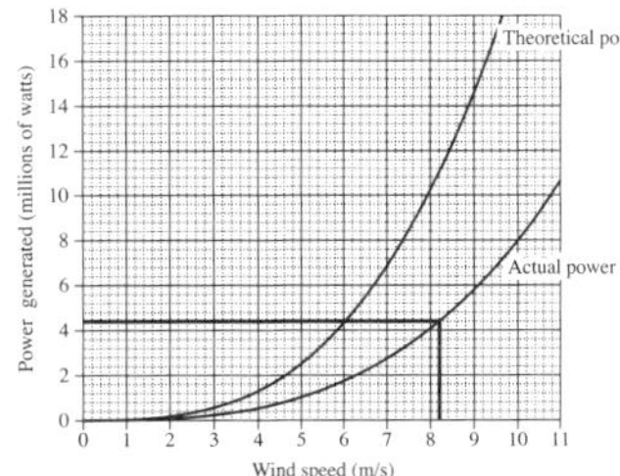

v $P = 0.61 \times \pi \times r^2 \times w^3$

When $r = 43$ and $w = 8$,

$$P = 0.61 \times \pi \times 43^2 \times 8^3$$
$$= 1\,814\,205.92\ldots$$ ✓

If w increases by 10%,

$$\text{new } w = 1.1 \times 8$$
$$= 8.8$$

$$P = 0.61 \times \pi \times 43^2 \times 8.8^3$$
$$= 2\,414\,708.08\ldots$$ ✓

Increase

$$= 2\,414\,708.08\ldots - 1\,814\,05.92\ldots$$
$$= 600\,502.1596\ldots$$

% increase

$$= \frac{600\,502.1596\ldots}{1\,814\,205.92\ldots} \times 100\%$$
$$= 33.1\%$$

The percentage increase in power would be 33.1%. ✓

(3 marks)

41

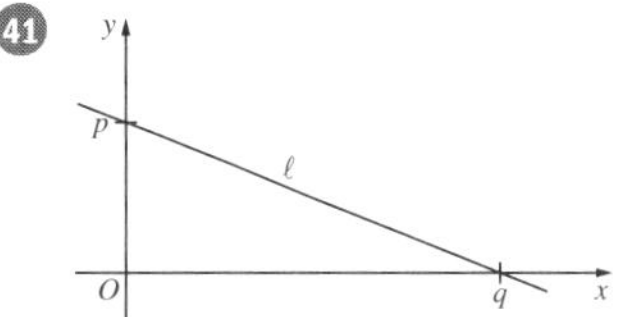

gradient

$$= \frac{\text{vertical change in position}}{\text{horizontal change in position}}$$
$$= \frac{-p}{q}$$
$$= -\frac{p}{q}$$

Answer A

42 $y - 2x = 3$

$\therefore y = 2x + 3$

When $x = 0, y = 3$

$\therefore$ y-intercept of $y - 2x = 3$ is $(0, 3)$.

When $y = 0$, $0 = 2x + 3$

$$2x = -3$$
$$x = -\frac{3}{2}$$

$\therefore$ x-intercept of $y - 2x = 3$

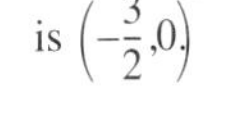

is $\left(-\frac{3}{2}, 0\right)$ ✓

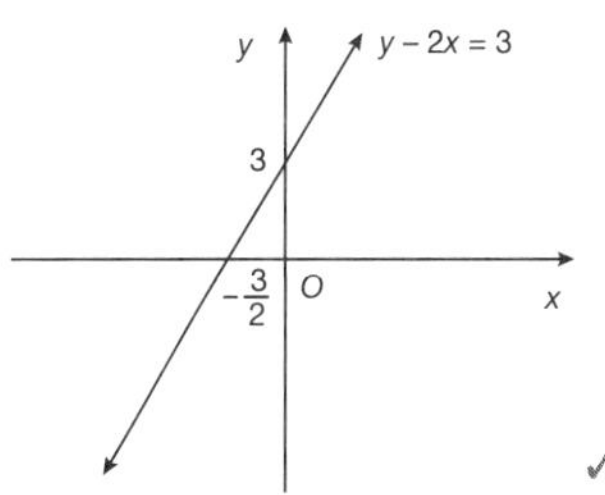

✓

(2 marks)

43 $x^2 - (k+4)x + (k+7) = 0$

$\Delta = b^2 - 4ac$
$= [-(k+4)]^2 - 4(1)(k+7)$
$= k^2 + 8k + 16 - 4k - 28$
$= k^2 + 4k - 12$ ✓

For equal roots, $\Delta = 0$.
$\therefore k^2 + 4k - 12 = 0$
$(k+6)(k-2) = 0$
$\therefore k = -6$ or $k = 2$. ✓✓

(3 marks)

44

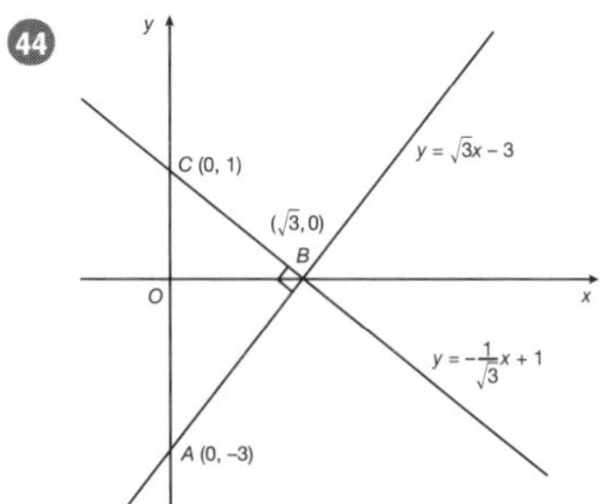

i Equation of line AB is
$y = \sqrt{3}x - 3$
$\therefore$ Gradient of $AB = \sqrt{3}$

$\therefore$ Gradient of $BC = -\dfrac{1}{\sqrt{3}}$ ✓
$(AB \perp BC)$

Line AB cuts the x-axis at B where $y = 0$.
$\therefore 0 = \sqrt{3}x - 3$
$x = \dfrac{3}{\sqrt{3}}$
$= \sqrt{3}$

$\therefore$ Coordinates of B are $(\sqrt{3}, 0)$.

Equation of line BC, with gradient $= -\dfrac{1}{\sqrt{3}}$, passing through $B(\sqrt{3}, 0)$ is given by
$y - y_1 = m(x - x_1)$
$y - 0 = -\dfrac{1}{\sqrt{3}}(x - \sqrt{3})$
$y = -\dfrac{1}{\sqrt{3}}x + 1$ ✓

(2 marks)

ii Line AB cuts the y-axis at A where $x = 0$,
$\therefore y = \sqrt{3}(0) - 3$
$= -3$
$\therefore$ Coordinates of A are $(0, -3)$.

Line BC cuts the y-axis at C where $x = 0$

$\therefore y = -\dfrac{1}{\sqrt{3}}(0) + 1$
$= 1$

$\therefore C$ has coordinates $(0, 1)$. ✓

Area of triangle ABC
$= \dfrac{1}{2} \times b \times h$
$= \dfrac{1}{2} \times AC \times OB$
$= \dfrac{1}{2} \times 4 \times \sqrt{3}$
$= 2\sqrt{3}$ units^2 ✓

(2 marks)

45 $M = \left(\dfrac{x_1 + x_2}{2}, \dfrac{y_1 + y_2}{2}\right)$
$= \left(\dfrac{-1+5}{2}, \dfrac{4+8}{2}\right)$
$= (2, 6)$ ✓

Equation of the line with gradient $= -\dfrac{1}{2}$, through the point $(2, 6)$ is given by

$y - 6 = -\dfrac{1}{2}(x - 2)$
$2y - 12 = -x + 2$
i.e. $x + 2y - 14 = 0$ ✓

(2 marks)

46

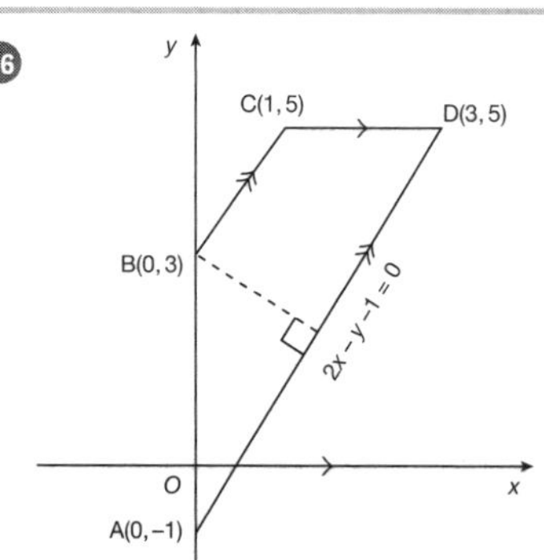

i Gradient of $BC = \dfrac{y_2 - y_1}{x_2 - x_1}$
$= \dfrac{5-3}{1-0}$
$= 2$

Gradient of $AD = \dfrac{-A}{B}$
$= \dfrac{-2}{-1}$
$= 2$ ✓

Since $m_{BC} = m_{AD}$, $BC \parallel AD$
$\therefore ABCD$ is a trapezium with a pair of opposite sides parallel. ✓

(2 marks)

ii Since CD is parallel to the x-axis, at D, $y = 5$.
Now D lies on $2x - y - 1 = 0$
$\therefore 2x - 5 - 1 = 0$
$2x - 6 = 0$
$2x = 6$
$\therefore x = 3$

$\therefore D$ is the point $(3,5)$. ✓

(1 mark)

47 $2x + y + 4 = 0$
$y = -2x - 4$
$\therefore m_1 = -2$

Now $m_1 \times m_2 = -1$
$\therefore -2 \times m_2 = -1$
$\therefore m_2 = \dfrac{1}{2}$ ✓

$\therefore$ Equation of the line passing through $(-1, 3)$ and perpendicular to $2x + y + 4 = 0$ is given by
$y - y_1 = m(x - x_1)$
$\therefore y - 3 = \dfrac{1}{2}(x - (-1))$
$y - 3 = \dfrac{1}{2}x + \dfrac{1}{2}$
$y = \dfrac{1}{2}x + 3\dfrac{1}{2}$
$2y = x + 7$
$\therefore x - 2y + 7 = 0$ ✓

(2 marks)

48

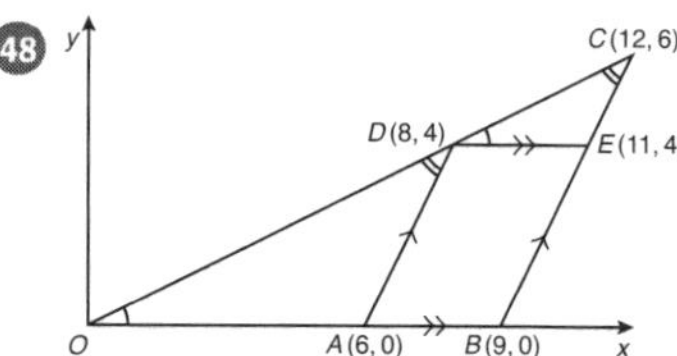

i Gradient of $BC = \dfrac{y_2 - y_1}{x_2 - x_1}$
$= \dfrac{6-0}{12-9}$
$= \dfrac{6}{3}$
$= 2$ ✓

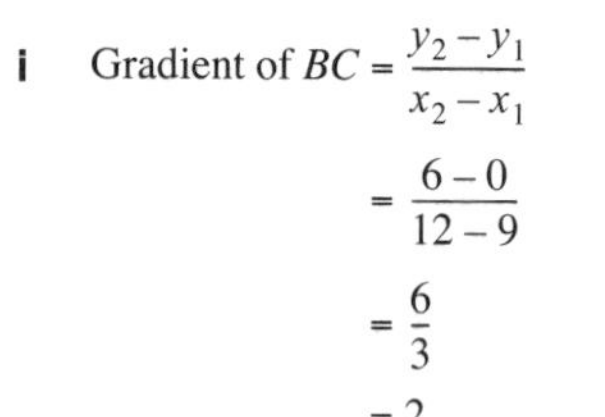

Now $m_{AD} = m_{BC}$
$\therefore m_{AD} = 2$

$\therefore$ Equation of line AD with gradient $= 2$, passing through $A(6, 0)$ is given by
$y - 0 = 2(x - 6)$
$\therefore y = 2x - 12$ ✓

(2 marks)

ii D is the point of intersection of

$x - 2y = 0 \quad \ldots (1)$

$y = 2x - 12 \quad \ldots (2)$

Solving simultaneously:

Substituting (2) into (1),

$x - 2(2x - 12) = 0$

$x - 4x + 24 = 0$

$-3x = -24$

$\therefore x = 8$ ✓

$\therefore y = 2(8) - 12$

$= 4$

$\therefore$ Coordinates of D are $(8, 4)$. ✓

(2 marks)

iii Coordinates of E are $(11, 4)$.

(Since $AB = 3$, $DE = 3$) ✓

(1 mark)

49 Let C represent cost and L represent length.

$\therefore C = kL^2$

$L = 30$ mm, $C = 130$

$L = 40$ mm, $C = ?$

$\therefore 130 = k(30)^2$

$\therefore \frac{130}{30^2} = k$ ✓

If $L = 40$ mm,

$C = \frac{130}{30^2} \times 40^2$

$= 231.111\ldots$

$= 231$ pesos to nearest peso. ✓

(2 marks)

50 Gradient $= 2$,

y-intercept is 3,

x-intercept is $-1\frac{1}{2}$. ✓

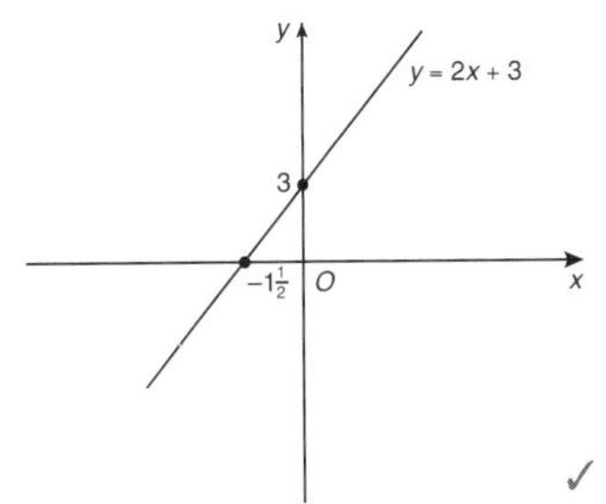

✓

(2 marks)

51 **i** $\Delta = b^2 - 4ac$

$= (2)^2 - 4(3)(k)$

$= 4 - 12k$ ✓

ii Real roots occur when

$\Delta \geqslant 0$

$\therefore 4 - 12k \geqslant 0$

$4 \geqslant 12k$

$k \leqslant \frac{1}{3}$ ✓

(2 marks)

(Total mark allocation only in exam)

52 Slope of line AB, $m_1 = \frac{3 - 1}{2 - 0}$

$= \frac{2}{2}$

$= 1$

(No mark allocation in exam)

53 **a**

At A, $y = 0$

$\therefore x + 0 + 8 = 0$

$\therefore x = -8$

$\therefore$ Coordinates of A are $(-8, 0)$ ✓

(1 mark)

b Line k: $y = -\frac{1}{2}x + 6$

Gradient $= m_1 = -\frac{1}{2}$

Line l: $x + 2y + 8 = 0$

i.e. $2y = -x - 8$

$y = -\frac{1}{2}x - 4$

Gradient $= m_2 = -\frac{1}{2}$

Since $m_1 = m_2$, the lines are parallel. ✓

(1 mark)

c

$y = -\frac{1}{2}x + 6$ ✓✓

$\therefore m_1 = -\frac{1}{2}$, y-intercept is 6.

(2 marks)

54 $6x - ky = 4$

Substituting $(3, 2)$:

$6(3) - k(2) = 4$

$18 - 2k = 4$

$-2k = -14$

$k = 7$

(No mark allocation in exam)

1 A student believes that the time it takes for an ice cube to melt (M minutes) varies inversely with the room temperature (T °C). The student observes that at a room temperature of 15 °C it takes 12 minutes for an ice cube to melt.

a Find the equation relating M and T. *(2 marks)* Easy

b By first completing this table of values, graph the relationship between temperature and time from $T = 5$ °C to $T = 30$ °C.

T	5	15	30
M			

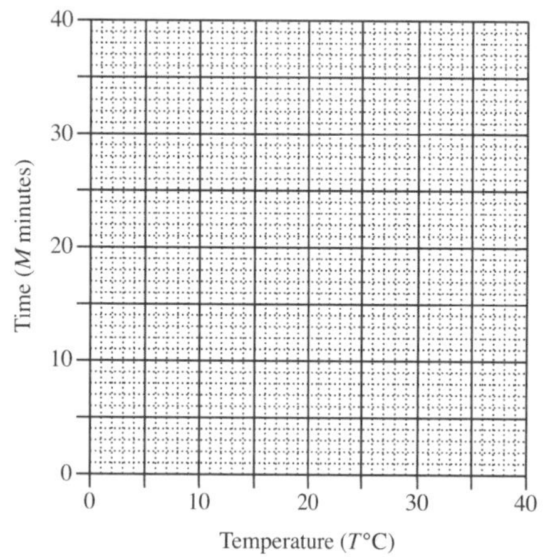

(2 marks) Easy

(Q12, **2022 HSC**)

2 The graph of $y = f(x)$ is shown.

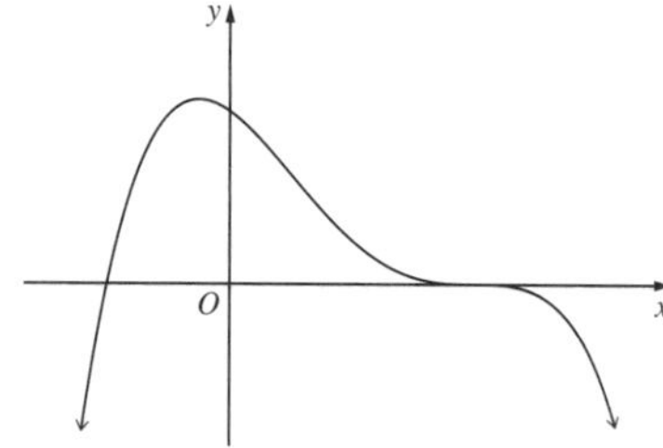

Which of the following could be the equation of this graph?

A $y = (1 - x)(2 + x)^3$ **B** $y = (x + 1)(x - 2)^3$
C $y = (x + 1)(2 - x)^3$ **D** $y = (x - 1)(2 + x)^3$

(Q8, **2021 HSC**) Medium

3 Logan's dentist determined the number of cavities in his patient's mouth each year is inversely proportional to the total number of seconds spent brushing during each session.

Logan developed 6 cavities during the year in which he spent 20 seconds brushing each time.

How many cavities will Logan develop annually if he increases his brushing time to a minute each time? *(2 marks)*

Bonus question (see page iv) Easy

4 Find the centre and radius of the circle $x^2 + y^2 - 8x + 4y - 16 = 0$. *(2 marks)*

Bonus question Easy

5

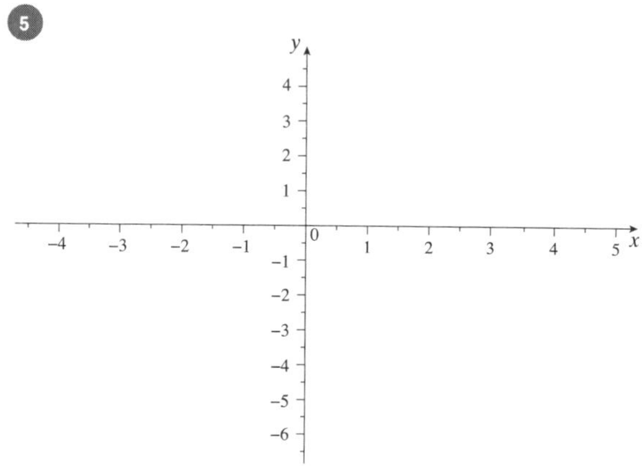

Line l cuts the axes at the points $(-2, 0)$ and $(0, 4)$.

The line m passes through $A(0, -6)$ and is parallel to line l.

The point $B(3, 0)$ lies on the line m.

Line n also passes through B and is perpendicular to line m.

The point C is the point of intersection of lines l and n.

i What are the coordinates of C? *(3 marks)* Medium

ii A circle is drawn with centre B and passing through C. The circle cuts line n again at D. Solve a pair of simultaneous equations to find the coordinates of D. *(3 marks)* Medium

Bonus question

6 The coordinates of A, B and C are $(4, -6)$, $(-18, 0)$ and $(0, 6)$.

D is the midpoint of AB.

i Plot the points on a number plane, and show that D has coordinates $(-7, -3)$. *(1 mark)* Easy

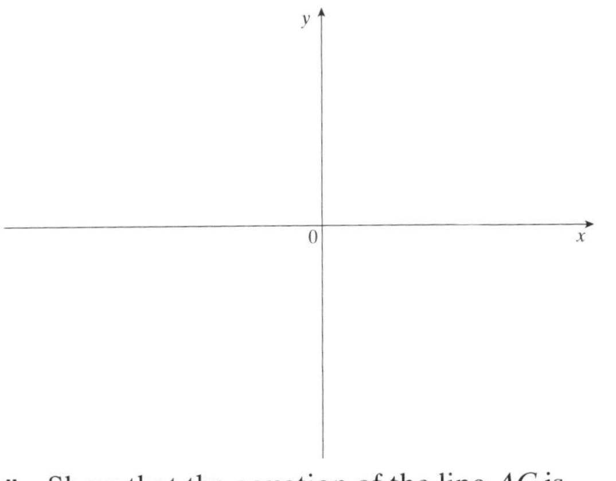

ii Show that the equation of the line AC is $3x + y - 6 = 0$. *(2 marks)* Easy

iii Show that the line BC is perpendicular to the line AC. *(1 mark)* Easy

iv Find the equation of the circle with diameter AB. *(2 marks)* Medium

Bonus question

---

7 The volume V of gas varies inversely to the pressure p.

The volume of a gas is 160 cm^3 under pressure of 36 kg/cm^2. What will be its volume under pressure of 48 kg/cm^2? *(2 marks)*

Bonus question Easy

---

8 The time it takes to fly from Chicago to London varies inversely with the speed of the plane. If the trip takes 9 hours at an average speed of 800 km/h, how long would it take at an average speed of 850 km/h?

Express your answer to the nearest minute. *(2 marks)*

Bonus question Easy

---

9 The number of hours N that it takes for a block of ice to melt varies inversely with the temperature t. At 32 °C it takes 6 hours for a block of ice to melt.

i Find an equation for the graph in terms of N and t in the form $N = \frac{k}{t}$, where k is a constant. *(1 mark)* Easy

ii How much longer will it take the same size block of ice to melt at 10 °C compared to 25 °C? *(2 marks)* Easy

Bonus question

---

10 The time to prepare a paddock for planting is inversely proportional to the number of people who are working. A large paddock can be prepared by 3 workers in 16 days. In order to finish the paddock sooner, the farmer plans to hire additional workers. How many workers are needed to finish the paddock in 6 days? *(2 marks)*

Bonus question Medium

---

11 A scientist is studying the relationship between the pressure, measured in atm (atmospheres) and the volume of a gas, measured in litres. The following is a sample of the scientist's results.

Pressure P (atm)	Volume V (L)
2.4	5.2

The data can be modelled in the form $P = \frac{k}{V}$, where k is a constant.

i Using this data find the value of k. *(1 mark)* Easy

ii Predict the volume when the pressure of the gas is 3 atm. *(2 marks)* Easy

Bonus question

---

12 i Sketch the graph of $y = |x - 1|$ for $-4 \leq x \leq 4$. *(1 mark)* Medium

ii Using the sketch from part **i**, or otherwise, solve $|x - 1| = 2x + 4$. *(2 marks)* Medium

(Q13e, **2019 HSC**)

---

13 When people walk in snow, the depth (D cm) of each footprint depends on both the area (A cm^2) of the shoe sole and the weight of the person.

The graph shows the relationship between the area of the shoe sole and the depth of the footprint in snow, for a group of people of the same weight.

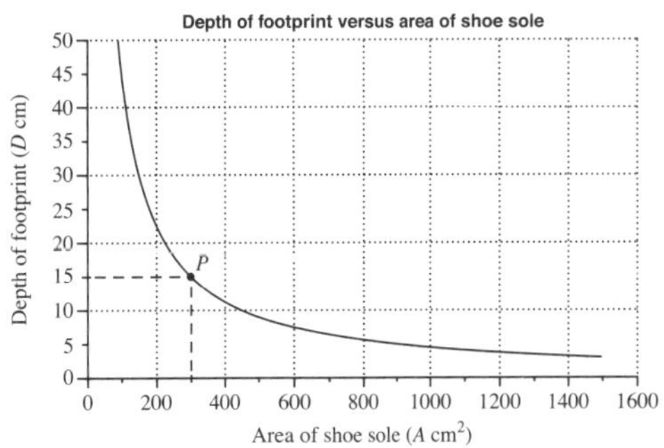

i The graph is a hyperbola because D is inversely proportional to A. The point P lies on the hyperbola.

Find the equation relating D and A. *(2 marks)* Medium

ii A man from this group walks in snow and the depth of his footprint is 4 cm.
Use your equation from part **i** to calculate the area of his shoe sole. *(1 mark)* **Medium**

(Q29c, **2018 MS2 HSC**)

14 Solve $|3x - 1| = 2$. *(2 marks)*

(Q11g, **2017 HSC**) **Easy**

15 Sketch the graph of $(x - 3)^2 + (y + 2)^2 = 4$. *(2 marks)*

(Q11a, **2016 HSC**) **Easy**

16 Solve $|x - 2| \leq 3$. *(2 marks)*

(Q11c, **2016 HSC**) **Easy**

17 The cost of hiring an open space for a music festival is \$120 000. The cost will be shared equally by the people attending the festival, so that C (in dollars) is the cost per person when n people attend the festival.

i Complete the table below by filling in the THREE missing values. *(1 mark)* **Easy**

Number of people (n)	500	1000	1500	2000	2500	3000
Cost per person (C)				60	48	40

ii Using the values from the table, draw the graph showing the relationship between n and C. *(2 marks)* **Easy**

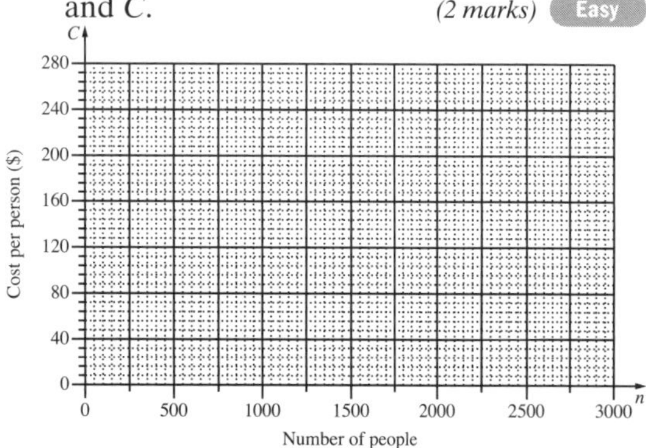

iii What equation represents the relationship between n and C? *(1 mark)* **Easy**

iv Give ONE limitation of this equation in relation to this context. *(1 mark)* **Easy**

v Is it possible for the cost per person to be \$94? Support your answer with appropriate calculations. *(1 mark)* **Easy**

(Q29a, **2014 MS2 HSC**)

18 The time taken to complete a journey varies inversely with the speed of a car. A car takes 6 hours to complete a journey when travelling at 60 km/h.

How long would the same journey take if the car were travelling at 100 km/h?

A 36 minutes
B 1 hour and 40 minutes
C 3 hours and 6 minutes
D 3 hours and 36 minutes *(1 mark)*

(Q15, **2012 MS2 HSC**) **Easy**

19 The air pressure, P, in a bubble varies inversely with the volume, V, of the bubble.

i Write an equation relating P, V and a, where a is a constant. *(1 mark)* **Medium**

ii It is known that $P = 3$ when $V = 2$.
By finding the value of the constant, a, find the value of P when $V = 4$. *(2 marks)* **Medium**

iii Sketch a graph to show how P varies for different values of V.
Use the horizontal axis to represent volume and the vertical axis to represent air pressure. *(2 marks)* **Medium**

(Q28a, **2011 MS2 HSC**)

20 Write down the equation of the circle with centre $(-1, 2)$ and radius 5. *(1 mark)*

(Q1c, **2010 HSC**) **Easy**

21 Solve $|2x + 3| = 9$. *(2 marks)*

(Q1d, **2010 HSC**) **Easy**

22 The number of hours that it takes for a block of ice to melt varies inversely with the temperature. At 30°C it takes 8 hours for a block of ice to melt.

How long will it take the same size block of ice to melt at 12°C?

A 3.2 hours **B** 20 hours
C 26 hours **D** 45 hours

(Q13, **2010 MS2 HSC**) **Medium**

23 Solve $|x + 1| = 5$. *(2 marks)*

(Q1c, **2009 HSC**) **Easy**

24 The time for a car to travel a certain distance varies inversely with its speed.

Which of the following graphs shows this relationship?

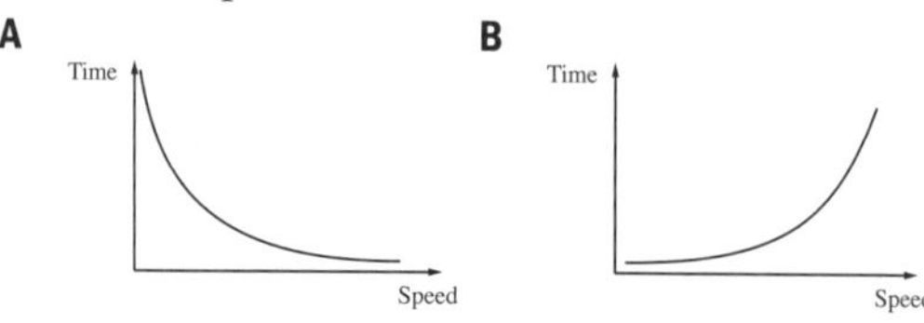

C

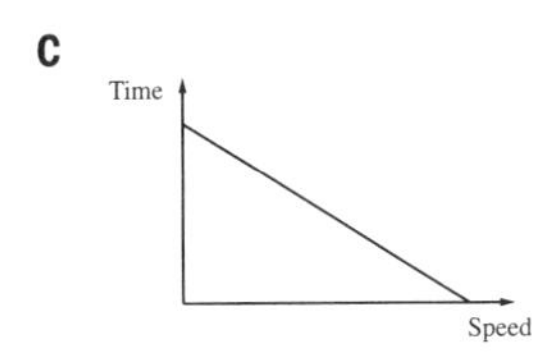

D

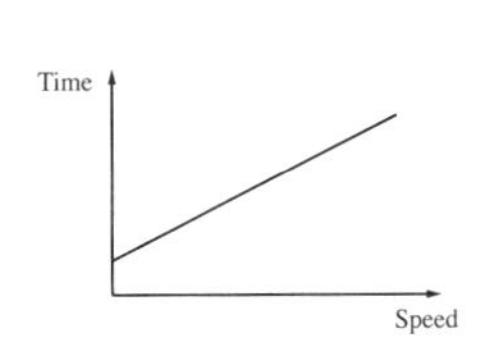

(1 mark)

(Q16, **2009 MS2 HSC**) Easy

---

25 Solve $|4x - 3| = 7$. *(2 marks)*

(Q1d, **2008 HSC**) Easy

---

26 The height of a particular termite mound is directly proportional to the square root of the number of termites.

The height of this mound is 35 cm when the number of termites is 2000.

What is the height of this mound, in centimetres, when there are 10 000 termites?

A 16 **B** 78
C 175 **D** 875 *(1 mark)*

(Q19, **2008 MS2 HSC**) Medium

---

27 If pressure (p) varies inversely with volume (V), which formula correctly expresses p in terms of V and k, where k is a constant?

A $p = \frac{k}{V}$ **B** $p = \frac{V}{k}$
C $p = kV$ **D** $p = k + V$ *(1 mark)*

(Q15, **2007 MS2 HSC**) Easy

---

28 Sketch the graph of $y = |x + 4|$. *(2 marks)*

(Q1c, **2006 HSC**) Easy

---

29 Find the values of x for which $|x - 3| \leq 1$. *(2 marks)*

(Q1e, **2005 HSC**) Easy

---

30 Find the values of x for which $|x + 1| \leq 5$. *(2 marks)*

(Q1f, **2004 HSC**) Easy

---

31 The time (t) taken to clean a house varies inversely with the number (n) of people cleaning the house.

Which graph represents this relationship?

A

B

C

D

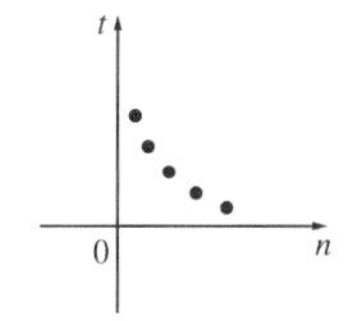

(1 mark)

(Q21, **2004 MS2 HSC**) Medium

---

32 Solve $|x - 3| = 7$. *(2 marks)*

(Q1f, **2003 HSC**) Easy

---

33 If an electrical current varies inversely with resistance, what is the effect on the current when the resistance is doubled?

A The current is doubled.
B The current is exactly the same.
C The current is halved.
D The current is squared. *(1 mark)*

(Q17, **2003 MS2 HSC**) Easy

---

34 Solve $|x - 1| \geq 3$ and graph your solution on the number line. *(2 marks)*

(Q4a, **2002 HSC**) Easy

---

35 Sketch the graph of $y = \sqrt{4 - x^2}$, and state the range. *(2 marks)*

(Q6a, **2002 HSC**) Easy

---

36 Solve $|x + 3| < 2$.
Graph your solution on a number line. *(2 marks)*

(Q1b, **2001 HSC**) Easy

---

37 Assume that the surface area S of a human satisfies the equation

$$S = kM^{\frac{2}{3}}$$

where M is the body mass in kilograms, and k is the constant of proportionality. A human with body mass 70 kg has surface area 18 600 cm^2.

Find the value of k, and hence find the surface area of a human with body mass 60 kg. *(2 marks)*

(Q3b, **2001 HSC**) Medium

---

38 State the domain and range of the function $y = 2\sqrt{25 - x^2}$. *(3 marks)*

(Q5a, **2001 HSC**) Medium

---

39 Joe uses a microwave oven to heat lasagne. The time taken for heating is inversely proportional to the power setting (in watts). It takes ten minutes at a power setting of 240 watts to heat the lasagne.

How long would it take at a power setting of 500 watts? *(3 marks)*

(Q28b, **2001 MS2 HSC**) **Medium**

40 Solve $|x - 5| = 3$. *(2 marks)*

(Q1f, **2000 HSC**) **Medium**

41 The diagram shows the points $P(0, 2)$ and $Q(4, 0)$. The point M is the midpoint of PQ. The line MN is perpendicular to PQ and meets the x axis at G and the y axis at N.

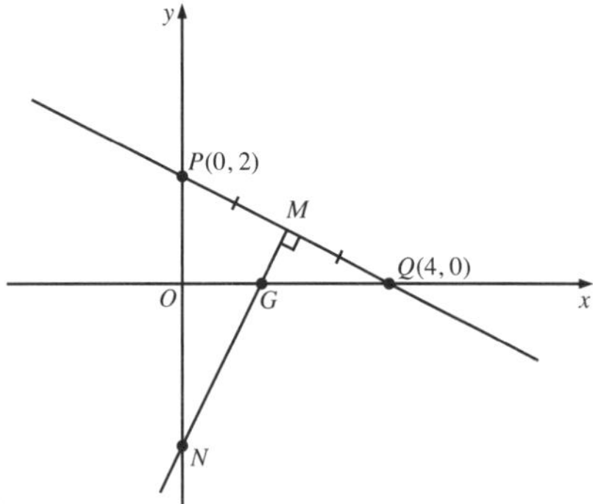

a Show that the gradient of PQ is $-\frac{1}{2}$. *(1 mark)* **Easy**

b Find the coordinates of M. *(2 marks)* **Easy**

c Find the equation of the line MN. *(2 marks)* **Easy**

d Show that N has coordinates $(0, -3)$. *(1 mark)* **Easy**

e Find the distance NQ. *(1 mark)* **Easy**

f Find the equation of the circle with centre N and radius NQ. *(2 marks)* **Easy**

g Hence show that the circle in part **f** passes through the point P. *(1 mark)* **Easy**

h The point R lies in the first quadrant, and $PNQR$ is a rhombus. Find the coordinates of R. *(2 marks)* **Easy**

(Q2, **2000 HSC**)

42 Write down the equation of the vertical asymptote of $y = \dfrac{4x}{(x-3)}$. *(1 mark)*

(Q1d, **1999 ME1 HSC**) **Medium**

43 Simplify $|-5| - |8|$. *(1 mark)*

(Q1b, **1998 HSC**) **Easy**

44 The line l cuts the x axis at $L(-4, 0)$ and the y axis at $M(0, 3)$ as shown. N is a point on the line l, and P is the point $(0, 8)$.

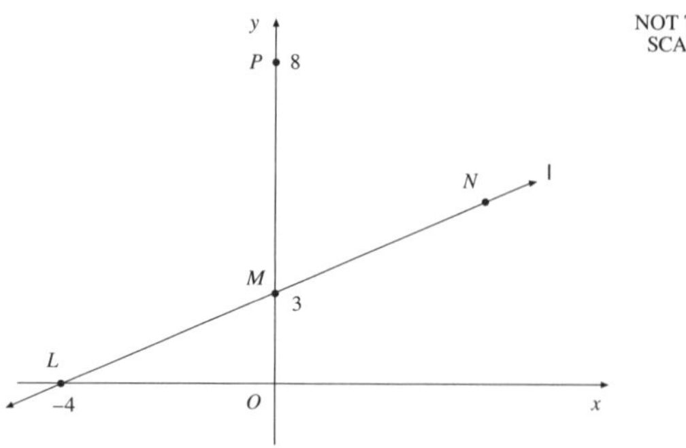

Copy the diagram into your Writing Booklet.

a Find the equation of the line l. *(2 marks)* **Easy**

b Show that the point $(16, 15)$ lies on the line l. *(1 mark)* **Easy**

c By considering the lengths of ML and MP, show that ΔLMP is isosceles. *(2 marks)* **Easy**

d Calculate the gradient of the line PL. *(1 mark)* **Easy**

e M is the midpoint of the interval LN. Find the coordinates of the point N. *(2 marks)* **Easy**

f Show that $\angle NPL$ is a right angle. *(2 marks)* **Easy**

g Find the equation of the circle that passes through the points N, P, and L. *(2 marks)* **Easy**

(Q2, **1995 HSC**)

45 **i** Draw the graphs of $y = |x|$ and $y = x + 4$ on the same set of axes. **Medium**

ii Find the coordinates of the point of intersection of these two graphs. *(4 marks)* **Medium**

(Q4b, **1995 HSC**)

46 Solve $|x - 1| = 4$.

(Q1d, **1994 HSC**) **Easy**

47 **i** Verify that $x = \frac{1}{3}$ and $x = 2$ satisfy the equation $7 - 3x = \frac{2}{x}$. **Easy**

ii On the same set of axes, sketch the graphs of $y = 7 - 3x$ and $y = \frac{2}{x}$. **Easy**

iii Using part **ii**, or otherwise, write down all values of x for which $7 - 3x < \frac{2}{x}$. **Hard**

(Q2c, **1994 ME1 HSC**)

1 **a** $M = \frac{k}{T}$ ✓

$12 = \frac{k}{15}$

$k = 180$ ✓

$\therefore M = \frac{180}{T}$

(2 marks)

b

T	5	15	30
M	36	12	6

✓

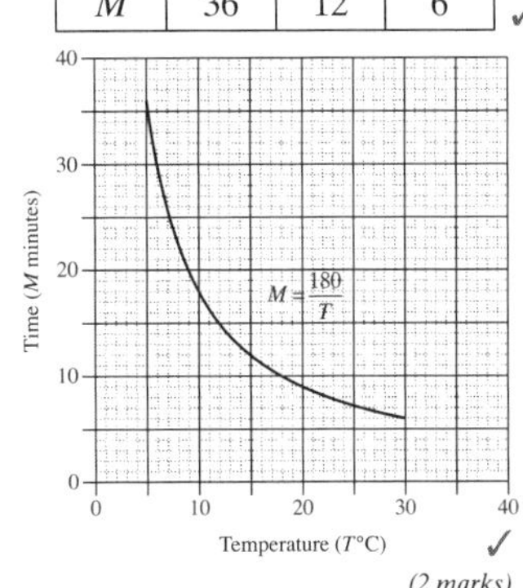

✓

(2 marks)

2 Considering the options, there is a single root at $x = -1$ and a triple root at $x = 2$.

This means the correct answer could be B or C.

The graph shows that as $x > 2$, then $y < 0$.

The equation of the graph is $y = (x + 1)(2 - x)^3$.

Answer C

3 Let C = cavities and t = time in seconds:

$C = \frac{k}{t}$

Substitute $C = 6, t = 20$:

$6 = \frac{k}{20}$

$k = 120$

$C = \frac{120}{t}$ ✓

Substitute $t = 60$:

$C = \frac{120}{60}$

$C = 2$

Logan would have 2 cavities. ✓

(2 marks)

4 $x^2 + y^2 - 8x + 4y - 16 = 0$

$x^2 - 8x + y^2 + 4y = 16$

$x^2 - 8x + 16 + y^2 + 4y + 4 = 16 + 16 + 4$

$(x - 4)^2 + (y + 2)^2 = 36$ ✓

Centre $(4, -2)$ and radius 6 units. ✓

(2 marks)

5

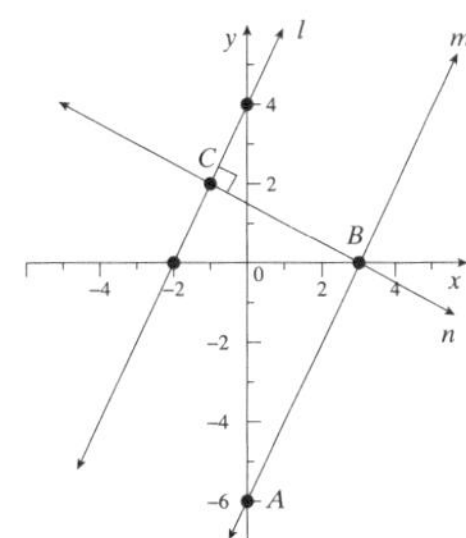

i Line l: gradient $= \frac{4}{2} = 2$ and

y-intercept = 4

$\therefore y = 2x + 4$ ✓

Line m: gradient = 2 and y-intercept = −6

$\therefore y = 2x - 6$

Line n: gradient $= -\frac{1}{2}$

and passes through $(3, 0)$

$y - 0 = -\frac{1}{2}(x - 3)$

$2y = -x + 3$

$x + 2y - 3 = 0$ ✓

Coordinates of C is point of intersection of $y = 2x + 4$ and $x + 2y - 3 = 0$.

$\therefore x + 2(2x + 4) - 3 = 0$

$x + 4x + 8 - 3 = 0$

$5x + 5 = 0$

$5x = -5$

$x = -1$

Subs. in $y = 2x + 4$:

$y = 2(-1) + 4$

$= 2$

$\therefore C(-1, 2)$ ✓

(3 marks)

ii Length BC

$= \sqrt{(3-(-1))^2 + (0-2)^2}$

$= \sqrt{20}$

Circle with centre $(3, 0)$ and radius $\sqrt{20}$ is $(x - 3)^2 + y^2 = 20$. ✓

Coordinates of D is point of intersection of $x + 2y - 3 = 0$ and $(x - 3)^2 + y^2 = 20$.

Substitute $x = 3 - 2y$ in $(x - 3)^2 + y^2 = 20$:

$(3 - 2y - 3)^2 + y^2 = 20$

$4y^2 + y^2 = 20$

$5y^2 = 20$

$y = \pm 2$ ✓

For point D, substitute $y = -2$ in $x + 2y - 3 = 0$:

$x + 2(-2) - 3 = 0$

$x = 7$

$\therefore D(7, -2)$ ✓

(3 marks)

6 **i**

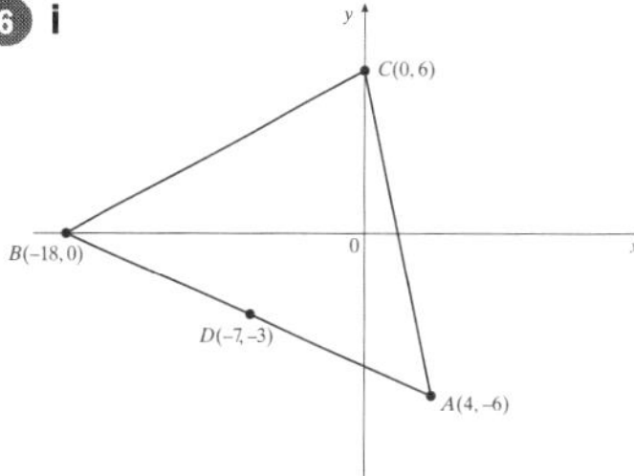

Midpoint of AB

$= \left(\frac{-18+4}{2}, \frac{0-6}{2}\right)$

$= (-7, -3)$

$\therefore D$ has coordinates $(-7, -3)$. ✓

(1 mark)

ii gradient $AC = \frac{6+6}{0-4}$

$= -3$ ✓

$\therefore$ gradient = −3 and y-intercept = 6:

$y = -3x + 6$

$\therefore 3x + y - 6 = 0$ ✓

(2 marks)

iii gradient $BC = \frac{6-0}{0+18}$

$= \frac{1}{3}$

As $-3 \times \frac{1}{3} = -1$, then BC is perpendicular to AC. ✓

(1 mark)

iv Length AD

$= \sqrt{(-7-4)^2 + (-3+6)^2}$

$= \sqrt{130}$ ✓

Circle has centre $(-7, -3)$ and radius $\sqrt{130}$:

Equation is $(x + 7)^2 + (y + 3)^2 = 130$. ✓

(2 marks)

7 $V = \frac{k}{p}$

Substitute $V = 160, p = 36$:

$160 = \frac{k}{36}$

$k = 5760$

$V = \frac{5760}{p}$ ✓

Substitute $p = 48$:

$V = \frac{5760}{48}$

$= 120$

The volume is 120 cm^3. ✓

(2 marks)

8 Let T = time and s = speed:

$T = \frac{k}{s}$

Substitute $T = 9, s = 800$:

$9 = \frac{k}{800}$

$k = 7200$

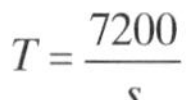

$T = \dfrac{7200}{s}$ ✓

Substitute $s = 850$:

$T = \dfrac{7200}{850}$

$= 8.470\,588\,235\ldots$

$= 8$ h 28 m (nearest minute)

The time would be about 8 hours 28 minutes. ✓

(2 marks)

9 **i** Substitute $N = 6, t = 32$:

$6 = \dfrac{k}{32}$

$k = 192$

$\therefore N = \dfrac{192}{t}$ ✓ *(1 mark)*

ii Substitute $t = 10$ and $t = 25$:

Difference $= \dfrac{192}{10} - \dfrac{192}{25}$

$= 19.2 - 7.68$ ✓

$= 11.52$

It will take 11.52 hours longer. ✓

(2 marks)

10 Let T = time and w = workers:

$T = \dfrac{k}{w}$

Substitute $T = 16, w = 3$:

$16 = \dfrac{k}{3}$

$k = 48$

$T = \dfrac{48}{w}$ ✓

Substitute $T = 6$:

$6 = \dfrac{48}{w}$

$6w = 48$

$= 8$

The farmer would need 8 workers. ✓

(2 marks)

11 **i** $P = \dfrac{k}{V}$

Substitute $P = 2.4, V = 5.2$:

$2.4 = \dfrac{k}{5.2}$

$k = 12.48$ ✓

(1 mark)

ii $P = \dfrac{12.48}{V}$

Substitute $P = 3$:

$3 = \dfrac{12.48}{V}$ ✓

$3V = 12.48$

$V = 4.16$

The volume is 4.16 L. ✓

(2 marks)

12 **i, ii**

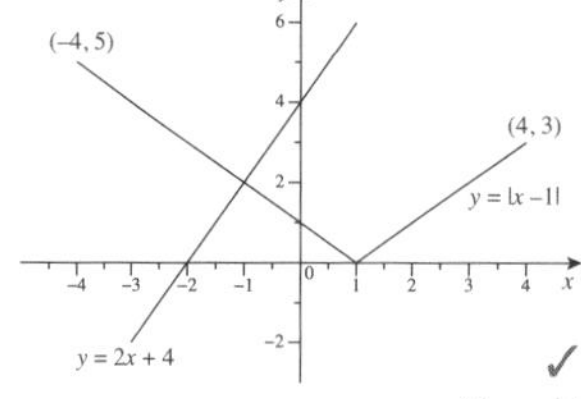

✓ *(1 mark)*

Point of intersection at $(-1, 2)$

Hence, solution is $x = -1$. ✓

Alternative solution:

$|x - 1| = 2x + 4$

Two cases:

$x - 1 = 2x + 4$

$x = -5$

$-(x - 1) = 2x + 4$

$-x + 1 = 2x + 4$

$3x = -3$

$x = -1$

Check solutions:

$x = -5$: $|-5 - 1| = 2(5) + 4 \ldots$ no!

$x = -1$: $|-1 - 1| = 2(-1) + 4 \ldots$ yes!

$\therefore$ solution is $x = -1$.

(2 marks)

13 **i** D is inversely proportional to A.

$\therefore D = \dfrac{k}{A}$ where k is a constant. ✓

At P, $A = 300$ and $D = 15$

$15 = \dfrac{k}{300}$

$k = 4500$

$\therefore D = \dfrac{4500}{A}$ ✓

(2 marks)

ii $D = \dfrac{4500}{A}$

When $D = 4$,

$4 = \dfrac{4500}{A}$

$4A = 4500$

$A = 1125$

The area of the shoe sole is 1125 cm^2. ✓

(1 mark)

14 $3x - 1 = 2$ $\quad$ $3x - 1 = -2$

$3x = 2 + 1$ $\quad$ $3x = -2 + 1$

$3x = 3$ $\quad$ $3x = -1$

$x = 1$ $\quad$ $x = -\dfrac{1}{3}$

$\therefore x = 1, -\dfrac{1}{3}$ ✓✓ *(2 marks)*

15 Circle, centre $(3, -2)$ and radius 2. ✓

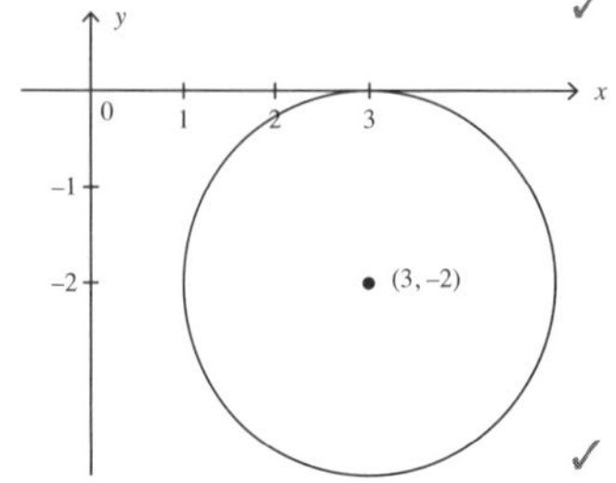

✓ *(2 marks)*

16 $|x - 2| \le 3$

$x - 2 \le 3$ $\quad$ $x - 2 \ge -3$

$x \le 5$ $\quad$ $x \ge -1$

$\therefore\ -1 \le x \le 5$ ✓✓

[or: $-3 \le x - 2 \le 3$

$-1 \le x \le 5$]

(2 marks)

17 **i** $120\,000 \div 500 = 240$

$120\,000 \div 1000 = 120$

$120\,000 \div 1500 = 80$

(n)	500	1000	1500	2000	2500	3000
(C)	240	120	80	60	48	40

✓ *(1 mark)*

ii

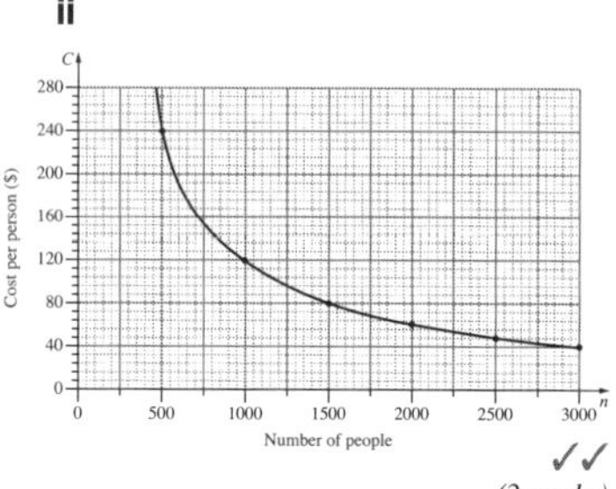

✓✓ *(2 marks)*

iii $C = \dfrac{120\,000}{n}$ ✓ *(1 mark)*

iv [Possible answers]

- n can only have positive integral values.
- There must be a limit to the number of people that can attend.
- There must be a limit to the price or no-one will choose to attend.
- The equation would be difficult to put into practice. People would not commit to attend without knowing the cost, but the cost cannot be calculated without knowing the number of people. ✓

(1 mark)

v If $C = 94$,

$94 = \dfrac{120\,000}{n}$

$94n = 120\,000$

$n = 1276.595\,74\ldots$

So as n is not an exact number, the cost per person cannot be exactly \$94.

[But if 1277 people attend, the cost per person will be \$93.97 which could be rounded to \$94.] ✓ *(1 mark)*

18 $T = \dfrac{D}{S}$

$6 = \dfrac{D}{60}$

$D = 6 \times 60$
$= 360$

So $T = \dfrac{360}{S}$

When $S = 100$,

$T = \dfrac{360}{100}$
$= 3.6$

So the journey takes 3.6 hours or 3 hours and 36 minutes.

Answer D

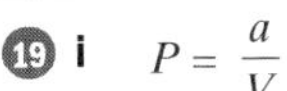

19 **i** $P = \dfrac{a}{V}$ ✓ *(1 mark)*

ii $P = 3$ when $V = 2$

$P = \dfrac{a}{V}$

$3 = \dfrac{a}{2}$

$a = 3 \times 2$
$= 6$ ✓

So $P = \dfrac{6}{V}$

When $V = 4$,

$P = \dfrac{6}{4}$
$= 1.5$ ✓ *(2 marks)*

iii

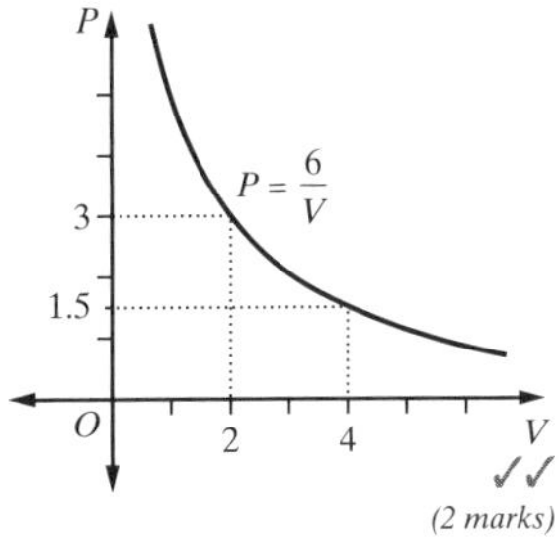

✓✓ *(2 marks)*

20 Using the formula
$(x - h)^2 + (y - k)^2 = r^2$,
with centre (h, k) and radius r:
Now, $C(-1, 2)$ and $r = 5$:
$(x - (-1))^2 + (y - 2)^2 = 5^2$
$(x + 1)^2 + (y - 2)^2 = 25$ ✓ *(1 mark)*

21 $|2x + 3| = 9$.
Two cases:

$2x + 3 = 9$ or $-(2x + 3) = 9$
$2x = 9 - 3$ or $-2x - 3 = 9$
$2x = 6$ or $-2x = 9 + 3$
$x = 3$ or $-2x = 12$
$x = -6$

$\therefore x = 3$ or $x = -6$ ✓✓ *(2 marks)*

22 Let h be the number of hours and $T°$ the temperature.

Now $h \propto \dfrac{1}{T}$

So $h = \dfrac{k}{T}$ for some constant k

When $T = 30$, $h = 8$

$8 = \dfrac{k}{30}$
$k = 8 \times 30$
$k = 240$

So $h = \dfrac{240}{T}$

When $T = 12$,

$h = \dfrac{240}{12}$
$= 20$

It would take 20 hours for the ice to melt at 12°C.

Answer B

23 $|x + 1| = 5$
$\therefore x + 1 = \pm 5$
$\therefore x + 1 = 5$ or $x + 1 = -5$
$\therefore x = 4$ or $x = -6$ ✓✓ *(2 marks)*

24 Time varying inversely with speed means that
$t = \dfrac{d}{s}$ where d is a certain distance.
This equation can be represented by a hyperbola, noting that as speed becomes larger, the time becomes smaller.
Answer A

25 $|4x - 3| = 7$

$\therefore 4x - 3 = 7$ or $4x - 3 = -7$
$4x = 10$ $\qquad 4x = -4$
$x = \dfrac{10}{4}$ $\qquad \therefore x = -1$

$\therefore x = 2\frac{1}{2}$

Test

$\left|4 \times 2\frac{1}{2} - 3\right| = 7$ $\qquad |4 \times -1 - 3| = 7$
$|10 - 3| = 7$ $\qquad |-4 - 3| = 7$

$\therefore x = 2\frac{1}{2}$ or $x = -1$ ✓✓ *(2 marks)*

26 $H = k\sqrt{n}$

When $H = 35$, $n = 2000$

$\therefore 35 = k \times \sqrt{2000}$

$\therefore k = \dfrac{35}{\sqrt{2000}}$

When $n = 10\,000$,

$H = \dfrac{35}{\sqrt{2000}} \times \sqrt{10\,000}$
$= 78.2623\ldots$
$\doteqdot 78$

Answer B

27 $p = \dfrac{k}{V}$

(N.B. If p varies inversely with V, then as V increases, p decreases.)
Answer A

28

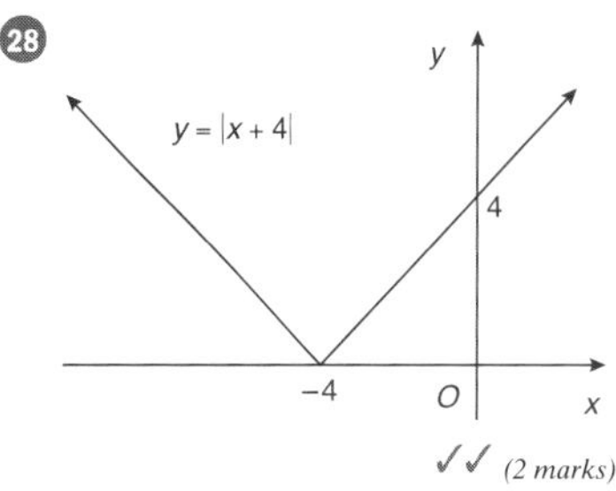

✓✓ *(2 marks)*

29 $|x - 3| \leq 1$
$\therefore -1 \leq x - 3 \leq 1$
$\therefore 2 \leq x \leq 4$. ✓✓ *(2 marks)*

30 $|x + 1| \leq 5$
$\therefore -5 \leq x + 1 \leq 5$
$\therefore -6 \leq x \leq 4$. ✓✓ *(2 marks)*

31 $t = \dfrac{k}{n}$

This represents a hyperbola.
As the number of people (n) increases, the time (t) it takes to clean the house will decrease but not at a constant rate.
Answer D

32 $|x - 3| = 7$
$\therefore x - 3 = \pm 7$
$\therefore x - 3 = 7$ or $x - 3 = -7$
$\therefore x = 10$ or $x = -4$ ✓✓

(2 marks)

33 $E = \dfrac{k}{R}$

Where E is electric current and R is resistance.

Let $R = 10$ and $k = 20$

$\therefore E = \dfrac{20}{10} = 2$

If R is doubled, then

$E = \dfrac{20}{20} = 1$

$\therefore E$ is halved

Answer C

34 $|x - 1| \geq 3$
$\therefore x - 1 \geq 3$ or $-(x - 1) \geq 3$
$\therefore x \geq 4$ or $-x + 1 \geq 3$
$\therefore x \geq 4$ or $x \leq -2$ ✓

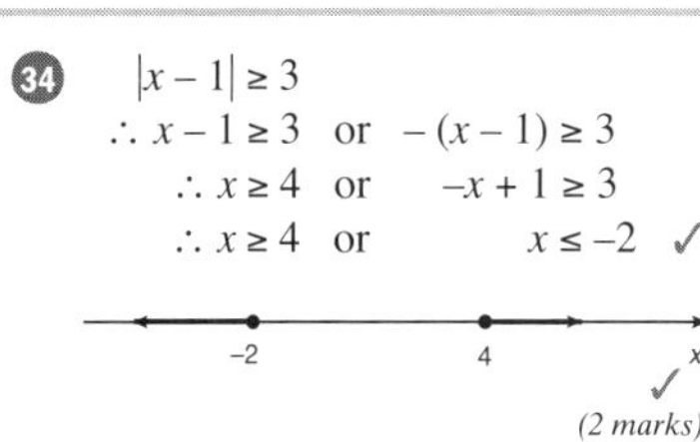

✓ *(2 marks)*

35

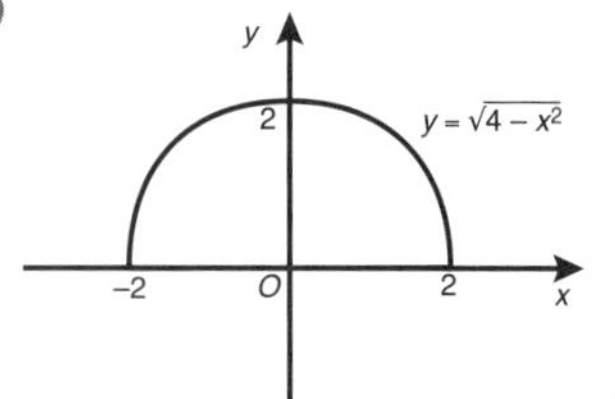

✓

Range is $0 \leqslant y \leqslant 2$. ✓

(2 marks)

36 $|x + 3| < 2$

$\therefore -2 < x + 3 < 2$

$\therefore -5 < x < -1$. ✓

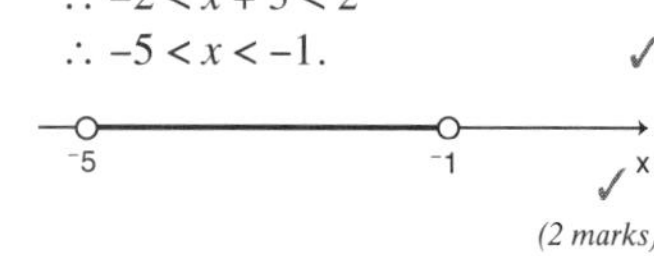

✓

(2 marks)

37 $S = kM^{\frac{2}{3}}$

When $M = 70$, $S = 18\,600$

$\therefore 18\,600 = k(70)^{\frac{2}{3}}$

$k = \dfrac{18\,600}{70^{\frac{2}{3}}}$ ✓

$= 1095.0843\ldots$

$\therefore S = 1095.0843\ldots \times M^{\frac{2}{3}}$

When $M = 60$ kg,

$S = 1095.0843\ldots \times 60^{\frac{2}{3}}$

$= 16\,783.469\ldots$

$\doteqdot 16\,783 \text{ cm}^2$ ✓

(2 marks)

38 Domain: $25 - x^2 \geq 0$

$25 \geq x^2$ ✓

$\therefore -5 \leq x \leq 5$ ✓

Range: $0 \leq y \leq 10$ ✓

(3 marks)

39 $T = \dfrac{k}{P}$

If $T = 10$, $P = 240$

$\therefore 10 = \dfrac{k}{240}$ ✓

$\therefore k = 2400$

$\therefore T = \dfrac{2400}{P}$ ✓

Now if $P = 500$,

$T = \dfrac{2400}{500}$

$= 4.8$ minutes ✓

(3 marks)

40 $|x - 5| = 3$

$\therefore x - 5 = \pm 3$ ✓

$\therefore x - 5 = 3$ or $x - 5 = -3$

$\therefore x = 8$ or $x = 2$ ✓

(2 marks)

41

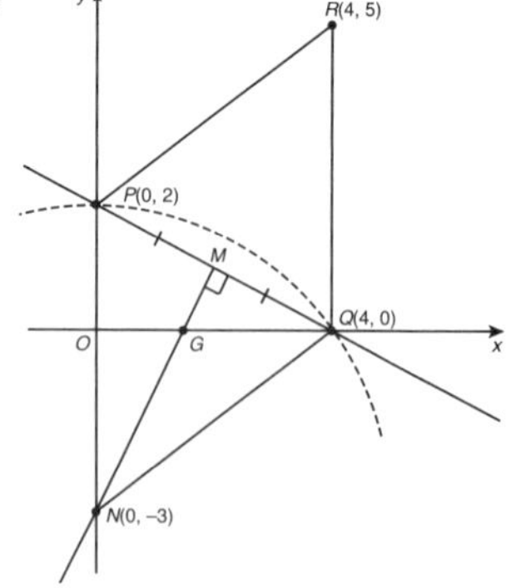

a Gradient of $PQ = \dfrac{y_2 - y_1}{x_2 - x_1}$

$= \dfrac{0 - 2}{4 - 0}$

$= \dfrac{-2}{4}$

$= -\dfrac{1}{2}$ ✓

(1 mark)

b Coordinates of M are

$\left(\dfrac{0 + 4}{2}, \dfrac{2 + 0}{2}\right)$ ✓

i.e. $(2, 1)$. ✓

(2 marks)

c Gradient of $PQ = -\frac{1}{2}$

$\therefore$ Gradient of $MN = 2$ ($MN \perp PQ$) ✓

$\therefore$ Equation of MN with gradient 2, passing through $M(2, 1)$ is given by

$y - y_1 = m(x - x_1)$

$y - 1 = 2(x - 2)$

$y - 1 = 2x - 4$

$y = 2x - 3$. ✓

(2 marks)

d Line MN cuts y-axis at N where $x = 0$.

$\therefore y = 2(0) - 3$

$= -3$

$\therefore N$ has coordinates $(0, -3)$. ✓

(1 mark)

e $NQ = \sqrt{(4 - 0)^2 + (0 - (-3))^2}$

$= \sqrt{16 + 9}$

$= \sqrt{25}$

$= 5$ units. ✓

(1 mark)

f Equation of circle with centre $(0, -3)$ and radius, $NQ = 5$ units is given by:

$(x - 0)^2 + (y - (-3))^2 = 5^2$ ✓

$x^2 + (y + 3)^2 = 25$

$x^2 + y^2 + 6y + 9 = 25$

$x^2 + y^2 + 6y - 16 = 0$ ✓

(2 marks)

g Substituting $P(0, 2)$ into $x^2 + y^2 + 6y - 16 = 0$

LHS $= 0^2 + 2^2 + 6(2) - 16$

$= 4 + 12 - 16$

$= 0$

RHS $= 0$

$\therefore$ Circle passes through P. ✓

(1 mark)

h If $PNQR$ is a rhombus, $M(2, 1)$ is the midpoint of $R(x, y)$ and $N(0, -3)$

$\therefore \dfrac{x + 0}{2} = 2$ and $\dfrac{y - 3}{2} = 1$ ✓

$\therefore x = 4$ and $y = 5$

$\therefore$ Coordinates of R are $(4, 5)$. ✓

(2 marks)

42 The vertical asymptote of $y = \dfrac{4x}{x - 3}$ is the line with equation $x - 3 = 0$, i.e. $x = 3$. ✓ *(1 mark)*

43 $|-5| - |8| = 5 - 8$

$= -3$ ✓ *(1 mark)*

44

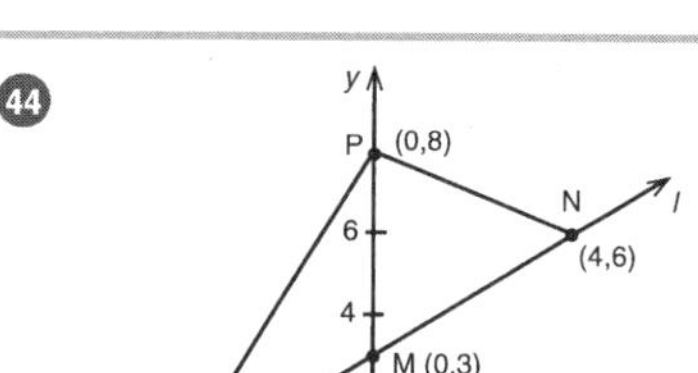

a Equation of line l passing through $(-4, 0)$ and $(0, 3)$ is given by

$\dfrac{y - 0}{x - (-4)} = \dfrac{3 - 0}{0 - (-4)}$ ✓

$\dfrac{y}{x + 4} = \dfrac{3}{4}$

$4y = 3x + 12$

i.e. $3x - 4y + 12 = 0$ ✓

(2 marks)

b Substituting $(16, 15)$ into equation:

LHS $= 3(16) - 4(15) + 12$

$= 48 - 60 + 12$

$= 0$

$=$ RHS

$\therefore (16, 15)$ lies on line l. ✓

(1 mark)

c $MP = 8 - 3$

$= 5$ units

$ML = \sqrt{(-4 - 0)^2 + (0 - 3)^2}$

$= \sqrt{16 + 9}$ ✓

$= \sqrt{25}$

$= 5$ units

$\therefore ML = MP$

$\therefore \triangle LMP$ is isosceles. ✓

(2 marks)

d Gradient of

$$PL = m_1 = \frac{8-0}{0-(-4)} = \frac{8}{4} = 2$$ ✓

(1 mark)

e If M is the midpoint of $L(-4, 0)$ and $N(x_1, y_1)$,

$$\therefore \frac{x_1 + (-4)}{2} = 0$$

$$\therefore \frac{x_1 - 4}{2} = 0$$

$$\therefore x_1 - 4 = 0$$

$$\therefore x_1 = 4$$ ✓

and $\frac{y_1 + 0}{2} = 3$

$$\therefore y_1 = 6$$

$\therefore$ Coordinates of N are (4, 6). ✓

(2 marks)

f Gradient of NP

$$= m_2 = \frac{8-6}{0-4} = \frac{2}{-4}$$ ✓✓

$$= -\frac{1}{2}$$ ✓

Gradient of $PL = m_1 = 2$ (from **d**)

Now $m_1 \times m_2 = 2 \times -\frac{1}{2} = -1$

$\therefore PL \perp NP$

$\therefore \angle NPL = 90°$. ✓

(2 marks)

g Since $\angle NPL = 90°$, LN is a diameter, $M(0,3)$ is the centre of the circle as M is the midpoint of LN, and $MP = 5$ units is the radius. (OR Since $ML = MP = MN = 5$, M is the centre of the circle of radius 5 units.) ✓

$\therefore$ Equation of circle with centre (0, 3) and radius = 5 units is given by

$$(x-0)^2 + (y-3)^2 = 5^2$$

$$x^2 + y^2 - 6y + 9 = 25$$

$$x^2 + y^2 - 6y - 16 = 0.$$ ✓

(2 marks)

45 **i**

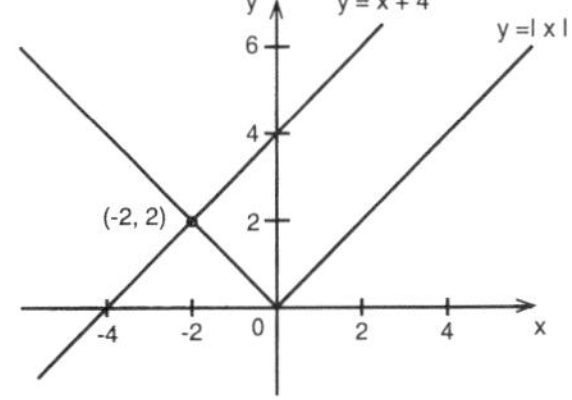

ii By definition,
$|x| = -x$ for $x < 0$
Solving simultaneously for $x < 0$

$$y = -x$$

$$y = x + 4$$

$$\therefore -x = x + 4$$

$$-2x = 4$$

$$x = -2$$ ✓

$$\therefore y = 2$$

$\therefore$ The point of intersection is (−2, 2). ✓

(4 marks)
(Total mark allocation only in exam)

46 $|x-1| = 4$

$\therefore (x-1) = \pm 4$

$\therefore x - 1 = 4$ or $x - 1 = -4$

$x = 5$ or $x = -3$

(No mark allocation in exam)

47 **i** For $x = \frac{1}{3}$:

$$\text{LHS} = 7 - 3\left(\tfrac{1}{3}\right) = 7 - 1 = 6$$

$$\text{RHS} = \frac{2}{\frac{1}{3}} = 2 \times 3 = 6$$

$\therefore x = \frac{1}{3}$ satisfies the equation.

For $x = 2$:

$$\text{LHS} = 7 - 3(2) = 7 - 6 = 1 \qquad \text{RHS} = \frac{2}{2} = 1$$

$\therefore x = 2$ satisfies the equation.

ii

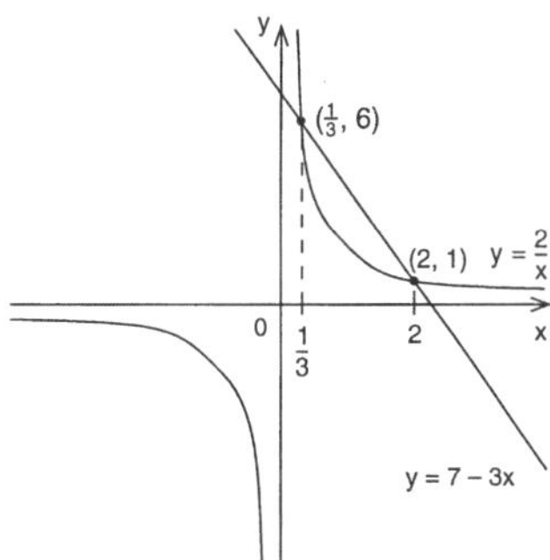

iii From the diagram:

$$7 - 3x < \frac{2}{x}$$

for $0 < x < \frac{1}{3}$ and $x > 2$.

(No mark allocation in exam)

1 A tower BT has height h metres.

From point A, the angle of elevation to the top of the tower is 26° as shown.

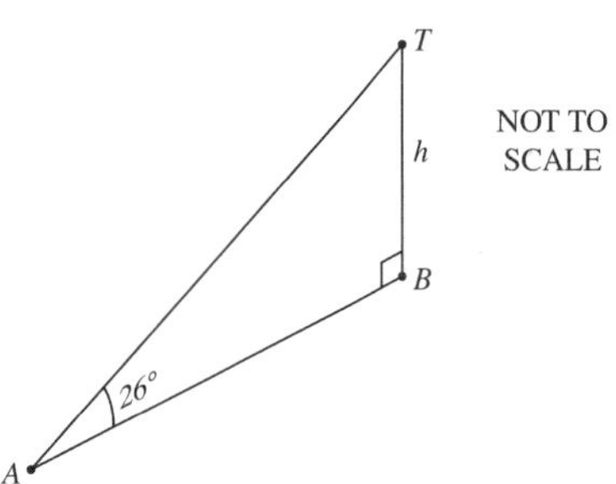

Which of the following is the correct expression for the length of AB?

A $h \tan 26°$ **B** $h \cot 26°$

C $h \sin 26°$ **D** $h \operatorname{cosec} 26°$ *(1 mark)*

(Q3, **2022 HSC**) Easy

2 A right-angled triangle XYZ is cut out from a semicircle with centre O. The length of the diameter XZ is 16 cm and $\angle YXZ = 30°$, as shown on the diagram.

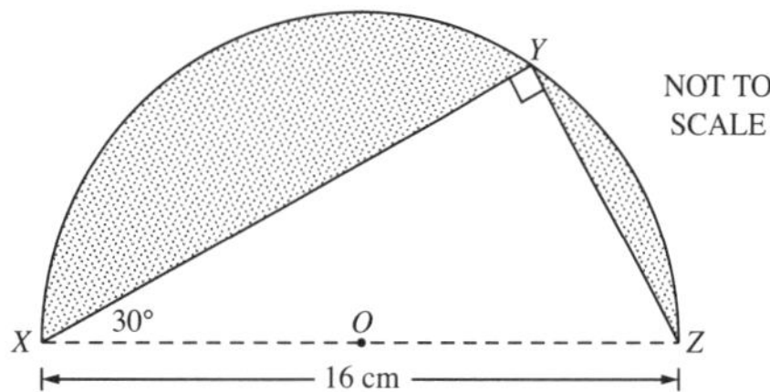

a Find the length of XY in centimetres, correct to two decimal places. *(2 marks)* Easy

b Hence, find the area of the shaded region in square centimetres, correct to one decimal place. *(3 marks)* Medium

(Q12, **2021 HSC**)

3 The diagram shows a triangle ABC where $AC = 25$ cm, $BC = 16$ cm, $\angle BAC = 28°$ and angle ABC is obtuse.

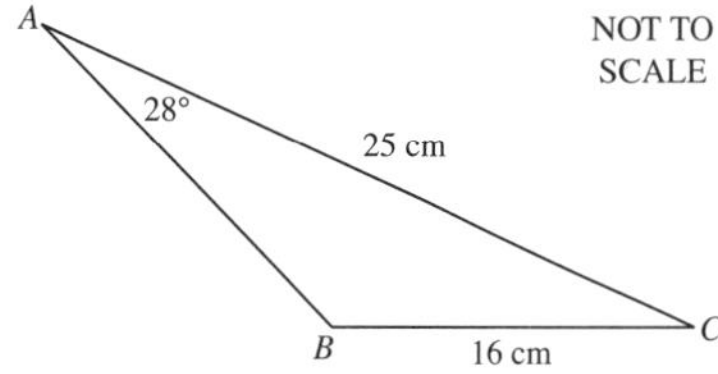

Find the size of the obtuse angle ABC correct to the nearest degree. *(3 marks)*

(Q18, **2021 HSC**) Medium

4 Mr Ali, Ms Brown and a group of students were camping at the site located at P. Mr Ali walked with some of the students on a bearing of 035° for 7 km to location A. Ms Brown, with the rest of the students, walked on a bearing of 100° for 9 km to location B.

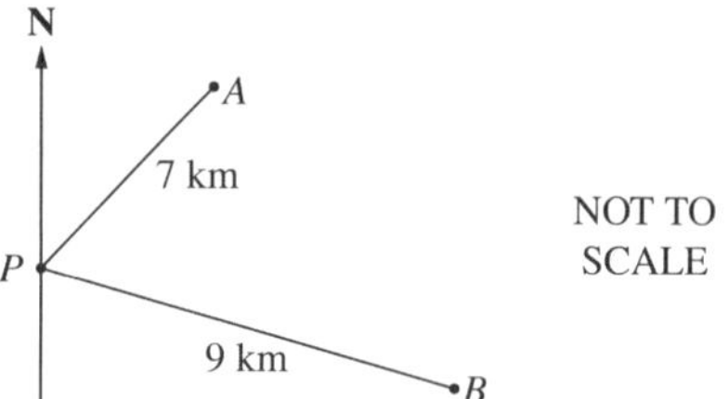

a Show that the angle APB is 65°. *(1 mark)* Easy

b Find the distance AB. *(2 marks)* Easy

c Find the bearing of Ms Brown's group from Mr Ali's group. Give your answer correct to the nearest degree. *(2 marks)* Medium

(Q15, **2020 HSC**)

5 The diagram shows a regular decagon (ten-sided shape with all sides equal and all interior angles equal). The decagon has centre O.

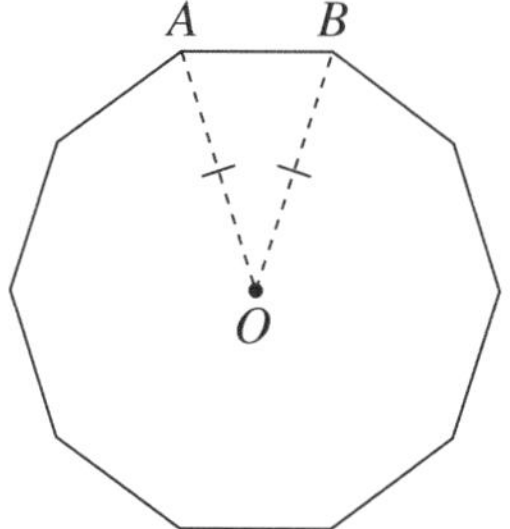

The perimeter of the shape is 80 cm.

By considering triangle OAB, calculate the area of the ten-sided shape. Give your answer in square centimetres correct to one decimal place. *(4 marks)*

(Q22, **2020 HSC**)

6 An 8-metre-high vertical flagpole ST is situated on flat ground.

The angle of elevation from A to the top of the tower S is 30°.

The angle of elevation from B to the top of the tower S is 45°.

B is south of the tower, and A is west of the tower.

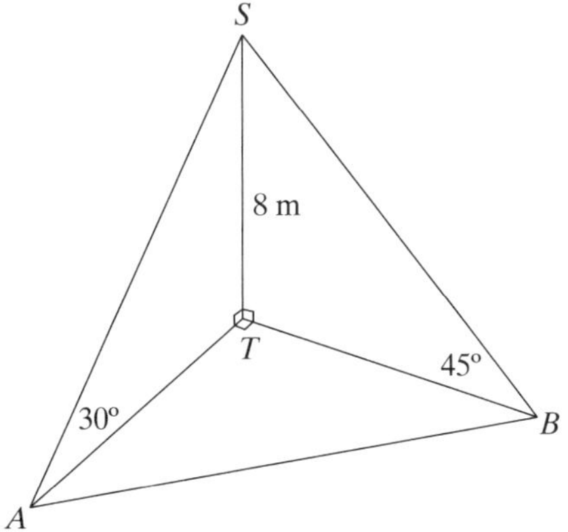

Find the distance from A to B. *(3 marks)*

Bonus question (see page iv) **Hard**

7 Using the sine rule, find the value of x correct to one decimal place.

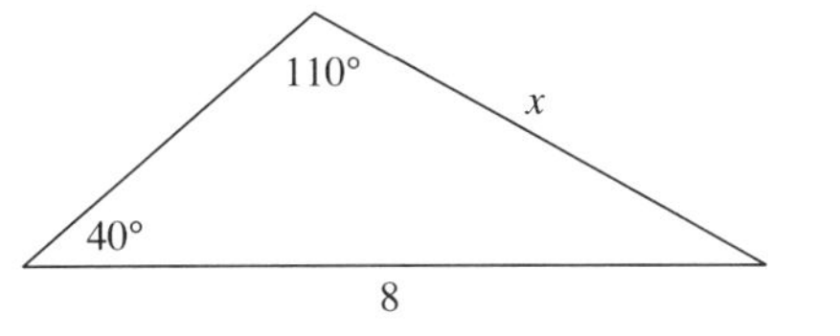

(2 marks)

(Q11a, **2019 HSC**) **Easy**

8 The regular hexagon $ABCDEF$ has sides of length 1. The diagonal AE and the side CD are produced to meet at the point X.

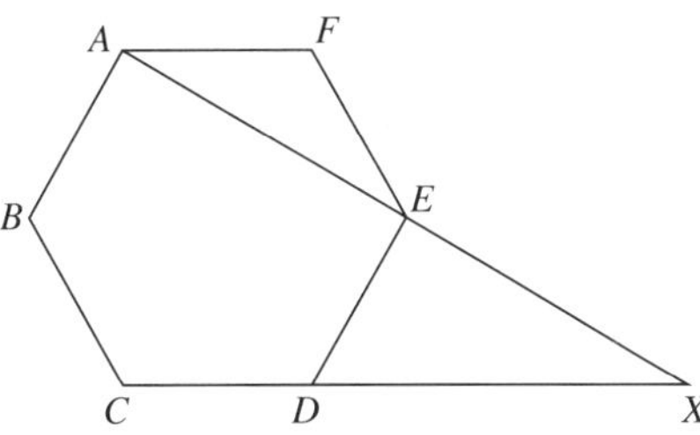

Copy or trace the diagram into your writing booklet.

Find the exact length of the line segment EX, justifying your answer. *(3 marks)*

(Q14c, **2019 HSC**) **Medium**

9 Which compass bearing is the same as a true bearing of 110°?

A S20°E **B** S20°W
C S70°E **D** S70°W *(1 mark)*

(Q4, **2019 MS2 HSC**) **Easy**

10 An owl is 7 metres above ground level, in a tree. The owl sees a mouse on the ground at an angle of depression of 32°. How far must the owl fly in a straight line to catch the mouse, assuming the mouse does not move?

A 3.7 m **B** 5.9 m
C 8.3 m **D** 13.2 m *(1 mark)*

(Q12, **2019 MS2 HSC**) **Medium**

11 The diagram shows a triangle with sides of length x cm, 11 cm and 13 cm and an angle of 80°.

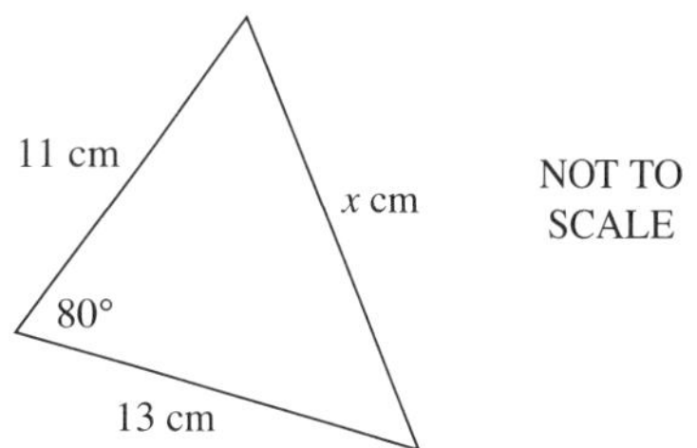

Use the cosine rule to calculate the value of x, correct to two significant figures. *(3 marks)*

(Q17, **2019 MS2 HSC**) **Medium**

12 Two right-angled triangles, ABC and ADC, are shown.

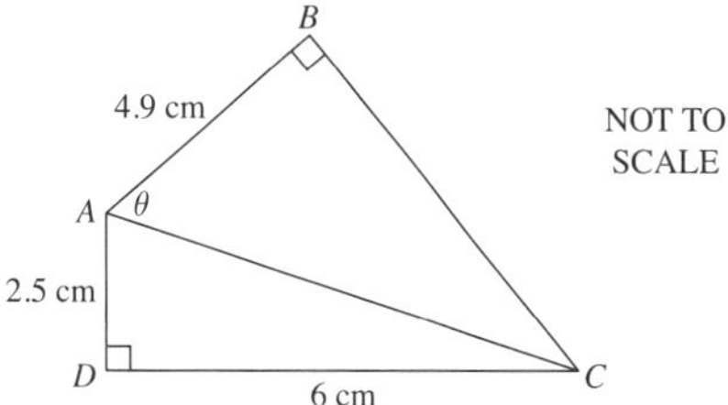

Calculate the size of angle θ, correct to the nearest minute. *(3 marks)*

(Q22, **2019 MS2 HSC**) **Medium**

13 A compass radial survey shows the positions of four towns A, B, C and D relative to point O.

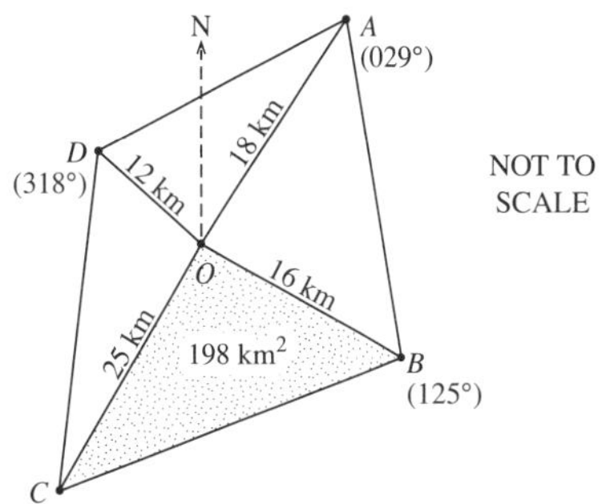

The area of triangle BOC is 198 km^2. Calculate the bearing of town C from point O, correct to the nearest degree. *(3 marks)*

(Q35, **2019 MS2 HSC**) **Medium**

14 A ship travels from Port A on a bearing of 050° for 320 km to Port B. It then travels on a bearing of 120° for 190 km to Port C.

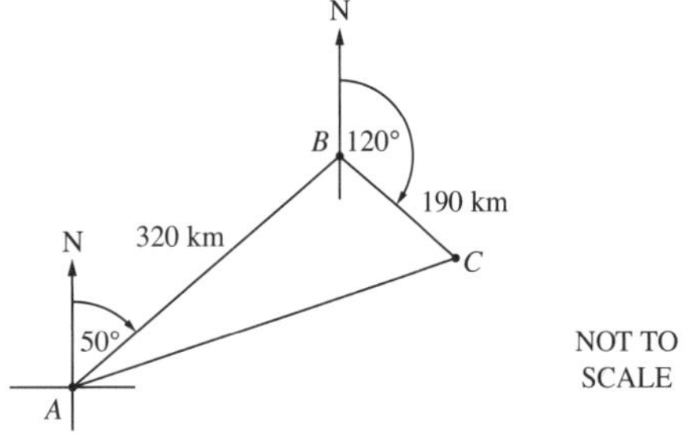

i What is the size of ABC? *(1 mark)* **Easy**

ii What is the distance from Port A to Port C? Answer to the nearest 10 kilometres. *(2 marks)* **Easy**

(Q12a, **2018 HSC**)

---

15 In ΔKLM, KL has length 3, LM has length 6 and KLM is 60°. The point N is chosen on side KM so that LN bisects KLM. The length LN is x.

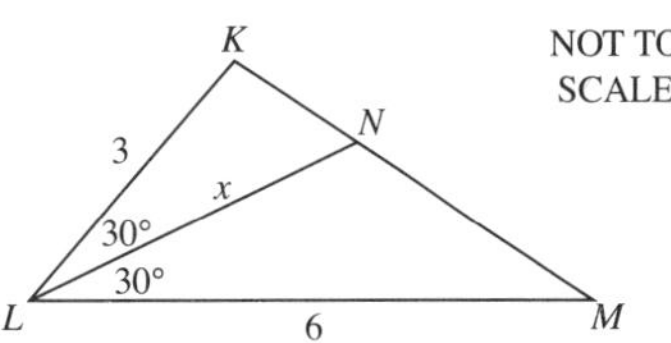

i Find the exact value of the area of ΔKLM. *(1 mark)* **Easy**

ii Hence, or otherwise, find the exact value of x. *(2 marks)* **Medium**

(Q14a, **2018 HSC**)

---

16 The diagram shows two triangles.

Triangle ABC is right-angled, with $AB = 13$ cm and $\angle ABC = 62°$.

In triangle ACD, $AD = x$ cm and $\angle DAC = 40°$. The area of triangle ACD is 30 cm^2.

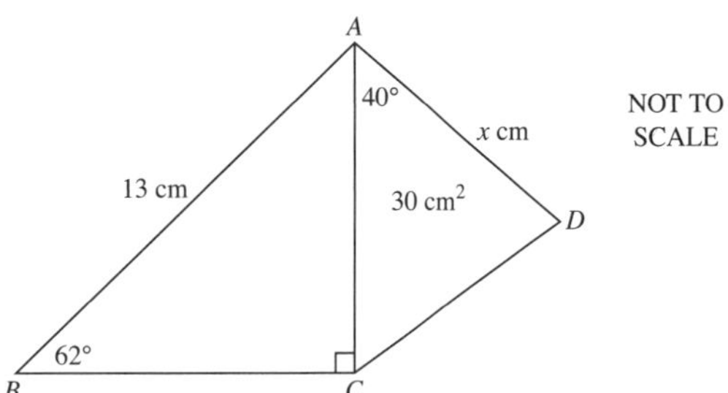

What is the value of x, correct to one decimal place? *(3 marks)*

(Q30c, **2018 MS2 HSC**) **Medium**

---

17 Using the cosine rule, find the value of x in the following diagram.

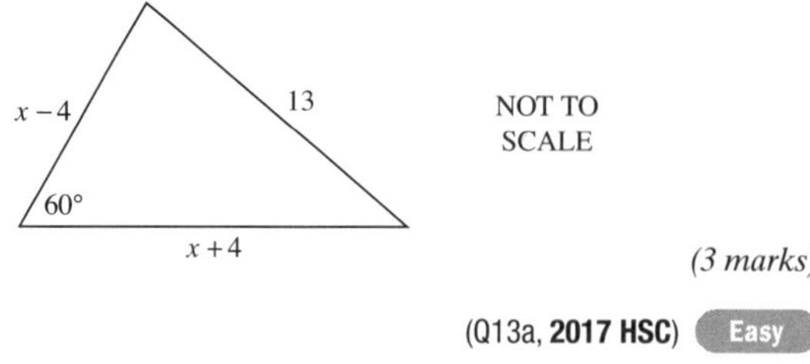

(3 marks)

(Q13a, **2017 HSC**) **Easy**

---

18 The diagram shows the location of three schools. School A is 5 km due north of school B, school C is 13 km from school B and $\angle ABC$ is 135°.

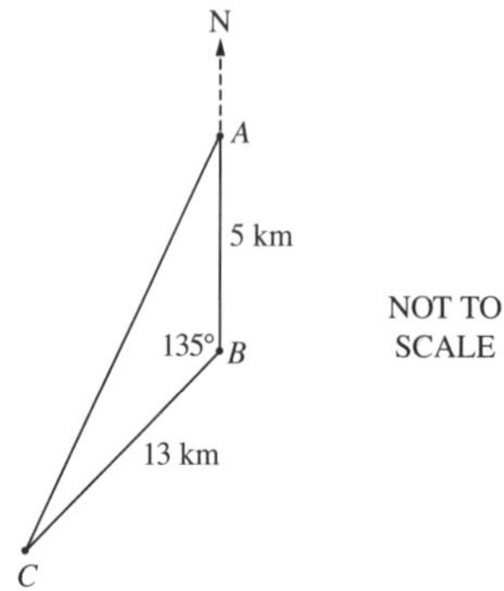

i Calculate the shortest distance from school A to school C, to the nearest kilometre. *(2 marks)* **Easy**

ii Determine the bearing of school C from school A, to the nearest degree. *(3 marks)* **Medium**

(Q30c, **2017 MS2 HSC**)

---

19 For the angle θ, $\sin\theta = \dfrac{7}{25}$ and $\cos\theta = -\dfrac{24}{25}$.

Which diagram best shows the angle θ?

A

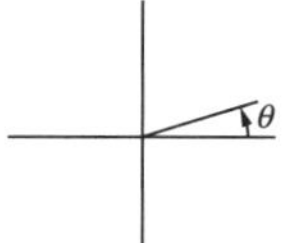

B

C

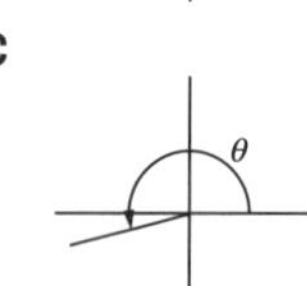

D

(1 mark)

(Q1, **2016 HSC**) **Easy**

---

20 Square tiles of side length 20 cm are being used to tile a bathroom.

The tiler needs to drill a hole in one of the tiles at a point P which is 8 cm from one corner and 15 cm from an adjacent corner.

To locate the point P the tiler needs to know the size of the angle θ shown in the diagram.

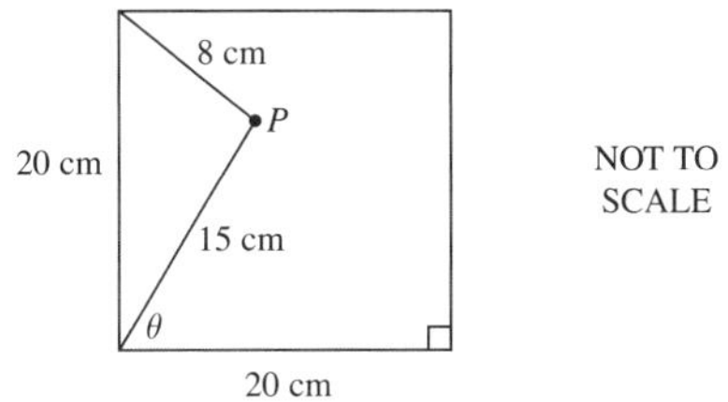

Find the size of the angle θ to the nearest degree. *(3 marks)*

(Q12c, **2016 HSC**) Easy

21 The diagram shows towns A, B and C. Town B is 40 km due north of town A. The distance from B to C is 18 km and the bearing of C from A is 025°. It is known that $\angle BCA$ is obtuse.

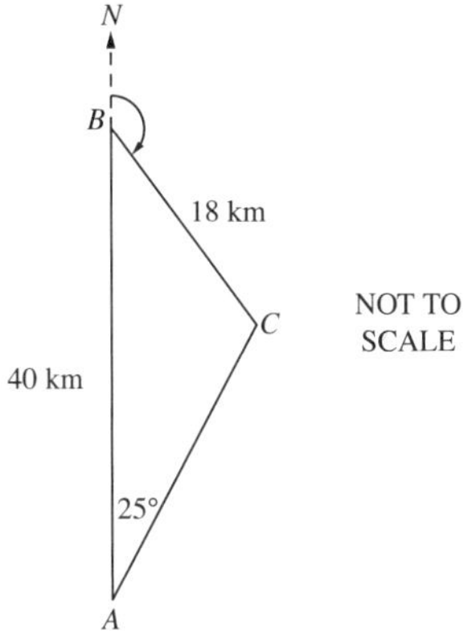

What is the bearing of C from B?

A 070° **B** 095°
C 110° **D** 135° *(1 mark)*

(Q25, **2016 MS2 HSC**) Medium

22 The diagram shows ΔABC with sides $AB = 6$ cm, $BC = 4$ cm and $AC = 8$ cm.

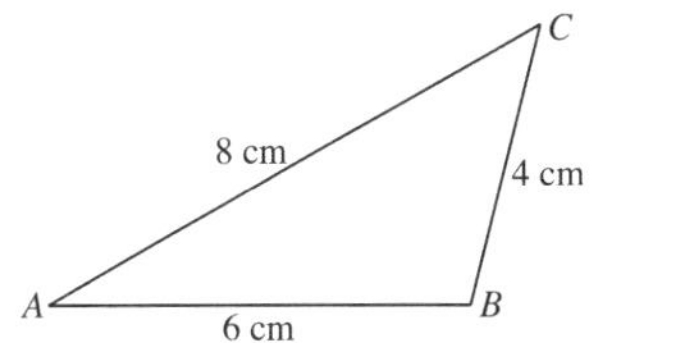

i Show that $\cos A = \dfrac{7}{8}$. *(1 mark)* Easy

ii By finding the exact value of $\sin A$, determine the exact value of the area of ΔABC. *(2 marks)* Medium

(Q13a, **2015 HSC**)

23 From point S, which is 1.8 m above the ground, a pulley at P is used to lift a flat object F. The lengths SP and PF are 5.4 m and 2.1 m respectively. The angle PSC is 108°.

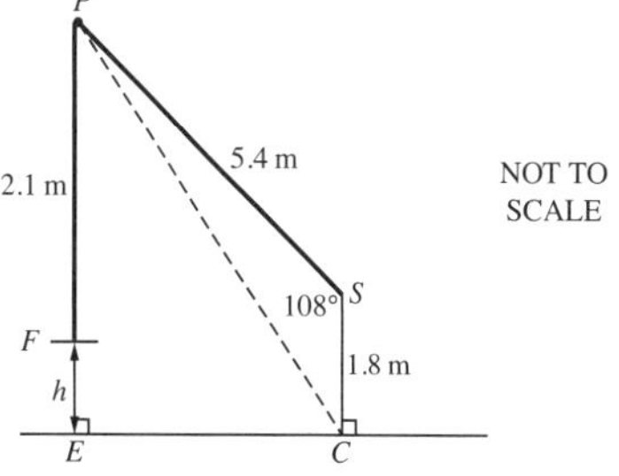

i Show that the length PC is 6.197 m, correct to 3 decimal places. *(1 mark)* Easy

ii Calculate h, the height of the object above the ground. *(4 marks)* Hard

(Q30e, **2015 MS2 HSC**)

24 Chris leaves island A in a boat and sails 142 km on a bearing of 078° to island B. Chris then sails on a bearing of 191° for 220 km to island C, as shown in the diagram.

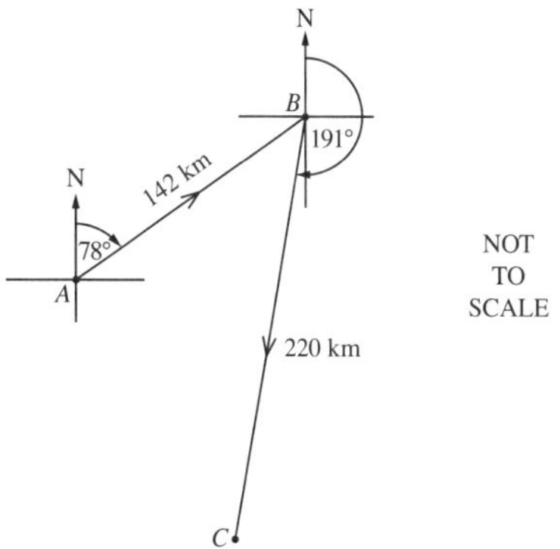

i Show that the distance from island C to island A is approximately 210 km. *(2 marks)* Easy

ii Chris wants to sail from island C directly to island A. On what bearing should Chris sail? Give your answer correct to the nearest degree. *(3 marks)* Medium

(Q13d, **2014 HSC**)

25 The following information is given about the locations of three towns X, Y and Z:

- X is due east of Z
- X is on a bearing of 145° from Y
- Y is on a bearing of 060° from Z.

Which diagram best represents this information?

A

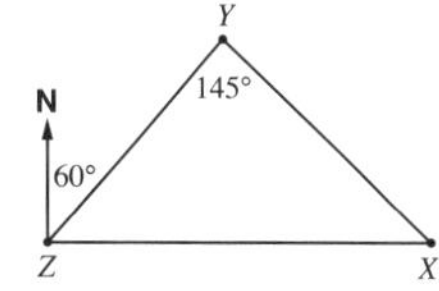

B

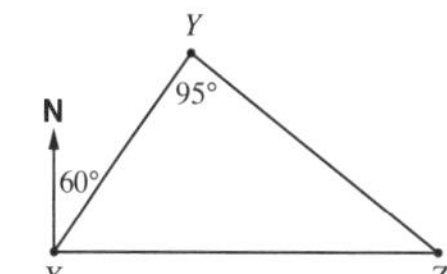

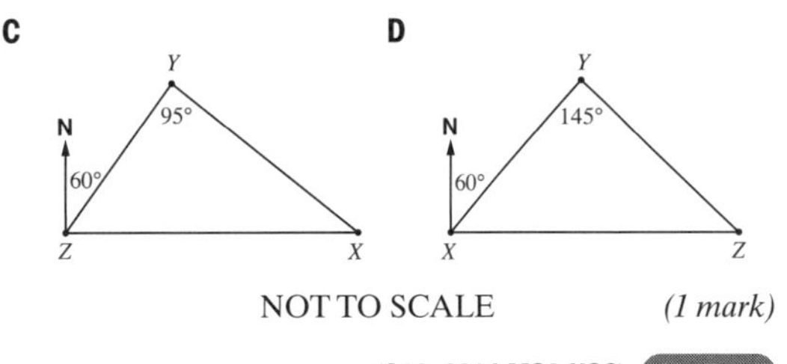

NOT TO SCALE (1 mark)

(Q23, **2014 MS2 HSC**) Medium

26 The diagram shows a triangle ABC. The line $2x + y = 8$ meets the x and y axes at the points A and B respectively. The point C has coordinates (7, 4).

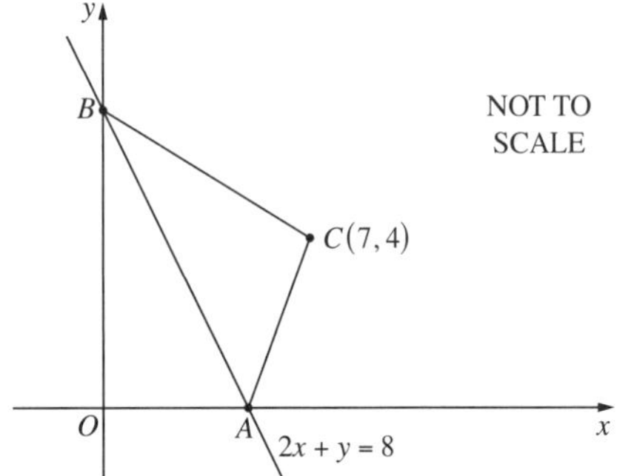

NOT TO SCALE

i Calculate the distance AB. *(2 marks)* Easy

ii It is known that $AC = 5$ and $BC = \sqrt{65}$. (Do NOT prove this.)
Calculate the size of $\angle ABC$ to the nearest degree. *(2 marks)* Medium

iii The point N lies on AB such that CN is perpendicular to AB.
Find the coordinates of N. *(3 marks)* Medium

(Q13a, **2012 HSC**)

27 Raj cycles around a course. The course starts at E, passes through F, G and H and finishes at E. The distances EH and GH are equal.

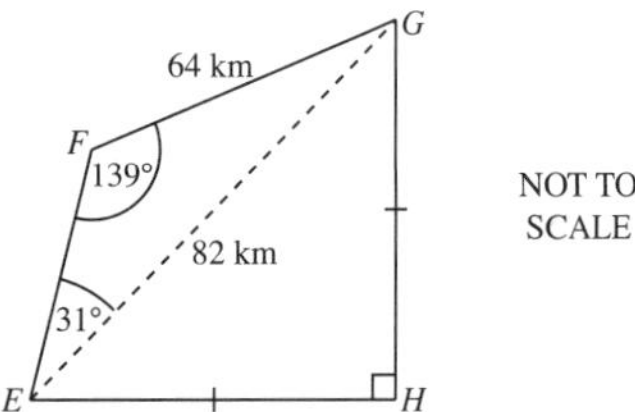

NOT TO SCALE

i What is the length of EF, to the nearest kilometre? *(2 marks)* Easy

ii What is the total distance that Raj cycles, to the nearest kilometre? *(3 marks)* Medium

(Q29c, **2012 MS2 HSC**)

28 In the diagram, the shop at S is 20 kilometres across the bay from the post office at P. The distance from the shop to the lighthouse at L is 22 kilometres and $\angle SPL$ is 60°.

Let the distance PL be x kilometres.

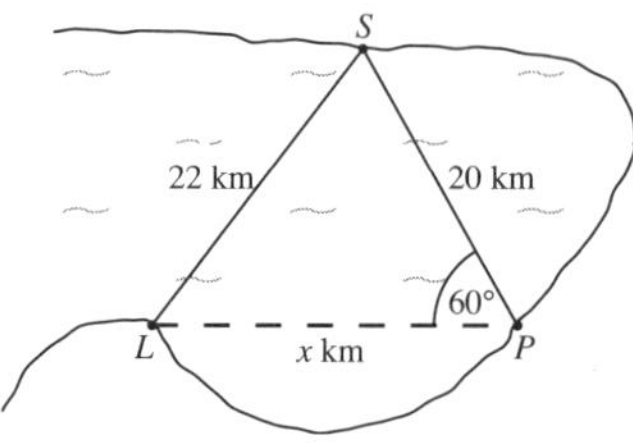

i Use the cosine rule to show that $x^2 - 20x - 84 = 0$. *(1 mark)* Medium

ii Hence, find the distance from the post office to the lighthouse. Give your answer correct to the nearest kilometre. *(2 marks)* Medium

(Q8a, **2011 HSC**)

29 A ship sails 6 km from A to B on a bearing of 121°. It then sails 9 km to C. The size of angle ABC is 114°.

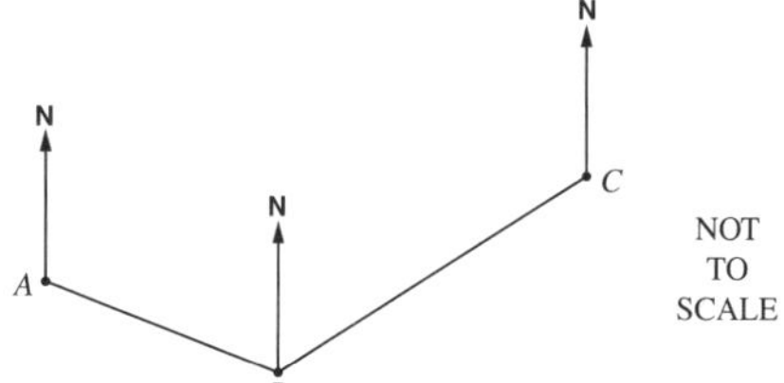

NOT TO SCALE

Copy the diagram into your writing booklet and show all the information on it.

i What is the bearing of C from B? *(1 mark)* Medium

ii Find the distance AC. Give your answer correct to the nearest kilometre. *(2 marks)* Medium

iii What is the bearing of A from C? Give your answer correct to the nearest degree. *(3 marks)* Hard

(Q24c, **2011 MS2 HSC**)

30 A boat is sailing due north from a point A towards a point P on the shore line. The shore line runs from west to east.

In the diagram, T represents a tree on a cliff vertically above P, and L represents a landmark on the shore. The distance PL is 1 km.

From A the point L is on a bearing of 020°, and the angle of elevation to T is 3°.

After sailing for some time the boat reaches a point B, from which the angle of elevation to T is 30°.

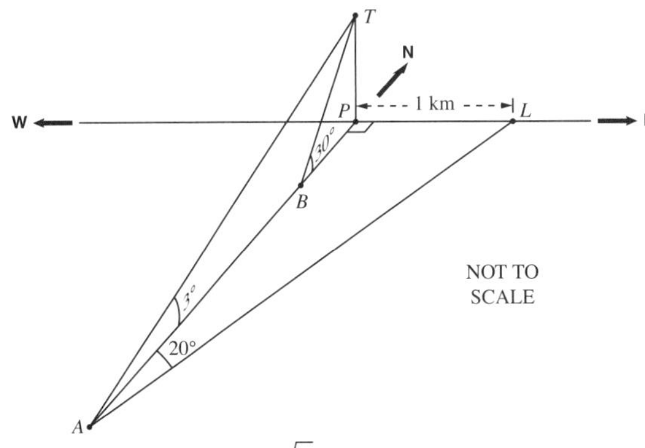

i Show that $BP = \dfrac{\sqrt{3}\tan 3°}{\tan 20°}$. *(3 marks)* Medium

ii Find the distance AB. *(1 mark)* Medium

(Q5a, **2010 ME1 HSC**)

31 The base of a lighthouse, D, is at the top of a cliff 168 metres above sea level. The angle of depression from D to a boat at C is 28°. The boat heads towards the base of the cliff, A, and stops at B. The distance AB is 126 metres.

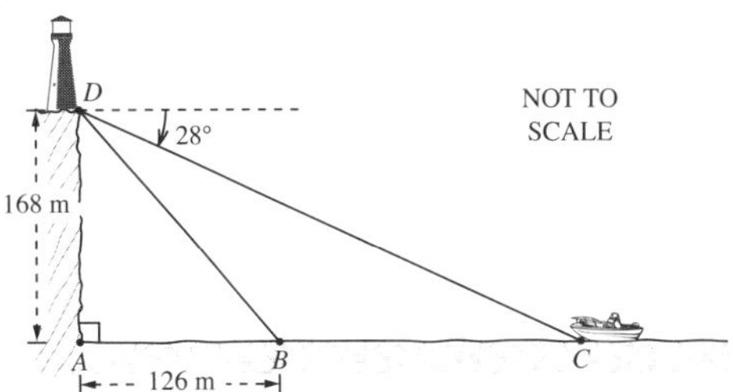

i What is the angle of depression from D to B, correct to the nearest minute? *(3 marks)* Medium

ii How far did the boat travel from C to B, correct to the nearest metre? *(2 marks)* Medium

(Q24d, **2010 MS2 HSC**)

32 Find the area of triangle ABC, correct to the nearest square metre.

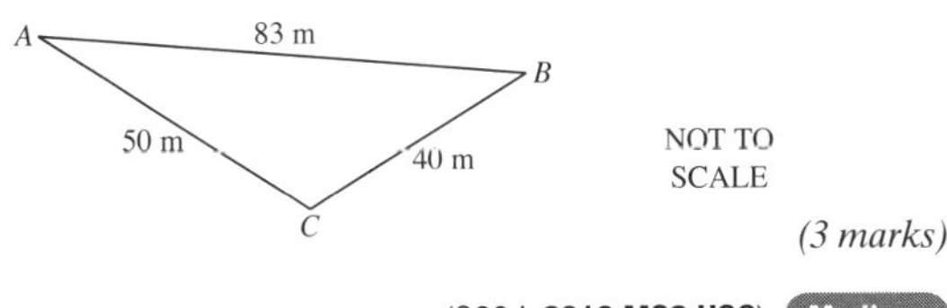

(3 marks)

(Q26d, **2010 MS2 HSC**) Medium

33 In the diagram, AD and DC are equal to 30 cm.

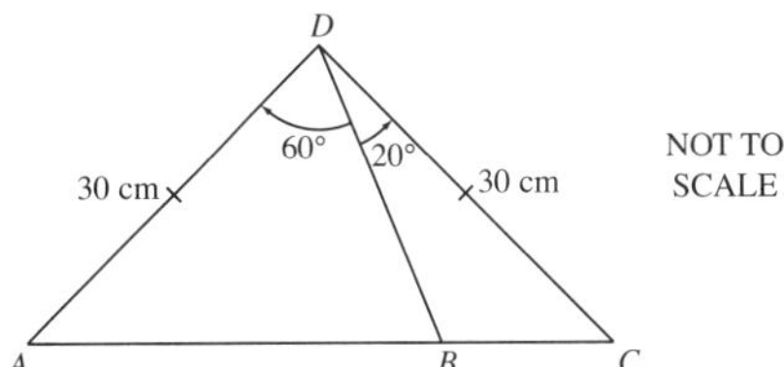

What is the length of AB to the nearest centimetre?

A 28 cm **B** 31 cm
C 34 cm **D** 39 cm *(1 mark)*

(Q22, **2009 MS2 HSC**) Medium

34 The point A is 25 m from the base of a building. The angle of elevation from A to the top of the building is 38°.

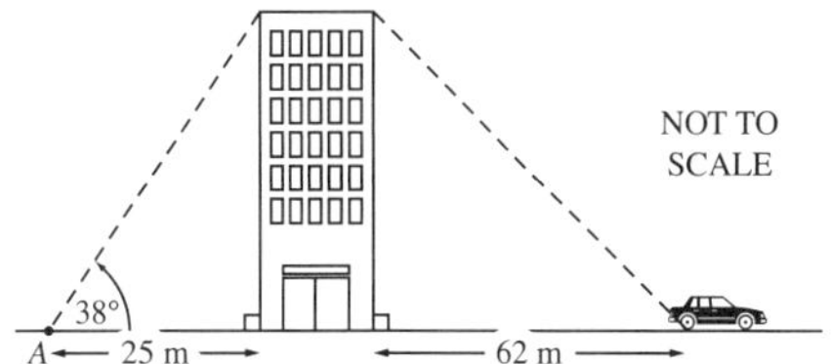

i Show that the height of the building is approximately 19.5 m. *(1 mark)* Easy

ii A car is parked 62 m from the base of the building.
What is the angle of depression from the top of the building to the car? Give your answer to the nearest degree. *(2 marks)* Medium

(Q23a, **2009 MS2 HSC**)

35 A yacht race follows the triangular course shown in the diagram. The course from P to Q is 1.8 km on a true bearing of 058°. At Q the course changes direction. The course from Q to R is 2.7 km and $\angle PQR = 74°$.

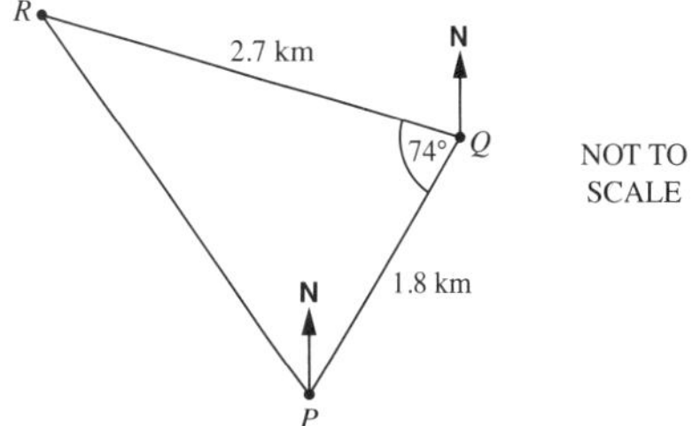

i What is the bearing of R from Q? *(1 mark)* Easy

ii What is the distance from R to P? *(2 marks)* Medium

iii The area inside this triangular course is set as a 'no-go' zone for other boats while the race is on.
What is the area of this 'no-go' zone? *(1 mark)* Medium

(Q27b, **2009 MS2 HSC**)

36 From a point A due south of a tower, the angle of elevation of the top of the tower T, is 23°. From another point B, on a bearing of 120° from the tower, the angle of elevation of T is 32°. The distance AB is 200 metres.

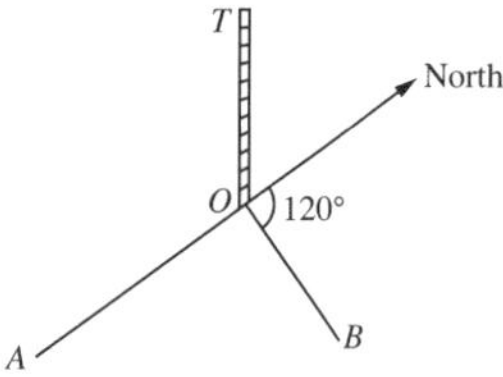

i Copy or trace the diagram into your writing booklet, adding the given information to your diagram. *(1 mark)* Easy

ii Hence find the height of the tower. *(3 marks)* Hard

(Q6a, **2008 ME1 HSC**)

37 The angle of depression from J to M is 75°. The length of JK is 20 m and the length of MK is 18 m.

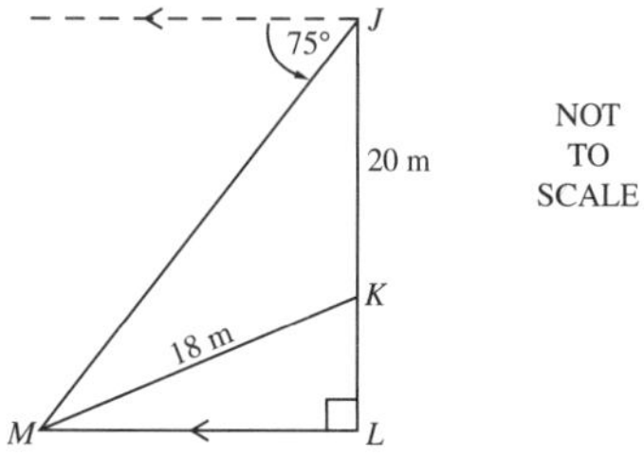

Copy or trace this diagram into your writing booklet and calculate the angle of elevation from M to K. Give your answer to the nearest degree. *(3 marks)*

(Q25b, **2007 MS2 HSC**) Medium

38 The diagram shows information about the locations of towns A, B and Q.

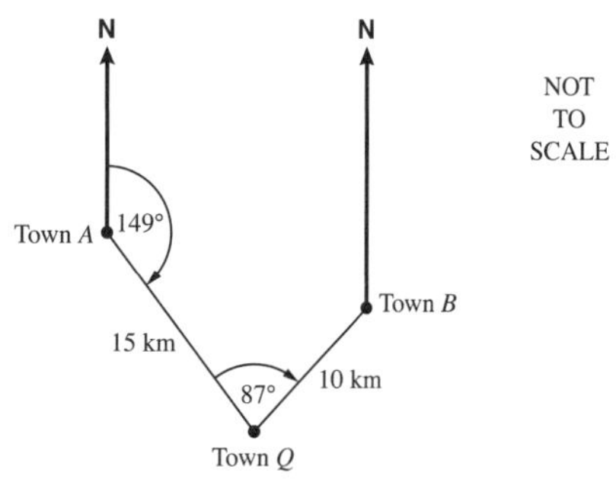

i It takes Elina 2 hours and 48 minutes to walk directly from Town A to Town Q. Calculate her walking speed correct to the nearest km/h. *(1 mark)* Medium

ii Elina decides, instead, to walk to Town B from Town A and then to Town Q. Find the distance from Town A to Town B. Give your answer to the nearest km. *(2 marks)* Easy

iii Calculate the bearing of Town Q from Town B. *(1 mark)* Medium

(Q26a, **2007 MS2 HSC**)

39 Find the value of θ in the diagram. Give your answer to the nearest degree.

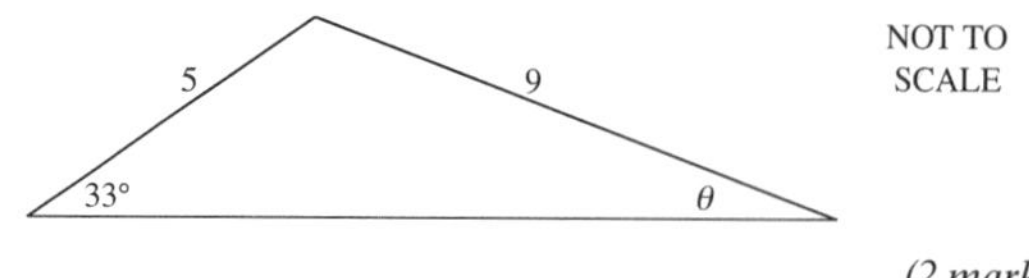

(2 marks)

(Q1d, **2006 HSC**) Easy

40 The lengths of the sides of a triangle are 7 cm, 8 cm and 13 cm.

i Find the size of the angle opposite the longest side. *(2 marks)* Easy

ii Find the area of the triangle. *(1 mark)* Easy

(Q3b, **2005 HSC**)

41 The diagram shows a point P which is 30 km due west of the point Q.

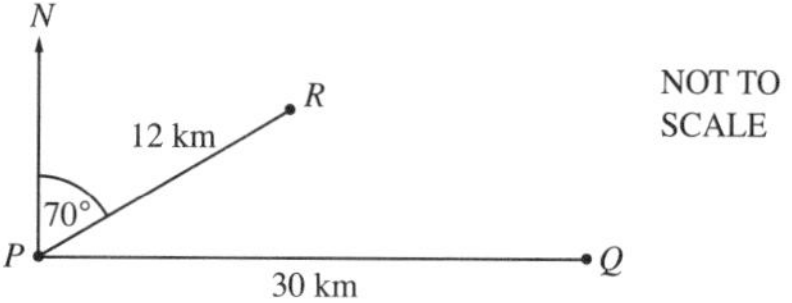

The point R is 12 km from P and has a bearing from P of 070°.

i Find the distance of R from Q. *(2 marks)* Medium

ii Find the bearing of R from Q. *(2 marks)* Medium

(Q3c, **2004 HSC**)

42 The length of each edge of the cube $ABCDEFGH$ is 2 metres. A circle is drawn on the face $ABCD$ so that it touches all four edges of the face. The centre of the circle is O and the diagonal AC meets the circle at X and Y.

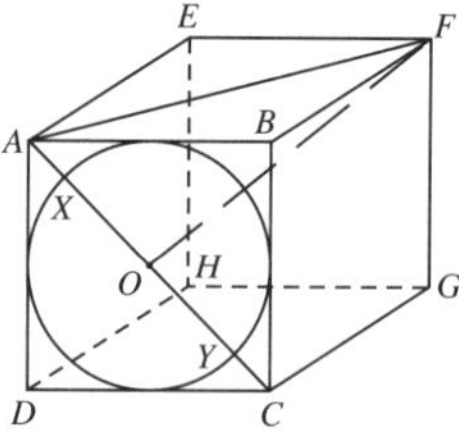

i Explain why $\angle FAC = 60°$. *(1 mark)* Easy

ii Show that $FO = \sqrt{6}$ metres. *(1 mark)* Medium

iii Calculate the size of $\angle XFY$ to the nearest degree. *(1 mark)* Medium

(Q3d, **2004 ME1 HSC**)

43 In the diagram, the point Q is due east of P. The point R is 38 km from P and 20 km from Q. The bearing of R from Q is 325°.

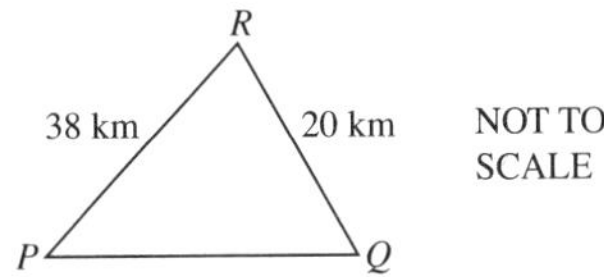

i What is the size of $\angle PQR$? *(1 mark)* Easy

ii What is the bearing of R from P? *(3 marks)* Medium

(Q4a, **2003 HSC**)

44 David is in a life raft and Anna is in a cabin cruiser searching for him. They are in contact by mobile telephone. David tells Anna that he can see Mt Hope. From David's position the mountain has a bearing of 109°, and the angle of elevation to the top of the mountain is 16°.

Anna can also see Mt Hope. From her position it has a bearing of 139°, and the top of the mountain has an angle of elevation of 23°.

The top of Mt Hope is 1500 m above sea level.

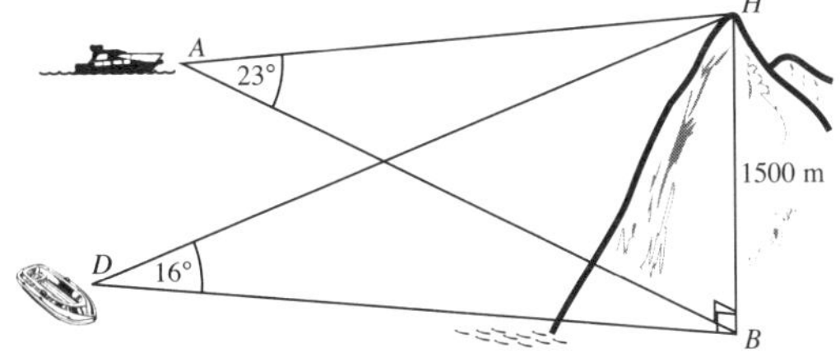

Find the distance and bearing of the life raft from Anna's position. *(4 marks)*

(Q7a, **2003 ME1 HSC**) Hard

45

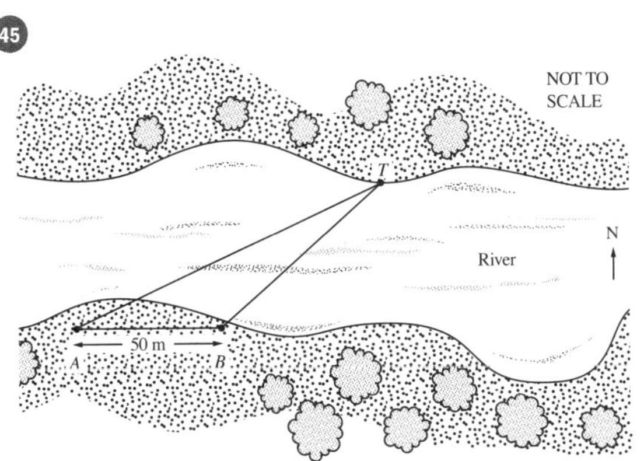

In the diagram above, the following measurements are given:

$\angle TAB = 30°$.
B is 50 m due east of A.

The bearing of T from B is 020°.

Copy or trace ΔABT into your answer booklet.

i Explain why $\angle ABT$ is 110°. *(1 mark)* Easy

ii Calculate the distance BT (to the nearest metre). *(3 marks)* Easy

(Q26b, **2003 MS2 HSC**)

46 In the diagram, XYZ is a triangle where $\angle ZYX = 45°$ and $\angle ZXY = 60°$.

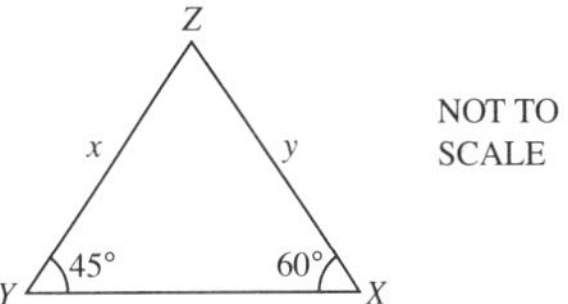

Find the exact value for the ratio $\frac{x}{y}$. *(3 marks)*

(Q2c, **2002 HSC**) Medium

47 In the diagram, LMN is a triangle where $LM = 5.2$ metres, $LN = 8.9$ metres and angle $MLN = 110°$.

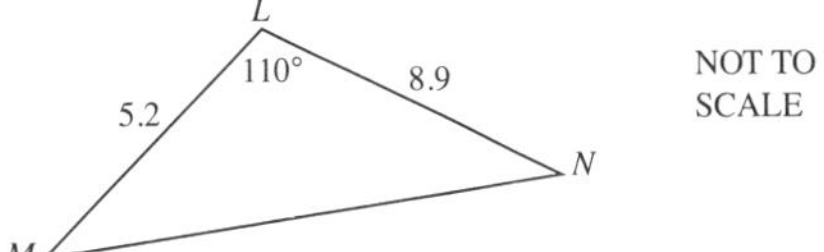

i Find the length of MN. *(2 marks)* Easy

ii Calculate the area of triangle LMN. *(2 marks)* Easy

(Q4c, **2002 HSC**)

48 The diagram shows a triangle with sides 7 cm, 13 cm and x cm, and an angle of 60° as marked.

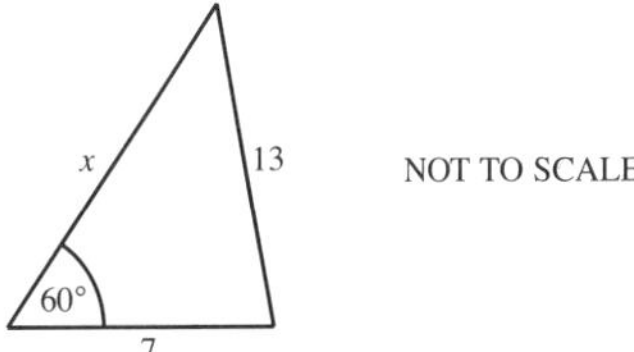

Use the cosine rule to show that $x^2 - 7x = 120$, and hence find the exact value of x. *(4 marks)*

(Q3d, **2001 HSC**) Hard

49 A surveyor stands at a point A, which is due south of a tower OT of height h m. The angle of elevation of the top of the tower from A is 45°. The surveyor then walks 100 m due east to point B, from where she measures the angle of elevation of the top of the tower to be 30°.

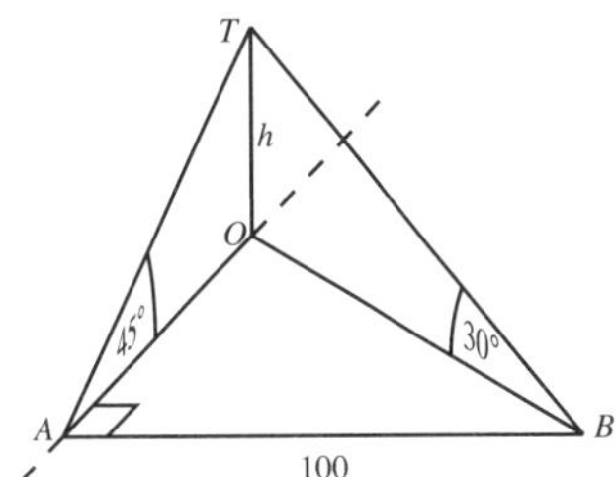

i Express the length of OB in terms of h. Easy

ii Show that $h = 50\sqrt{2}$. Hard

iii Calculate the bearing of B from the base of the tower. Hard

(5 marks) (Q3c, **2000 ME1 HSC**)

50 In the diagram, AC is parallel to DB, AB is 6 cm, BC is 3 cm and AC is 7 cm.

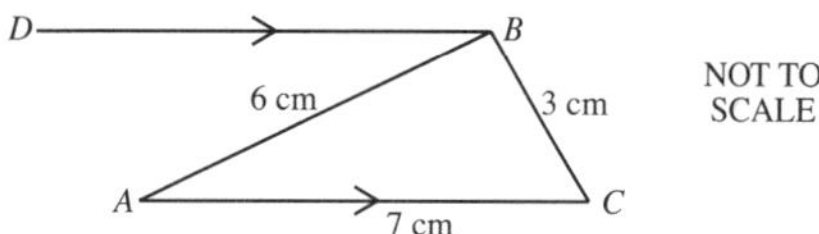

i Use the cosine rule to find the size of $\angle ACB$, to the nearest degree. Easy

ii Hence find the size of $\angle DBC$, giving reasons for your answer. Easy

(4 marks) (Q3c, **1999 HSC**)

51 The figure shows the net of an oblique pyramid with a rectangular base. In this figure, $PXZYR$ is a straight line, $PX = 15$ cm, $RY = 20$ cm, $AB = 25$ cm, and $BC = 10$ cm. Further, $AP = PD$ and $BR = RC$.

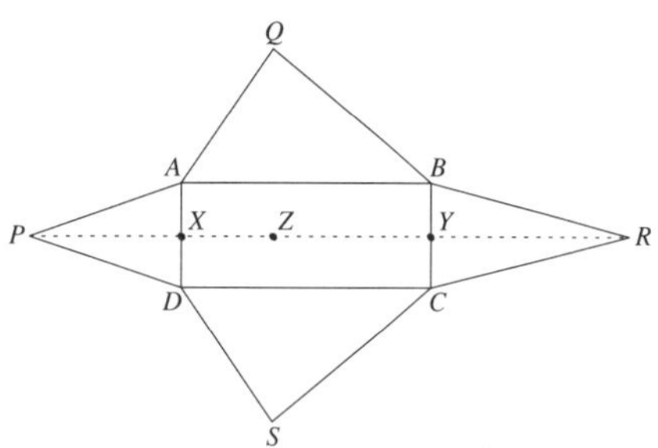

When the net is folded, points P, Q, R, and S all meet at the apex T, which lies vertically above the point Z in the horizontal base, as shown below.

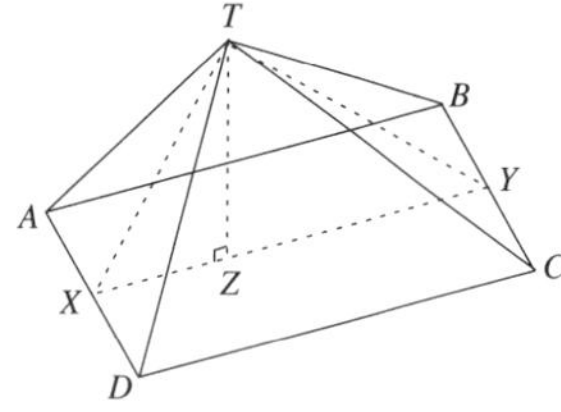

i Show that triangle TXY is right-angled. Medium

ii Hence show that T is 12 cm above the base. Hard

iii Hence find the angle that the face DCT makes with the base. Hard

(5 marks) (Q4c, **1996 ME1 HSC**)

52 A horizontal bridge is built between points A and B. The bridge is supported by cables SP and SR, which are attached to the top of a vertical pylon ST.

The section of the pylon, SQ, above the bridge is 8 metres long and $\angle SRQ = 52°$.

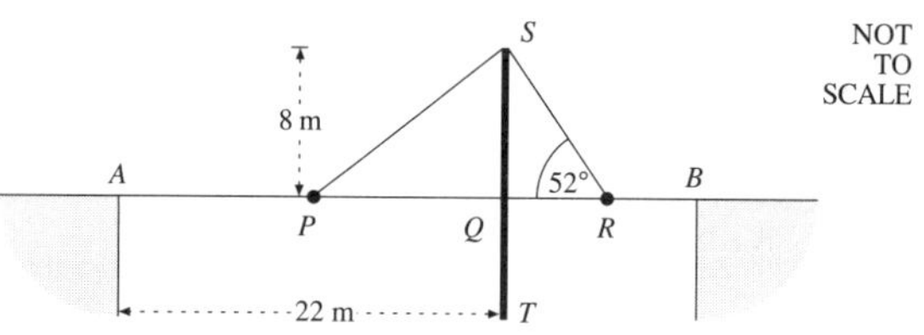

The distance AQ is 22 metres, and P is the midpoint of AQ.

i Find the length of the cable SR. Medium

ii Find the size of $\angle SPQ$ to the nearest degree. Medium

(4 marks) (Q3b, **1995 HSC**)

53 The figure shows the side view of a bridge opened to let boats pass underneath. When the equal arms of the bridge PA and QB are lowered, they meet exactly to form the straight roadway PQ, which is 50 m long. When the arms PA and QB are raised through an angle θ as shown, the 'corridor' AB is 12 m wide.

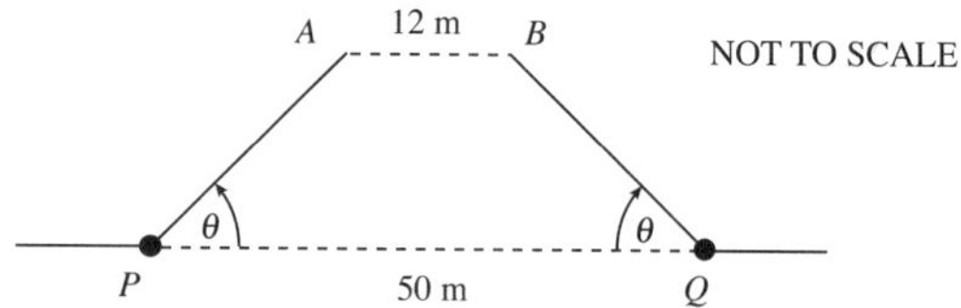

Calculate the size of angle θ.

(Q5b, **1994 HSC**) Medium

54 From position P, Anne finds that the angle of elevation of the top A of the rock pillar AB is 8°. After walking 300 m directly towards the pillar to the point Q she finds that the angle of elevation of A is 21°.

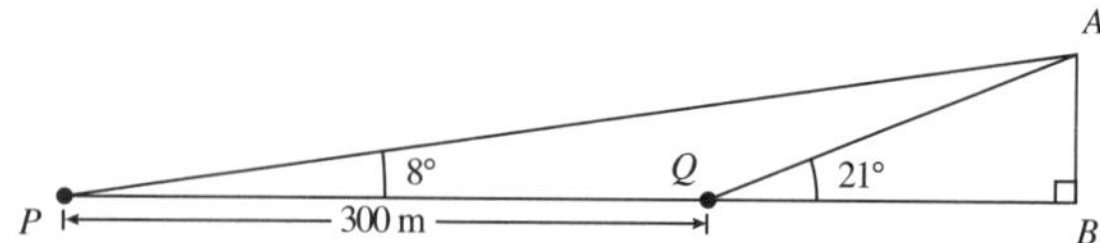

i Copy the diagram into your Writing Booklet and find $\angle PAQ$. Easy

ii Calculate the length AQ. Easy

iii Find the height of the rock pillar AB. Easy

(Q4a, **1993 HSC**)

Year 11 Trigonometry—Worked answers

1 $\dfrac{h}{AB} = \tan 26^o$

$\therefore AB = \dfrac{h}{\tan 26^o}$

$= h \cot 26^o$

Answer B

2 **a** $\dfrac{XY}{16} = \cos 30^o$ ✓

$XY = 16 \times \cos 30^o$

$= 13.856\,406\,46\ldots$

$= 13.86$ (2 dec. pl.)

$\therefore$ the length is 13.86 cm. ✓

(2 marks)

b Area of shaded region

= Area of semicircle − Area of triangle

$= \dfrac{1}{2} \times \pi \times 8^2 - \dfrac{1}{2} \times 13.86 \times 16 \times \sin 30^o$ ✓✓

$= 45.090\,964\,91\ldots$

$= 45.1$ (1 dec. pl.)

$\therefore$ the area is 45.1 cm^2. ✓

Alternatively, the area of the triangle can be found by first using Pythagoras' theorem.

(3 marks)

3 Using the sine rule:

$\dfrac{\sin \angle ABC}{25} = \dfrac{\sin 28^\circ}{16}$ ✓

$\sin \angle ABC = \dfrac{25 \times \sin 28^\circ}{16}$

$\sin \angle ABC = 0.733\,549\,316\ldots$

$\angle ABC = 47^\circ$

or 133° (nearest degree) ✓

But $\angle ABC$ is obtuse, hence,

$\angle ABC = 133^\circ$. ✓

(3 marks)

4 **a** As $100 - 35 = 65$, then angle $APB = 65^\circ$.

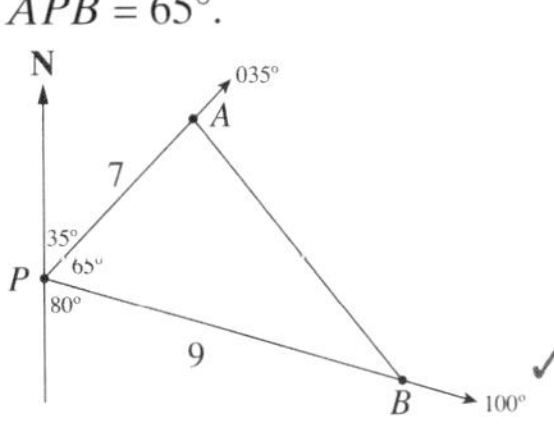

✓

(1 mark)

b Using cosine rule,

$AB^2 = 7^2 + 9^2 - 2(7)(9) \cos 65^\circ$ ✓

$= 76.750\,099\,02\ldots$

$AB = 8.760\,713\,385\ldots$

$= 8.76$ (2 dec. pl.)

$\therefore$ the distance is 8.76 km. ✓

(2 marks)

c

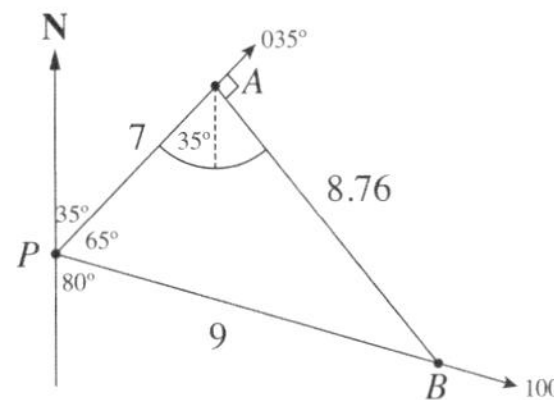

Using cosine rule,

$\cos \angle PAB = \dfrac{7^2 + 8.76^2 - 9^2}{2(7)(8.76)}$

$= 0.364\,787\,997\ldots$

$\angle PAB = 68.605\,463\,71\ldots$

$= 69$ ✓ (nearest whole)

As $69 - 35 = 34$, and $180 - 34 = 146$, the bearing of B from A is 146°. ✓

(2 marks)

5 As $80 \div 10 = 8$, then $AB = 8$ cm.

As $360 \div 10 = 36$, angle $AOB = 36^\circ$.

As ΔAOB is isosceles, then OM is perpendicular to AB and BM is 4 cm. ✓

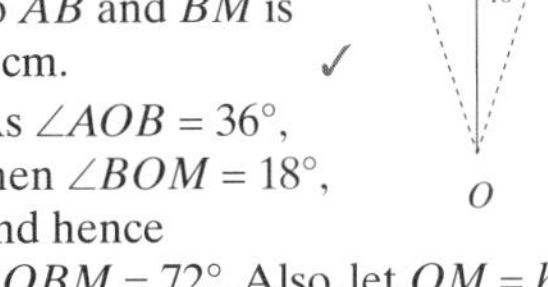

As $\angle AOB = 36^\circ$, then $\angle BOM = 18^\circ$, and hence $\angle OBM = 72^\circ$. Also, let $OM = h$.

$\dfrac{h}{4} = \tan 72^\circ$

$h = 4 \tan 72^\circ$ ✓

Area $\Delta OAB = \dfrac{1}{2} \times AB \times h$

$= \dfrac{1}{2} \times 8 \times 4 \tan 72^\circ$

$= 16 \tan 72^\circ$ ✓

Area of decagon

$= 10 \times 16 \tan 72^\circ$

$= 492.429\,3659\ldots$

$= 492.4$ (1 dec. pl.)

$\therefore$ the area is 492.4 cm^2. ✓ *(4 marks)*

6

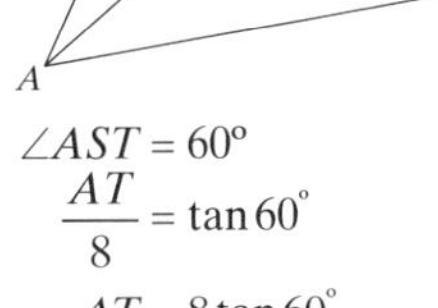

$\angle AST = 60^o$

$\dfrac{AT}{8} = \tan 60^\circ$

$AT = 8 \tan 60^\circ$

$= 8\sqrt{3}$ ✓

As ΔSBT is isosceles, $BT = 8$. ✓

Let $AB = x$.

$x^2 = 8^2 + (8\sqrt{3})^2$ (by Pythagoras)

$= 64 + 192$

$= 256$

$x = 16$

The distance is 16 metres. ✓

(3 marks)

7 $\dfrac{x}{\sin 40^\circ} = \dfrac{8}{\sin 110^\circ}$

$x = \dfrac{8 \times \sin 40^\circ}{\sin 110^\circ}$ ✓

$= 5.472\,322\,293\ldots$

$= 5.5$ (1 dec. pl.) ✓ *(2 marks)*

8 Angle sum of hexagon = 720°

Internal angle of regular hexagon = 120°

$\angle AFE = 120^o$

$\therefore \angle FEA = 30^o$ (base angles of isosceles triangle AFE equal)

$\angle AED = 120^o - 30^o$

$= 90^o$

$\therefore \angle DEX = 90^o$ (straight angle) ✓

$\angle EDC = 60^o$ (exterior angle of regular hexagon) ✓

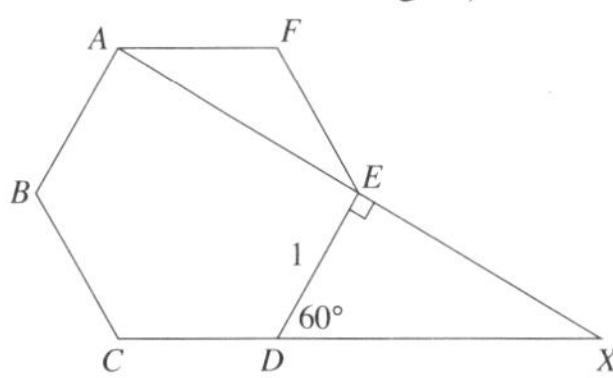

Using triangle EDX:

$\dfrac{x}{1} = \tan 60^o$

$x = \sqrt{3}$

$\therefore$ the length of EX is $\sqrt{3}$ units ✓

(3 marks)

9 $110^o = 180^o - 70^o$

Compass bearing is S70°E

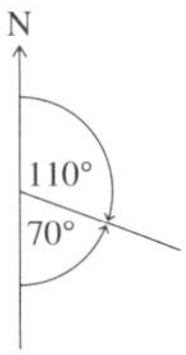

Answer C

10 Let x m be the distance the owl must fly.

$\sin 32^\circ = \dfrac{7}{x}$

$x = \dfrac{7}{\sin 32^\circ}$

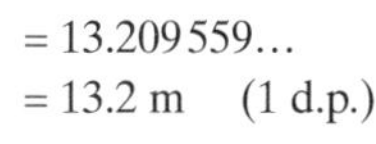

$= 13.209\,559\ldots$

$= 13.2$ m (1 d.p.)

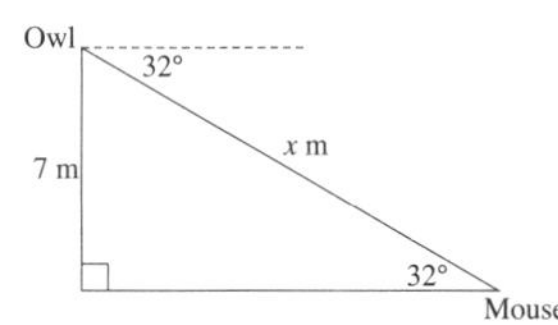

Answer D

11 $c^2 = a^2 + b^2 - 2ab\cos C$

$x^2 = 11^2 + 13^2 - 2 \times 11 \times 13 \times \cos 80°$ ✓

$= 240.336\,621\ldots$

$x = \sqrt{240.336\,621\ldots}$ $(x > 0)$

$= 15.502\,7939\ldots$ ✓

$= 16$ (2 sig. figs) ✓

(3 marks)

12 In ΔADC,

$AC^2 = 2.5^2 + 6^2$

$= 42.25$

So $AC = 6.5$ cm ✓

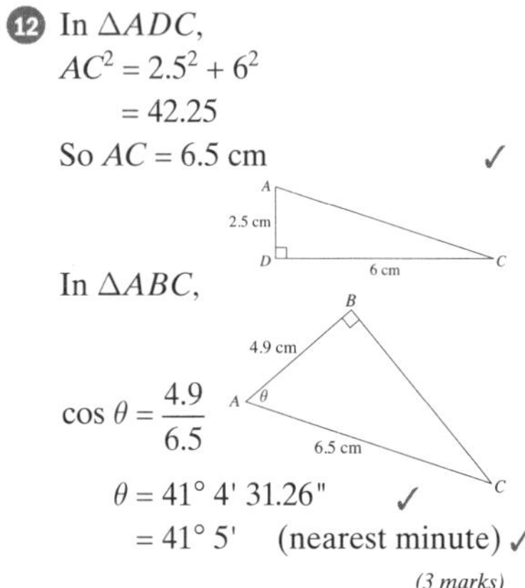

In ΔABC,

$\cos\theta = \dfrac{4.9}{6.5}$

$\theta = 41°\,4'\,31.26''$ ✓

$= 41°\,5'$ (nearest minute) ✓

(3 marks)

13 Let θ be the size of $\angle BOC$.

$A = \dfrac{1}{2}ab\sin C$

So $\dfrac{1}{2} \times 16 \times 25 \times \sin\theta = 198$ ✓

$\sin\theta = \dfrac{198}{200}$

$\theta = 81.890\,3855\ldots°$

$= 82°$ (nearest degree) ✓

The bearing of C from O is $(125 + 82)°$ or 207° to the nearest degree. ✓

(3 marks)

14 **i** $\angle ABC = 50° + (180° - 120°)$

$\angle ABC = 50° + 60°$

$= 110°$ ✓

(1 mark)

ii Using the cosine rule:

$AC^2 = 320^2 + 190^2 - 2(320)(190)\cos 110°$

$= 180\,089.6494\ldots$ ✓

$AC = 424.369\,7084\ldots$

$= 420$ (nearest 10)

$\therefore$ the distance is 420 km. ✓

(2 marks)

15 **i** Using Area $= \dfrac{1}{2}ab\sin C$

$= \dfrac{1}{2} \times 3 \times 6 \times \sin 60°$

$= \dfrac{1}{2} \times 3 \times 6 \times \dfrac{\sqrt{3}}{2}$

$= \dfrac{9\sqrt{3}}{2}$

$\therefore$ the area is $\dfrac{9\sqrt{3}}{2}u^2$. ✓

(1 mark)

ii Area of ΔKLN + Area of ΔNLM = Area of ΔKLM

$\dfrac{1}{2} \times 3 \times x \times \sin 30° + \dfrac{1}{2} \times x \times 6 \times \sin 30°$

$= \dfrac{9\sqrt{3}}{2}$

$\dfrac{1}{2} \times 3 \times x \times \dfrac{1}{2} + \dfrac{1}{2} \times x \times 6 \times \dfrac{1}{2}$

$= \dfrac{9\sqrt{3}}{2}$

$\therefore \dfrac{3x}{4} + \dfrac{3x}{2} = \dfrac{9\sqrt{3}}{2}$

$3x + 6x = 18\sqrt{3}$

$9x = 18\sqrt{3}$

$x = 2\sqrt{3}$ ✓

(2 marks)

16 In ΔABC,

$\sin 62° = \dfrac{AC}{13}$

$AC = 13\sin 62°$ ✓

$= 11.478\,3187\ldots$ cm

In ΔACD,

Area $= \dfrac{1}{2}cd\sin A$

$30 = \dfrac{1}{2} \times x \times 11.478\,3187\ldots \times \sin 40°$ ✓

$= x \times 3.689\,060\,52\ldots$

$x = 30 \div 3.689\,060\,52\ldots$

$= 8.132\,151\,75\ldots$

$= 8.1$ (1 d.p.) ✓

(3 marks)

17 $c^2 = a^2 + b^2 - 2ab\cos C$

$13^2 = (x+4)^2 + (x-4)^2 - 2(x+4)(x-4)\cos 60°$ ✓

$169 = x^2 + 8x + 16 + x^2 - 8x + 16 - 2(x^2 - 16) \times \dfrac{1}{2}$

$169 = 2x^2 + 32 - x^2 + 16$ ✓

$x^2 = 169 - 48$

$x^2 = 121$

$x = 11$ $(x > 4)$ ✓

(3 marks)

18 **i** By the cosine rule in ΔABC:

$b^2 = a^2 + c^2 - 2ac\cos B$

$= 13^2 + 5^2 - 2 \times 13 \times 5 \times \cos 135°$ ✓

$= 285.923\,88\ldots$

$b = \sqrt{285.923\,88\ldots}$ $(b > 0)$

$= 16.909\,2838\ldots$

$= 17$ (nearest unit)

The shortest distance from A to C is 17 km, to the nearest kilometre. ✓

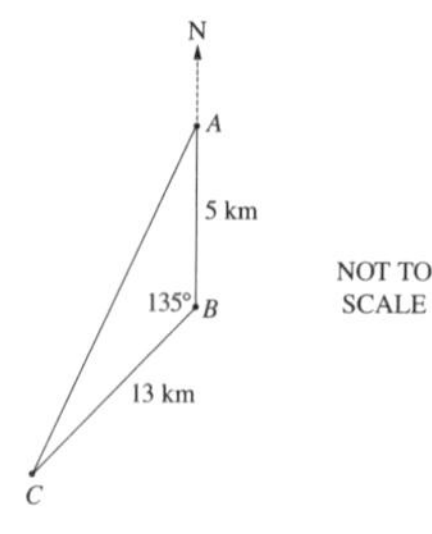

(2 marks)

ii By the sine rule in ΔABC:

$\dfrac{\sin A}{a} = \dfrac{\sin B}{b}$

$\dfrac{\sin A}{13} = \dfrac{\sin 135°}{16.909\,2838\ldots}$ ✓

$\sin A = \dfrac{13\sin 135°}{16.909\,2838\ldots}$

$A = \sin^{-1}\left(\dfrac{13\sin 135°}{16.909\,2838\ldots}\right)$

$= 32.931\,0687\ldots°$

$= 33°$ (nearest degree) ✓

Now A is due north of B.

So the bearing of B from A is 180°.

The bearing of C from $A = 180° + 33°$

$= 213°$.

The bearing of school C from school A is 213° to the nearest degree. ✓

(3 marks)

19 $\sin\theta > 0$ and $\cos\theta < 0$.

$\therefore$ 2nd quadrant

Answer B

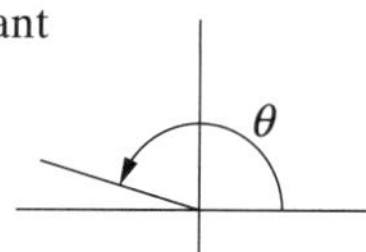

20 Consider α. Using the cosine rule:

$\cos\alpha = \dfrac{20^2 + 15^2 - 8^2}{2(20)(15)}$

$= 0.935$ ✓

$\alpha = 20.771\,855\,05\ldots$

$= 21$ (nearest whole)

$\therefore$ $\theta = 90 - 21$ ($\angle$ in square) ✓

$= 69$

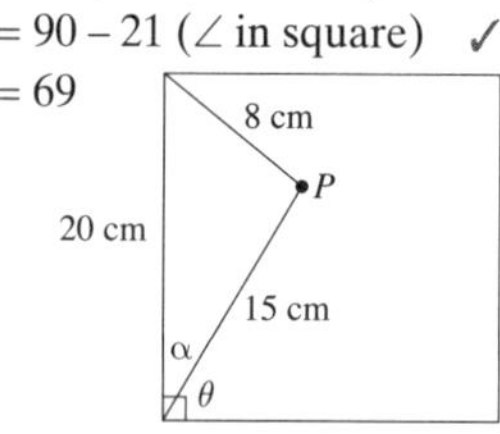

The size of the angle is 69° to the nearest degree. ✓

(3 marks)

21 $\dfrac{\sin C}{c} = \dfrac{\sin A}{a}$

$\dfrac{\sin C}{40} = \dfrac{\sin 25°}{18}$

$\sin C = \dfrac{40\sin 25°}{18}$

$= 0.939\,1516\ldots$

$C = 70°$ or $110°$ (nearest degree)

But the angle is obtuse so $\angle BCA = 110°$.

So, using the exterior angle of the triangle, the bearing of C from B is $110° + 25° = 135°$

Answer D

22 **i** 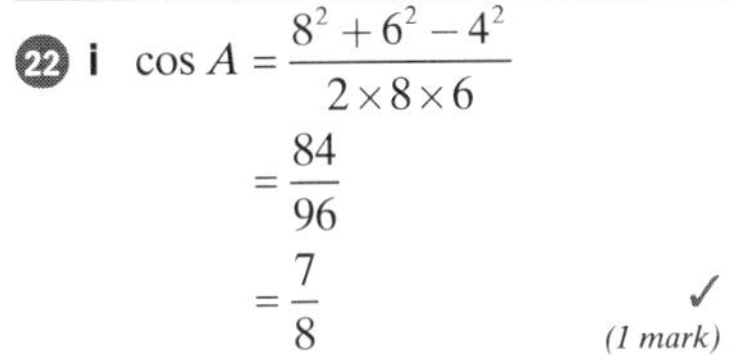

$\cos A = \dfrac{8^2 + 6^2 - 4^2}{2 \times 8 \times 6}$

$= \dfrac{84}{96}$

$= \dfrac{7}{8}$ ✓ *(1 mark)*

ii Draw a triangle with $\cos A = \dfrac{7}{8}$:

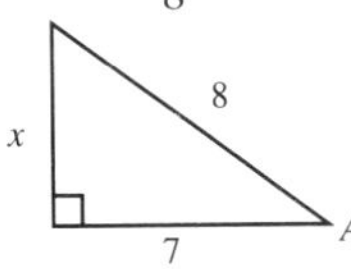

Using Pythagoras' theorem, and letting unknown side be x:

$x^2 = 8^2 - 7^2$

$= 15$

$x = \sqrt{15} \quad (x > 0)$

$\therefore \sin A = \dfrac{\sqrt{15}}{8}$ ✓

Now, area of

$\Delta ABC = \dfrac{1}{2}bc\sin A$

$= \dfrac{1}{2} \times 8 \times 6 \times \dfrac{\sqrt{15}}{8}$

$= 3\sqrt{15}$

$\therefore$ the area is $3\sqrt{15}$ cm^2. ✓ *(2 marks)*

23 **i** $c^2 = a^2 + b^2 - 2ab\cos C$

In ΔPSC,

$PC^2 = 1.8^2 + 5.4^2 - 2 \times 1.8 \times 5.4 \times \cos 108°$

$= 38.407\,29\ldots$

$PC = \sqrt{38.407\,29\ldots} \quad (PC > 0)$

$= 6.197\,361\,56\ldots$

$= 6.197$ m (3 d.p.) ✓ *(1 mark)*

ii Let $\angle PCS = \theta$

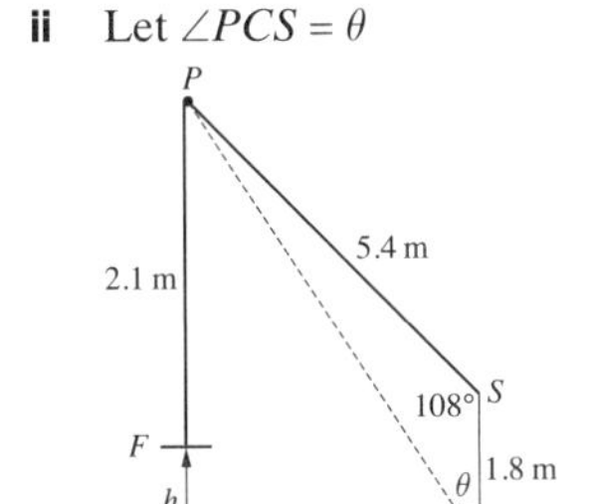

By the sine rule in ΔPSC,

$\dfrac{\sin\theta}{5.4} = \dfrac{\sin 108°}{6.197\ldots}$

$\sin\theta = 0.828\,69\ldots$

$\theta = 55.9646\ldots°$ ✓

$\angle PCE = 90° - 55.9646\ldots°$

$= 34.035\ldots°$ ✓

In ΔPEC,

$\sin 34.035\ldots° = \dfrac{PE}{6.197\ldots}$

$PE = 3.468\,6917\ldots$ ✓

$PE = 3.47$ m (2 d.p.)

$h = 3.47 - 2.1$

$= 1.37$ m

The object is 1.37 m above the ground, to the nearest centimetre. ✓ *(4 marks)*

24 **i** $270 - (12 + 191) = 67$

$\therefore \angle ABC = 67°$ ✓

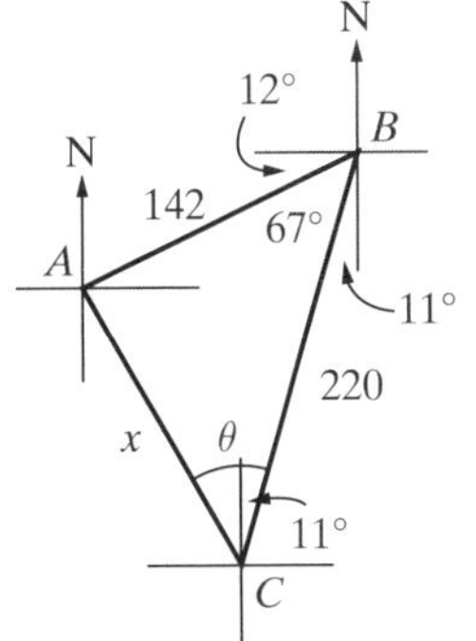

Let $AC = x$:

$x^2 = 142^2 + 220^2 - 2(142)(220)\cos 67°$

$= 44\,151.119\,09\ldots$

$x = 210.121\,6769\ldots$

$= 210$ (nearest whole)

$\therefore$ distance is approx. 210 km. ✓ *(2 marks)*

ii Let $\angle ACB = \theta$:

$\cos\theta = \dfrac{210^2 + 220^2 - 142^2}{2(210)(220)}$

$= 0.782\,857\,142\ldots$

$\theta = 38.477\,077\,75\ldots$

$= 38$ (nearest whole) ✓

Also, $191 - 180 = 11$, ✓

$38 - 11 = 27$ and

$360 - 27 = 333$.

$\therefore$ bearing is 333°. ✓ *(3 marks)*

25 X is due east of Z.

[So options B and D are eliminated.]

Let P be a point due north of Y.

$\angle PYX = 145°$

$\angle PYZ = 180° - 60°$ (co-interior angles, parallel lines)

$= 120°$

$\angle XYZ = 360° - (120° + 145°)$

$= 95°$

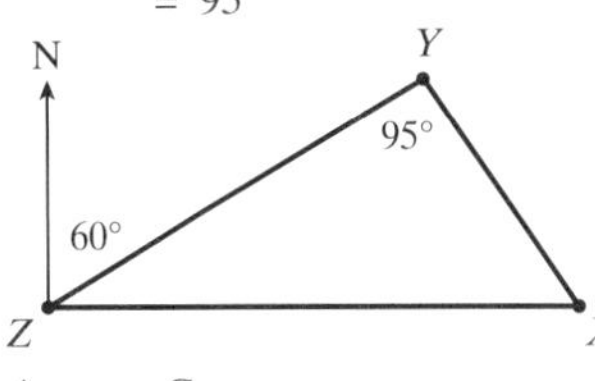

Answer C

26 **i** $A(x, 0)$: subs $y = 0$ in $2x + y = 8$

$2x = 8$

$x = 4$

$B(0, y)$: subs $x = 0$ in $2x + y = 8$

$y = 8$

$\therefore A(4, 0), B(0, 8)$ ✓

$AB^2 = 4^2 + 8^2$

$= 16 + 64$

$= 80$

$AB = \sqrt{80}$

$\therefore$ distance is $4\sqrt{5}$ units ✓ *(2 marks)*

ii

Let $\angle ABC = \theta$

Using $\cos\theta$

$= \dfrac{(\sqrt{65})^2 + (\sqrt{80})^2 - 5^2}{2 \times \sqrt{65} \times \sqrt{80}}$ ✓

$= \dfrac{120}{144.222\,051\ldots}$

$= 0.832\,05\ldots$

$\theta = 33.69006751...$

$\therefore \angle ABC = 34°$ ✓ (nearest degree)

(2 marks)

iii Gradient of $2x + y = 8$ is -2

$\therefore$ gradient of perpendicular $= \frac{1}{2}$

equation of CN:

$y - y_1 = m(x - x_1)$

$y - 4 = \frac{1}{2}(x - 7)$

$2y - 8 = x - 7$

$x - 2y = -1$ ✓

Solve simultaneously:

$2x + y = 8 \quad ... \mathbf{1}$

$x - 2y = -1 \quad ... \mathbf{2}$

$2 \times \mathbf{2}: 2x - 4y = -2 \quad ... \mathbf{3}$

$\mathbf{1} - \mathbf{3}: 5y = 10$

$y = 2$ ✓

Subs in **1**:

$2x + 2 = 8$

$2x = 8 - 2$

$2x = 6$

$x = 3$

$\therefore N(3, 2)$ ✓

(3 marks)

27

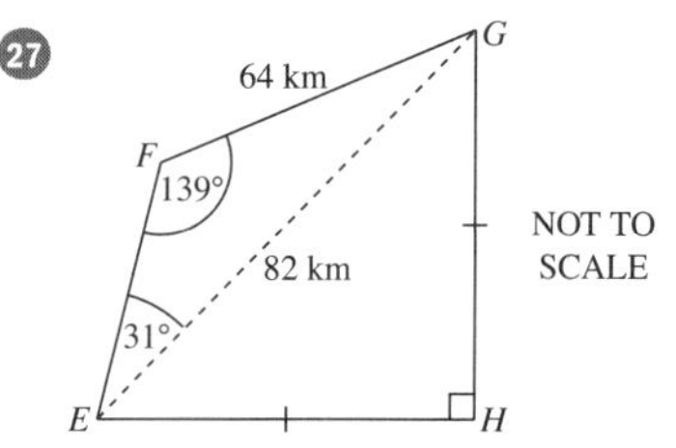

i $\therefore FGE = 180° - (139 + 31)°$
$= 10°$

By the sine rule in ΔEFG,

$\frac{EF}{\sin 10°} = \frac{82}{\sin 139°}$ ✓

$EF = \frac{82 \sin 10°}{\sin 139°}$

$= 21.704069...$

$= 22$ km [nearest km] ✓

{Or: $\frac{EF}{\sin 10°} = \frac{64}{\sin 31°}$

$EF = \frac{64 \sin 10°}{\sin 31°}$

$= 21.577984...$

$= 22$ km [nearest km]}

(2 marks)

Or: By the cosine rule in ΔEFG,

$EF^2 = 64^2 + 82^2 - 2 \times 64 \times 82 \times \cos 10°$

$= 483.45782...$

$EF = 21.987674... \quad (EF > 0)$

$= 22$ km [nearest km]}

(2 marks)

ii Let the length of GH be x km.

$EH = GH = x$ km

By Pythagoras' theorem in ΔGEH,

$x^2 + x^2 = 82^2$

$2x^2 = 6724$

$x^2 = 3362$

$x = 57.982756...$ ✓ $[x > 0]$

The length of GH is 58 km to the nearest kilometre.

Total distance

$= (2 \times 57.982... + 64 + 21.704...)$ km ✓

$= 201.66958...$ km

$= 202$ km ✓ [nearest kilometre]

(3 marks)

28 **i** $22^2 = x^2 + 20^2 - 2 \times x \times 20 \times \cos 60°$

$484 = x^2 + 400 - 20x$

$x^2 - 20x - 84 = 0$ ✓

(1 mark)

ii $x = \frac{20 \pm \sqrt{400 - 4 \times 1 \times (-84)}}{2 \times 1}$

$= \frac{20 \pm \sqrt{736}}{2}$ ✓

$= 23.56465..., -3.56465...$

$\therefore$ the distance is 24 km. ✓

(2 marks)

29

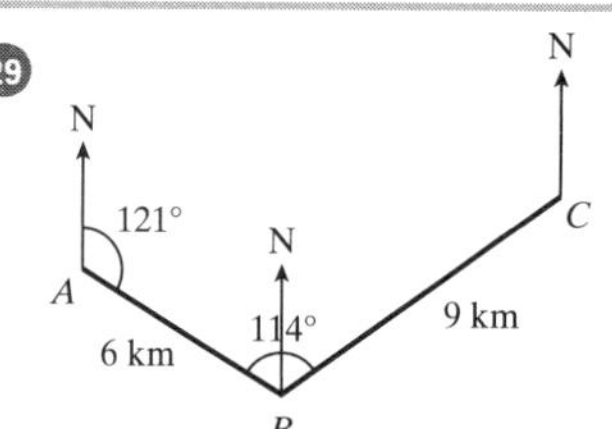

i Let X be a point due North of B.

$\angle ABX = 180° - 121°$ (co-interior angles, parallel lines)

$= 59°$

$\angle XBC = 114° - 59°$

$= 55°$

$\therefore$ the bearing of C from B is 055°. ✓

(1 mark)

ii Using the cosine rule in ΔABC,

$b^2 = a^2 + c^2 - 2ac \cos B$

$= 9^2 + 6^2 - 2 \times 9 \times 6 \times \cos 114°$

$= 160.92755...$ ✓

$b = \sqrt{160.92755...} \quad (b > 0)$

$= 12.685722...$

$= 13$ [nearest whole number]

The distance AC is 13 km to the nearest kilometre. ✓

(2 marks)

iii Using the sine rule in ΔABC,

$\frac{\sin C}{c} = \frac{\sin B}{b}$

$\frac{\sin C}{6} = \frac{\sin 114°}{12.685722...}$

$\sin C = \frac{6 \times \sin 114°}{12.685722...}$

$= 0.432082028...$

$C = 25.59976364...°$

$= 26°$ [nearest degree] ✓

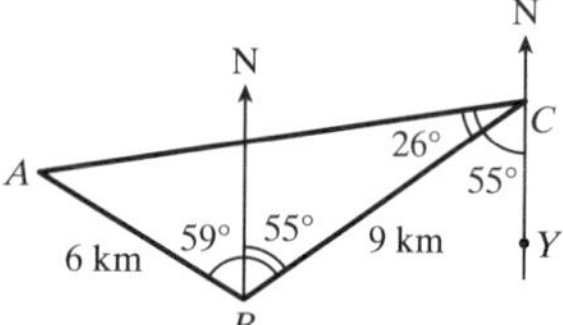

Let Y be a point due south of C.

$\angle BCY = 55°$ (alternate angles, parallel lines) ✓

Required bearing

$= (180 + 55 + 26)°$

$= 261°$

The bearing of A from C is 261°. ✓

(3 marks)

30

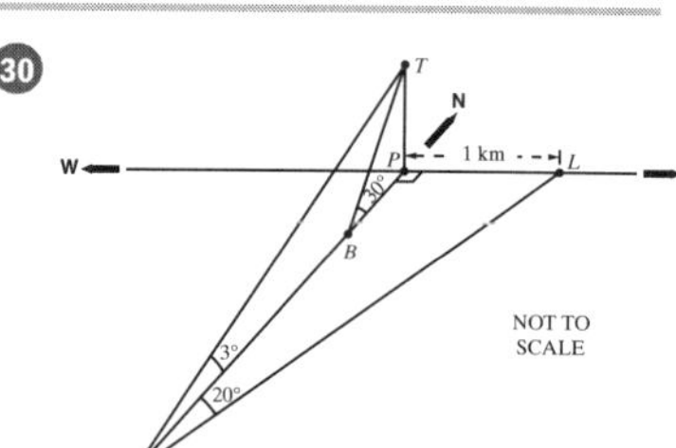

i In ΔAPL, $\tan 20° = \frac{1}{AP}$

$\therefore AP = \frac{1}{\tan 20°}$ ✓

In ΔATP,

$\tan 3° = \frac{TP}{AP}$

$\therefore TP = AP \tan 3°$

$= \frac{\tan 3°}{\tan 20°}$ ✓

In ΔTPB,

$\tan 30° = \frac{TP}{BP}$

$\frac{1}{\sqrt{3}} = \frac{TP}{BP}$

$BP = \sqrt{3}\, TP$

$= \frac{\sqrt{3} \tan 3°}{\tan 20°}$ ✓

(3 marks)

ii $AB = AP - BP$

$= \dfrac{1}{\tan 20°} - \dfrac{\sqrt{3}\tan 3°}{\tan 20°}$

$= 2.498\,080\,826\ldots$

$= 2.5$ km ✓ [1 decimal place]

(1 mark)

31

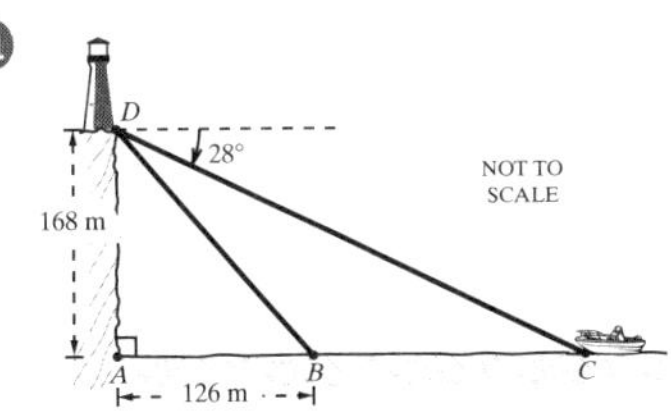

i The angle of depression from D to $B = \angle DBA$ (alternate angles, parallel lines)

In ΔDBA,

$\tan B = \dfrac{168}{126}$ ✓

$B = 53.130\,102\,35\ldots°$ ✓

$= 53° 8'$ [nearest minute]

So the angle of depression is 53° 8′ correct to the nearest minute. ✓

(3 marks)

ii $\angle ADC = 90° - 28°$

$= 62°$

In ΔADC,

$\tan 62° = \dfrac{AC}{168}$

$AC = 168 \times \tan 62°$

$= 315.962\,04\ldots$ m ✓

$BC = 315.962\,04\ldots - 126$

$= 189.962\,04\ldots$

$= 190$ m ✓ [nearest metre]

(2 marks)

32

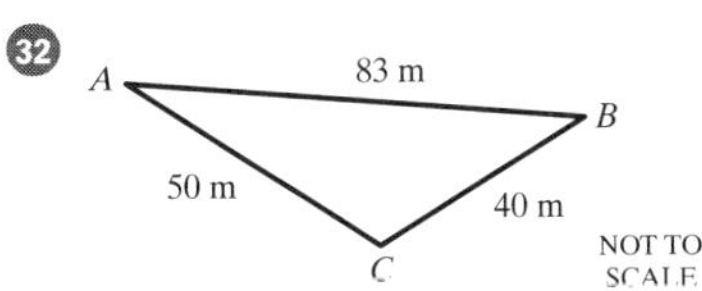

$\text{Cos } C = \dfrac{a^2 + b^2 - c^2}{2ab}$

$= \dfrac{40^2 + 50^2 - 83^2}{2 \times 40 \times 50}$

$= -0.697\,25$

$C = 134.206\,78\ldots°$ ✓

$A = \dfrac{1}{2}ab \sin C$

$= \dfrac{1}{2} \times 40 \times 50 \times \sin 134.206\,78\ldots°$ ✓

$= 716.828\,039\ldots$

$= 717 \text{ m}^2$ [nearest square metre]

The area is 717 m^2 to the nearest square metre. ✓

(3 marks)

33 $\angle DAC = \angle DCA$, since ΔADC is isosceles

$\therefore \angle DAC = \angle DCA = \dfrac{180° - 80°}{2}$

(Angle sum of $\Delta ADC = 180°$)

$= 50°$

$\therefore \angle DBA = 180° - 60° - 50°$ (angle sum ΔDBA)

$= 70°$

Using the sine rule,

$\dfrac{AB}{\sin 60°} = \dfrac{30}{\sin 70°}$

$\therefore AB = \dfrac{30 \sin 60°}{\sin 70°}$

$= 27.6481\ldots$ cm

$= 28$ cm to nearest cm

Answer A

34

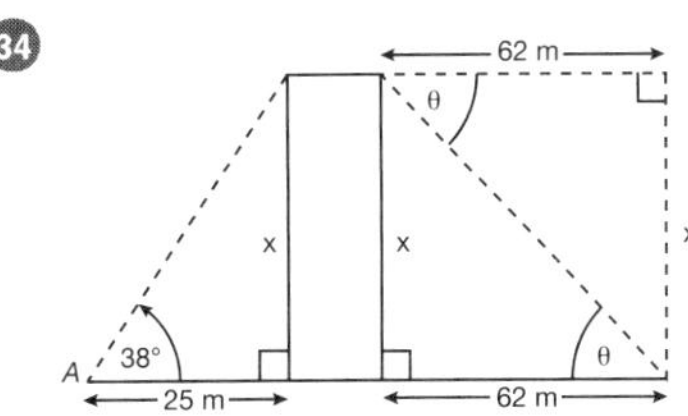

i Let x be the height of the building.

$\therefore \tan 38° = \dfrac{x}{25}$

$x = 25 \tan 38°$

$= 19.5321\ldots$ m

$= 19.5$ m ✓

(1 mark)

ii Let θ be the angle of depression which is equal to the angle of elevation.

$\therefore \tan \theta = \dfrac{x}{62}$ ✓

$= \dfrac{19.5321\ldots}{62}$

$\therefore \theta = 17.48\ldots°$

$= 17°$ ✓ to nearest degree.

(2 marks)

35 i

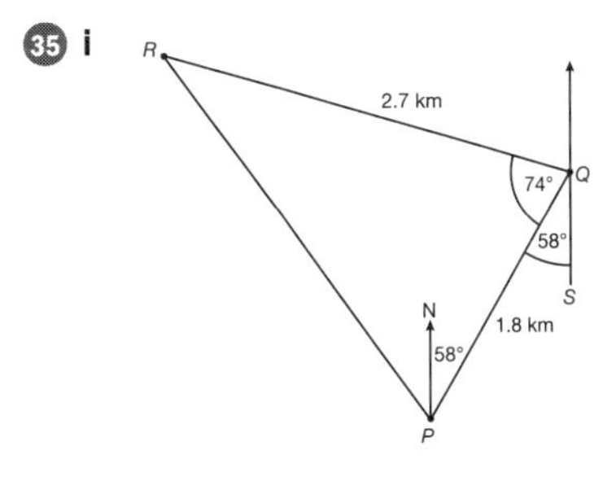

$\angle NPQ = \angle PQS = 58°$

$\therefore$ Bearing of R from Q is $(180 + 58 + 74)° = 312°$. ✓

(1 mark)

ii Using the cosine rule,

$RP^2 = 2.7^2 + 1.8^2 - 2(2.7)(1.8)\cos 74°$ ✓

$= 7.8508\ldots$

$RP = 2.8019\ldots$

$\doteqdot 2.8$ km ✓

(2 marks)

iii $A = \dfrac{1}{2}ab \sin C$

$= \dfrac{1}{2} \times 2.7 \times 1.8 \times \sin 74°$

$= 2.3358\ldots$

$\doteqdot 2.3 \text{ km}^2$. ✓

(1 mark)

36 i

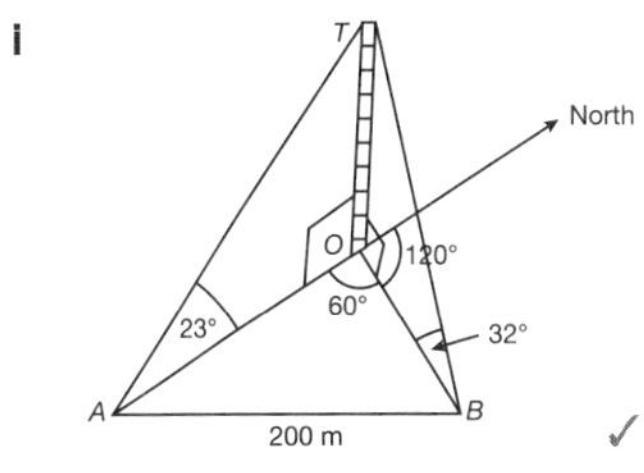

✓

(1 mark)

ii By the cosine rule, in ΔAOB,

$200^2 = OA^2 + OB^2 - 2(OA)(OB)\cos 60° \quad \ldots(1)$

In ΔAOT, $\tan 23° = \dfrac{OT}{OA}$

$\therefore OA = \dfrac{OT}{\tan 23°}$

In ΔBOT, $\tan 32° = \dfrac{OT}{OB}$

$\therefore OB = \dfrac{OT}{\tan 32°}$ ✓

Substituting for OA and OB in (1):

$200^2 = \dfrac{OT^2}{\tan^2 23°} + \dfrac{OT^2}{\tan^2 32°} - 2\left(\dfrac{OT}{\tan 23°}\right)\left(\dfrac{OT}{\tan 32°}\right)\cos 60°$

$200^2 = OT^2\left(\dfrac{1}{\tan^2 23°} + \dfrac{1}{\tan^2 32°} - \dfrac{1}{\tan 23° \tan 32°}\right)$ ✓

$40\,000 = OT^2 \times 4.3409\ldots$

$OT^2 = 40\,000 \div 4.3409\ldots$

$= 9214.5535$

$\therefore OT = 95.9924\ldots$

$= 96$ m to the nearest m ✓

(3 marks)

37

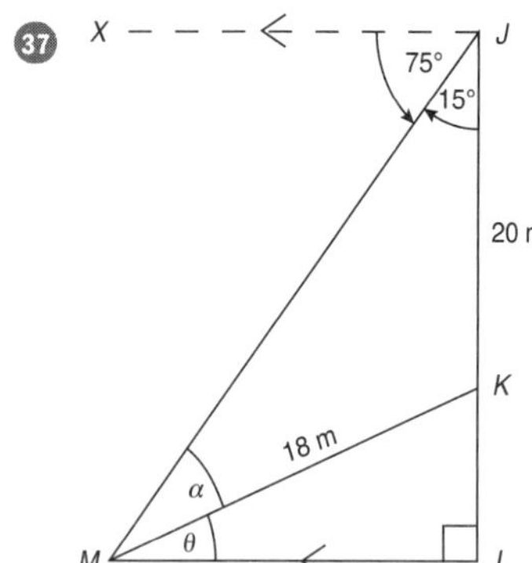

Let θ = angle of elevation from M to K and $\alpha = \angle JMK$

Now $\angle JML = 75° = \angle XJM$

$\therefore \quad \angle MJK = 90° - 75°$
$= 15°$ ✓

$\therefore$ By the sine rule in ΔMJK,

$$\frac{\sin\alpha}{20} = \frac{\sin 15°}{18}$$

$$\sin\alpha = \frac{20\sin 15°}{18}$$
$= 0.2875\ldots$

$\therefore \alpha = 17°$ to nearest degree ✓

$\therefore \theta = 75° - 17°$
$= 58°$ ✓

(3 marks)

38

i speed $= \dfrac{\text{distance}}{\text{time}}$

$= \dfrac{15 \text{ km}}{2 \text{ h } 48 \text{ min}}$

$= 5.3571\ldots$ km/h

$= 5$ km/h to the nearest km/h

N.B. To enter 2 h 48 min into calculator, use the degrees and minutes button. ✓ *(1 mark)*

ii Let x = distance from A to B.

$\therefore$ Using the cosine rule in ΔABQ,

$x^2 = 15^2 + 10^2 - 2 \times 15 \times 10 \times \cos 87°$ ✓

$= 309.299\ldots$

$\therefore x = 17.5869\ldots$

$= 18$ km to nearest km ✓

(2 marks)

iii $\angle AQN_1 = 180° - 149°$
$= 31°$

$\therefore \angle N_1QB = 87° - 31°$
$= 56°$

$\therefore \angle QBN = 180° - 56°$
$= 124°$

$\therefore$ The bearing of Q from B is $360° - 124° = 236°$. ✓

(1 mark)

39 By the sine rule:

$$\frac{\sin\theta}{5} = \frac{\sin 33°}{9}$$

$$\therefore \sin\theta = \frac{5\sin 33°}{9}$$

$= 0.3025\ldots$ ✓

$\therefore \theta = 17.6124\ldots°$

$= 18°$ to the nearest degree. ✓

(2 marks)

40 **i**

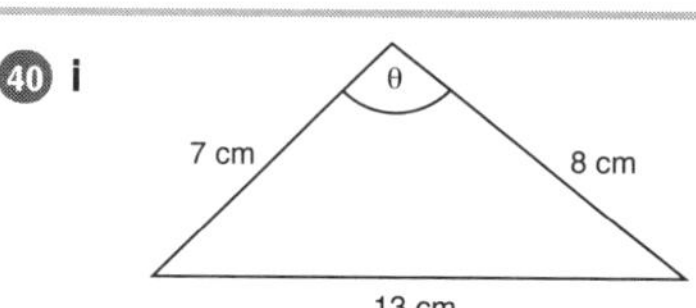

By the cosine rule:

$$\cos\theta = \frac{7^2 + 8^2 - 13^2}{2 \times 7 \times 8}$$

$= \dfrac{-56}{112}$

$= -0.5$ ✓

$\therefore \theta = 180° - 60°$

(N.B. angle is obtuse)

$= 120°$. ✓

(2 marks)

ii $A = \dfrac{1}{2}ab\sin C$

$= \dfrac{1}{2} \times 7 \times 8 \times \sin 120°$

$= 28 \times \dfrac{\sqrt{3}}{2}$

$= 14\sqrt{3}$

$\therefore$ Area $= 14\sqrt{3}$ cm^2. ✓

(1 mark)

41

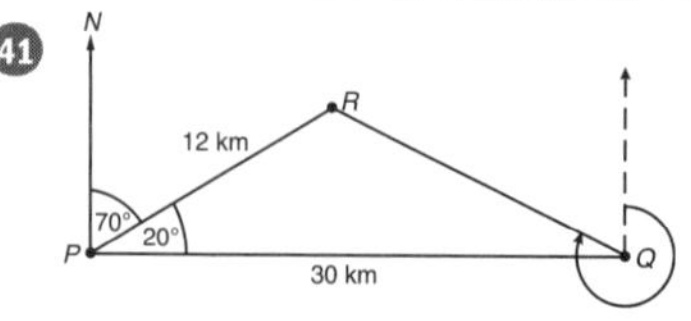

i By the cosine rule in ΔPRQ:

$RQ^2 = PR^2 + PQ^2 - 2 . PR . PQ . \cos \angle RPQ$

$= 12^2 + 30^2 - 2(12)(30)\cos 20°$ ✓

$= 367.4213\ldots$

$\therefore RQ = 19.1682\ldots$

$\doteqdot 19.2$ km to 1 decimal place. ✓

(2 marks)

ii By the sine rule in ΔPRQ:

$$\frac{\sin\angle RQP}{12} = \frac{\sin 20°}{RQ}$$

$$\sin\angle RQP = \frac{12\sin 20°}{19.1682\ldots}$$

$= 0.2141\ldots$

$\therefore \angle RQP = 12.3637\ldots°$ ✓

$\therefore$ The bearing of R from Q

$= 270° + \angle RQP$

$= 270° + 12.3637\ldots°$

$= 282.3637\ldots°$

$= 282° 22'$

$= 282°$ to the nearest degree. ✓

(2 marks)

42 **i** Now $AF = AC = FC$ (diagonals of congruent squares)

$\therefore \Delta AFC$ is equilateral and $\angle FAC = 60°$. ✓ *(1 mark)*

ii

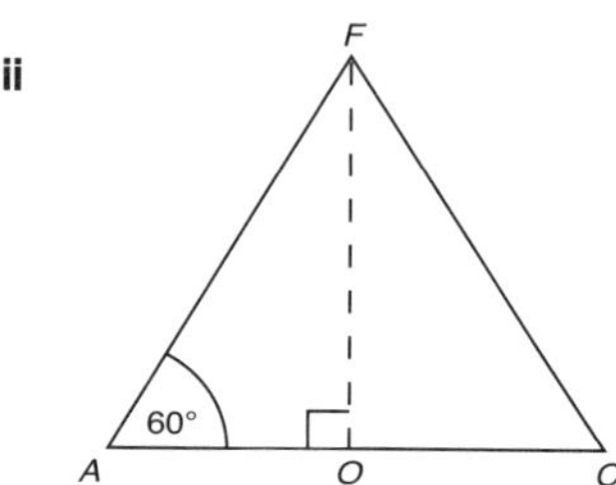

$\sin 60° = \dfrac{FO}{AF}$

Now $AF^2 = AE^2 + EF^2$ (In ΔAEF)

$= 2^2 + 2^2$

$= 8$

$\therefore AF = \sqrt{8}$

$\therefore \sin 60° = \dfrac{FO}{\sqrt{8}}$

$\dfrac{\sqrt{3}}{2} = \dfrac{FO}{\sqrt{8}}$

$\therefore FO = \dfrac{\sqrt{3} \times \sqrt{8}}{2}$

$= \dfrac{\sqrt{3} \times 2\sqrt{2}}{2}$

$= \sqrt{6}$ metres ✓

(1 mark)

iii

$\tan\angle XFO = \dfrac{OX}{FO}$

$= \dfrac{1}{\sqrt{6}}$

($OX = 1$, radius of circle)

$\therefore \angle XFO = 22.207\ldots°$

and $\angle XFY = 2 \times \angle XFO$

(since $\triangle FXY$ is isosceles)

$= 2 \times 22.207\ldots°$

$= 44.415\ldots°$

$= 44°$ to nearest degree. ✓

(1 mark)

43

N, R, P, Q, θ, 38 km, 20 km, 55°, 325°

i $\angle PQR = 325° - 270°$

$= 55°$ ✓

(1 mark)

ii By the sine rule in $\triangle PQR$:

$$\frac{\sin\theta}{20} = \frac{\sin 55°}{38}$$

$$\sin\theta = \frac{20 \times \sin 55°}{38}$$

$= 0.4311\ldots$

$\therefore \theta = 25.539\ldots°$

$= 25°32'$ ✓

$\therefore \angle NPR = 90° - 25°32'$

$= 64°28'$ ✓

$\therefore$ Bearing of R from

$P = \angle NPR$

$\doteqdot 064°$ ✓

(3 marks)

44 In the right-angled $\triangle DBH$,

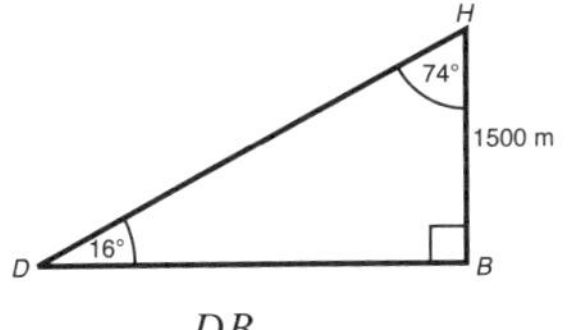

$$\tan 74° = \frac{DB}{1500}$$

$\therefore DB = 1500 \tan 74°$

$= 5231.121\ldots$

In the right-angled $\triangle ABH$.

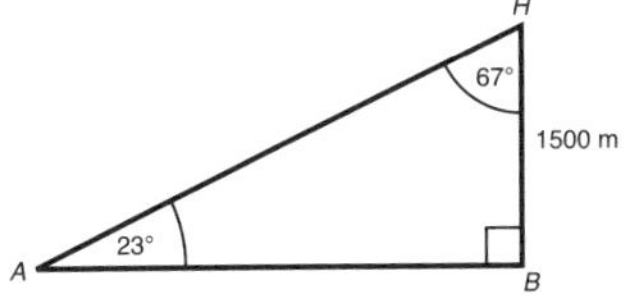

$$\tan 67° = \frac{AB}{1500}$$

$AB = 1500 \tan 67°$

$= 3533.778\ldots$ ✓

Y, A, 139°, 41°, S, N, D, 109°, 71°, T, 3533.778... m, 5231.121... m, 71°, 30°, 41°, Z, B

Now $\angle NDB = 109°$ (given)

and $\angle YAB = 139°$ (given)

$\therefore \angle SAB = 180° - 139°$

$= 41°$

and $\angle TDB = 180° - 109°$

$= 71°$

Also, $\angle SAB = \angle ZBA = 41°$

(alternate angles, $SY \parallel BZ$)

and $\angle TDB = \angle ZBD = 71°$

(alternate angles, $TN \parallel BZ$)

$\therefore \angle DBA = 71° - 41°$

$= 30°$ ✓

In $\triangle DAB$,

$AD^2 = AB^2 + DB^2 - 2(AB)(DB)\cos 30°$

$= (3533.778\ldots)^2 + (5231.121\ldots)^2 - 2(3533.778\ldots)(5231.121\ldots)\cos 30°$

$= 7\,834\,182.089\ldots$

$\therefore AD = 2798.960\ldots$ m

$\therefore$ The distance of the life raft from Anna'sposition is approximately 2799 m. ✓

In $\triangle DAB$,

$$\cos \angle DAB = \frac{AD^2 + AB^2 - DB^2}{2(AD)(AB)}$$

$$= \frac{(2798.960\ldots)^2 + (3533.778\ldots)^2 - (5231.121\ldots)^2}{2(2798.960\ldots)(3533.778\ldots)}$$

$= -0.35602\ldots$

$\therefore \angle DAB = 110.856\ldots°$

$\doteqdot 111°$

$\therefore$ The bearing of D (life raft) from A (Anna's position) is $139° + 111° = 250°$. ✓

(4 marks)

45

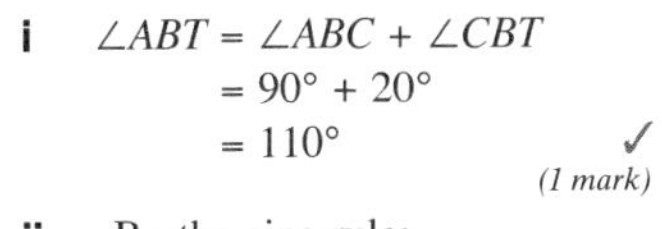

i $\angle ABT = \angle ABC + \angle CBT$

$= 90° + 20°$

$= 110°$ ✓

(1 mark)

ii By the sine rule:

$$\frac{BT}{\sin A} = \frac{AB}{\sin T}$$

Now $\angle BTA$

$= 180° - 30° - 110°$

$= 40°$ ✓

$$\therefore \frac{BT}{\sin 30°} = \frac{50}{\sin 40°}$$

$$\therefore BT = \frac{50 \sin 30°}{\sin 40°}$$ ✓

$= 38.893\ldots$ m

$= 39$ m to the nearest m. ✓

(3 marks)

46

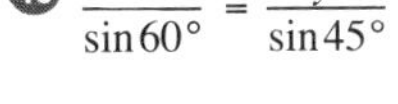

$$\frac{x}{\sin 60°} = \frac{y}{\sin 45°}$$

$$\therefore \frac{x}{y} = \frac{\sin 60°}{\sin 45°}$$ ✓

$$= \frac{\frac{\sqrt{3}}{2}}{\frac{1}{\sqrt{2}}}$$ ✓

$$= \frac{\sqrt{3}}{2} \times \frac{\sqrt{2}}{1}$$

$$= \frac{\sqrt{6}}{2}$$ ✓

(3 marks)

47 **i** By the cosine rule:

$MN^2 = LM^2 + LN^2 - 2(LM)(LN)\cos \angle MLN$

$= 5.2^2 + 8.9^2 - 2(5.2)(8.9)\cos 110°$ ✓

$= 137.9073\ldots$

$\therefore MN = 11.7433\ldots$

$\doteqdot 11.7$ m ✓

(2 marks)

ii Area ΔLMN

$= \frac{1}{2}(LM)(LN)\sin \angle MLN$

$= \frac{1}{2} \times 5.2 \times 8.9 \times \sin 110°$ ✓

$= 21.7444\ldots$

$\doteqdot 21.7\text{ m}^2$ ✓

(2 marks)

48 $13^2 = x^2 + 7^2 - 2(x)(7)\cos 60°$

$169 = x^2 + 49 - 2(x)(7)(\frac{1}{2})$ ✓

$169 = x^2 + 49 - 7x$

$120 = x^2 - 7x$

i.e. $x^2 - 7x - 120 = 0$ ✓

$(x - 15)(x + 8) = 0$

$\therefore x = 15$ or -8 ✓

$\therefore x = 15$

(a length must be positive) ✓

(4 marks)

49 **i**

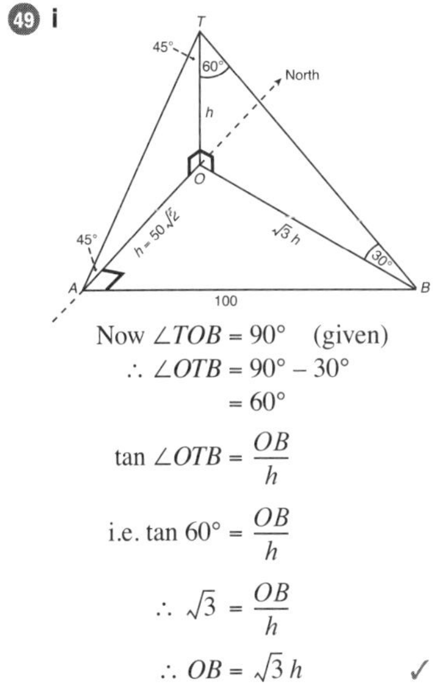

Now $\angle TOB = 90°$ (given)

$\therefore \angle OTB = 90° - 30°$

$= 60°$

$$\tan \angle OTB = \frac{OB}{h}$$

$$\text{i.e. } \tan 60° = \frac{OB}{h}$$

$$\therefore \sqrt{3} = \frac{OB}{h}$$

$\therefore OB = \sqrt{3}\,h$ ✓

ii Now $\angle TOA = 90°$ (given)
$\therefore \angle OTA = 90° - 45° = 45°$

$\tan \angle OTA = \frac{OA}{OT}$

i.e. $\tan 45° = \frac{OA}{h}$

$\therefore 1 = \frac{OA}{h}$

$\therefore OA = h$

Now $OB^2 = OA^2 + 100^2$ (Pythagoras' theorem)

i.e. $(\sqrt{3}\,h)^2 = h^2 + 100^2$

$\therefore 3h^2 = h^2 + 100^2$ ✓

$\therefore 2h^2 = 100^2$

$\therefore \sqrt{2}\,h = 100$

$h = \frac{100}{\sqrt{2}} = \frac{100\sqrt{2}}{2} = 50\sqrt{2}$ ✓

iii Now $OA = h = 50\sqrt{2}$

$\therefore \angle\tan AOB = \frac{AB}{OA} = \frac{100}{50\sqrt{2}} = \frac{100\sqrt{2}}{100} = \sqrt{2}$ ✓

$\therefore \angle AOB = \tan^{-1}\sqrt{2} = 55°$ (to nearest degree).

The bearing of B from the base of the tower is $180° - 55° = 125°$. ✓

(5 marks)
(Total mark allocation only in exam)

50 **i** $\cos \angle ACB = \frac{7^2 + 3^2 - 6^2}{2 \times 7 \times 3} = \frac{49 + 9 - 36}{42}$ ✓ $= \frac{22}{42}$

$\therefore \angle ACB \doteqdot 58°25'$
$= 58°$ to the nearest ✓ degree.

ii $\angle DBC = 180° - 58°25'$ (co-interior angles) ✓
$= 121°35'$
$= 122°$
to the nearest degree. ✓

(4 marks)
(Total mark allocation only in exam)

51

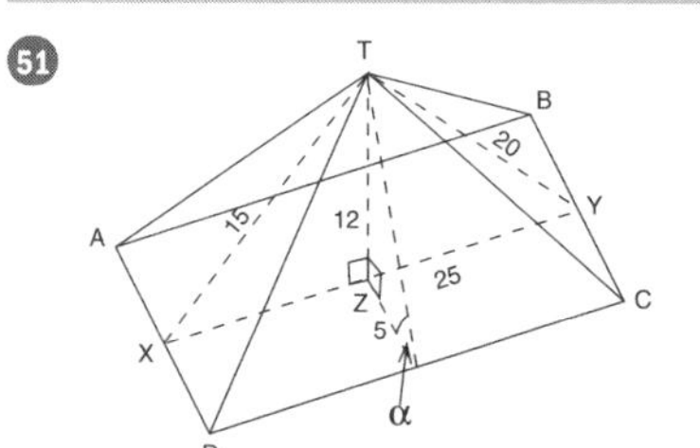

From the data: $TX = PX = 15$ cm and $TY = RY = 20$ cm

i Now P is on the perpendicular bisector of AD since $AP = PD$ (given).
$\therefore P$ is also on the perpendicular bisector of BC since $ABCD$ is a rectangle (given).
And R is on the perpendicular bisector of BC since $BR = RC$ (given)
$\therefore$ The straight line $PXZYR$ is the perpendicular bisector of BC and AD.
$\therefore XY = AB$ ($ABYX$ is a rectangle)
$\therefore XY = 25$ cm ✓

Now $TX^2 + TY^2 = 15^2 + 20^2 = 625 = 25^2 = XY^2$

$\therefore \Delta TXY$ is right-angled (Pythagoras' Theorem). ✓

ii $\Delta TXZ\ |||\ \Delta YXT$
(both are right-angled and have a common angle at X)

$\therefore \frac{TZ}{TY} = \frac{TX}{XY}$

$\therefore \frac{TZ}{20} = \frac{15}{25}$

$\therefore TZ = \frac{20 \times 15}{25} = 12$ cm. ✓

iii Let α be the angle that the face DCT makes with the base. The perpendicular distrance from Z to $DC = YC = 5$ cm (since Y is the ✓ midpoint of BC as shown in **i**).

$\therefore \tan \alpha = \frac{12}{5}$

$\therefore \alpha \doteqdot 67°$ ✓

(5 marks)
(Total mark allocation only in exam)

52

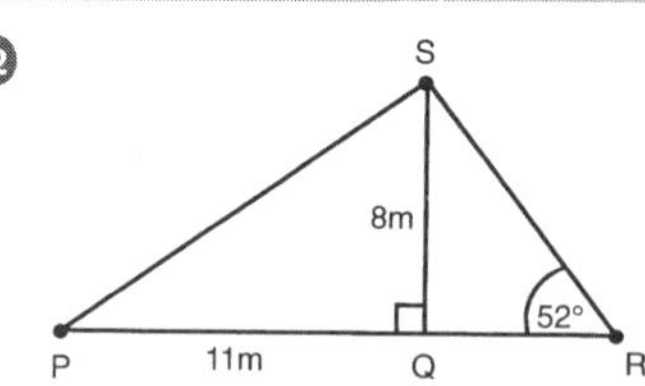

i In ΔSQR,

$\sin 52° = \frac{8}{SR}$ ✓

$SR = \frac{8}{\sin 52°} = 10.1521\ldots$ m $\doteqdot 10.2$ m. ✓

ii $\tan \angle SPQ = \frac{8}{11}$ ✓

$\angle SPQ = 36.0273\ldots° \doteqdot 36°$ ✓

(4 marks)
(Total mark allocation only in exam)

53

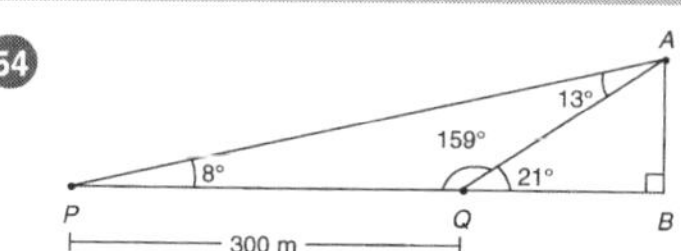

$DC = AB = 12$ m (opposite sides of rectangle)

$\therefore PD = QC = (50 - 12) \div 2 = 19$ m (congruent triangles)

Also $PA = QB = 50 \div 2 = 25$ m

$\cos\theta = \frac{19}{25}$

$\therefore \theta = 40.5358\ldots° \doteqdot 41°$

(No mark allocation in exam)

54

i $\angle AQP = 180° - 21°$ (BQP a straight line) $= 159°$

$\angle PAQ = 180° - 8° - 159°$ (angle sum PAQ) $= 13°$

ii By the sine rule in ΔPAQ,

$\frac{300}{\sin 13°} = \frac{AQ}{\sin 8°}$

$AQ = \frac{300 \sin 8°}{\sin 13°} = 185.6045\ldots = 185.6$ m to 1 decimal place.

iii In ΔQAB:

$\sin 21° = \frac{AB}{AQ}$

$\therefore AB = AQ \times \sin 21°$ (using calc. memory for AQ) $= 66.5147\ldots = 66.5$ m to 1 decimal place.

(No mark allocation in exam)

1 Which of these is the exact value of $\cos\left(\frac{\pi}{4}\right) + \cos\left(\frac{2\pi}{3}\right)$?

A $\frac{1}{2\sqrt{2}}$ **B** $\frac{2-\sqrt{2}}{2\sqrt{2}}$

C $\frac{2-\sqrt{3}}{2\sqrt{2}}$ **D** $\frac{\sqrt{3}-2}{2\sqrt{2}}$ *(1 mark)*

Bonus question (see page iv) Easy

2 The diagram shows a circle with centre O and radius 20 cm.

The points A and B lie on the circle such that $\angle AOB = 70°$.

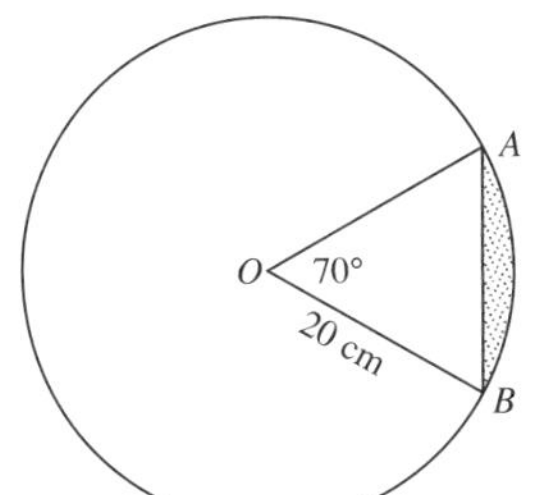

Find the perimeter of the shaded segment, giving your answer correct to one decimal place. *(3 marks)*

(Q13b, **2019 HSC**) Easy

3 In the diagram, OAB is a sector of the circle with centre O and radius 6 cm, where $\angle AOB = 30°$.

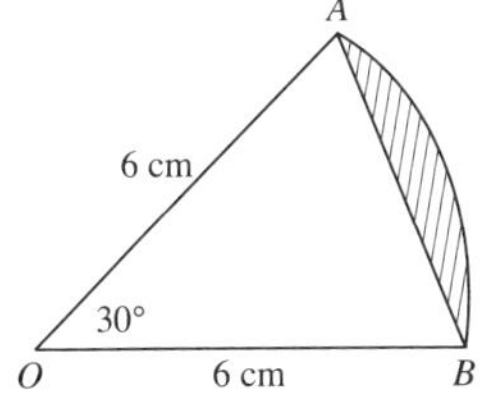

i Find the exact value of the area of the triangle OAB. *(1 mark)* Easy

ii Find the exact value of the area of the shaded segment. *(1 mark)* Medium

(Q11e, **2017 HSC**)

4 The circle centred at O has radius 5. Arc AB has length 7 as shown in the diagram.

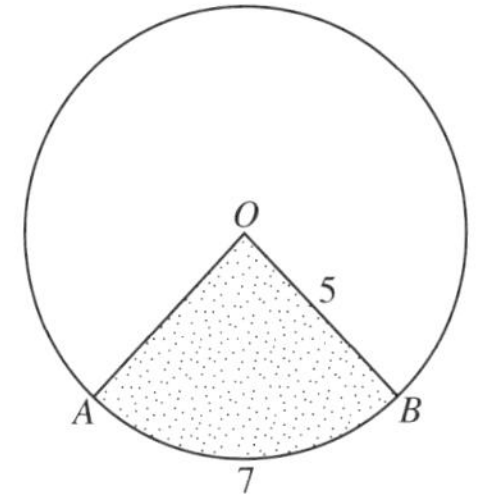

What is the area of the shaded sector OAB?

A $\frac{35}{2}$ **B** $\frac{35}{2}\pi$

C $\frac{125}{14}$ **D** $\frac{125}{14}\pi$ *(1 mark)*

(Q7, **2016 HSC**) Easy

5 A school playground consists of part of a circle, with centre O, and a rectangle as shown in the diagram. The radius OB of the circle is 45 m, the width BC of the rectangle is 20 m and $\angle AOB$ is 100°.

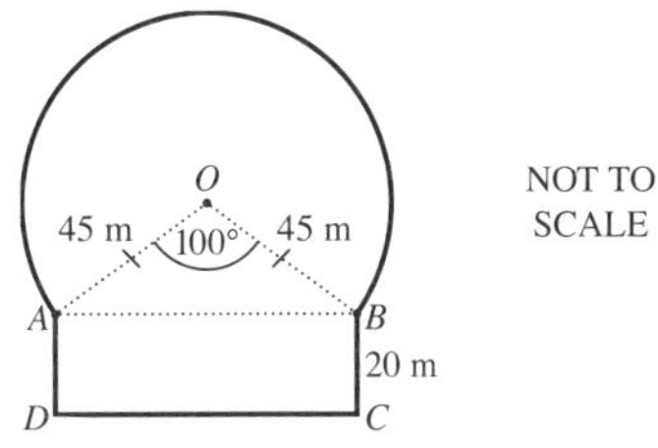

What is the area of the whole playground, correct to the nearest square metre? *(5 marks)*

(Q30c, **2016 MS2 HSC**) Medium

6 The angle of a sector in a circle of radius 8 cm is $\frac{\pi}{7}$ radians, as shown in the diagram.

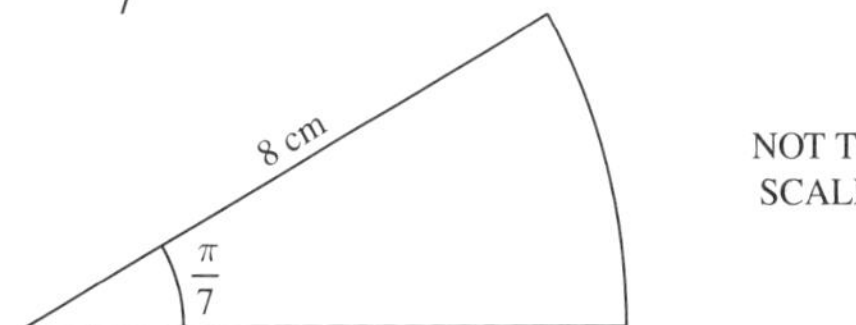

Find the exact value of the perimeter of the sector. *(2 marks)*

(Q11g, **2014 HSC**) Easy

7 The region ABC is a sector of a circle with radius 30 cm, centred at C. The angle of the sector is θ. The arc DE lies on a circle also centred at C, as shown in the diagram.

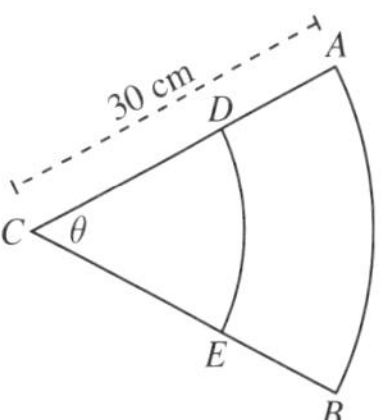

The arc DE divides the sector ABC into two regions of equal area.

Find the exact length of the interval CD. *(2 marks)*

(Q13c, **2013 HSC**) Medium

8 The right-angled triangle ABC has hypotenuse $AB = 13$. The point D is on AC such that $DC = 4$, $\angle DBC = \frac{\pi}{6}$ and $\angle ABD = x$.

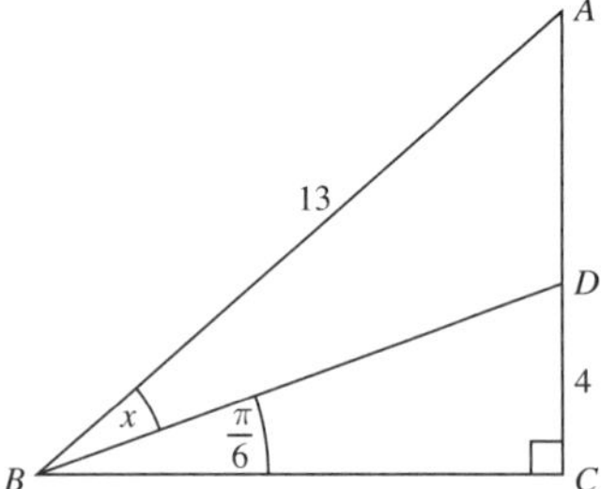

NOT TO SCALE

Using the sine rule, or otherwise, find the exact value of $\sin x$. *(3 marks)*

(Q14c, **2013 HSC**) **Hard**

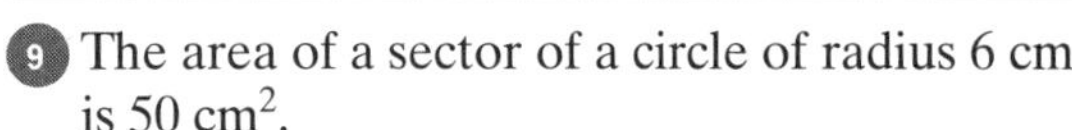

9 The area of a sector of a circle of radius 6 cm is 50 cm^2.

Find the length of the arc of the sector. *(2 marks)*

(Q11f, **2012 HSC**) **Easy**

---

10 The diagram shows a circle with centre O and radius 5 cm.

The length of the arc PQ is 9 cm. Lines drawn perpendicular to OP and OQ at P and Q respectively meet at T.

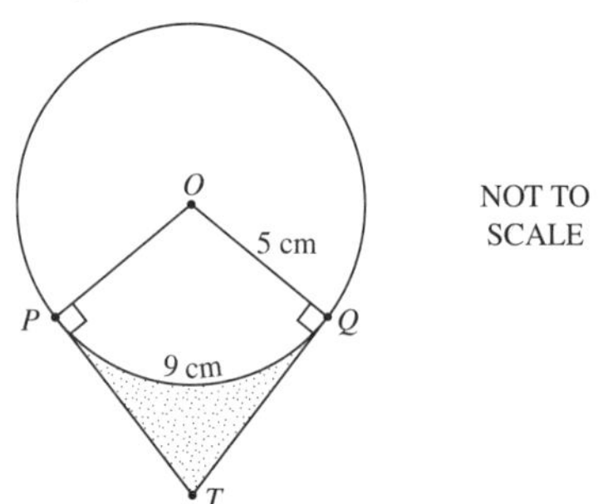

NOT TO SCALE

i Find $\angle POQ$ in radians. *(1 mark)* **Medium**

ii [Prove that ΔOPT is congruent to ΔOQT.]

iii Find the length of PT. *(1 mark)* **Hard**

iv Find the area of the shaded region. *(2 marks)* **Hard**

(Q6b, **2010 HSC**)

---

11 The diagram shows a circle with centre O and radius 2 centimetres. The points A and B lie on the circumference of the circle and $\angle AOB = \theta$.

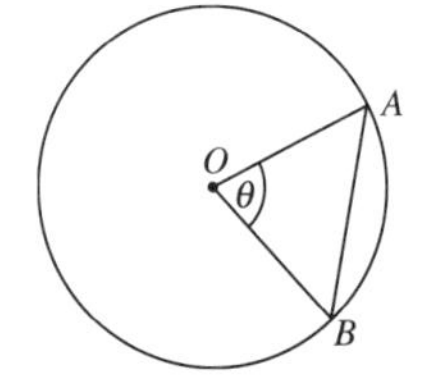

NOT TO SCALE

i There are two possible values of θ for which the area of ΔAOB is $\sqrt{3}$ square centimetres. One value is $\frac{\pi}{3}$. Find the other value. *(2 marks)* **Medium**

ii Suppose that $\theta = \frac{\pi}{3}$.

1 Find the area of the sector AOB. *(1 mark)* **Easy**

2 Find the exact length of the perimeter of the minor segment bounded by the chord AB and the arc AB. *(2 marks)* **Medium**

(Q5c, **2009 HSC**)

---

12 Evaluate $2 \cos \frac{\pi}{5}$ correct to three significant figures. *(2 marks)*

(Q1a, **2008 HSC**) **Easy**

---

13 The diagram shows a sector with radius r and angle θ where $0 < \theta \leq 2\pi$.

The arc length is $\frac{10\pi}{3}$.

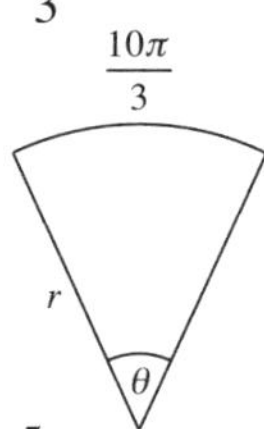

i Show that $r \geq \frac{5}{3}$. *(2 marks)* **Medium**

ii Calculate the area of the sector when $r = 4$. *(2 marks)* **Medium**

(Q7b, **2008 HSC**)

---

14 An advertising logo is formed from two circles, which intersect as shown in the diagram.

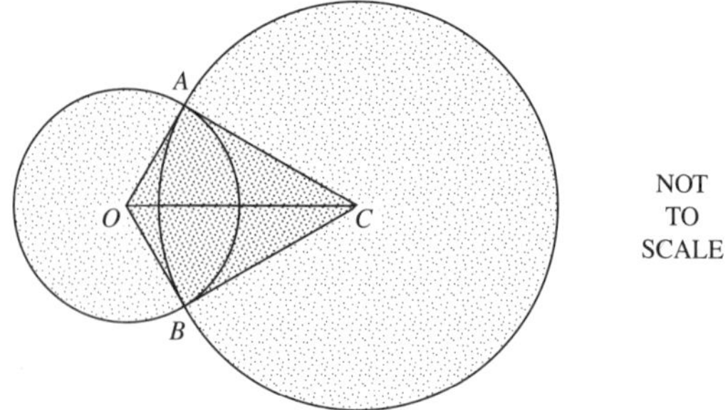

NOT TO SCALE

The circles intersect at A and B and have centres at O and C.

The radius of the circle centred at O is 1 metre and the radius of the circle centred at C is $\sqrt{3}$ metres. The length of OC is 2 metres.

i Use Pythagoras' theorem to show that $\angle OAC = \frac{\pi}{2}$. *(1 mark)* **Easy**

ii Find $\angle ACO$ and $\angle AOC$. *(2 marks)* **Easy**

iii Find the area of the quadrilateral $AOBC$. *(1 mark)* **Easy**

iv Find the area of the major sector ACB. *(1 mark)* **Hard**

v Find the total area of the logo (the sum of all the shaded areas). *(2 marks)* **Hard**

(Q4c, **2007 HSC**)

15 In the diagram, $ABCD$ represents a garden. The sector BCD has centre B and $\angle DBC = \frac{5\pi}{6}$.

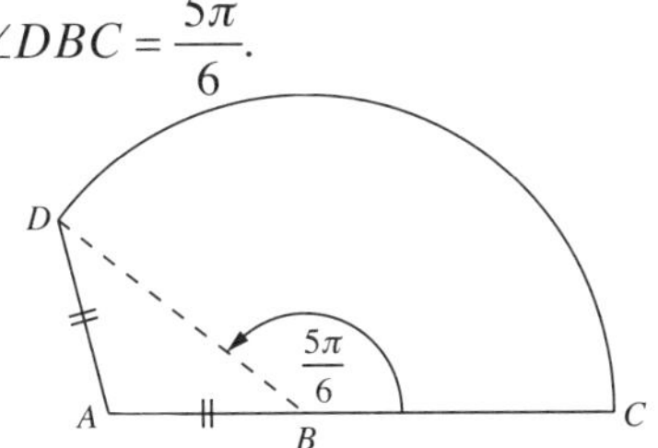

The points A, B and C lie on a straight line and $AB = AD = 3$ metres.

Copy or trace the diagram into your writing booklet.

i Show that $\angle DAB = \frac{2\pi}{3}$. *(1 mark)* **Easy**

ii Find the length of BD. *(2 marks)* **Easy**

iii Find the area of the garden $ABCD$. *(2 marks)* **Medium**

(Q4a, **2006 HSC**)

16 A pendulum is 90 cm long and swings through an angle of 0.6 radians. The extreme positions of the pendulum are indicated by the points A and B in the diagram.

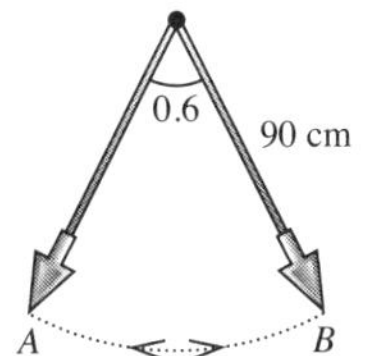

i Find the length of the arc AB. *(1 mark)* **Easy**

ii Find the straight-line distance between the extreme positions of the pendulum. *(2 marks)* **Easy**

iii Find the area of the sector swept out by the pendulum. *(1 mark)* **Easy**

(Q4a, **2005 HSC**)

17 AOB is a sector of a circle, centre O and radius 6 cm. The length of the arc AB is 5π cm.

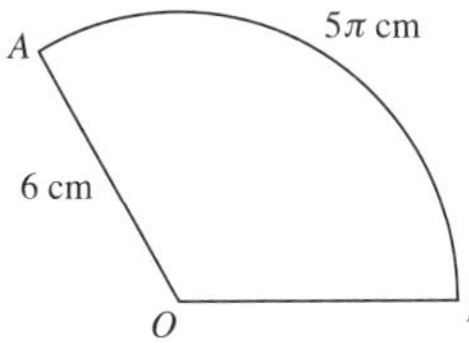

Calculate the exact area of the sector AOB. *(2 marks)*

(Q4a, **2004 HSC**) **Medium**

18 An arc of length 5 units subtends an angle θ at the centre of a circle of radius 3 units. Find the value of θ to the nearest degree. *(2 marks)*

(Q1c, **2003 HSC**) **Medium**

19 The point A lies on the circle $\mathscr{C}_1$ with centre O and radius r. The circle $\mathscr{C}_2$ with centre A and radius $r\sqrt{3}$ intersects $\mathscr{C}_1$ at the points B and C, as shown in the diagram.

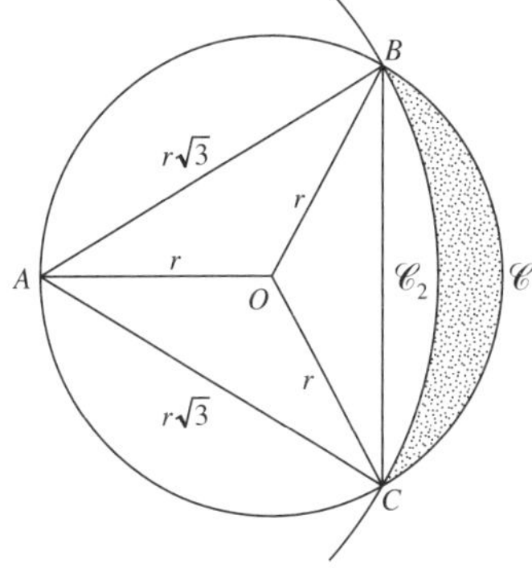

i Using the cosine rule (or otherwise) find $\angle BAO$. *(1 mark)* **Medium**

ii Find the area of the sector BAC of $\mathscr{C}_2$. *(1 mark)* **Medium**

iii Find the area of the sector BOC of $\mathscr{C}_1$. *(1 mark)* **Hard**

iv Hence (or otherwise) find the area of the shaded region. *(2 marks)* **Hard**

(Q9b, **2003 HSC**)

20 The length of the arc between two spokes on a car's steering wheel is 38 cm. Each spoke is 20 cm in length.

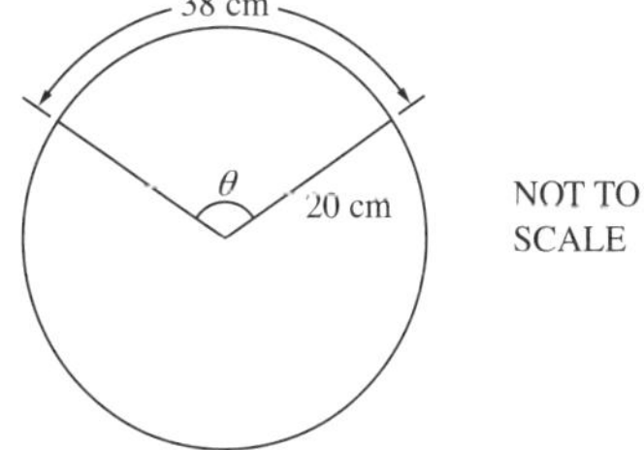

Calculate the angle θ between the two spokes. Give your answer correct to the nearest degree. *(3 marks)*

(Q5b, **2002 HSC**) **Medium**

21 Find the length of the radius of the sector of the circle shown in the diagram.

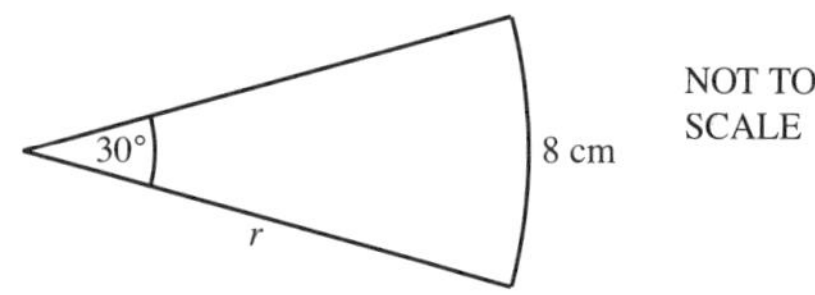

Give your answer correct to the nearest mm.

(2 marks)

(Q5c, **2001 HSC**) Easy

22 What is the exact value of $\cos \dfrac{\pi}{6}$? *(1 mark)*

(Q1c, **2000 HSC**) Easy

23 In the diagram, AB is an arc of a circle with centre O. The radius OA is 20 cm.

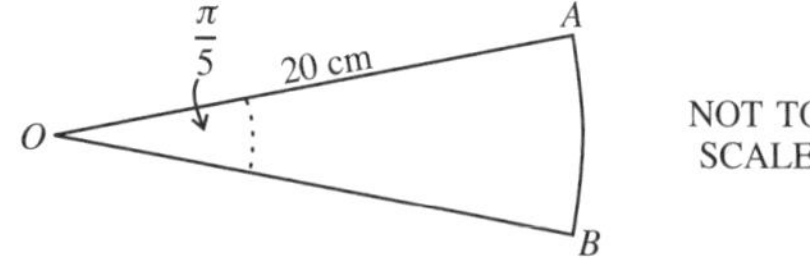

The angle AOB is $\dfrac{\pi}{5}$ radians. Find the area of the sector AOB. *(2 marks)*

(Q1f, **1999 HSC**) Medium

24 The diagram shows a hanging basket in the shape of a hemisphere with radius 10 cm. Let O be the centre of the sphere and let OA be the central axis. Two vertical wire supports, AB and AC, are shown on the diagram. The length of the arc BC is 4 cm.

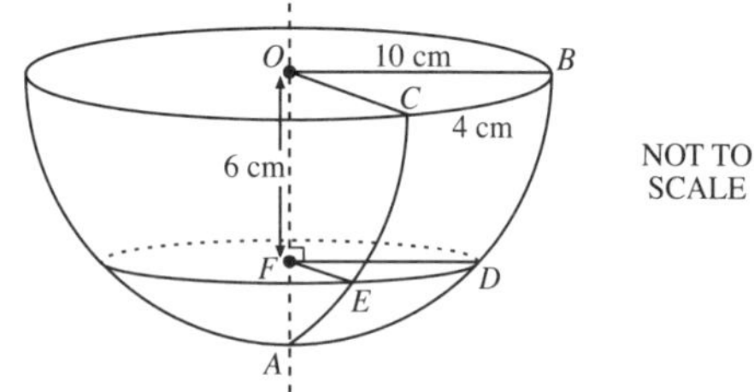

A horizontal wire support is placed around the surface of the basket. This wire meets AB at D and AC at E. The plane through DE parallel to the plane OBC cuts OA at F. The length OF is 6 cm. Note that $\angle BOC = \angle DFE$.

i Show that the length of FD is 8 cm. Easy

ii Find $\angle DFE$ in radians. Medium

iii Find the size of the angle $\angle DOE$ in radians, correct to 3 decimal places. Hard

(6 marks) (Q6b, **1999 ME1 HSC**)

25 Find the exact value of

$\sin\left(\dfrac{\pi}{4}\right) + \sin\left(\dfrac{2\pi}{3}\right)$. *(2 marks)*

(Q1e, **1998 HSC**) Medium

26 Write down the exact value of 135° in radians.

(1 mark)

(Q1c, **1997 HSC**) Easy

27 In the diagram, PQ is an arc of a circle with centre O. The radius $OP = 3$ cm and the angle POQ is $\dfrac{5\pi}{6}$ radians.

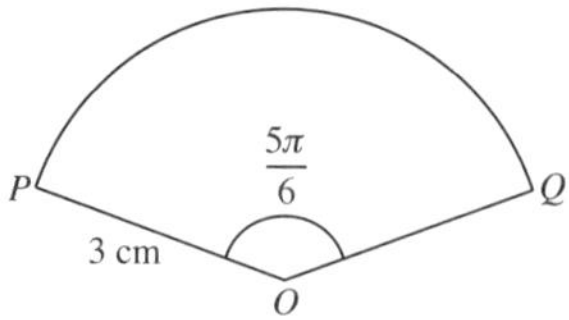

Find the length of the arc PQ. *(1 mark)*

(Q1d, **1997 HSC**) Easy

28 Convert $\dfrac{3\pi}{5}$ radians to degrees. *(1 mark)*

(Q1c, **1995 HSC**) Easy

29 A table-top is in the shape of a circle with a small segment removed as shown. The circle has centre O and radius 0·6 metres. The length of the straight edge AB is also 0·6 metres.

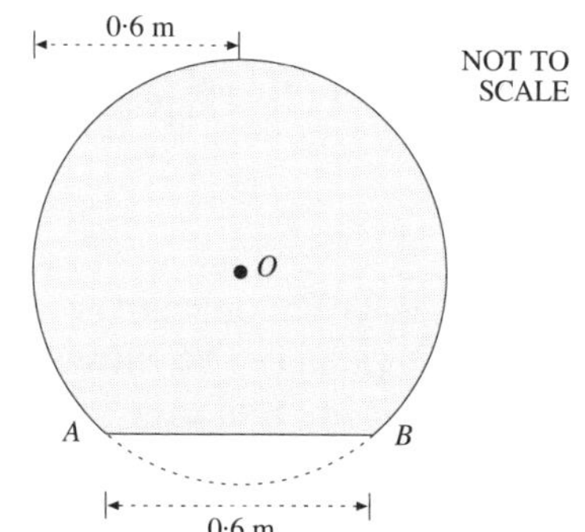

i Explain why $\angle AOB = \dfrac{\pi}{3}$.

ii Find the area of the table-top. *(4 marks)*

(Q5a, **1995 HSC**) Easy

30 A car windscreen wiper traces out the area $RSTU$ where RS and UT are arcs of circles centre O, radii 50 cm and 20 cm respectively, as shown in the figure.

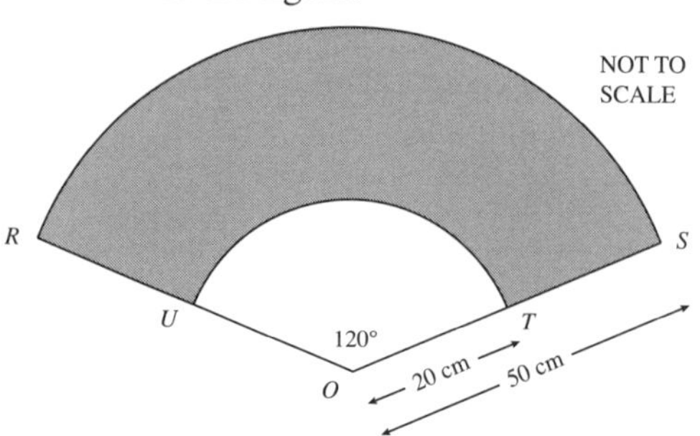

Calculate the area $RSTU$.

(Q4b, **1994 HSC**) Medium

31 The diagram shows the first quadrant of the circle $x^2 + y^2 = 1$. The point A has coordinates $\left(\frac{1}{2}, 0\right)$ and AB is perpendicular to the x axis.

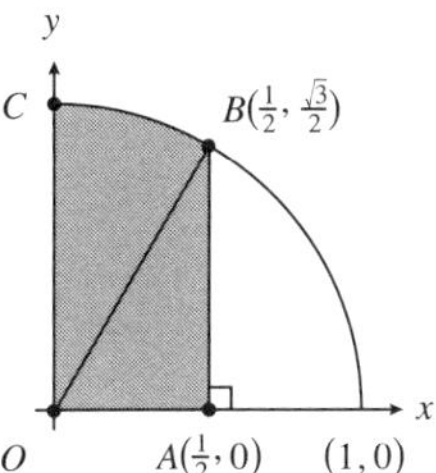

i What is the exact value of $\angle COB$? **Medium**

ii Show that the exact value of the shaded area $OABC$ is $\dfrac{2\pi + 3\sqrt{3}}{24}$. **Medium**

(Q7a, **1993 HSC**)

32 The diagram shows the sector of a circle.

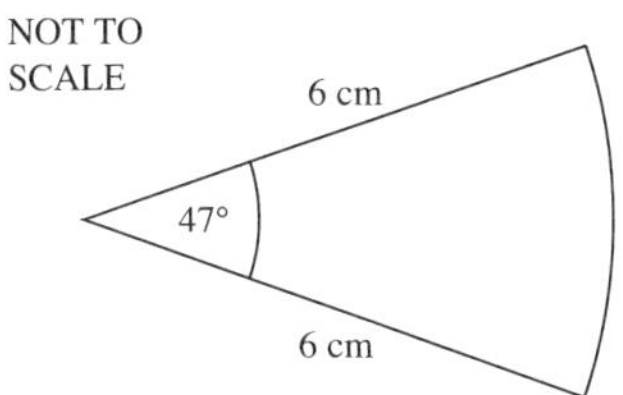

Find the area of this sector.

Give your answer to the nearest square centimetre.

(Q5a, **1992 HSC**) **Easy**

Year 11 Radians—Worked answers

1 $\cos\left(\frac{\pi}{4}\right)+\cos\left(\frac{2\pi}{3}\right)$

$=\frac{1}{\sqrt{2}}-\frac{1}{2}$

$=\frac{2-\sqrt{2}}{2\sqrt{2}}$

Answer B

2 $\theta = 70° = \frac{7\pi}{18}$ radians

$\ell = r\theta$

$= 20 \times \frac{7\pi}{18}$

$= \frac{70\pi}{9}$ ✓

Using cosine rule,

$AB^2 = 20^2 + 20^2 - 2(20)(20)\cos 70°$

$= 526.3838853\ldots$

$AB = 22.94305745\ldots$

$= 22.943$ (3 dec. pl.) ✓

Perimeter $= 22.943 + \frac{70\pi}{9}$

$= 47.37766698\ldots$

$= 47.4$ (1 dec. pl.) ✓

$\therefore$ perimeter is 47.4 cm.

(3 marks)

3 **i** Area of triangle

$= \frac{1}{2}ab\sin C$

$= \frac{1}{2} \times 6 \times 6 \times \sin 30°$

$= 9$

$\therefore$ the area is 9 cm^2. ✓

(1 mark)

ii Area

= Area of sector − Area of triangle

$= \frac{1}{2}r^2\theta - 9$

$= \frac{1}{2} \times 6^2 \times \frac{\pi}{6} - 9$

$= 3\pi - 9$

$\therefore$ the area is $(3\pi - 9)$ cm^2. ✓

(1 mark)

4 Use $\ell = r\theta$:

$7 = 5\theta$

$\theta = \frac{7}{5}$

Use $A = \frac{1}{2}r^2\theta$:

$= \frac{1}{2} \times 5^2 \times \frac{7}{5}$

$= \frac{35}{2}$

Answer A

5 By the cosine rule in ΔAOB:

$c^2 = a^2 + b^2 - 2ab\cos C$

$AB^2 = 45^2 + 45^2 - 2 \times 45 \times 45 \times \cos 100°$

$= 4753.2751\ldots$

$AB = 68.9439998\ldots$ m ✓

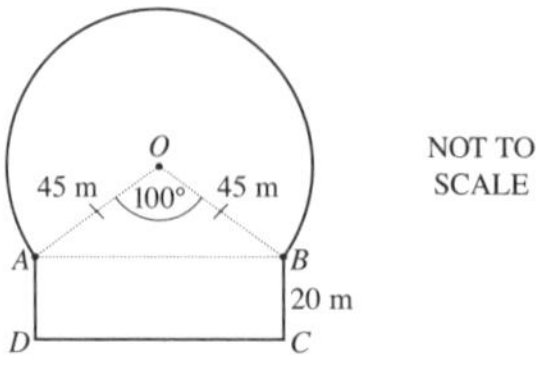

Area of rectangle $ABCD$

$= 68.9439998\ldots \times 20$

$= 1378.87999\ldots$ m^2 ✓

Area of triangle AOB:

$A = \frac{1}{2}ab\sin C$

$= \frac{1}{2} \times 45 \times 45 \times \sin 100°$

$= 997.117849\ldots$ m^2 ✓

Angle in major sector

$= 360° - 100° = 260°$

Area of major sector:

$A = \frac{\theta}{360}\pi r^2$

$= \frac{260}{360} \times \pi \times 45^2$

$= 4594.57925\ldots$ m^2 ✓

Total area

$= 1378.87999\ldots + 997.117849\ldots + 4594.57925\ldots$

$= 6970.577\ldots$

$= 6971$ m^2 (nearest square metre) ✓

(5 marks)

6 Arc length $= r\theta$

$= 8 \times \frac{\pi}{7}$

$= \frac{8\pi}{7}$ ✓

Perimeter $= 8 + 8 + \frac{8\pi}{7}$

$= 16 + \frac{8\pi}{7}$

$\therefore \left(16 + \frac{8\pi}{7}\right)$ cm. ✓

(2 marks)

7 Area of sector ACB

$= \frac{1}{2}r^2\theta$

$= \frac{1}{2} \times 30^2 \times \theta$

$= 450\theta$

Area of sector DCE

$= \frac{1}{2} \times 450\theta$

$= 225\theta$ ✓

$\therefore \frac{1}{2} \times r^2 \times \theta = 225\theta$

$\therefore r^2 = 450$

$r = \sqrt{450}$

$= 15\sqrt{2}$

$\therefore$ length of CD is $15\sqrt{2}$ cm ✓

(2 marks)

8

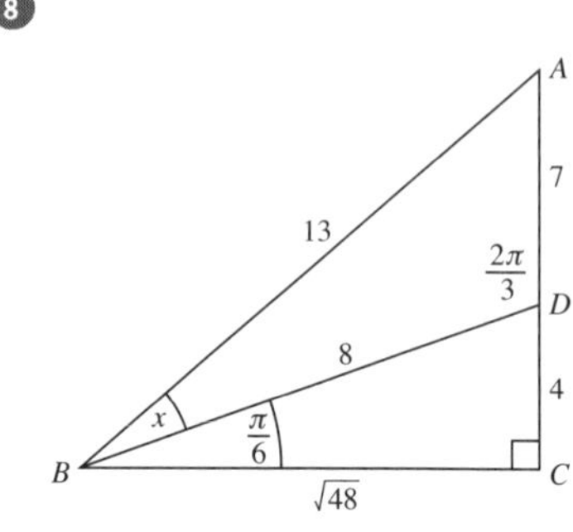

$\frac{4}{BD} = \sin\frac{\pi}{6}$

$\frac{4}{BD} = \frac{1}{2}$

$\therefore BD = 8$ ✓

By Pythagoras,

$BC^2 = 8^2 - 4^2$

$= 64 - 16$

$= 48$

$BC = \sqrt{48}$

By Pythagoras,

$AC^2 = 13^2 - (\sqrt{48})^2$

$= 121$

$AC = 11$

$\therefore AD = 7$

Also, $\angle ADB = \frac{\pi}{6} + \frac{\pi}{2}$

$= \frac{2\pi}{3}$ ✓

$\therefore \frac{\sin x}{7} = \frac{\sin\frac{2\pi}{3}}{13}$

$\sin x = \frac{7\sin\frac{2\pi}{3}}{13}$

$= \frac{7 \times \frac{\sqrt{3}}{2}}{13}$

$= \frac{7\sqrt{3}}{26}$ ✓

(3 marks)

9 Area $= \frac{1}{2}r^2\theta$

$50 = \frac{1}{2} \times 6^2 \times \theta$

$\therefore 50 = 18\theta$

$\therefore \theta = \frac{25}{9}$ ✓

Now, $\ell = r\theta$

$= 6 \times \frac{25}{9}$

$= \frac{50}{3}$

$\therefore$ length is $\frac{50}{3}$ cm ✓

(2 marks)

10 **i** Using $l = 9, r = 5$,

$l = r\theta$

$9 = 5\theta$

$\theta = \frac{9}{5}$

$= 1.8$

$\therefore \angle POQ$ is 1.8 radians. ✓

(1 mark)

iii [From **ii** $\angle POT = \angle QOT$ (matching $\angle$s of cong Δs)]

Also, $\angle POQ$

$= 1.8$ radians (from **i**),

then $\angle POT$

$= \frac{1}{2} \times 1.8$ radians

$= 0.9$ radians

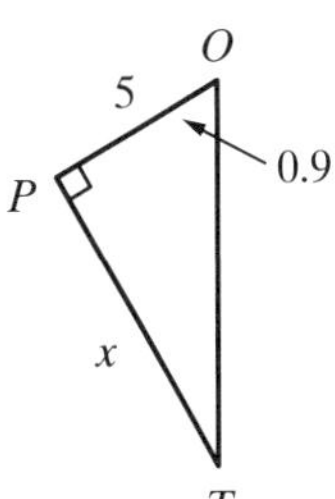

$\frac{x}{5} = \tan 0.9$

$x = 5 \times \tan 0.9$

$= 6.300\ldots$

$= 6.3$ (corr. to 1 dec. pl.)

$\therefore$ length of PT is 6.3 cm ✓

(1 mark)

iv Area $OPTQ$

$= 2 \times$ Area of ΔOPT

$= 2 \times \frac{1}{2} \times 5 \times 6.300..$

$= 31.503\ldots$

$= 31.5$ (corr. to 1 dec. pl.)

$\therefore$ The area of $OPTQ$ is 31.5 cm^2. ✓

Shaded area

= Area $OPTQ$ – Area sector

$= 31.503\ldots - \frac{1}{2} \times 5^2 \times 1.8$

$= 9.003\ldots$

$= 9$ (nearest whole)

$\therefore$ The shaded area is 9 cm^2.

✓ *(2 marks)*

11 **i** Area of ΔAOB

$= \frac{1}{2} \times AO \times OB \times \sin\theta$

$(0 < \theta < \pi)$

$\therefore \sqrt{3} = \frac{1}{2} \times 2 \times 2 \times \sin\theta$

$\sqrt{3} = 2\sin\theta$

$\sin\theta = \frac{\sqrt{3}}{2}$ ✓

$\therefore \theta = \frac{\pi}{3}, \pi - \frac{\pi}{3}$

$= \frac{\pi}{3}, \frac{2\pi}{3}$

$\therefore$ The other value is $\frac{2\pi}{3}$. ✓

(2 marks)

ii **1** $A = \frac{1}{2}r^2\theta$

where $\theta = \frac{\pi}{3}$ and $r = 2$.

$\therefore A = \frac{1}{2} \times 2^2 \times \frac{\pi}{3}$

$= \frac{2\pi}{3}$

$\therefore$ Area of sector AOB

$= \frac{2\pi}{3}$ cm^2. ✓

(1 mark)

2 $l = r\theta$

where $\theta = \frac{\pi}{3}$ and $r = 2$

$\therefore l = 2 \times \frac{\pi}{3}$

$\therefore l = \frac{2\pi}{3}$ ✓

By the cosine rule,

$AB^2 = 2^2 + 2^2 - 2 \times 2^2 \times \cos\frac{\pi}{3}$

$= 4 + 4 - 8 \times \frac{1}{2}$

$= 4$

$\therefore AB = 2$

$\therefore$ Perimeter

$= \left(2 + \frac{2\pi}{3}\right)$ cm. ✓

(2 marks)

12 $2\cos\frac{\pi}{5} = 1.6180\ldots$ ✓

$= 1.62$ to 3 significant figures.

✓ *(2 marks)*

13 **i** $l = r\theta$ where $0 < \theta \leqslant 2\pi$

$\therefore \frac{10\pi}{3} = r\theta$

$\therefore \theta = \frac{10\pi}{3r}$ ✓

Now $0 < \theta \leqslant 2\pi$

$\therefore 0 < \frac{10\pi}{3r} \leqslant 2\pi$

$\therefore 0 < 10\pi \leqslant 6\pi r$

$\therefore 0 < \frac{10\pi}{6\pi} \leqslant r$

$\therefore 0 < \frac{5}{3} \leqslant r$

$\therefore r \geqslant \frac{5}{3}$ ✓

(2 marks)

ii $A = \frac{1}{2}r^2\theta$

where $\theta = \frac{10\pi}{3r}$ and $r = 4$

$\therefore A = \frac{1}{2} \times 4^2 \times \frac{10\pi}{3 \times 4}$ ✓

$= \frac{1}{2} \times 16 \times \frac{10\pi}{12}$

$= \frac{80\pi}{12}$

$= \frac{20\pi}{3}$

$\therefore$ Area of sector is $\frac{20\pi}{3}$ $units^2$. ✓

(2 marks)

14

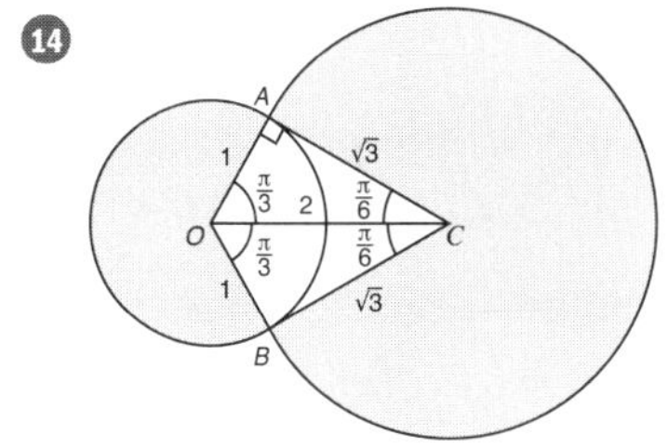

i In ΔOAC,

$OA = 1$

$AC = \sqrt{3}$

By Pythagoras' theorem:

$OC^2 = OA^2 + AC^2$

$= 1^2 + (\sqrt{3})^2$

$= 4$

$\therefore OC = \sqrt{4}$

$= 2$

$\therefore \Delta OAC$ is right angled

$\therefore \angle OAC = \frac{\pi}{2}$ ✓

(1 mark)

ii In ΔOAC,

$\tan\angle ACO = \frac{OA}{AC}$

$= \frac{1}{\sqrt{3}}$

$\therefore \angle ACO = \frac{\pi}{6}$ ✓

$\therefore \angle AOC = \pi - \frac{\pi}{2} - \frac{\pi}{6}$

(angle sum of ΔOAC)

$= \frac{\pi}{3}$ ✓

(2 marks)

iii Area of $\Delta AOC = \frac{1}{2}(OA)(AC)$

$= \frac{1}{2} \times 1 \times \sqrt{3}$

$= \frac{\sqrt{3}}{2}$ m^2.

$\therefore$ Area of quadrilateral $AOBC$

$= \frac{\sqrt{3}}{2} \times 2$

$= \sqrt{3}$ m^2 ✓

(1 mark)

iv Now ΔOAC and ΔOBC are congruent (SSS),

$\therefore \angle AOC = \angle BOC = \frac{\pi}{3}$

and $\angle ACO = \angle BCO = \frac{\pi}{6}$

$\therefore \angle ACB = \frac{\pi}{6} + \frac{\pi}{6} = \frac{\pi}{3}$

Area of major sector ACB

$= \frac{1}{2}r^2\theta$ where $\theta = (2\pi - \angle ACB)$

$= \frac{1}{2}(\sqrt{3})^2\left(2\pi - \frac{\pi}{3}\right)$

$= \frac{1}{2} \times 3 \times \frac{5\pi}{3}$

$= \frac{5\pi}{2}$ m^2 ✓

(1 mark)

v Now $\angle AOB = \angle AOC + \angle BOC$

$= \frac{\pi}{3} + \frac{\pi}{3}$

$= \frac{2\pi}{3}$

Area of major sector AOB

$= \frac{1}{2}r^2\theta$ where $\theta = (2\pi - \angle AOB)$

$= \frac{1}{2}r^2\left(2\pi - \frac{2\pi}{3}\right)$

$= \frac{1}{2} \times 1 \times \frac{4\pi}{3}$

$= \frac{2\pi}{3}$ m^2 ✓

Total area of the logo
= area of quadrilateral $AOBC$
+ area of major sector ACB
+ area of major sector AOB

$= \sqrt{3} + \frac{5\pi}{2} + \frac{2\pi}{3}$

$= \frac{6\sqrt{3} + 15\pi + 4\pi}{6}$

$= \left(\frac{6\sqrt{3} + 19\pi}{6}\right)$ m^2 ✓

(2 marks)

15 **i**

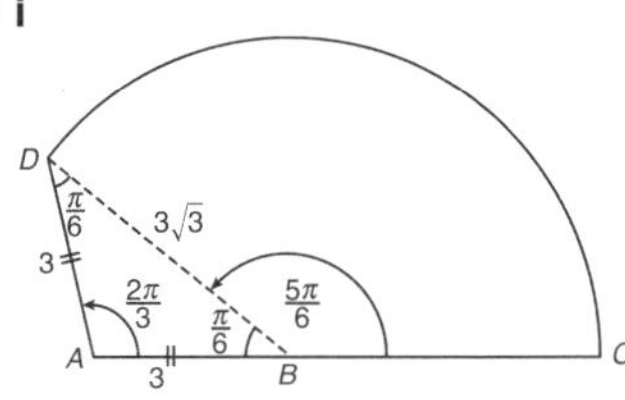

In ΔABD,

$\angle ABD = \pi - \angle CBD$
(ABC is a straight angle, π)

$= \pi - \frac{5\pi}{6}$

$= \frac{\pi}{6}$

Also $\angle ABD = \angle ADB$
(angles opp. equal sides; $AD = AB$)

$= \frac{\pi}{6}$

Now
$\angle DAB + \angle ABD + \angle ADB = \pi$
(angle sum of ΔABD is π)

$\therefore \angle DAB + \frac{\pi}{6} + \frac{\pi}{6} = \pi$

$\therefore \angle DAB = \pi - \frac{2\pi}{6}$

$= \frac{2\pi}{3}$ ✓

(1 mark)

ii By the sine rule:

$$\frac{BD}{\sin\frac{2\pi}{3}} = \frac{3}{\sin\frac{\pi}{6}}$$

$$\therefore BD = \frac{3 \times \sin\frac{2\pi}{3}}{\sin\frac{\pi}{6}}$$ ✓

$$= \frac{3 \times \frac{\sqrt{3}}{2}}{\frac{1}{2}}$$

$$= \frac{6 \times \sqrt{3}}{2}$$

$= 3\sqrt{3}$ m ✓

(2 marks)

iii Area of Garden $ABCD$
= Area of ΔABD
+ Area of Sector BCD:

Area of ΔABD

$= \frac{1}{2}bd\sin A$

$= \frac{1}{2} \times 3^2 \times \sin\frac{2\pi}{3}$

$= \frac{1}{2} \times 9 \times \frac{\sqrt{3}}{2}$

$= \frac{9\sqrt{3}}{4}$ m^2

Area of Sector BCD

$= \frac{1}{2}r^2\theta$

$= \frac{1}{2} \times (3\sqrt{3})^2 \times \frac{5\pi}{6}$

$= \frac{1}{2} \times 27 \times \frac{5\pi}{6}$

$= \frac{135\pi}{12}$

$= \frac{45\pi}{4}$ m^2 ✓

Area of Garden
= Area of ΔABD
+ Area of Sector BCD

$= \frac{9\sqrt{3}}{4} + \frac{45\pi}{4}$

$= \frac{9\sqrt{3} + 45\pi}{4}$ m^2 ✓

(2 marks)

16 **i** $l = r\theta$

$= 90 \times 0.6$

$= 54$ cm ✓

(1 mark)

ii By the cosine rule:

$AB^2 = 90^2 + 90^2 - 2(90)(90)\cos 0.6$ ✓

$= 2829.5630\ldots$

$\therefore AB = 53.1936\ldots$

$\doteqdot 53.2$ cm ✓

(2 marks)

iii $A = \frac{1}{2}r^2\theta$

$= \frac{1}{2}(90)^2(0.6)$

$= 2430$

$\therefore$ Area $= 2430$ cm^2 ✓

(1 mark)

17 $l = r\theta$

$\therefore 5\pi = 6 \times \theta$

$\therefore \theta = \frac{5\pi}{6}$ ✓

Area of sector AOB

$= \frac{1}{2}r^2\theta$

$= \frac{1}{2} \times (6)^2 \times \frac{5\pi}{6}$

$= 15\pi$ cm^2 ✓

(2 marks)

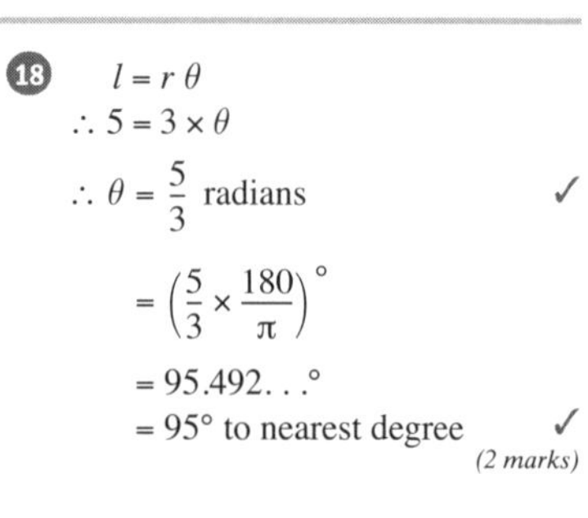

18 $l = r\theta$

$\therefore 5 = 3 \times \theta$

$\therefore \theta = \frac{5}{3}$ radians ✓

$= \left(\frac{5}{3} \times \frac{180}{\pi}\right)^\circ$

$= 95.492\ldots^\circ$

$= 95^\circ$ to nearest degree ✓

(2 marks)

19 **i** $\cos \angle BAO = \dfrac{r^2 + (r\sqrt{3})^2 - r^2}{2(r)(r\sqrt{3})}$

$= \dfrac{r^2 + 3r^2 - r^2}{2\sqrt{3}r^2}$

$= \dfrac{3r^2}{2\sqrt{3}r^2}$

$= \dfrac{3}{2\sqrt{3}} \times \dfrac{\sqrt{3}}{\sqrt{3}}$

$= \dfrac{3\sqrt{3}}{6}$

i.e. $\cos \angle BAO = \dfrac{\sqrt{3}}{2}$

$\therefore \angle BAO = \dfrac{\pi}{6}$ ✓

(1 mark)

ii $A = \dfrac{1}{2} r^2 \theta$

$r = r\sqrt{3}$

$\theta = \angle BAC = \dfrac{\pi}{6} + \dfrac{\pi}{6} = \dfrac{2\pi}{6} = \dfrac{\pi}{3}$

($\angle BAC = \angle BAO + \angle CAO$

and $\angle CAO = \dfrac{\pi}{6}$ as

$\Delta BAO \equiv \Delta CAO$)

$\therefore A = \dfrac{1}{2}(r\sqrt{3})^2 \left(\dfrac{\pi}{3}\right)$

$= \dfrac{1}{2} . 3r^2 . \dfrac{\pi}{3}$

$= \dfrac{\pi r^2}{2}$ ✓

(1 mark)

iii $A = \dfrac{1}{2} r^2 \theta$

$r = r$

$\theta = \angle BOC = 120° = \dfrac{2\pi}{3}$

(angle at centre is twice angle at circumference)

$\therefore A = \dfrac{1}{2} r^2 \left(\dfrac{2\pi}{3}\right)$

$= \dfrac{\pi r^2}{3}$ ✓

(1 mark)

iv Now triangles AOB, AOC and BOC are all congruent and hence have the same area:

$A = \dfrac{1}{2} ab \sin C$

$= \dfrac{1}{2} (r)(r) \sin 120°$

$= \dfrac{1}{2} r^2 \left(\dfrac{\sqrt{3}}{2}\right)$

$= \dfrac{\sqrt{3}}{4} r^2$ ✓

Area of shaded region
= Area of Sector BOC – (Area of Sector BAC – Area Δ's AOB and AOC)

$= \dfrac{\pi r^2}{3} - \left[\dfrac{\pi r^2}{2} - 2\left(\dfrac{\sqrt{3}r^2}{4}\right)\right]$

$= \dfrac{\pi r^2}{3} - \left(\dfrac{\pi r^2}{2} - \dfrac{\sqrt{3}r^2}{2}\right)$

$= \dfrac{\pi r^2}{3} - \dfrac{\pi r^2}{2} + \dfrac{\sqrt{3}r^2}{2}$

$= r^2 \left(\dfrac{\pi}{3} - \dfrac{\pi}{2} + \dfrac{\sqrt{3}}{2}\right)$

$= r^2 \left(\dfrac{\sqrt{3}}{2} - \dfrac{\pi}{6}\right)$ units^2 ✓

(2 marks)

20 $l = r\theta$

$\therefore 38 = 20 \times \theta$

$\theta = \dfrac{38}{20}$

$= 1.9$ radians

Now 1.9 radians ✓

$= \left(1.9 \times \dfrac{180}{\pi}\right)^{\circ}$

$= 108.8619\ldots°$ ✓

$= 109°$ to nearest degree ✓

(3 marks)

21 $30° = \dfrac{30°}{180°} \times \pi$

$= \dfrac{\pi}{6}$.

$l = r\theta$

$\therefore 8 = r \times \dfrac{\pi}{6}$ ✓

$r = 8 \times \dfrac{6}{\pi}$

$= \dfrac{48}{\pi}$

$= 15.2788\ldots$ cm

$= 153$ mm to nearest mm ✓

(2 marks)

22 $\cos \dfrac{\pi}{6} = \dfrac{\sqrt{3}}{2}$ ✓

(1 mark)

23 Area of sector AOB

$= \dfrac{1}{2} r^2 \theta$

$= \dfrac{1}{2} \times (20)^2 \times \dfrac{\pi}{5}$ ✓

$= \dfrac{1}{2} \times 400 \times \dfrac{\pi}{5}$

$= 40\pi$ cm^2 ✓

(2 marks)

24 **i**

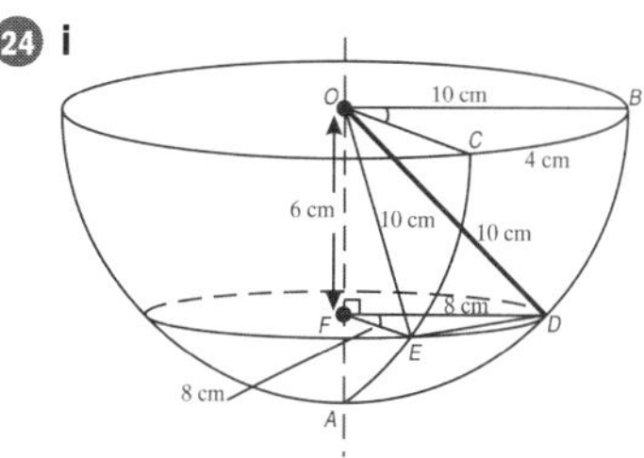

OFD is a right-angle triangle with hypotenuse $OD = 10$ cm because OD is a radius of the sphere.

In the ΔOFD,

$FD^2 + OF^2 = OD^2$

$\therefore FD^2 + 6^2 = 10^2$

$FD^2 = 100 - 36$

$= 64$

$\therefore FD = 8$ cm ✓

ii Let $\angle BOC = \theta$

In sector BOC:

$l = r\theta$

$4 = 10\theta$ ✓

$\theta = 0.4$

So $\angle BOC = 0.4$ radians ✓

Now $\angle DFE = \angle BOC$ (given)

$\therefore \angle DFE = 0.4$ radians

iii In ΔEFD,

$ED^2 = FD^2 + FE^2 - 2FD.FE.\cos \frac{2}{5}$

$= 8^2 + 8^2 - 2(8)(8) \cos \frac{2}{5}$

(note FD and FE are equal radii)

$= 64 + 64 - 128 \cos \frac{2}{5}$

$= 128 - 128 \cos \frac{2}{5}$ ✓

In ΔDOE,

$ED^2 = OE^2 + OD^2 - 2OE.OD \cos \angle DOE$

$= 10^2 + 10^2 - 2(10)(10) \cos \angle DOE$

$= 100 + 100 - 200 \cos \angle DOE$

$= 200 - 200 \cos \angle DOE$ ✓

$\therefore$ Equating the above:

$200 - 200 \cos \angle DOE$

$= 128 - 128 \cos \frac{2}{5}$

$200 \cos \angle DOE$

$= 200 - 128 + 128 \cos \frac{2}{5}$

$\cos \angle DOE = \dfrac{72 + 128 \cos \frac{2}{5}}{200}$

$= 0.9494\ldots$

$\therefore \angle DOE = 0.319$ radians to 3 decimal places. ✓

(6 marks)
(Total mark allocation only in exam)

25 $\sin\left(\dfrac{\pi}{4}\right) + \sin\left(\dfrac{2\pi}{3}\right) = \dfrac{1}{\sqrt{2}} + \dfrac{\sqrt{3}}{2}$ ✓

$= \dfrac{\sqrt{2} + \sqrt{3}}{2}$ ✓

(2 marks)

26 $135° = \left(135 \times \frac{\pi}{180}\right)$ radians

$= \frac{3\pi}{4}$ radians ✓

(1 mark)

27 $l = r \times \theta$

$= 3 \times \frac{5\pi}{6}$

$= \frac{5\pi}{2}$ cm ✓

(1 mark)

28 $\frac{3\pi}{5}$ radians $= \frac{3 \times 180°}{5}$

$= 108°$ ✓

(1 mark)

29

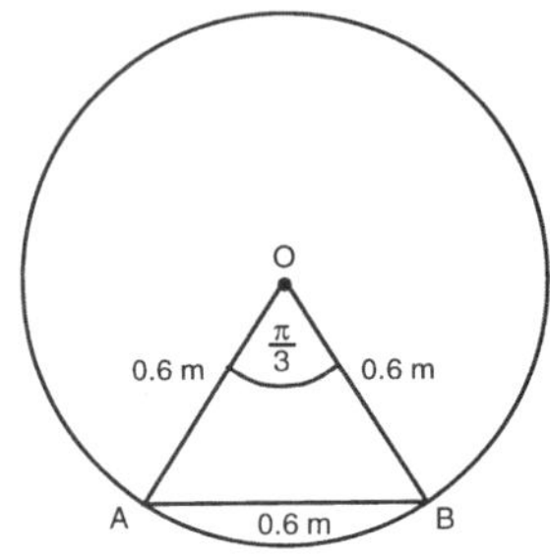

i $\angle AOB = \frac{\pi}{3}(= 60°)$ because because ΔAOB is ✓ equilateral, since the side AB equals each of the two radii OA and OB. ✓

ii

Area of table-top

= Area of circle − Area of minor segment AB

$= \pi r^2 - \frac{1}{2}r^2(\theta - \sin\theta)$

$= \pi(0.6)^2 - \frac{1}{2}(0.6)^2\left(\frac{\pi}{3} - \sin\frac{\pi}{3}\right)$ ✓

$= 0.36\pi - 0.18\left(\frac{\pi}{3} - \frac{\sqrt{3}}{2}\right)$

$= 0.36\pi - 0.06\pi + 0.09\sqrt{3}$

$= (0.3\pi + 0.09\sqrt{3})$ m^2 ✓

$\doteqdot 1.1$ m^2.

(4 marks)
(Total mark allocation only in exam)

30 Let $r_1 = OT$ and $r_2 = OS$

$Area = \frac{1}{2}r_2^2\theta - \frac{1}{2}r_1^2\theta$

$= \frac{1}{2}\theta(r_2^2 - r_1^2)$

$= \frac{1}{2}\cdot\frac{2\pi}{3}(50^2 - 20^2)$

$= \frac{\pi}{3}(2500 - 400)$

$= \frac{\pi}{3} \times 2100$

$= 700\pi$

$= 2199.1148\ldots$

$\doteqdot 2199$ cm^2

(No mark allocation in exam)

31 **i** In ΔBOA,

$\tan\angle BOA = \dfrac{\frac{\sqrt{3}}{2}}{\frac{1}{2}}$

$= \frac{\sqrt{3}}{2} \times \frac{2}{1}$

$= \sqrt{3}$

$\therefore \angle BOA = 60°$

$\therefore \angle COB = 90° - 60°$
$(\angle COA = 90°)$
$= 30°$

$\left(\therefore \angle COB = 30° \times \frac{\pi}{180°}\right.$
$\left.= \frac{\pi}{6} \text{ radians}\right)$

ii Area of $OABC$

= Area of ΔOAB
+ Area of sector COB

$= \frac{1}{2}bh + \frac{1}{2}r^2\theta$

$= \frac{1}{2}\left(\frac{1}{2}\right)\left(\frac{\sqrt{3}}{2}\right) + \frac{1}{2}(1)^2\left(\frac{\pi}{6}\right)$

$= \frac{\sqrt{3}}{8} + \frac{\pi}{12}$

$= \frac{3\sqrt{3}}{24} + \frac{2\pi}{24}$

$= \frac{2\pi + 3\sqrt{3}}{24}$ units ^{2}

(No mark allocation in exam)

32 $A = \frac{1}{2}r^2\theta$

$= \frac{1}{2} \times 6^2 \times \frac{47\pi}{180}$

$= 14.7654\ldots$

$= 15$ cm^2 to the nearest square centimetre

(No mark allocation in exam)

1 Which of the following is equivalent to $\sin^2 5x$?

A $1 + \cos^2 5x$ **B** $1 - \cos^2 5x$

C $-1 + \cos^2 5x$ **D** $-1 - \cos^2 5x$ *(1 mark)*

(Q1, **2021 HSC**) Easy

2 Prove that $\sec\theta - \cos\theta = \sin\theta\tan\theta$.

(2 marks)

(Q19, **2020 HSC**) Medium

3 Which is the graph of $y = \cot x$ for values of x between $-\pi$ and π?

A **B**

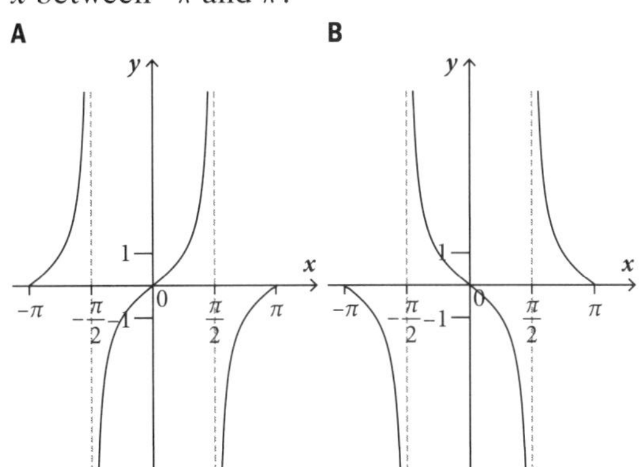

C **D**

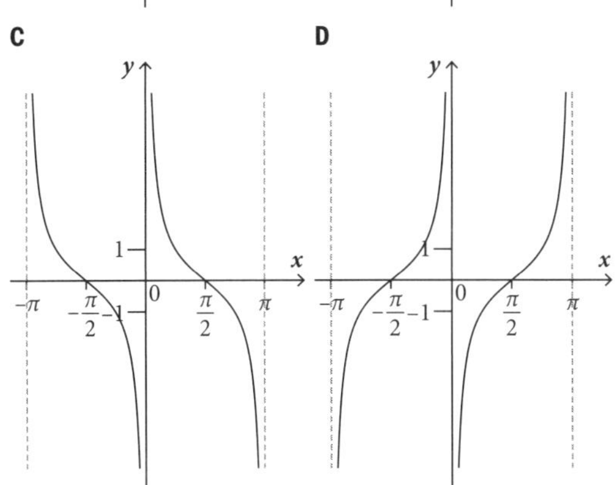

(1 mark)

Bonus question (see page iv) Easy

4 Show that $\text{cosec}^2 x + \text{cosec}\, x \cot x = \dfrac{1}{1 - \cos x}$.

(2 marks)

Bonus question Medium

5 Declan walks 1600 metres due north along a road from point A to point B.

The point A is due east of a hill XY, where X is the top of the hill. The point Y is directly below point X and is on the same horizontal plane as the road.

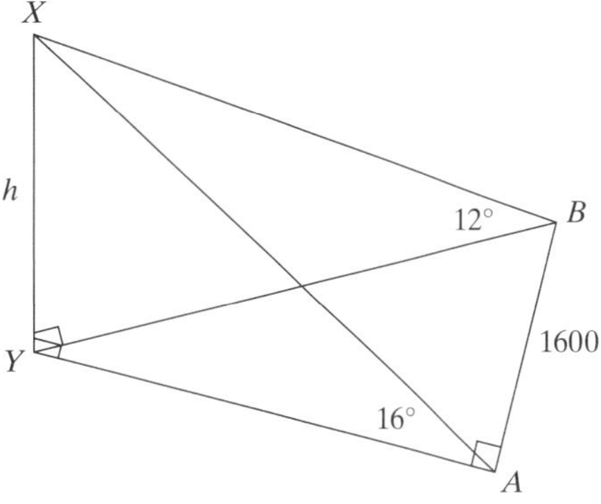

Let the height of the hill above point Y be h metres.

From point A, the angle of elevation to the top of the hill is 16°.

From point B, the angle of elevation to the top of the hill is 12°.

a Show that $BY = h\cot 12°$. *(1 mark)* Easy

b Hence, find the value of h. *(3 marks)* Hard

Bonus question

6 Solve $2\sin x\cos x = \sin x$ for $0 \leq x \leq 2\pi$.

(3 marks)

(Q13a, **2019 HSC**) Medium

7 Which expression is equivalent to $\tan\theta + \cot\theta$?

A $\text{cosec}\,\theta + \sec\theta$ **B** $\sec\theta\,\text{cosec}\,\theta$

C 2 **D** 1 *(1 mark)*

(Q7, **2017 HSC**) Medium

8 How many solutions does the equation $|\cos(2x)| = 1$ have for $0 \leq x \leq 2\pi$?

A 1 **B** 3

C 4 **D** 5 *(1 mark)*

(Q8, **2016 HSC**) Hard

9 Solve $\sin\left(\dfrac{x}{2}\right) = \dfrac{1}{2}$ for $0 \leq x \leq 2\pi$. *(2 marks)*

(Q11g, **2016 HSC**) Medium

10 Find the solutions of $2\sin\theta = 1$ for $0 \leq \theta \leq 2\pi$. *(2 marks)*

(Q12a, **2015 HSC**) Easy

11 A person walks 2000 metres due north along a road from point A to point B. The point A is due east of a mountain OM, where M is the top of the mountain. The point O is directly below point M and is on the same horizontal plane as the road. The height of the mountain above point O is h metres.

From point A, the angle of elevation to the top of the mountain is 15°.

From point B, the angle of elevation to the top of the mountain is 13°.

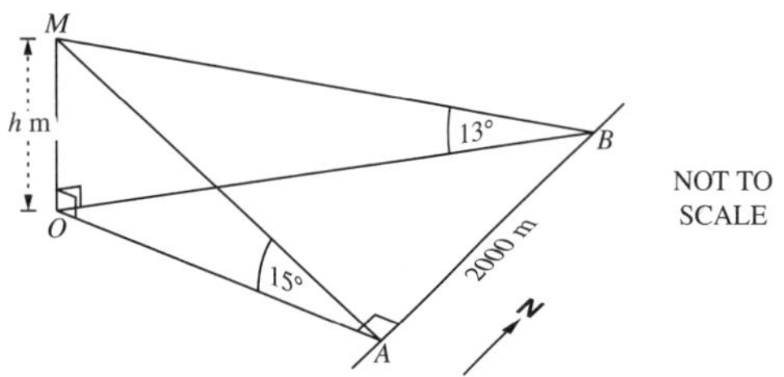

i Show that $OA = h \cot 15°$. *(1 mark)* Easy

ii Hence, find the value of h. *(2 marks)* Hard

(Q12c, **2015 ME1 HSC**)

---

12 How many solutions of the equation $(\sin x - 1)(\tan x + 2) = 0$ lie between 0 and 2π?

A 1 **B** 2

C 3 **D** 4 *(1 mark)*

(Q7, **2014 HSC**) Hard

---

13 Find all solutions of $2 \sin^2 x + \cos x - 2 = 0$, where $0 \leq x \leq 2\pi$. *(3 marks)*

(Q15a, **2014 HSC**) Hard

---

14 What are the solutions of $\sqrt{3} \tan x = -1$ for $0 \leq x \leq 2\pi$?

A $\frac{2\pi}{3}$ and $\frac{4\pi}{3}$ **B** $\frac{2\pi}{3}$ and $\frac{5\pi}{3}$

C $\frac{5\pi}{6}$ and $\frac{7\pi}{6}$ **D** $\frac{5\pi}{6}$ and $\frac{11\pi}{6}$

(1 mark)

(Q6, **2012 HSC**) Easy

---

15 Find the exact values of x such that $2 \sin x = -\sqrt{3}$, where $0 \leq x \leq 2\pi$. *(2 marks)*

(Q2b, **2011 HSC**) Easy

---

16 Find the exact value of θ such that $2 \cos \theta = 1$, where $0 \leq \theta \leq \frac{\pi}{2}$. *(2 marks)*

(Q1e, **2009 HSC**) Easy

---

17 Solve $2 \sin^2 \frac{x}{3} = 1$ for $-\pi \leq x \leq \pi$. *(3 marks)*

(Q6a, **2008 HSC**) Medium

---

18 Solve $\sqrt{2} \sin x = 1$ for $0 \leq x \leq 2\pi$. *(2 marks)*

(Q4a, **2007 HSC**) Easy

---

19 Solve $\cos \theta = \frac{1}{\sqrt{2}}$ for $0 \leq \theta \leq 2\pi$. *(2 marks)*

(Q2a, **2005 HSC**) Easy

---

20 **i** Show that $\cos\theta \tan\theta = \sin\theta$. *(1 mark)* Easy

ii Hence solve $8\sin\theta \cos\theta \tan\theta = \operatorname{cosec}\theta$ for $0 \leq \theta \leq 2\pi$. *(2 marks)* Easy

(Q8a, **2004 HSC**) Easy

---

21 Solve $2 \sin^2 x - 3 \sin x - 2 = 0$ for $0 \leq x \leq 2\pi$. *(2 marks)*

(Q9a, **2003 HSC**) Medium

---

22 Find all values of θ, where $0° \leq \theta \leq 360°$, that satisfy the equation

$$\cos \theta - \frac{2}{5} = 0.$$

Give your answer(s) to the nearest degree. *(2 marks)*

(Q4b, **2002 HSC**) Easy

---

23 Solve $\tan x = 2$ for $0 < x < 2\pi$.

Express your answer in radian measure correct to two decimal places. *(2 marks)*

(Q5a, **2000 HSC**) Medium

---

24 The diagram shows a square $ABCD$ of side x cm, with a point P within the square, such that $PC = 6$ cm, $PB = 2$ cm and $AP = 2\sqrt{5}$ cm.

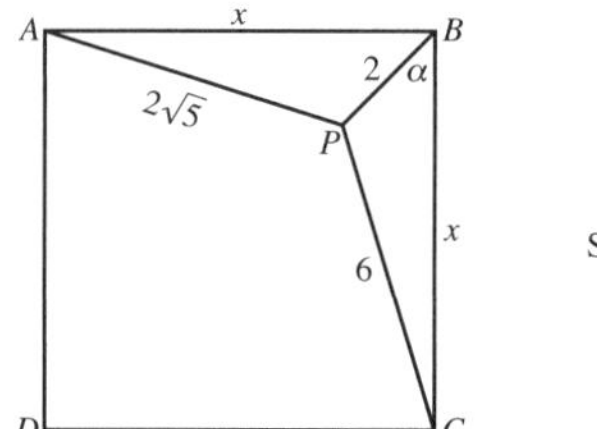

Let $\angle PBC = \alpha$.

i Using the cosine rule in triangle PBC, show that $\cos \alpha = \frac{x^2 - 32}{4x}$. Medium

ii By considering triangle PBA, show that $\sin \alpha = \frac{x^2 - 16}{4x}$. Medium

iii Hence, or otherwise, show that the value of x is a solution of $x^4 - 56x^2 + 640 = 0$. Hard

iv Find x. Give reasons for your answer. Hard

(7 marks) (Q9c, **2000 HSC**)

---

25 By expressing $\sec \theta$ and $\tan \theta$ in terms of $\sin \theta$ and $\cos \theta$, show that

$\sec^2\theta - \tan^2\theta = 1$ *(2 marks)*

(Q7a, **1997 HSC**) Medium

26 A triangle ABC is right-angled at B. D is the point on AC such that BD is perpendicular to AC. Let $\angle BAC = \theta$.

i In your Writing Booklet, draw a diagram showing this information. Easy

You are given that $6AD + BC = 5AC$.

ii Show that $6 \cos \theta + \tan \theta = 5 \sec \theta$. Medium

iii Deduce that $6 \sin^2\theta - \sin \theta - 1 = 0$. Medium

iv Find θ. Medium

(Q10b, **1994 HSC**)

Year 11 Trigonometric functions and identities—Worked answers

1 As $\cos^2 5x + \sin^2 5x = 1$, then

$$\sin^2 5x = 1 - \cos^2 5x$$

Answer B

2
$$\begin{aligned} LHS &= \sec\theta - \cos\theta \\ &= \frac{1}{\cos\theta} - \cos\theta \\ &= \frac{1-\cos^2\theta}{\cos\theta} \\ &= \frac{\sin^2\theta}{\cos\theta} \quad ✓ \\ &= \sin\theta \times \frac{\sin\theta}{\cos\theta} \\ &= \sin\theta\tan\theta \\ &= RHS \quad ✓ \end{aligned}$$

(2 marks)

3 $\cot x = \dfrac{1}{\tan x}$

So cot x is undefined whenever $\tan x = 0$. So there are asymptotes at $x = 0$ and $x = \pm\pi$.

Also cot x is positive whenever tan x is positive (the first and third quadrants).

Answer C

4
$$\begin{aligned} &\text{cosec}^2 x + \text{cosec}\, x \cot x \\ &= \frac{1}{\sin^2 x} + \frac{1}{\sin x}\cdot\frac{\cos x}{\sin x} \\ &= \frac{1+\cos x}{\sin^2 x} \\ &= \frac{1+\cos x}{1-\cos^2 x} \quad ✓ \\ &= \frac{1+\cos x}{(1+\cos x)(1-\cos x)} \\ &= \frac{1}{1-\cos x} \quad ✓ \end{aligned}$$

(2 marks)

5 **a**
$$\begin{aligned} \frac{h}{BY} &= \tan 12° \\ \therefore \frac{BY}{h} &= \cot 12° \\ BY &= h\cot 12° \quad ✓ \end{aligned}$$

(1 mark)

b Similarly, $AY = h\cot 16°$

Using Pythagoras' theorem for triangle BAY:

$$\begin{aligned} &(h\cot 12°)^2 \\ &= (h\cot 16°)^2 + 1600^2 \quad ✓ \\ &h^2(\cot^2 12°) \\ &= h^2(\cot^2 16°) + 1600^2 \\ &h^2(\cot^2 12°) - h^2(\cot^2 16°) \\ &= 1600^2 \\ &h^2[\cot^2 12° - \cot^2 16°] \\ &= 1600^2 \\ h^2 &= \frac{1600^2}{\cot^2 12° - \cot^2 16°} \quad ✓ \\ h &= \frac{1600}{\sqrt{\cot^2 12° - \cot^2 16°}} \\ h &= 506.687\,3518\ldots \\ &= 507 \text{ (nearest whole)} \quad ✓ \end{aligned}$$

(3 marks)

6
$$\begin{aligned} 2\sin x\cos x &= \sin x \\ 2\sin x\cos x - \sin x &= 0 \\ \sin x(2\cos x - 1) &= 0 \\ \sin x = 0 \quad & 2\cos x = 1 \quad ✓ \\ x = 0, \pi, 2\pi \quad & \cos x = \frac{1}{2} \\ & x = \frac{\pi}{3}, \frac{5\pi}{3} \\ \therefore x = 0, \pi, 2\pi, \frac{\pi}{3}, \frac{5\pi}{3} & \quad ✓✓ \end{aligned}$$

(3 marks)

7
$$\begin{aligned} \tan\theta + \cot\theta &= \frac{\sin\theta}{\cos\theta} + \frac{\cos\theta}{\sin\theta} \\ &= \frac{\sin^2\theta + \cos^2\theta}{\cos\theta\sin\theta} \\ &= \frac{1}{\cos\theta\sin\theta} \\ &= \sec\theta\ \text{cosec}\,\theta \end{aligned}$$

Answer B

8

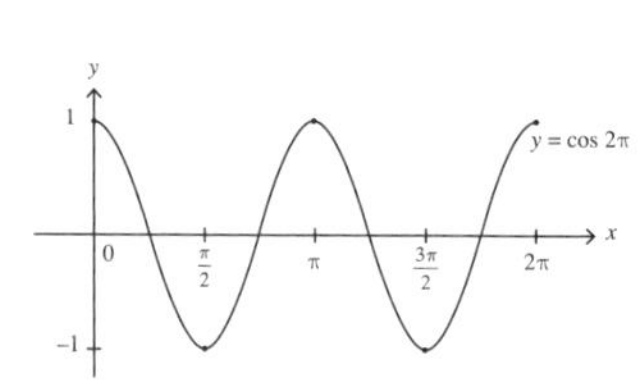

cos $2x = 1$ three times and cos $2x = -1$ twice.

$\therefore$ a total of 5 times.

Answer D

9
$$\begin{aligned} \sin\frac{x}{2} &= \frac{1}{2} \\ \therefore \frac{x}{2} &= \frac{\pi}{6}, \frac{5\pi}{6} \quad ✓ \\ \therefore x &= \frac{\pi}{3}, \frac{5\pi}{3} \quad ✓ \end{aligned}$$

(2 marks

10
$$\begin{aligned} 2\sin\theta &= 1 \\ \sin\theta &= \frac{1}{2} \\ \theta &= \frac{\pi}{6} \text{ or } \frac{5\pi}{6}. \quad ✓✓ \end{aligned}$$

(2 marks)

11

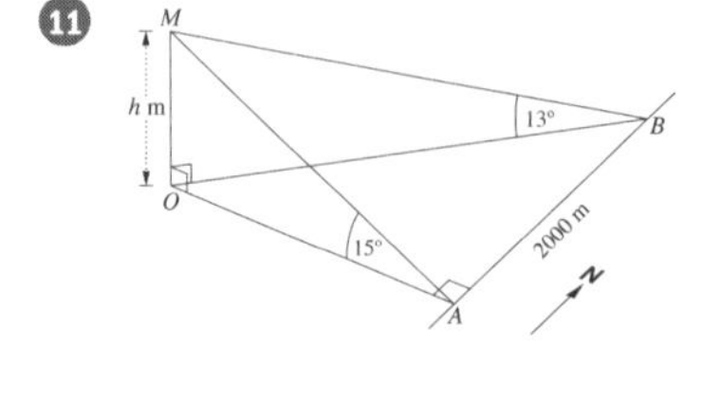

i In ΔOMA,

$$\begin{aligned} \tan 15° &= \frac{h}{OA} \\ OA &= \frac{h}{\tan 15°} \\ &= h\cot 15° \quad ✓ \text{ (1 mark)} \end{aligned}$$

ii In ΔOMB,

$$\begin{aligned} \tan 13° &= \frac{h}{OB} \\ OB &= \frac{h}{\tan 13°} \\ &= h\cot 13° \end{aligned}$$

In ΔOBA,

$$\begin{aligned} OB^2 &= OA^2 + AB^2 \\ (h\cot 13°)^2 &= (h\cot 15°)^2 + 2000^2 \\ h^2\cot^2 13° &= h^2\cot^2 15° + 2000^2 \quad ✓ \\ &h^2\cot^2 13° - h^2\cot^2 15° \\ &= 2000^2 \\ &h^2(\cot^2 13° - \cot^2 15°) \\ &= 2000^2 \\ h^2 &= \frac{2000^2}{\cot^2 13° - \cot^2 15°} \\ h &= \frac{2000}{\sqrt{\cot^2 13° - \cot^2 15°}} \quad (h > 0) \\ &= 909.703\,84\ldots \\ &= 910 \quad \text{(nearest unit)} \quad ✓ \end{aligned}$$

(2 marks)

12
$$\begin{aligned} (\sin x - 1)(\tan x + 2) &= 0 \\ \sin x - 1 &= 0 \\ \sin x &= 1 \\ x &= \frac{\pi}{2} \\ \tan x + 2 &= 0 \end{aligned}$$

$\tan x = -2$ has two solutions.

But $x = \dfrac{\pi}{2}$ is not a solution of $\tan x + 2 = 0$, as $\tan\dfrac{\pi}{2}$ is undefined.

$\therefore$ there are only 2 solutions.

Answer B

13
$$\begin{aligned} 2\sin^2 x + \cos x - 2 &= 0 \\ 2(1-\cos^2 x) + \cos x - 2 &= 0 \\ 2 - 2\cos^2 x + \cos x - 2 &= 0 \\ 2\cos^2 x - \cos x &= 0 \quad ✓ \\ \cos x(2\cos x - 1) &= 0 \\ \cos x = 0 \qquad \cos x &= \frac{1}{2} \\ \therefore x = \frac{\pi}{2}, \frac{3\pi}{2}, \frac{\pi}{3}, \frac{5\pi}{3} & \quad ✓✓ \end{aligned}$$

(3 marks)

14 $\sqrt{3}\tan x = -1$

$\tan x = \dfrac{-1}{\sqrt{3}}$

$x = \dfrac{5\pi}{6}, \dfrac{11\pi}{6}$

Answer D

15 $2\sin x = -\sqrt{3}$

$\sin x = -\dfrac{\sqrt{3}}{2}$

$x = \dfrac{4\pi}{3}, \dfrac{5\pi}{3}$ ✓✓ *(2 marks)*

16 $2\cos\theta = 1 \quad 0 \leq \theta \leq \dfrac{\pi}{2}$

$\cos\theta = \dfrac{1}{2}$ ✓

$\therefore \theta = \dfrac{\pi}{3}.$ ✓ *(2 marks)*

17 $2\sin^2\dfrac{x}{3} = 1$ for $-\pi \leq x \leq \pi$

$\therefore \sin^2\dfrac{x}{3} = \dfrac{1}{2}$ for $-\dfrac{\pi}{3} \leq \dfrac{x}{3} \leq \dfrac{\pi}{3}$

$\therefore \sin\dfrac{x}{3} = \pm\dfrac{1}{\sqrt{2}}$ ✓

$\therefore \dfrac{x}{3} = \dfrac{\pi}{4}, \pi - \dfrac{\pi}{4}, -\dfrac{\pi}{4}, -\pi + \dfrac{\pi}{4}$

$= \dfrac{\pi}{4}, \dfrac{3\pi}{4}, -\dfrac{\pi}{4}, -\dfrac{3\pi}{4}$

$\therefore x = \dfrac{3\pi}{4}, \dfrac{9\pi}{4}, -\dfrac{3\pi}{4}, -\dfrac{9\pi}{4}$

$\therefore x = -\dfrac{3\pi}{4}, \dfrac{3\pi}{4}$ (by domain). ✓✓ *(3 marks)*

18 $\sqrt{2}\sin x = 1$ for $0 \leq x \leq 2\pi$

$\sin x = \dfrac{1}{\sqrt{2}}$

$\therefore x = \dfrac{\pi}{4}$ or $\pi - \dfrac{\pi}{4}$

$\therefore x = \dfrac{\pi}{4}$ or $\dfrac{3\pi}{4}$ ✓✓ *(2 marks)*

19 $\cos\theta = \dfrac{1}{\sqrt{2}}$ for $0 \leq \theta \leq 2\pi$

$\therefore \theta = \dfrac{\pi}{4}$ or $2\pi - \dfrac{\pi}{4}$

$= \dfrac{\pi}{4}$ or $\dfrac{7\pi}{4}$ ✓✓ *(2 marks)*

20 **i** $\cos\theta\tan\theta = \sin\theta$

LHS $= \cos\theta\tan\theta$

$= \cos\theta\left(\dfrac{\sin\theta}{\cos\theta}\right)$

$= \sin\theta$

$=$ RHS

$\therefore \cos\theta\tan\theta = \sin\theta$ ✓ *(1 mark)*

ii $8\sin\theta\cos\theta\tan\theta$

$= \operatorname{cosec}\theta \quad 0 \leq \theta \leq 2\pi$

$\therefore 8\sin\theta(\sin\theta) = \operatorname{cosec}\theta$ (from **i**)

$8\sin^2\theta = \dfrac{1}{\sin\theta}$

$8\sin^3\theta = 1$ ✓

$\sin^3\theta = \dfrac{1}{8}$

$\sin\theta = \dfrac{1}{2}$

$\therefore \theta = \dfrac{\pi}{6}, \dfrac{5\pi}{6}$ ✓ *(2 marks)*

21 $2\sin^2 x - 3\sin x - 2 = 0$

$0 \leq x \leq 2\pi$

$(2\sin x + 1)(\sin x - 2) = 0$

$\therefore 2\sin x + 1 = 0$

or $\sin x - 2 = 0$

$\therefore \sin x = -\dfrac{1}{2}$ or $\sin x = 2$ ✓

$\sin x = 2$ has no real solutions since $-1 \leq \sin x \leq 1$.

$\therefore \sin x = -\dfrac{1}{2}$

$x = \left(\pi + \dfrac{\pi}{6}\right)$ or $\left(2\pi - \dfrac{\pi}{6}\right)$

$= \dfrac{7\pi}{6}$ or $\dfrac{11\pi}{6}$ ✓ *(2 marks)*

22 $\cos\theta - \dfrac{2}{5} = 0 \quad 0° \leq \theta \leq 360°$

$\therefore \cos\theta = \dfrac{2}{5}$

$\therefore \theta = 66.421\ldots°$

or $360° - 66.421\ldots°$

$\therefore \theta = 66°$ or $293.578\ldots°$

$= 66°$ or $294°$ ✓✓

to nearest degree *(2 marks)*

23 $\tan x = 2$ for $0 < x < 2\pi$

$\therefore x = 1.1071\ldots$ or $\pi + 1.1071\ldots$

$= 1.1071\ldots$ or $4.2487\ldots$

$= 1.11$ or 4.25 radians ✓✓

(to 2 decimal places) *(2 marks)*

24 **i** In $\triangle PBC$,

$\cos\alpha = \dfrac{x^2 + 2^2 - 6^2}{2 \times x \times 2}$

$= \dfrac{x^2 + 4 - 36}{4x}$

$= \dfrac{x^2 - 32}{4x}$ ✓

ii In $\triangle PBA$,

$\angle ABP = 90° - \alpha$

$\cos(90° - \alpha) = \dfrac{x^2 + 2^2 - (2\sqrt{5})^2}{2 \times x \times 2}$

$= \dfrac{x^2 + 4 - 20}{4x}$

$= \dfrac{x^2 - 16}{4x}$ ✓

Now $\cos(90° - \alpha) = \sin\alpha$

$\therefore \sin\alpha = \dfrac{x^2 - 16}{4x}$ ✓

iii Now $\cos^2\alpha + \sin^2\alpha = 1$

$\therefore \left(\dfrac{x^2 - 32}{4x}\right)^2 + \left(\dfrac{x^2 - 16}{4x}\right)^2 = 1$

$\dfrac{x^4 - 64x^2 + 1024}{16x^2} + \dfrac{x^4 - 32x^2 + 256}{16x^2} = 1$ ✓

$2x^4 - 96x^2 + 1280 = 16x^2$

$2x^4 - 112x^2 + 1280 = 0$

$x^4 - 56x^2 + 640 = 0$ ✓

iv $x^4 - 56x^2 + 640 = 0$

Let $X = x^2$

$\therefore X^2 - 56X + 640 = 0$

$(X - 40)(X - 16) = 0$

$\therefore X = 40$ or 16

$\therefore x^2 = 40$ or $x^2 = 16$

$\therefore x = \pm\sqrt{40}$ or $x = \pm 4$

Since x is a length,

$x = +\sqrt{40}$ or 4

$= 2\sqrt{10}$ or 4

$\doteqdot 6.3$ or 4 ✓

If $x = 4$, $\sin\alpha = \dfrac{16 - 16}{16}$

$= 0$

$\therefore \alpha = 0°$

and $\cos\alpha = \dfrac{16 - 32}{16}$

$= -1$

$\therefore \alpha = 180°$

If $x = 2\sqrt{10}$, $\sin\alpha = \dfrac{40 - 16}{8\sqrt{10}}$

$= \dfrac{24}{8\sqrt{10}}$

$= \dfrac{3}{\sqrt{10}}$

$\therefore \alpha = 71°34'$

and $\cos\alpha = \dfrac{40 - 32}{8\sqrt{10}}$

$= \dfrac{8}{8\sqrt{10}}$

$= \dfrac{1}{\sqrt{10}}$

$\therefore \alpha = 71°34'$

$\therefore x = 2\sqrt{10}$, as this is the only value of x which gives the same angle α when substituted into the

expressions for both $\sin \alpha$ and $\cos \alpha$.
(Alternately, the length of one side of a triangle must be greater than the difference of the other two sides, i.e. greater than $6 - 2 = 4. \therefore x = 2\sqrt{10}$ ($\doteqdot 6.3$).) ✓

(7 marks)
(Total mark allocation only in exam)

25 Now $\sec\theta = \dfrac{1}{\cos\theta}$

and $\tan\theta = \dfrac{\sin\theta}{\cos\theta}$

$$\begin{aligned} \text{LHS} &= \sec^2\theta - \tan^2\theta \\ &= \frac{1}{\cos^2\theta} - \frac{\sin^2\theta}{\cos^2\theta} \quad ✓ \\ &= \frac{1 - \sin^2\theta}{\cos^2\theta} \\ &= \frac{\cos^2\theta}{\cos^2\theta} \\ &= 1 = \text{RHS} \end{aligned}$$

$\therefore \sec^2\theta - \tan^2\theta = 1.$ ✓ *(2 marks)*

26 **i**

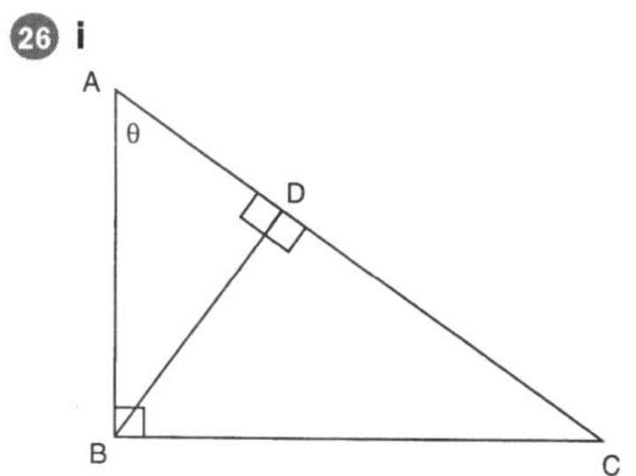

ii In $\triangle ABC$:

$$\tan\theta = \frac{BC}{AB}$$
$$\therefore BC = AB\tan\theta$$
$$\sec\theta = \frac{AC}{AB}$$
$$\therefore AC = AB\sec\theta$$

In $\triangle ADB$:

$$\cos\theta = \frac{AD}{AB}$$
$$\therefore AD = AB\cos\theta$$

Now $6AD + BC = 5AC$
$\therefore 6AB\cos\theta + AB\tan\theta = 5AB\sec\theta$
$\therefore 6\cos\theta + \tan\theta = 5\sec\theta$

iii Now $6\cos\theta + \tan\theta = 5\sec\theta$

$$\therefore 6\cos\theta + \frac{\sin\theta}{\cos\theta} = \frac{5}{\cos\theta}$$
$$\therefore 6\cos^2\theta + \sin\theta = 5$$
$$6(1 - \sin^2\theta) + \sin\theta = 5$$
$$6 - 6\sin^2\theta + \sin\theta = 5$$
$$1 - 6\sin^2\theta + \sin\theta = 0$$
$$6\sin^2\theta - \sin\theta - 1 = 0$$

iv In ΔABC, $\angle ABC = 90°$,
$\therefore 0 < \theta < 90°$.

If $6\sin^2\theta - \sin\theta - 1 = 0$
$(3\sin\theta + 1)(2\sin\theta - 1) = 0$
$\therefore \sin\theta = \frac{1}{2}$ or $-\frac{1}{3}$
$\therefore \theta = 30°$

($\sin\theta = -\frac{1}{3}$ is ignored as $0 < \theta < 90°$).

(No mark allocation in exam)

1 Which graph shows a function that is discontinuous?

A

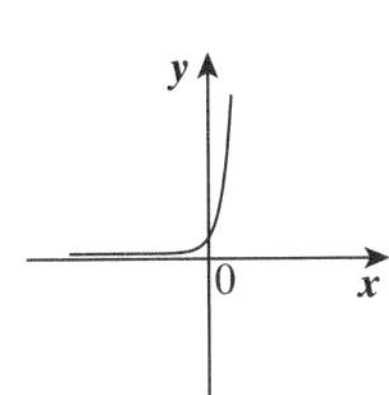

B

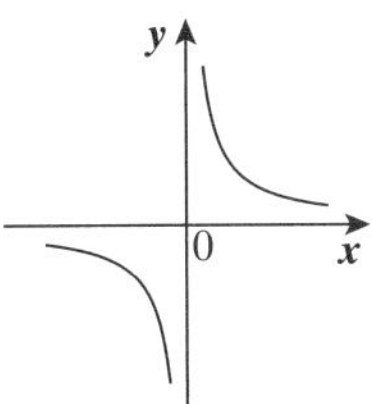

C

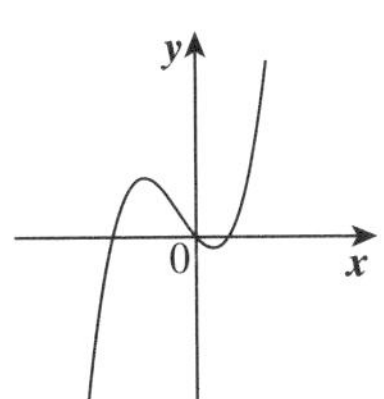

D

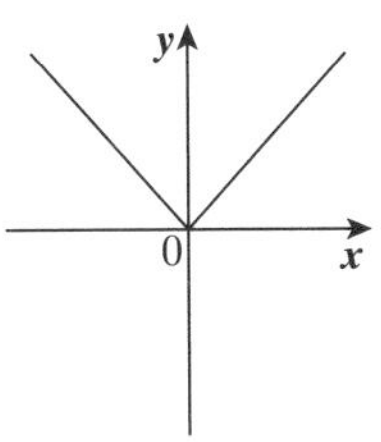

(1 mark)

Bonus question (see page iv) 

2 The points A and B have coordinates $(1, 0)$ and $(7, 4)$ as shown in the diagram. The angle between the line AB and the x-axis is θ.

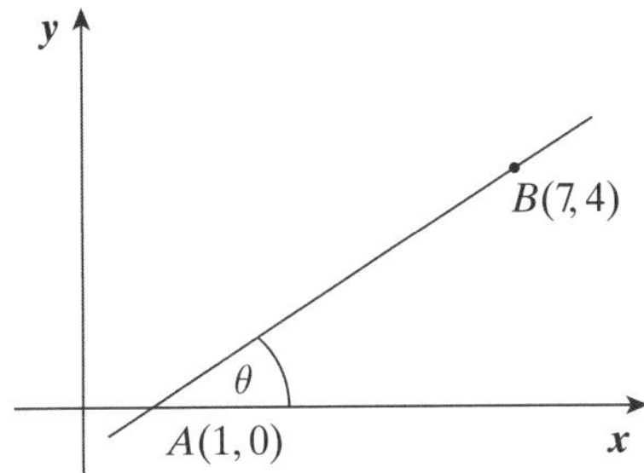

i Find the gradient of the line AB. *(1 mark)* Easy

ii Find the size of angle θ to the nearest degree. *(1 mark)* Easy

Bonus question

3 **i** Sketch the function given by

$$f(x) = \begin{cases} x^2 & \text{for } x \geq -1 \\ \dfrac{x+3}{2} & \text{for } x < -1 \end{cases}$$

(2 marks) Medium

ii Is the curve continuous? *(1 mark)* Easy

iii Is the curve smooth? *(1 mark)* Easy

Bonus question

4 The diagram shows a curve $y = f(x)$.

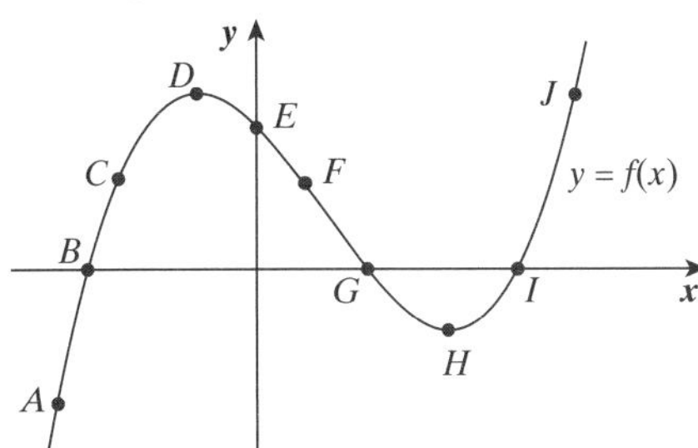

i Where is the curve decreasing? *(1 mark)* Easy

ii At what points is the curve increasing at an increasing rate? *(1 mark)* Easy

iii What feature will the gradient of the tangent have at B? *(1 mark)* Easy

Bonus question

5 A secant to the curve $y = \dfrac{24}{x}$ passes through the points where $x = 2$ and $x = 8$.
Find its gradient. *(2 marks)*

Bonus question Medium

6 A function is defined as

$$f(x) = \begin{cases} x^2 + 3 & \text{for } x < 1 \\ p - x^2 & \text{for } x \geq 1 \end{cases}.$$

Find the value of p if $f(x)$ is a continuous function. *(2 marks)*

Bonus question Medium

7 **i** Show that $y = x^2 - 2$ and $y = 6x - 11$ meet at just one point and find the coordinates of that point. *(2 marks)* Medium

ii Briefly explain why the gradient of the tangent to the curve $y = x^2 - 2$ at $x = 3$ must be equal to 6. *(1 mark)* Easy

iii Show that the gradient of the secant to the curve $y = x^2 - 2$ through the points at $x = 2$ and $x = 4$ is equal to the gradient of the tangent at $x = 3$. *(2 marks)* Medium

Bonus question

8 A particle moves in a straight line. At time t seconds it is x m to the right of its starting point, O.

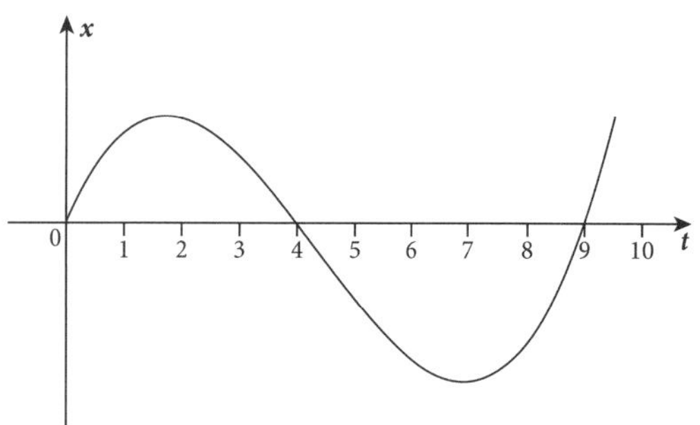

i Explain the significance of a negative value of x. *(1 mark)* Easy

ii What does it mean if the velocity of the particle is zero. *(1 mark)* Easy

iii At what approximate times is the particle stationary? *(1 mark)* Easy

iv Briefly describe the motion of the particle. *(2 marks)* Medium

Bonus question

---

9 The amount of input resources, R tonnes, used in a manufacturing process is given by the formula $R = 120t + \dfrac{30}{t+1}$, where t is the time in hours that the process has been in operation. Find the average rate of resource use in the first 9 hours of operation. *(2 marks)* Medium

Bonus question

---

10 **i** Show that for all values of m, the line $y = mx - 3m^2$ touches the parabola $x^2 = 12y$. *(2 marks)* Medium

ii Find the values of m for which this line passes through the point $(5, 2)$. *(2 marks)* Easy

iii Hence determine the equations of the two tangents to the parabola $x^2 = 12y$ from the point $(5, 2)$. *(1 mark)* Medium

(Q8d, **2003 HSC**)

---

11 The diagram shows the points $A(-2, 0)$, $B(3, 5)$ and the point C which lies on the x axis. The point D also lies on the x axis such that BD is perpendicular to AC.

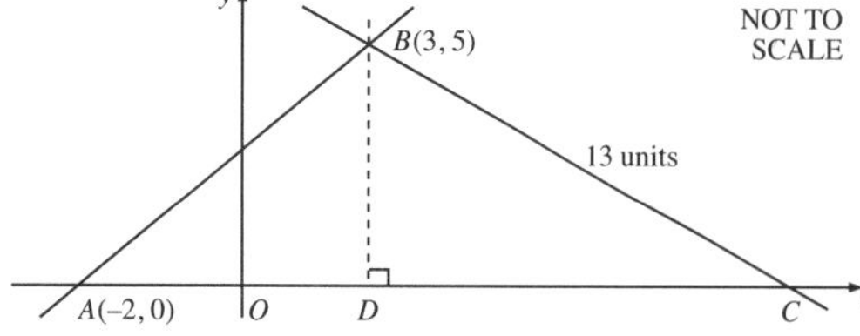

i Show that the gradient of AB is 1.

ii Find the equation of the line AB.

iii What is the size of $\angle BAC$?

(Q2b, **1999 HSC**) Easy

Year 11 Introduction to differentiation—Worked answers

1 The hyperbola is undefined when $x = 0$.

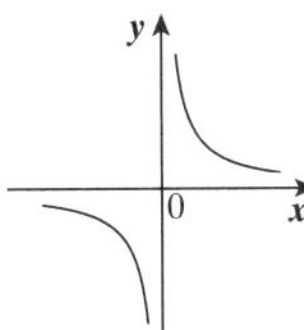

So, it is discontinuous.
Answer B

2 **i** $A(1, 0) \quad B(7, 4)$

$$m = \frac{y_2 - y_1}{x_2 - x_1} = \frac{4-0}{7-1} = \frac{2}{3}$$ ✓ *(1 mark)*

ii $\tan\theta = \frac{2}{3}$

$$\theta = \tan^{-1}\left(\frac{2}{3}\right) = 33.6900675\ldots° = 34° \text{ (nearest degree)}$$ ✓ *(1 mark)*

3 **i** $f(x) = \begin{cases} x^2 & \text{for } x \geq -1 \\ \dfrac{x+3}{2} & \text{for } x < -1 \end{cases}$

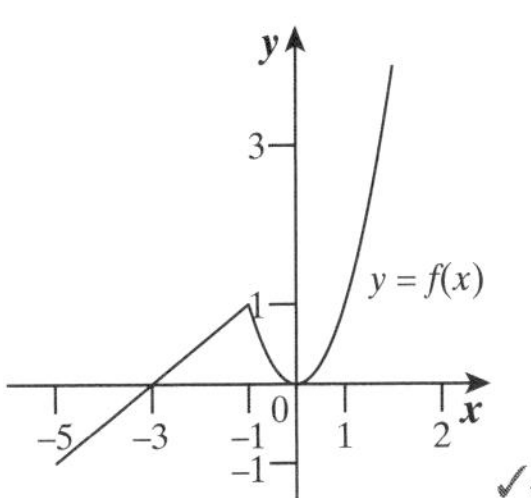

✓✓ *(2 marks)*

ii Yes, the curve is continuous. ✓ *(1 mark)*

iii No, the curve is not smooth. ✓ *(1 mark)*

4 **i** Between D and H ✓ *(1 mark)*

ii At I and at J ✓ *(1 mark)*

iii The curve is increasing so the gradient of the tangent will be positive. ✓ *(1 mark)*

5 $y = \dfrac{24}{x} \quad x = 2$ and $x = 8$

$$m = \frac{f(x_2) - f(x_1)}{x_2 - x_1} = \frac{\frac{24}{8} - \frac{24}{2}}{8-2} = -\frac{3}{2}$$ ✓

The gradient of the secant is -1.5.
✓ *(2 marks)*

6 $f(x) = \begin{cases} x^2 + 3 & \text{for } x < 1 \\ p - x^2 & \text{for } x \geq 1 \end{cases}$

$f(1) = p - 1^2$
$= p - 1$

When $x = 1$, $x^2 + 3 = 1^2 + 3$
$= 4$ ✓

If the function is continuous
$p - 1 = 4$
$p = 5$ ✓ *(2 marks)*

7 **i** At any point of intersection

$$x^2 - 2 = 6x - 11$$
$$x^2 - 6x + 9 = 0$$
$$(x-3)^2 = 0$$
$$x = 3$$ ✓

When $x = 3$, $y = 3^2 - 2 = 7$.
So, $y = x^2 - 2$ and $y = 6x - 11$ meet at just one point, $(3, 7)$. ✓ *(2 marks)*

ii As the line $y = 6x - 11$ meets the curve in just one point, it must be a tangent at that point. The gradient of the line is 6 so the gradient of the tangent is 6. ✓ *(1 mark)*

iii $y = x^2 - 2 \quad x = 2$ and $x = 4$

$$m = \frac{f(x_2) - f(x_1)}{x_2 - x_1} = \frac{(4^2-2)-(2^2-2)}{4-2} = 6$$ ✓

So the gradient of the secant is equal to the gradient of the tangent. ✓ *(2 marks)*

8 **i** When $x < 0$, the particle will be to the left of the starting point. ✓ *(1 mark)*

ii If the velocity is zero, the particle will not be moving. It will be stationary. ✓ *(1 mark)*

iii The velocity is zero when the gradient of the tangent is horizontal. After approximately 2 seconds and again after about 7 seconds. ✓ *(1 mark)*

iv The particle begins at O and moves to the right. It stops after 2 seconds and then moves to the left, passing ✓ through O at $t = 4$ seconds. It stops again after about 7 seconds and moves back to the right, passing through O again at $t = 9$ seconds and continues for another second. ✓ *(2 marks)*

9 $R = 120t + \dfrac{30}{t+1}$,

Average rate

$$= \frac{R(9) - R(0)}{9 - 0}$$ ✓

$$= \frac{120 \times 9 + \frac{30}{9+1} - \left(120 \times 0 + \frac{30}{0+1}\right)}{9}$$

$= 117$ tonnes per hour ✓ *(2 marks)*

10 **i** $y = mx - 3m^2$

$$y = \frac{x^2}{12}$$

Solving simultaneously:

$$\frac{x^2}{12} = mx - 3m^2$$
$$x^2 = 12mx - 36m^2$$
$$x^2 - 12mx + 36m^2 = 0$$ ✓

If the line touches the curve, then $x^2 - 12mx + 36m^2 = 0$ should have equal roots, i.e. $b^2 - 4ac = 0$.

Now $b^2 - 4ac$
$= (-12m)^2 - 4(1)(36m^2)$
$= 144m^2 - 144m^2$
$= 0$

Hence, for all values of m,
$b^2 - 4ac = 0$
$\therefore y = mx - 3m^2$ touches the parabola $x^2 = 12y$. ✓ *(2 marks)*

ii If $(5, 2)$ satisfies the equation of the line,

$$y = mx - 3m^2$$
$$\therefore 2 = m(5) - 3m^2$$
$$2 = 5m - 3m^2$$
$$3m^2 - 5m + 2 = 0$$
$$(3m - 2)(m - 1) = 0$$
$$\therefore m = \frac{2}{3} \text{ or } m = 1$$ ✓✓ *(2 marks)*

iii For $m = 1$, $y = x - 3(1)^2$
i.e. $y = x - 3$

For $m = \frac{2}{3}$, $y = \frac{2}{3}x - 3\left(\frac{2}{3}\right)^2$

$$= \frac{2}{3}x - 3\left(\frac{4}{9}\right)$$
$$= \frac{2}{3}x - \frac{4}{3}$$

i.e. $3y = 2x - 4$ ✓ *(1 mark)*

11 **i** Gradient of $AB = \dfrac{5-0}{3-(-2)}$

$$= \frac{5}{5} = 1$$

ii Equation of line AB through $B(3, 5)$ with gradient = 1 is

$y - 5 = 1(x - 3)$

$y - 5 = x - 3$

$y = x + 2$

iii $\tan \angle BAC$ = slope of AB

$\therefore \tan \angle BAC = 1$

$\therefore \angle BAC = 45°$.

(No mark allocation in exam)

1 Let $h(x) = f(g(x))$ where the function $f(x)$ is an odd function and the function $g(x)$ is an even function.

The tangent to $y = h(x)$ at $x = k$, where $k > 0$, has the equation $y = mx + c$.

What is the equation of the tangent to $y = h(x)$ at $x = -k$?

A $y = mx + c$ **B** $y = -mx + c$

C $y = mx - c$ **D** $y = -mx - c$ *(1 mark)*

(Q9, **2021 HSC**) Medium

2 The derivative of a function is defined as $f'(x) = \lim_{h \to 0} \frac{f(x+h) - f(x)}{h}$.

Differentiate $f(x) = 3x - x^2$ from first principles. *(2 marks)*

Bonus question (see page iv) Medium

3 The diagram shows the graph of a function $y = f(x)$.

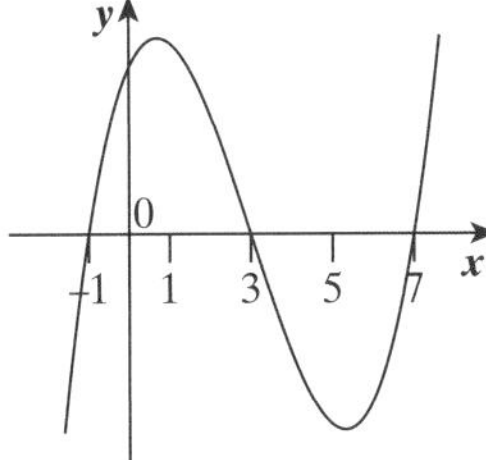

i What feature will the gradient function have when $x = 5$? *(1 mark)* Easy

ii For what values of x is the derivative $f'(x)$ positive? *(1 mark)* Easy

iii What, if anything, can be said about the graph of $y = f'(x)$ when $x = 2$? *(1 mark)* Easy

Bonus question

4 **i** Use differentiation by first principles to find the gradient function y' if $y = 2x^2 - 8x + 3$. *(2 marks)* Medium

ii What is the gradient of the tangent at $x = 3$? *(1 mark)* Medium

iii At what point on the curve is the tangent parallel to the x-axis? *(2 marks)* Medium

Bonus question

5 The diagram shows the graph of the gradient function of the function $f(x)$.

For what value of x does $f(x)$ have a local maximum? Justify your answer. *(2 marks)*

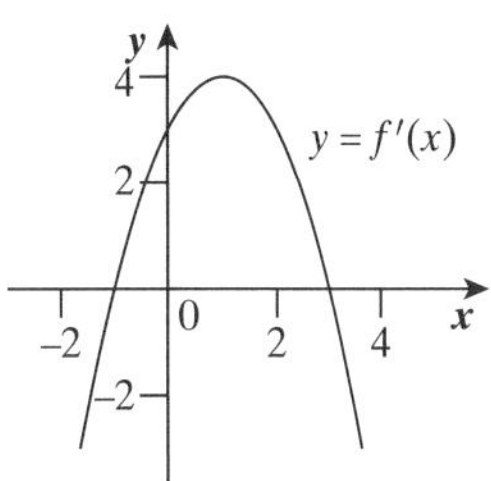

Bonus question Medium

6 Use differentiation by first principles to find $\frac{dx}{dt}$ if $x = t^3 + t$. *(2 marks)*

Bonus question Medium

7 The diagram shows the graph of a function $y = f(x)$.

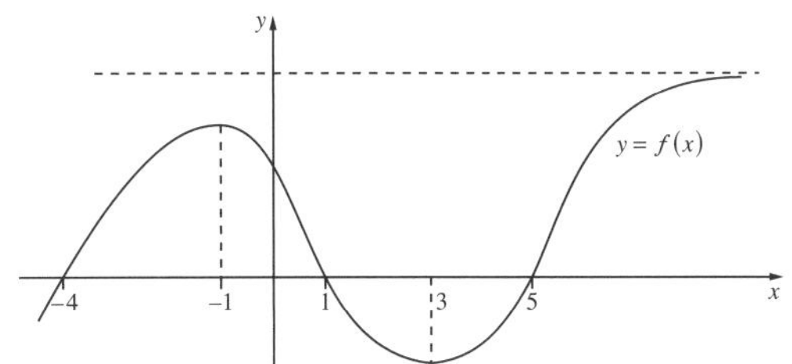

i For which values of x is the derivative, $f'(x)$, negative? *(1 mark)* Easy

ii What happens to $f'(x)$ for large values of x? *(1 mark)* Easy

iii Sketch the graph $y = f'(x)$. *(2 marks)* Medium

(Q8a, **2009 HSC**)

8 Use the definition of the derivative, $f'(x) = \lim_{h \to 0} \frac{f(x+h) - f(x)}{h}$, to find $f'(x)$ when $f(x) = x^2 + 5x$. *(2 marks)*

(Q3c, **2005 ME1 HSC**) Medium

9

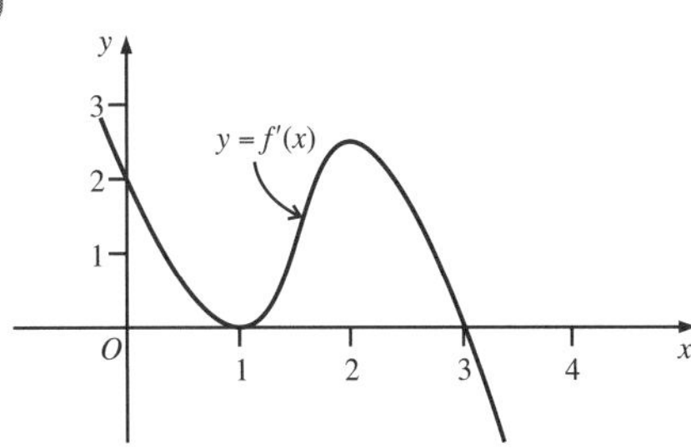

The above diagram shows a sketch of the gradient function of the curve $y = f(x)$.

In your Writing Booklet, draw a sketch of the function $y = f(x)$ given that $f(0) = 0$. *(2 marks)*

(Q9b, **2000 HSC**) Medium

10 The diagram shows the graph of a certain function $f(x)$.

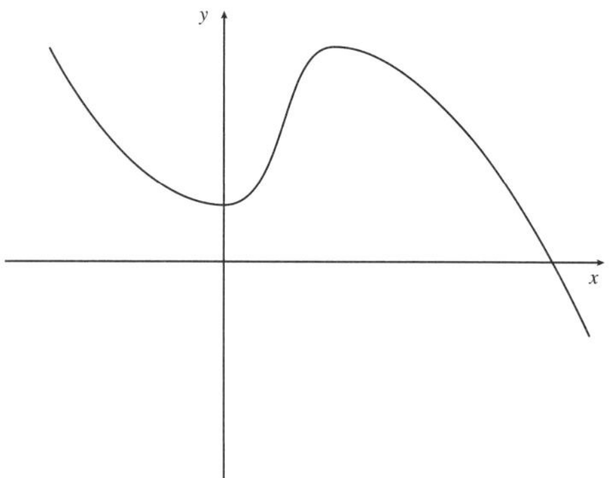

i Copy this graph into your Writing Booklet. Easy

ii On the *same* set of axes, draw a sketch of the derivative $f'(x)$ of the function. Medium

(Q6b, **1994 HSC**)

Year 11 The derivative function and its graph—Worked answers

1 Consider odd function $f(x) = x^3$ and even function $g(x) = x^2$.

$h(x) = f(g(x))$
$= (x^2)^3$
$= x^6$, which is an even function.

This means if $f(x)$ is an odd function and $g(x)$ is an even function then $h(x)$ is an even function.

As an even function is symmetrical about the y-axis, then the equation of the tangent at $x = -k$ is $y = -mx + c$.

Answer B

2
$$f'(x) = \lim_{h \to 0} \frac{3(x+h)-(x+h)^2-(3x-x^2)}{h}$$
$$= \lim_{h \to 0} \frac{3x+3h-(x^2+2xh+h^2)-3x+x^2}{h}$$
$$= \lim_{h \to 0} \frac{3x+3h-x^2-2xh-h^2-3x+x^2}{h}$$
$$= \lim_{h \to 0} \frac{3h-2xh-h^2}{h} \quad ✓$$
$$= \lim_{h \to 0} \frac{h(3-2x-h)}{h}$$
$$= \lim_{h \to 0} (3-2x-h)$$
$$= 3-2x \quad ✓$$

(2 marks)

3 **i** It will be zero. ✓ *(1 mark)*

ii $x < 1$ or $x > 5$ ✓ *(1 mark)*

iii It will be negative. ✓ *(1 mark)*

4 **i**
$$y = 2x^2 - 8x + 3$$
$$y' = \lim_{h \to 0} \frac{f(x+h)-f(x)}{h}$$
$$= \lim_{h \to 0} \frac{2(x+h)^2-8(x+h)+3-(2x^2-8x+3)}{h}$$
$$= \lim_{h \to 0} \frac{2x^2+4xh+2h^2-8x-8h+3-2x^2+8x-3}{h}$$
$$= \lim_{h \to 0} \frac{4xh+2h^2-8h}{h} \quad ✓$$
$$= \lim_{h \to 0} \frac{h(4x+2h-8)}{h}$$
$$= \lim_{h \to 0} (4x+2h-8)$$
$$= 4x-8 \quad ✓$$

(2 marks)

ii The gradient function is $4x - 8$.

When $x = 3$, $y' = 4 \times 3 - 8$
$= 4$

The gradient of the tangent at $x = 3$ is 4. ✓ *(1 mark)*

iii The tangent will be parallel to the x-axis when $y' = 0$.

So, $4x - 8 = 0$ ✓
$4x = 8$
$x = 2$

When $x = 2$, $y = 2 \times 2^2 - 8 \times 2 + 3$
$= -5$

So the tangent at the point $(2, -5)$ will be parallel to the x-axis. ✓

(2 marks)

5 $x = 3$ ✓

$f'(x) = 0$ when $x = -1$ and $x = 3$ so $f(x)$ has stationary values at both of those values of x.

The gradient is positive before $x = 3$ and negative afterwards, so that will be a local maximum.

(The gradient is negative before $x = -1$ and positive afterwards so that will mean it has a local minimum at $x = -1$.) ✓

(2 marks)

6 **i**
$$x = t^3 + t$$
$$\frac{dx}{dt} = \lim_{h \to 0} \frac{x(t+h)-x(t)}{h}$$
$$= \lim_{h \to 0} \frac{(t+h)^3+t+h-(t^3+t)}{h}$$
$$= \lim_{h \to 0} \frac{t^3+3t^2h+3th^2+h^3+t+h-t^3-t}{h}$$
$$= \lim_{h \to 0} \frac{3t^2h+3th^2+h^3+h}{h} \quad ✓$$
$$= \lim_{h \to 0} \frac{h(3t^2+3th+h^2+1)}{h}$$
$$= \lim_{h \to 0} (3t^2+3th+h^2+1)$$
$$= 3t^2 + 1 \quad ✓$$

(2 marks)

7 **i** $-1 < x < 3$ ✓ *(1 mark)*

ii As $x \to \infty, f'(x) \to 0$ ✓ *(1 mark)*

iii

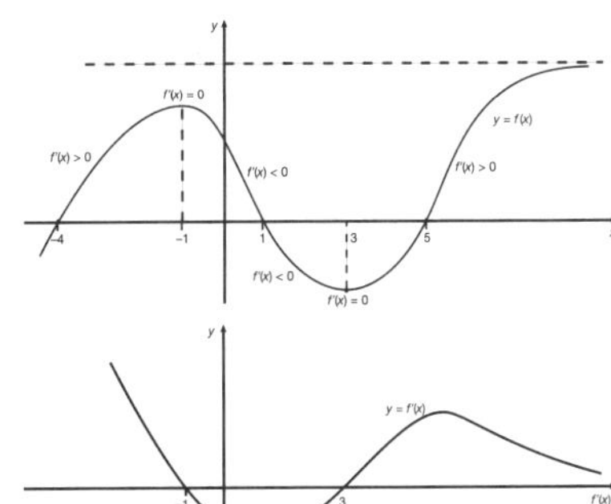

✓

For $x < -1, f'(x) > 0$
At $x = -1, f'(x) = 0$
For $-1 < x < 3, f'(x) < 0$
At $x = 3, f'(x) = 0$
For $x > 3, f'(x) > 0$
As x approaches large values, $f'(x)$ approaches 0. ✓

(2 marks)

8
$$f(x) = x^2 + 5x$$
$$f'(x) = \lim_{h \to 0} \frac{f(x+h)-f(x)}{h}$$
$$= \lim_{h \to 0} \frac{(x+h)^2+5(x+h)-(x^2+5x)}{h}$$
$$= \lim_{h \to 0} \frac{x^2+2xh+h^2+5x+5h-x^2-5x}{h}$$
$$= \lim_{h \to 0} \frac{2xh+h^2+5h}{h} \quad ✓$$

$$= \lim_{h \to 0} 2x + h + 5$$
$$= 2x + 5.$$ ✓

(2 marks)

9

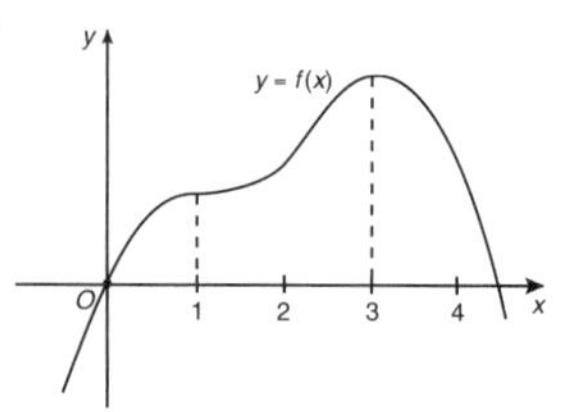

✓

Explanation:
$f'(1) = f'(3) = 0$
$\therefore$ Stationary points occur at $x = 1$ and $x = 3$.
Now $f'(1 - \varepsilon) > 0$
and $f'(1 + \varepsilon) > 0$
$\therefore$ A horizontal point of inflexion at $x = 1$.
Also, $f'(3 - \varepsilon) > 0$
and $f'(3 + \varepsilon) < 0$
$\therefore$ A maximum turning point at $x = 3$.
For $x < 3$, the gradient of $y = f(x)$ is always positive except at the above stationary points. ✓

(2 marks)

10 **i** and **ii**

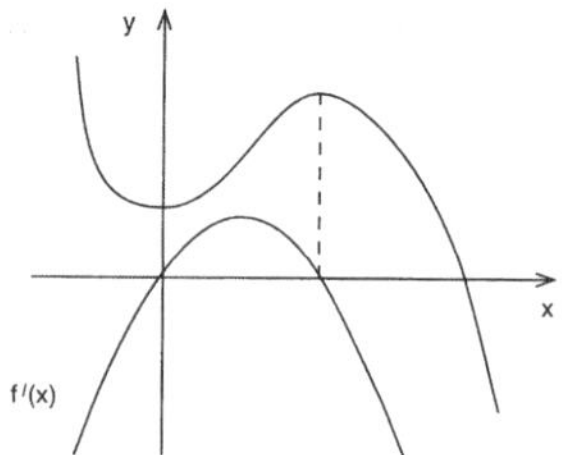

(No mark allocation in exam)

1 Let $h(x) = \dfrac{f(x)}{g(x)}$, where

$f(1) = 2 \qquad f'(1) = 4$
$g(1) = 8 \qquad g'(1) = 12.$

What is the gradient of the tangent to the graph of $y = h(x)$ at $x = 1$?

A -8 **B** 8

C $-\dfrac{1}{8}$ **D** $\dfrac{1}{8}$ *(1 mark)*

(Q5, **2022 HSC**) Easy

2 For what values of x is $f(x) = x^2 - 2x^3$ increasing? *(3 marks)*

(Q16, **2021 HSC**) Medium

3 By considering the equation of the tangent to $y = x^2 - 1$ at the point $(a, a^2 - 1)$, find the equations of the two tangents to $y = x^2 - 1$ which pass through $(3, -8)$. *(4 marks)*

(Q31, **2021 HSC**) Medium

4 Differentiate $\dfrac{2x+1}{x+5}$. *(2 marks)*

(Q11c, **2019 HSC**) Easy

5 The equation of the tangent to the curve $y = x^3 + ax^2 + bx + 4$ at the point where $x = 2$ is $y = x - 4$.

Find the values of a and b. *(3 marks)*

(Q14d, **2019 HSC**) Medium

6 A particle is moving along a straight line.

The graph shows the velocity, v, of the particle for time $t \geq 0$.

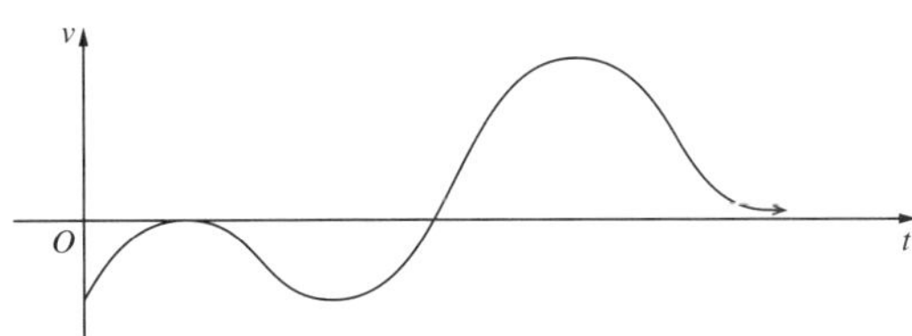

How many times does the particle change direction?

A 1 **B** 2

C 3 **D** 4 *(1 mark)*

(Q10, **2017 HSC**) Hard

7 Find the equation of the tangent to the curve $y = x^2 + 4x - 7$ at the point $(1, -2)$. *(2 marks)*

(Q12a, **2017 HSC**) Easy

8 Differentiate $\dfrac{x+2}{3x-4}$. *(2 marks)*

(Q11b, **2016 HSC**) Easy

9 Find $f'(x)$, where $f(x) = \dfrac{x^2+3}{x-1}$. *(2 marks)*

(Q12c, **2015 HSC**) Easy

10 Differentiate $\dfrac{x^3}{x+1}$. *(2 marks)*

(Q11c, **2014 HSC**) Easy

11 Find the equation of the tangent to the curve $y = x^2$ at the point where $x = 3$. *(2 marks)*

(Q11c, **2012 HSC**) Easy

12 Find the equation of the tangent to the curve $y = (2x+1)^4$ at the point where $x = -1$. *(3 marks)*

(Q2c, **2011 HSC**) Medium

13 The parabola shown in the diagram is the graph $y = x^2$. The points $A\ (-1, 1)$ and $B\ (2, 4)$ are on the parabola.

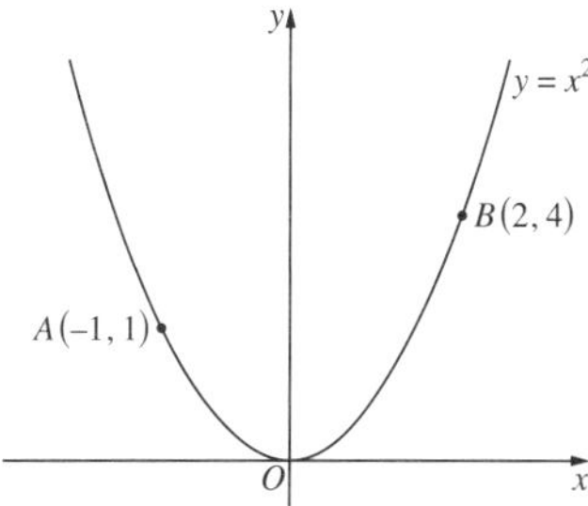

i Find the equation of the tangent to the parabola at A. *(2 marks)* Easy

ii Let M be the midpoint of AB.
There is a point C on the parabola such that the tangent at C is parallel to AB.
Show that the line MC is vertical. *(2 marks)* Hard

iii The tangent at A meets the line MC at T.
Show that the line BT is a tangent to the parabola. *(2 marks)* Hard

(Q7b, **2010 HSC**)

14 Find the gradient of the tangent to the curve $y = x^4 - 3x$ at the point $(1, -2)$. *(2 marks)*

(Q1d, **2009 HSC**) Easy

15 The diagram illustrates the design for part of a roller-coaster track. The section RO is a straight line with slope 1.2, and the section PQ is a straight line with slope -1.8. The section OP is a parabola $y = ax^2 + bx$. The horizontal distance from the y-axis to P is 30 m.

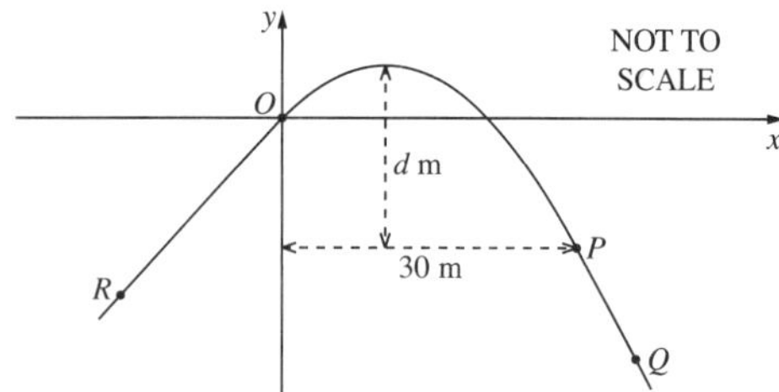

In order that the ride is smooth, the straight line sections must be tangent to the parabola at O and at P.

i Find the values of a and b so that the ride is smooth. *(3 marks)* **Medium**

ii Find the distance d, from the vertex of the parabola to the horizontal line through P, as shown on the diagram. *(2 marks)* **Hard**

(Q6c, **2009 HSC**)

16 Differentiate with respect to x:
$(x^2 + 3)^9$ *(2 marks)*

(Q2a i, **2008 HSC**) Easy

17 Differentiate with respect to x:
$\dfrac{x^2}{x-1}$. *(2 marks)*

(Q2b ii, **2005 HSC**) Easy

18 Differentiate $x^4 + 5x^{-1}$ with respect to x. *(2 marks)*

(Q1b, **2004 HSC**) Easy

19 Differentiate $x^3 + 2$. *(2 marks)*

(Q1b, **2002 HSC**) Easy

20 Find the equation of the tangent to the curve $y = x^2 + 3x$ at the point $(1, 4)$. *(3 marks)*

(Q2a, **2001 HSC**) Easy

21 Find the equation of the normal to the curve $y = \sqrt{x+2}$ at the point $(7, 3)$. *(4 marks)*

(Q3b, **1999 HSC**) Medium

22 Differentiate the following functions:
$(3x^2 + 4)^5$

(Q2a i, **1998 HSC**) Easy

23 Differentiate the following functions:
$(x^2 + 5)^3$

(Q3a i, **1997 HSC**) Easy

24 Differentiate: $\sqrt{x}$

(Q3a i, **1996 HSC**) Easy

25 Differentiate:
$2x^{-3}$

(Q3a i, **1995 HSC**) Easy

26 Differentiate $3x^5 - 4x + 7$.

(Q1c, **1994 HSC**) Easy

27 Find:
$\dfrac{1}{x^4}$

(Q3a i, **1994 HSC**) Easy

28 Differentiate:
$\dfrac{1}{x^3}$

(Q3a ii, **1993 HSC**) Easy

29 The line $y = mx + b$ is a tangent to the curve $y = x^3 - 3x + 1$ at the point $(-2, -1)$.

Find m and b.

(Q5a, **1993 HSC**) Medium

30 Differentiate $\dfrac{1}{\sqrt{1+x^2}}$.

(Q1b, **1992 ME1 HSC**) Medium

1 Using quotient rule:

$$\frac{dy}{dx}=\frac{v\frac{du}{dx}-u\frac{dv}{dx}}{v^2}$$

$$h'(x)=\frac{g(x)f'(x)-f(x)g'(x)}{[g(x)]^2}$$

At $x=1$:

$$h'(1)=\frac{g(1)\times f'(1)-f(1)\times g'(1)}{[g(1)]^2}$$

$$=\frac{8\times 4-2\times 12}{8^2}$$

$$=\frac{1}{8}$$

Answer D

2 $f(x)=x^2-2x^3$

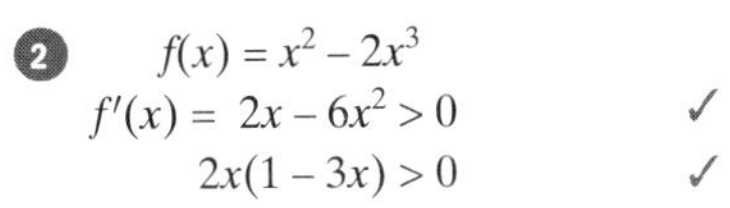

$f'(x)=2x-6x^2>0$ ✓

$2x(1-3x)>0$ ✓

Consider $y=2x(1-3x)$ and when $y>0$:

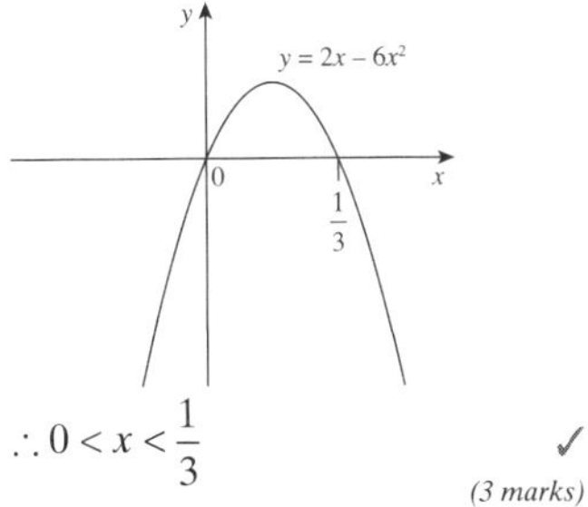

$\therefore 0<x<\frac{1}{3}$ ✓

(3 marks)

3 $y=x^2-1$

$y'=2x$

Substitute $x=a$:

$y'(a)=2a$ ✓

Equation of tangent:

$y-y_1=m(x-x_1)$

$y-(a^2-1)=2a(x-a)$ … * ✓

Substitute the point $(3,-8)$:

$-8-(a^2-1)=2a(3-a)$

$-8-a^2+1=6a-2a^2$

$a^2-6a-7=0$

$(a-7)(a+1)=0$

$a=7,-1$ ✓

Substitute $a=7$ into *:

$y-(7^2-1)=2(7)(x-7)$

$y-48=14(x-7)$

$y-48=14x-98$

$y=14x-50$

Substitute $a=-1$ into *:

$y-((-1)^2-1)=2(-1)(x-(-1))$

$y=-2(x+1)$

$y=-2x-2$

$\therefore$ the two tangents are $y=14x-50$ and $y=-2x-2$. ✓

(4 marks)

4 Using the quotient rule,

Let $u=2x+1$ $\quad\frac{du}{dx}=2$

Let $v=x+5$ $\quad\frac{dv}{dx}=1$

$$\frac{d}{dx}\left[\frac{2x+1}{x+5}\right]$$

$$=\frac{v\frac{du}{dx}-u\frac{dv}{dx}}{v^2}$$

$$=\frac{(x+5).2-(2x+1).1}{(x+5)^2}$$ ✓

$$=\frac{2x+10-2x-1}{(x+5)^2}$$

$$=\frac{9}{(x+5)^2}$$ ✓

(2 marks)

5 The curve and the tangent have the same gradient at the point of contact.

$y=x-4$ has gradient 1.

For $y=x^3+ax^2+bx+4$,

$\frac{dy}{dx}=3x^2+2ax+b$ ✓

Substitute $x=2$, $\frac{dy}{dx}=1$:

$1=3(2)^2+2a(2)+b$

$1=12+4a+b$

$4a+b=-11$ … ①

Also, substitute $x=2$ into equation of tangent:

$y=x-4$

$=2-4$

$=-2$

$\therefore$ point of contact is at $(2,-2)$. ✓

Substitute $x=2$, $y=-2$:

$y=x^3+ax^2+bx+4$

$-2=2^3+a(2)^2+b(2)+4$

$-2=8+4a+2b+4$

$4a+2b=-14$ … ②

② – ①: $b=-3$

Substitute into ①:

$4a-3=-11$

$4a=-8$

$a=-2$

$\therefore a=-2, b=-3$ ✓

(3 marks)

6 Let the times when the particle stops ($v=0$) be t_1 and t_2.

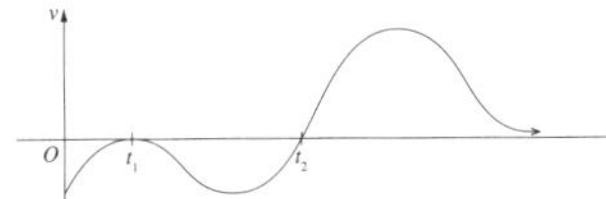

The particle changes direction when the particle's velocity changes sign. This occurs only once at t_2 seconds. (At time t_1 the particle stops but continues in the same direction: moving to the left.)

Answer A

7 $\frac{dy}{dx}=2x+4$

$\frac{dy}{dx}(1)=2(1)+4$

$=6$ ✓

Using $(1,-2)$ and $m=6$,

$y-y_1=m(x-x_1)$

$y+2=6(x-1)$

$y+2=6x-6$

$6x-y-8=0$ ✓

(2 marks)

8 Using the quotient rule,

Let $u=x+2$ $\quad\frac{du}{dx}=1$

Let $v=3x-4$ $\quad\frac{dv}{dx}=3$

$$\frac{d}{dx}\left[\frac{x+2}{3x-4}\right]=\frac{v\frac{du}{dx}-u\frac{dv}{dx}}{v^2}$$

$$=\frac{(3x-4).1-(x+2).3}{(3x-4)^2}$$ ✓

$$=\frac{3x-4-3x-6}{(3x-4)^2}$$

$$=\frac{-10}{(3x-4)^2}$$ ✓

(2 marks)

9 $f(x)=\frac{x^2+3}{x-1}$

Using the quotient rule:

Let $u=x^2+3$ $\quad u'=2x$

Let $v=x-1$ $\quad v'=1$

$$\frac{dy}{dx}=\frac{vu'-uv'}{v^2}$$

$$=\frac{(x-1).2x-(x^2+3).1}{(x-1)^2}$$ ✓

$$=\frac{2x^2-2x-x^2-3}{(x-1)^2}$$

$$=\frac{x^2-2x-3}{(x-1)^2}$$ ✓

(2 marks)

10 $\frac{x^3}{x+1}$

Using the quotient rule:

Let $u=x^3$, $u'=3x^2$

Let $v=x+1$, $v'=1$

$$\frac{dy}{dx}=\frac{vu'-uv'}{v^2}$$

$$=\frac{(x+1).3x^2-x^3.1}{(x+1)^2}$$ ✓

$$=\frac{3x^3+3x^2-x^3}{(x+1)^2}$$

$$=\frac{2x^3+3x^2}{(x+1)^2}$$ ✓

(2 marks)

11 $y=x^2$

$\frac{dy}{dx}=2x$

At $x = 3$, gradient $m = 2 \times 3$
$= 6$ ✓

At $x = 3$, $y = (3)^2$
$= 9 \quad \therefore (3, 9)$

Using $y - y_1 = m(x - x_1)$
$y - 9 = 6(x - 3)$
$y - 9 = 6x - 18$
$\therefore 6x - y - 9 = 0$ ✓

(2 marks)

12 $y = (2x + 1)^4$

Using the function of a function rule:

$$\frac{dy}{dx} = 4(2x+1)^3 . \frac{d}{dx}(2x+1)$$
$$= 4(2x+1)^3 . 2$$
$$= 8(2x+1)^3$$ ✓

At $x = -1$,
gradient $m = 8(2(-1) + 1)^3$
$= 8 \times -1$
$= -8$

At $x = -1$, $y = (2(-1) + 1)^4$
$= 1 \quad \therefore (-1, 1)$ ✓

Using $y - y_1 = m(x - x_1)$
$y - 1 = -8(x - (-1))$
$y - 1 = -8(x + 1)$
$y - 1 = -8x - 8$
$8x + y + 7 = 0$ ✓

(3 marks)

13

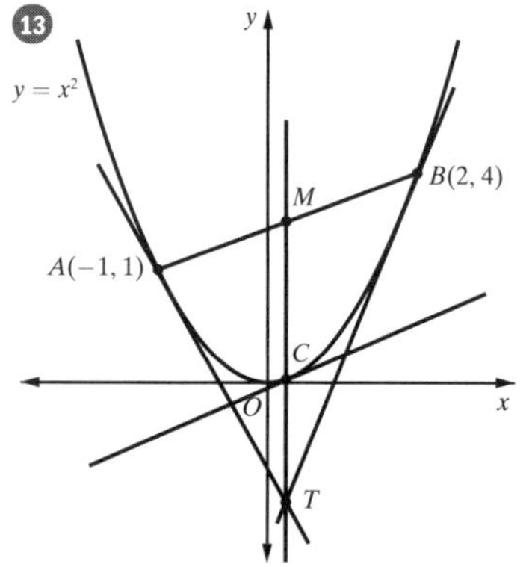

i $y = x^2$
$$\frac{dy}{dx} = 2x$$
At $x = -1$, $\frac{dy}{dx} = -2$ ✓

using $(-1, 1)$ and $m = -2$,
$y - y_1 = m(x - x_1)$
$y - 1 = -2(x - (-1))$
$y - 1 = -2(x + 1)$
$y - 1 = -2x - 2$
$2x + y + 1 = 0$

$\therefore$ The equation of the tangent is $2x + y + 1 = 0$. ✓

(2 marks)

ii Using $A(-1, 1)$ and $B(2, 4)$,

$$MP = \left(\frac{x_1 + x_2}{2}, \frac{y_1 + y_2}{2}\right)$$
$$= \left(\frac{-1+2}{2}, \frac{1+4}{2}\right)$$
$$= \left(\frac{1}{2}, \frac{5}{2}\right)$$
$$= (0.5, 2.5)$$

$\therefore$ The co-ordinates of M are $(0.5, 2.5)$. ✓

Now, using $A(-1, 1)$ and $B(2, 4)$,
$$m = \frac{y_2 - y_1}{x_2 - x_1}$$
$$= \frac{4-1}{2+1}$$
$$= \frac{3}{3}$$
$$= 1$$

$\therefore$ The gradient of the tangent at C is 1.

Let $\frac{dy}{dx} = 2x = 1$
$$\therefore x = \frac{1}{2}$$

$\therefore$ As M and C both have an x-value of $\frac{1}{2}$, then MC is vertical. ✓

(2 marks)

iii Tangent at A is $2x + y + 1 = 0$:

At T, subs $x = \frac{1}{2}$:
$$2 \times \frac{1}{2} + y + 1 = 0$$
$$1 + y + 1 = 0$$
$$y = -2$$
$$\therefore T\left(\frac{1}{2}, -2\right)$$ ✓

Now, using $T\left(\frac{1}{2}, -2\right)$ and $B(2, 4)$,

$$m = \frac{y_2 - y_1}{x_2 - x_1}$$
$$= \frac{4+2}{2-\frac{1}{2}}$$
$$= \frac{6}{1\frac{1}{2}}$$
$$= 4$$

$\therefore$ The gradient of BT is 4.

But the gradient of the tangent at B:
$$y = x^2$$
$$\frac{dy}{dx} = 2x$$
At $x = 2$, $\frac{dy}{dx} = 4$

As both have a gradient of 4, then BT is a tangent to the parabola. ✓

(2 marks)

14
$$y = x^4 - 3x$$
$$\frac{dy}{dx} = 4x^3 - 3$$ ✓

At $x = 1$, $\frac{dy}{dx} = 4 - 3$
$= 1$

$\therefore$ Gradient of tangent is 1. ✓

(2 marks)

15 i
$$y = ax^2 + bx$$
$$\frac{dy}{dx} = 2ax + b$$ ✓

At O, $x = 0$ and $\frac{dy}{dx} = 1.2$
$\therefore 2a(0) + b = 1.2$
$\therefore b = 1.2$ ✓

At P, $x = 30$ and $\frac{dy}{dx} = -1.8$

$\therefore 2ax + b = -1.8$
$\therefore 2a(30) + 1.2 = -1.8$
$60a = -3$
$\therefore a = -0.05.$ ✓

(3 marks)

ii
$$y = ax^2 + bx$$
$$\therefore y = -0.05x^2 + 1.2x$$
$$\frac{dy}{dx} = -0.1x + 1.2$$

At the vertex,
$$\frac{dy}{dx} = 0$$
$-0.1x + 1.2 = 0$
$0.1x = 1.2$
$x = 12$ ✓

OR

At the vertex, $x = \frac{-b}{2a}$
$$= \frac{-1.2}{2(-0.05)}$$
$$= \frac{-1.2}{-0.1}$$
$$= 12$$

When $x = 12$,
$y = -0.05(12)^2 + 1.2(12)$
$= 7.2$

$\therefore$ Coordinates of vertex are $(12, 7.2)$

At P, $x = 30$
$\therefore y = -0.05(30)^2 + 1.2(30)$
$= -9.$

$\therefore$ Coordinates of P are $(30, -9)$
$\therefore d = 7.2 + 9$
$= 16.2$

The distance, d, is 16.2 m. ✓

(2 marks)

16 $\frac{d}{dx}\left[(x^2+3)^9\right]$

$= 9(x^2+3)^8 \cdot \frac{d}{dx}(x^2+3)$ ✓

$= 9(x^2+3)^8 \cdot 2x$

$= 18x(x^2+3)^8$ ✓

(2 marks)

17 $\frac{d}{dx}\left(\frac{x^2}{x-1}\right)$

$= \frac{(x-1)\cdot\frac{d}{dx}(x^2) - x^2\cdot\frac{d}{dx}(x-1)}{(x-1)^2}$

$= \frac{(x-1)\cdot 2x - x^2 \cdot 1}{(x-1)^2}$ ✓

$= \frac{2x^2-2x-x^2}{(x-1)^2}$

$= \frac{x^2-2x}{(x-1)^2}$

$= \frac{x(x-2)}{(x-1)^2}$ ✓

(2 marks)

18 $\frac{d}{dx}(x^4+5x^{-1}) = 4x^3 - 5x^{-2}$ ✓✓

(2 marks)

19 $\frac{d}{dx}(x^3+2) = 3x^2$ ✓✓

(2 marks)

20 $y = x^2 + 3x$

$\frac{dy}{dx} = 2x + 3$ ✓

At $x = 1$, $\frac{dy}{dx} = 2(1) + 3$

$= 5$ ✓

Equation of tangent at $(1, 4)$ with gradient $= 5$ is

$y - 4 = 5(x-1)$
$y - 4 = 5x - 5$
$y = 5x - 1$ ✓

(3 marks)

21 $y = (x+2)^{\frac{1}{2}}$

$\frac{dy}{dx} = \frac{1}{2}(x+2)^{-\frac{1}{2}} \cdot 1$

$= \frac{1}{2\sqrt{x+2}}$ ✓

At $x = 7$, $\frac{dy}{dx} = \frac{1}{2\sqrt{9}}$

$= \frac{1}{6}$ ✓

$\therefore$ Gradient of normal $= -\frac{1}{\frac{1}{6}}$

$= -6$ ✓

$\therefore$ Equation of normal through $(7, 3)$ with gradient $= -6$ is

$y - 3 = -6(x-7)$
$y - 3 = -6x + 42$
$6x + y - 45 = 0$ ✓

(4 marks)

22 $\frac{d}{dx}[(3x^2+4)^5]$

$= 5(3x^2+4)^4 \cdot \frac{d}{dx}(3x^2+4)$
$= 5(3x^2+4)^4(6x)$
$= 30x(3x^2+4)^4.$

(No mark allocation in exam)

23 $\frac{d}{dx}(x^2+5)^3$

$= 3(x^2+5)^2 \cdot \frac{d}{dx}(x^2+5)$
$= 3(x^2+5)^2 . 2x$
$= 6x(x^2+5)^2.$

(No mark allocation in exam)

24 $\frac{d}{dx}(\sqrt{x}) = \frac{d}{dx}(x^{\frac{1}{2}})$

$= \frac{1}{2}x^{-\frac{1}{2}}$

$= \frac{1}{2x^{\frac{1}{2}}}$

$= \frac{1}{2\sqrt{x}}.$

(No mark allocation in exam)

25 $\frac{d}{dx}(2x^{-3}) = -6x^{-4}.$

(No mark allocation in exam)

26 $\frac{d}{dx}(3x^5 - 4x + 7)$

$= 15x^4 - 4$

(No mark allocation in exam)

27 $\frac{d}{dx}(x^{-4}) = -4x^{-5}$

$= \frac{-4}{x^5}$

(No mark allocation in exam)

28 $\frac{d}{dx}\left(\frac{1}{x^3}\right) = \frac{d}{dx}(x^{-3})$

$= -3x^{-4}$

$= \frac{-3}{x^4}$

(No mark allocation in exam)

29 $(-2, -1)$ lies on the line $y = mx + b$

$\therefore -1 = m(-2) + b$
$-1 = -2m + b \ldots (1)$

Now $y = x^3 - 3x + 1$

$\frac{dy}{dx} = 3x^2 - 3$

$= 3(-2)^2 - 3$ when $x = -2$
$= 12 - 3$
$= 9$

Since $y = mx + b$ is a tangent at $(-2, -1)$

$m = \frac{dy}{dx} = 9$

Substituting $m = 9$ into (1):
$-1 = -2(9) + b$
$-1 = -18 + b$
$b = 17$

(No mark allocation in exam)

30 $\frac{d}{dx}\left(\frac{1}{\sqrt{1+x^2}}\right)$

$= \frac{d}{dx}(1+x^2)^{-\frac{1}{2}}$

$= -\frac{1}{2}(1+x^2)^{-\frac{3}{2}} \cdot \frac{d}{dx}(1+x^2)$

$= -\frac{1}{2}(1+x^2)^{-\frac{3}{2}} . 2x$

$= \frac{-x}{\sqrt{(1+x^2)^3}}$

$= \frac{-x}{(1+x^2)\sqrt{1+x^2}}$

(No mark allocation in exam)

1 Which expression is equivalent to $x = \log_3 7$?

A $3 = 7^x$ **B** $3 = x^7$

C $7 = 3^x$ **D** $7 = x^3$ *(1 mark)*

Bonus question (see page iv) Easy

2 What is the value of $\log_4 16 - \log_2 16$? *(1 mark)*

Bonus question Medium

3 Which expression is equivalent to $y = p^5$?

A $5 = \log_y p$ **B** $5 = \log_p y$

C $p = \log_y 5$ **D** $p = \log_5 y$ *(1 mark)*

Bonus question Easy

4 Find the value of x if $\log_x 9 = 2$. Briefly explain why there is only one value of x. *(2 marks)*

Bonus question Easy

5 Which graph best represents $y = 10^{-x}$?

A

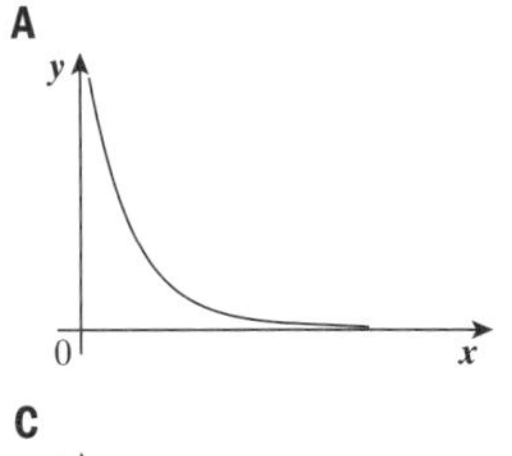

B

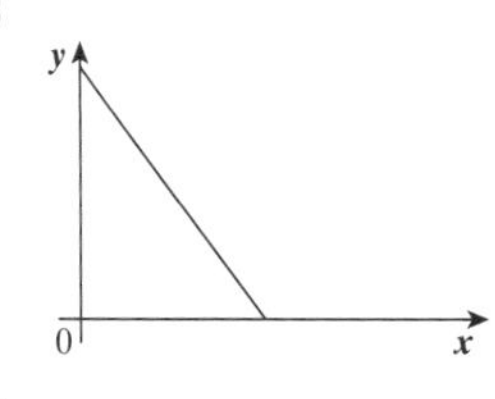

C

D

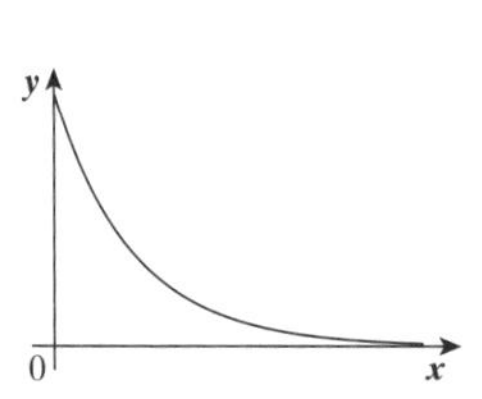

(1 mark)

Bonus question Easy

6 Which is the graph of $y = \log_2 x$?

A

B

C

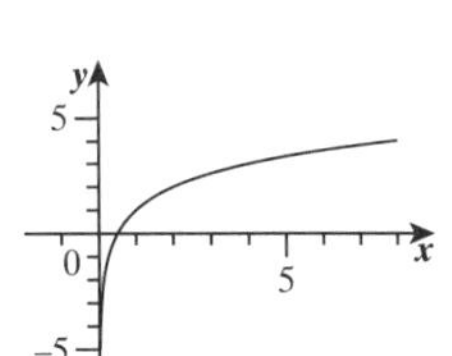

D

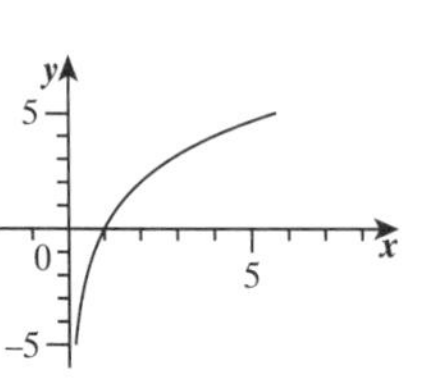

(1 mark)

Bonus question Medium

7 Find the value of x for which $10^x = 7$. Give the answer to 4 decimal places. *(2 marks)*

Bonus question Easy

8 Find x if $\log_9 x = 4.5$. *(1 mark)*

Bonus question Easy

9 Which of the following graphs best represents the equation $y = a^x$, where a is a positive number greater than 1?

A

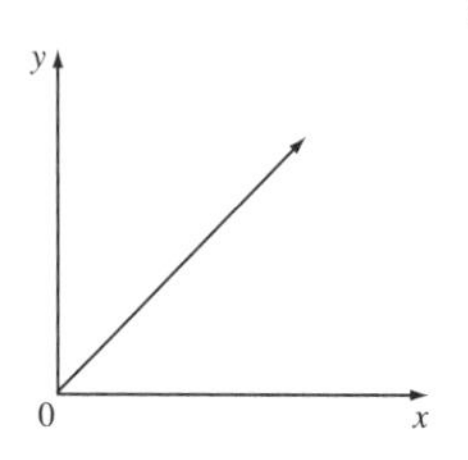

B

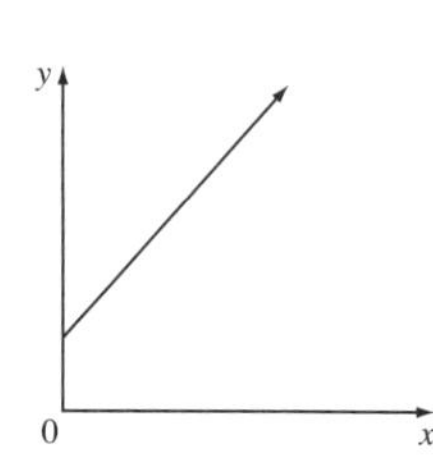

C

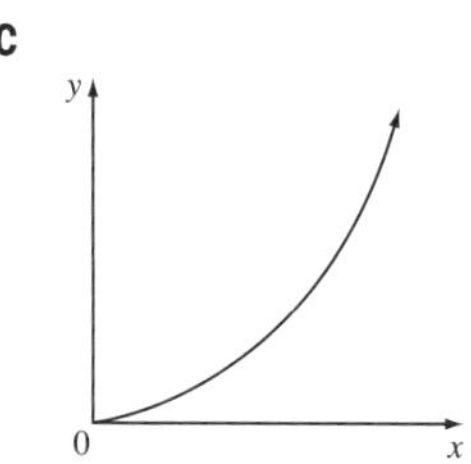

D

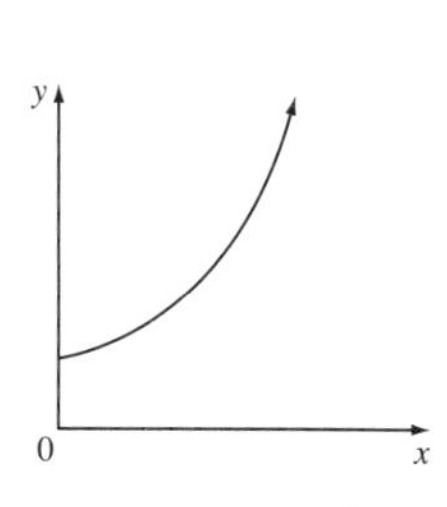

(1 mark)

(Q6, **2011 MS2 HSC**) Medium

10 Which graph best represents $y = 3^x$?

A

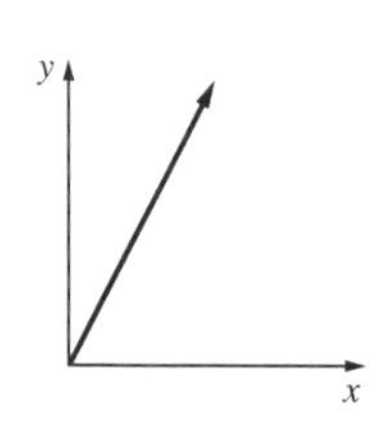

B

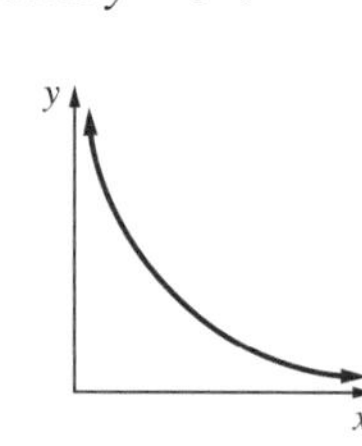

C

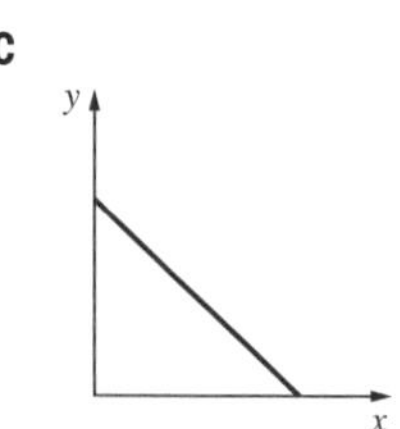

D

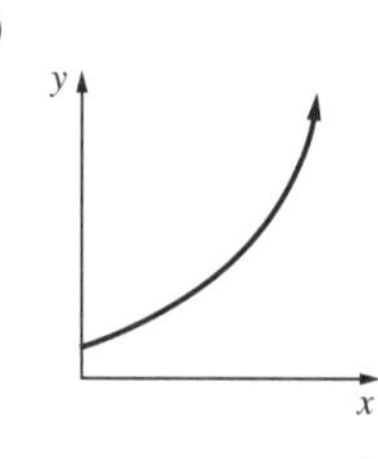

(1 mark)

(Q4, **2008 MS2 HSC**) Medium

Year 11 Introducing logarithms—Worked answers

1 $y = a^x$ is equivalent to $x = \log_a y$.
So $7 = 3^x$ is equivalent to
$x = \log_3 7$.
Answer C

2 $4^2 = 16$ and $2^4 = 16$
So $\log_4 16 - \log_2 16 = 2 - 4$
$= -2$ ✓
(1 mark)

3 $y = a^x$ is equivalent to $x = \log_a y$.
So $y = p^5$ is equivalent to
$5 = \log_p y$
Answer B

4 $\log_x 9 = 2$
So, $x^2 = 9$
$x = 3$ ✓
By definition, the base must be a positive number so $x = -3$ is not a solution. ✓
There is only one value of x $(x > 0)$ for which $x^2 = 9$.
(2 marks)

5 $y = 10^{-x}$ is an exponential curve.
When $x = 0, y = 1$.
As x gets large, y gets closer to 0.
So the x-axis is an asymptote of the curve.
Answer D

6 $y = \log_2 x$
When $x = 1, y = \log_2 1 = 0$.
When $x = 2, y = \log_2 2 = 1$.
When $x = 8, y = \log_2 8 = 3$.
Answer A

7 $10^x = 7$
$\therefore x = \log_{10} 7$ ✓
$= 0.845\,0980\ldots$
$= 0.8451$ ✓
(4 decimal places)
(2 marks)

8 $\log_9 x = 4.5$
$\therefore x = 9^{4.5}$
$= 19\,683$ ✓
(1 mark)

9 The graph of $y = a^x$ is an exponential curve (not a straight line). When $x = 0, y = 1$.
The graph that best represents $y = a^x$ is [D].

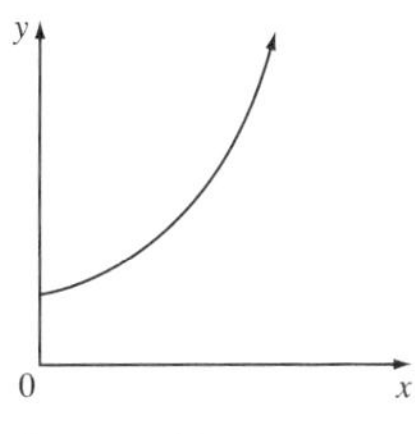

Answer D

10 Using GDC, graph $y = 3^x$.
or
$y = 3^x$ is not linear so correct answer is B or D.
When $x = 1, y = 3^1 = 3$
When $x = 2, y = 3^2 = 9$
Answer D

1 On the decibel scale, the loudness L of a sound of intensity S is given by $L = 10\log_{10}\left(\frac{S}{S_0}\right)$, where S_0 is a reference intensity used for comparisons.

How much louder is a sound of 47 decibels than a sound of 27 decibels?

A 2 times **B** 20 times
C 100 times **D** 200 times *(1 mark)*

Bonus question (see page iv) Medium

2 On the Richter scale, the magnitude R of an earthquake of intensity I is given by the formula $R = \log_{10}\left(\frac{I}{I_0}\right)$, where I_0 is a reference intensity used for comparisons.

The 1989 Newcastle Earthquake measured a magnitude of 5.6 on the Richter scale.

Which of these is closest to $\left(\frac{I}{I_0}\right)$, which is the number of times this earthquake is more intense than the reference intensity?

A 400 times **B** 4000 times
C 40 000 times **D** 400 000 times

(1 mark)

Bonus question Medium

3 The pH value of a solution is given by the formula $\text{pH} = -\log_{10}[\text{H}^+]$, where $[\text{H}^+]$ is the concentration of hydrogen ions in moles per litre.

Tylah tested a quantity of apple juice and determined it had a $\text{H}^+ = 3 \times 10^{-4}$.

Sean tested a quantity of ammonia and determined it had a $\text{H}^+ = 1.3 \times 10^{-9}$.

Which of these is closest to the difference in their pH?

A 1.7 **B** 4.3
C 4.8 **D** 5.4 *(1 mark)*

Bonus question Medium

4 On the decibel scale, the loudness L of a sound of intensity S is given by $L = 10\log_{10}\left(\frac{S}{S_0}\right)$, where S_0 is a reference intensity used for comparisons.

How many times more intense is the sound in the front row of a concert of a heavy metal band (110 decibels) than a normal conversation (60 decibels)? *(2 marks)*

Bonus question Medium

5 Solve $\log_4 x + \log_4 (x - 12) = 3$. *(2 marks)*

Bonus question Medium

6 Which of the following is equal to $\frac{\log_2 9}{\log_2 3}$?

A 2 **B** 3
C $\log_2 3$ **D** $\log_2 6$ *(1 mark)*

(Q5, **2019 HSC**) Medium

7 Solve $\log_2 5 + \log_2(x - 2) = 3$. *(2 marks)*

(Q11b, **2018 ME1 HSC**) Medium

8 It is given that $\log_a 8 = 1.893$, correct to 3 decimal places.

What is the value of $\log_a 4$, correct to 2 decimal places?

A 0.95 **B** 1.26
C 1.53 **D** 2.84 *(1 mark)*

(Q2, **2017 ME1 HSC**) Medium

9 Which expression is equivalent to $4 + \log_2 x$?

A $\log_2(2x)$ **B** $\log_2(16 + x)$
C $4\log_2(2x)$ **D** $\log_2(16x)$ *(1 mark)*

(Q10, **2016 HSC**) Medium

10 What is the solution to the equation $\log_2(x - 1) = 8$?

A 4 **B** 17
C 65 **D** 257 *(1 mark)*

(Q3, **2014 HSC**) Medium

11 Use the change of base formula to evaluate $\log_3 7$, correct to two decimal places. *(1 mark)*

(Q5a, **2005 HSC**) Easy

12 Solve $\log_2(3x - 4) = 5$. *(2 marks)*

(Q6a, **2003 HSC**) Easy

13 Solve $2^x = 3$.

Express your answer correct to two decimal places. *(2 marks)*

(Q2a, **2002 ME1 HSC**) Easy

14 i Find $\log_{10}(2^{1000})$ correct to 3 decimal places. *(2 marks)* Easy

ii We know that $2^{10} = 1024$, so that 2^{10} can be represented by a 4 digit numeral. How many digits are there in 2^{1000} when written as a numeral? *(1 mark)* Medium

(Q5b, **2001 HSC**)

---

15 Let $\log_a 2 = x$ and $\log_a 3 = y$.

Find an expression for $\log_a 12$ in terms of x and y. *(2 marks)*

(Q5b, **1999 HSC**) Medium

---

16 Given that $\log_2 7 = 2{\cdot}807$ (to three decimal places), find $\log_2 14$. *(1 mark)*

(Q1d, **1998 ME1 HSC**) Medium

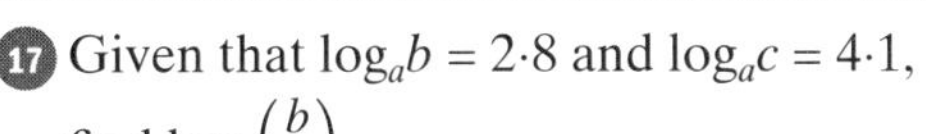

17 Given that $\log_a b = 2{\cdot}8$ and $\log_a c = 4{\cdot}1$, find $\log_a\left(\frac{b}{c}\right)$. *(1 mark)*

(Q1c, **1997 ME1 HSC**)  Easy

---

18 A layer of plastic cuts out 15% of the light and lets through the remaining 85%.

i Show that two layers of the plastic let through 72·25% of the light. Easy

ii How many layers of the plastic are required to cut out at least 90% of the light? Hard

(3 marks)

(Q3b, **1996 HSC**)

---

19 Given that $\log_a b = 2{\cdot}75$ and $\log_a c = 0{\cdot}25$, find the value of:

i $\log_a\left(\frac{b}{c}\right)$ Easy

ii $\log_a(bc)^2$. *(3 marks)* Medium

(Q7d, **1995 HSC**)

---

20 Solve $2\log_5 3 = \log_5 x - \log_5 6$.

(Q7b, **1994 HSC**) Medium

Year 11 Logarithmic laws and applications—Worked answers

1 Difference in decibels = 47 – 27
= 20

$$20 = 10\log_{10}\left(\frac{S}{S_0}\right)$$

$$\log_{10}\left(\frac{S}{S_0}\right) = 2$$

$$\frac{S}{S_0} = 10^2 = 100$$

The sound is 100 times louder.
Answer C

2 $5.6 = \log_{10}\left(\frac{I}{I_0}\right)$,

$$\frac{I}{I_0} = 10^{5.6}$$
= 398 107.1706...
= 400 000 (nearest hundred thousand)

Answer D

3 Apple juice:
$\text{pH} = -\log_{10}[3 \times 10^{-4}]$
= 3.522 878 745...
= 3.5 (1 dec. pl.)

Ammonia:
$\text{pH} = -\log_{10}[1.3 \times 10^{-9}]$
= 8.886 095 6648...
= 8.9 (1 dec. pl.)

Difference = 8.9 – 3.5
= 5.4

Answer D

4 Difference in decibels = 110 – 60
= 50

$$L = 10\log_{10}\left(\frac{S}{S_0}\right)$$

$$50 = 10\log_{10}\left(\frac{S}{S_0}\right)$$

$$\log_{10}\left(\frac{S}{S_0}\right) = 5 \quad ✓$$

$$\frac{S}{S_0} = 10^5 = 100\,000 \quad ✓$$

It is 100 000 times more intense.
(2 marks)

5 $\log_4 x + \log_4(x - 12) = 3$

$\log_4[x(x - 12)] = 3$

$x(x - 12) = 4^3$

$x^2 - 12x - 64 = 0$ ✓

$(x - 16)(x + 4) = 0$

$x = 16$ ✓

(as $x > 0$)
(2 marks)

6 $\frac{\log_2 9}{\log_2 3} = \log_3 9$
(by change of base formula)
$= \log_3 3^2$
$= 2\log_3 3$
$= 2 \times 1$
$= 2$

Answer A

7 $\log_2 5 + \log_2(x - 2) = 3$

$\log_2(5x - 10) = 3$

$\therefore 5x - 10 = 2^3$ ✓

$5x - 10 = 8$

$5x = 18$

$x = 3.6$ ✓

(2 marks)

8 $\log_a 8 = 1.893$

Now $\log_a 8 = \log_a 2^3$
$= 3\log_a 2$

$\log_a 4 = \log_a 2^2$
$= 2\log_a 2$

$\therefore \log_a 4 = \frac{2}{3}\log_a 8$
= 1.26 (2 d.p.)

Answer B

9 Use $4 = \log_2 2^4$.

$4 + \log_2 x = \log_2 2^4 + \log_2 x$
$= \log_2 16 + \log_2 x$
$= \log_2 16x$

Answer D

10 $\log_2(x - 1) = 8$

$x - 1 = 2^8$

$x - 1 = 256$

$x = 257$

Answer D

11 $\log_3 7 = \frac{\log_{10} 7}{\log_{10} 3}$
= 1.7712 . . .
= 1.77 to 2 decimal places. ✓

(1 mark)

12 $\log_2(3x - 4) = 5$

$3x - 4 = 2^5$ ✓

$3x - 4 = 32$

$3x = 36$

$\therefore x = 12$ ✓

(2 marks)

13 $2^x = 3$

$\therefore \log 2^x = \log 3$

$\therefore x\log 2 = \log 3$

$\therefore x = \frac{\log 3}{\log 2}$ ✓

= 1.5849 . . .
= 1.58 ✓
to 2 decimal places.

(2 marks)

14 **i** $\log_{10}(2^{1000})$
$= 1000\log_{10} 2$ ✓
$= 1000 \times 0.301\,029\,9...$
$= 301.0299...$
= 301.030 to 3 decimal places. ✓
(2 marks)

ii Now $\log_{10} 2^{10}$
$= 10\log_{10} 2$
$= 10 \times 0.301\,029\,9...$
$= 3.0102...$

$\therefore 2^{10} = 10^{3.0102...}$
= 1024 (4 digits)

Also $\log_{10} 2^{20}$
$= 20\log_{10} 2$
$= 20 \times 0.301\,029\,9...$
$= 6.0205...$

$\therefore 2^{20} = 10^{6.0205...}$
= 1 048 576 (7 digits)

$\therefore 2^{1000} = 10^{301.0299...}$

$\therefore$ 302 digits ✓

(1 mark)

15 $\log_a 12 = \log_a(4 \times 3)$
$= \log_a 4 + \log_a 3$
$= \log_a 2^2 + \log_a 3$ ✓
$= 2\log_a 2 + \log_a 3$
$= 2x + y$ ✓
(2 marks)

16 $\log_2 14 = \log_2(2 \times 7)$
$= \log_2 2 + \log_2 7$
$= 1 + 2.807$
$= 3.807$ ✓ *(1 mark)*

17 $\log_a b = 2.8$

$\log_a c = 4.1$

$\therefore \log_a\left(\frac{b}{c}\right) = \log_a b - \log_a c$
$= 2.8 - 4.1$
$= -1.3$ ✓

(1 mark)

18 **i** Two layers let through
85% × 85%
$= \frac{85}{100} \times \frac{85}{100}$
= 0.7225
= 72.25% ✓

ii Cut out at least 90% means let through 10% or less

$\therefore \left(\frac{85}{100}\right)^n \leqslant 10\%$

$\left(\frac{85}{100}\right)^n \leqslant 0.1$

$n\log(0.85) \leqslant \log(0.1)$ ✓

$n \geqslant \frac{\log(0.1)}{\log(0.85)}$

$n \geqslant 14.1681 . . .$

$\therefore n = 15$

$\therefore$ 15 layers are required. ✓

(3 marks)
(Total mark allocation only in exam)

19 **i** $\log_a \left(\frac{b}{c}\right) = \log_a b - \log_a c$
$= 2.75 - 0.25$
$= 2.5$ ✓

ii $\log_a (bc)^2 = 2 \log_a (bc)$
$= 2(\log_a b + \log_a c)$
$= 2(2.75 + 0.25)$ ✓
$= 2 \times 3$
$= 6$ ✓

(3 marks)
(Total mark allocation only in exam)

20 $2 \log_5 3 = \log_5 x - \log_5 6$
$\log_5 3^2 = \log_5 \left(\frac{x}{6}\right)$
$9 = \frac{x}{6}$
$\therefore x = 54$

(No mark allocation in exam)

1 Which of the following represents the domain of the function $f(x) = \ln(1 - x)$?

A $[1, \infty)$ **B** $(1, \infty)$
C $(-\infty, 1]$ **D** $(-\infty, 1)$ *(1 mark)*

(Q3, **2021 HSC**) Easy

2 Solve $e^{2\ln x} = x + 6$. *(2 marks)*

(Q15a, **2019 HSC**) Medium

3 What is the domain of the function $f(x) = \ln(4 - x)$?

A $x < 4$ **B** $x \leq 4$
C $x > 4$ **D** $x \geq 4$ *(1 mark)*

(Q1, **2019 ME1 HSC**) Medium

4 It is given that $\ln a = \ln b - \ln c$, where $a, b, c > 0$.

Which statement is true?

A $a = b - c$ **B** $a = \dfrac{b}{c}$
C $\ln a = \dfrac{b}{c}$ **D** $\ln a = \dfrac{\ln b}{\ln c}$ *(1 mark)*

(Q5, **2017 HSC**) Medium

5 The line $y = mx$ is a tangent to the curve $y = e^{2x}$ at a point P.

i Sketch the line and the curve on one diagram. *(1 mark)* Easy
ii Find the coordinates of P. *(3 marks)* Hard
iii Find the value of m. *(1 mark)* Hard

(Q15c, **2014 HSC**)

6 Evaluate ln 3 correct to three significant figures. *(1 mark)*

(Q11a, **2013 HSC**) Easy

7 Let $a = e^x$.

Which expression is equal to $\log_e(a^2)$?

A e^{2x} **B** e^{x^2}
C $2x$ **D** x^2 *(1 mark)*

(Q7, **2012 HSC**) Easy

8 What is the range of the function $f(x) = \ln(x^2 + e)$? *(1 mark)*

(Q1f, **2011 ME1 HSC**) Hard

9 Let $f(x) = 1 + e^x$.

Show that $f(x) \times f(-x) = f(x) + f(-x)$ *(2 marks)*

(Q4d, **2010 HSC**) Medium

10 Solve $\ln(x + 6) = 2\ln x$. *(3 marks)*

(Q1c, **2010 ME1 HSC**) Medium

11 Solve the equation $\ln x = 2$. Give your answer correct to four decimal places. *(2 marks)*

(Q1f, **2009 HSC**) Medium

12 Let $f(x) = \ln(x - 3)$. What is the domain of $f(x)$? *(1 mark)*

(Q1b, **2009 ME1 HSC**) Easy

13 Solve $\log_e x - \dfrac{3}{\log_e x} = 2$. *(3 marks)*

(Q7a, **2008 HSC**) Hard

14 Solve the following equation for x: $2e^{2x} - e^x = 0$. *(2 marks)*

(Q6a, **2007 HSC**) Medium

15 Evaluate $e^{-0.5}$ correct to three decimal places. *(2 marks)*

(Q1a, **2006 HSC**) Easy

16 Solve the following equation for x: $e^{2x} + 3e^x - 10 = 0$. *(2 marks)*

(Q6a, **2004 HSC**) Medium

17 Evaluate $e^{-3.5}$ correct to 3 significant figures. *(2 marks)*

(Q1a, **2003 HSC**) Easy

18 Find α so that the equation $P = 100(1.23)^t$ can be rewritten as $P = 100e^{\alpha t}$.

Give your answer in decimal form. *(2 marks)*

(Q6b, **2001 HSC**) Medium

19 Find the value of $\log_e 8$ correct to two decimal places. *(2 marks)*

(Q1a, **2000 HSC**) Easy

20 Find the value of e^3, correct to three significant figures. *(2 marks)*

(Q1b, **1999 HSC**) Easy

21 Solve $\ln(7x - 12) = 2\ln x$. *(2 marks)*

(Q9a, **1998 HSC**) Medium

22 Solve the equation $2\ln x = \ln(5 + 4x)$.

(Q9c, **1993 HSC**) Medium

Year 11 Exponential function and natural logarithms—Worked answers

1 As domain of $f(x) = \ln x$ is $x > 0$, domain of $f(x) = \ln(1 - x)$ is $1 - x > 0$

$\therefore x < 1 \;\therefore (-\infty, 1)$

Answer D

2

$$e^{2\ln x} = x + 6$$
$$e^{\ln x^2} = x + 6$$
$$x^2 = x + 6 \quad ✓$$
$$x^2 - x - 6 = 0$$
$$(x - 3)(x + 2) = 0$$
$$x = 3 \text{ (as } x > 0) \quad ✓$$

(2 marks)

3 $f(x) = \ln(4 - x)$

Now $4 - x > 0$

$\therefore x < 4$

Answer A

4 $\ln a = \ln b - \ln c$

$\ln a = \ln \dfrac{b}{c}$

$a = \dfrac{b}{c}$

Answer B

5 **i**

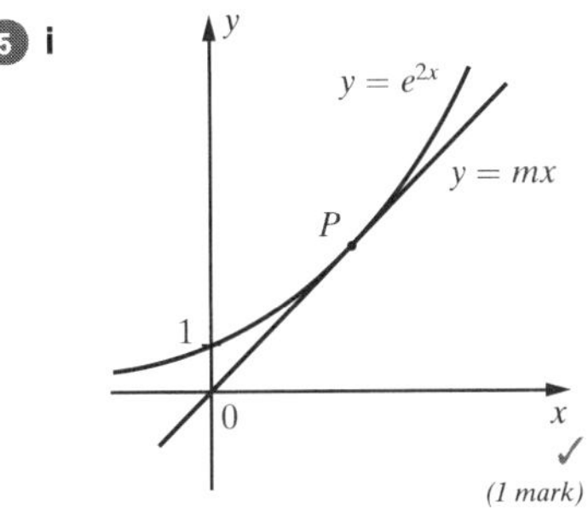

✓ *(1 mark)*

ii For common point, equate y:

$mx = e^{2x}$ ....①

For common gradient: equate y':

$m = 2e^{2x}$ ....②

Subs. ② in ①:

$$2e^{2x}x = e^{2x} \quad ✓$$
$$2e^{2x}x - e^{2x} = 0$$
$$e^{2x}(2x - 1) = 0$$
$$x = \frac{1}{2} \text{ (as } e^{2x} \neq 0) \quad ✓$$

Subs. in $y = e^{2x}$:

$$y = e^{2\left(\frac{1}{2}\right)}$$
$$= e$$

$$\therefore P\left(\frac{1}{2}, e\right) \quad ✓$$

(3 marks)

iii Subs. $P\left(\frac{1}{2}, e\right)$ in $y = mx$:

$$e = m\left(\frac{1}{2}\right)$$
$$m = 2e \quad ✓$$

(1 mark)

6 $\ln 3 = 1.0986 \ldots$

$= 1.10$ (3 sig. figs) ✓

(1 mark)

7

$$a = e^x$$
$$a^2 = (e^x)^2$$
$$= e^{2x}$$

Taking logs of both sides:

$$\log_e(a^2) = \log_e(e^{2x})$$
$$= 2x \log_e e$$
$$= 2x$$

Answer C

8 $f(x) = \ln(x^2 + e)$

Now $x^2 \geq 0$

So $x^2 + e \geq e$

$\ln e = 1$

So the range is $y \geq 1$ ✓ *(1 mark)*

9

$$f(x) = 1 + e^x$$
$$\text{LHS} = f(x) \times f(-x)$$
$$= (1 + e^x)(1 + e^{-x})$$
$$= 1 + e^{-x} + e^x + e^0$$
$$= 2 + e^{-x} + e^x \quad ✓$$
$$\text{RHS} = f(x) + f(-x)$$
$$= 1 + e^x + 1 + e^{-x}$$
$$= 2 + e^{-x} + e^x \quad ✓$$

$\therefore$ LHS = RHS

(2 marks)

10

$$\ln(x + 6) = 2\ln x$$
$$\ln(x + 6) = \ln x^2$$
$$\therefore x + 6 = x^2$$
$$x^2 - x - 6 = 0 \quad ✓$$
$$(x - 3)(x + 2) = 0$$
$$x = 3 \text{ or } x = -2$$

But $x > 0$ ✓

$\therefore x = 3$ ✓

(3 marks)

11 $\ln x = 2$

$\therefore x = e^2$ ✓

$= 7.389\,05 \ldots$

$= 7.3891$ to 4 decimal places. ✓

(2 marks)

12 Domain: $x - 3 > 0$

$\therefore x > 3$ ✓

(1 mark)

13 $\log_e x - \dfrac{3}{\log_e x} = 2$

Let $m = \log_e x$

$$\therefore m - \frac{3}{m} = 2$$
$$m^2 - 3 = 2m$$
$$m^2 - 2m - 3 = 0 \quad ✓$$
$$(m - 3)(m + 1) = 0$$

$\therefore m - 3 = 0$ *or* $m + 1 = 0$

$\therefore m = 3$ *or* $m = -1$

$\therefore \log_e x = 3$ *or* $\log_e x = -1$

$\therefore x = e^3$ ✓ $\quad x = e^{-1}$

$\therefore x = \dfrac{1}{e}$ ✓

(3 marks)

14

$$2e^{2x} - e^x = 0$$
$$\therefore 2(e^x)^2 - e^x = 0$$

Let $m = e^x$,

$$\therefore 2m^2 - m = 0$$
$$m(2m - 1) = 0$$

$\therefore m = 0$ or $2m - 1 = 0$

no solution (as $e^x > 0$) $\quad m = \dfrac{1}{2}$ ✓

$$\therefore e^x = \frac{1}{2}$$
$$\therefore \ln e^x = \ln\left(\frac{1}{2}\right)$$
$$\therefore x = \ln\frac{1}{2}$$
$$\therefore x = -\ln 2 \quad ✓$$

(2 marks)

15 $e^{-0.5} = 0.6065\ldots$ ✓

$= 0.607$ to 3 decimal places. ✓

(2 marks)

16

$$e^{2x} + 3e^x - 10 = 0$$
$$\therefore (e^x)^2 + 3e^x - 10 = 0$$

Let $m = e^x$,

$$\therefore m^2 + 3m - 10 = 0$$
$$(m + 5)(m - 2) = 0$$

$\therefore m + 5 = 0$ or $m - 2 = 0$

$m = -5$ or $m = 2$

$\therefore e^x = -5$ or $\therefore e^x = 2$ ✓

no solution ($e^x > 0$) $\quad \ln e^x = \ln 2$

$\therefore x = \ln 2$

$\therefore x = \ln 2.$ ✓

(2 marks)

17 $e^{-3.5} = 0.03019\ldots$ ✓

$= 0.0302$ to 3 sig. figures. ✓

(2 marks)

18

$P = 100(1.23)^t$

and $P = 100e^{\alpha t}$

$\therefore (1.23)^t = e^{\alpha t}$

$\therefore (1.23)^t = (e^{\alpha})^t$

$\therefore e^{\alpha} = 1.23$ ✓

$\ln e^{\alpha} = \ln 1.23$

$\alpha \ln e = \ln 1.23$

$\therefore \alpha = \ln 1.23$

$\doteqdot 0.21$ ✓

(2 marks)

19 $\log_e 8 = 2.0794\ldots$ ✓

$= 2.08$ to 2 decimal places. ✓

(2 marks)

20 $e^3 \doteqdot 20.0855\ldots$ ✓

$= 20.1$ to 3 sig. figures ✓

(2 marks)

21

$\ln(7x - 12) = 2\ln x$

$\ln(7x - 12) = \ln x^2$

$\therefore 7x - 12 = x^2$

$x^2 - 7x + 12 = 0$ ✓

$(x - 3)(x - 4) = 0$

$\therefore x = 3$ or 4 ✓

(2 marks)

22

$2\ln x = \ln(5 + 4x)$

$\ln x^2 = \ln(5 + 4x)$

$x^2 = 5 + 4x$

$x^2 - 4x - 5 = 0$

$(x - 5)(x + 1) = 0$

$\therefore x = 5$

($x = -1$ is ignored as $\ln(-1)$ does not exist.)

(No mark allocation in exam)

1 Which of the following best represents the graph of $y = 10(0.8)^x$?

A

B

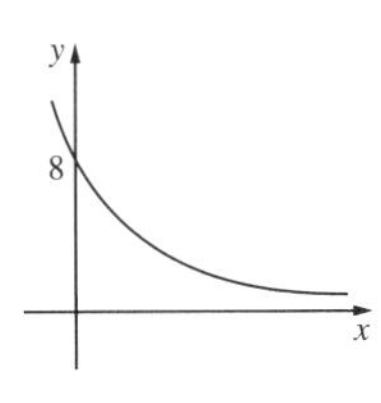

C

D

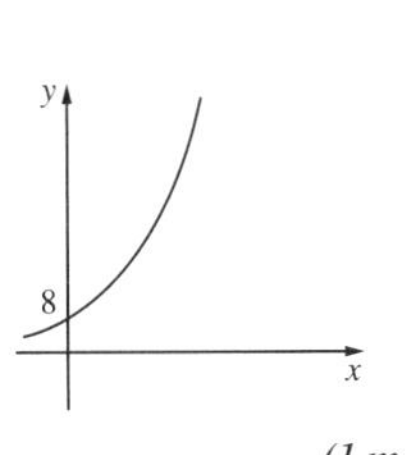

(1 mark)

CQ (Q5, **2021 HSC**) Easy

2 Find the value(s) of k for which the equation $2^{kx-8} = 4^{x^2}$ has exactly one solution. *(2 marks)*

Bonus question Medium

3 Solve $2^{2x+3} - 4.2^x = 0$ *(3 marks)*

Bonus question Medium

4 The population of a country grew exponentially between 1910 and 2010. This population can be modelled by the equation $P(t) = 92e^{kt}$, where $P(t)$ is the population of the country in millions, t is the time in years after 1910 and k is a positive constant. The population of the country in 1960 was 184 million.

i Show that the value of k is 0.0139, correct to 4 decimal places. *(2 marks)* Easy

ii Assuming that this model continues to be valid after 2010, estimate the population of the country in 2020 to the nearest million. *(2 marks)* Easy

(Q13c, **2018 HSC**)

5 A radioactive isotope of Curium has a half-life of 163 days. Initially there are 10 mg of Curium in a container.

The mass $M(t)$ in milligrams of Curium, after t days, is given by $M(t) = Ae^{-kt}$,

where A and k are constants.

i State the value of A. *(1 mark)* Easy

ii Given that after 163 days only 5 mg of Curium remain, find the value of k. *(2 marks)* Medium

(Q13c, **2016 HSC**)

6 The mass M kg of a baby pig at age x days is given by $M = A(1.1)^x$ where A is a constant. The graph of this equation is shown.

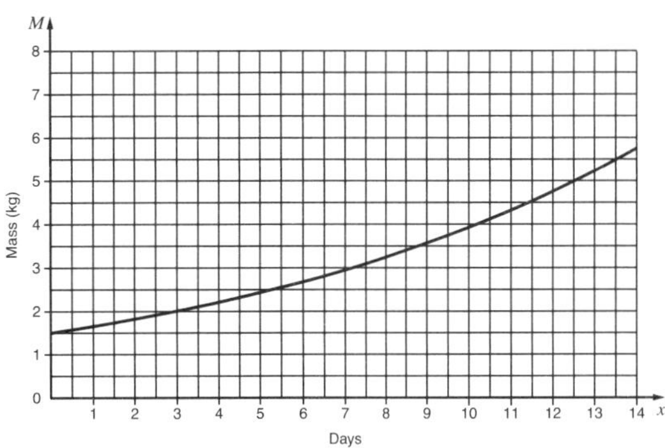

i What is the value of A? *(1 mark)* Medium

ii What is the daily growth rate of the pig's mass? Write your answer as a percentage. *(1 mark)* Hard

(Q29b, **2016 MS2 HSC**)

7 What is the solution of $5^x = 4$?

A $x = \dfrac{\log_e 4}{5}$ **B** $x = \dfrac{4}{\log_e 5}$

C $x = \dfrac{\log_e 4}{\log_e 5}$ **D** $x = \log_e\left(\dfrac{4}{5}\right)$ *(1 mark)*

(Q9, **2013 HSC**) Easy

8 Professor Smith has a colony of bacteria. Initially there are 1000 bacteria. The number of bacteria, $N(t)$, after t minutes is given by

$$N(t) = 1000e^{kt}.$$

i After 20 minutes there are 2000 bacteria. Show $k = 0.0347$ correct to four decimal places. *(1 mark)* Easy

ii How many bacteria are there when $t = 120$? *(1 mark)* Easy

iii What is the rate of change of the number of bacteria per minute, when $t = 120$? *(1 mark)* Medium

iv How long does it take for the number of bacteria to increase from 1000 to 100 000? *(2 marks)* Medium

(Q14c, **2012 HSC**)

9 In 2010, the city of Thagoras modelled the predicted population of the city using the equation

$$P = A(1.04)^n.$$

That year, the city introduced a policy to slow its population growth. The new predicted population was modelled using the equation

$$P = A(b)^n.$$

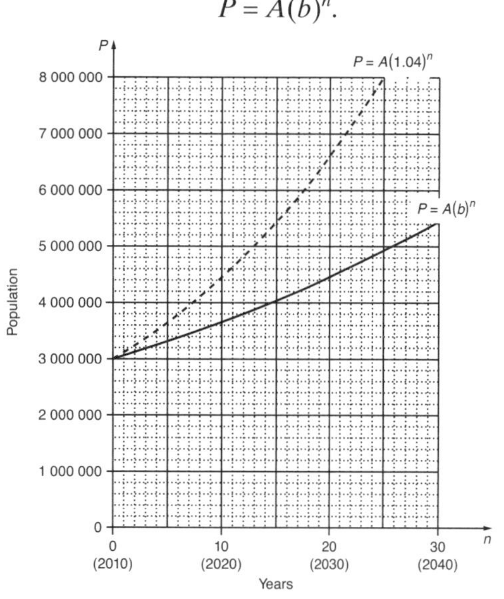

In both equations, P is the predicted population and n is the number of years after 2010. The graph shows the two predicted populations.

i Use the graph to find the predicted population of Thagoras in 2030 if the population policy had NOT been introduced. *(1 mark)* Easy

ii In each of the two equations given, the value of A is 3000000.
What does A represent? *(1 mark)* Easy

iii The guess-and-check method is to be used to find the value of b, in $P = A(b)^n$.

1 Explain, with or without calculations, why 1.05 is not a suitable first estimate for b. *(1 mark)* Easy

2 With $n = 20$ and $P = 4460000$, use the guess-and-check method and the equation $P = A(b)^n$ to estimate the value of b to two decimal places. Show at least TWO estimate values for b, including calculations and conclusions. *(2 marks)* Easy

iv The city of Thagoras was aiming to have a population under 7000000 in 2050. Does the model indicate that the city will achieve this aim?
Justify your answer with suitable calculations. *(2 marks)* Easy

(Q30c, **2012 MS2 HSC**)

10 The intensity I, measured in watt/m^2, of a sound is given by

$I = 10^{-12} \times e^{0.1L}$,

where L is the loudness of the sound in decibels.

i If the loudness of a sound at a concert is 110 decibels, find the intensity of the sound. Give your answer in scientific notation. *(1 mark)* Easy

ii Ear damage occurs if the intensity of a sound is greater than 8.1×10^{-9} watt/m^2. What is the maximum loudness of a sound so that no ear damage occurs? *(2 marks)* Medium

iii By how much will the loudness of a sound have increased if its intensity has doubled? *(2 marks)* Hard

(Q10a, **2011 HSC**)

11 The number of penguins, P, after t years in a new colony can be found using the following formula.

$$P = a \times 2^t$$

i If there are 24 penguins after two years, find the value of a. *(2 marks)* Easy

ii How many years will it take for the number of penguins to first exceed 1500? *(2 marks)* Medium

(Q25a, **2008 MS2 HSC**)

12 In 2004 there were 13.5 million registered motor vehicles in Australia. The number of registered motor vehicles is increasing at a rate of 2.3% per year.

Which expression represents the number (in millions) of registered motor vehicles, if y represents the number of years after 2004?

A $13.5 \times (1.023)^y$ **B** $13.5 \times (0.023)^y$
C $13.5 \times (1.023) \times y$ **D** $13.5 \times (0.023) \times y$

(1 mark)

(Q14, **2006 MS2 HSC**) Easy

13 **i** The number of bacteria in a culture grows from 100 to 114 in one hour.
What is the percentage increase in the number of bacteria? *(1 mark)* Easy

ii The bacteria continue to grow according to the formula $n = 100(1.14)^t$, where n is the number of bacteria after t hours. What is the number of bacteria after 15 hours? *(1 mark)* Easy

Time in hours (t)	0	5	10	15
Number of bacteria (n)	100	193	371	?

iii Use the values of n from $t = 0$ to $t = 15$ to draw a graph of $n = 100(1.14)^t$.
Use about half a page for your graph and mark a scale on each axis. *(4 marks)* Medium

iv Using your graph or otherwise, estimate the time in hours for the number of bacteria to reach 300. *(1 mark)* Easy

(Q26a, **2004 MS2 HSC**)

14 On the island of Wupetoi the unit of currency is the clam. The rate of inflation on Wupetoi has been constant for many years. Assuming the rate of inflation remains constant, the price of a surfboard will increase as shown in the graph.

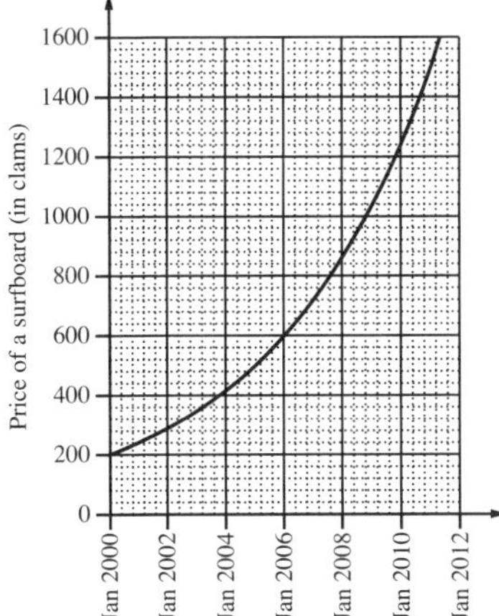

i The formula used to draw this graph was $P = A \times (1.2)^t$,
where P = price of a surfboard
and t = number of years after January 2000.

1. What is the value of A, and what does it represent? *(2 marks)* Easy
2. What annual rate of inflation has been assumed? *(1 mark)* Easy

ii In January 2000, Tana started saving a fixed number of clams each month, in order to buy a surfboard.
The straight line on the graph below represents Tana's savings.

1. During which year will Tana first be able to afford a surfboard? Explain your answer. *(2 marks)* Easy
2. If Tana uses his savings to buy a surfboard in January 2006, how many clams will he have left? *(1 mark)* Easy
3. Write an equation that describes the relationship between Tana's savings in clams (c) and the number of months (n) after January 2000. *(2 marks)* Easy

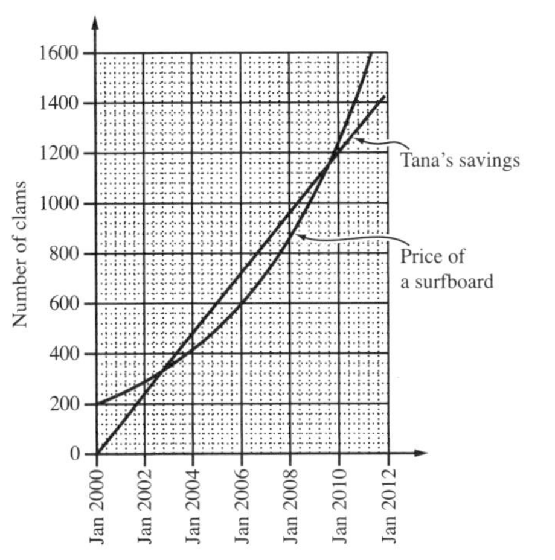

(Q26b, **2001 MS2 HSC**)

15 **i** Without using calculus, sketch $y = \log_e x$. Easy

ii On the same sketch, find, graphically, the number of solutions of the equation $\log_e x - x = -2$. Medium

(3 marks) (Q9a, **2000 HSC**)

16 A Geiger counter is taken into a region after a nuclear accident and gives a reading of 10 000. One year later, the same Geiger counter gives a reading of 9000. It is known that the reading T is given by the formula

$$T = T_0 e^{-kt},$$

where T_0 and k are constants and t is the time measured in years.

i Evaluate the constants T_0 and k. Easy

ii It is known that the region will become safe when the reading reaches 40. After how many years will the region become safe? Medium

(Q7c, **1994 HSC**)

17 Ten kilograms of sugar is placed in a container of water and begins to dissolve. After t hours the amount A kg of undissolved sugar is given by

$$A = 10e^{-kt}.$$

i Calculate k, given that $A = 3{\cdot}2$ when $t = 4$. Easy

ii After how many hours does 1 kg of sugar remain undissolved? Medium

(Q4c, **1992 HSC**)

18 Solve the equation $3^{2x} + 2.3^x - 15 = 0$.

(Q7a, **1992 HSC**) Medium

1 Substitute $x = 0$: $y = 10(0.8)^0$
$= 10$

This means either A or C is the correct graph.
As $0.8 < 1$ and $x > 0$, the graph of $y = 10(0.8)^x$ is decreasing.
This means the correct graph is A.

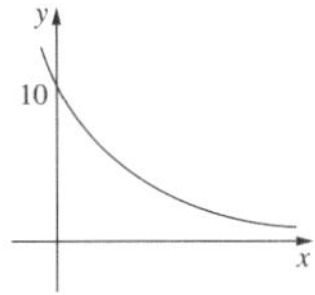

Answer A

2
$$2^{kx-8} = 4^{x^2}$$
$$2^{kx-8} = 2^{2x^2}$$
$$kx - 8 = 2x^2$$
$$2x^2 - kx + 8 = 0 \quad ✓$$
For equal roots,
$$\Delta = (-k)^2 - 4(2)(8) = 0$$
$$k^2 - 64 = 0$$
$$k = \pm 8 \quad ✓$$
(2 marks)

3
$$2^{2x+3} - 2^2.2^x = 0$$
$$2^{2x+3} - 2^{x+2} = 0 \quad ✓$$
$$2^{x+2}(2^{x+1} - 1) = 0$$
$$2^{x+1} - 1 = 0 \text{ (as } 2^{x+2} \neq 0) \quad ✓$$
$$2^{x+1} = 1$$
$$2^{x+1} = 2^0$$
$$x = -1 \quad ✓$$
Another method:
$$2^{2x+3} - 4.2^x = 0$$
$$8.2^{2x} - 4.2^x = 0$$
$$2.2^{2x} - 2^x = 0$$
Let $m = 2^x$:
$$2m^2 - m = 0$$
$$m(2m - 1) = 0$$
$$m = 0, \frac{1}{2}$$
$$\therefore 2^x = 0, \quad 2^x = \frac{1}{2}$$
$$\therefore x = -1 \text{ (as } 2^x \neq 0)$$
(3 marks)

4 i $P = 92e^{kt}$
Substitute $P = 184$, $t = 50$:
$$184 = 92e^{k(50)}$$
$$e^{50k} = 2$$
$$50k = \ln 2$$
$$k = \frac{\ln 2}{50} \quad ✓$$
$$= 0.01386943\ldots$$
$$= 0.0139 \text{ (4 dec. pl.)} \quad ✓$$
(2 marks)

ii Let $t = 110$:
$$P = 92e^{k(110)} \quad ✓$$
$$= 422.7209946\ldots$$
$$= 423 \text{ (nearest whole)}$$
$\therefore$ the population will be 423 million. ✓

(or, if using $k = 0.0139$, then $P = 424$ million)
(2 marks)

5 i Consider $M = Ae^{-kt}$
Let $t = 0$, $M = 10$:
$$\therefore 10 = Ae^{-k(0)}$$
$$\therefore A = 10 \quad ✓$$
(1 mark)

ii $M = 10e^{-kt}$
Let $t = 163$, $M = 5$:
$$\therefore 5 = 10e^{-k(163)}$$
$$e^{-163k} = 0.5 \quad ✓$$
$$-163k = \ln 0.5$$
$$k = -\frac{\ln 0.5}{163} \left(\text{or } k = \frac{\ln 2}{163}\right)$$
✓ *(2 marks)*

6 i $A = 1.5$ ✓
[the initial value from the graph]
(1 mark)

ii Growth rate = 10% ✓
[From the equation: 1.1 = 110% so the rate of increase is 10%.]
(1 mark)

7
$$5^x = 4$$
$$\log_e 5^x = \log_e 4$$
$$x \log_e 5 = \log_e 4$$
$$x = \frac{\log_e 4}{\log_e 5}$$
Answer C

8 i $N = 1000e^{kt}$, $t = 20$, $N = 2000$
$$2000 = 1000e^{20k}$$
$$e^{20k} = 2$$
$$\log_e e^{20k} = \log_e 2$$
$$20k = \log_e 2$$
$$k = \frac{\log_e 2}{20}$$
$$= 0.034657359\ldots$$
$$= 0.0347 \quad ✓$$
(to 4 dec. pl.)
(1 mark)

ii $N = 1000e^{kt}$, $t = 120$,
$$= 1000e^{k(120)}$$
$$= 64000$$
$\therefore$ 64 000 bacteria ✓
[or, using $k = 0.0347$, $N = 64328$]
(1 mark)

iii $N = 1000e^{kt}$
$$\frac{dN}{dt} = k.1000e^{kt}$$
$$= k.64000$$
$$= 64000k$$
$$= 2218.070978\ldots$$
$$= 2218 \text{ (nearest whole)}$$
$\therefore$ 2218 bacteria per minute
[or, using $k = 0.0347$, 2232 per minute] ✓
(1 mark)

iv $N = 1000e^{kt}$, $N = 100000$
$$100000 = 1000e^{kt}$$
$$e^{kt} = 100$$
$$\log_e e^{kt} = \log_e 100$$
$$kt = \log_e 100$$
$$t = \frac{\log_e 100}{k} \quad ✓$$
$$= 132.8771238\ldots$$
$$= 133 \text{ (nearest whole)}$$
$\therefore$ 2h 13 min ✓
(2 marks)

9 i In 2030, $n = 20$
From the graph, when $n = 20$, $P = 6600000$
If the policy is not introduced, the predicted population in 2030 is 6 600 000. ✓ *(1 mark)*

ii A = 3 000 000 represents the population of Thagoras in 2010. ✓
(1 mark)

iii 1 The population growth was slower using $P = A(b)^n$. So the value of b must be smaller than in the original equation in order for P to be smaller. That is $b < 1.04$. So a value of 1.05 is not suitable. ✓
(1 mark)

2 When $n = 20$,
$$P = 4460000$$
$$P = 3000000(b)^{20}$$
Try $b = 1.03$
$$P = 3000000(1.03)^{20}$$
$$= 5418333.704\ldots.$$
This value of P is too high so b needs to be smaller.
Try $b = 1.01$
$$P = 3000000(1.01)^{20}$$
$$= 3660570.12\ldots$$
This value of P is too small so b needs to be larger. ✓
Try $b = 1.02$
$$P = 3000000(1.02)^{20}$$
$$= 4457842.188\ldots$$
$$= 4460000$$
[3 significant figures]

So the value of b should be 1.02 to 2 decimal places. ✓ *(2 marks)*

iv In 2050, $n = 40$

$P = 3\,000\,000(1.02)^{40}$

$= 6\,624\,118.991\ldots$ ✓

This is less than 7 000 000 so the model does suggest that the city will achieve its goal. ✓

(2 marks)

10 i $I = 10^{-12} \times e^{0.1L}$

Subs $L = 110$,

$I = 10^{-12} \times e^{0.1(110)}$

$= 0.000\,000\,059\ldots$

$= 6 \times 10^{-8}$ (1 sig. fig.)

✓ *(1 mark)*

ii Let $I = 8.1 \times 10^{-9}$

$8.1 \times 10^{-9} = 10^{-12} \times e^{0.1L}$

$e^{0.1L} = (8.1 \times 10^{-9}) \div 10^{-12}$

$= 8100$

$0.1\,L \log_e e = \log_e 8100$

$L = \dfrac{\log_e 8100}{0.1}$ ✓

$= 89.996\,19\ldots$

$= 90$ (nearest whole)

$\therefore$ max loudness is 90 decibels.

✓ *(2 marks)*

iii From **i**,

$I = 6 \times 10^{-8}$, when $L = 110$.

We can double the value of I:

$I = 12 \times 10^{-8}$

$= 1.2 \times 10^{-7}$

$1.2 \times 10^{-7} = 10^{-12} \times e^{0.1L}$

$e^{0.1L} = (1.2 \times 10^{-7}) \div 10^{-12}$

$= 120\,000$ ✓

$0.1\,L \log_e e = \log_e 120\,000$

$= 116.952\,47\ldots$

$= 117$

(nearest whole)

$\therefore$ As $117 - 110 = 7$, the loudness has increased by 7 decibels. ✓

(2 marks)

11 i $P = a \times 2^t$

If $t = 2$ and $P = 24$,

$24 = a \times 2^2$ ✓

$24 = 4a$

$\therefore a = 6$ ✓

(2 marks)

ii $P = 1500, a = 6$

$\therefore 1500 = 6 \times 2^t$

$\dfrac{1500}{6} = 2^t$

$250 = 2^t$ ✓

By guess and check:

$2^7 = 128$

$2^8 = 256$ ✓

$\therefore$ It will take nearly 8 years.

(2 marks)

12 Each year, the number of vehicles is 2.3% more than the year before. This is similar to compound interest.

Using the formula,

$A = P(1 + r)^n$

$r = 2.3\% = 0.023$

$\therefore A = 13.5 \times (1 + 0.023)^y$

$= 13.5 \times (1.023)^y$

Answer A

13 i Percentage increase

$= \dfrac{114 - 100}{100} \times 100$

$= 14\%$ ✓

(1 mark)

ii $n = 100(1.14)^t$

When $t = 15$,

$n = 100(1.14)^{15}$

$= 713.793\ldots$

$\doteqdot 714$ bacteria. ✓

(1 mark)

iii

✓✓✓✓ *(4 marks)*

iv Using the equation,

$300 = 100(1.14)^t$

$3 = 1.14^t$

By guess and check, and using the graph,

when $t = 8.4$

$1.14^{8.4} = 3.006\ldots$

$\therefore$ It takes about 8.4 hours for the number of bacteria to reach 300. ✓

(1 mark)

14 i **1** $A = 200$ ✓

It is the price of a surfboard in Jan 2000, i.e. at the beginning (when $t = 0$). ✓

(2 marks)

2 20%

($P = A \times (1.2)^t$ corresponds to the compound interest formula,

$A = P(1 + r)^n$

where

$1 + r = 1.2$

$\therefore r = 0.2$

$= 0.2 \times 100\%$

$= 20\%$) ✓

(1 mark)

ii 1 During 2002, Tana will have just saved enough for a surfboard. This ✓ can be seen by the intersection of the two graphs. ✓

(2 marks)

2 $720 - 600 = 120$ clams. ✓

(1 mark)

3 $y = mx + b$ (straight line)

$\therefore c = mn + b$

When $n = 0, c = 0$ $\therefore b = 0$

$\therefore c = mn$

When $n = 24$ months (2 years),

$c = 240$ ✓

$\therefore 240 = m \times 24$

$\therefore m = 240 \div 24$

$= 10$

$\therefore c = 10n$ ✓

(2 marks)

15 i

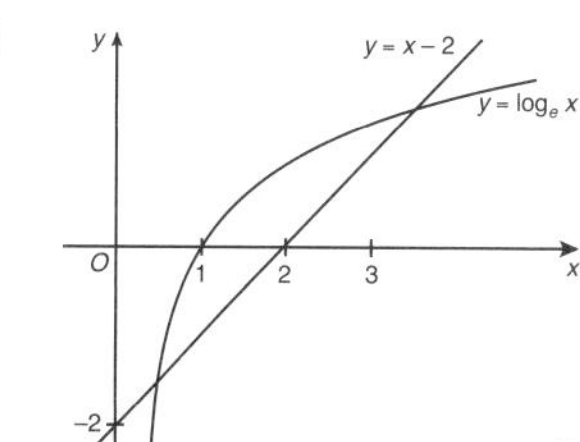

✓✓

ii $\log_e x - x = -2$

$\therefore \log_e x = x - 2$

The solution of the equation is found by sketching the graphs of $y = \log_e x$ and $y = x - 2$.

From the sketch there are 2 solutions. ✓

(3 marks)

(Total mark allocation only in exam)

16 i $T = T_0 e^{-kt}$

When $t = 0, T = 10\,000$

$\therefore 10\,000 = T_0 e^0$

$\therefore T_0 = 10\,000$

$\therefore T = 10\,000e^{-kt}$

When $t = 1, T = 9000$

$\therefore 9000 = 10\,000e^{-k(1)}$

$\dfrac{9000}{10\,000} = e^{-k}$

$\dfrac{9}{10} = e^{-k}$

$\ln 0.9 = \ln e^{-k}$

$$\ln 0.9 = -k \ln e$$
$$\ln 0.9 = -k$$
$$k = -\ln 0.9$$
$$\doteqdot 0.1054$$

ii $T = 10\,000e^{-(-\ln 0.9)t}$
$= 10\,000e^{(\ln 0.9)t}$

When $T = 40$,
$$40 = 10\,000e^{(\ln 0.9)t}$$
$$\frac{40}{10\,000} = e^{(\ln 0.9)t}$$
$$\frac{4}{1000} = e^{(\ln 0.9)t}$$
$$0.004 = e^{(\ln 0.9)t}$$
$$\ln 0.004 = \ln 0.9 \times t$$
$$t = \frac{\ln 0.004}{\ln 0.9}$$
$$= 52.4054\ldots \text{ years}$$

∴ The region will be safe after approximately 52.4 years.

(No mark allocation in exam)

---

17 **i**
$$A = 10e^{-kt}$$
$$3.2 = 10e^{-4k}$$
$$0.32 = e^{-4k}$$
$$\log_e 0.32 = -4k$$
$$k = \frac{\log_e 0.32}{-4}$$
$$\doteqdot 0.28485857$$

ii
$$1 = 10e^{-kt}$$
$$0.1 = e^{-kt}$$
$$\log_e 0.1 = -kt$$
$$t = \frac{\log_e 0.1}{-k}$$
(using calc. memory for k)
$= 8.0832\ldots$
$= 8.1$ hours to 1 decimal place.

(No mark allocation in exam)

18
$$3^{2x} + 2.3^x - 15 = 0$$
$$(3^x)^2 + 2.3^x - 15 = 0$$

Let $m = 3^x$:

$$m^2 + 2m - 15 = 0$$
$$(m + 5)(m - 3) = 0$$
$\therefore m = 3$ or $m = -5$
$\therefore 3^x = 3$ or $3^x = -5$
$\therefore x = 1$
($3^x = -5$ has no real solutions).

(No mark allocation in exam)

1 Liam is playing two games. He is equally likely to win each game. The probability that Liam will win at least one of the games is 80%.

Which of the following is closest to the probability that Liam will win both games?

A 31% **B** 40%
C 55% **D** 64% *(1 mark)*

(Q9, **2022 HSC**) Medium

2 In a bag there are 3 six-sided dice. Two of the dice have faces marked 1, 2, 3, 4, 5, 6. The other is a special die with faces marked 1, 2, 3, 5, 5, 5.

One die is randomly selected and tossed.

a What is the probability that the die shows a 5? *(1 mark)* Easy

b Given that the die shows a 5, what is the probability that it is the special die? *(1 mark)* Medium

(Q15, **2022 HSC**)

3 There are 8 chocolates in a box. Three have peppermint centres (P) and five have caramel centres (C).

Kim randomly chooses a chocolate from the box and eats it. Sam then randomly chooses and eats one of the remaining chocolates.

A partially completed probability tree is shown.

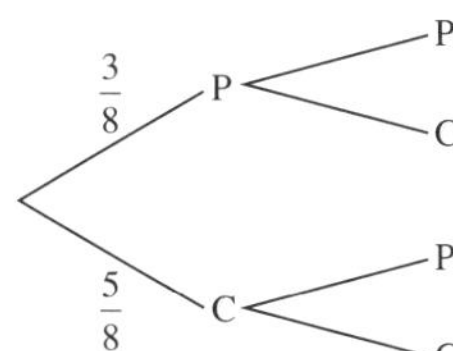

What is the probability that Kim and Sam choose chocolates with different centres?

A $\frac{15}{64}$ **B** $\frac{15}{56}$
C $\frac{15}{32}$ **D** $\frac{15}{28}$ *(1 mark)*

CQ (Q6, **2021 HSC**) Easy

4 History and Geography are two of the subjects students may decide to study. For a group of 40 students, the following is known.

- 7 students study neither History nor Geography
- 20 students study History
- 18 students study Geography

a A student is chosen at random. By using a Venn diagram, or otherwise, find the probability that the student studies both History and Geography. *(2 marks)* Easy

b A student is chosen at random. Given that the student studies Geography, what is the probability that the student does NOT study History? *(1 mark)* Medium

c Two different students are chosen at random, one after the other. What is the probability that the first student studies History and the second student does NOT study History? *(2 marks)* Medium

(Q14, **2020 HSC**)

5 The top of a rectangular table is divided into 8 equal sections as shown.

1	2	3	4
5	6	7	8

A standard die with faces labelled 1 to 6 is rolled onto the table. The die is equally likely to land in any of the 8 sections of the table. If the die does not land entirely in one section of the table, it is rolled again. A score is calculated by multiplying the value shown on the top face of the die by the number shown in the section of the table where the die lands.

What is the probability of getting a score of 6?

A $\frac{1}{48}$ **B** $\frac{1}{12}$
C $\frac{1}{8}$ **D** $\frac{1}{6}$ *(1 mark)*

(Q15, **2020 MS2 HSC**) Easy

6 For events A and B from a sample space,

$P(A|B) = \frac{2}{3}$ and $P(B) = \frac{1}{4}$.

i Calculate $P(A \cap B)$. *(1 mark)* Easy

ii Calculate $P(\overline{A} \cap B)$. *(1 mark)* Easy

iii If the events A and B are independent, calculate $P(A \cup B)$. *(1 mark)* Easy

Bonus question (see page iv)

7 For events A and B from a sample space,

$P(A|B) = \frac{1}{4}$ and $P(B|A) = \frac{1}{3}$.

Let $P(A \cap B) = p$.

i Find $P(B)$ in terms of p. *(1 mark)* Medium

ii Find $P(\overline{B} \cap \overline{A})$ in terms of p. *(2 marks)* Medium

Bonus question

8 For events A and B, $P(A \cap B) = p$, $P(\bar{A} \cap B) = 2p - \frac{1}{4}$ and $P(A \cap \bar{B}) = \frac{p}{3}$.
Given that A and B are independent, find the value of p. *(3 marks)*

Bonus question Medium

9 Two events, A and B are independent, where $P(B) = 2P(A)$ and $P(A \cup B) = \frac{5}{8}$.
What is the value of $P(A)$? *(3 marks)*

Bonus question Medium

10 A and B are events of a sample space S. $P(A \cap B) = \frac{2}{5}$ and $P(A \cap \bar{B}) = \frac{3}{8}$
What is the value of $P(\bar{B}|A)$? *(2 marks)*

Bonus question Medium

11 If X is a random variable such that $P(X > 2) = c$ and $P(X > 5) = d$, then what is $P(X \le 2 | X \le 5)$?

A $\frac{c}{d}$ **B** $\frac{1-d}{1-c}$

C $\frac{1-c}{1-d}$ **D** $\frac{3-c}{d}$ *(1 mark)*

Bonus question Medium

12 A and B are events of a sample space.
Given that $P(B|A) = \sqrt{p}$, $P(A) = p$ and $P(B) = p^2$, which of these is an expression for $P(A|B)$?

A $p^{\frac{3}{2}}$ **B** $\sqrt{p}$

C $\frac{1}{p}$ **D** $\frac{1}{\sqrt{p}}$ *(1 mark)*

Bonus question Medium

13 Two events, A and B, are such that $P(A) = \frac{2}{7}$ and $P(B) = \frac{3}{5}$.

i If $P(A \cup B) = \frac{4}{5}$, find the value of $P(\bar{A} \cap B)$. *(2 marks)* Medium

ii When A and B are mutually exclusive, find the value of $P(A \cap \bar{B})$. *(1 mark)* Medium

Bonus question

14 A game is played by tossing an ordinary 6-sided die and an ordinary coin at the same time. The game is won if the uppermost face of the die shows an even number or the uppermost face of the coin shows a tail (or both).

What is the probability of winning this game?

A $\frac{1}{4}$ **B** $\frac{1}{2}$

C $\frac{3}{4}$ **D** 1 *(1 mark)*

(Q6, **2019 HSC**) Medium

15 A bag contains 5 green beads and 7 purple beads. Two beads are selected at random, without replacement.

What is the probability that the two beads are the same colour? *(2 marks)*

(Q11f, **2019 HSC**) Easy

16 The probability that a person chosen at random has red hair is 0.02 .

i Two people are chosen at random. What is the probability that at least ONE has red hair? *(2 marks)* Medium

ii What is the smallest number of people that can be chosen at random so that the probability that at least ONE has red hair is greater than 0.4? *(2 marks)* Hard

(Q15d, **2019 HSC**)

17 A roulette wheel has the numbers $0, 1, 2, \ldots, 36$ where each of the 37 numbers is equally likely to be spun.

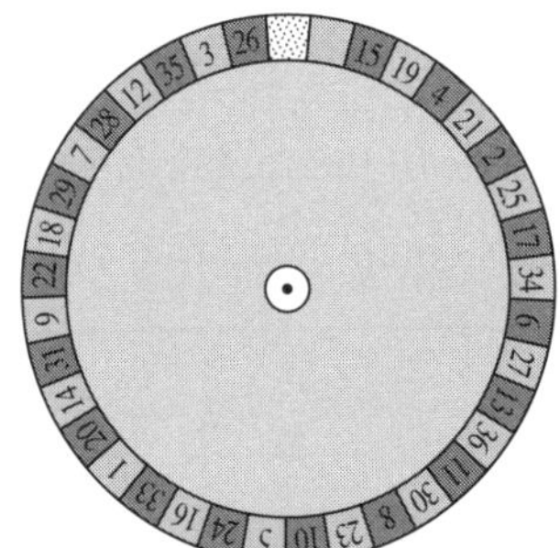

If the wheel is spun 18 500 times, calculate the expected frequency of spinning the number 8. *(2 marks)*

(Q20, **2019 MS2 HSC**) Easy

18 A bowl of fruit contains 17 apples of which 9 are red and 8 are green.

Dennis takes one apple at random and eats it. Margaret also takes an apple at random and eats it.

By drawing a probability tree diagram, or otherwise, find the probability that Dennis and Margaret eat apples of the same colour. *(3 marks)*

(Q25, **2019 MS2 HSC**) Medium

19 A runner has four different pairs of shoes.

If two shoes are selected at random, what is the probability that they will be a matching pair?

A $\frac{1}{56}$ **B** $\frac{1}{16}$

C $\frac{1}{7}$ **D** $\frac{1}{4}$ *(1 mark)*

(Q6, **2018 HSC**) Hard

20 Two machines, A and B, produce pens. It is known that 10% of the pens produced by machine A are faulty and that 5% of the pens produced by machine B are faulty.

i One pen is chosen at random from each machine.
What is the probability that at least one of the pens is faulty? *(1 mark)* Medium

ii A coin is tossed to select one of the two machines. Two pens are chosen at random from the selected machine. What is the probability that neither pen is faulty? *(2 marks)* Medium

(Q14e, **2018 HSC**)

21 A game involves rolling two six-sided dice, followed by rolling a third six-sided die. To win the game, the number rolled on the third die must lie between the two numbers rolled previously. For example, if the first two dice show 1 and 4, the game can only be won by rolling a 2 or 3 with the third die.

i What is the probability that a player has no chance of winning before rolling the third die? *(2 marks)* Medium

ii What is the probability that a player wins the game? *(2 marks)* Hard

(Q16b, **2018 HSC**)

22 Jeremy rolled a biased 6-sided die a number of times. He recorded the results in a table.

Number	1	2	3	4	5	6
Frequency	23	19	48	20	21	19

What is the relative frequency of rolling a 3? *(1 mark)*

(Q26a, **2018 MS2 HSC**) Easy

23 A spinner is marked with the numbers 1, 2, 3, 4 and 5. When it is spun, each of the five numbers is equally likely to occur.

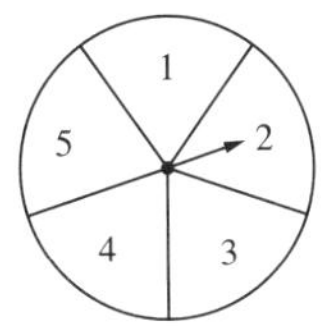

The spinner is spun three times.

i What is the probability that an even number occurs on the first spin? *(1 mark)* Easy

ii What is the probability that an even number occurs on at least one of the three spins? *(1 mark)* Medium

iii What is the probability that an even number occurs on the first spin and odd numbers occur on the second and third spins? *(1 mark)* Easy

iv What is the probability that an even number occurs on exactly one of the three spins? *(1 mark)* Medium

(Q12e, **2017 HSC**)

24 A deck of 52 playing cards contains 12 picture cards. Two cards from the deck are drawn at random and placed on a table.

What is the probability, correct to four decimal places, that exactly one picture card is on the table?

A 0.0498 **B** 0.1810

C 0.3550 **D** 0.3620 *(1 mark)*

(Q24, **2017 MS2 HSC**) Easy

25 In a raffle, 30 tickets are sold and there is one prize to be won.

What is the probability that someone buying 6 tickets wins the prize?

A $\frac{1}{30}$ **B** $\frac{1}{6}$

C $\frac{1}{5}$ **D** $\frac{1}{4}$ *(1 mark)*

(Q2, **2016 HSC**) Easy

26 The probability that Mel's soccer team wins this weekend is $\frac{5}{7}$.

The probability that Mel's rugby league team wins this weekend is $\frac{2}{3}$.

What is the probability that neither team wins this weekend?

A $\frac{2}{21}$ **B** $\frac{10}{21}$

C $\frac{13}{21}$ **D** $\frac{19}{21}$ *(1 mark)*

(Q4, **2015 HSC**) Easy

27 Weather records for a town suggest that:

- if a particular day is wet (W), the probability of the next day being dry is $\frac{5}{6}$.
- if a particular day is dry (D), the probability of the next day being dry is $\frac{1}{2}$.

In a specific week Thursday is dry. The tree diagram shows the possible outcomes for the next three days: Friday, Saturday and Sunday.

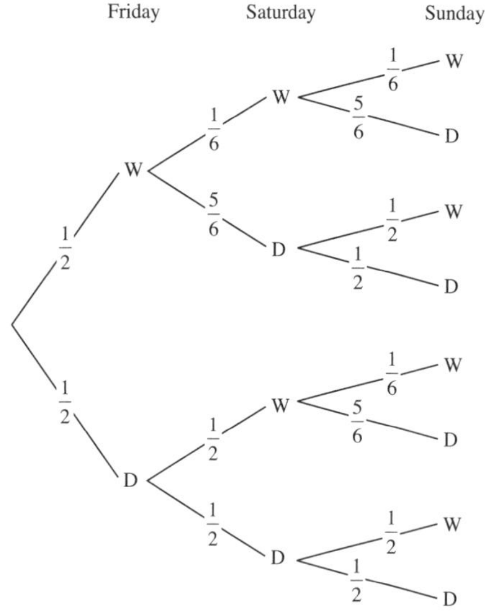

i Show that the probability of Saturday being dry is $\frac{2}{3}$. *(1 mark)* Easy

ii What is the probability of both Saturday and Sunday being wet? *(2 marks)* Easy

iii What is the probability of at least one of Saturday and Sunday being dry? *(1 mark)* Easy

(Q14b, **2015 HSC**)

---

28 Three runners compete in a race. The probabilities that the three runners finish the race in under 10 seconds are $\frac{1}{4}$, $\frac{1}{6}$ and $\frac{2}{5}$ respectively.

What is the probability that at least one of the three runners will finish the race in under 10 seconds?

A $\frac{1}{60}$ **B** $\frac{37}{60}$

C $\frac{3}{8}$ **D** $\frac{5}{8}$ *(1 mark)*

(Q10, **2014 HSC**) Medium

---

29 A packet of lollies contains 5 red lollies and 14 green lollies. Two lollies are selected at random without replacement.

i Draw a tree diagram to show the possible outcomes. Include the probability on each branch. *(2 marks)* Easy

ii What is the probability that the two lollies are of different colours? *(1 mark)* Easy

(Q12c, **2014 HSC**)

---

30 In Mathsville, there are on average eight rainy days in October.

Which expression could be used to find a value for the probability that it will rain on two consecutive days in October in Mathsville?

A $\frac{8}{31} \times \frac{7}{30}$ **B** $\frac{8}{31} \times \frac{7}{31}$

C $\frac{8}{31} \times \frac{8}{30}$ **D** $\frac{8}{31} \times \frac{8}{31}$ *(1 mark)*

(Q16, **2014 MS2 HSC**) Easy

---

31 A bag contains 4 red marbles and 6 blue marbles. Three marbles are selected at random without replacement.

What is the probability that at least one of the marbles selected is red?

A $\frac{1}{6}$ **B** $\frac{1}{2}$

C $\frac{5}{6}$ **D** $\frac{29}{30}$ *(1 mark)*

(Q5, **2013 HSC**) Easy

---

32 Two unbiased dice, each with faces numbered 1, 2, 3, 4, 5 and 6, are rolled.

What is the probability of obtaining a sum of 6?

A $\frac{1}{6}$ **B** $\frac{1}{12}$

C $\frac{5}{12}$ **D** $\frac{5}{36}$ *(1 mark)*

(Q18, **2013 MS2 HSC**) Easy

---

33 Two buckets each contain red marbles and white marbles. Bucket A contains 3 red and 2 white marbles. Bucket B contains 3 red and 4 white marbles.

Chris randomly chooses one marble from each bucket.

i What is the probability that both marbles are red? *(1 mark)* Easy

ii What is the probability that at least one of the marbles is white? *(1 mark)* Easy

iii What is the probability that both marbles are the same colour? *(2 marks)* Easy

(Q13c, **2012 HSC**)

---

34 Two unbiased dice, each with faces numbered 1, 2, 3, 4, 5 and 6, are rolled.

What is the probability of a 6 appearing on at least one of the dice?

A $\frac{1}{6}$ **B** $\frac{11}{36}$

C $\frac{25}{36}$ **D** $\frac{5}{6}$ *(1 mark)*

(Q12, **2012 MS2 HSC**) Medium

35 A batch of 800 items is examined. The probability that an item from this batch is defective is 0.02. How many items from this batch are defective? *(1 mark)*

(Q1g, **2011 HSC**) Easy

36 Kim has three red shirts and two yellow shirts. On each of the three days, Monday, Tuesday and Wednesday, she selects one shirt at random to wear. Kim wears each shirt that she selects only once.

i What is the probability that Kim wears a red shirt on Monday? *(1 mark)* Easy

ii What is the probability that Kim wears a shirt of the same colour on all three days? *(1 mark)* Easy

iii What is the probability that Kim does not wear a shirt of the same colour on consecutive days? *(2 marks)* Easy

(Q5b, **2011 HSC**)

37 There are twelve chocolates in a box. Four of the chocolates have mint centres, four have caramel centres and four have strawberry centres. Ali randomly selects two chocolates and eats them.

i What is the probability that the two chocolates have mint centres? *(1 mark)* Easy

ii What is the probability that the two chocolates have the same centre? *(1 mark)* Easy

iii What is the probability that the two chocolates have different centres? *(1 mark)* Easy

(Q4c, **2010 HSC**)

38 Two identical biased coins are tossed together, and the outcome is recorded. After a large number of trials it is observed that the probability that both coins land showing heads is 0.36.

What is the probability that both coins land showing tails? *(2 marks)*

(Q8b, **2010 HSC**) Hard

39 Lou and Ali are on a fitness program for one month. The probability that Lou will finish the program successfully is 0.7 while the probability that Ali will finish successfully is 0.6. The probability tree shows this information.

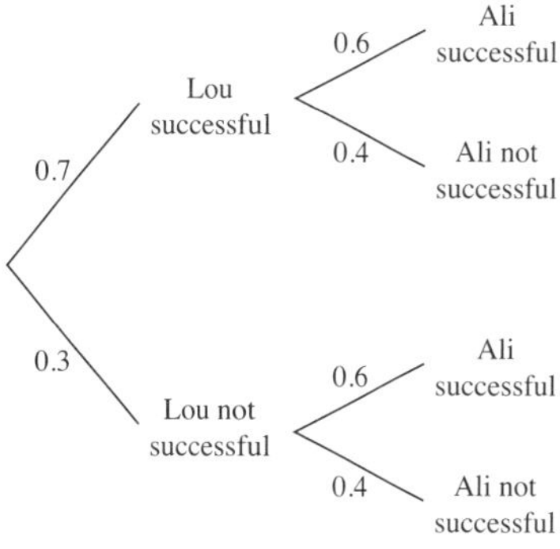

What is the probability that only one of them will be successful?

A 0.18 **B** 0.28
C 0.42 **D** 0.46 *(1 mark)*

(Q20, **2010 MS2 HSC**) Easy

40 On each working day James parks his car in a parking station which has three levels. He parks his car on a randomly chosen level. He always forgets where he has parked, so when he leaves work he chooses a level at random and searches for his car. If his car is not on that level, he chooses a different level and continues in this way until he finds his car.

i What is the probability that his car is on the first level he searches? *(1 mark)* Easy

ii What is the probability that he must search all three levels before he finds his car? *(1 mark)* Hard

iii What is the probability that on every one of the five working days in a week, his car is not on the first level he searches? *(1 mark)* Hard

(Q5b, **2009 HSC**)

41 Each week Van and Marie take part in a raffle at their respective workplaces.

The probability that Van wins a prize in his raffle is $\frac{1}{9}$. The probability that Marie wins a prize in her raffle is $\frac{1}{16}$.

What is the probability that, during the next three weeks, at least one of them wins a prize? *(2 marks)*

(Q9a, **2009 HSC**) Medium

42 In each of three raffles, 100 tickets are sold and one prize is awarded.

Mary buys two tickets in one raffle. Jane buys one ticket in each of the other two raffles.

Determine who has the better chance of winning at least one prize. Justify your response using probability calculations. *(4 marks)*

(Q27c, **2009 MS2 HSC**) Medium

43 In an experiment, two unbiased dice, with faces numbered 1, 2, 3, 4, 5, 6, are rolled 18 times. The difference between the numbers on their uppermost faces is recorded each time. Juan performs this experiment twice and his results are shown in the tables.

Experiment 1

Difference	*Frequency*
0	3
1	3
2	2
3	4
4	3
5	3

Experiment 2

Difference	*Frequency*
0	4
1	4
2	3
3	3
4	2
5	2

Juan states that Experiment 2 has given results that are closer to what he expected than the results given by Experiment 1.

Is he correct? Explain your answer by finding the sample space for the dice differences and using theoretical probability. *(4 marks)*

(Q28d, **2009 MS2 HSC**) **Medium**

44 Xena and Gabrielle compete in a series of games. The series finishes when one player has won two games. In any game, the probability that Xena wins is $\frac{2}{3}$ and the probability that Gabrielle wins is $\frac{1}{3}$.

Part of the tree diagram for this series of games is shown.

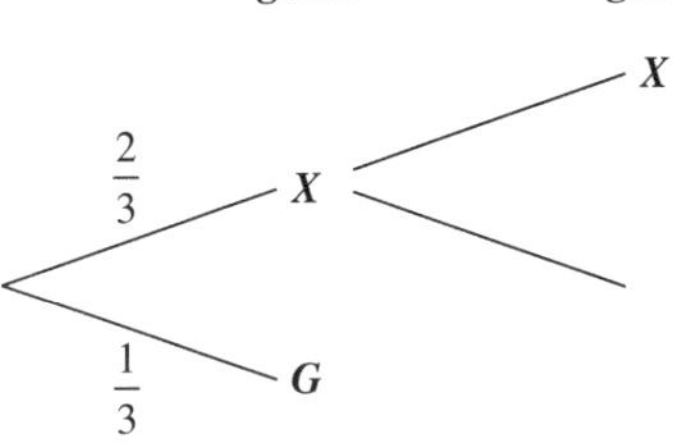

i Copy and complete the tree diagram showing the possible outcomes. *(1 mark)* **Easy**

ii What is the probability that Gabrielle wins the series? *(2 marks)* **Medium**

iii What is the probability that three games are played in the series? *(2 marks)* **Hard**

(Q7c, **2008 HSC**)

45 It is estimated that 85% of students in Australia own a mobile phone.

i Two students are selected at random. What is the probability that neither of them owns a mobile phone? *(2 marks)* **Medium**

ii Based on a recent survey, 20% of the students who own a mobile phone have used their mobile phone during class time. A student is selected at random. What is the probability that the student owns a mobile phone and has used it during class time? *(1 mark)* **Easy**

(Q9a, **2008 HSC**)

46 Two ordinary dice are rolled. The score is the sum of the numbers on the top faces.

i What is the probability that the score is 10? *(2 marks)* **Medium**

ii What is the probability that the score is not 10? *(1 mark)* **Easy**

(Q4b, **2007 HSC**)

47 A pack of 52 cards consists of four suits with 13 cards in each suit.

i One card is drawn from the pack and kept on the table. A second card is drawn and placed beside it on the table. What is the probability that the second card is from a different suit to the first? *(1 mark)* **Medium**

ii The two cards are replaced and the pack shuffled. Four cards are chosen from the pack and placed side by side on the table. What is the probability that these four cards are all from different suits? *(2 marks)* **Medium**

(Q9b, **2007 HSC**)

48 A chessboard has 32 black squares and 32 white squares. Tanya chooses three different squares at random.

i What is the probability that Tanya chooses three white squares? *(2 marks)* **Medium**

ii What is the probability that the three squares Tanya chooses are the same colour? *(1 mark)* **Medium**

iii What is the probability that the three squares Tanya chooses are not the same colour? *(1 mark)* **Easy**

(Q4c, **2006 HSC**)

49 A total of 300 tickets are sold in a raffle which has three prizes. There are 100 red, 100 green and 100 blue tickets.

At the drawing of the raffle, winning tickets are NOT replaced before the next draw.

i What is the probability that each of the three winning tickets is red? *(2 marks)* **Medium**

ii What is the probability that at least one of the winning tickets is not red? *(1 mark)* **Medium**

iii What is the probability that there is one winning ticket of each colour? *(2 marks)* **Medium**

(Q5d, **2005 HSC**)

50 Xuan and Yvette would like to meet at a cafe on Monday. They each agree to come to the cafe sometime between 12 noon and 1 pm, wait for 15 minutes, and then leave if they have not seen the other person.

Their arrival times can be represented by the point (x, y) in the Cartesian plane, where x represents the fraction of an hour after 12 noon that Xuan arrives, and y represents the fraction of an hour after 12 noon that Yvette arrives.

Thus $\left(\frac{1}{3}, \frac{2}{5}\right)$ represents Xuan arriving at 12:20 pm and Yvette arriving at 12:24 pm. Note that the point (x, y) lies somewhere in the unit square $0 \le x \le 1$ and $0 \le y \le 1$ as shown in the diagram.

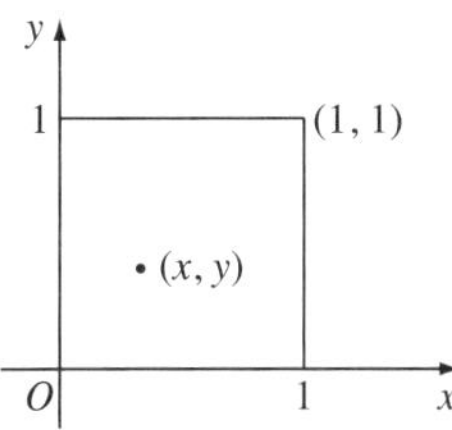

i Explain why Xuan and Yvette will meet if $x - y \le \frac{1}{4}$ or $y - x \le \frac{1}{4}$. *(1 mark)* **Medium**

ii The probability that they will meet is equal to the area of the part of the region given by the inequalities in part **i** that lies within the unit square $0 \le x \le 1$ and $0 \le y \le 1$. Find the probability that they will meet. *(2 marks)* **Medium**

iii Xuan and Yvette agree to try to meet again on Tuesday. They agree to arrive between 12 noon and 1 pm, but on this occasion they agree to wait for t minutes before leaving.
For what value of t do they have a 50% chance of meeting? *(2 marks)* **Hard**

(Q10b, **2005 HSC**)

51 A packet contains 12 red, 8 green, 7 yellow and 3 black jellybeans. One jellybean is selected from the packet at random.

What is the probability that the selected jellybean is red or yellow? *(2 marks)*

(Q1e, **2004 HSC**) **Easy**

52 In a game, a turn involves rolling two dice, each with faces marked 0, 1, 2, 3, 4 and 5. The score for each turn is calculated by multiplying the two numbers uppermost on the dice.

i What is the probability of scoring zero on the first turn? *(2 marks)* **Medium**

ii What is the probability of scoring 16 or more on the first turn? *(1 mark)* **Easy**

iii What is the probability that the sum of the scores in the first two turns is less than 45? *(2 marks)* **Hard**

(Q6c, **2004 HSC**)

53 The diagram shows two spinners which are spun simultaneously.

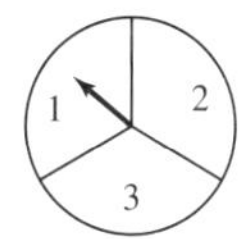

Each of the three outcomes on the first spinner are equally likely, and each of the four outcomes on the second spinner are equally likely.

i What is the probability that both spinners stop on the same number? *(2 marks)* **Medium**

ii What is the probability that at least one of the spinners stops on a 3? *(1 mark)* **Medium**

(Q4b, **2003 HSC**)

54 Chris has four pairs of socks in a drawer, each pair a different colour.

He selects socks one at a time and at random from the drawer.

i The probability that he does NOT have a matching pair after selecting the second sock is $\frac{6}{7}$. Explain why this is so. *(1 mark)* **Medium**

ii Find the probability that he does NOT have a matching pair after selecting the third sock. *(2 marks)* **Medium**

iii What is the probability that the first three socks include a matching pair? *(1 mark)* **Medium**

(Q7c, **2002 HSC**)

55 Rob, Alex and Tan plan a swimming race against each other. Rob and Alex are each twice as likely as Tan to win the race.

What is the probability that Tan will win the race?

A $\frac{1}{6}$ **B** $\frac{1}{5}$

C $\frac{1}{4}$ **D** $\frac{1}{3}$ *(1 mark)*

(Q20, **2002 MS2 HSC**) Medium

56 Onslo tries to connect to his internet service provider. The probability that he connects on any single attempt is 0.75.

i What is the probability that he connects for the first time on his second attempt? *(2 marks)* Easy

ii What is the probability that he is still not connected after his third attempt? *(1 mark)* Easy

(Q7b, **2001 HSC**)

57 Five candidates, A, B, C, D and E, are standing for an election. Their names are written on pieces of cardboard that are placed in a barrel and are drawn out randomly to determine their positions on the ballot paper.

i What is the probability that A is drawn first? *(1 mark)* Easy

ii What is the probability that the order of the names on the ballot paper is that shown below? *(2 marks)* Medium

A
B
C
D
E

(Q8b, **2001 HSC**)

58 Sonia has written letters to four of her friends and sealed the letters in envelopes. Now she does not know which envelope contains which letter.

If Sonia addresses the envelopes to her four friends at random, what is the probability that each envelope contains the correct letter?

A $\frac{1}{256}$ **B** $\frac{1}{24}$

C $\frac{1}{16}$ **D** $\frac{1}{4}$ *(1 mark)*

(Q22, **2001 MS2 HSC**) Medium

59 A bag contains red marbles and blue marbles in the ratio 2:3.

A marble is selected at random.

What is the probability that the marble is blue? *(1 mark)*

(Q1d, **2000 HSC**) Easy

60 Four white (W) balls and two red (R) balls are placed in a bag. One ball is selected at random, removed and replaced by a ball of the other colour. The bag is then shaken and another ball is randomly selected.

i Copy the tree diagram into your Writing Booklet. Complete the tree diagram, showing the probability on each branch. Easy

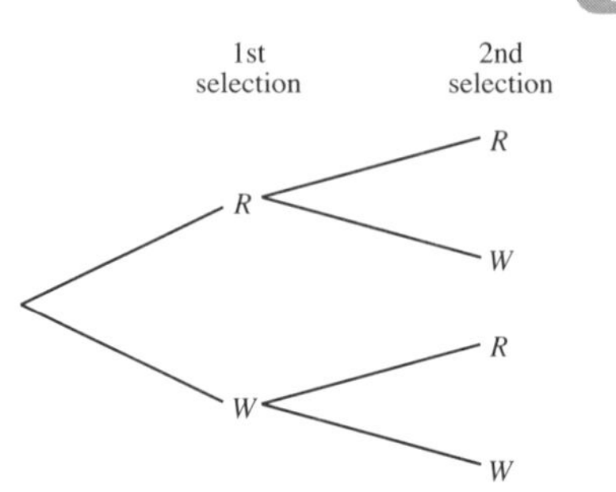

ii Find the probability that both balls selected are white. Easy

iii Find the probability that the second ball selected is white. Medium

(5 marks)

(Q5b, **2000 HSC**)

61 A box contains five cards. Each card is labelled with a number. The numbers on the cards are 0, 3, 3, 5, 5.

Cameron draws one card at random from the box and then draws a second card at random *without* replacing the first card drawn.

i What is the probability that he draws a '5', then a '3'? Easy

ii What is the probability that the sum of the two numbers drawn is at least 8? Easy

iii What is the probability that the second card drawn is labelled '3'? Medium

(5 marks)

(Q8b, **1999 HSC**)

62 A coin is tossed three times. What is the probability that 'heads' appears every time? *(2 marks)*

(Q1c, **1998 HSC**) Easy

63 A game is played in which two coloured dice are thrown once. The six faces of the blue die are numbered 4, 6, 8, 9, 10 and 12. The six faces of the pink die are numbered 2, 3, 5, 7, 11 and 13. The player wins if the number on the pink die is larger than the number on the blue die.

i By drawing up a table of possible outcomes, or otherwise, calculate the probability of the player winning a game.

ii Calculate the probability that the player wins at least once in two successive games. *(5 marks)*

(Q10a, **1998 HSC**) **Medium**

64 Forty-five balls, numbered 1 to 45, are placed in a barrel, and one ball is drawn at random.

What is the probability that the number on the ball drawn is even? *(2 marks)*

(Q1f, **1997 HSC**) **Easy**

65 A bag contains two red balls, one black ball, and one white ball. Andrew selects one ball from the bag and keeps it hidden. He then selects a second ball, also keeping it hidden.

i Draw a tree diagram to show all the possible outcomes. **Easy**

ii Find the probability that both the selected balls are red. **Easy**

iii Find the probability that at least one of the selected balls is red. **Medium**

iv Andrew drops one of the selected balls and we can see that it is red. What is the probability that the ball that is still hidden is also red? **Medium**

(5 marks)

(Q9a, **1997 HSC**)

66 Students studying at least one of the languages, French and Japanese, attend a meeting. Of the 28 students present, 18 study French and 22 study Japanese.

i What is the probability that a randomly chosen student studies French? **Easy**

ii What is the probability that two randomly chosen students both study French? **Medium**

iii What is the probability that a randomly chosen student studies both languages? **Easy**

(4 marks)

(Q8a, **1996 HSC**)

67 A factory assembles torches. Each torch requires one battery and one bulb. It is known that 6% of all batteries and 4% of all bulbs are defective.

Find the probability that, in a torch selected at random, both the battery and the bulb are NOT defective. Give your answer in exact form. *(2 marks)*

(Q7a, **1995 HSC**) **Medium**

68 Greg and Jack are playing in a golf tournament. They will play two rounds and each has an equal chance of winning the first round. If Greg wins the first round, his probability of winning the second round is increased to 0·6.

If Greg loses the first round, his probability of winning the second round is reduced to 0·3.

i Draw a tree diagram for the two-round sequence. Label each branch of the diagram with the appropriate probability. **Medium**

ii Find the probability that Greg wins exactly one round. **Easy**

(4 marks)

(Q8a, **1995 HSC**)

69 'A bag contains green, black, and red jellybeans. Therefore, if I choose one jellybean at random from the bag, the probability that it is black is $\frac{1}{3}$.'

Is this statement true or false? Explain why, in no more than one sentence.

(Q6a, **1994 HSC**) **Easy**

70 The die used in a new game has 20 faces. Each face has a different letter of the alphabet marked on it, however the letters Q, U, V, X, Y, and Z have not been used.

i The die is rolled twice. What is the probability that the same letter appears on the upper face twice?

ii The die is rolled three times. What is the probability that the letter E appears on the upper face exactly twice?

(Q7c, **1993 HSC**) **Medium**

71 Pat and Chris each throw a die.

i Find the probability that they throw the same number. **Easy**

ii Find the probability that the number thrown by Chris is greater than the number thrown by Pat. **Medium**

(Q4b, **1992 HSC**)

72 A box contains twelve chocolates all of exactly the same appearance. Four of the chocolates are hard and eight are soft. Kim eats three chocolates chosen randomly from the box. Using a tree diagram, or otherwise, find the probability that:

i the first chocolate Kim eats is hard; **Easy**

ii Kim eats three hard chocolates; **Easy**

iii Kim eats exactly one hard chocolate. **Medium**

(Q8b, **1992 HSC**)

Year 11 Probability and Venn diagrams—Worked answers

1 $P(\text{win at least 1 game}) = 0.8$

$P(\text{lose both games}) = P(LL)$
$= 1 - 0.8$
$= 0.2$

$\therefore P(L) = \sqrt{0.2}$

$\therefore P(W) = 1 - \sqrt{0.2}$
$= 0.5527$ (4 dec. pl.)

$P(WW) = 0.5527 \times 0.5527$
$= 0.31$ (2 dec. pl.)

$\therefore$ the probability of winning both games is about 31%.

Answer A

2 **a** $P(\text{choosing normal dice}) = \frac{2}{3}$

$P(\text{special dice}) = \frac{1}{3}$

$P(\text{choosing a 5})$

$= \frac{2}{3} \times \frac{1}{6} + \frac{1}{3} \times \frac{1}{2}$

$= \frac{5}{18}$ ✓ *(1 mark)*

b $P(\text{special dice}|\text{shows 5})$

$= \dfrac{P(\text{special dice} \cap \text{shows 5})}{P(\text{shows 5})}$

$= \dfrac{\frac{1}{3} \times \frac{1}{2}}{\frac{5}{18}}$

$= \frac{3}{5}$ ✓ *(1 mark)*

3

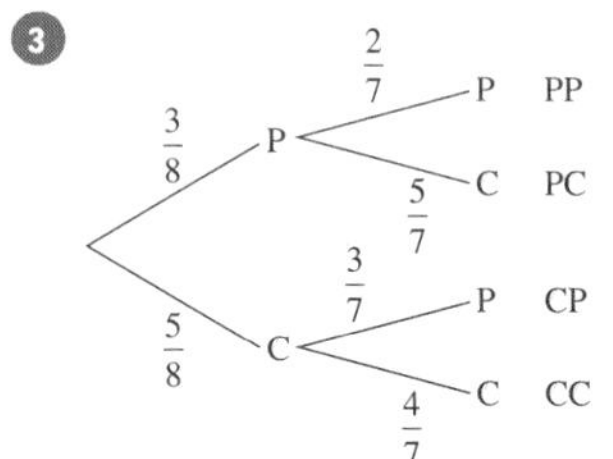

$P(\text{different centres})$

$= P(\text{PC}) + P(\text{CP})$

$= \frac{3}{8} \times \frac{5}{7} + \frac{5}{8} \times \frac{3}{7}$

$= \frac{15}{28}$

Answer D

4 **a** Let H = History, and G = Geography.

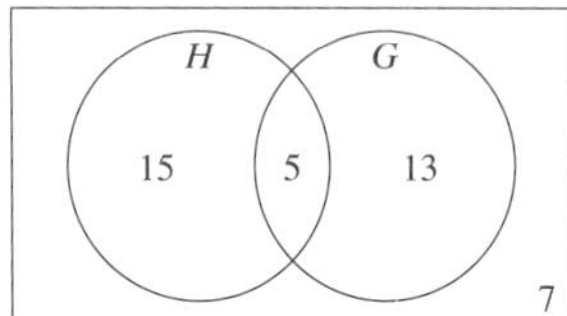

As $40 - 7 = 33$, then 33 students study History, or Geography, or both.

As $20 + 18 - 33 = 5$, then 5 students study both. ✓

$P(\text{student studies both})$

$= P(H \cap G)$

$= \frac{5}{40}$

$= \frac{1}{8}$ ✓ *(2 marks)*

b $P(\text{Not History}|\text{Geography})$

$= P(\bar{H}|G)$

$= \dfrac{P(\bar{H} \cap G)}{P(G)}$

$= \dfrac{\frac{13}{40}}{\frac{18}{40}}$

$= \frac{13}{18}$ ✓ *(1 mark)*

c $P(\text{History then not History})$

$= P(H) \times P(\bar{H}|H)$

$= \frac{20}{40} \times \frac{20}{39}$ ✓

$= \frac{10}{39}$ ✓ *(2 marks)*

5 Ways of getting a score of 6:

- rolling 6 in section 1
- rolling 3 in section 2
- rolling 2 in section 3
- rolling 1 in section 6.

So required probability

$= 4 \times \frac{1}{6} \times \frac{1}{8}$

$= \frac{1}{12}$

Answer B

6 **i** $\text{P}(A|B) = \dfrac{\text{P}(A \cap B)}{\text{P}(B)}$

$\frac{2}{3} = \dfrac{\text{P}(A \cap B)}{\frac{1}{4}}$

$\text{P}(A \cap B) = \frac{2}{3} \times \frac{1}{4}$

$= \frac{1}{6}$ ✓ *(1 mark)*

ii $\text{P}(A \cap B) + \text{P}(A' \cap B) = \text{P}(B)$

$\frac{1}{6} + \text{P}(A' \cap B) = \frac{1}{4}$

$\text{P}(A' \cap B) = \frac{1}{4} - \frac{1}{6}$

$= \frac{1}{12}$ ✓ *(1 mark)*

iii If A and B are independent, then $\text{P}(A|B) = \text{P}(A) = \frac{2}{3}$.

$\text{P}(A \cup B)$

$= \text{P}(A) + \text{P}(B) - \text{P}(A \cap B)$

$= \frac{2}{3} + \frac{1}{4} - \frac{1}{6}$

$= \frac{3}{4}$ ✓ *(1 mark)*

7 **i** $P(A|B) = \frac{1}{4}, P(A \cap B) = p$

$P(A|B) = \dfrac{P(A \cap B)}{P(B)}$

$\frac{1}{4} = \dfrac{p}{P(B)}$

$P(B) = 4p$ ✓ *(1 mark)*

ii $P(B|A) = \frac{1}{3}, P(A \cap B) = p$

$P(B|A) = \dfrac{P(B \cap A)}{P(A)}$

$\frac{1}{3} = \dfrac{p}{P(A)}$

$P(A) = 3p$ ✓

Now,

$P(B' \cap A')$

$= P((B \cup A)')$

$= 1 - P(B \cup A)$

$= 1 - [P(B) + P(A) - P(B \cap A)]$

$= 1 - [4p + 3p - p]$

$= 1 - 6p$ ✓ *(2 marks)*

8 $P(A \cap B) = p, P(A' \cap B) = 2p - \frac{1}{4}, P(A \cap B') = \frac{p}{3}.$

$P(A) = \text{P}(A \cap B) + \text{P}(A \cap B')$

$= p + \frac{p}{3}$

$= \frac{4p}{3}$ ✓

$P(B) = P(B \cap A) + P(B \cap A')$

$= p + 2p - \frac{1}{4}$

$= 3p - \frac{1}{4}$ ✓

If A and B are independent then $P(A).P(B) = \text{P}(A \cap B)$

$\frac{4p}{3}(3p - \frac{1}{4}) = p$

$12p^2 - p = 3p$

$12p^2 - 4p = 0$

$4p(3p - 1) = 0$

$p = \frac{1}{3}$ (as $p > \frac{1}{8}$) ✓ *(3 marks)*

9 $P(A \cup B)$

$= P(A) + P(B) - P(A \cap B)$

$= P(A) + P(B) - P(A).P(B)$

$\frac{5}{8} = P(A) + 2P(A) - P(A).2P(A)$ ✓

$\frac{5}{8} = 3P(A) - 2[P(A)]^2$

$2[P(A)]^2 - 3P(A) + \frac{5}{8} = 0$

Let $m = P(A)$:

$2m^2 - 3m + \frac{5}{8} = 0$

$16m^2 - 24m + 5 = 0$ ✓

$(4m-5)(4m-1)=0$

$m=\frac{5}{4},\frac{1}{4}$

$\therefore P(A)=\frac{1}{4}$ (as $0 \le P(A) \le 1$) ✓

(3 marks)

10 $P(B'|A)=\frac{P(B' \cap A)}{P(A)}$

$=\frac{P(B' \cap A)}{P(B' \cap A)+P(B \cap A)}$

$=\frac{\frac{3}{8}}{\frac{3}{8}+\frac{2}{5}}$ ✓

$=\frac{15}{31}$ ✓ *(2 marks)*

11 As $P(X>2)=c$, then $P(X \le 2)=1-c$, and similarly, $P(X \le 5)=1-d$.

$P(X \le 2|X \le 5)$

$=\frac{P(X \le 2 \cap X \le 5)}{P(X \le 5)}$

$=\frac{P(X \le 2)}{P(X \le 5)}$

$=\frac{1-c}{1-d}$

Answer C

12 As $P(B|A)=\frac{P(B \cap A)}{P(A)}$

$\sqrt{p}=\frac{P(B \cap A)}{p}$

$P(B \cap A)=p^{\frac{3}{2}}$

As $P(A|B)=\frac{P(A \cap B)}{P(B)}$

$=\frac{p^{\frac{3}{2}}}{p^2}$

$=p^{-\frac{1}{2}}$

$=\frac{1}{\sqrt{p}}$

Answer D

13 **i** $P(A')=\frac{5}{7}$ and $P(B)=\frac{3}{5}$

As $P(A \cup B)$

$=P(A)+P(B)-P(A \cap B)$

then, $P(A \cap B)$

$=P(A)+P(B)-P(A \cup B)$

Also, $P(B)$

$=P(A \cap B)+P(A' \cap B)$ ✓

Hence, $P(A' \cap B)$

$=P(B)-P(A \cap B)$

$P(A' \cap B)=P(B)-[P(A)+P(B)-P(A \cup B)]$

$=P(A \cup B)-P(A)$

$=\frac{4}{5}-\frac{2}{7}$

$=\frac{18}{35}$ ✓

(2 marks)

ii If A and B are mutually exclusive then $P(A \cap B)=0$.

$P(A \cap B')=P(A)$

$=\frac{2}{7}$ ✓

(1 mark)

14 $P(A \cup B)$

$=P(A)+P(B)-P(A \cap B)$

P(winning) = P(even number) + P(tail) − P(even number and tail)

$=\frac{1}{2}+\frac{1}{2}-\frac{1}{2}\times\frac{1}{2}$

$=\frac{3}{4}$

Alternative solution:

P(winning) = 1 − P(losing)

= 1 − P(odd number and heads)

$=1-\frac{1}{2}\times\frac{1}{2}$

$=\frac{3}{4}$

Answer C

15 P(same colour)

= P(GG) + P(PP) ✓

$=\frac{5}{12}\times\frac{4}{11}+\frac{7}{12}\times\frac{6}{11}$

$=\frac{31}{66}$ ✓

(2 marks)

16 **i** P(red hair) = 0.02

∴ P(not red hair) = 0.98 ✓

P(at least one red hair)

= 1 − P(both not red hair)

$=1-0.98^2$

= 0.0396 ✓

(2 marks)

ii $1-0.98^n>0.4$

$0.98^n<0.6$

$n \ln 0.98<\ln 0.6$

$n>\frac{\ln 0.6}{\ln 0.98}$ ✓

(as $\ln 0.98<0$)

$n>25.28500838\ldots$

∴ smallest number of people is 26. ✓

(2 marks)

17 $P(\text{spinning } 8)=\frac{1}{37}$ ✓

Expected frequency of 8

$=\frac{1}{37}\times 18500$

$=500$ ✓

(2 marks)

18 P(same colour)

$=P(RR)+P(GG)$ ✓

$=\frac{9}{17}\times\frac{8}{16}+\frac{8}{17}\times\frac{7}{16}$ ✓

$=\frac{8}{17}$ ✓

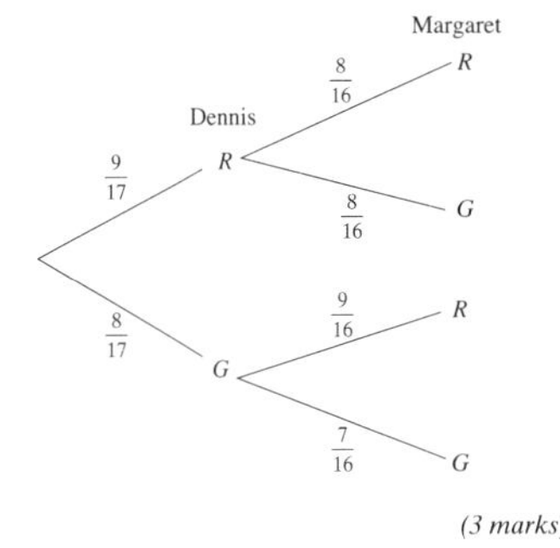

(3 marks)

19 Once a shoe is chosen, the probability that the next shoe matches it is 1 shoe out of the remaining 7 shoes.

P(matching pair) $=\frac{1}{7}$

Answer C

20 **i** P(pen from A not faulty) = 0.9, and P(pen from B not faulty) = 0.95

P(at least one faulty)

= 1 − P(none faulty)

= 1 − P(A not faulty and B not faulty)

$=1-(0.9\times 0.95)$

= 0.145 ✓

(1 mark)

ii P(neither faulty)

= P(A chosen and neither faulty) + P(B chosen and neither faulty)

$=0.5\times 0.9^2+0.5\times 0.95^2$ ✓

= 0.856 25 ✓

(2 marks)

21 **i** The table shows the number of results on the third die that lie between the results on the first two dice. The dash (–) means no chance of winning before rolling the third die.

		First die					
Second die		1	2	3	4	5	6
	1	–	–	1	2	3	4
	2	–	–	–	1	2	3
	3	1	–	–	–	1	2
	4	2	1	–	–	–	1
	5	3	2	1	–	–	–
	6	4	3	2	1	–	–

P(no chance of winning before rolling the third die) ✓

$=\frac{16}{36}$

$=\frac{4}{9}$ ✓

(2 marks)

ii Total possible results on third die that produce win = 40 ✓

P(player wins the game)

$= \frac{40}{6^3}$

$= \frac{5}{27}$ ✓ *(2 marks)*

22 Total frequency

$= 23 + 19 + 48 + 20 + 21 + 19$

$= 150$

Relative frequency of 3

$= \frac{48}{150}$

$= \frac{8}{25}$ (or 0.32) ✓ *(1 mark)*

23 **i** P(even number on first spin)

$= \frac{2}{5}$ ✓ *(1 mark)*

ii P(at least one even number)

$= 1 - \text{P(no even numbers)}$

$= 1 - \frac{3}{5} \times \frac{3}{5} \times \frac{3}{5}$

$= \frac{98}{125}$ ✓ *(1 mark)*

iii P(even, then odd, then odd)

$= \frac{2}{5} \times \frac{3}{5} \times \frac{3}{5}$

$= \frac{18}{125}$ ✓ *(1 mark)*

iv P(even, then odd, then odd) + P(odd, then even, then odd) + P(odd, then odd, then even)

$= 3 \times \frac{2}{5} \times \frac{3}{5} \times \frac{3}{5}$

$= \frac{54}{125}$ ✓ *(1 mark)*

24 *P*(exactly one picture card)

$= P(\text{picture, not picture}) + P(\text{not picture, picture})$

$= \frac{12}{52} \times \frac{40}{51} + \frac{40}{52} \times \frac{12}{51}$

$= \frac{80}{221}$

$= 0.361\,990\,95\ldots$

$= 0.3620$ (4 d.p.)

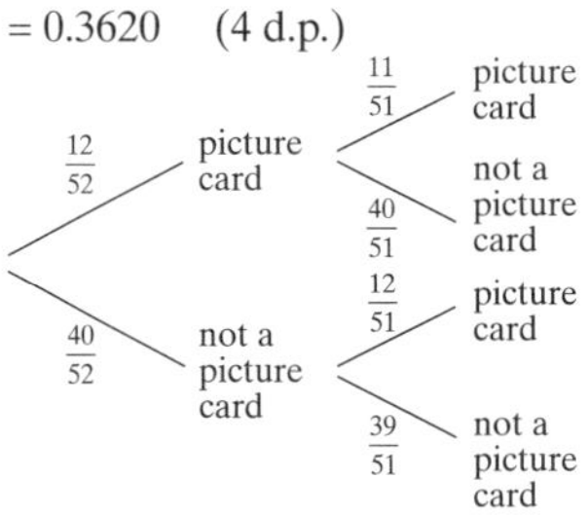

$\frac{12}{52}$ picture card — $\frac{11}{51}$ picture card; $\frac{40}{51}$ not a picture card

$\frac{40}{52}$ not a picture card — $\frac{12}{51}$ picture card; $\frac{39}{51}$ not a picture card

Answer D

25 $\text{P(wins)} = \frac{6}{30}$

$= \frac{1}{5}$

Answer C

26 P(Soccer team does not win)

$= 1 - \frac{5}{7} = \frac{2}{7}$

P(Rugby League team does not win)

$= 1 - \frac{2}{3} = \frac{1}{3}$

P(neither team wins)

$= \frac{2}{7} \times \frac{1}{3}$

$= \frac{2}{21}$

Answer A

27 **i** P(Sat. dry)

$= P(WD) + P(DD)$

$= \frac{1}{2} \times \frac{5}{6} + \frac{1}{2} \times \frac{1}{2}$

$= \frac{2}{3}$ ✓ *(1 mark)*

ii P(Sat. and Sun. wet)

$= P(WWW) + P(DWW)$ ✓

$= \frac{1}{2} \times \frac{1}{6} \times \frac{1}{6} + \frac{1}{2} \times \frac{1}{2} \times \frac{1}{6}$

$= \frac{1}{18}$

∴ the probability that Saturday and Sunday are both wet is $\frac{1}{18}$. ✓ *(2 marks)*

iii P (at least one of Sat. and Sun. dry)

$= 1 - \text{P(Sat. and Sun. wet)}$

$= 1 - \frac{1}{18}$ (from part **ii**)

$= \frac{17}{18}$

∴ the probability that at least one of Saturday and Sunday being dry is $\frac{17}{18}$. ✓ *(1 mark)*

28 P(at least one is under 10 s)

$= 1 - \text{P (none under 10 s)}$

$= 1 - \left[\frac{3}{4} \times \frac{5}{6} \times \frac{3}{5}\right]$

$= \frac{5}{8}$

Answer D

29 **i**

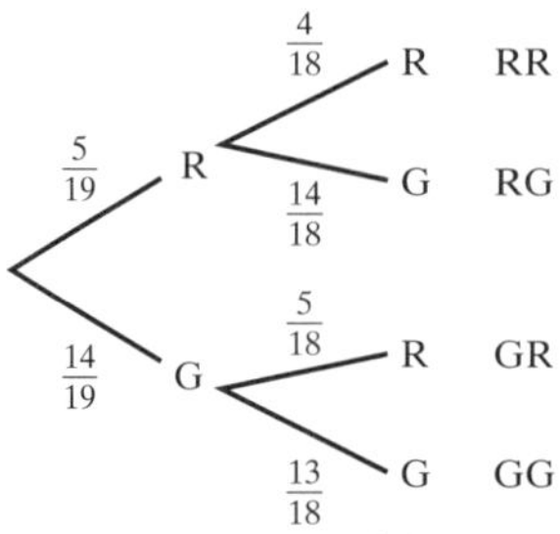

$\frac{5}{19}$ R — $\frac{4}{18}$ R RR; $\frac{14}{18}$ G RG

$\frac{14}{19}$ G — $\frac{5}{18}$ R GR; $\frac{13}{18}$ G GG

✓✓ *(2 marks)*

ii P(different colours)

$= P(RG) + P(GR)$

$= \frac{5}{19} \times \frac{14}{18} + \frac{14}{19} \times \frac{5}{18}$

$= \frac{70}{171}$ ✓ *(1 mark)*

30 Based on averages, the probability of rain on any given day in October is $\frac{8}{31}$.

Assuming the events are independent, the probability of rain on two consecutive days could be given by $\frac{8}{31} \times \frac{8}{31}$.

Answer D

31 P(at least one is red)

$= 1 + \text{P(none red)}$

$= 1 + \frac{6}{10} \times \frac{5}{9} \times \frac{4}{8}$

$= \frac{5}{6}$

Answer C

32 Number of outcomes $= 6 \times 6$

$= 36$

Outcomes of a sum of 6 are $1 + 5, 2 + 4, 3 + 3, 4 + 2$, and $5 + 1$

So there are 5 favourable outcomes.

$\text{P(sum of 6)} = \frac{5}{36}$

Answer D

33 **i** $\text{P(RR)} = \frac{3}{5} \times \frac{3}{7}$

$= \frac{9}{35}$ ✓ *(1 mark)*

ii P(at least one white)

$= 1 - \text{P(RR)}$

$= 1 - \frac{9}{35}$

$= \frac{26}{35}$ ✓ *(1 mark)*

iii P(both same colour)

$= \text{P(RR)} + \text{P(WW)}$

$= \frac{3}{5} \times \frac{3}{7} + \frac{2}{5} \times \frac{4}{7}$ ✓

$= \frac{17}{35}$ ✓ *(2 marks)*

34 On one die:

$P(\text{rolling a 6}) = \frac{1}{6}$

$P(\text{not rolling a 6}) = \frac{5}{6}$

On two dice:

$P(\text{neither showing 6}) = \frac{5}{6} \times \frac{5}{6}$

$= \frac{25}{36}$

$P(\text{at least one 6}) = 1 - \frac{25}{36}$

$= \frac{11}{36}$

Answer B

35 Defective = 0.02×800

$= 16$

∴ there are 16 items defective. ✓

(1 mark)

36 **i** $P(\text{red}) = \frac{3}{5}$ ✓

(1 mark)

ii P(same colour)

$= P(\text{red, red, red})$

$= \frac{3}{5} \times \frac{2}{4} \times \frac{1}{3}$

$= \frac{1}{10}$ ✓

(1 mark)

iii P(does not wear shirt of same colour over consecutive days)

$= P(\text{RYR or YRY})$ ✓

$= \frac{3}{5} \times \frac{2}{4} \times \frac{2}{3} + \frac{2}{5} \times \frac{3}{4} \times \frac{1}{3}$

$= \frac{3}{10}$ ✓

(2 marks)

37 4M, 4C, 4S

i $P(\text{MM}) = \frac{4}{12} \times \frac{3}{11}$

$= \frac{1}{11}$ ✓

(1 mark)

ii P(MM or CC or SS)

$= \frac{1}{11} + \frac{1}{11} + \frac{1}{11}$

$= \frac{3}{11}$ ✓

(1 mark)

iii Method 1:

$P(\text{different}) = 1 - P(\text{same})$

$= 1 - \frac{3}{11}$

$= \frac{8}{11}$

Method 2:

P(MC or MS or CM or CS or SM or SC)

$= 6 \times \frac{4}{12} \times \frac{4}{11}$

$= \frac{8}{11}$ ✓

(1 mark)

38 $P(\text{both heads}) = 0.36$

$\therefore P(\text{head}) = \sqrt{0.36}$

$= 0.6$ ✓

$\therefore P(\text{tail}) = 1 - P(\text{head})$

$= 1 - 0.6$

$= 0.4$

$\therefore P(\text{both tails}) = 0.4 \times 0.4$

$= 0.16$

(2 marks)

39 *P*(only one successful)

= *P*(Lou is successful, Ali is not) + *P*(Lou is not successful, Ali is)

$= 0.7 \times 0.4 + 0.3 \times 0.6$

$= 0.28 + 0.18$

$= 0.46$

Answer D

40 **i** P(Correct level chosen)

$= \frac{1}{3}$ ✓

(1 mark)

ii P(All levels searched)

$= \frac{2}{3} \times \frac{1}{2} \times 1$

$= \frac{1}{3}$ ✓

(1 mark)

iii P(Not on first level)

$= 1 - \frac{1}{3}$

$= \frac{2}{3}$

∴ P(Not on first level each day)

$= \left(\frac{2}{3}\right)^5$

$= \frac{32}{243}$ ✓

(1 mark)

41

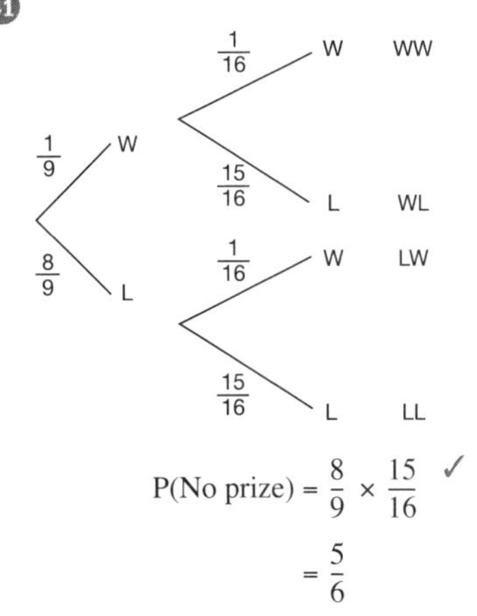

$P(\text{No prize}) = \frac{8}{9} \times \frac{15}{16}$ ✓

$= \frac{5}{6}$

$\therefore P(\text{No prize over 3 weeks}) = \left(\frac{5}{6}\right)^3$

$= \frac{125}{216}$

$\therefore P(\text{At least one prize}) = 1 - \frac{125}{216}$

$= \frac{91}{216}$ ✓

(2 marks)

42

2 Mary
98 Other

RAFFLE A

$P(\text{Mary wins a prize}) = \frac{2}{100}$

$= \frac{1}{50}$

$= 0.02$ ✓

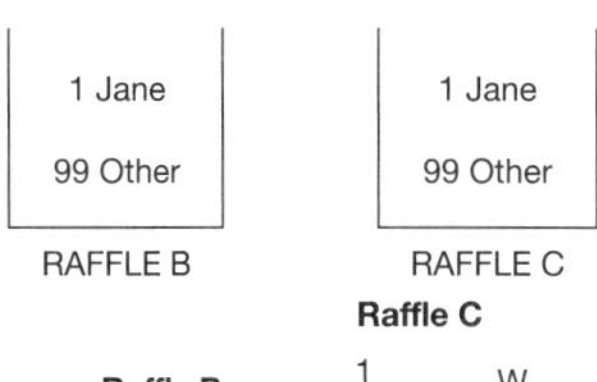

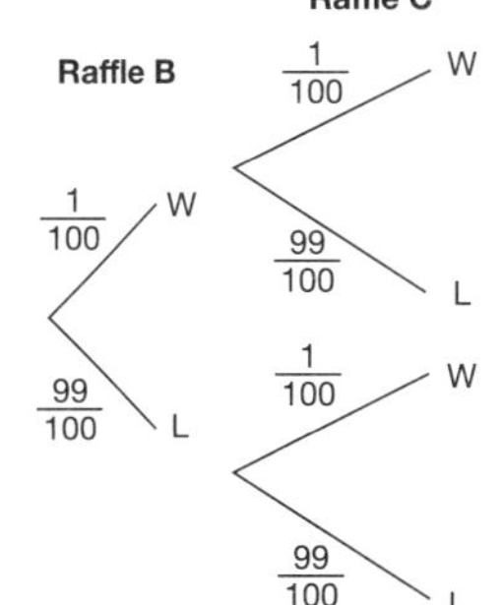

✓

P(Jane wins at least 1 prize)

= P(wins B, not C) or P(not B, wins C) or P(wins both)

$= \frac{1}{100} \times \frac{99}{100} + \frac{99}{100} \times \frac{1}{100} + \frac{1}{100} \times \frac{1}{100}$

$= 0.0199$ ✓

∴ Mary has a better chance of winning a prize by buying 2 tickets in the one raffle. ✓

(4 marks)

43 The sample space for the dice differences is:

		Die 1					
		1	2	3	4	5	6
	1	0	1	2	3	4	5
	2	1	0	1	2	3	4
Die 2	3	2	1	0	1	2	3
	4	3	2	1	0	1	2
	5	4	3	2	1	0	1
	6	5	4	3	2	1	0

✓

Theoretically the results should be

Difference	*Frequency*
0	6
1	10
2	8
3	6
4	4
5	2
	36 ✓

If the dice are rolled 18 times, you would expect half these frequencies:

Difference	*Frequency*
0	3
1	5
2	4
3	3
4	2
5	1 ✓

∴ Experiment 2 has given results closer to what is expected and Juan is correct. ✓

(4 marks)

44 **i**

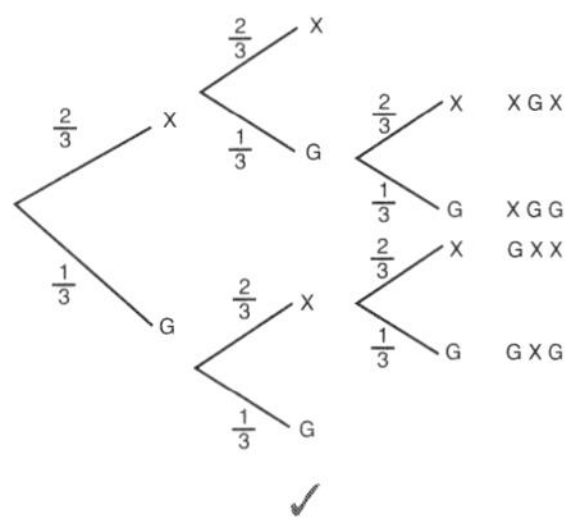

✓

(1 mark)

ii P(Gabrielle wins series)

$= P(XGG) + P(GXG) + P(GG)$

$= P\left(\frac{2}{3} \times \frac{1}{3} \times \frac{1}{3}\right) + P\left(\frac{1}{3} \times \frac{2}{3} \times \frac{1}{3}\right) + P\left(\frac{1}{3} \times \frac{1}{3}\right)$ ✓

$= \frac{2}{27} + \frac{2}{27} + \frac{1}{9}$

$= \frac{7}{27}$ ✓

(2 marks)

iii P(three games are played)

$= 1 - P(XX) - P(GG)$

$= 1 - P\left(\frac{2}{3} \times \frac{2}{3}\right) - P\left(\frac{1}{3} \times \frac{1}{3}\right)$ ✓

$= 1 - \frac{4}{9} - \frac{1}{9}$

$= \frac{4}{9}$ ✓

(2 marks)

45 **i**

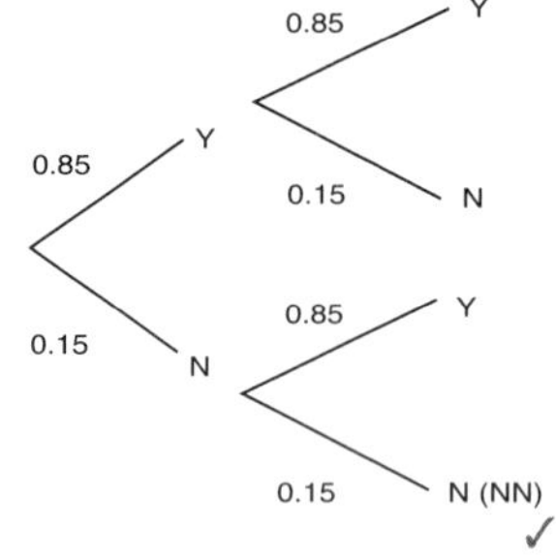

✓

P(no phone) = P(NN)

$= 0.15 \times 0.15$

$= 0.0225$

$= 2.25\%$ ✓

(2 marks)

ii

0.85 Y — 0.2 On (Y, On); 0.8 Off

0.15 N

Required probability

= P(Y, On)

$= 0.85 \times 0.20$

$= 0.17$

$= 17\%$ ✓

(1 mark)

46 **i**

Die 1

Die 2	1	2	3	4	5	6
1	2	3	4	5	6	7
2	3	4	5	6	7	8
3	4	5	6	7	8	9
4	5	6	7	8	9	10*
5	6	7	8	9	10*	11
6	7	8	9	10*	11	12

✓

$P(\text{score} = 10) = \frac{3}{36} = \frac{1}{12}$ ✓

(2 marks)

ii $P(\text{score} \neq 10)$

$= 1 - P(\text{score} = 10)$

$= 1 - \frac{1}{12}$

$= \frac{11}{12}$ ✓

(1 mark)

47 **i** $P = \frac{52}{52} \times \frac{39}{51}$

$= \frac{13}{17}$

$= 0.7647\ldots$

$= 0.765$ to 3 decimal places ✓

(1 mark)

ii $P = \frac{52}{52} \times \frac{39}{51} \times \frac{26}{50} \times \frac{13}{49}$ ✓

$= \frac{13\,182}{124\,950}$

$= 0.1054\ldots$

$= 0.105$ to 3 decimal places ✓

(2 marks)

48 **i** P(three white squares)

= P(WWW)

$= \frac{32}{64} \times \frac{31}{63} \times \frac{30}{62}$ ✓

$= \frac{29\,760}{249\,984}$

$= \frac{5}{42}$

$= 0.11904\ldots$

$\doteqdot 0.1190.$ ✓ *(2 marks)*

ii P(three squares are the same colour)

= P(WWW) + P(BBB)

$= \frac{29\,760}{249\,984} + \frac{29\,760}{249\,984}$

$= \frac{5}{42} + \frac{5}{42}$

$= \frac{5}{21}$

$= 0.23809\ldots$

$\doteqdot 0.2381$ ✓ *(1 mark)*

iii P(three squares not the same colour)

= 1 – (P(WWW) + P(BBB))

$= 1 - \frac{5}{21}$

$= \frac{16}{21}$

$= 0.76190\ldots$

$\doteqdot 0.7619$ ✓

(1 mark)

49 **i** P(each ticket red)

= P(RRR)

$= \frac{100}{300} \times \frac{99}{299} \times \frac{98}{298}$ ✓

$= \frac{1617}{44\,551}$

$= 0.03629\ldots$

$\doteqdot 0.0363$ ✓

(2 marks)

ii P(at least 1 not red)

= 1 – P(RRR)

$= 1 - \frac{1617}{44\,551}$

$= 0.96370\ldots$

$\doteqdot 0.9637$ ✓ *(1 mark)*

iii The following outcomes contain one ticket of each colour:

RGB GRB BRG
RBG GBR BGR

∴ There are 6 favourable outcomes.

∴ P(One ticket of each colour)

$= 6 \times \frac{100}{300} \times \frac{100}{299} \times \frac{100}{298}$ ✓

$= 0.22446\ldots$

$\doteqdot 0.2245$ ✓

(2 marks)

50 **i** Both Xuan and Yvette wait for a quarter of an hour so if the difference between their arrival times ($x - y$ or $y - x$) is less than or equal to a quarter of an hour, then they will both be there at the same time and hence meet. ✓

(1 mark)

ii *Sketch* $x - y = \frac{1}{4}$

i.e. $y = x - \frac{1}{4}$

When $y = 0, x = \frac{1}{4}$

$\therefore$ x intercept of $x - y = \frac{1}{4}$ is $\left(\frac{1}{4}, 0\right)$.

Also when $y = 1, x = \frac{3}{4}$.

Sketch $y - x = \frac{1}{4}$

i.e. $y = \frac{1}{4} + x$

When $x = 0, y = \frac{1}{4}$

$\therefore$ y intercept of $y - x = \frac{1}{4}$ is $\left(0, \frac{1}{4}\right)$.

Also when $x = 1, y = \frac{3}{4}$.

Determine the region where each inequality holds:

Test $(0, 0)$ in $x - y \leqslant \frac{1}{4}$:

$0 - 0 \leqslant \frac{1}{4}$

$\therefore$ Region is above the line $x - y = 4$.

Test $(0, 0)$ in $y - x \leqslant \frac{1}{4}$:

$0 - 0 \leqslant \frac{1}{4}$

$\therefore$ Region is below the line $y - x = \frac{1}{4}$.

Also, $0 \leqslant x \leqslant 1$ and $0 \leqslant y \leqslant 1$.

$\therefore$ Shaded region is region where all 4 inequalities hold.

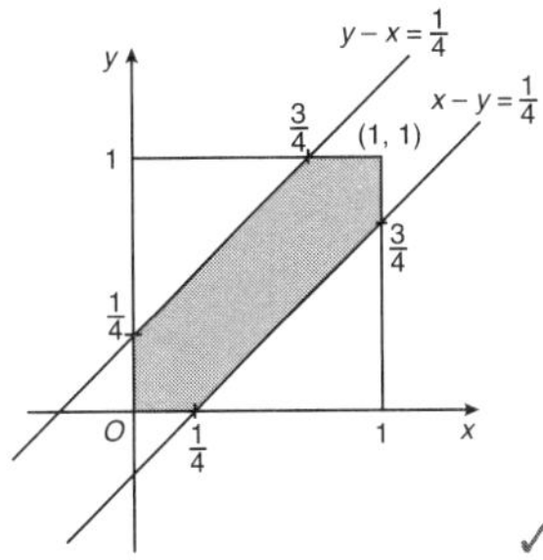

✓

Area shaded
= Area of square –
2 × unshaded triangles

$= 1 - 2 \times \frac{1}{2} \times \frac{3}{4} \times \frac{3}{4}$

$= 1 - \frac{9}{16}$

$= \frac{7}{16}$ units^2

$\therefore$ P(They will meet) $= \frac{7}{16}$. ✓

(2 marks)

iii For the probability of meeting to be 50%, half of the square must be shaded. Therefore half of the square must be unshaded, i.e.

unshaded area $= \frac{1}{2}$.

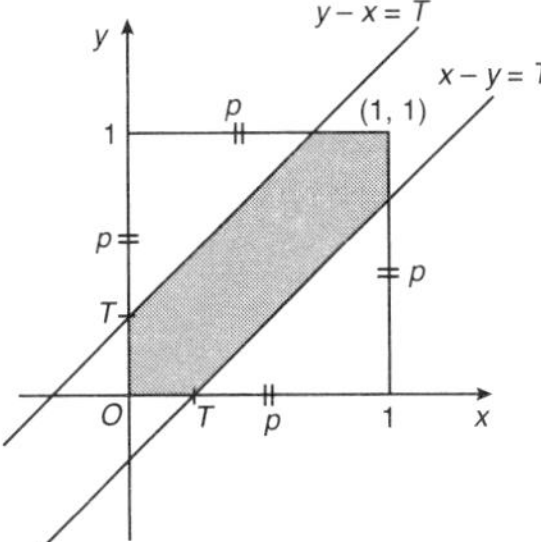

Let p be the distance as shown on the diagram.

Let $T = \frac{t}{60}$ represent the fraction of an hour they wait before leaving.

$\therefore$ Unshaded area

$= 2 \times \frac{b \times h}{2} = \frac{1}{2}$

$= 2 \times \frac{p^2}{2} = \frac{1}{2}$

$\therefore p^2 = \frac{1}{2}$

$\therefore p = \pm\frac{1}{\sqrt{2}}$

But $p > 0$

$\therefore p = \frac{1}{\sqrt{2}}$ ✓

Now $T = 1 - p$

$= 1 - \frac{1}{\sqrt{2}}$

$\therefore \frac{t}{60} = 1 - \frac{1}{\sqrt{2}}$

$t = 60\left(1 - \frac{1}{\sqrt{2}}\right)$

$= 17.5735\ldots$

$\doteqdot 18$ minutes

$\therefore$ They have a 50% chance of meeting if $t \doteqdot 18$ minutes. ✓

(2 marks)

51 P(red or yellow)

= P(red) + P(yellow)

$= \frac{12}{30} + \frac{7}{30}$ ✓

$= \frac{19}{30}$ ✓

(2 marks)

52 **i**

		Die 1					
		0	1	2	3	4	5
	0	0	0	0	0	0	0
	1	0	1	2	3	4	5
Die	2	0	2	4	6	8	10
2	3	0	3	6	9	12	15
	4	0	4	8	12	16	20
	5	0	5	10	15	20	25

P(0) $= \frac{11}{36}$ ✓ ✓

(2 marks)

ii $P(\geqslant 16) = \frac{4}{36} = \frac{1}{9}$ ✓

(1 mark)

iii $P(\text{Sum} < 45) = 1 - P(\text{Sum} \geqslant 45)$

A sum of scores $\geqslant 45$ can be obtained in the following ways:

- A score of 20 followed by a score of 25

  $P(20, 25) = \frac{2}{36} \times \frac{1}{36}$

  $= \frac{2}{1296}$

- A score of 25 followed by a score of 20

  $P(25, 20) = \frac{1}{36} \times \frac{2}{36}$

  $= \frac{2}{1296}$

- A score of 25 followed by a score of 25

  $P(25, 25) = \frac{1}{36} \times \frac{1}{36}$

  $= \frac{1}{1296}$ ✓

$\therefore P(\text{Sum} \geqslant 45)$

$= \frac{2}{1296} + \frac{2}{1296} + \frac{1}{1296}$

$= \frac{5}{1296}$

$\therefore P(\text{sum} < 45) = 1 - \frac{5}{1296}$

$= \frac{1291}{1296}$ ✓

(2 marks)

53 **i**

		SPINNER 2			
		1	3	5	7
SPINNER 1	1	(11)	13*	15	17
	2	21	23*	25	27
	3	31*	(33*)	35*	37*

P(same no.) $= \frac{2}{12}$ ✓

$= \frac{1}{6}$ ✓

(2 marks)

ii P(at least one 3) $= \frac{6}{12}$

$= \frac{1}{2}$ ✓

(1 mark)

54 **i** The first sock selected can be any colour. The second sock must not be the same colour as the first drawn.

$\therefore$ Probability $= \frac{8}{8} \times \frac{6}{7}$

$= \frac{6}{7}$ ✓

(1 mark)

ii The third sock cannot be the same colour as the first two colours drawn.

$\therefore$ Required probability

$= \frac{8}{8} \times \frac{6}{7} \times \frac{4}{6}$ ✓

$= \frac{4}{7}$ ✓

(2 marks)

iii P(matching pair) $= 1 - \frac{4}{7}$

$= \frac{3}{7}$ ✓

(1 mark)

55 Let $P(\text{Tan wins}) = \frac{1}{x}$

$\therefore P(\text{Alex or Rob wins}) = \frac{2}{x}$

Someone must win

$\therefore \frac{1}{x} + \frac{2}{x} + \frac{2}{x} = 1$

$\frac{5}{x} = 1$

$\therefore x = 5$

$\therefore P(\text{Tan wins}) = \frac{1}{5}$

Answer B

56 **i**

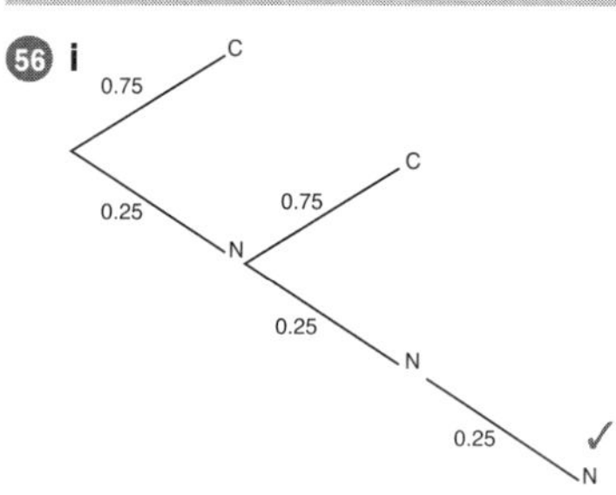

✓

i P(NC) $= 0.25 \times 0.75$
$= 0.1875$ ✓ *(2 marks)*

ii P(NNN) $= 0.25 \times 0.25 \times 0.25$
$= 0.015\,625$ ✓

(1 mark)

57 **i** P(A first) $= \frac{1}{5}$ ✓

(1 mark)

ii P(ABCDE)

$= \frac{1}{5} \times \frac{1}{4} \times \frac{1}{3} \times \frac{1}{2} \times \frac{1}{1}$ ✓

$= \frac{1}{120}$ ✓

(2 marks)

58 No. of ways of placing the letters
$= 4 \times 3 \times 2 \times 1$
$= 24$

$\therefore P(\text{correct letter}) = \frac{1}{24}$

Answer B

59 P(Blue) $= \frac{3}{5}$ ✓

(1 mark)

60 **i**

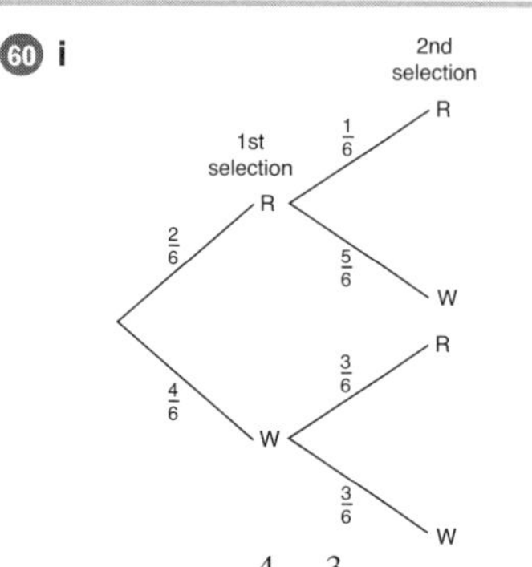

✓

ii $P(WW) = \frac{4}{6} \times \frac{3}{6}$ ✓

$= \frac{12}{36}$

$= \frac{1}{3}$ ✓

iii $P(WW \text{ or } RW)$

$= \frac{4}{6} \times \frac{3}{6} + \frac{2}{6} \times \frac{5}{6}$ ✓

$= \frac{12}{36} + \frac{10}{36}$

$= \frac{22}{36}$

$= \frac{11}{18}$ ✓

(5 marks)
(Total mark allocation only in exam)

61 **i**

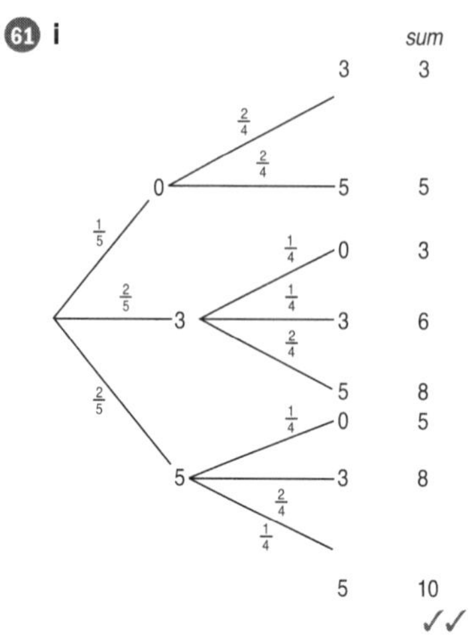

✓✓

P(5 then 3) $= \frac{2}{5} \times \frac{2}{4}$

$= \frac{4}{20}$

$= \frac{1}{5}$ ✓

ii P(Sum > 8)

$= (\frac{2}{5} \times \frac{2}{4}) + (\frac{2}{5} \times \frac{2}{4}) + (\frac{2}{5} \times \frac{1}{4})$

$= \frac{4}{20} + \frac{4}{20} + \frac{2}{20}$

$= \frac{10}{20}$

$= \frac{1}{2}$ ✓

iii P(second card 3)

$= (\frac{1}{5} \times \frac{2}{4}) + (\frac{2}{5} \times \frac{1}{4}) + (\frac{2}{5} \times \frac{2}{4})$

$= \frac{2}{20} + \frac{2}{20} + \frac{4}{20}$

$= \frac{8}{20}$

$= \frac{2}{5}$ ✓

(5 marks)
(Total mark allocation only in exam)

62

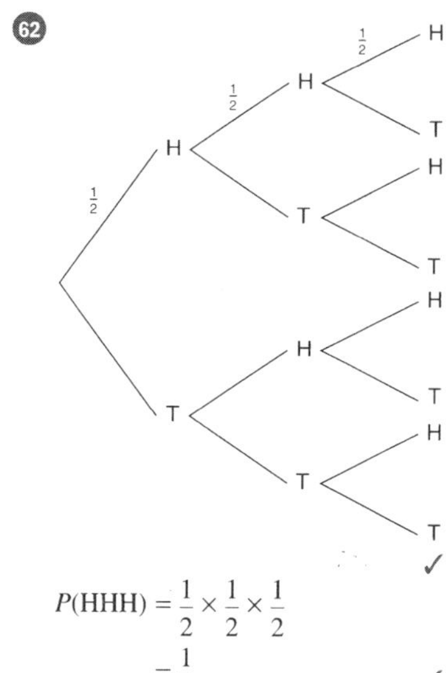

✓

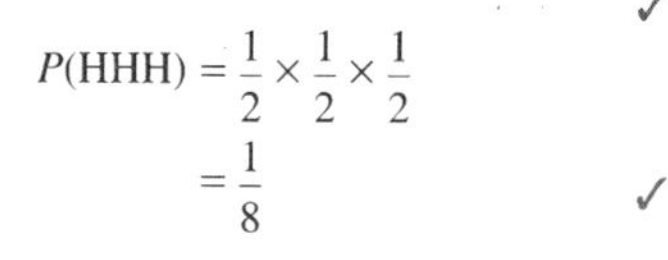

$P(HHH) = \frac{1}{2} \times \frac{1}{2} \times \frac{1}{2}$

$= \frac{1}{8}$ ✓

(2 marks)

63 **i**

BLUE \ PINK DIE	2	3	5	7	11	13
4			✓	✓	✓	✓
6				✓	✓	✓
8					✓	✓
9					✓	✓
10					✓	✓
12						✓

✓

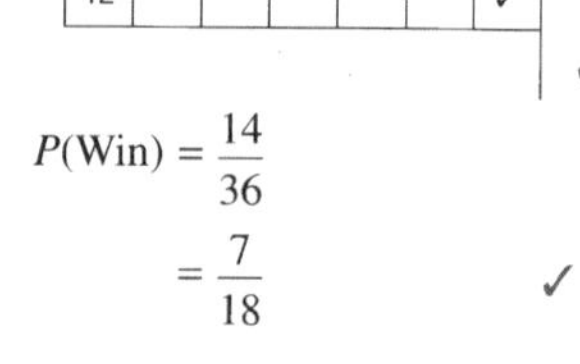

$P(\text{Win}) = \frac{14}{36}$

$= \frac{7}{18}$ ✓

ii $P(\text{Loss}) = 1 - \dfrac{7}{18}$

$= \dfrac{11}{18}$ ✓

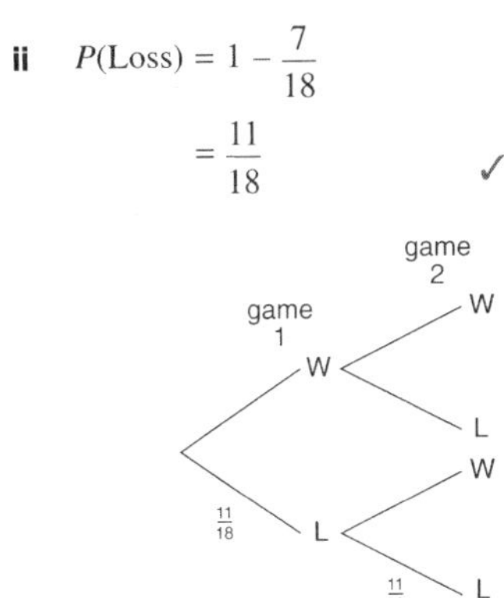

$P(\text{At least one win})$

$= 1 - P(LL)$

$= 1 - \left(\dfrac{11}{18} \times \dfrac{11}{18}\right)$ ✓

$= 1 - \dfrac{121}{324}$

$= \dfrac{203}{324}.$ ✓

(5 marks)
(Total mark allocation only in exam)

64 $P(\text{Even}) = \dfrac{22}{45}.$ ✓✓

(2 marks)

65 **i**

Branch	Outcome	Probability
2/4 R → 1/3 R	RR	$\frac{2}{12}$
2/4 R → 1/3 B	RB	$\frac{2}{12}$
2/4 R → 1/3 W	RW	$\frac{2}{12}$
1/4 B → 2/3 R	BR	$\frac{2}{12}$
1/4 B → 1/3 W	BW	$\frac{1}{12}$
1/4 W → 2/3 R	WR	$\frac{2}{12}$
1/4 W → 1/3 B	WB	$\frac{1}{12}$
		$\frac{12}{12}$

✓✓

ii $P(RR) = \dfrac{2}{12}$

$= \dfrac{1}{6}$ ✓

iii $P(\text{At least 1 red})$

$= 1 - [P(BW + P(WB)]$

$= 1 - (\frac{1}{12} + \frac{1}{12})$

$= 1 - \frac{2}{12}$

$= \frac{10}{12}$

$= \frac{5}{6}$ ✓

iv If one of the selected balls is red, there are 5 possible outcomes only one of which will yield another red: <u>RR</u>, RB, RW, BR and WR

$\therefore P(\text{Hidden Red}) = \dfrac{1}{5}$ ✓

(5 marks)
(Total mark allocation only in exam)

66 Let x be the number of students studying both French and Japanese.

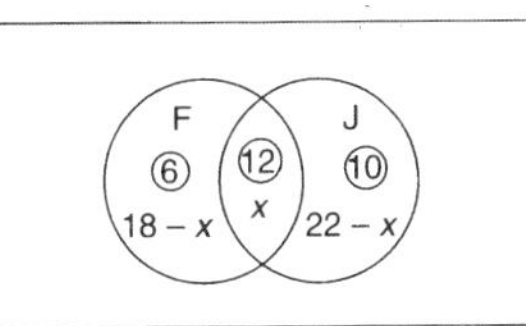

Now $18 - x + x + 22 - x = 28$

$40 - x = 28$

$x = 12$ ✓

i $P(F) = \dfrac{18}{28}$

$= \dfrac{9}{14}$ ✓

ii $P(FF) = \dfrac{18}{28} \times \dfrac{17}{27}$

$= \dfrac{17}{42}$ ✓

iii $P(\text{Both}) = \dfrac{12}{28}$

$= \dfrac{3}{7}$ ✓

(4 marks)
(Total mark allocation only in exam)

67 $P(\text{both not defective})$

$= 0.94 \times 0.96$ ✓

$= 0.9024$ ✓

(2 marks)

68 **i**

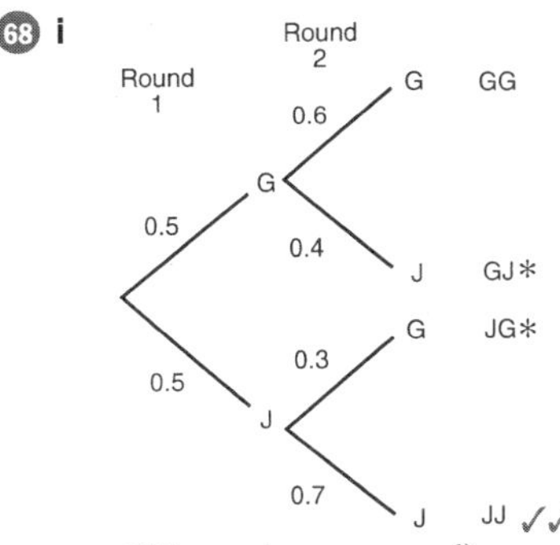

ii $P(\text{Greg wins one round})$

$= P(GJ) + P(JG)$

$= 0.5 \times 0.4 + 0.5 \times 0.3$ ✓

$= 0.2 + 0.15$

$= 0.35$ ✓

(4 marks)
(Total mark allocation only in exam)

69 The statement is false unless there are equal numbers of green, black and red jelly beans.

(No mark allocation in exam)

70 **i** Prob. of rolling a particular letter twice

$= \frac{1}{20} \times \frac{1}{20}$

$= \frac{1}{400}$

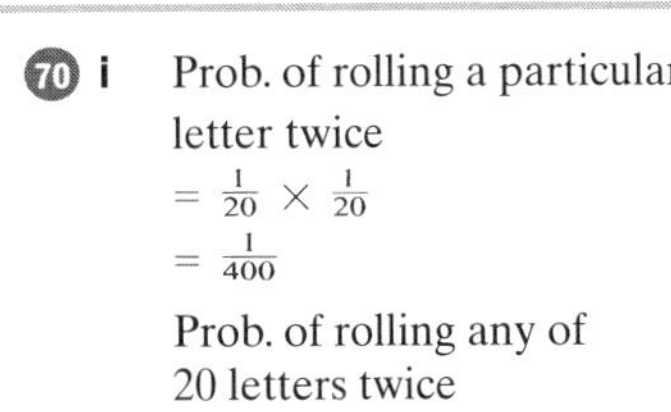

Prob. of rolling any of 20 letters twice

$= \frac{1}{400} \times 20$

$= \frac{20}{400}$

$= \frac{1}{20}$

ii

$P(\text{E appearing twice})$

$= P(EE\tilde{E}) + P(E\tilde{E}E) + P(\tilde{E}EE)$

$= (\frac{1}{20} \times \frac{1}{20} \times \frac{19}{20}) +$

$(\frac{1}{20} \times \frac{19}{20} \times \frac{1}{20}) + (\frac{19}{20} \times \frac{1}{20} \times \frac{1}{20})$

$= \frac{19}{8000} + \frac{19}{8000} + \frac{19}{8000}$

$= \frac{57}{8000}$

(No mark allocation in exam)

71

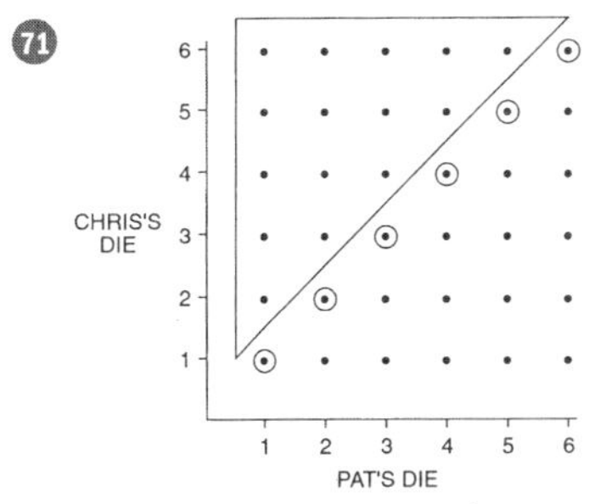

i $P(\text{same number}) = \frac{6}{36}$ (⊙)

$= \frac{1}{6}$

ii Required probability

$= \frac{15}{36}$ (◸)

$= \frac{5}{12}$

(No mark allocation in exam)

72

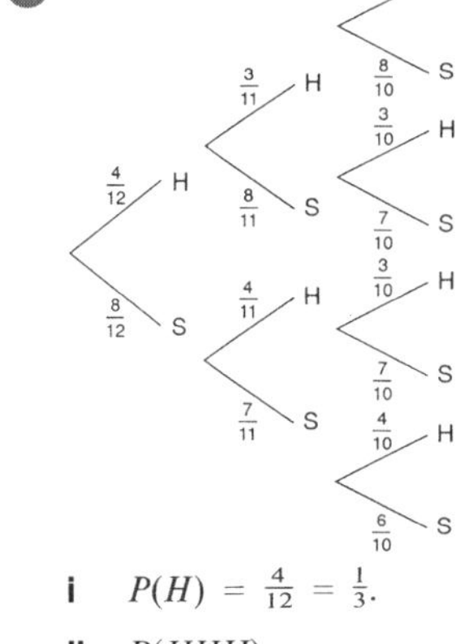

i $P(H) = \frac{4}{12} = \frac{1}{3}.$

ii $P(HHH)$

$= \frac{4}{12} \times \frac{3}{11} \times \frac{2}{10}$

$= \frac{1}{55}$

iii

$P(\text{exactly 1 hard chocolate})$

$= P(HSS) + P(SHS) + P(SSH)$

$= (\frac{4}{12} \times \frac{8}{11} \times \frac{7}{10})$

$+ (\frac{8}{12} \times \frac{4}{11} \times \frac{7}{10})$

$+ (\frac{8}{12} \times \frac{7}{11} \times \frac{4}{10})$

$= \frac{28}{165} + \frac{28}{165} + \frac{28}{165}$

$= \frac{84}{165}$

$= \frac{28}{55}$

(No mark allocation in exam)

1 The probability distribution table for a discrete random variable X is shown.

x	$P(X = x)$
1	0.6
2	0.3
3	0.1

What is the expected value of X?

A 0.6 **B** 1.0
C 1.5 **D** 2.0 *(1 mark)*

(Q2, **2021 HSC**) Easy

---

2 The discrete random variable X can only take the values 0, 1, 2, 3, or 4.

The probability distribution is described by the following statements:

- $P(X = 0) = P(X = 2) = a$
- $P(X = 1) = P(X = 3) = b$
- $P(X = 4) = 2a + b$
- $b = 2a$

Show that $Var(X) = 2.04$ *(3 marks)*

Bonus question (see page iv) Medium

---

3 The probability distribution of a discrete random variable, X, is given by the table below.

x	0	1	2	3
$P(X = x)$	0.1	$0.6p^2$	0.3	$1 - p$

i Show that $p = \frac{2}{3}$ or $p = 1$. *(2 marks)* Easy

ii Let $p = \frac{2}{3}$.

Find correct to 2 decimal places,

α $E(X)$ *(1 mark)* Easy

β $P(X \leq E(X))$. *(1 mark)* Medium

Bonus question

---

4 For the random variable, X, it is known that $E(X) = 2.6$.

If the expected value $E(X^2) = 8.4$, which of these is closest to the standard deviation σ?

A 1.28 **B** 2.41
C 7.99 **D** 8.24 *(1 mark)*

Bonus question Easy

---

5 On any given day, the number X of phone messages that Ella receives as she commutes to and from work is a random variable with a probability distribution given in the table below.

x	0	1	2	3	4	5
$P(X = x)$	0.2	0.3	0.2	0.15	0.1	0.05

What is the mean of X?

A 0.3 **B** 1.5
C 1.8 **D** 2.5 *(1 mark)*

Bonus question Easy

---

6 In a game a six-sided dice is rolled twice.

The difference between the two results is recorded.

The following table is produced. Some of the results are missing.

		Second roll					
		1	**2**	**3**	**4**	**5**	**6**
First roll	**1**	0	1	2	3		5
	2	1	0	1	2	3	
	3	2	1			2	3
	4		2	1	0	1	
	5	4	3	2	1	0	1
	6	5	4	3	2		0

i Complete the table. *(1 mark)* Easy

ii A probability distribution table is drawn to summarise the results. Complete the table. *(2 marks)* Medium

x	0	1	2	3	4	5
$P(X = x)$			$\frac{2}{9}$			$\frac{1}{18}$

iii Find the expected value, $E(X)$, and the standard deviation σ. *(3 marks)* Medium

Bonus question

---

7 Consider the following discrete probability distribution for the random variable X.

x	0	1	2	5
$P(X = x)$	p	$2p$	$3p$	$4p$

What is the mean of the distribution?

(2 marks)

Bonus question Easy

Year 11 Discrete probability distribution—Worked answers

1 $E(X) = 1(0.6) + 2(0.3) + 3(0.1)$
$= 0.6 + 0.6 + 0.3$
$= 1.5$
Answer C

2 The sum of the probabilites = 1
$2a + 2b + 2a + b = 1$
$4a + 3b = 1$
Substitute $b = 2a$ into
$4a + 3b = 1$:
$4a + 3(2a) = 1$
$10a = 1$
$a = 0.1$, and hence $b = 0.2$. ✓

x	0	1	2	3	4
$P(X = x)$	0.1	0.2	0.1	0.2	0.4

$E(X) = 0(0.1) + 1(0.2) + 2(0.1) + 3(0.2) + 4(0.4)$
$= 2.6$ ✓
$\therefore \quad \mu^2 = 2.6^2$
$= 6.76$
$E(X^2) = 0^2(0.1) + 1^2(0.2) + 2^2(0.1) + 3^2(0.2) + 4^2(0.4)$
$= 8.8$
$\text{Var}(X) = 8.8 - 6.76$
$= 2.04$ ✓

(3 marks)

3 **i** $0.1 + 0.6p^2 + 0.3 + 1 - p = 1$
$0.6p^2 - p + 0.4 = 0$
$3p^2 - 5p + 2 = 0$ ✓
$(3p - 2)(p - 1) = 0$
$p = \frac{2}{3}, 1$ ✓

(2 marks)

ii α $E(X)$
$= 0(0.1) + 1(0.6)\left(\frac{2}{3}\right)^2 + 2(0.3) + 3\left(1 - \frac{2}{3}\right)$
$= 1.86666\ldots$
$= 1.87$ (2 dec. pl.) ✓

(1 mark)

β $P(X \leq 1.87)$
$= P(X = 0) + P(X = 1)$
$= 0.1 + (0.6)\left(\frac{2}{3}\right)^2$
$= 0.36666\ldots$
$= 0.37$ (2 dec. pl.) ✓

(1 mark)

4 $E(X) = \mu = 2.6$
Hence, $\mu^2 = 6.76$
$E(X^2) = 8.4$
$Var(X) = E(X^2) - \mu^2$
$= 8.4 - 6.76$
$= 1.64$
$\sigma = \sqrt{1.64}$
$= 1.28$ (2 dec. pl.)
Answer A

5 $E(X) = 0(0.2) + 1(0.3) + 2(0.2) + 3(0.15) + 4(0.1) + 5(0.05)$
$= 1.8$
Answer C

6 **i**

		Second roll					
		1	2	3	4	5	6
First roll	1	0	1	2	3	4	5
	2	1	0	1	2	3	4
	3	2	1	0	1	2	3
	4	3	2	1	0	1	2
	5	4	3	2	1	0	1
	6	5	4	3	2	1	0

✓ *(1 mark)*

ii

x	0	1	2	3	4	5
$P(X = x)$	$\frac{1}{6}$	$\frac{5}{18}$	$\frac{2}{9}$	$\frac{1}{6}$	$\frac{1}{9}$	$\frac{1}{18}$

✓✓ *(2 marks)*

iii $E(X)$
$= 0\left(\frac{1}{6}\right) + 1\left(\frac{5}{18}\right) + 2\left(\frac{2}{9}\right) + 3\left(\frac{1}{6}\right) + 4\left(\frac{1}{9}\right) + 5\left(\frac{1}{18}\right)$
$= 1\frac{17}{18}$ ✓
$E(X^2)$
$= 0^2\left(\frac{1}{6}\right) + 1^2\left(\frac{5}{18}\right) + 2^2\left(\frac{2}{9}\right) + 3^2\left(\frac{1}{6}\right) + 4^2\left(\frac{1}{9}\right) + 5^2\left(\frac{1}{18}\right)$
$= 5\frac{5}{6}$ ✓
As $E(X) = \mu = 1\frac{17}{18}$, then
$\mu^2 = 3\frac{253}{324}$:
$Var(X) = E(X^2) - \mu^2$
$= 5\frac{5}{6} - 3\frac{253}{324}$
$= 2.052\,469\,136\ldots$
$\sigma = 1.43$ (2 dec. pl.) ✓

(3 marks)

7 $p + 2p + 3p + 4p = 1$
$10p = 1$
$p = \frac{1}{10}$ ✓
$E(X) = 0\left(\frac{1}{10}\right) + 1\left(\frac{2}{10}\right) + 2\left(\frac{3}{10}\right) + 5\left(\frac{4}{10}\right)$
$= 2.8$ ✓

(2 marks)

1 The graph of the function $f(x) = x^2$ is translated m units to the right, dilated vertically by a scale factor of k and then translated 5 units down. The equation of the transformed function is $g(x) = 3x^2 - 12x + 7$.

Find the values of m and k. *(3 marks)*

(Q19, **2022 HSC**) Easy

2 For what values of x is $f(x) = x^2 - 2x^3$ increasing? *(3 marks)*

(Q16, **2021 HSC**) Medium

3 Without using calculus, sketch the graph of $y = 2 + \dfrac{1}{x+4}$, showing the asymptotes and the x and y intercepts. *(3 marks)*

(Q19, **2021 HSC**) Medium

4 Consider the graph of $y = f(x)$ as shown.

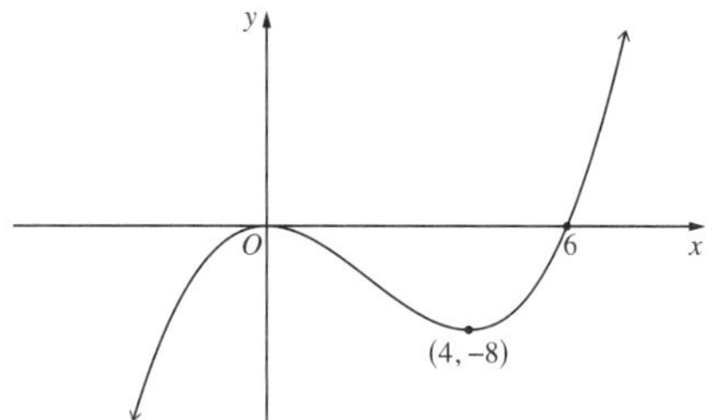

NOT TO SCALE

Sketch the graph of $y = 4f(2x)$ showing the x-intercepts and the coordinates of the turning points. *(2 marks)*

(Q21, **2021 HSC**) Medium

5 The function $f(x) = x^3$ is transformed to $g(x) = (x-2)^3 + 5$ by a horizontal translation of 2 units followed by a vertical translation of 5 units.

Which row of the table shows the directions of the translations?

	Horizontal translation of 2 units	*Vertical translation of 5 units*
A	Left	Up
B	Right	Up
C	Left	Down
D	Right	Down

(1 mark)

(Q2, **2020 HSC**) Easy

6 The circle $x^2 - 6x + y^2 + 4y - 3 = 0$ is reflected in the x-axis.

Sketch the reflected circle, showing the coordinates of the centre and the radius. *(3 marks)*

(Q24, **2020 HSC**) Medium

7 The graph of $y = f(x)$ is shown below.

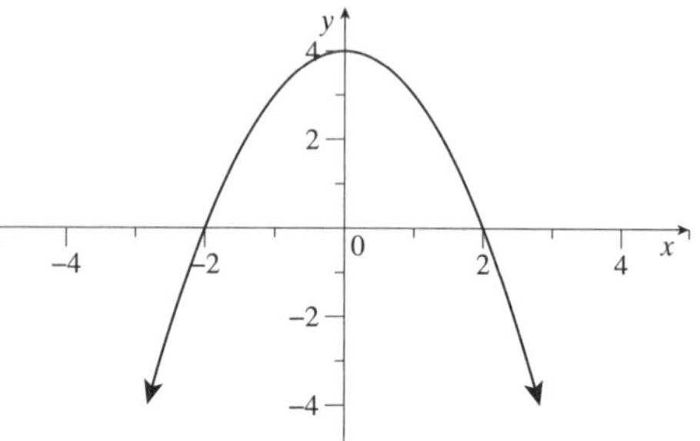

Which of the following is the graph of $y = -f(x+1)$?

A **B**

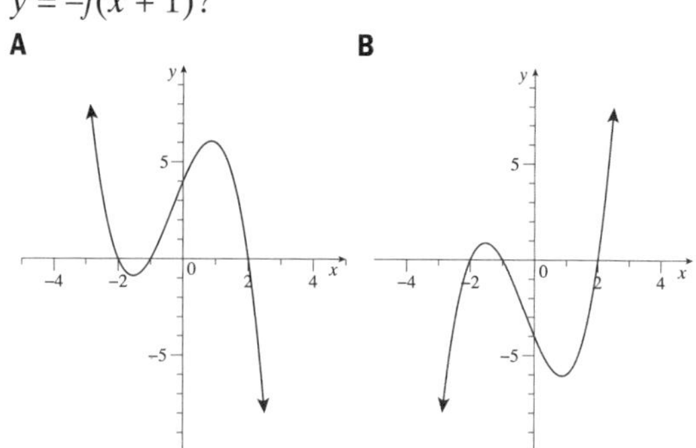

C **D**

(1 mark)

Bonus question (see page iv) Easy

8

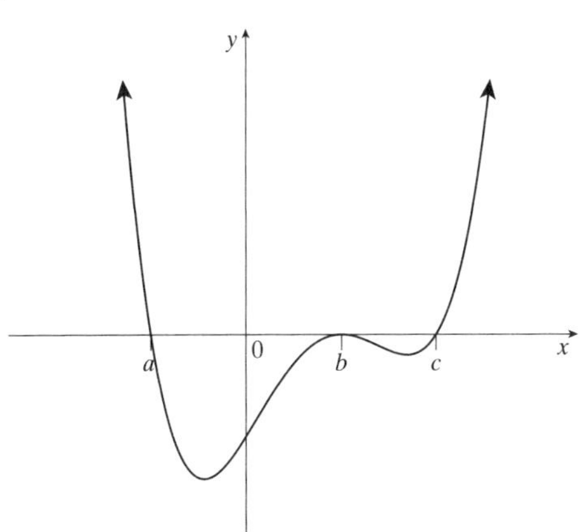

The rule for a function with the graph above could be

A $y = -2(x - a)(x - b)^2(x - c)$

B $y = 2(x - a)(x - b)^2(x - c)$

C $y = 2(x + a)(x - b)^2(x - c)$

D $y = 2(x - a)(x - b)(x - c)$ *(1 mark)*

Bonus question Easy

9 The function $y = (1 - x)(x + 1)(x - 2)^2$ is to be graphed.

Which of these is the graph of the function?

A

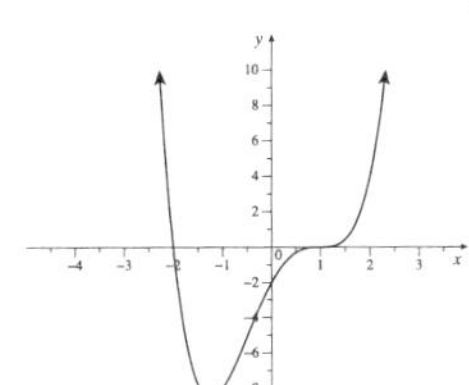

B

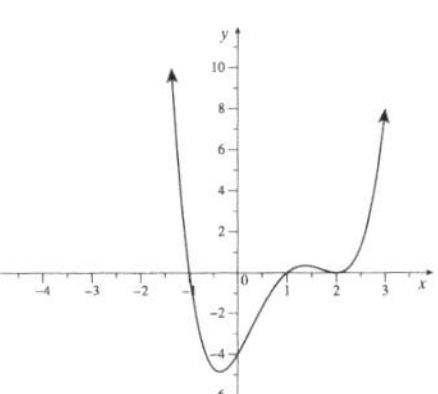

C

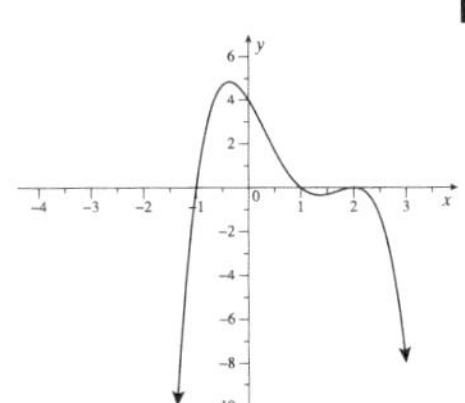

D

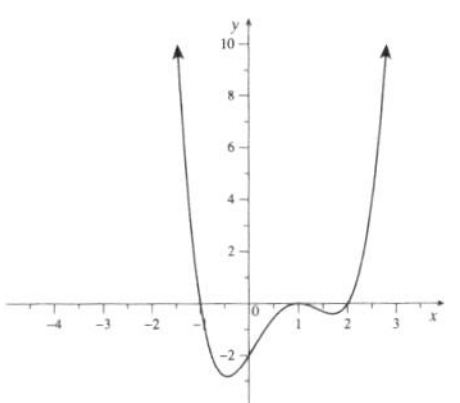

(1 mark)

Bonus question Easy

10 Let $f(x) = \dfrac{x}{x-1}$.

i Show $\dfrac{x}{x-1} = 1 + \dfrac{1}{x-1}$. *(2 marks)* Easy

ii Hence, or otherwise, on the same number plane, draw sketches of $y = \dfrac{x}{x-1}$ and $y = \dfrac{1}{x-1}$. *(2 marks)* Medium

Bonus question

11 The graph of $y = a\cos(x + b) + c$, over the domain $0 \le x \le 2\pi$, has been drawn on the number plane below.

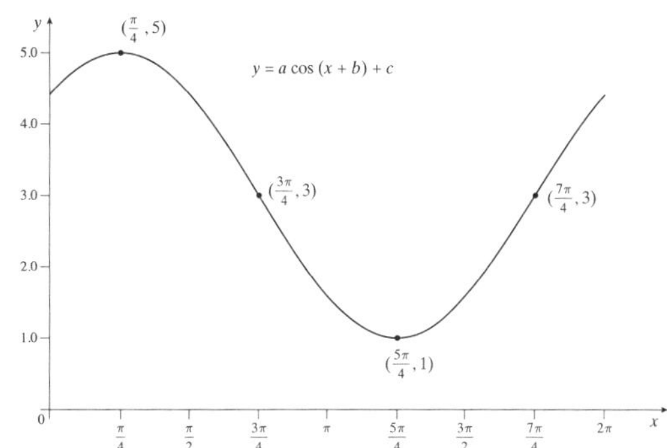

What are the values of a, b and c? *(3 marks)*

Bonus question Medium

12 The function $y = e^x$ is translated right 2 units and up by 1 unit to give the function $f(x)$.

i Write down the equation of the function $f(x)$. *(1 mark)* Easy

ii Sketch $y = f(x)$ on the number plane below.

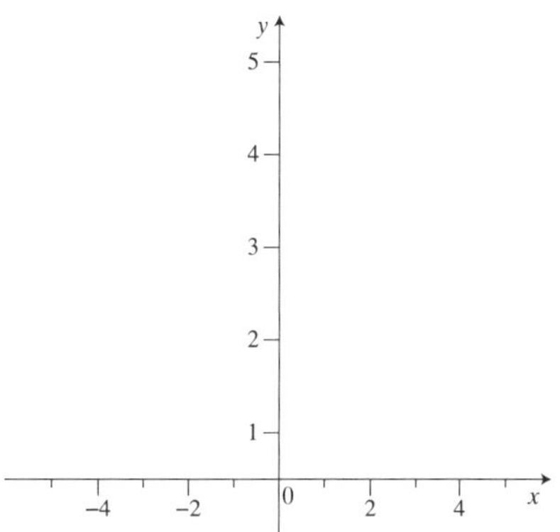

(1 mark) Easy

Bonus question

13 i Carefully sketch the graphs of $y = |x + 1|$ and $y = 3 - |x - 2|$ on the same axes, showing all intercepts. *(3 marks)* Medium

ii Using the graphs from part **i**, or otherwise, find the range of values of x for which $|x + 1| + |x - 2| = 3$. *(1 mark)* Hard

(Q12b, **2017 ME1 HSC**)

14 Which diagram best shows the graph of the parabola $y = 3 - (x - 2)^2$?

A

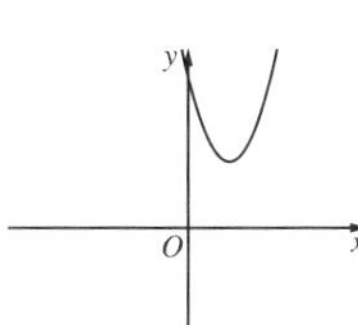

B

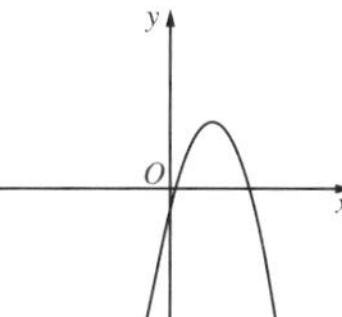

C 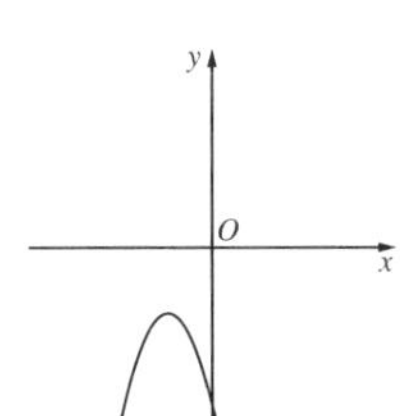

D 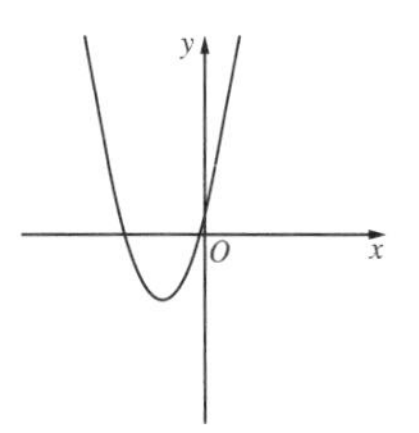

(1 mark)

(Q3, **2016 HSC**) Easy

---

15 i Sketch the graph $y = |2x - 3|$. *(1 mark)* Medium

ii Using the graph from part **i**, or otherwise, find all values of m for which the equation $|2x - 3| = mx + 1$ has exactly one solution. *(2 marks)* Hard

(Q15c, **2013 HSC**)

---

16 Solve the inequality $x^2 - x - 12 < 0$. *(2 marks)*

(Q2b, **2010 HSC**) Easy

---

17 The circle $x^2 + (y - c)^2 = r^2$, where $c > 0$ and $r > 0$, lies inside the parabola $y = x^2$. The circle touches the parabola at exactly two points located symmetrically on opposite sides of the y-axis, as shown in the diagram.

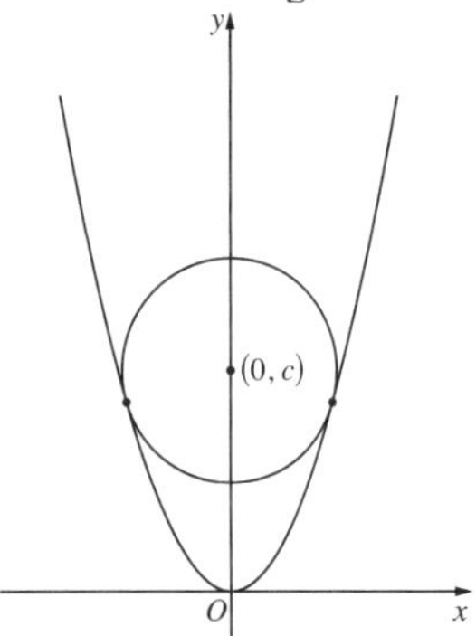

i Show that $4c = 1 + 4r^2$. *(2 marks)* Medium

ii Deduce that $c > \frac{1}{2}$. *(1 mark)* Hard

(Q16c, **2012 HSC**)

---

18 Let $f(x) = \log_e[(x - 3)(5 - x)]$

What is the domain of $f(x)$? *(2 marks)*

(Q1f, **2008 ME1 HSC**) Medium

---

19 i Sketch the graph of $y = |2x - 1|$. *(1 mark)* Easy

ii Hence, or otherwise, solve $|2x - 1| \leq |x - 3|$. *(3 marks)* Hard

(Q3a, **2008 ME1 HSC**)

---

20 i Sketch the graph of $y = x^2 - 6$, and label all intercepts with the axes.

ii On the same set of axes, carefully sketch the graph of $y = |x|$.

iii Find the x coordinates of the two points where the graphs intersect.

iv Hence solve the inequality $x^2 - 6 \leq |x|$.

(6 marks)

(Q4b, **1997 HSC**) Easy

---

21 Find the values of m for which $12 + 4m - m^2 > 0$. *(1 mark)*

(Q5a, **1994 HSC**) Medium

---

22 i Verify that $x = \frac{1}{3}$ and $x = 2$ satisfy the equation $7 - 3x = \frac{2}{x}$. Easy

ii On the same set of axes, sketch the graphs of $y = 7 - 3x$ and $y = \frac{2}{x}$. Easy

iii Using part **ii**, or otherwise, write down all values of x for which $7 - 3x < \frac{2}{x}$. Medium

(Q2c, **1994 ME1 HSC**)

---

23 i On the same diagram, sketch the graphs of $y = x$ and $y = |2x - 1|$. Medium

ii By using **i** or otherwise, determine for what value of c the equation

$$|2x - 1| = x + c$$

has exactly two solutions. Hard

(Q2b, **1993 ME1 HSC**)

Year 12 Graphing techniques—Worked answers

1 $g(x) = 3x^2 - 12x + 7$
$= 3(x^2 - 4x) + 7$
$= 3(x^2 - 4x + 4) - 12 + 7$
$= 3(x - 2)^2 - 5$ ✓

As $g(x) = 3(x - 2)^2 - 5$, then $f(x) = x^2$ has been translated 2 units to the right, dilated vertically by factor of 3 and then translated 5 units down.

$\therefore m = 2$ and $k = 3$ ✓✓

(3 marks)

2 $f(x) = x^2 - 2x^3$
$f'(x) = 2x - 6x^2 > 0$ ✓
$2x(1 - 3x) > 0$ ✓

Consider $y = 2x(1 - 3x)$ and when $y > 0$:

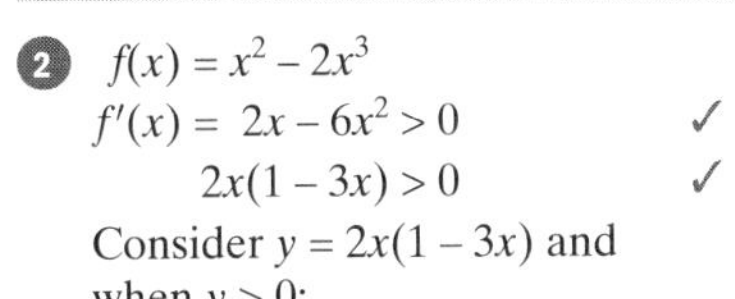
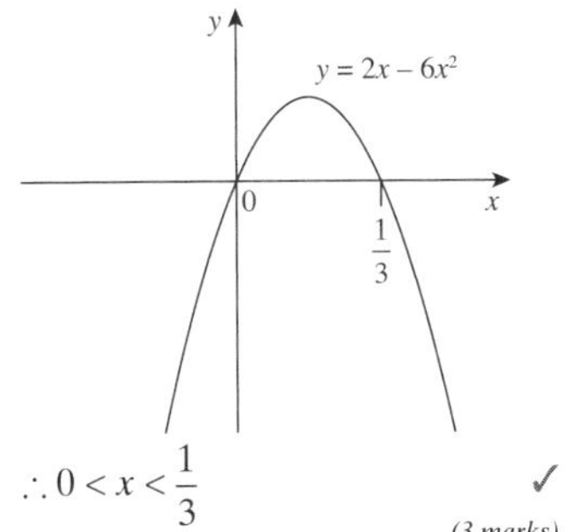

$\therefore 0 < x < \frac{1}{3}$ ✓

(3 marks)

3 The hyperbola $y = \frac{1}{x}$ is translated 4 units to the left and then 2 units vertically upwards.

Substitute $y = 0$:

$0 = 2 + \frac{1}{x+4}$

$\frac{1}{x+4} = -2$

$1 = -2x - 8$

$2x = -9$

$x = -\frac{9}{2}$

$\therefore$ x-intercept at $x = -\frac{9}{2}$.

Substitute $x = 0$:

$y = 2 + \frac{1}{0+4}$

$= 2\frac{1}{4}$

$\therefore$ y-intercept at $y = 2\frac{1}{4}$. ✓

Discontinuous at $x = -4$

$\therefore$ vertical asymptote at $x = -4$.

As $x \to \infty$, $\frac{1}{x+4} \to 0$, and

so $y = 2 + \frac{1}{x+4} \to 2$. ✓

$\therefore$ horizontal asymptote at $y = 2$.

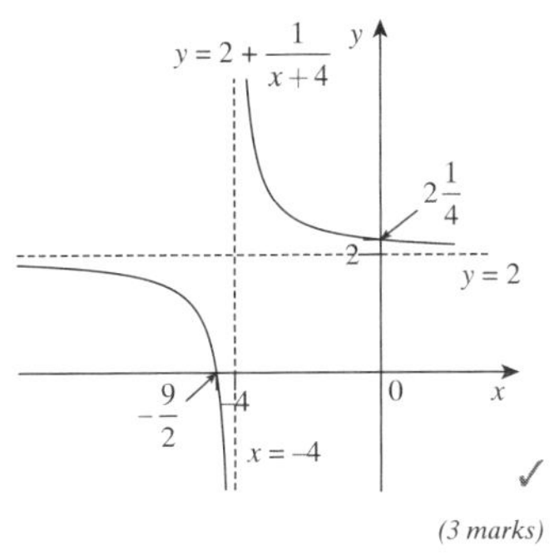

✓

(3 marks)

4 $y = f(x)$ is dilated horizontally by a factor of $\frac{1}{2}$ and dilated vertically by a factor of 4.

As $y = f(x)$ has x-intercepts 0 and 6, then $y = 4f(2x)$ has x-intercepts of 0 and 3.

As $y = f(x)$ has turning points $(0, 0)$ and $(4, -8)$, then $y = 4f(2x)$ has turning points $(0, 0)$ and $(2, -32)$. ✓

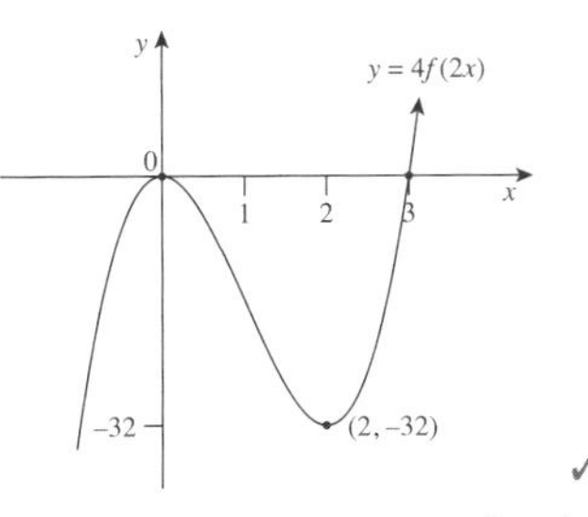

✓

(2 marks)

5 From $y = x^3$ to $y = (x - 2)^3 + 5$ the graph has been translated 2 units to the right and 5 units up.

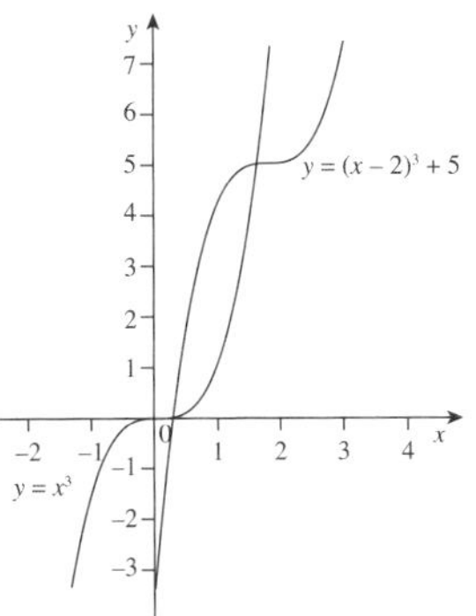

Answer B

6 $x^2 - 6x + y^2 + 4y - 3 = 0$
$x^2 - 6x + y^2 + 4y = 3$
$x^2 - 6x + 9 + y^2 + 4y + 4$
$= 3 + 9 + 4$
$(x - 3)^2 + (y + 2)^2 = 16$ ✓

Circle centre $(3, -2)$, radius 4 units. ✓

Reflecting about x-axis means new circle has centre at $(3, 2)$ and radius 4 units.

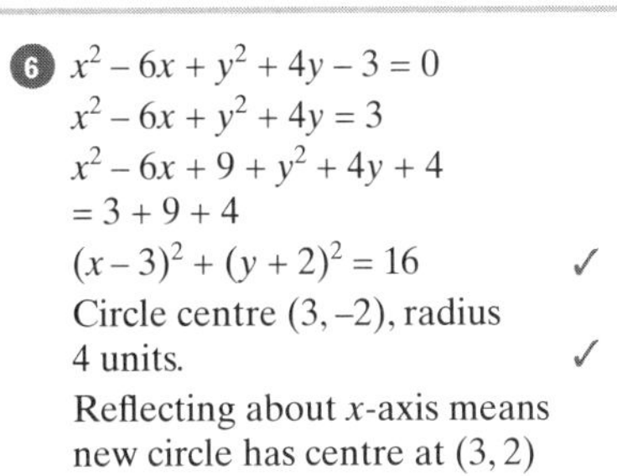
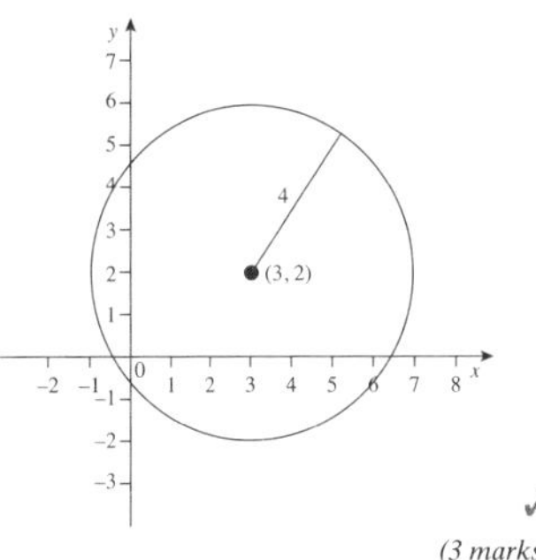

✓

(3 marks)

7

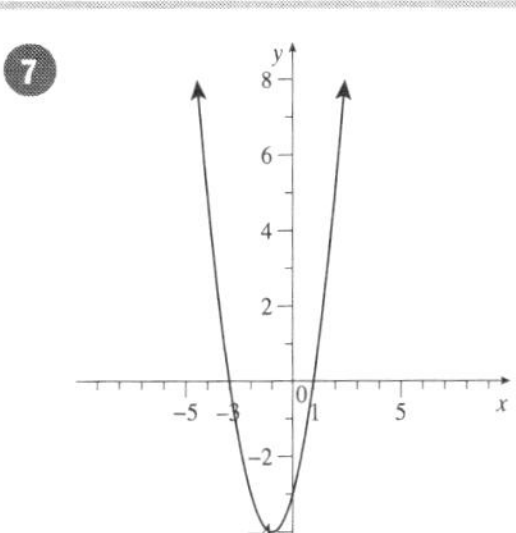

Answer D

8 The function has zeros (x-intercepts) at $x = a$, $x = b$ (twice) and $x = c$.

Also, for $x > c$, $y > 0$.

The rule for the function is $y = 2(x - a)(x - b)^2(x - c)$

Answer B

9 The function has zeros (x-intercepts) at $x = -1$, $x = 1$ and $x = 2$ (twice).

Substitute $x = 0$:

$y(0) = (1 - 0)(0 + 1)(0 - 2)^2$
$= 4$

The curve passes through $(0, 4)$.

The correct graph is

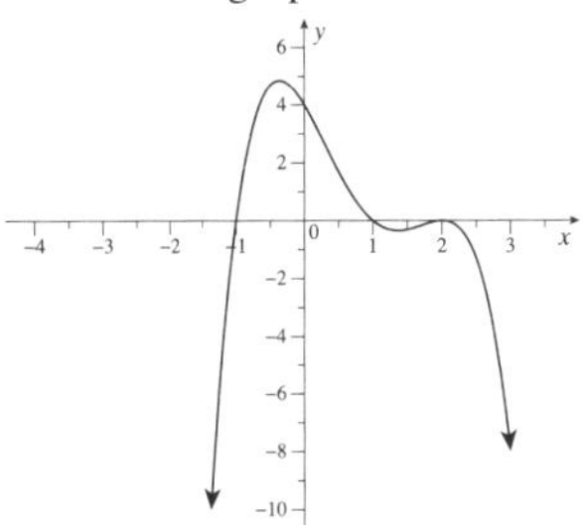

Answer C

10 **i** RHS $= 1 + \frac{1}{x-1}$

$= \frac{x-1+1}{x-1}$ ✓

$= \frac{x}{x-1}$

$=$ LHS

$\therefore \frac{x}{x-1} = 1 + \frac{1}{x-1}$ ✓ *(2 marks)*

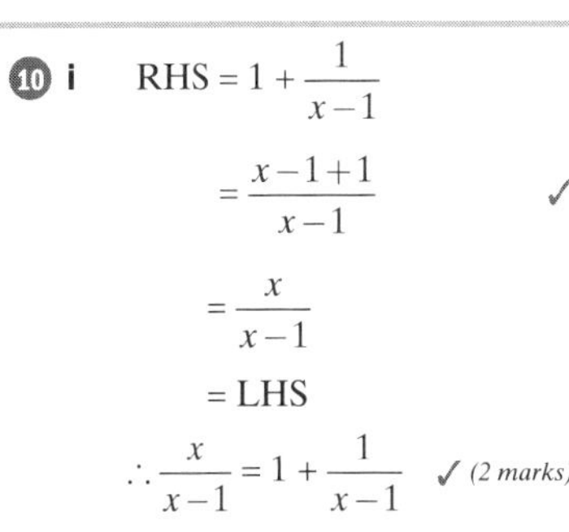

ii

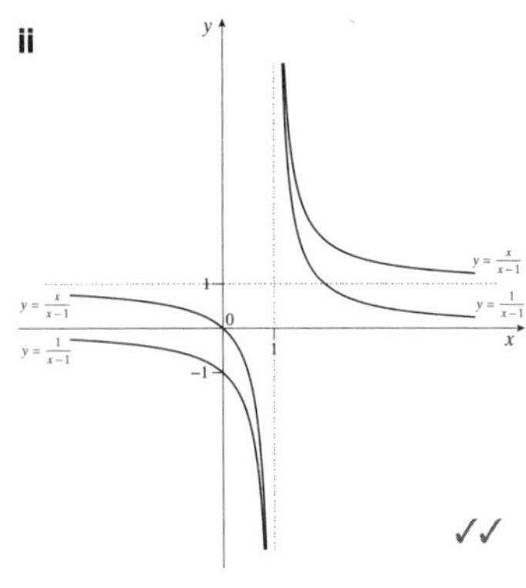

✓✓

(2 marks)

11 Amplitude $= \dfrac{5-1}{2} = 2$

$\therefore a = 2$ ✓

$\therefore y = 2\cos(x+b)+c$

As the centre of motion is at $y = 3$, then $c = 3$. ✓

$\therefore y = 2\cos(x+b)+3$

Now, compared to $y = \cos x$, the graph has been shifted to the right $\dfrac{\pi}{4}$ units.

$\therefore b = -\dfrac{\pi}{4}$ ✓

The equation of the curve is $y = 2\cos\left(x - \dfrac{\pi}{4}\right) + 3$, and hence $a = 2, b = -\dfrac{\pi}{4}$ and $c = 3$.

(3 marks)

12 **i** $f(x) = 1 + e^{x-2}$ ✓

(1 mark)

ii

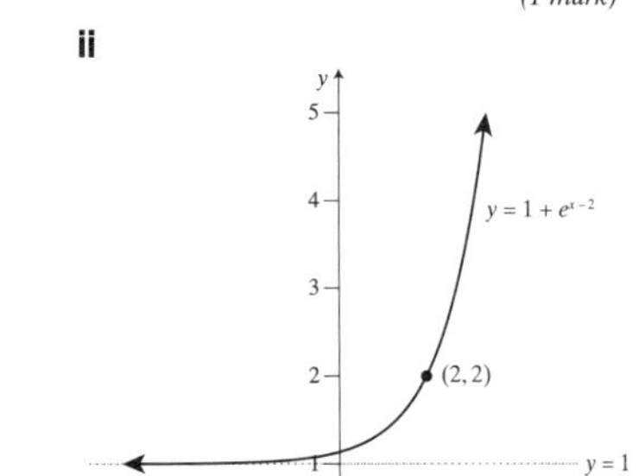

✓

(1 mark)

13 **i**

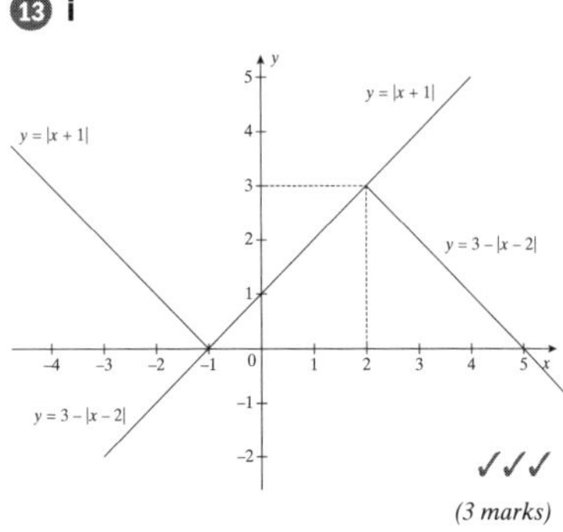

✓✓✓

(3 marks)

ii $|x+1| + |x-2| = 3$

$|x+1| = 3 - |x-2|$

From the graph $-1 \le x \le 2$

✓ *(1 mark)*

14 $y = (x-2)^2$ has an axis of symmetry $x = 2$, and $y = 3 - (x-2)^2$ is concave down.

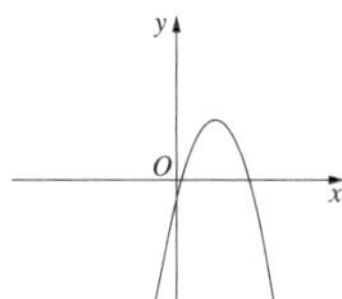

Answer B

15 **i**

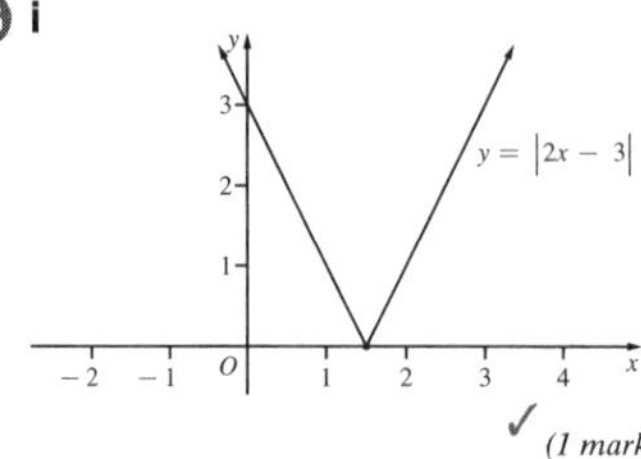

✓ *(1 mark)*

ii Gradient of $y = 2x - 3$ is 2, and gradient of $y = -(2x-3)$ is -2.

If only one point of intersection, then $m < -2$, $m = -\dfrac{2}{3}$ or $m \ge 2$. ✓✓

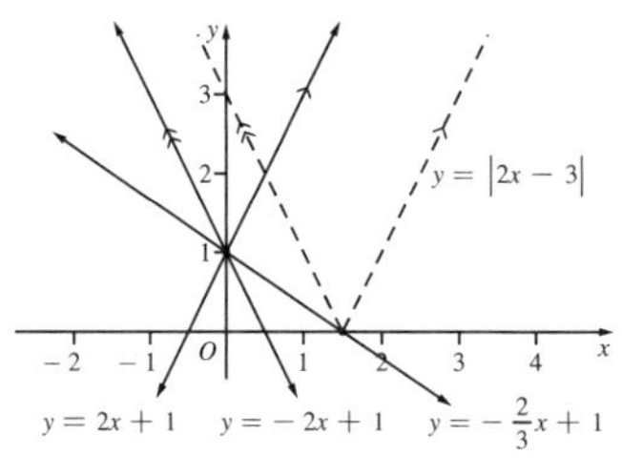

(2 marks)

16 $x^2 - x - 12 < 0$

$(x-4)(x+3) < 0$

Method 1:

Let $y = x^2 - x - 12$, so we need to consider values of x for which $y < 0$:

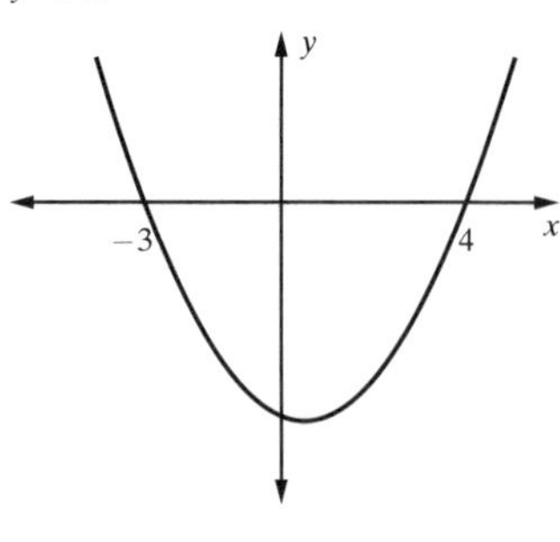

$\therefore -3 < x < 4$ ✓✓

Method 2:

Let $(x-4)(x+3) = 0$ gives $x = 4, -3$:

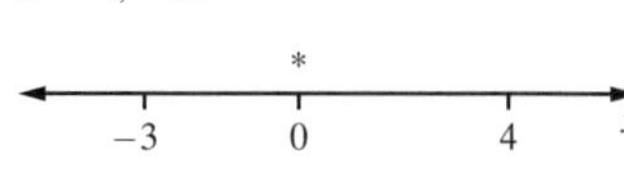

Choose 0: subs in

$(x-4)(x+3) < 0$

LHS $= (0-4)(0+3)$

$= -12 < 0$, $\therefore$ true

$\therefore -3 < x < 4$

(2 marks)

17 **i** $y = x^2$ **1**

$x^2 + (y-c)^2 = r^2$ **2**

$\therefore$ Subs **1** in **2**:

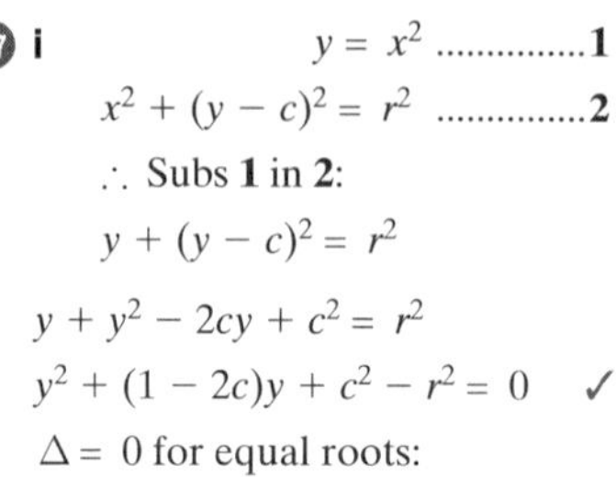

$y + (y-c)^2 = r^2$

$y + y^2 - 2cy + c^2 = r^2$

$y^2 + (1-2c)y + c^2 - r^2 = 0$ ✓

$\Delta = 0$ for equal roots:

$\therefore \Delta = (1-2c)^2 - 4 \times 1 \times (c^2 - r^2)$

$= 1 - 4c + 4c^2 - 4c^2 + 4r^2 = 0$

$4r^2 - 4c + 1 = 0$

$4c = 1 + 4r^2$ ✓

(2 marks)

ii From diagram, $c > r$

$\therefore 4c > 4r$

$16c^2 > 16r^2$

$16c^2 > 4(4r^2)$

$16c^2 > 4(4c - 1)$ (using part **i**)

$16c^2 > 16c - 4$

$16c^2 - 16c + 4 > 0$

$4c^2 - 4c + 1 > 0$

$(2c-1)^2 > 0$

$2c - 1 > 0$ (since $c > 0$)

$2c > 1$

$c > \dfrac{1}{2}$ ✓

(1 mark)

18 $f(x) = \log_e[(x-3)(5-x)]$

Domain: $(x-3)(5-x) > 0$ ✓

$\therefore 3 < x < 5$ ✓

(2 marks)

19 **i**

✓ *(1 mark)*

ii On the axes above, sketch $y = |x - 3|$.

By observation,
$|2x - 1| \leqslant |x - 3|$
when x is between the two points of intersection:

$$2x - 1 = -x + 3$$
$$3x = 4$$
$$x = \frac{4}{3} \quad ✓$$

$$-2x + 1 = -x + 3$$
$$-x = 2$$
$$x = -2 \quad ✓$$

$\therefore$ Solution is $-2 \leqslant x \leqslant \frac{4}{3}$. ✓

(3 marks)

20 **i and ii**

$$y = x^2 - 6$$
$$= (x + \sqrt{6})(x - \sqrt{6})$$

$\therefore$ x intercepts are $\pm\sqrt{6}$

y intercept occurs when $x = 0$

$$\therefore y = 0 - 6$$
$$= -6 \quad ✓$$

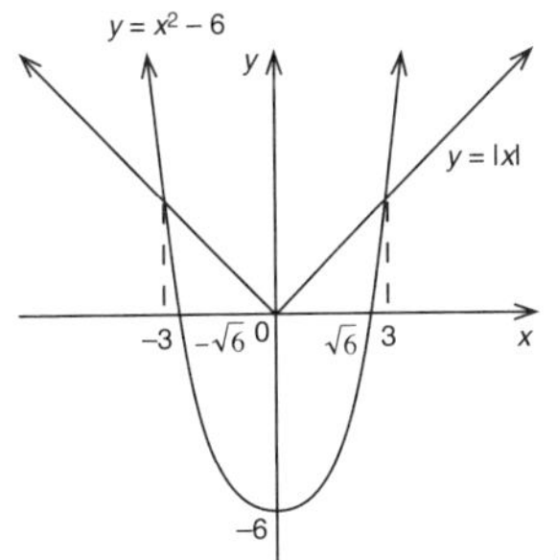

✓✓

iii Graphs intersect when

$$x^2 - 6 = |x|$$
$$(x^2 - 6)^2 = x^2$$
$$x^4 - 12x^2 + 36 = x^2$$
$$x^4 - 13x^2 + 36 = 0$$
$$(x^2 - 9)(x^2 - 4) = 0$$
$$(x + 3)(x - 3)$$
$$(x + 2)(x - 2) = 0$$
$$\therefore x = \pm 3 \text{ or } \pm 2 \quad ✓$$

From the graph $x = \pm 3$. ✓

iv $x^2 - 6 \leqslant |x|$
when $-3 \leqslant x \leqslant 3$
✓(from graph).

(6 marks)
(Total mark allocation only in exam)

21 $12 + 4m - m^2 > 0$

$(6 - m)(2 + m) > 0$

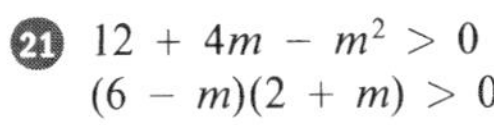

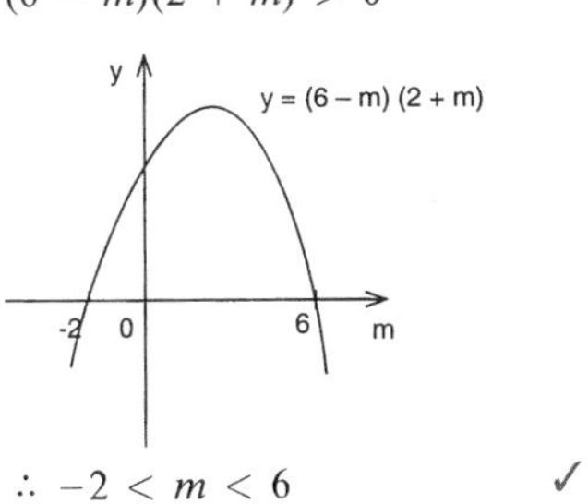

$\therefore -2 < m < 6$ ✓

(1 mark)

22 **i** For $x = \frac{1}{3}$:

$$\text{LHS} = 7 - 3\left(\tfrac{1}{3}\right)$$
$$= 7 - 1$$
$$= 6$$

$$\text{RHS} = \frac{2}{\frac{1}{3}}$$
$$= 2 \times 3$$
$$= 6$$

$\therefore$ $x = \frac{1}{3}$ satisfies the equation.

For $x = 2$:

$$\text{LHS} = 7 - 3(2)$$
$$= 7 - 6$$
$$= 1$$

$$\text{RHS} = \frac{2}{2}$$
$$= 1$$

$\therefore$ $x = 2$ satisfies the equation.

ii

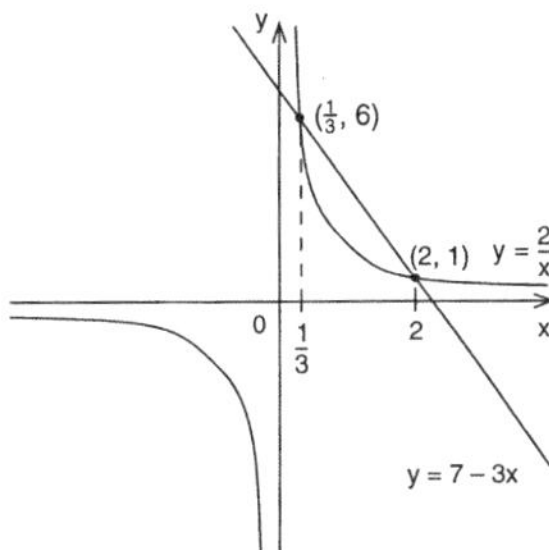

ii From the diagram:

$$7 - 3x < \frac{2}{x}$$

for $0 < x < \frac{1}{3}$ and $x > 2$.

(No mark allocation in exam)

23 **i**

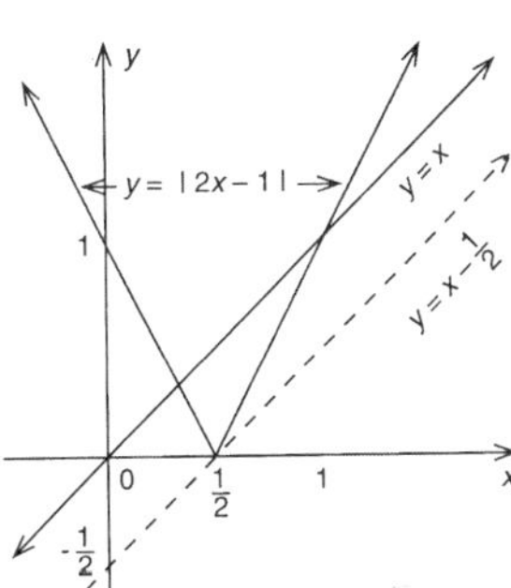

ii The equation $|2x - 1| = x + c$ has exactly two solutions for $c > -\frac{1}{2}$.

From the graph:

If $c > -\frac{1}{2}$, the line $y = x + c$ will intersect $y = |2x - 1|$ at 2 different points.

If $c = -\frac{1}{2}$, they will have 1 common point $(\frac{1}{2}, 0)$.

If $c < -\frac{1}{2}$, there are no common points.

(No mark allocation in exam)

1 The graph of $y = k \sin (ax)$ is shown.

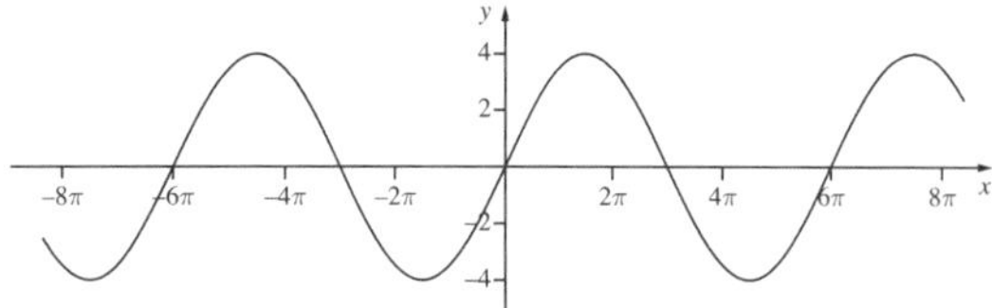

What are the values of k and a? *(2 marks)*

(Q14, **2022 HSC**) Easy

2 The depth of water in a bay rises and falls with the tide. On a particular day the depth of the water, d metres, can be modelled by the equation

$$d = 1.3 - 0.6\cos\left(\frac{4\pi}{25}t\right),$$

where t is the time in hours since low tide.

a Find the depth of water at low tide and at high tide. *(2 marks)* Easy

b What is the time interval, in hours, between two successive low tides? *(1 mark)* Easy

c For how long between successive low tides will the depth of water be at least 1 metre? *(3 marks)* Medium

(Q23, **2022 HSC**)

3 The line $y = mx$ is a tangent to the curve $y = \cos x$ at the point where $x = a$, as shown in the diagram.

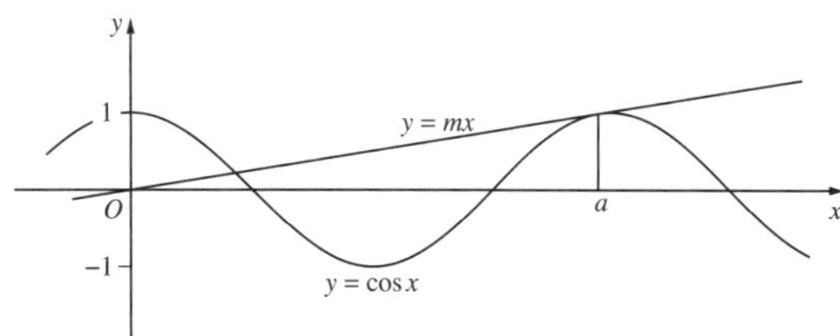

Which of the following statements is true?

A $m < \frac{1}{a} < \frac{1}{2\pi}$ **B** $\frac{1}{2\pi} < m < \frac{1}{a}$

C $\frac{1}{2\pi} < \frac{1}{a} < m$ **D** $m < \frac{1}{2\pi} < \frac{1}{a}$ *(1 mark)*

(Q10, **2021 HSC**) Medium

4 For what values of x, in the interval $0 \le x \le \frac{\pi}{4}$, does the line $y = 1$ intersect the graph of $y = 2\sin 4x$? *(2 marks)*

(Q20, **2021 HSC**) Hard

5 Which interval gives the range of the function $y = 5 + 2\cos 3x$?

A $[2, 8]$ **B** $[3, 7]$

C $[4, 6]$ **D** $[5, 9]$ *(1 mark)*

(Q6, **2020 HSC**) Easy

6 **i** Write down the range of $f(x) = 2\cos x$. *(1 mark)* Easy

ii Sketch $y = f\left(x + \frac{\pi}{3}\right) + 1$, for $0 \le x \le 2\pi$, on the number plane below. *(2 marks)* Medium

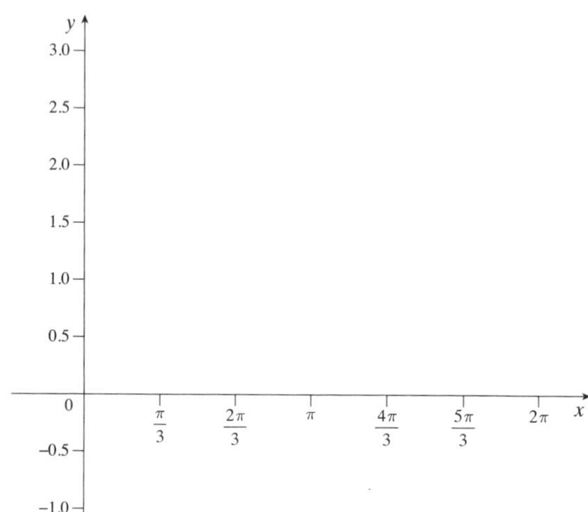

Bonus question (see page vi)

7 The diagram shows part of the graph of $y = a\sin(bx) + 4$.

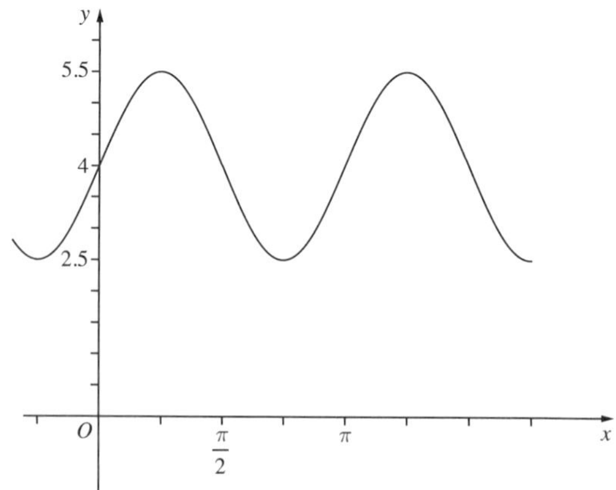

What are the values of a and b?

A $a = 3 \quad b = \frac{1}{2}$ **B** $a = 3 \quad b = 2$

C $a = 1.5 \quad b = \frac{1}{2}$ **D** $a = 1.5 \quad b = 2$

(1 mark)

(Q7, **2019 HSC**) Medium

8 The length of daylight, $L(t)$, is defined as the number of hours from sunrise to sunset, and can be modelled by the equation

$$L(t) = 12 + 2\cos\left(\frac{2\pi t}{366}\right),$$

where t is the number of days after 21 December 2015, for $0 \leq t \leq 366$.

i Find the length of daylight on 21 December 2015. *(1 mark)* Easy

ii What is the shortest length of daylight? *(1 mark)* Medium

iii What are the two values of t for which the length of daylight is 11? *(2 marks)* Medium

(Q15a, **2018 HSC**)

9 Sketch the curve $y = 4 + 3\sin 2x$ for $0 \leq x \leq 2\pi$. *(3 marks)*

(Q14a, **2017 HSC**) Easy

10 What is the period of the function $f(x) = \tan(3x)$?

A $\dfrac{\pi}{3}$ **B** $\dfrac{2\pi}{3}$

C 3π **D** 6π *(1 mark)*

(Q6, **2016 HSC**) Medium

11 Which diagram shows the graph $y = \sin\left(2x + \dfrac{\pi}{3}\right)$?

A

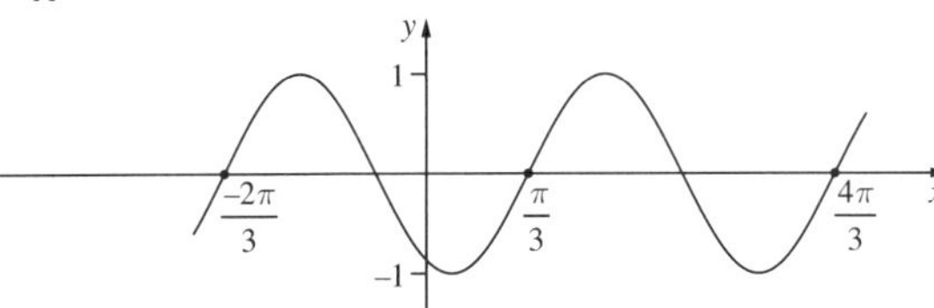

B

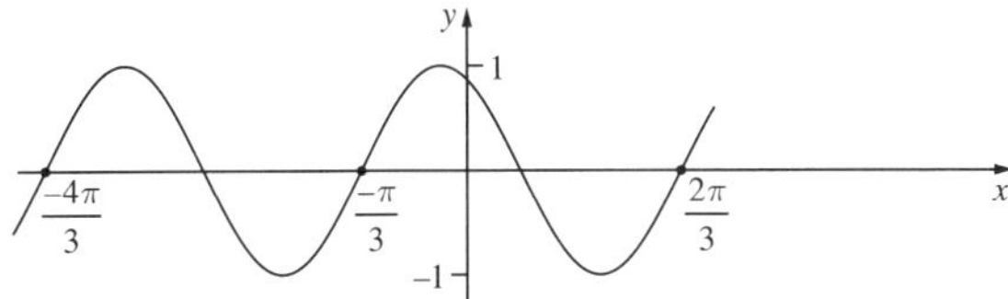

C

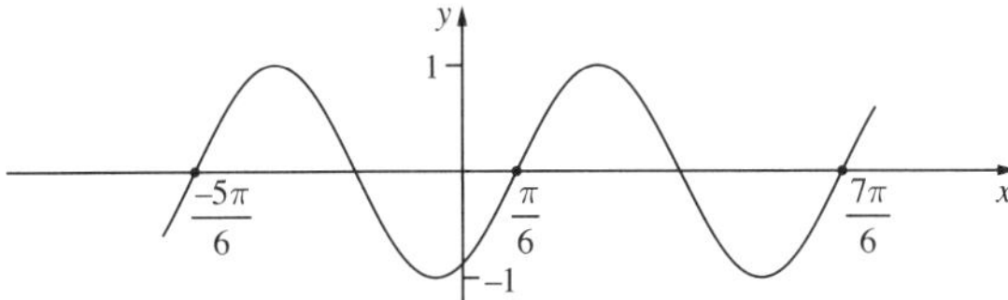

D

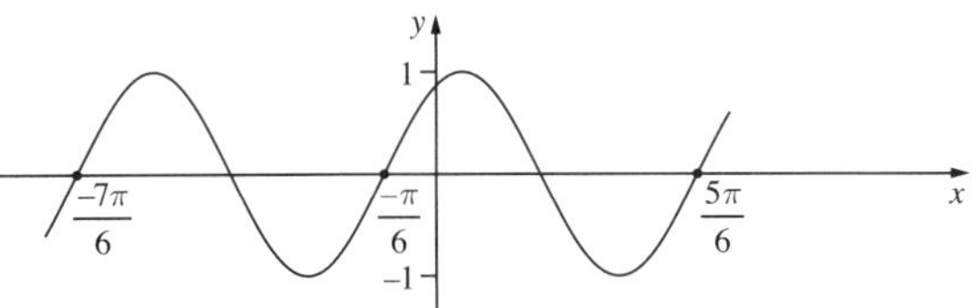

(1 mark)

(Q6, **2013 HSC**) Hard

12 The population of a herd of wild horses is given by

$$P(t) = 400 + 50\cos\left(\frac{\pi}{6}t\right),$$

where t is time in months.

i Find all times during the first 12 months when the population equals 375 horses. *(2 marks)* Medium

ii Sketch the graph of $P(t)$ for $0 \leq t \leq 12$. *(2 marks)* Hard

(Q13a, **2013 HSC**)

13 The graph shown is $y = A\sin bx$.

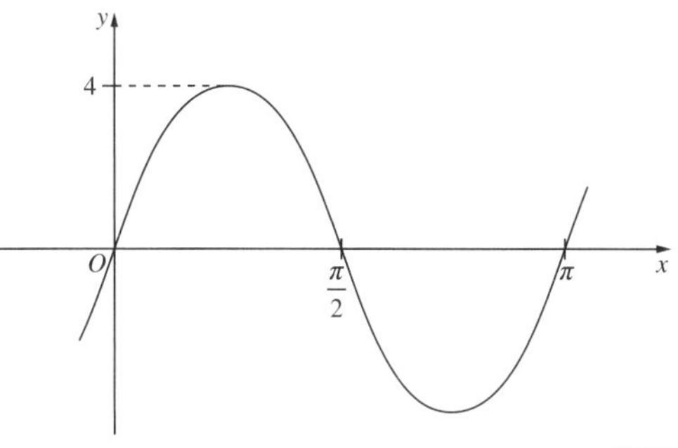

i Write down the value of A. *(1 mark)* Easy

ii Find the value of b. *(1 mark)* Medium

iii Copy or trace the graph into your writing booklet.
On the same set of axes, draw the graph $y = 3\sin x + 1$, for $0 \leq x \leq \pi$. *(2 marks)* Medium

(Q8c, **2010 HSC**)

14 Between 5 am and 5 pm on 3 March 2009, the height, h, of the tide in a harbour was given by

$$h = 1 + 0.7\sin\frac{\pi}{6}t \text{ for } 0 \leq t \leq 12,$$

where h is in metres and t is in hours, with $t = 0$ at 5 am.

i What is the period of the function h? *(1 mark)* Medium

ii What was the value of h at low tide, and at what time did low tide occur? *(2 marks)* Hard

iii A ship is able to enter the harbour only if the height of the tide is at least 1.35 m. Find all times between 5 am and 5 pm on 3 March 2009 during which the ship was able to enter the harbour. *(3 marks)* Hard

(Q7b, **2009 HSC**)

15 Sketch the curve $y = 1 - \sin 2x$ for $0 \leq x \leq \pi$. *(3 marks)*

(Q6a, **2000 HSC**) Medium

16 **i** Show that $x = \dfrac{\pi}{3}$ is a solution of $\sin x = \dfrac{1}{2}\tan x$.

ii On the same set of axes, sketch the graphs of the functions $y = \sin x$ and $y = \frac{1}{2}\tan x$ for $-\pi \leq x \leq \pi$.

iii Hence find all solutions of $\sin x = \frac{1}{2}\tan x$ for $-\frac{\pi}{2} < x < \frac{\pi}{2}$.

iv Use your graphs to solve $\sin x \leq \frac{1}{2}\tan x$ for $-\frac{\pi}{2} < x < \frac{\pi}{2}$. *(6 marks)*

(Q10a, **1999 HSC**) Medium

---

17 **i** On the same set of axes, sketch the graphs of $y = 2\sin\theta$ and $y = \theta$ for $-\pi \leq \theta \leq \pi$.

ii Use your sketch to find the number of solutions of the equation $2\sin\theta = \theta$ for $-\pi \leq \theta \leq \pi$. *(3 marks)*

(Q3a, **1997 ME1 HSC**) Medium

---

18 **i** Sketch the graph of $y = 2\cos x$ for $0 \leq x \leq 2\pi$. Easy

ii On the same set of axes, sketch the graph of $y = 2\cos x - 1$ for $0 \leq x \leq 2\pi$. Easy

iii Find the exact values of the x coordinates of the points where the graph of $y = 2\cos x - 1$ crosses the x axis in the domain $0 \leq x \leq 2\pi$. *(5 marks)* Medium

(Q7a, **1996 HSC**)

---

19 **i** Draw the graphs of $y = 4\cos x$ and $y = 2 - x$ on the same set of axes for $-2\pi \leq x \leq 2\pi$.

ii Explain why all the solutions of the equation $4\cos x = 2 - x$ must lie between $x = -2$ and $x = 6$. *(3 marks)*

(Q10a, **1995 HSC**) Medium

20

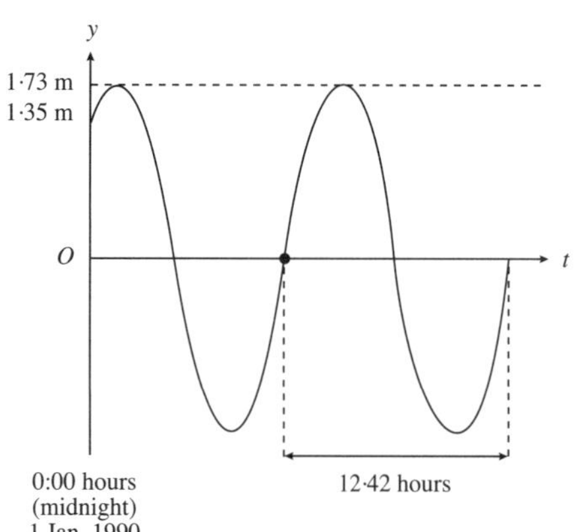

The diagram shows the tidal effect due to the Moon at Port Hedland on 1 January 1900. The water level can be approximated by a sine curve of the form

$$y = A\sin(at + b)$$

where y is the water level in metres measured as on the diagram and t is the time in hours after 0:00 hours.

i Find the amplitude A.

ii Estimate b by letting $t = 0$.

iii Estimate a.

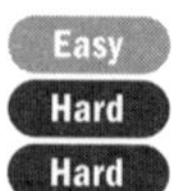

(Q9b, **1993 HSC**)

---

21 Let m be a negative number. Show that the equation

$$\sin x = mx$$

has $x = 0$ as its only solution satisfying $-\pi \leq x \leq \pi$.

(Q9c, **1992 HSC**) Medium

Year 12 Trigonometric functions and graphs—Worked answers

1 Amplitude = 4, hence $k = 4$. ✓

Period = 6π. Hence $\frac{2\pi}{a} = 6\pi$.

$$a = \frac{2\pi}{6\pi}$$
$$= \frac{1}{3} \quad ✓$$

$\therefore k = 4$ and $a = \frac{1}{3}$ *(2 marks)*

2 a Use $-1 \le \cos f(x) \le 1$.

Let $\cos\left(\frac{4\pi}{25}t\right) = -1$:

$\therefore \quad d = 1.3 - 0.6(-1)$
$= 1.9$ ✓

Let $\cos\left(\frac{4\pi}{25}t\right) = 1$:

$\therefore \quad d = 1.3 - 0.6(1)$
$= 0.7$ ✓

$\therefore$ low tide = 0.7 m,
high tide = 1.9 m *(2 marks)*

b For $\cos f(x)$,

$$\text{period} = \frac{2\pi}{n}$$
$$= 2\pi \div \frac{4\pi}{25}$$
$$= 12.5 \quad ✓$$

$\therefore$ 12.5 hours *(1 mark)*

c Let $d =$

$$1.3 - 0.6\cos\left(\frac{4\pi}{25}t\right) = 1$$
$$0.6\cos\left(\frac{4\pi}{25}t\right) = 0.3$$
$$\cos\left(\frac{4\pi}{25}t\right) = 0.5$$
$$\frac{4\pi}{25}t = \frac{\pi}{3}, \frac{5\pi}{3} \quad ✓$$

$\frac{4\pi}{25}t = \frac{\pi}{3} \qquad \frac{4\pi}{25}t = \frac{5\pi}{3}$

$t = \frac{\pi}{3} \times \frac{25}{4\pi} \qquad t = \frac{5\pi}{3} \times \frac{25}{4\pi}$

$t = \frac{25}{12} \qquad t = \frac{125}{12}$ ✓

As $\frac{125}{12} - \frac{25}{12} = \frac{100}{12} = 8\frac{1}{3}$, the time is $8\frac{1}{3}$ hours, or 8 hours 20 minutes. ✓ *(3 marks)*

3

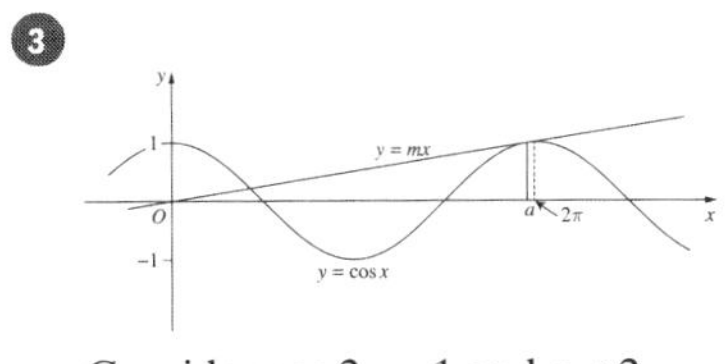

Consider $\cos 2\pi = 1$ and $a < 2\pi$.

Consider the lines $y = \frac{1}{a}x$, $y = mx$ and $y = \frac{1}{2\pi}x$.

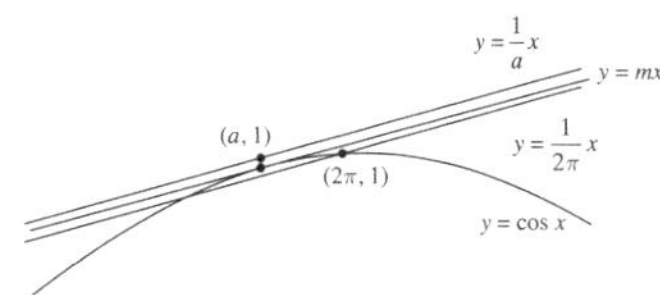

The gradients in increasing steepness are $\frac{1}{2\pi}$, m and $\frac{1}{a}$. This means $\frac{1}{2\pi} < m < \frac{1}{a}$.

Answer B

4 $2\sin 4x = 1$

$$\sin 4x = \frac{1}{2}$$
$$4x = \frac{\pi}{6}, \frac{5\pi}{6} \quad ✓$$
$$x = \frac{\pi}{24}, \frac{5\pi}{24} \quad ✓$$

(2 marks)

5 The range of $y = \cos x$ is $[-1, 1]$.

$\therefore -1 \le \cos 3x \le 1$
$-2 \le 2\cos 3x \le 2$
$3 \le 5 + 2\cos 3x \le 7$

$\therefore$ the range is $[3, 7]$.

Answer B

6 i For $f(x) = 2\cos x$,
range = $-2 \le f(x) \le 2$. ✓ *(1 mark)*

ii $f\left(x + \frac{\pi}{3}\right) + 1$
$= 2\cos\left(x + \frac{\pi}{3}\right) + 1.$

The new range will be
$-1 \le f(x) \le 3$. ✓

To sketch $y = 2\cos\left(x + \frac{\pi}{3}\right) + 1$, substitute values for x.

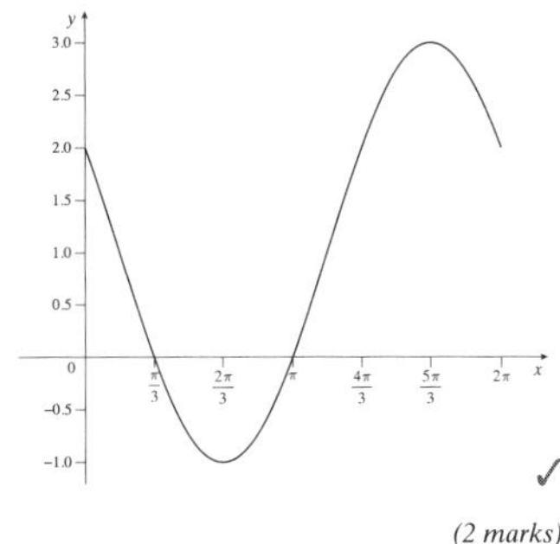

✓ *(2 marks)*

7 Amplitude $a = 5.5 - 4$
$= 1.5$

Period = π

$\therefore$ the equation is
$y = 1.5\sin(2x) + 4$

$\therefore a = 1.5, b = 2$

Answer D

8 i Substitute $t = 0$:

$$L(0) = 12 + 2\cos\left(\frac{2\pi(0)}{366}\right)$$
$$= 12 + 2\cos 0$$
$$= 14$$

$\therefore$ 14 hours daylight ✓ *(1 mark)*

ii As $-1 \le \cos\theta \le 1$, then minimum value of $\cos\left(\frac{2\pi t}{366}\right)$ is -1.

Minimum value of L
$= 12 + 2(-1)$
$= 10$

$\therefore$ 10 hours ✓ *(1 mark)*

iii $12 + 2\cos\left(\frac{2\pi t}{366}\right) = 11$

$$2\cos\left(\frac{2\pi t}{366}\right) = -1$$
$$\cos\left(\frac{2\pi t}{366}\right) = -\frac{1}{2} \quad ✓$$
$$\frac{2\pi t}{366} = \frac{2\pi}{3}, \frac{4\pi}{3}$$
$$t = 122, 244 \quad ✓$$

(2 marks)

9

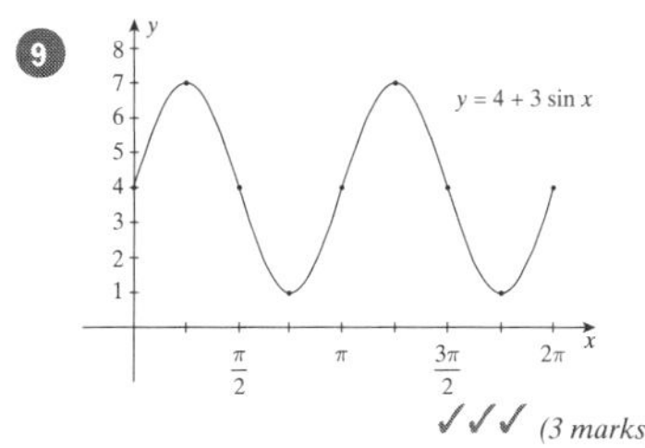

✓✓✓ *(3 marks)*

10

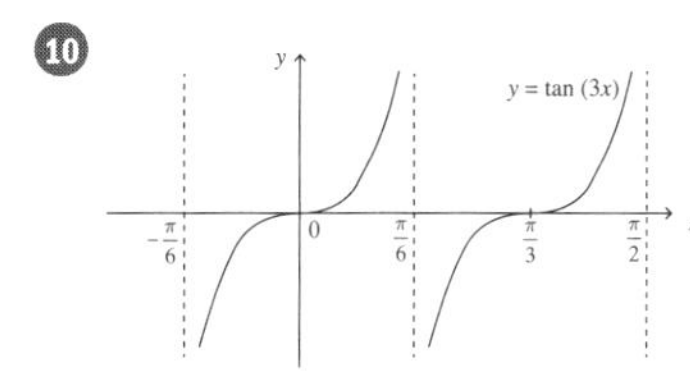

The period is $\frac{\pi}{3}$.

Answer A

11 Solve $\sin\left(2x + \frac{\pi}{3}\right) = 0$

$$2x + \frac{\pi}{3} = 0, \pi, 2\pi \ldots -\pi, -2\pi \ldots$$
$$2x = -\frac{\pi}{3}, \frac{2\pi}{3}, \frac{5\pi}{3} \ldots \frac{-4\pi}{3}, \frac{-7\pi}{3} \ldots$$
$$x = -\frac{\pi}{6}, \frac{\pi}{3}, \frac{5\pi}{6}, \frac{-2\pi}{3}, \frac{-7\pi}{6} \ldots$$

Answer D

12 $P(t) = 400 + 50\cos\left(\frac{\pi}{6}t\right)$

i $400 + 50\cos\left(\frac{\pi}{6}t\right) = 375$

$50\cos\left(\frac{\pi}{6}t\right) = 375 - 400$

$50\cos\left(\frac{\pi}{6}t\right) = -25$

$\cos\left(\frac{\pi}{6}t\right) = -\frac{1}{2}$ ✓

$\frac{\pi}{6}t = \frac{2\pi}{3}, \frac{4\pi}{3}, \frac{8\pi}{3}\ldots$

$t = 4, 8, 16 \ldots$

After 4 months and 8 months. ✓

(2 marks)

ii $P(0)$

$= 400 + 50\cos\left(\frac{\pi}{6} \times 0\right)$

$= 400 + 50\cos 0$

$= 450$

$P(12)$

$= 400 + 50\cos\left(\frac{\pi}{6} \times 12\right)$

$= 400 + 50\cos 2\pi$

$= 450$

Also, $P(6)$

$= 400 + 50\cos\left(\frac{\pi}{6} \times 6\right)$

$= 400 + 50\cos\pi$

$= 350$ ✓

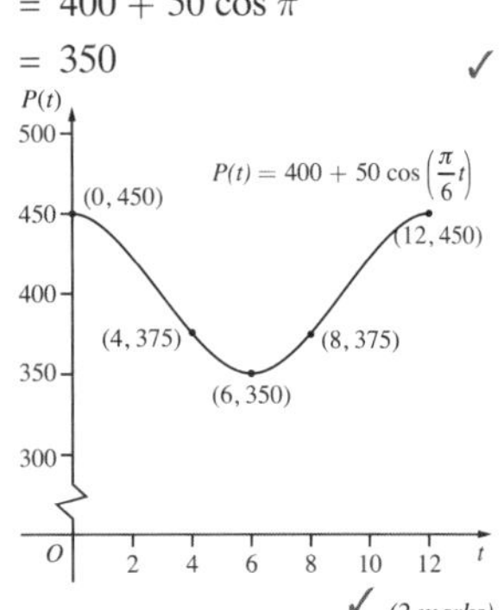

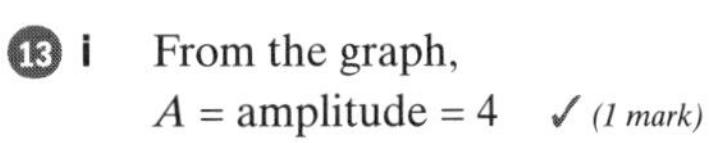

✓ *(2 marks)*

13 **i** From the graph,
A = amplitude = 4 ✓ *(1 mark)*

ii From the graph $\left(\frac{\pi}{4}, 4\right)$ lies on the curve.

Subs. into $y = 4\sin bx$:

$4 = 4\sin\frac{\pi . b}{4}$

$\sin\frac{\pi b}{4} = 1$

$\frac{\pi b}{4} = \frac{\pi}{2}$

$b = \frac{\pi}{2} \div \frac{\pi}{4}$

$= 2$ ✓ *(1 mark)*

iii

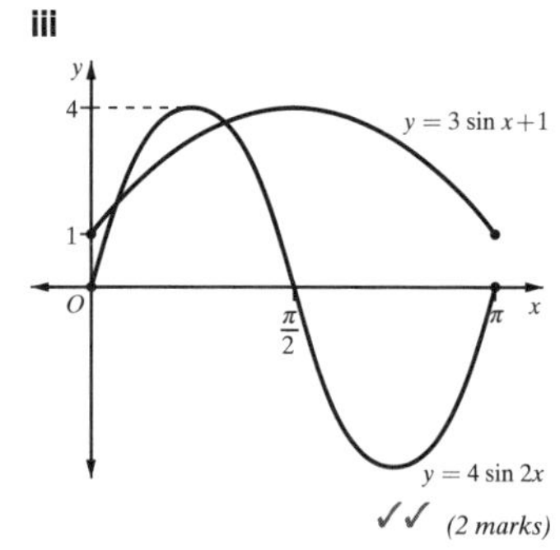

✓✓ *(2 marks)*

14 **i** Period $= \frac{2\pi}{n}$ where $n = \frac{\pi}{6}$

$= \frac{2\pi}{\frac{\pi}{6}}$

$= 12$ ✓ *(1 mark)*

ii The minimum value of h will occur when $\sin\frac{\pi}{6}t$ is a minimum.

The minimum value of $\sin\frac{\pi}{6}t$ is -1,

as $-1 \leqslant \sin\frac{\pi}{6}t \leqslant 1$.

$h = 1 + 0.7\sin\frac{\pi}{6}t$ for $0 \leqslant t \leqslant 12$.

$\therefore h = 1 + 0.7(-1)$

$= 1 - 0.7$

$= 0.3$ ✓

$\therefore$ At low tide, the value of h is 0.3.

$\therefore 0.3 = 1 + 0.7\sin\frac{\pi}{6}t$

$-0.7 = 0.7\sin\frac{\pi}{6}t$ ✓

$\sin\frac{\pi}{6}t = -1$

$\frac{\pi}{6}t = \frac{3\pi}{2}$

$t = \frac{3\pi}{2} \times \frac{6}{\pi}$

$= 9$

$\therefore$ Low tide occurs after 9 hours, that is, 5 a.m. + 9 hours.

$\therefore$ Low tide occurs at 2 p.m. ✓

(2 marks)

iii The ship may enter the harbour when $h \geqslant 1.35$.

Solving for $h = 1.35$:

$1 + 0.7\sin\frac{\pi}{6}t = 1.35$

$0.7\sin\frac{\pi}{6}t = 0.35$

$\sin\frac{\pi}{6}t = \frac{1}{2}$ ✓

$\frac{\pi}{6}t = \frac{\pi}{6}, \pi - \frac{\pi}{6}$

$= \frac{\pi}{6}, \frac{5\pi}{6}$

$\therefore t = 1, 5$ ✓

By drawing the graphs of $y = \sin\frac{\pi}{6}t$ and $y = \frac{1}{2}$, it can be seen that the ship may enter the harbour between $t = 1$ and $t = 5$, i.e. $\sin\frac{\pi}{6}t \geqslant \frac{1}{2}$.

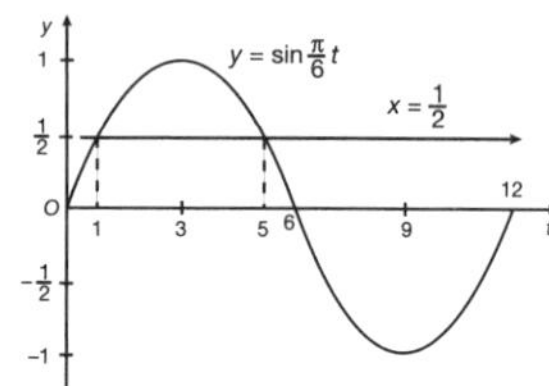

$\therefore$ The ship may enter the harbour between 6 a.m. and 10 a.m. as $t = 0$ at 5 a.m. ✓

(3 marks)

15

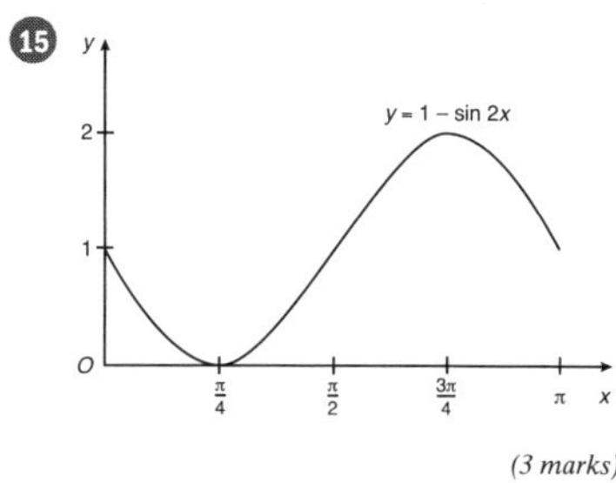

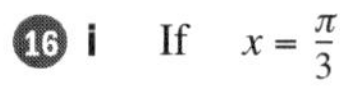

(3 marks)

16 **i** If $x = \frac{\pi}{3}$

LHS $= \sin\frac{\pi}{3}$

$= \frac{\sqrt{3}}{2}$

RHS $= \frac{1}{2}\tan\frac{\pi}{3}$

$= \frac{1}{2} \times \sqrt{3}$

$= \frac{\sqrt{3}}{2}$

$\therefore x = \frac{\pi}{3}$ is a solution ✓

ii

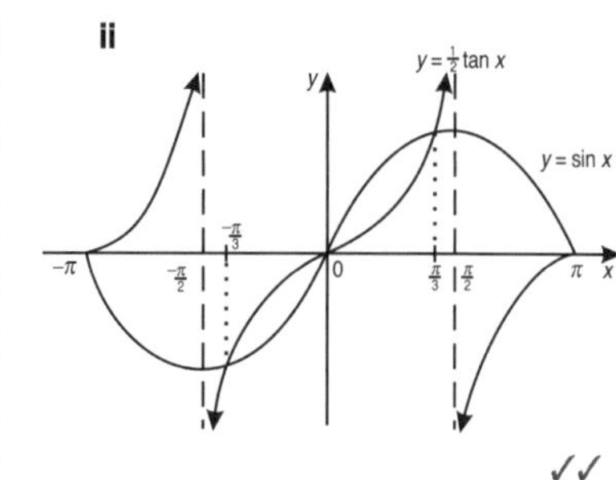

✓✓

iii From the graph, solutions are

$x = -\frac{\pi}{3}, 0$ and $\frac{\pi}{3}$ ✓

iv $\sin x \leqslant \frac{1}{2}\tan x \quad -\frac{\pi}{2} < x < \frac{\pi}{2}$

when $\frac{\pi}{3} \leqslant x < \frac{\pi}{2}$ ✓

and $-\frac{\pi}{3} \leqslant x \leqslant 0$ ✓

(6 marks)
(Total mark allocation only in exam)

17 **i**

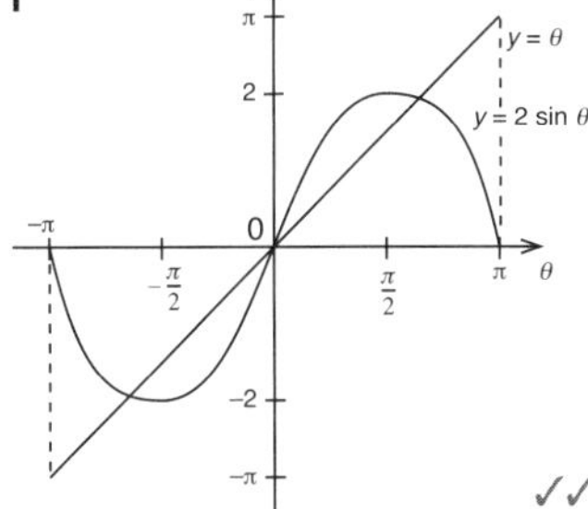

✓✓

ii From the graph, there are 3 solutions. ✓
(One is 0, one is between $-\pi$ and $-\frac{\pi}{2}$, and the other is between $\frac{\pi}{2}$ and π.)

(3 marks)
(Total mark allocation only in exam)

18 **i and ii**

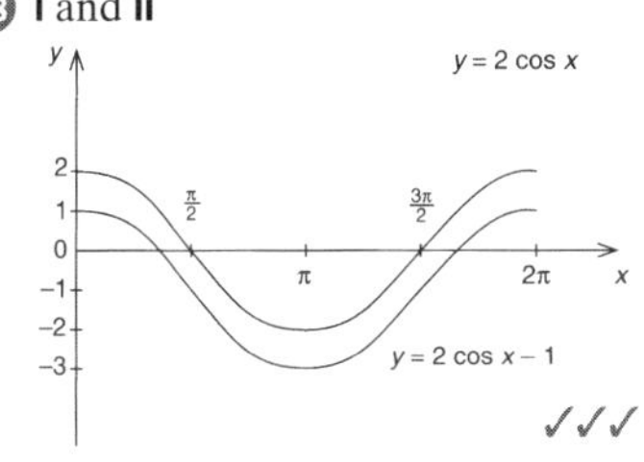

✓✓✓

iii $\therefore 2\cos x - 1 = 0$
$0 \leqslant x \leqslant 2\pi.$

$2\cos x = 1$

$\cos x = \frac{1}{2}$

$\therefore x = \frac{\pi}{3}$ or $\left(2\pi - \frac{\pi}{3}\right)$

$= \frac{\pi}{3}$ or $\frac{5\pi}{3}$ ✓✓

(5 marks)
(Total mark allocation only in exam)

19 **i**

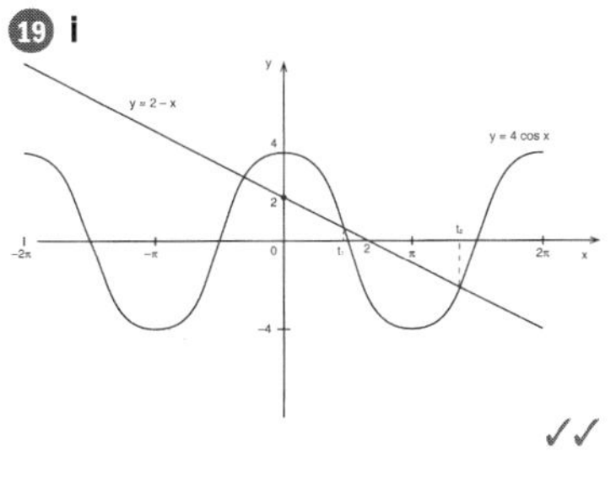

✓✓

ii Since the maximum and minimum valeus of $y = 4\cos x$ are 4 and –4 respectively, it can be seen from the graph that solutions to the equation $4\cos x = 2 - x$ must lie between the points where $2 - x = 4$ and $2 - x = -4$, ✓
i.e. between $x = -2$ and $x = 6$

(3 marks)
(Total mark allocation only in exam)

20 **i** $A = 1.73$ (from graph)

ii $y = 1.73\sin(at + b)$

When $t = 0, y = 1.35$

$\therefore 1.35 = 1.73\sin b$

$\sin b = \frac{1.35}{1.73}$

$\therefore b = 0.8952\ldots$

$= 0.90$ radians to 2 decimal places.

iii $T = \frac{2\pi}{a}$

$a = \frac{2\pi}{T}$

$= \frac{2\pi}{12.42}$

$= 0.5058\ldots$

$= 0.51$ to 2 decimal places.

(No mark allocation in exam)

21

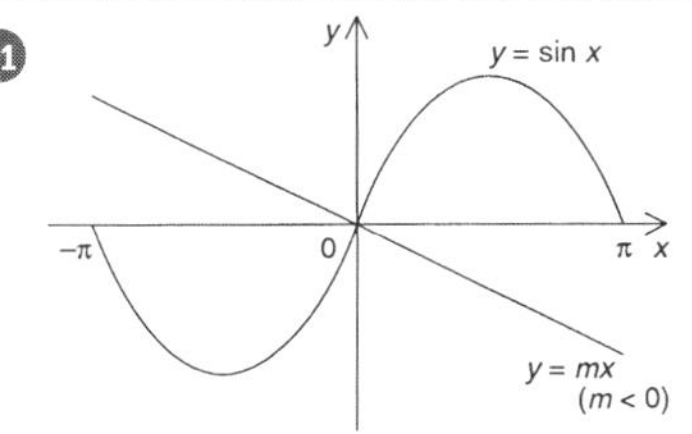

The solution to the equation can be found by considering the intersection of the curve $y = \sin x$ and the line $y = mx$ $(m < 0, -\pi \leq x \leq \pi)$.
From the sketch, there is only 1 possible point of intersection – (0, 0).

(No mark allocation in exam)

1 $\frac{d}{dx}(\log_a x) =$

A $\frac{x}{\ln a}$ **B** $\frac{1}{x \ln a}$

C $\frac{\ln a}{x}$ **D** $\frac{1}{\ln(ax)}$ *(1 mark)*

Bonus question (see page vi) Easy

2 $\frac{d}{dx}(7^x) =$

A 7^x **B** $\frac{7^x}{\ln 7}$

C $(\ln 7)\, 7^x$ **D** $\log_7 x$ *(1 mark)*

Bonus question Easy

3 What is the derivative of $\cos x - \sin x$?

A $-\cos x - \sin x$ **B** $-\cos x + \sin x$

C $\cos x - \sin x$ **D** $\cos x + \sin x$ *(1 mark)*

Bonus question Medium

4 Find the gradient of the tangent to $y = \sin x$ at the point where $x = \frac{\pi}{3}$. *(2 marks)*

Bonus question Easy

5 Find $f'(1)$ if $f(x) = 5^x$. *(2 marks)*

Bonus question Easy

6 Find the exact value of $f'\left(\frac{2\pi}{3}\right)$ if $f(x) = 3 \sin x + 4 \cos x$. *(2 marks)*

Bonus question Medium

7 **i** Use the logarithmic laws to simplify $6 \log_e x^3$. *(1 mark)* Medium

ii Hence find $f'(9)$ if $f(x) = 6 \log_e x^3$. *(2 marks)* Medium

Bonus question

8 The displacement, x metres at time t seconds, of a moving particle is given by $x = 2 - \cos t$.

i Find the velocity after $\frac{\pi}{6}$ seconds. *(2 marks)* Easy

ii Find the acceleration after π seconds. *(2 marks)* Easy

Bonus question

9 Differentiate $7x - 2 \sin x$ with respect to x. *(1 mark)*

Bonus question Easy

10 **i** Find the equation of the tangent to $y = e^x$ at $x = 0$. *(2 marks)* Medium

ii Find the equation of the tangent to $y = \ln x$ at $x = 1$. *(2 marks)* Medium

iii What conclusion can be made about the two tangents? *(1 mark)* Medium

Bonus question

11 Two particles, A and B, move along a straight line so that their displacements from the origin at time t ($t \geq 0$) are given respectively by:

$x_A = \sin t$ and $x_B = \cos t$.

Which particle is moving faster when $t = \frac{\pi}{3}$? *(3 marks)*

Bonus question Medium

12 Maxine is using the formula $f(x) = 1000 - 65 \ln x$ to model the number of a particular type of animal in a conservation park. x is the number of years after the start of the program.

i Find the approximate number of animals after 10 years and after 20 years. *(1 mark)* Medium

ii Find $f'(10)$ and $f'(20)$. *(2 marks)* Medium

iii Briefly explain what conclusion can be reached about the number of those animals in the park over that time. *(1 mark)* Medium

Bonus question

1 $\frac{d}{dx}(\log_a x) = \frac{1}{x \ln a}$
Answer B

2 $\frac{d}{dx}(7^x) = (\ln 7)\, 7^x$
Answer C

3 $y = \cos x - \sin x$
$\frac{dy}{dx} = -\sin x - \cos x$
$= -\cos x - \sin x$
Answer A

4 $y = \sin x$
$\frac{dy}{dx} = \cos x$ ✓
When $x = \frac{\pi}{3}$,
$\frac{dy}{dx} = \cos \frac{\pi}{3}$
$= \frac{1}{2}$
The gradient of the tangent is $\frac{1}{2}$.
✓ *(2 marks)*

5 $f(x) = 5^x$
$f'(x) = (\ln 5)\, 5^x$ ✓
$f'(1) = (\ln 5)\, 5^1$
$= 5 \ln 5$ ✓
(2 marks)

6 $f(x) = 3 \sin x + 4 \cos x$
$f'(x) = 3 \cos x - 4 \sin x$ ✓
$f'\left(\frac{2\pi}{3}\right) = 3 \cos \frac{2\pi}{3} - 4 \sin \frac{2\pi}{3}$
$= 3 \times -\frac{1}{2} - 4 \times \frac{\sqrt{3}}{2}$
$= \frac{-3 - 4\sqrt{3}}{2}$ ✓
(2 marks)

7 i $6 \log_e x^3 = 18 \log_e x$ ✓
(1 mark)

ii $f(x) = 6 \log_e x^3$
$= 18 \log_e x$
$f'(x) = 18 \times \frac{1}{x}$ ✓
$f'(9) = \frac{18}{9}$
$= 2$ ✓
(2 marks)

8 i $x = 2 - \cos t$
$v = \sin t$ ✓
When $t = \frac{\pi}{6}$,
$v = \sin \frac{\pi}{6}$
$= \frac{1}{2}$
The velocity is 0.5 ms^{-1}. ✓
(2 marks)

ii $v = \sin t$
$a = \cos t$ ✓
When $t = \pi$,
$a = \cos \pi$
$= -1$
The acceleration is -1 ms^{-2}.
✓ *(2 marks)*

9 $y = 7x - 2 \sin x$
$y' = 7 - 2 \cos x$ ✓
(1 mark)

10 i $y = e^x$
When $x = 0$, $y = e^0 = 1$
$\frac{dy}{dx} = e^x$ ✓
When $x = 0$, $\frac{dy}{dx} = e^0 = 1$
So $m = 1$, $(0, 1)$
$y - y_1 = m(x - x_1)$
$y - 1 = 1(x - 0)$
$y - 1 = x$
$y = x + 1$ ✓
(2 marks)

ii $y = \ln x$
When $x = 1$, $y = \ln 1 = 0$
$\frac{dy}{dx} = \frac{1}{x}$ ✓
When $x = 1$, $\frac{dy}{dx} = \frac{1}{1} = 1$
$y - y_1 = m(x - x_1)$
$y - 0 = 1(x - 1)$
$y = x - 1$ ✓
(2 marks)

iii The tangents are parallel. ✓
(1 mark)

11 $x_A = \sin t$
$v_A = \cos t$ ✓
When $t = \frac{\pi}{3}$,
$v_A = \cos \frac{\pi}{3}$
$= \frac{1}{2}$
$x_B = \cos t$
$v_B = -\sin t$ ✓
When $t = \frac{\pi}{3}$,
$v_B = -\sin \frac{\pi}{3}$
$= -\frac{\sqrt{3}}{2}$
So particle B is moving faster than particle A but in the opposite direction. ✓
3 marks

12 i $f(x) = 1000 - 65 \ln x$
$f(10) = 1000 - 65 \ln 10$
≈ 850
$f(20) = 1000 - 65 \ln 20$
≈ 805
There are approximately 850 animals after 10 years and 805 after 20 years. ✓
(1 mark)

ii $f'(x) = -\frac{65}{x}$
$f'(10) = -\frac{65}{10}$
$= -6.5$ ✓
$f'(20) = -\frac{65}{20}$
$= -3.25$ ✓
(2 marks)

iii The number of animals in the park is decreasing but at a decreasing rate. ✓
(1 mark)

1 Find the exact gradient of the tangent to the curve $y = x\tan x$ at the point where $x = \frac{\pi}{3}$. *(3 marks)*

(Q13, **2021 HSC**) Medium

2 Which of these is the derivative of $\log_2(x+1)$?

A $\frac{2}{x+1}$ **B** $\frac{1}{2(x+1)}$

C $\frac{1}{(x+1)\ln 2}$ **D** $\frac{1}{\ln(2x+1)}$ *(1 mark)*

Bonus question (see page vi) Medium

3 Differentiate $\log_2(x^2+1)$. *(2 marks)*

Bonus question Medium

4 Differentiate $x^2 \sin x$. *(2 marks)*

(Q11b, **2019 HSC**) Easy

5 What is the derivative of $\sin(\ln x)$, where $x > 0$?

A $\cos\left(\frac{1}{x}\right)$ **B** $\cos(\ln x)$

C $\cos\left(\frac{\ln x}{x}\right)$ **D** $\frac{\cos(\ln x)}{x}$ *(1 mark)*

(Q5, **2018 HSC**) Easy

6 Differentiate $x^2 \tan x$. *(2 marks)*

(Q11f, **2018 HSC**) Easy

7 Differentiate $\frac{e^x}{x+1}$. *(2 marks)*

(Q11g, **2018 HSC**) Easy

8 What is the derivative of e^{x^2}?

A $x^2e^{x^2-1}$ **B** $2xe^{2x}$

C $2xe^{x^2}$ **D** $2e^{x^2}$ *(1 mark)*

(Q3, **2017 HSC**) Easy

9 Differentiate $\frac{\sin x}{x}$. *(2 marks)*

(Q11c, **2017 HSC**) Easy

10 Differentiate $x^3 \ln x$. *(2 marks)*

(Q11d, **2017 HSC**) Easy

11 What is the derivative of $\ln(\cos x)$?

A $-\sec x$ **B** $-\tan x$

C $\sec x$ **D** $\tan x$ *(1 mark)*

(Q5, **2016 HSC**) Easy

12 What is the value of the derivative of $y = 2\sin 3x - 3\tan x$ at $x = 0$?

A -1 **B** 0

C 3 **D** -9 *(1 mark)*

(Q6, **2015 HSC**) Easy

13 Differentiate $(e^x + x)^5$. *(2 marks)*

(Q11e, **2015 HSC**) Easy

14 Differentiate $y = (x+4)\ln x$. *(2 marks)*

(Q11f, **2015 HSC**) Easy

15 Differentiate $\frac{e^x \ln x}{x}$. *(2 marks)*

(Q11f, **2014 ME1 HSC**) Medium

16 What is the derivative of $\frac{x}{\cos x}$?

A $\frac{\cos x + x\sin x}{\cos^2 x}$ **B** $\frac{\cos x - x\sin x}{\cos^2 x}$

C $\frac{x\sin x - \cos x}{\cos^2 x}$ **D** $\frac{-x\sin x - \cos x}{\cos^2 x}$

(1 mark)

(Q4, **2013 HSC**) Easy

17 Differentiate $(\sin x - 1)^8$. *(2 marks)*

(Q11c, **2013 HSC**) Easy

18 Differentiate $(3 + e^{2x})^5$. *(2 marks)*

(Q11d, **2012 HSC**) Easy

19 Differentiate with respect to x.

i $(x-1)\log_e x$ *(2 marks)* Easy

ii $\frac{\cos x}{x^2}$ *(2 marks)* Easy

(Q12a, **2012 HSC**)

20 Differentiate $x^2 \tan x$ with respect to x. *(2 marks)*

(Q11b, **2012 ME1 HSC**) Easy

21 Differentiate $\ln(5x+2)$ with respect to x. *(2 marks)*

(Q1d, **2011 HSC**) Easy

22 Find the derivative of $y = x^2e^x$ with respect to x. *(2 marks)*

(Q2d, **2011 HSC**) Easy

23 Differentiate $\frac{x}{\sin x}$ with respect to x. *(2 marks)*

(Q4a, **2011 HSC**) Easy

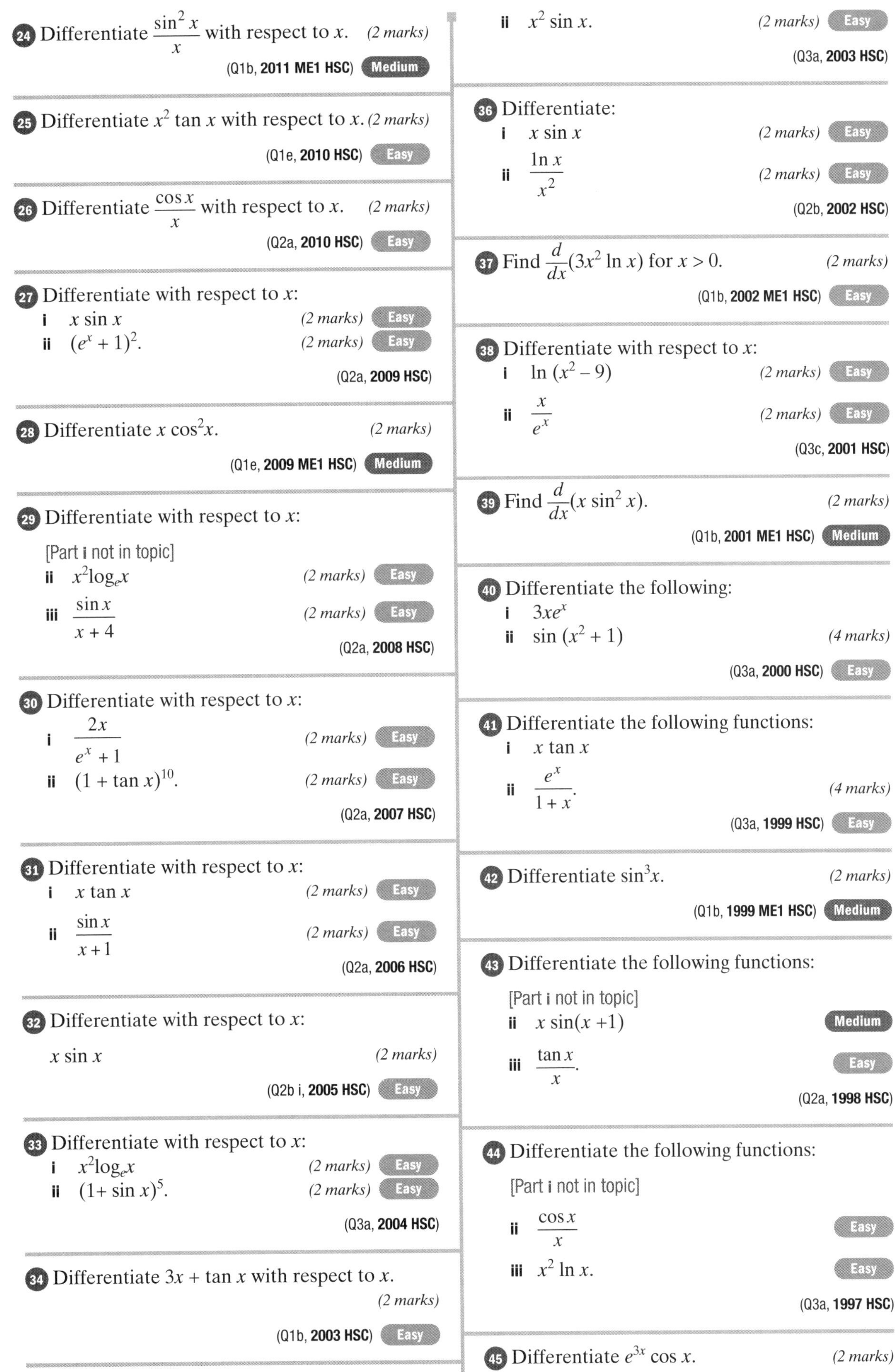

24 Differentiate $\dfrac{\sin^2 x}{x}$ with respect to x. *(2 marks)*

(Q1b, **2011 ME1 HSC**) **Medium**

25 Differentiate $x^2 \tan x$ with respect to x. *(2 marks)*

(Q1e, **2010 HSC**) **Easy**

26 Differentiate $\dfrac{\cos x}{x}$ with respect to x. *(2 marks)*

(Q2a, **2010 HSC**) **Easy**

27 Differentiate with respect to x:

i $x \sin x$ *(2 marks)* **Easy**

ii $(e^x + 1)^2$. *(2 marks)* **Easy**

(Q2a, **2009 HSC**)

28 Differentiate $x \cos^2 x$. *(2 marks)*

(Q1e, **2009 ME1 HSC**) **Medium**

29 Differentiate with respect to x:

[Part **i** not in topic]

ii $x^2 \log_e x$ *(2 marks)* **Easy**

iii $\dfrac{\sin x}{x + 4}$ *(2 marks)* **Easy**

(Q2a, **2008 HSC**)

30 Differentiate with respect to x:

i $\dfrac{2x}{e^x + 1}$ *(2 marks)* **Easy**

ii $(1 + \tan x)^{10}$. *(2 marks)* **Easy**

(Q2a, **2007 HSC**)

31 Differentiate with respect to x:

i $x \tan x$ *(2 marks)* **Easy**

ii $\dfrac{\sin x}{x + 1}$ *(2 marks)* **Easy**

(Q2a, **2006 HSC**)

32 Differentiate with respect to x:

$x \sin x$ *(2 marks)*

(Q2b i, **2005 HSC**) **Easy**

33 Differentiate with respect to x:

i $x^2 \log_e x$ *(2 marks)* **Easy**

ii $(1 + \sin x)^5$. *(2 marks)* **Easy**

(Q3a, **2004 HSC**)

34 Differentiate $3x + \tan x$ with respect to x. *(2 marks)*

(Q1b, **2003 HSC**) **Easy**

35 Differentiate with respect to x:

i $(2e^x - 4)^9$ *(2 marks)* **Easy**

ii $x^2 \sin x$. *(2 marks)* **Easy**

(Q3a, **2003 HSC**)

36 Differentiate:

i $x \sin x$ *(2 marks)* **Easy**

ii $\dfrac{\ln x}{x^2}$ *(2 marks)* **Easy**

(Q2b, **2002 HSC**)

37 Find $\dfrac{d}{dx}(3x^2 \ln x)$ for $x > 0$. *(2 marks)*

(Q1b, **2002 ME1 HSC**) **Easy**

38 Differentiate with respect to x:

i $\ln (x^2 - 9)$ *(2 marks)* **Easy**

ii $\dfrac{x}{e^x}$ *(2 marks)* **Easy**

(Q3c, **2001 HSC**)

39 Find $\dfrac{d}{dx}(x \sin^2 x)$. *(2 marks)*

(Q1b, **2001 ME1 HSC**) **Medium**

40 Differentiate the following:

i $3xe^x$

ii $\sin (x^2 + 1)$ *(4 marks)*

(Q3a, **2000 HSC**) **Easy**

41 Differentiate the following functions:

i $x \tan x$

ii $\dfrac{e^x}{1 + x}$. *(4 marks)*

(Q3a, **1999 HSC**) **Easy**

42 Differentiate $\sin^3 x$. *(2 marks)*

(Q1b, **1999 ME1 HSC**) **Medium**

43 Differentiate the following functions:

[Part **i** not in topic]

ii $x \sin(x + 1)$ **Medium**

iii $\dfrac{\tan x}{x}$. **Easy**

(Q2a, **1998 HSC**)

44 Differentiate the following functions:

[Part **i** not in topic]

ii $\dfrac{\cos x}{x}$ **Easy**

iii $x^2 \ln x$. **Easy**

(Q3a, **1997 HSC**)

45 Differentiate $e^{3x} \cos x$. *(2 marks)*

(Q1a, **1997 ME1 HSC**) **Easy**

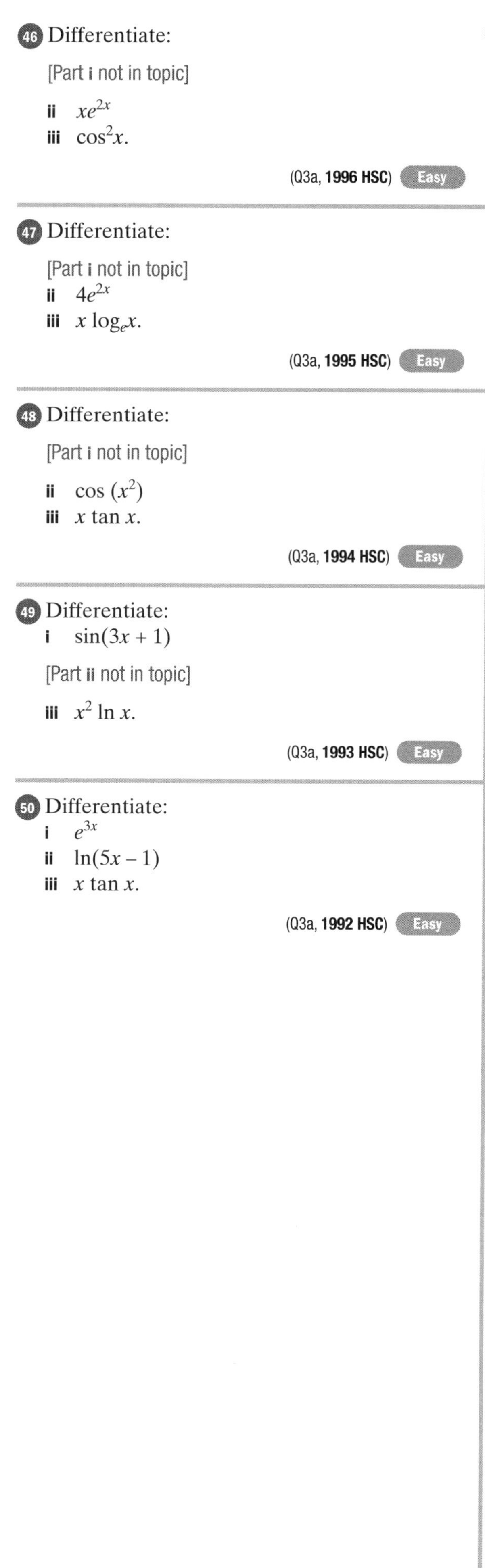

46 Differentiate:

[Part **i** not in topic]

ii xe^{2x}

iii $\cos^2 x$.

(Q3a, **1996 HSC**) Easy

47 Differentiate:

[Part **i** not in topic]

ii $4e^{2x}$

iii $x \log_e x$.

(Q3a, **1995 HSC**) Easy

48 Differentiate:

[Part **i** not in topic]

ii $\cos(x^2)$

iii $x \tan x$.

(Q3a, **1994 HSC**) Easy

49 Differentiate:

i $\sin(3x + 1)$

[Part **ii** not in topic]

iii $x^2 \ln x$.

(Q3a, **1993 HSC**) Easy

50 Differentiate:

i e^{3x}

ii $\ln(5x - 1)$

iii $x \tan x$.

(Q3a, **1992 HSC**) Easy

Year 12 Rules of differentiation—Worked answers

1 $y = x \tan x$

Using the product rule:

Let $u = x$ $\quad \dfrac{du}{dx} = 1$

Let $v = \tan x$ $\quad \dfrac{dv}{dx} = \sec^2 x$

$$\frac{dy}{dx} = u\frac{dv}{dx} + v\frac{du}{dx}$$
$$= x.\sec^2 x + \tan x.1$$
$$= x \sec^2 x + \tan x \quad ✓$$
$$\frac{dy}{dx}\left(\frac{\pi}{3}\right) = \frac{\pi}{3}\sec^2\frac{\pi}{3} + \tan\frac{\pi}{3}$$
$$= \frac{\pi}{3} \times \frac{1}{\cos^2\frac{\pi}{3}} + \tan\frac{\pi}{3}$$
$$= \frac{\pi}{3} \times \frac{1}{\left(\frac{1}{2}\right)^2} + \sqrt{3} \quad ✓$$
$$= \frac{\pi}{3} \times 4 + \sqrt{3}$$
$$= \sqrt{3} + \frac{4\pi}{3} \quad ✓$$

(3 marks)

2 $\dfrac{d}{dx}(\log_2(x+1))$

$$= \frac{d}{dx}\left(\frac{\ln(x+1)}{\ln 2}\right)$$
$$= \frac{d}{dx}\left(\frac{1}{\ln 2}\ln(x+1)\right)$$
$$= \frac{1}{\ln 2}.\frac{1}{x+1}$$
$$= \frac{1}{(x+1)\ln 2}$$

Answer C

3 Firstly $\log_2(x^2+1) = \dfrac{\ln(x^2+1)}{\ln 2}$. ✓

$$\therefore \frac{d}{dx}\left(\frac{\ln(x^2+1)}{\ln 2}\right) = \frac{2x}{(x^2+1)\ln 2} \quad ✓$$

(2 marks)

4 Using the product rule:

Let $u = x^2$ $\quad \dfrac{du}{dx} = 2x$

Let $v = \sin x$ $\quad \dfrac{dv}{dx} = \cos x$ ✓

$$\frac{d}{dx}[x^2 \sin x] = u\frac{dv}{dx} + v\frac{du}{dx}$$
$$= x^2 \cos x + 2x \sin x$$

✓ *(2 marks)*

5 $\dfrac{d}{dx}[\sin(\ln x)] = \dfrac{1}{x}\cos(\ln x)$

$$= \frac{\cos(\ln x)}{x}$$

Answer D

6 Using the product rule:

Let $u = x^2$ $\quad \dfrac{du}{dx} = 2x$

Let $v = \tan x$ $\quad \dfrac{dv}{dx} = \sec^2 x$

$$\frac{d}{dx}\left[x^2 \tan x\right] = u\frac{dv}{dx} + v\frac{du}{dx}$$
$$= x^2.\sec^2 x + \tan x.2x$$
$$= x^2 \sec^2 x + 2x \tan x$$

✓✓ *(2 marks)*

7 Using the quotient rule:

Let $u = e^x$ $\quad \dfrac{du}{dx} = e^x$

Let $v = x + 1$ $\quad \dfrac{dv}{dx} = 1$

$$\frac{d}{dx}\left[\frac{e^x}{x+1}\right] = \frac{v\frac{du}{dx} - u\frac{dv}{dx}}{v^2}$$
$$= \frac{(x+1).e^x - e^x.1}{(x+1)^2} \quad ✓$$
$$= \frac{xe^x + e^x - e^x}{(x+1)^2}$$
$$= \frac{xe^x}{(x+1)^2} \quad ✓$$

(2 marks)

8 $\dfrac{d}{dx}\left[e^{x^2}\right] = 2xe^{x^2}$

Answer C

9 Using the quotient rule:

Let $u = \sin x$ $\quad \dfrac{du}{dx} = \cos x$

Let $v = x$ $\quad \dfrac{dv}{dx} = 1$ ✓

$$\frac{d}{dx}\left[\frac{\sin x}{x}\right] = \frac{v\frac{du}{dx} - u\frac{dv}{dx}}{v^2}$$
$$= \frac{x.\cos x - \sin x.1}{x^2}$$
$$= \frac{x\cos x - \sin x}{x^2} \quad ✓$$

(2 marks)

10 Using the product rule:

Let $u = x^3$ $\quad \dfrac{du}{dx} = 3x^2$

Let $v = \ln x$ $\quad \dfrac{dv}{dx} = \dfrac{1}{x}$ ✓

$$\frac{d}{dx}\left[x^3 \ln x\right] = u\frac{dv}{dx} + v\frac{du}{dx}$$
$$= x^3.\frac{1}{x} + \ln x.3x^2$$
$$= x^2 + 3x^2 \ln x \quad ✓$$

(2 marks)

11 $\dfrac{d}{dx}[\ln(\cos x)] = \dfrac{1}{\cos x}.-\sin x$

$$= -\tan x$$

Answer B

12
$$y = 2\sin 3x - 3\tan x$$
$$y' = 6\cos 3x - 3\sec^2 x$$
$$= 6\cos 3x - \frac{3}{\cos^2 x}$$
$$y'(0) = 6\cos 3(0) - \frac{3}{\cos^2(0)}$$
$$= 6 \times 1 - \frac{3}{1}$$
$$= 3$$

Answer C

13 Using the function of a function rule:

$$\frac{d}{dx}[(e^x + x)^5] = 5(e^x + x)^4.(e^x + 1)$$

✓✓ *(2 marks)*

14 $y = (x + 4)\ln x$

Using the product rule:

Let $u = x + 4$ $\quad u' = 1$

Let $v = \ln x$ $\quad v' = \dfrac{1}{x}$ ✓

$$\frac{dy}{dx} = u'v + v'u$$
$$= 1.\ln x + \frac{1}{x}(x + 4)$$
$$= \ln x + \frac{x+4}{x} \quad ✓$$

(2 marks)

15 $y = \dfrac{e^x \ln x}{x}$

$$\frac{dy}{dx} = \frac{x(e^x \times \frac{1}{x} + (\ln x) \times e^x) - (e^x \ln x) \times 1}{x^2} \quad ✓$$
$$= \frac{e^x + xe^x \ln x - e^x \ln x}{x^2}$$
$$= \frac{e^x}{x^2}(1 + x\ln x - \ln x) \quad ✓$$

(2 marks)

16 Using the quotient rule:

Let $u = x,$ $\quad u' = 1$

Let $v = \cos x,$ $\quad v' = -\sin x$

$$\frac{d}{dx}\left[\frac{x}{\cos x}\right] = \frac{v.u' - u.v'}{v^2}$$
$$= \frac{\cos x.1 - x.(-\sin x)}{(\cos x)^2}$$
$$= \frac{\cos x + x\sin x}{\cos^2 x}$$

Answer A

17 $(\sin x - 1)^8$

Using the function of a function rule:

$$\frac{d}{dx}[(\sin x - 1)^8]$$
$$= 8(\sin x - 1)^7.\frac{d}{dx}(\sin x - 1) \quad ✓$$
$$= 8(\sin x - 1)^7.\cos x$$
$$= 8\cos x(\sin x - 1)^7 \quad ✓$$

(2 marks)

18 $y = (3 + e^{2x})^5$

Using the function of a function rule:

$$\frac{dy}{dx} = 5(3 + e^{2x})^4.\frac{d}{dx}(3 + e^{2x}) \quad ✓$$

$= 5(3 + e^{2x})^4 . 2e^{2x}$

$= 10e^{2x}(3 + e^{2x})^4$ ✓

(2 marks)

19 **i** $(x - 1)\log_e x$

Using the product rule:

Let $u = x - 1,\ u' = 1$

Let $v = \log_e x,\ v' = \frac{1}{x}$ ✓

$\frac{dy}{dx} = u'.v + v'.u$

$= 1.\log_e x + \frac{1}{x}.(x - 1)$

$= \log_e x + \frac{x - 1}{x}$

$\left[\text{or } \frac{x \log_e x + x - 1}{x}\right]$ ✓ *(2 marks)*

ii $\frac{\cos x}{x^2}$

Using the quotient rule:

Let $u = \cos x,\quad u' = -\sin x$

Let $v = x^2,\quad v' = 2x$

$\frac{dy}{dx} = \frac{v.u' - u.v'}{v^2}$

$= \frac{x^2.(-\sin x) - \cos x.2x}{(x^2)^2}$ ✓

$= \frac{-x^2 \sin x - 2x \cos x}{x^4}$

$= \frac{-x \sin x - 2\cos x}{x^3}$ ✓

(2 marks)

20 $y = x^2 \tan x$

$\frac{dy}{dx} = x^2 \times \sec^2 x + (\tan x) \times 2x$ ✓

$= x(x \sec^2 x + 2\tan x)$ ✓

(2 marks)

21 $\frac{d}{dx}[\ln(5x + 2)] = \frac{5}{5x + 2}$ ✓✓

(2 marks)

22 $y = x^2 e^x$

Using the product rule:

Let $u = x^2,\quad u' = 2x$

Let $v = e^x,\quad v' = e^x$ ✓

$\frac{dy}{dx} = u'.v + v'.u$

$= 2x.e^x + e^x.x^2$

$= x e^x(2 + x)$ ✓

(2 marks)

23 Let $y = \frac{x}{\sin x}$

Using the quotient rule:

Let $u = x,\quad u' = 1$

Let $v = \sin x,\quad v' = \cos x$ ✓

$\frac{dy}{dx} = \frac{v.u' - u.v'}{v^2}$

$= \frac{\sin x.1 - x.\cos x}{\sin^2 x}$

$= \frac{\sin x - x\cos x}{\sin^2 x}$ ✓

(2 marks)

24 $y = \frac{\sin^2 x}{x}$

$\frac{dy}{dx} = \frac{x \times 2\sin x\cos x - \sin^2 x \times 1}{x^2}$ ✓

$= \frac{\sin x(2x\cos x - \sin x)}{x^2}$ ✓

(2 marks)

25 Let $y = x^2 \tan x$.

Using the product rule:

Let $u = x^2, u' = 2x$

Let $v = \tan x, v' = \sec^2 x$ ✓

$\frac{dy}{dx} = u'.v + v'.u$

$= 2x.\tan x + \sec^2 x.x^2$

$= 2x\tan x + x^2 \sec^2 x$ ✓

(2 marks)

26 Let $y = \frac{\cos x}{x}$.

Using the quotient rule:

Let $u = \cos x, u' = -\sin x$

Let $v = x, v' = 1$ ✓

$\frac{dy}{dx} = \frac{v.u' - u.v'}{v^2}$

$= \frac{x.(-\sin x) - \cos x.1}{x^2}$

$= \frac{-x\sin x - \cos x}{x^2}$ ✓

(2 marks)

27 **i** $\frac{d}{dx}(x \sin x)$

$= x.\frac{d}{dx}(\sin x) + \sin x.\frac{d}{dx}(x)$ ✓

$= x.\cos x + \sin x.1$

$= x\cos x + \sin x$ ✓

(2 marks)

ii $\frac{d}{dx}(e^x + 1)^2$

$= 2(e^x + 1).\frac{d}{dx}(e^x + 1)$ ✓

$= 2(e^x + 1).e^x$

$= 2e^x(e^x + 1)$ ✓ *(2 marks)*

28 $\frac{d}{dx}(x\cos^2 x)$

$= x(2\cos x).-\sin x + (\cos^2 x).1$ ✓

$= -2x\cos x \sin x + \cos^2 x$

$= \cos x(\cos x - 2x\sin x)$ ✓

(2 marks)

29 **ii** $\frac{d}{dx}(x^2 \log_e x)$

$= x^2.\frac{d}{dx}(\log_e x) + \log_e x.\frac{d}{dx}(x^2)$

$= x^2.\frac{1}{x} + \log_e x.2x$ ✓

$= x + 2x\log_e x$

$= x(1 + 2\log_e x)$ ✓

(2 marks)

iii $\frac{d}{dx}\left(\frac{\sin x}{x + 4}\right)$

$= \frac{(x+4).\frac{d}{dx}(\sin x) - \sin x.\frac{d}{dx}(x+4)}{(x+4)^2}$ ✓

$= \frac{(x+4).\cos x - \sin x.1}{(x+4)^2}$

$= \frac{(x+4)\cos x - \sin x}{(x+4)^2}$ ✓

(2 marks)

30 **i** $\frac{d}{dx}\left(\frac{2x}{e^x + 1}\right)$

$= \frac{(e^x+1).\frac{d}{dx}(2x) - 2x.\frac{d}{dx}(e^x+1)}{(e^x+1)^2}$

$= \frac{(e^x+1).2 - 2x.e^x}{(e^x+1)^2}$ ✓

$= \frac{2e^x + 2 - 2xe^x}{(e^x+1)^2}$

$= \frac{2(e^x + 1 - xe^x)}{(e^x+1)^2}$ ✓

(2 marks)

ii $\frac{d}{dx}\left[(1 + \tan x)^{10}\right]$

$= 10(1 + \tan x)^9.\frac{d}{dx}(1 + \tan x)$ ✓

$= 10(1 + \tan x)^9.\sec^2 x$

$= 10\sec^2 x(1 + \tan x)^9$ ✓

(2 marks)

31 **i** $\frac{d}{dx}\ x\tan x)$

$= x.\frac{d}{dx}(\tan x) + \tan x.\frac{d}{dx}(x)$ ✓

$= x.\sec^2 x + \tan x.1$

$= x\sec^2 x + \tan x$ ✓

(2 marks)

ii $\frac{d}{dx}\left(\frac{\sin x}{x + 1}\right)$

$= \frac{(x+1).\frac{d}{dx}(\sin x) - \sin x.\frac{d}{dx}(x+1)}{(x+1)^2}$ ✓

$= \frac{(x+1).\cos x - \sin x.1}{(x+1)^2}$

$= \frac{(x+1).\cos x - \sin x}{(x+1)^2}$ ✓

(2 marks)

32 $\frac{d}{dx}(x \sin x)$

$= x \cdot \frac{d}{dx}(\sin x) + \sin x \cdot \frac{d}{dx}(x)$ ✓

$= x \cos x + \sin x \cdot 1$

$= x \cos x + \sin x$ ✓

(2 marks)

33 **i** $\frac{d}{dx}(x^2 \log_e x)$

$= x^2 \cdot \frac{d}{dx}(\log_e x) + \log_e x \cdot \frac{d}{dx}(x^2)$ ✓

$= x^2 \cdot \frac{1}{x} + \log_e x \cdot 2x$

$= x + 2x \log_e x$

$= x(1 + 2\log_e x)$ ✓

(2 marks)

ii $\frac{d}{dx}[(1 + \sin x)^5]$

$= 5(1 + \sin x)^4 \cdot \frac{d}{dx}(1 + \sin x)$ ✓

$= 5(1 + \sin x)^4 \cdot \cos x$

$= 5 \cos x (1 + \sin x)^4$ ✓

(2 marks)

34 $\frac{d}{dx}(3x + \tan x) = 3 + \sec^2 x$ ✓✓

(2 marks)

35 **i** $\frac{d}{dx}[(2e^x - 4)^9]$

$= 9(2e^x - 4)^8 \cdot \frac{d}{dx}(2e^x - 4)$ ✓

$= 9(2e^x - 4)^8 \cdot (2e^x)$

$= 18e^x(2e^x - 4)^8$ ✓

(2 marks)

ii $\frac{d}{dx}(x^2 \sin x)$

$= x^2 \cdot \frac{d}{dx}(\sin x) + \sin x \cdot \frac{d}{dx}(x^2)$ ✓

$= x^2 \cdot \cos x + \sin x \cdot 2x$

$= x^2 \cos x + 2x \sin x$ ✓ *(2 marks)*

36 **i** $\frac{d}{dx}(x \sin x)$

$= x \cdot \frac{d}{dx}(\sin x) + \sin x \cdot \frac{d}{dx}(x)$ ✓

$= x \cos x + \sin x \cdot 1$

$= x \cos x + \sin x$ ✓

(2 marks)

ii $\frac{d}{dx}\left(\frac{\ln x}{x^2}\right)$

$$= \frac{x^2 \cdot \frac{d}{dx}(\ln x) - \ln x \cdot \frac{d}{dx}(x^2)}{(x^2)^2}$$

$$= \frac{x^2 \cdot \frac{1}{x} - \ln x \cdot 2x}{x^4}$$ ✓

$$= \frac{x - 2x \ln x}{x^4}$$

$$= \frac{x(1 - 2\ln x)}{x^4}$$

$$= \frac{1 - 2\ln x}{x^3}$$ ✓

(2 marks)

37 $\frac{d}{dx}(3x^2 \ln x)$

$= 3x^2 \cdot \frac{d}{dx}(\ln x) + \ln x \cdot \frac{d}{dx}(3x^2)$ ✓

$= 3x^2 \cdot \frac{1}{x} + \ln x \cdot 6x$

$= 3x + 6x \ln x$ for $x > 0$ ✓

(2 marks)

38 **i** $\frac{d}{dx}\ln(x^2 - 9)$

$= \frac{1}{x^2 - 9} \cdot \frac{d}{dx}(x^2 - 9)$ ✓

$= \frac{1}{x^2 - 9} \cdot 2x$

$= \frac{2x}{x^2 - 9}$ ✓

(2 marks)

ii $\frac{d}{dx}\left(\frac{x}{e^x}\right)$

$$= \frac{e^x \cdot \frac{d}{dx}(x) - x \cdot \frac{d}{dx}(e^x)}{(e^x)^2}$$

$$= \frac{e^x \cdot 1 - x \cdot e^x}{(e^x)^2}$$ ✓

$$= \frac{e^x - xe^x}{(e^x)^2}$$

$$= \frac{e^x(1 - x)}{(e^x)^2}$$

$$= \frac{1 - x}{e^x}$$ ✓

(2 marks)

39 $\frac{d}{dx}(x \sin^2 x)$

$= x \cdot \frac{d}{dx}(\sin^2 x) + \sin^2 x \cdot \frac{d}{dx}(x)$ ✓

$= x(2 \sin x \cos x) + \sin^2 x \cdot 1$

$= x \sin 2x + \sin^2 x$ ✓

(2 marks)

40 **i** $\frac{d}{dx}(3xe^x)$

$= 3x \cdot \frac{d}{dx}(e^x) + e^x \cdot \frac{d}{dx}(3x)$ ✓

$= 3x \cdot e^x + e^x \cdot 3$

$= 3xe^x + 3e^x$ ✓

ii $\frac{d}{dx}[\sin(x^2 + 1)]$

$= \cos(x^2 + 1) \cdot \frac{d}{dx}(x^2 + 1)$ ✓

$= \cos(x^2 + 1) \cdot 2x$

$= 2x \cos(x^2 + 1)$. ✓

(4 marks)
(Total mark allocation only in exam)

41 **i** $y = x \tan x$

$\frac{dy}{dx} = x \cdot \frac{d}{dx}(\tan x) + \tan x \cdot \frac{d}{dx}(x)$ ✓

$= x \sec^2 x + \tan x \cdot 1$

$= x \sec^2 x + \tan x$ ✓

ii $y = \frac{e^x}{1 + x}$

$$\frac{dy}{dx} = \frac{(1+x) \cdot \frac{d}{dx}(e^x) - e^x \cdot \frac{d}{dx}(1+x)}{(1+x)^2}$$

$$= \frac{(1+x) \cdot e^x - e^x \cdot (1)}{(1+x)^2}$$ ✓

$$= \frac{e^x + xe^x - e^x}{(1+x)^2}$$

$$= \frac{xe^x}{(1+x)^2}$$ ✓

(4 marks)
(Total mark allocation only in exam)

42 $y = \sin^3 x$

$= (\sin x)^3$

$\frac{dy}{dx} = 3(\sin x)^2 \cdot \cos x$ ✓

$= 3 \sin^2 x \cos x$ ✓

(2 marks)

43 **ii** $\frac{d}{dx}[x \sin(x + 1)]$

$= x \cdot \frac{d}{dx}\sin(x + 1) + \sin(x + 1) \cdot \frac{d}{dx}(x)$

$= x \cos(x + 1) \cdot 1 + \sin(x + 1) \cdot 1$

$= x \cos(x + 1) + \sin(x + 1)$.

iii $\frac{d}{dx}\left(\frac{\tan x}{x}\right)$

$$= \frac{x \cdot \frac{d}{dx}(\tan x) - \tan x \cdot \frac{d}{dx}(x)}{x^2}$$

$$= \frac{x \sec^2 x - \tan x \cdot 1}{x^2}$$

$$= \frac{x \sec^2 x - \tan x}{x^2}$$

(No mark allocation in exam)

44 **ii** $\frac{d}{dx}\left(\frac{\cos x}{x}\right)$

$$= \frac{x \cdot \frac{d}{dx}(\cos x) - \cos x \cdot \frac{d}{dx}(x)}{x^2}$$

$$= \frac{x(-\sin x) - \cos x\,(1)}{x^2}$$

$$= \frac{-x \sin x - \cos x}{x^2}$$

iii $\frac{d}{dx}(x^2 \ln x)$

$= x^2 \cdot \frac{d}{dx}(\ln x) + \ln x \cdot \frac{d}{dx}(x^2)$

$= x^2\left(\frac{1}{x}\right) + \ln x\,(2x)$

$= x + 2x \ln x$.

(No mark allocation in exam)

45 $\frac{d}{dx}(e^{3x} \cos x)$

$= e^{3x}(-\sin x) + \cos x\,(3e^{3x})$ ✓

$= -e^{3x} \sin x + 3e^{3x} \cos x$

$= e^{3x}(-\sin x + 3 \cos x)$

$= e^{3x}(3 \cos x - \sin x)$ ✓

(2 marks)

46 **ii** $\frac{d}{dx}(xe^{2x})$

$= x.\frac{d}{dx}(e^{2x}) + e^{2x}.\frac{d}{dx}(x)$

$= x.2e^{2x} + e^{2x}.1$

$= 2xe^{2x} + e^{2x}$

iii $\frac{d}{dx}(\cos^2 x)$

$= \frac{d}{dx}[(\cos x)^2]$

$= 2(\cos x)^1.\frac{d}{dx}(\cos x)$

$= 2\cos x(-\sin x)$

$= -2\cos x \sin x$

(No mark allocation in exam)

iii $\frac{d}{dx}(x \tan x)$

$= x.\frac{d}{dx}(\tan x) + \tan x.\frac{d}{dx}(x)$

$= x.\sec^2 x + \tan x.1$

$= x\sec^2 x + \tan x$

(No mark allocation in exam)

---

47 **ii** $\frac{d}{dx}(4e^{2x}) = 4e^{2x}.\frac{d}{dx}(2x)$

$= 4e^{2x}.2$

$= 8e^{2x}$

iii $\frac{d}{dx}(x \log_e x)$

$= x.\frac{d}{dx}(\log_e x) + \log_e x.\frac{d}{dx}(x)$

$= x\left(\frac{1}{x}\right) + \log_e x.1$

$= 1 + \log_e x$

(No mark allocation in exam)

---

48 **ii** $\frac{d}{dx}\cos(x^2) = -\sin(x^2).\frac{d}{dx}(x^2)$

$= -\sin(x^2).2x$

$= -2x\sin(x^2)$

iii $\frac{d}{dx}(x \tan x)$

$= x.\frac{d}{dx}(\tan x) + \tan x.\frac{d}{dx}(x)$

$= x\sec^2 x + \tan x.1$

$= x\sec^2 x + \tan x$

(No mark allocation in exam)

---

49 **i** $\frac{d}{dx}\sin(3x + 1)$

$= \cos(3x + 1).\frac{d}{dx}(3x + 1)$

$= \cos(3x + 1).3$

$= 3\cos(3x + 1)$

iii $\frac{d}{dx}(x^2 \ln x)$

$= x^2.\frac{d}{dx}(\ln x) + \ln x.\frac{d}{dx}(x^2)$

$= x^2.\frac{1}{x} + \ln x.2x$

$= x + 2x\ln x$

$= x(1 + 2\ln x)$

(No mark allocation in exam)

---

50 **i** $\frac{d}{dx}(e^{3x}) = e^{3x}.\frac{d}{dx}(3x)$

$= e^{3x}.3$

$= 3e^{3x}$

ii $\frac{d}{dx}[\ln(5x-1)]$

$= \frac{1}{5x-1}.\frac{d}{dx}(5x-1)$

$= \frac{1}{5x-1}.5$

$= \frac{5}{5x-1}$

1 Let $f(x) = \sin(2x)$.

Find the value of x, for $0 < x < \pi$, for which $f'(x) = -\sqrt{3}$ AND $f''(x) = 2$. *(3 marks)*

(Q25, **2022 HSC**) Medium

2 The diagram shows part of $y = f(x)$ which has a local minimum at $x = -2$ and a local maximum at $x = 3$.

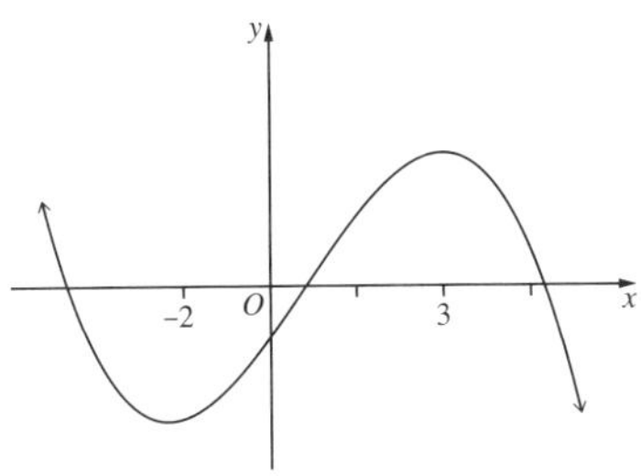

Which of the following shows the correct relationship between $f''(-2)$, $f(0)$ and $f'(3)$?

A $f(0) < f'(3) < f''(-2)$
B $f(0) < f''(-2) < f'(3)$
C $f''(-2) < f'(3) < f(0)$
D $f''(-2) < f(0) < f'(3)$ *(1 mark)*

(Q7, **2021 HSC**) Medium

3 The graph of $y = f(x)$ is shown.

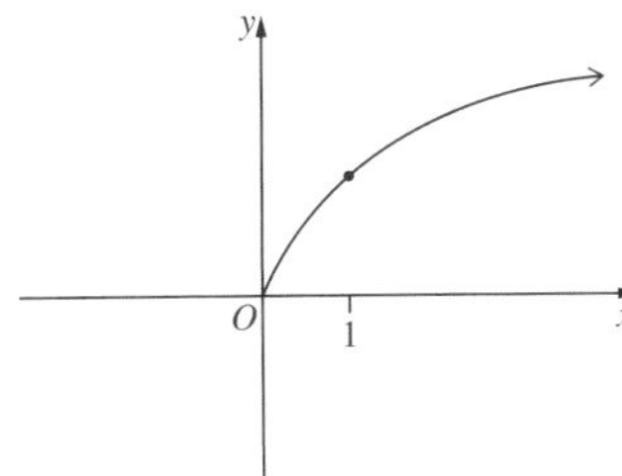

Which of the following inequalities is correct?

A $f''(1) < 0 < f'(1) < f(1)$
B $f''(1) < 0 < f(1) < f'(1)$
C $0 < f''(1) < f'(1) < f(1)$
D $0 < f''(1) < f(1) < f'(1)$ *(1 mark)*

(Q8, **2020 HSC**) Medium

4 The graph shows two functions $y = f(x)$ and $y = g(x)$.

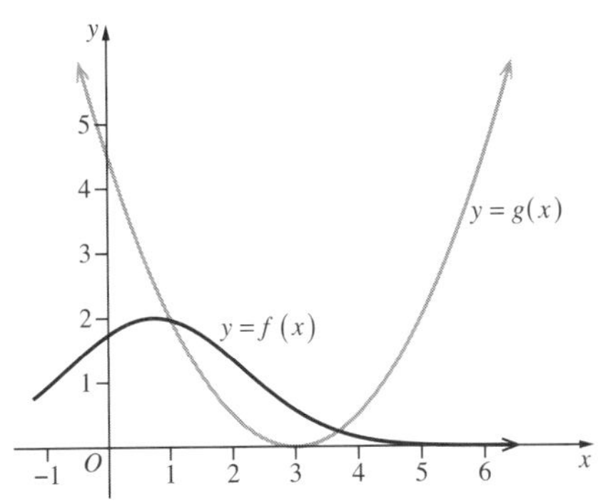

Define $h(x) = f(g(x))$.

How many stationary points does $y = h(x)$ have for $1 \leq x \leq 5$?

A 0 **B** 1
C 2 **D** 3 *(1 mark)*

(Q10, **2020 HSC**) Hard

5 Sketch the graph of the curve $y = -x^3 + 3x^2 - 1$, labelling the stationary points and point of inflection. Do NOT determine the x intercepts of the curve. *(4 marks)*

(Q16, **2020 HSC**) Easy

6 A particle is moving along a straight line. The graph shows the acceleration of the particle.

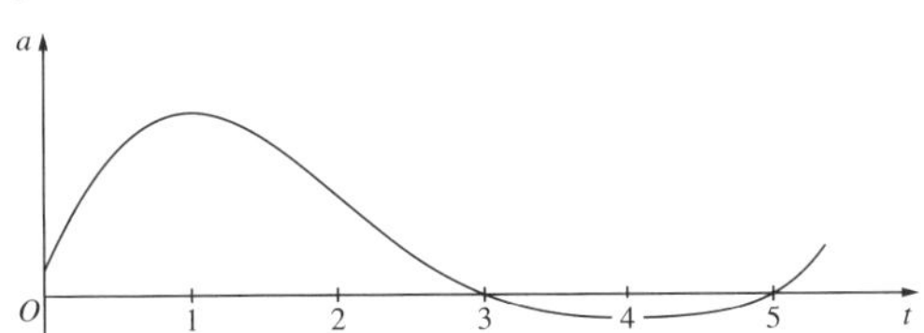

For what value of t is the velocity v a maximum?

A 1 **B** 2
C 3 **D** 5 *(1 mark)*

(Q8, **2019 HSC**) Hard

7 The diagram shows the graph of $f'(x)$, the derivative of a function.

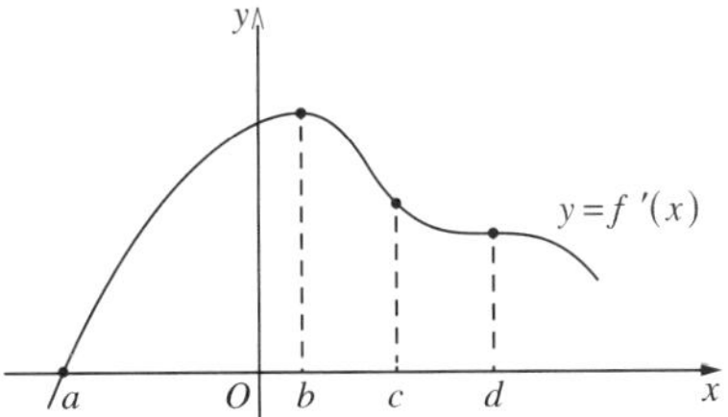

For what value of x does the graph of the function $f(x)$ have a point of inflexion?

A $x = a$ **B** $x = b$
C $x = c$ **D** $x = d$ *(1 mark)*

(Q9, **2018 HSC**) Medium

8 Let $f(x) = x^3 + kx^2 + 3x - 5$, where k is a constant.

Find the values of k for which $f(x)$ has NO stationary points. *(3 marks)*

(Q14c, **2018 HSC**) Easy

9 The function $f(x)$ is defined for $a \le x \le b$.

On this interval, $f'(x) > 0$ and $f''(x) < 0$.

Which graph best represents $y = f(x)$?

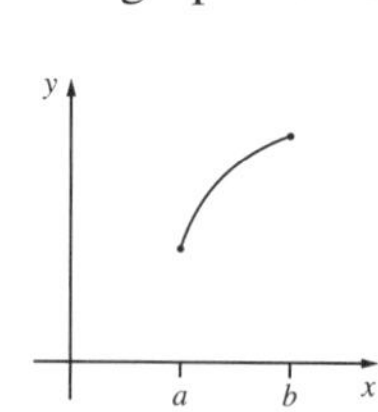

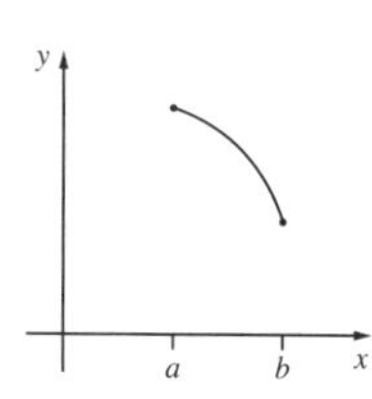

(1 mark)

(Q4, **2017 HSC**) Easy

10 The diagram shows the graph of $y = f(x)$.

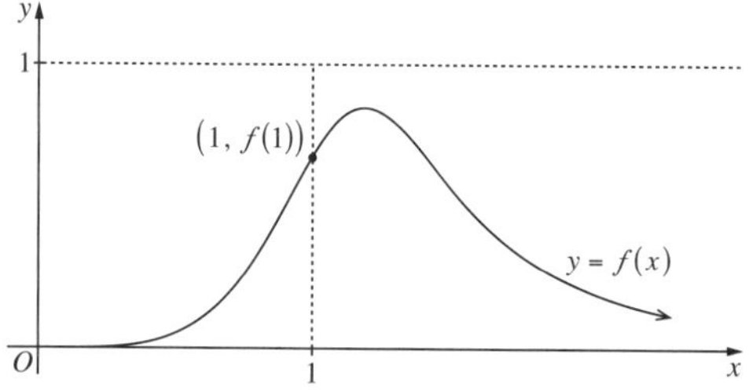

Which of the following is a correct statement?

A $f''(1) < f(1) < 1 < f'(1)$

B $f''(1) < f'(1) < f(1) < 1$

C $f(1) < 1 < f'(1) < f''(1)$

D $f'(1) < f(1) < 1 < f''(1)$ *(1 mark)*

(Q9, **2016 ME1 HSC**) Hard

11 The graph shows the displacement x of a particle moving along a straight line as a function of time t.

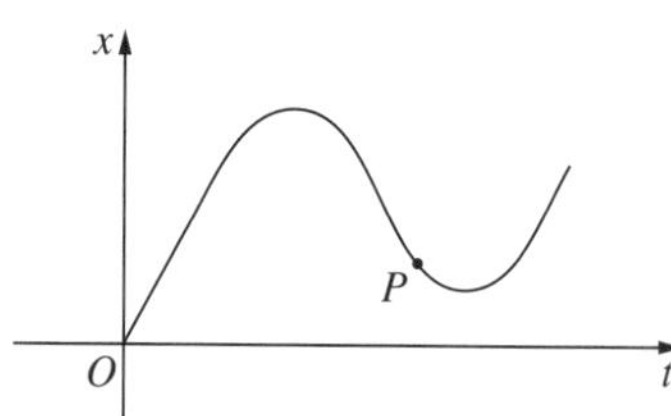

Which statement describes the motion of the particle at the point P?

A The velocity is negative and the acceleration is positive.

B The velocity is negative and the acceleration is negative.

C The velocity is positive and the acceleration is positive.

D The velocity is positive and the acceleration is negative. *(1 mark)*

(Q9, **2014 HSC**) Medium

12 The displacement of a particle moving along the x-axis is given by

$$x = t - \frac{1}{1+t},$$

where x is the displacement from the origin in metres, t is the time in seconds, and $t \ge 0$.

i Show that the acceleration of the particle is always negative. *(2 marks)* Medium

ii What value does the velocity approach as t increases indefinitely? *(1 mark)* Medium

(Q13c, **2014 HSC**)

13 The diagram shows the graph of a function $f(x)$.

The graph has a horizontal point of inflexion at A, a point of inflexion at B and a maximum turning point at C.

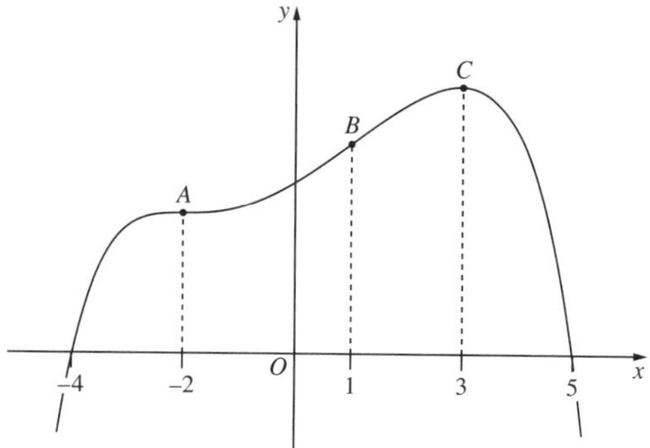

Sketch the graph of the derivative $f'(x)$. *(3 marks)*

(Q14e, **2014 HSC**) Medium

14 The diagram shows points A, B, C and D on the graph $y = f(x)$.

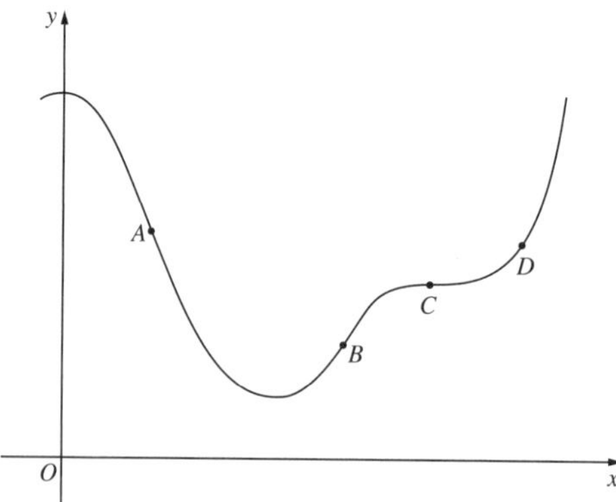

At which point is $f'(x) > 0$ and $f''(x) = 0$?

A A **B** B

C C **D** D *(1 mark)*

(Q8, **2013 HSC**) Medium

15 The cubic $y = ax^3 + bx^2 + cx + d$ has a point of inflexion at $x = p$.

Show that $p = -\dfrac{b}{3a}$. *(2 marks)*

(Q12a, **2013 HSC**) Easy

16 The diagram shows the graph $y = f(x)$.

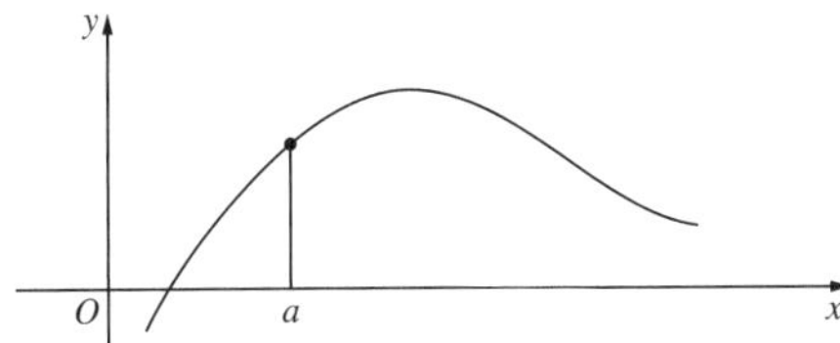

Which of the following statements is true?

A $f'(a) > 0$ and $f''(a) < 0$
B $f'(a) > 0$ and $f''(a) > 0$
C $f'(a) < 0$ and $f''(a) < 0$
D $f'(a) < 0$ and $f''(a) > 0$ *(1 mark)*

(Q4, **2012 HSC**) Easy

17 The graph $y = f(x)$ in the diagram has a stationary point when $x = 1$, a point of inflexion when $x = 3$, and a horizontal asymptote $y = -2$.

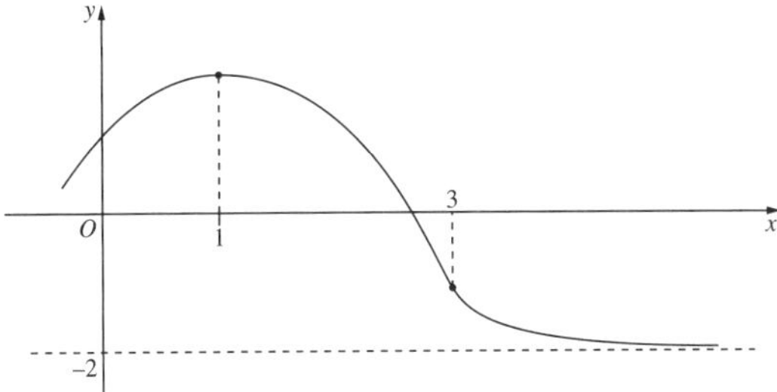

Sketch the graph $y = f'(x)$, clearly indicating its features at $x = 1$ and at $x = 3$, and the shape of the graph as $x \to \infty$. *(3 marks)*

(Q9c, **2011 HSC**) Medium

18 The graph of $y = x^3 + x^2 - x + 2$ is sketched below. The points A and B are the turning points.

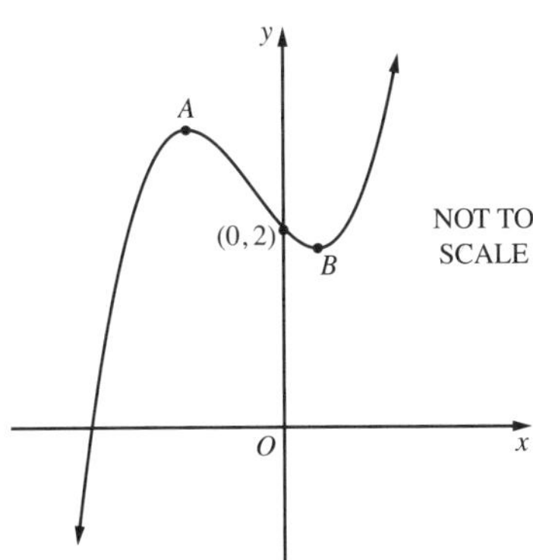

i Find the coordinates of A and B.
(3 marks) Medium

ii For what values of x is the curve concave up? Give reasons for your answer.
(2 marks) Easy

iii For what values of k has the equation $x^3 + x^2 - x + 2 = k$ three real solutions?
(1 mark) Medium

(Q6c, **2001 HSC**)

19 The diagram shows the graph of the gradient function of the curve $y = f(x)$.

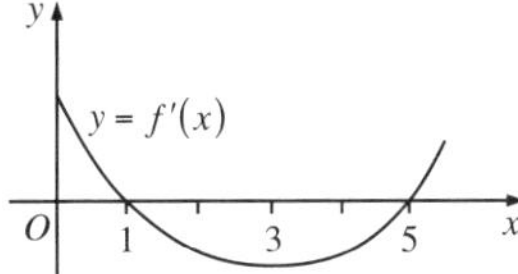

For what value of x does $f(x)$ have a local minimum? Justify your answer. *(2 marks)*

(Q8c, **1999 HSC**) Medium

20 A particle P moves along a straight line for 8 seconds, starting at the fixed point S at time $t = 0$. At time t seconds, P is $x(t)$ metres to the right of S. The graph of $x(t)$ is shown in the diagram.

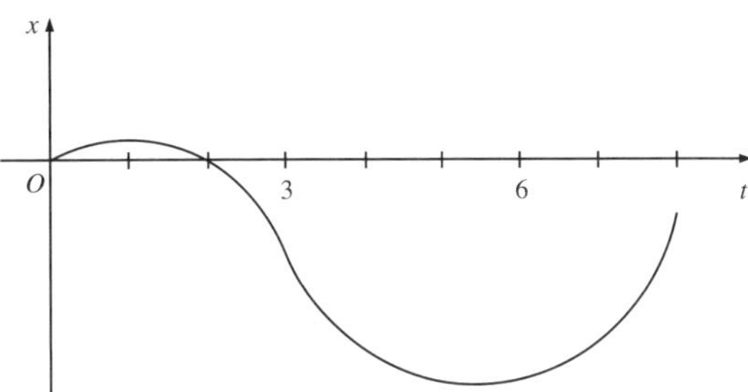

i At approximately what times is the velocity of the particle equal to 0?
ii At approximately what time is the acceleration of the particle equal to 0?
iii At approximately what time is the distance from S greatest?
iv At approximately what time is the particle moving with the greatest velocity?
(5 marks)

(Q6a, **1998 HSC**) Medium

21 The diagram shows a sketch of the curve $y = 6x^2 - x^3$. The curve cuts the x axis at L, and has a local maximum at M and a point of inflection at N.

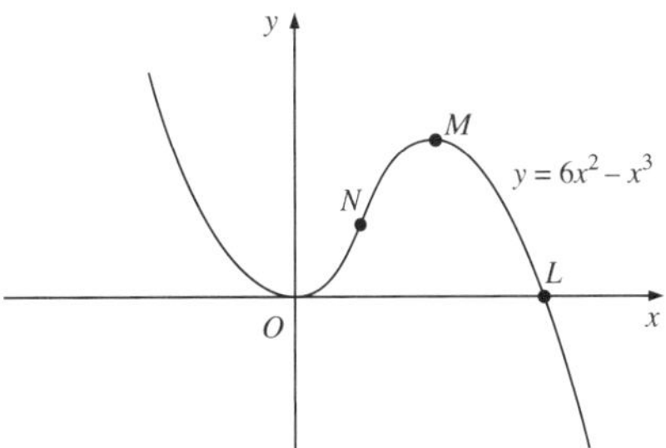

i Find the coordinates of L.
ii Find the coordinates of M.
iii Find the coordinates of N. *(6 marks)*

(Q2a, **1997 HSC**) Medium

22 For all x in the domain $0 \leq x \leq 4$, a function $g(x)$ satisfies $g'(x) < 0$ and $g''(x) < 0$.

Sketch a possible graph of $y = g(x)$ in this domain. *(2 marks)*

(Q5d, **1996 HSC**) **Medium**

23 The graph shows the levels of a pollutant in the atmosphere over the past 50 years.

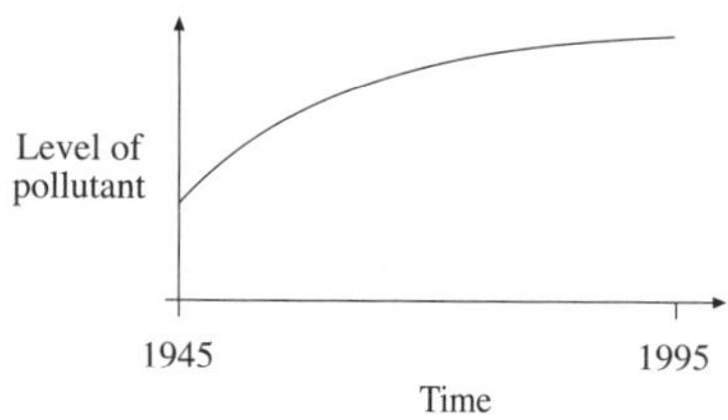

Describe briefly how the level of this pollutant has changed over this period of time. Include mention of the rate of change. *(2 marks)*

(Q7c, **1995 HSC**) **Medium**

24 Consider the curve given by $y = x^3 - 6x + 4$.

i Find the coordinates of the stationary points and determine their nature.

ii Find the coordinates of any point of inflexion.

iii Sketch the curve for the domain $-3 \leq x \leq 3$.

iv What is the maximum value of $x^3 - 6x + 4$ in the domain $-3 \leq x \leq 3$?

(Q6c, **1994 HSC**) **Medium**

25 The number of unemployed people u at time t was studied over a period of time. At the start of this period, the number of unemployed was 800000.

i Throughout the period, $\frac{du}{dt} < 0$.

What does this say about the number of people unemployed during the period?

ii It is also observed that, throughout the period, $\frac{d^2u}{dt^2} > 0$.

Sketch this graph of u against t.

(Q8a, **1994 HSC**) **Medium**

26 'Although the amount of pollution is increasing, the Government's policies to reduce pollution seem to be taking effect.'

Given that P is the amount of pollution, what does the above statement imply about $\frac{dP}{dt}$ and $\frac{d^2P}{dt^2}$?

(Q3b, **1993 HSC**) **Medium**

27 If $y = \cos(\ln x)$ find:

i $\frac{dy}{dx}$

ii $\frac{d^2y}{dx^2}$

(Q1d, **1993 ME1 HSC**) **Hard**

Year 12 The first and second derivative—Worked answers

1 $f(x) = \sin 2x$

$f'(x) = 2\cos 2x = -\sqrt{3}$

$\cos 2x = -\dfrac{\sqrt{3}}{2}$

$2x = \dfrac{5\pi}{6}, \dfrac{7\pi}{6}$

$x = \dfrac{5\pi}{12}, \dfrac{7\pi}{12}$ ✓

$f''(x) = -4\sin 2x = 2$

$\sin 2x = -\dfrac{1}{2}$

$2x = \dfrac{7\pi}{6}, \dfrac{11\pi}{6}$

$x = \dfrac{7\pi}{12}, \dfrac{11\pi}{12}$ ✓

$\therefore x = \dfrac{7\pi}{12}$ is the common solution. ✓ *(3 marks)*

2 A minimum at $x = -2$ means $f''(-2) > 0$.
As y-intercept is negative, then $f(0) < 0$.
A maximum at $x = 3$ means $f'(3) = 0$.
This means $f(0) < f'(3) < f''(-2)$.
Answer A

3 At $x = 1$, the graph is positive ($f(1) > 0$), increasing ($f'(1) > 0$) and concave down ($f''(1) < 0$).
The slope of the curve at $x = 1$ appears to be approximately 45°, which means $f'(x) = 1$.
At $x = 1$, the value of $f(1)$ appears to be greater than 1.
This means $f''(1) < 0 < f'(1) < f(1)$.
Answer A

4 $h(x) = f(g(x))$
Using the chain rule:
$h'(x) = g'(x)f'(g(x)) = 0$
For $g'(x) = 0$: from the graph, $g'(3) = 0$.
For $f'(g(x)) = 0$: from the graph, $f'(1) = 0$, and as $g(x) = 1$ twice, so $f'(g(x)) = 0$ occurs twice.
Hence there are three stationary points.
Answer D

5 $y = -x^3 + 3x^2 - 1$

$\dfrac{dy}{dx} = -3x^2 + 6x = 0$

$-3x(x-2) = 0$

$x = 0, 2$

Substitute into y:

$y(0) = -0^3 + 3(0)^2 - 1$
$= -1$

$y(2) = -(2)^3 + 3(2)^2 - 1$
$= 3$

$\therefore$ stationary points at $(0, -1)$ and $(2, 3)$.

$\dfrac{d^2y}{dx^2} = -6x + 6$

$\dfrac{d^2y}{dx^2}(0) = -6(0) + 6 > 0$

$\therefore$ minimum $(0, -1)$. ✓

$\dfrac{d^2y}{dx^2}(2) = -6(2) + 6 < 0$

$\therefore$ maximum $(2, 3)$. ✓

Also, $\dfrac{d^2y}{dx^2} = -6x + 6 = 0$

$6x = 6$

$x = 1$

Substitute into y:

$y(1) = -(1)^3 + 3(1)^2 - 1$
$= 1$

Possible point of inflection at $(1, 1)$
Consider neighbourhood of $x = 1$:

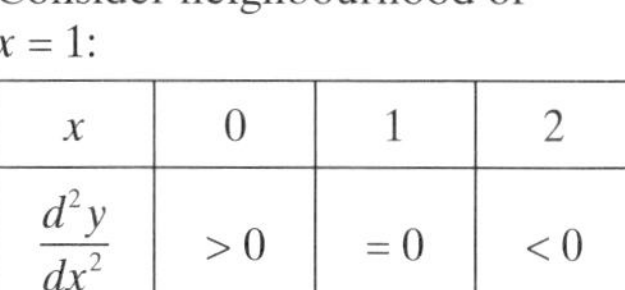

x	0	1	2
$\dfrac{d^2y}{dx^2}$	> 0	$= 0$	< 0

As sign change, then point of inflection at $(1, 1)$. ✓

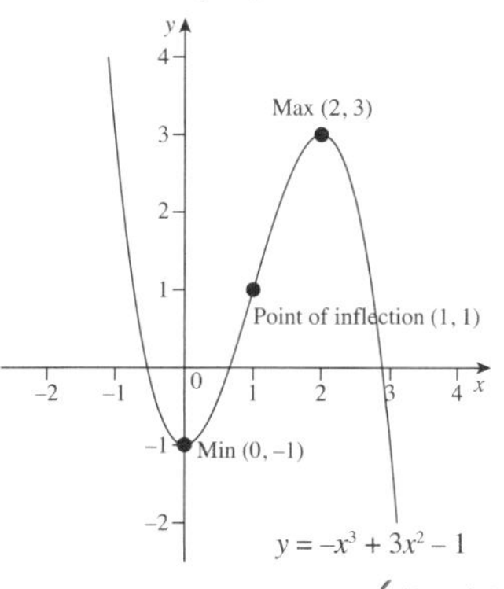

✓ *(4 marks)*

6 Velocity is maximum when acceleration = 0.
From the graph $t = 3$ and $t = 5$.
Now, consider behaviour of $\dfrac{dv}{dt}$ (i.e. a) near these values of t:
For maximum v, as t increases, then a changes from positive just before, then zero at max, then negative just after.
Hence $t = 3$.
Answer C

7 At a point of inflexion on a curve, $f''(x) = 0$. This means that at that same x-value, $f'(x)$ shows a turning point (a maximum or a minimum). Using the diagram, this occurs at $x = b$.
Answer B

8 $f(x) = x^3 + kx^2 + 3x - 5$

$f'(x) = 3x^2 + 2kx + 3$ ✓

If no stationary point, then $f'(x) \neq 0$.
Hence, $3x^2 + 2kx + 3$ has no real roots.

$\therefore \quad \Delta = b^2 - 4ac < 0$

$(2k)^2 - 4(3)(3) < 0$ ✓

$4k^2 - 36 < 0$

$k^2 - 9 < 0$

$(k-3)(k+3) < 0$

−3 0 3

$\therefore -3 < k < 3$ ✓ *(3 marks)*

9 As $f'(x) > 0$, function is increasing, and $f''(x) < 0$, function is concave down.

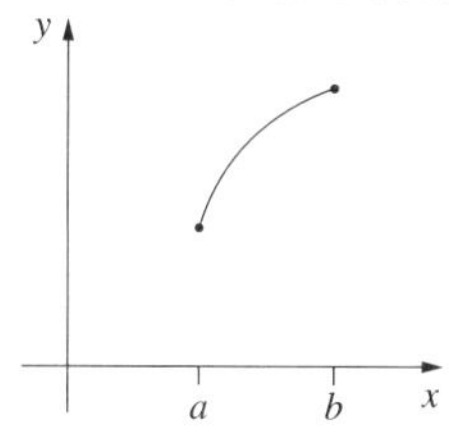

Answer A

10 When $x = 1$, the gradient is greater than 1.
So $f'(1) > 1$

At $x = 1$, the curve changes concavity
so $f''(1) = 0$
$\therefore f''(1) < f(1) < 1 < f'(1)$
Answer A

11 At P: decreasing ($f'(x) < 0$), and concave up ($f''(x) > 0$)
$\therefore$ velocity is negative and acceleration is positive.
Answer A

12 $x = t - \dfrac{1}{1+t}$

i $x = t - (1+t)^{-1}$

$\dot{x} = 1 + 1(1+t)^{-2}$

$= 1 + \dfrac{1}{(1+t)^2}$ ✓

$\ddot{x} = -2(1+t)^{-3}$

$= \dfrac{-2}{(1+t)^3}$

As $t > 0$, then $(1+t)^3 > 0$,

then $\ddot{x} < 0$ for all t. ✓ *(2 marks)*

ii Use $\dot{x} = 1 + \dfrac{1}{(1+t)^2}$.

As $t \to \infty$, $(1+t)^2 \to \infty$,

and $\dfrac{1}{(1+t)^2} \to 0$.

Hence $1 + \dfrac{1}{(1+t)^2} \to 1$.

$\therefore$ velocity approaches 1 m/s.

✓ *(1 mark)*

13

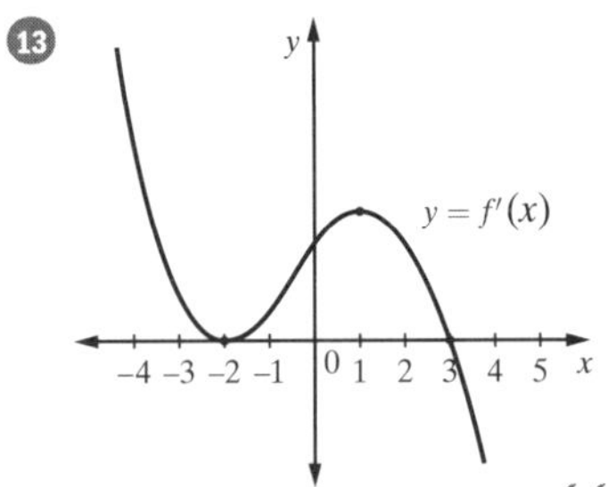

✓✓✓

[Note:

- $f(x)$ increasing when $x < -2$.

  $\therefore f'(x)$ is +ve.
- Hor. pt of inflexion at $x = -2$.

  $\therefore f'(-2) = 0$.
- $f(x)$ increasing when $-2 < x < 3$.

  $\therefore f'(x)$ is +ve.
- Maximum at $x = 3$.

  $\therefore f'(3) = 0$.
- $f(x)$ decreasing when $x > 3$.

  $\therefore f'(x)$ is −ve.
- Pt of inflexion at $x = 1$.

  $\therefore f'(1)$ is max value.] *(3 marks)*

14 At *B*: increasing ($f'(x)$. 0), and point of inflexion ($f''(x) = 0$)

[Note: At *A*, decreasing ($f'(x) < 0$)
At *C*, stationary ($f'(x) = 0$)
At *D*, concave up ($f''(x) > 0$)]

Answer B

15

$$y = ax^3 + bx^2 + cx + d$$
$$y' = 3ax^2 + 2bx + c$$
$$y'' = 6ax + 2b$$
$$y''(p) = 6ap + 2b = 0 \quad ✓$$
$$6ap = -2b$$
$$p = \frac{-2b}{6a} = \frac{-b}{3a} \quad ✓$$

(2 marks)

16 At $x = a$, the curve is *increasing* and *concave down.*

$\therefore f'(a) > 0$ and $f''(a) < 0$

Answer A

17

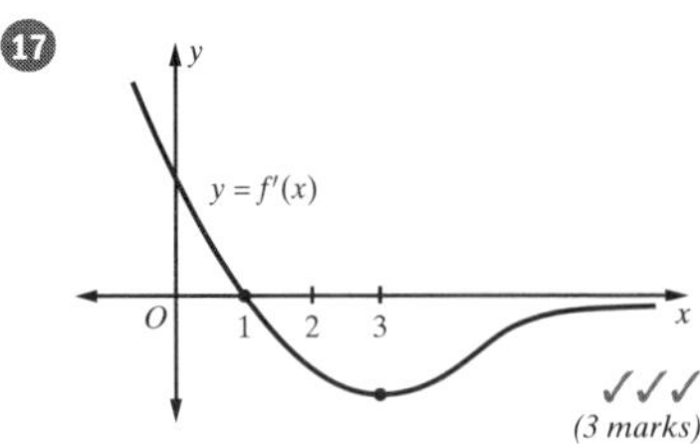

✓✓✓ *(3 marks)*

18 i

$$y = x^3 + x^2 - x + 2$$
$$\frac{dy}{dx} = 3x^2 + 2x - 1 = (3x-1)(x+1)$$

Turning points occur when

$$\frac{dy}{dx} = 0$$
$$(3x-1)(x+1) = 0 \quad ✓$$
$$x = \frac{1}{3} \text{ or } -1.$$

When $x = -1$,

$$y = (-1)^3 + (-1)^2 - (-1) + 2 = -1 + 1 + 1 + 2 = 3$$

$\therefore$ Coordinates of *A* are $(-1, 3)$ ✓

When $x = \dfrac{1}{3}$,

$$y = \left(\frac{1}{3}\right)^3 + \left(\frac{1}{3}\right)^2 - \left(\frac{1}{3}\right) + 2 = \frac{1}{27} + \frac{1}{9} - \frac{1}{3} + 2 = \frac{1}{27} + \frac{3}{27} - \frac{9}{27} + \frac{54}{27} = \frac{49}{27}.$$

$\therefore$ Coordinates of *B* are $\left(\dfrac{1}{3}, \dfrac{49}{27}\right)$. ✓

(3 marks)

ii $\dfrac{d^2y}{dx^2} = 6x + 2$

Curve is concave up when

$$\frac{d^2y}{dx^2} > 0 \quad ✓$$
$$6x + 2 > 0$$
$$6x > -2$$
$$x > -\frac{1}{3}. \quad ✓$$

(2 marks)

iii If $x^3 + x^2 - x + 2 = k$ has three real solutions, then the line $y = k$ intersects the curve $y = x^3 + x^2 - x + 2$ at 3 points. ✓

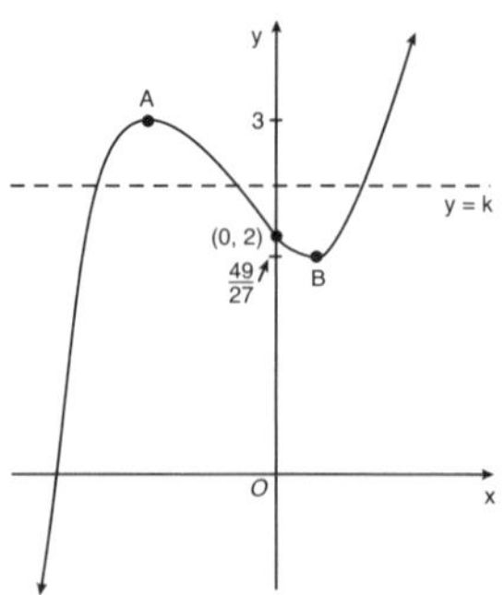

If *k* is between the *y*-coordinates of A and B, then the line $y = k$ will intersect the line at 3 points i.e. $\dfrac{49}{27} < k < 3$ *(1 mark)*

19 At $x = 5$, $f'(x) = 0$
and $f'(5 - \varepsilon) < 0$
and $f'(5 + \varepsilon) > 0$ ✓

$\therefore$ A local minimum at $x = 5$. ✓ *(2 marks)*

20
i 1 and $5\frac{1}{2}$ seconds. ✓✓
ii 3 seconds ✓
iii $5\frac{1}{2}$ seconds. ✓
iv 8 seconds ✓ *(5 marks)*
(Total mark allocation only in exam)

21 i

$$y = 6x^2 - x^3 = x^2(6 - x)$$

At *L*, $y = 0$

$$\therefore 0 = x^2(6 - x)$$
$$\therefore x = 0 \text{ or } 6$$

$\therefore$ Coordinates of *L* are $(6, 0)$. ✓

ii $\dfrac{dy}{dx} = 12x - 3x^2 = 3x(4 - x)$ ✓

At *M*, $\dfrac{dy}{dx} = 0$

$$\therefore 3x(4 - x) = 0$$
$$x = 0 \text{ or } 4 \quad ✓$$

When $x = 4$, $y = 6(4)^2 - (4)^3 = 96 - 64 = 32$

$\therefore$ Coordinates of *M* are $(4, 32)$. ✓

iii $\dfrac{d^2y}{dx^2} = 12 - 6x = 6(2 - x)$

At *N*, $\dfrac{d^2y}{dx^2} = 0$

$$\therefore 6(2 - x) = 0 \quad ✓$$
$$\therefore x = 2$$

When $x = 2$,

$$y = 6(2)^2 - (2)^3 = 24 - 8 = 16$$

$\therefore$ Coordinates of *N* are $(2, 16)$. ✓

(6 marks)
(Total mark allocation only in exam)

22

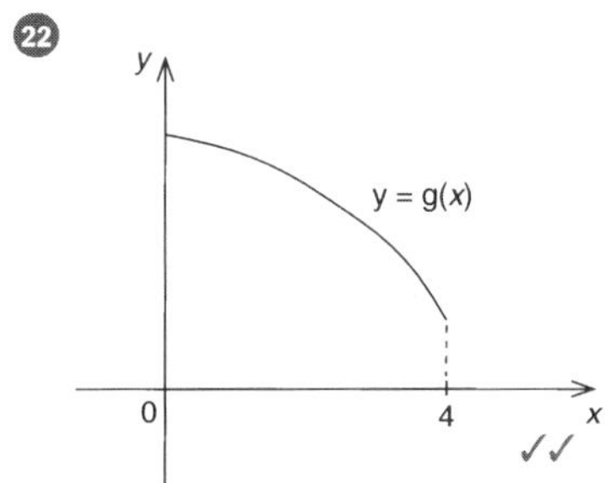

✓✓

(2 marks)

23 The level of pollutant has risen from 1945 to 1995 but the rate of increase has decreased to almost zero at 1995. ✓✓

(2 marks)

24 **i**

$$y = x^3 - 6x + 4$$

$$\frac{dy}{dx} = 3x^2 - 6$$

$$= 3(x^2 - 2)$$

$$= 3(x + \sqrt{2})(x - \sqrt{2})$$

$$\frac{d^2y}{dx^2} = 6x$$

Stationary points occur when

$$\frac{dy}{dx} = 0$$

$$3(x + \sqrt{2})(x - \sqrt{2}) = 0$$

$\therefore\ x = \pm\sqrt{2}$

When $x = \sqrt{2}$,

$$y = (\sqrt{2})^3 - 6(\sqrt{2}) + 4$$

$$= 2\sqrt{2} - 6\sqrt{2} + 4$$

$$= 4 - 4\sqrt{2}$$

and $\frac{d^2y}{dx^2} = 6(\sqrt{2}) > 0$

$\therefore$ A minimum turning point at $(\sqrt{2}, 4 - 4\sqrt{2})$.

When $x = -\sqrt{2}$,

$$y = (-\sqrt{2})^3 - 6(-\sqrt{2}) + 4$$

$$= -2\sqrt{2} + 6\sqrt{2} + 4$$

$$= 4 + 4\sqrt{2}$$

and $\frac{d^2y}{dx^2} = 6(-\sqrt{2}) < 0$

$\therefore$ A maximum turning point at $(-\sqrt{2}, 4 + 4\sqrt{2})$.

ii Points of inflexion may occur when

$$\frac{d^2y}{dx^2} = 0$$

i.e. $6x = 0$

$\therefore\ x = 0$

When $x < 0$, say $x = -\frac{1}{2}$,

$$\frac{d^2y}{dx^2} = 6(-\tfrac{1}{2}) < 0$$

When $x > 0$, say $x = \frac{1}{2}$

$$\frac{d^2y}{dx^2} = 6(\tfrac{1}{2}) > 0$$

$\therefore$ A point of inflexion at (0, 4)

iii Values at end points of domain:

At $x = -3$,

$$y = (-3)^3 - 6(-3) + 4$$

$$= -27 + 18 + 4$$

$$= -5$$

At $x = 3$,

$$= (3)^3 - 6(3) + 4$$

$$= 27 - 18 + 4$$

$$= 13$$

y intercept:

At $x = 0$, $y = 0 - 0 + 4$

$= 4$

i.e. (0, 4)

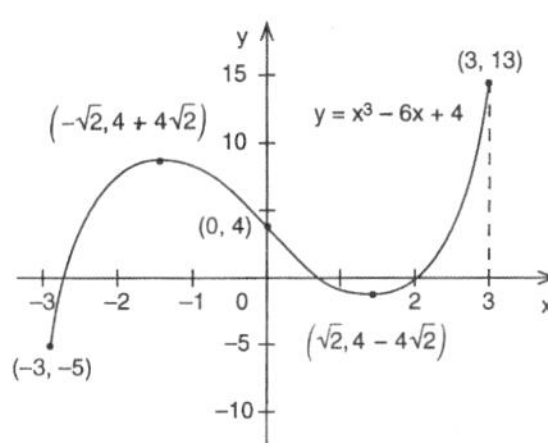

iv Maximum value is 13.

(No mark allocation in exam)

25 **i** The number of unemployed decreased during the period.

ii

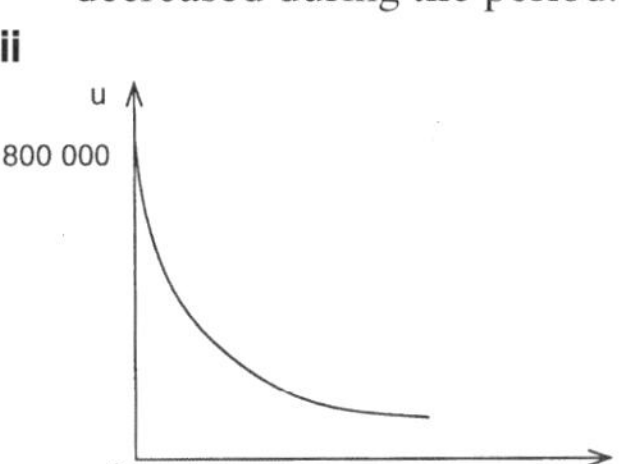

(No mark allocation in exam)

26

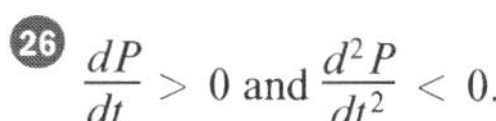

$\frac{dP}{dt} > 0$ and $\frac{d^2P}{dt^2} < 0$.

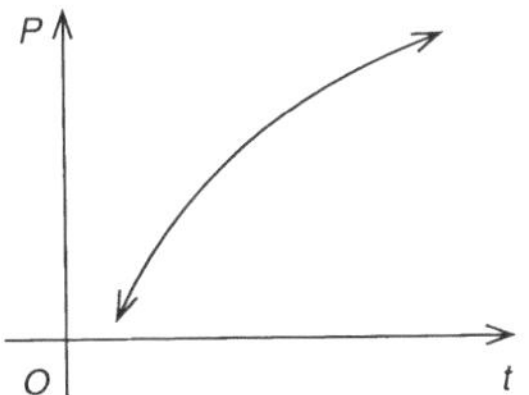

(No mark allocation in exam)

27 $y = \cos(\ln x)$

i

$$\frac{dy}{dx} = -\sin(\ln x).\frac{d}{dx}(\ln x)$$

$$= -\sin(\ln x).\frac{1}{x}$$

$$= -\frac{\sin(\ln x)}{x}$$

ii

$$\frac{d^2y}{dx^2}$$

$$= -\frac{x\cos(\ln x).\frac{1}{x} - 1.\sin(\ln x)}{x^2}$$

$$= \frac{\sin(\ln x) - \cos(\ln x)}{x^2}$$

(No mark allocation in exam)

1 A scientist is studying the growth of bacteria. The scientist models the number of bacteria, N, by the equation

$$N(t) = 200e^{0.013t},$$

where t is the number of hours after starting the experiment.

a What is the initial number of bacteria in the experiment? *(1 mark)* Easy

b What is the number of bacteria 24 hours after starting the experiment? *(1 mark)* Easy

c What is the rate of increase in the number of bacteria 24 hours after starting the experiment? *(2 marks)* Easy

(Q20, **2022 HSC**)

2 Find the global maximum and minimum values of $y = x^3 - 6x^2 + 8$, where $-1 \leq x \leq 7$. *(4 marks)*

(Q22, **2022 HSC**) Easy

3 Let $f(x) = xe^{-2x}$.

It is given that $f'(x) = e^{-2x} - 2xe^{-2x}$.

a Show that $f''(x) = 4(x - 1)\, e^{-2x}$. *(2 marks)* Easy

b Find any stationary points of $f(x)$ and determine their nature. *(2 marks)* Easy

c Sketch the curve $y = xe^{-2x}$, showing any stationary points, points of inflection and intercepts with the axes. *(3 marks)* Medium

(Q27, **2022 HSC**)

4 A line passes through the point $P(1, 2)$ and meets the axes at $X(x, 0)$ and $Y(0, y)$, where $x > 1$.

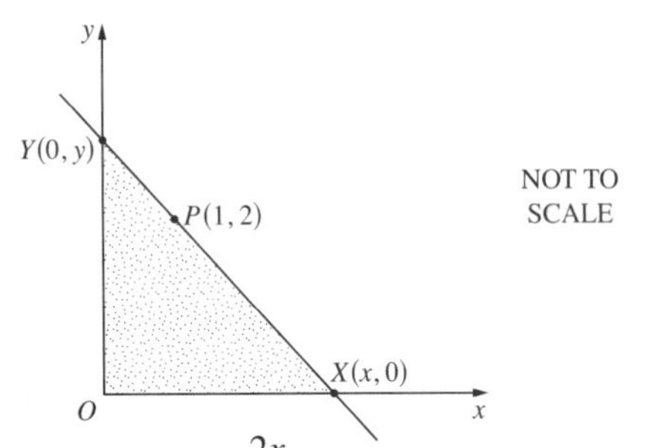

a Show that $y = \dfrac{2x}{x - 1}$. *(2 marks)* Easy

b Find the minimum value of the area of triangle XOY. *(4 marks)* Medium

(Q31, **2022 HSC**)

5 A population, P, which is initially 5000, varies according to the formula

$$P = 5000b^{\frac{-t}{10}},$$

where b is a positive constant and t is time in years, $t \geq 0$.

The population is 1250 after 20 years.

Find the value of t, correct to one decimal place, for which the instantaneous rate of decrease is 30 people per year. *(4 marks)*

(Q23, **2021 HSC**) Hard

6 A particle is shot vertically upwards from a point 100 metres above ground level. The position of the particle, y metres above the ground after t seconds, is given by $y(t) = -5t^2 + 70t + 100$.

a Find the maximum height above ground level reached by the particle. *(2 marks)* Medium

b Find the velocity of the particle, in metres per second, immediately before it hits the ground, leaving your answer in the form $a\sqrt{b}$, where a and b are integers. *(3 marks)* Medium

(Q26, **2021 HSC**)

7 Hot tea is poured into a cup. The temperature of tea can be modelled by $T = 25 + 70(1.5)^{-0.4t}$, where T is the temperature of the tea, in degrees Celsius, t minutes after it is poured.

a What is the temperature of the tea 4 minutes after it has been poured? *(1 mark)* Easy

b At what rate is the tea cooling 4 minutes after it has been poured? *(2 marks)* Medium

c How long after the tea is poured will it take for its temperature to reach 55°C? *(3 marks)* Easy

(Q21, **2020 HSC**)

8 A landscape gardener wants to build a garden bed in the shape of a rectangle attached to a quarter-circle. Let x and y be the dimensions of the rectangle in metres, as shown in the diagram.

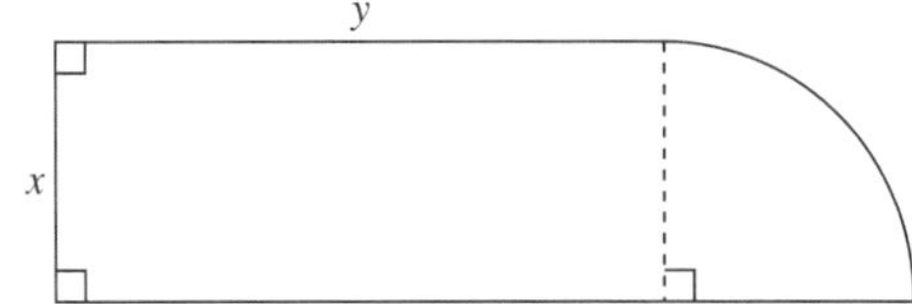

The garden bed is required to have an area of 36 m^2 and to have a perimeter which is as small as possible. Let P metres be the perimeter of the garden bed.

a Show that $P = 2x + \dfrac{72}{x}$. *(3 marks)* **Medium**

b Find the smallest possible perimeter of the garden bed, showing why this is the minimum perimeter. *(4 marks)* **Medium**

(Q25, **2020 HSC**)

9 The diagram shows the graph of $y = c \ln x$, $c > 0$.

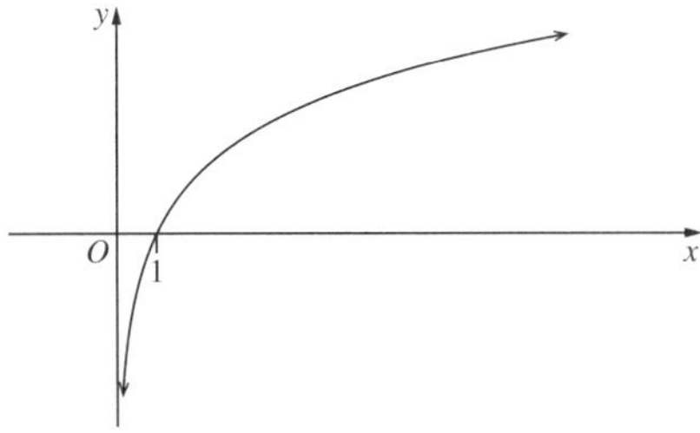

a Show that the equation of the tangent to $y = c \ln x$ at $x = p$, where $p > 0$, is $y = \dfrac{c}{p}x - c + c\ln p$. *(2 marks)* **Medium**

b Find the value of c such that the tangent from part **a** has a gradient of 1 and passes through the origin. *(2 marks)* **Hard**

(Q29, **2020 HSC**)

10 The population of mice on an isolated island can be modelled by the function

$$m(t) = a\sin\left(\frac{\pi}{26}t\right) + b,$$

where t is the time in weeks and $0 \le t \le 52$. The population of mice reaches a maximum of 35 000 when $t = 13$ and a minimum of 5000 when $t = 39$. The graph of $m(t)$ is shown.

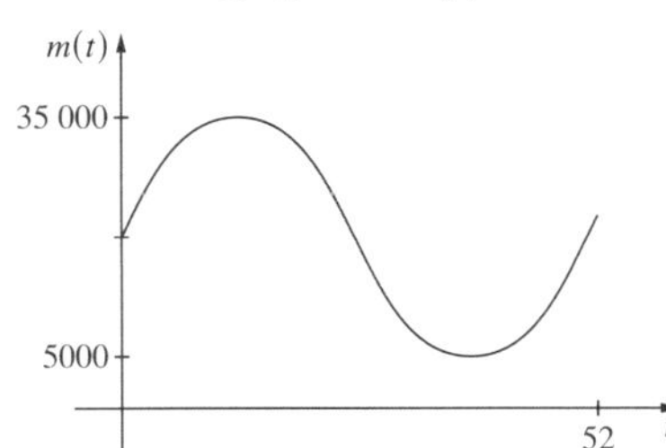

a What are the values of a and b? *(2 marks)* **Medium**

b On the same island, the population of cats can be modelled by the function

$$c(t) = -80\cos\left(\frac{\pi}{26}(t-10)\right) + 120.$$

Consider the graph of $m(t)$ and the graph of $c(t)$.

Find the values of t, $0 \le t \le 52$, for which both populations are increasing. *(3 marks)* **Hard**

c Find the rate of change of the mice population when the cat population reaches a maximum. *(2 marks)* **Hard**

(Q31, **2020 HSC**)

11 A particle is moving along a straight line with displacement x at time t.

The particle is stationary when $t = 11$ and when $t = 13$.

Which of the following MUST be true in this case?

A The particle changes direction at some time between $t = 11$ and $t = 13$.

B The displacement function of the particle has a stationary point at some time between $t = 11$ and $t = 13$.

C The acceleration of the particle is 0 at some time between $t = 11$ and $t = 13$.

D The acceleration function of the particle has a stationary point at some time between $t = 11$ and $t = 13$. *(1 mark)*

(Q10, **2019 HSC**) **Hard**

12 The number of leaves, $L(t)$, on a tree t days after the start of autumn can be modelled by $L(t) = 200000\, e^{-0.14t}$.

i What is the number of leaves on the tree when $t = 31$? *(1 mark)* **Easy**

ii What is the rate of change of the number of leaves on the tree when $t = 31$? *(2 marks)* **Easy**

iii For what value of t are there 100 leaves on the tree? *(2 marks)* **Easy**

(Q12c, **2019 HSC**)

13 The entry points, R and Q, to a national park can be reached via two straight access roads. The access roads meet the national park boundaries at right angles. The corner, P, of the national park is 8 km from R and 1 km from Q. The boundaries of the national park form a right angle at P.

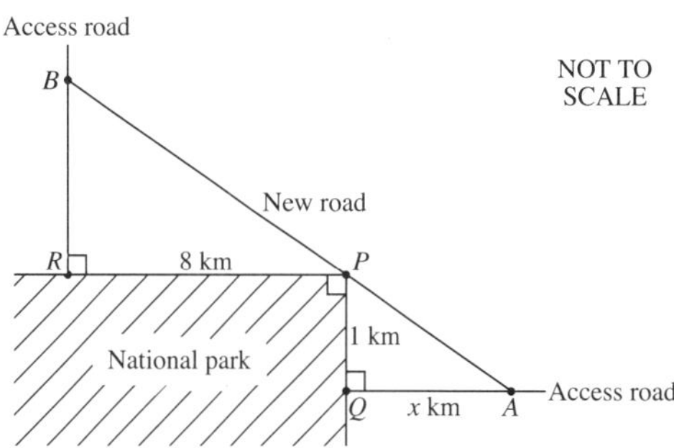

A new straight road is to be built joining these roads and passing through P.

Points A and B on the access roads are to be chosen to minimise the distance, D km, from A to B along the new road.

Let the distance QA be x km.

i Show that $D^2 = (x+8)^2 + \left(\frac{8}{x}+1\right)^2$. *(3 marks)* Medium

ii Show that $x = 2$ gives the minimum value of D^2. *(3 marks)* Hard

(Q15c, **2019 HSC**)

14 Find the equation of the tangent to the curve $y = \cos 2x$ at $x = \frac{\pi}{6}$. *(3 marks)*

(Q12b, **2018 HSC**) 

15 The displacement of a particle moving along the x-axis is given by

$$x = \frac{t^3}{3} - 2t^2 + 3t,$$

where x is the displacement from the origin in metres and t is the time in seconds, for $t \geq 0$.

i What is the initial velocity of the particle? *(1 mark)* Easy

ii At which times is the particle stationary? *(2 marks)* Easy

iii Find the position of the particle when the acceleration is zero. *(2 marks)* Easy

(Q12d, **2018 HSC**)

16 Consider the curve $y = 6x^2 - x^3$.

i Find the stationary points and determine their nature. *(3 marks)* Easy

ii Given that the point $(2, 16)$ lies on the curve, show that it is a point of inflexion. *(2 marks)* Easy

iii Sketch the curve, showing the stationary points, the point of inflexion and the x and y intercepts. *(2 marks)* Easy

(Q13a, **2018 HSC**)

17 A sector with radius 10 cm and angle θ is used to form the curved surface of a cone with base radius x cm, as shown in the diagram.

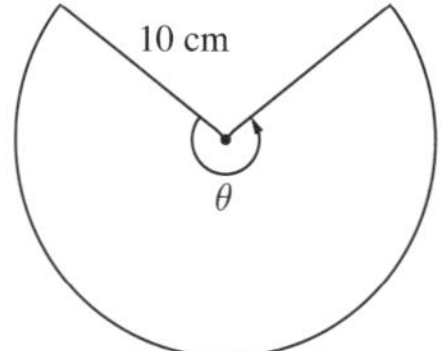

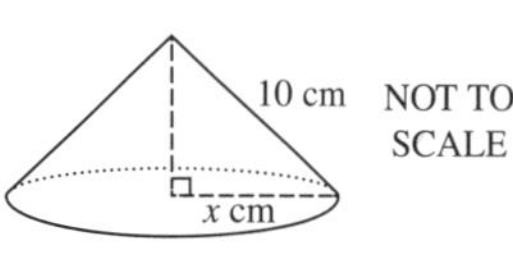

NOT TO SCALE

The volume of a cone of radius r and height h is given by $V = \frac{1}{3}\pi r^2 h$.

i Show that the volume, V cm^3, of the cone described above is given by $V = \frac{1}{3}\pi x^2\sqrt{100 - x^2}$. *(1 mark)* Easy

ii Show that $\frac{dV}{dx} = \frac{\pi x(200 - 3x^2)}{3\sqrt{100 - x^2}}$. *(2 marks)* Medium

iii Find the exact value of θ for which V is a maximum. *(3 marks)* Hard

(Q16a, **2018 HSC**)

18 Consider the curve $y = 2x^3 + 3x^2 - 12x + 7$.

i Find the stationary points of the curve and determine their nature. *(4 marks)* Easy

ii Sketch the curve, labelling the stationary points. *(2 marks)* Easy

iii Hence, or otherwise, find the values of x for which $\frac{dy}{dx}$ is positive. *(1 mark)* Medium

(Q13b, **2017 HSC**)

19 John's home is at point A and his school is at point B. A straight river runs nearby.

The point on the river closest to A is point C, which is 5 km from A.

The point on the river closest to B is point D, which is 7 km from B.

The distance from C to D is 9 km.

To get some exercise, John cycles from home directly to point E on the river, x km from C, before cycling directly to school at B, as shown in the diagram.

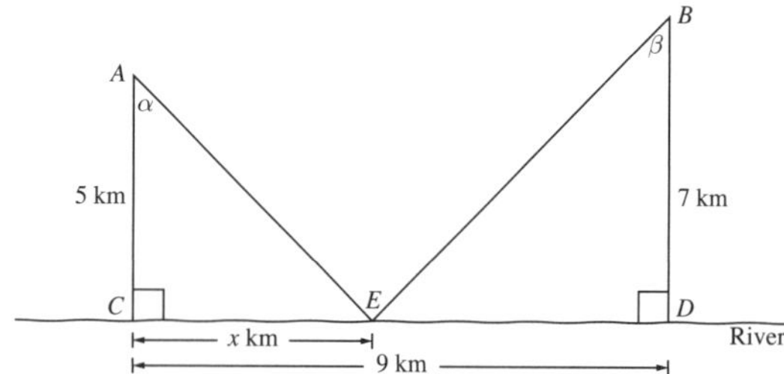

The total distance John cycles from home to school is L km.

i Show that $L = \sqrt{x^2 + 25} + \sqrt{49 + (9 - x)^2}$. *(1 mark)* Easy

ii Show that if $\frac{dL}{dx} = 0$, then $\sin\alpha = \sin\beta$. *(3 marks)* Medium

iii Find the value of x that makes $\sin\alpha = \sin\beta$. *(2 marks)* Medium

iv Explain why this value of x gives a minimum for L. *(1 mark)* Hard

(Q16a, **2017 HSC**)

20 Find the gradient of the tangent to the curve $y = \tan x$ at the point where $x = \frac{\pi}{8}$.
Give your answer correct to 3 significant figures. *(2 marks)*

(Q11f, **2016 HSC**) **Medium**

21 Consider the function $y = 4x^3 - x^4$.

i Find the two stationary points and determine their nature. *(4 marks)* **Easy**

ii Sketch the graph of the function, clearly showing the stationary points and the x and y intercepts. *(2 marks)* **Easy**

(Q13a, **2016 HSC**)

22 A farmer wishes to make a rectangular enclosure of area 720 m^2. She uses an existing straight boundary as one side of the enclosure. She uses wire fencing for the remaining three sides and also to divide the enclosure into four equal rectangular areas of width x m as shown.

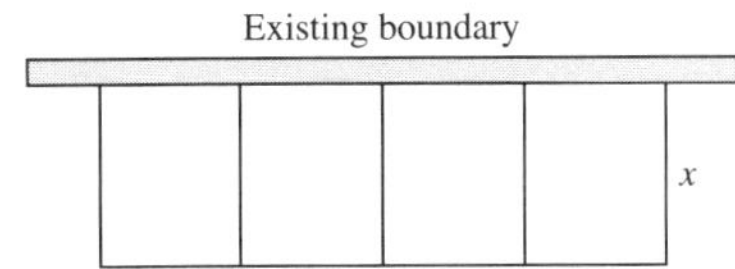

i Show that the total length, ℓ m, of the wire fencing is given by $\ell = 5x + \frac{720}{x}$. *(1 mark)*

ii Find the minimum length of wire fencing required, showing why this is the minimum length. *(3 marks)*

(Q14c, **2016 HSC**) **Medium**

23 Some yabbies are introduced into a small dam. The size of the population, y, of yabbies can be modelled by the function

$$y = \frac{200}{1 + 19e^{-0.5t}},$$

where t is the time in months after the yabbies are introduced into the dam.

i Show that the rate of growth of the size of the population is
$\frac{1900e^{-0.5t}}{(1 + 19e^{-0.5t})^2}$. *(2 marks)* **Medium**

ii Find the range of the function y, justifying your answer. *(2 marks)* **Hard**

iii Show that the rate of growth of the size of the population can be rewritten as

$\frac{y}{400}(200 - y)$. *(1 mark)* **Hard**

iv Hence, find the size of the population when it is growing at its fastest rate. *(2 marks)* **Hard**

(Q16b, **2016 HSC**)

24 Consider the curve $y = x^3 - x^2 - x + 3$.

i Find the stationary points and determine their nature. *(4 marks)* **Easy**

ii Given that the point $P\left(\frac{1}{3}, \frac{70}{27}\right)$ lies on the curve, prove that there is a point of inflexion at P. *(2 marks)* **Medium**

iii Sketch the curve, labelling the stationary points, point of inflexion and y-intercept. *(2 marks)* **Medium**

(Q13c, **2015 HSC**)

25 The diagram shows a cylinder of radius x and height y inscribed in a cone of radius R and height H, where R and H are constants.

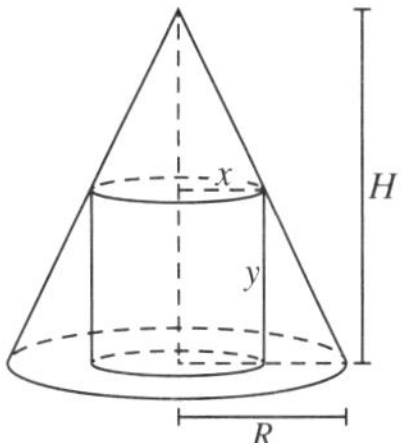

The volume of a cone of radius r and height h is $\frac{1}{3}\pi r^2 h$.

The volume of a cylinder of radius r and height h is $\pi r^2 h$.

i Show that the volume, V, of the cylinder can be written as $V = \frac{H}{R}\pi x^2(R - x)$. *(3 marks)* **Hard**

ii By considering the inscribed cylinder of maximum volume, show that the volume of any inscribed cylinder does not exceed $\frac{4}{9}$ of the volume of the cone. *(4 marks)* **Hard**

(Q16c, **2015 HSC**)

26 Find the coordinates of the stationary point on the graph $y = e^x - ex$, and determine its nature. *(3 marks)*

(Q14a, **2014 HSC**) **Medium**

27 The diagram shows a window consisting of two sections. The top section is a semicircle of diameter x m. The bottom section is a rectangle of width x m and height y m.

The entire frame of the window, including the piece that separates the two sections, is made using 10 m of thin metal.

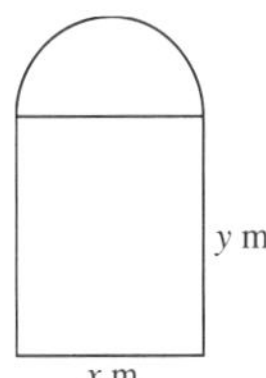

The semicircular section is made of coloured glass and the rectangular section is made of clear glass.

Under test conditions the amount of light coming through one square metre of the coloured glass is 1 unit and the amount of light coming through one square metre of the clear glass is 3 units.

The total amount of light coming through the window under test conditions is L units.

i Show that $y = 5 - x\left(1 + \frac{\pi}{4}\right)$.
(2 marks) **Medium**

ii Show that $L = 15x - x^2\left(3 + \frac{5\pi}{8}\right)$.
(2 marks) **Hard**

iii Find the values of x and y that maximise the amount of light coming through the window under test conditions.
(3 marks) **Hard**

(Q16c, **2014 HSC**)

28 A particle is moving along the x-axis. The displacement of the particle at time t seconds is x metres.

At a certain time, $\dot{x} = -3\text{ms}^{-1}$ and $\ddot{x} = 2\text{ms}^{-2}$.

Which statement describes the motion of the particle at that time?

A The particle is moving to the right with increasing speed.
B The particle is moving to the left with increasing speed.
C The particle is moving to the right with decreasing speed.
D The particle is moving to the left with decreasing speed. *(1 mark)*

(Q10, **2013 HSC**) **Hard**

29 Two straight roads meet at R at an angle of 60°. At time $t = 0$ car A leaves R on one road, and car B is 100 km from R on the other road. Car A travels away from R at a speed of 80 km/h, and car B travels towards R at a speed of 50 km/h.

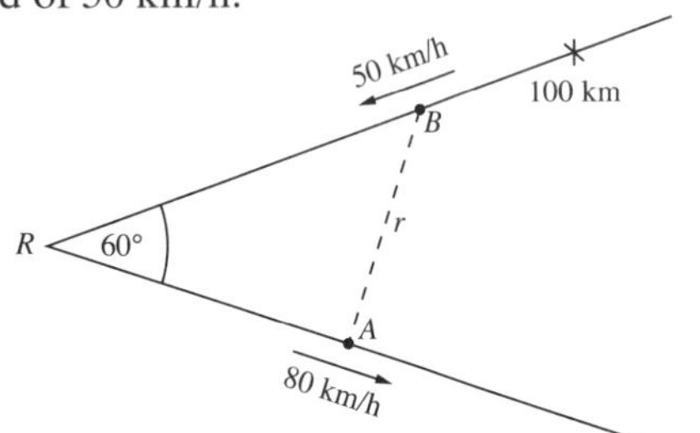

The distance between the cars at time t hours is r km.

i Show that $r^2 = 12900t^2 - 18000t + 10000$.
(2 marks) **Hard**

ii Find the minimum distance between the cars.
(3 marks) **Hard**

(Q14b, **2013 HSC**)

30 A function is given by $f(x) = 3x^4 + 4x^3 - 12x^2$.

i Find the coordinates of the stationary points $f(x)$ of and determine their nature.
(3 marks) **Easy**

ii Hence, sketch the graph $y = f(x)$ showing the stationary points. *(2 marks)* **Medium**

iii For what values of x is the function increasing? *(1 mark)* **Medium**

iv For what values of k will $3x^4 + 4x^3 - 12x^2 + k = 0$ have no solution?
(1 mark) **Hard**

(Q14a, **2012 HSC**)

31 The diagram shows a point T on the unit circle $x^2 + y^2 = 1$ at angle θ from the positive x-axis, where $0 < \theta < \frac{\pi}{2}$.

The tangent to the circle at T is perpendicular to OT, and intersects the x-axis at P, and the line $y = 1$ at Q. The line $y = 1$ intersects the y-axis at B.

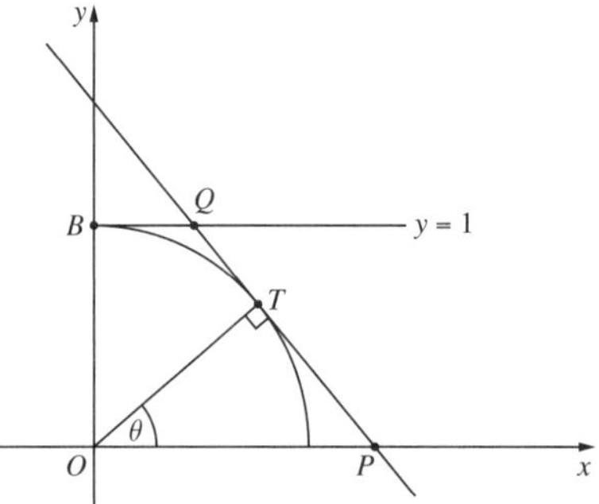

i Show that the equation of the line PT is $x\cos\theta + y\sin\theta = 1$. *(2 marks)* **Hard**

ii Find the length of BQ in terms of θ.
(1 mark) **Hard**

iii Show that the area, A, of the trapezium $OPQB$ is given by
$A = \frac{2 - \sin\theta}{2\cos\theta}$. *(2 marks)* **Hard**

iv Find the angle θ that gives the minimum area of the trapezium. *(3 marks)* **Hard**

(Q16b, **2012 HSC**)

32 Let $f(x) = x^3 - 3x + 2$.

i Find the coordinates of the stationary points of $y = f(x)$, and determine their nature. *(3 marks)* **Easy**

ii Hence, sketch the $y = f(x)$ graph showing all stationary points and the y-intercept.
(2 marks) **Easy**

(Q7a, **2011 HSC**)

33 The velocity of a particle moving along the x-axis is given by

$$\dot{x} = 8 - 8e^{-2t},$$

where t is the time in seconds and x is the displacement in metres.

- **i** Show that the particle is initially at rest. *(1 mark)* Easy
- **ii** Show that the acceleration of the particle is always positive. *(1 mark)* Medium
- **iii** Explain why the particle is moving in the positive direction for all $t > 0$. *(2 marks)* Hard
- **iv** As $t \to \infty$, the velocity of the particle approaches a constant. Find the value of this constant. *(1 mark)* Medium
- **v** Sketch the graph of the particle's velocity as a function of time. *(2 marks)* Medium

(Q7b, **2011 HSC**)

34 Consider the function $f(x) = e^{-x} - 2e^{-2x}$.

- **i** Find $f'(x)$. *(1 mark)* Easy
- **ii** The graph $y = f(x)$ has one maximum turning point. Find the coordinates of the maximum turning point. *(2 marks)* Medium
- **iii** Evaluate $f(\ln 2)$. *(1 mark)* Easy
- **iv** Describe the behaviour of $f(x)$ as $x \to \infty$. *(1 mark)* Easy
- **v** Find the y-intercept of the graph $y = f(x)$. *(1 mark)* Easy
- **vi** Sketch the graph $y = f(x)$ showing the features from parts **ii–v**. You are not required to find any points of inflexion. *(2 marks)* Medium

(Q4a, **2011 ME1 HSC**)

35 A farmer is fencing a paddock using P metres of fencing. The paddock is to be in the shape of a sector of a circle with radius r and sector angle θ in radians, as shown in the diagram.

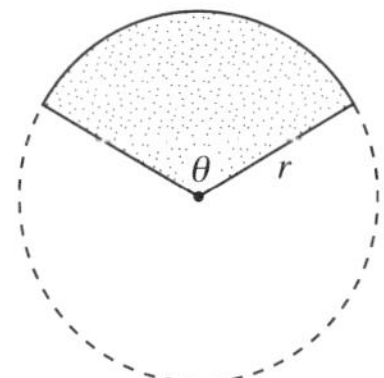

- **i** Show that the length of fencing required to fence the perimeter of the paddock is $P = r(\theta + 2)$. *(1 mark)* Easy
- **ii** Show that the area of the sector is $A = \frac{1}{2}Pr - r^2$. *(1 mark)* Medium
- **iii** Find the radius of the sector, in terms of P, that will maximise the area of the paddock. *(2 marks)* Hard
- **iv** Find the angle θ that gives the maximum area of the paddock. *(1 mark)* Hard
- **v** Explain why it is only possible to construct a paddock in the shape of a sector if $\frac{P}{2(\pi + 1)} < r < \frac{P}{2}$. *(2 marks)* Hard

(Q10b, **2011 HSC**)

36 Find the gradient of the tangent to the curve $y = \ln(3x)$ at the point where $x = 2$. *(2 marks)*

(Q2c, **2010 HSC**) Easy

37 A rainwater tank is to be designed in the shape of a cylinder with radius r metres and height h metres.

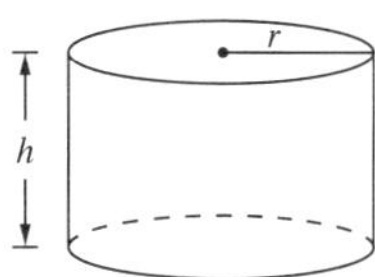

The volume of the tank is to be 10 cubic metres. Let A be the surface area of the tank, including its top and base, in square metres.

- **i** Given that $A = 2\pi r^2 + 2\pi rh$, show that $A = 2\pi r^2 + \frac{20}{r}$. *(2 marks)* Medium
- **ii** Show that A has a minimum value and find the value of r for which the minimum occurs. *(3 marks)* Medium

(Q5a, **2010 HSC**)

38 Let $f(x) = (x + 2)(x^2 + 4)$.

- **i** Show that the graph $y = f(x)$ has no stationary points. *(2 marks)* Easy
- **ii** Find the values of x for which the graph $y = f(x)$ is concave down, and the values for which it is concave up. *(2 marks)* Medium
- **iii** Sketch the graph $y = f(x)$, indicating the values of the x and y intercepts. *(2 marks)* Hard

(Q6a, **2010 HSC**)

39 An oil rig, S, is 3 km offshore. A power station, P, is on the shore. A cable is to be laid from P to S. It costs \$1000 per kilometre to lay the cable along the shore and \$2600 per kilometre to lay the cable underwater from the shore to S.

The point R is the point on the shore closest to S, and the distance PR is 5 km.

The point Q is on the shore, at a distance of x km from R, as shown in the diagram.

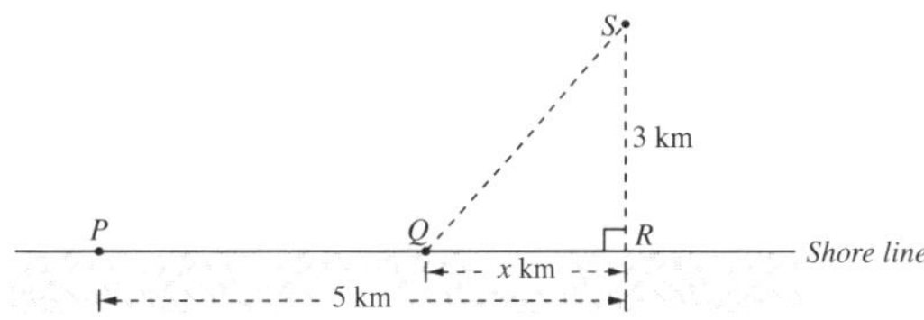

i Find the total cost of laying the cable in a straight line from P to R and then in a straight line from R to S. *(1 mark)* Easy

ii Find the cost of laying the cable in a straight line from P to S. *(1 mark)* Easy

iii Let $\$C$ be the total cost of laying the cable in a straight line from P to Q, and then in a straight line from Q to S.

Show that $C = 1000\left(5 - x + 2.6\sqrt{x^2+9}\right)$. *(2 marks)* Medium

iv Find the minimum cost of laying the cable. *(4 marks)* Hard

v New technology means that the cost of laying the cable underwater can be reduced to \$1100 per kilometre. Determine the path for laying the cable in order to minimise the cost in this case. *(2 marks)* Hard

(Q9b, **2009 HSC**)

40 Light intensity is measured in lux. The light intensity at the surface of a lake is 6000 lux. The light intensity, I lux, a distance s metres below the surface of the lake is given by

$$I = Ae^{-ks}$$

where A and k are constants.

i Write down the value of A. *(1 mark)* Easy

ii The light intensity 6 metres below the surface of the lake is 1000 lux. Find the value of k. *(2 marks)* Hard

iii At what rate, in lux per metre, is the light intensity decreasing 6 metres below the surface of the lake? *(2 marks)* Hard

(Q5c, **2008 HSC**)

41 The graph shows the velocity of a particle, v metres per second, as a function of time, t seconds.

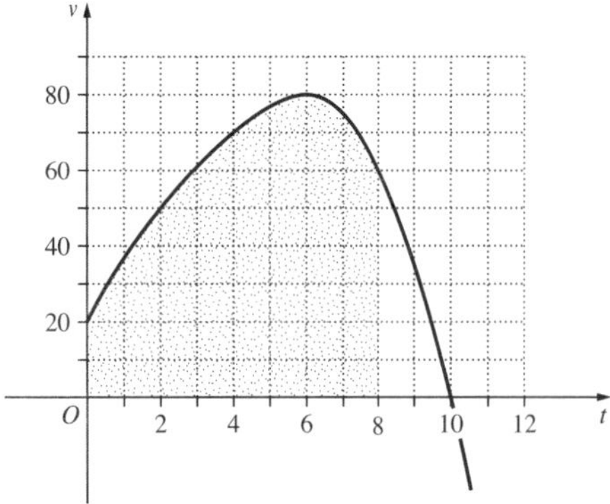

i What is the initial velocity of the particle? *(1 mark)* Easy

ii When is the velocity of the particle equal to zero? *(1 mark)* Easy

iii When is the acceleration of the particle equal to zero? *(1 mark)* Medium

(Q6b, **2008 HSC**)

42 Let $f(x) = x^4 - 8x^2$.

i Find the coordinates of the points where the graph of $y = f(x)$ crosses the axes. *(2 marks)* Easy

ii Show that $f(x)$ is an even function. *(1 mark)* Easy

iii Find the coordinates of the stationary points of $f(x)$ and determine their nature. *(4 marks)* Medium

iv Sketch the graph of $y = f(x)$. *(1 mark)* Easy

(Q8a, **2008 HSC**)

43 The diagram shows two parallel brick walls KJ and MN joined by a fence from J to M. The wall KJ is s metres long and $\angle KJM = \alpha$.

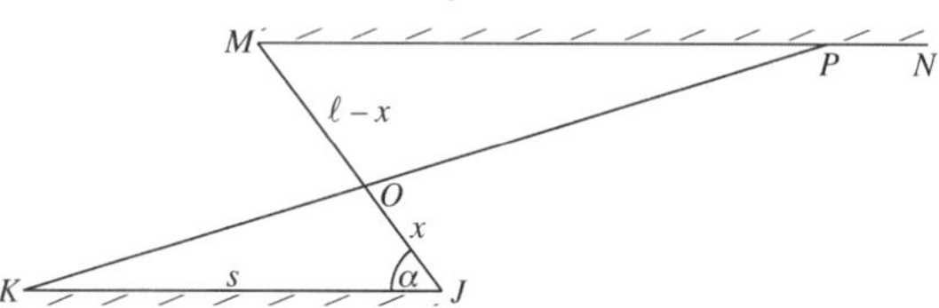

The fence JM is ℓ metres long.

A new fence is to be built from K to a point P somewhere on MN. The new fence KP will cross the original fence JM at O.

Let $OJ = x$ metres, where $0 < x < \ell$.

i Show that the total area, A square metres, enclosed by ΔOKJ and ΔOMP is given by

$A = s\left(x - \ell + \dfrac{\ell^2}{2x}\right)\sin\alpha$. *(3 marks)* Hard

ii Find the value of x that makes A as small as possible. Justify the fact that this value of x gives the minimum value for A. *(3 marks)* Medium

iii Hence, find the length of MP when A is as small as possible. *(1 mark)* Medium

(Q10b, **2008 HSC**)

44 The point $P(\pi, 0)$ lies on the curve $y = x\sin x$. Find the equation of the tangent to the curve at P. *(3 marks)*

(Q2c, **2007 HSC**) Medium

45 Let $f(x) = x^4 - 4x^3$.

i Find the coordinates of the points where the curve crosses the axes. *(2 marks)* Easy

ii Find the coordinates of the stationary points and determine their nature. *(4 marks)* Easy

iii Find the coordinates of the points of inflexion. *(1 mark)* Medium

iv Sketch the graph of $y = f(x)$, indicating clearly the intercepts, stationary points and points of inflexion. *(3 marks)* Medium

(Q6b, **2007 HSC**)

46 The graphs of the functions $y = kx^n$ and $y = \log_e x$ have a common tangent at $x = a$, as shown in the diagram.

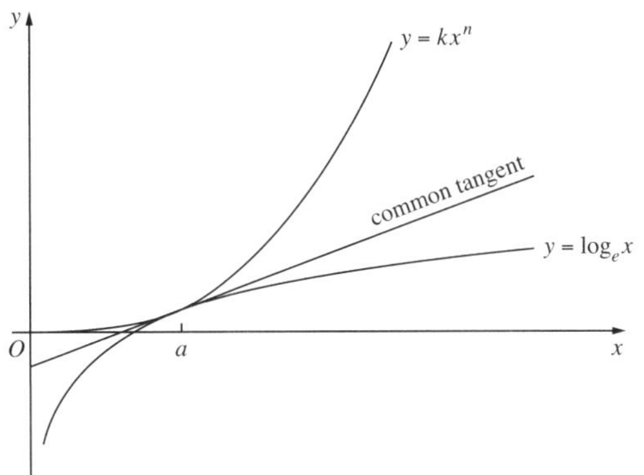

i By considering gradients, show that $a^n = \dfrac{1}{nk}$. *(1 mark)* **Hard**

ii Express k as a function of n by eliminating a. *(2 marks)* **Hard**

(Q7a, **2007 ME1 HSC**)

47 The noise level, N, at a distance d metres from a single sound source of loudness L is given by the formula

$$N = \frac{L}{d^2}.$$

Two sound sources, of loudness L_1 and L_2 are placed m metres apart.

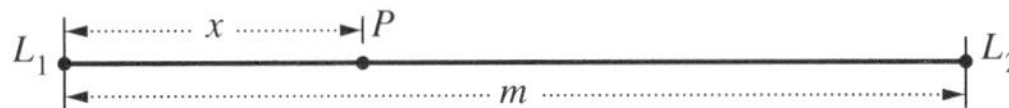

The point P lies on the line between the sound sources and is x metres from the sound source with loudness L_1.

i Write down a formula for the sum of the noise levels at P in terms of x. *(1 mark)* **Medium**

ii There is a point on the line between the sound sources where the sum of the noise levels is a minimum.
Find an expression for x in terms of m, L_1 and L_2 if P is chosen to be this point. *(4 marks)* **Hard**

(Q10b, **2007 HSC**)

48 Find the equation of the tangent to the curve $y = \cos 2x$ at the point whose x-coordinate is $\dfrac{\pi}{6}$. *(3 marks)*

(Q2c, **2006 HSC**) **Medium**

49 A function $f(x)$ is defined by $f(x) = 2x^2(3 - x)$.

i Find the coordinates of the turning points of $y = f(x)$ and determine their nature. *(3 marks)* **Medium**

ii Find the coordinates of the point of inflexion. *(1 mark)* **Medium**

iii Hence sketch the graph of $y = f(x)$, showing the turning points, the point of inflexion and the points where the curve meets the x-axis. *(3 marks)* **Medium**

iv What is the minimum value of $f(x)$ for $-1 \leq x \leq 4$? *(1 mark)* **Medium**

(Q5a, **2006 HSC**)

50 A rare species of bird lives only on a remote island. A mathematical model predicts that the bird population, P, is given by

$$P = 150 + 300e^{-0.05t}$$

where t is the number of years after observations began.

i According to the model, how many birds were there when observations began? *(1 mark)* **Medium**

ii According to the model, what will be the rate of change in the bird population ten years after observations began? *(2 marks)* **Medium**

iii What does the model predict will be the limiting value of the bird population? *(1 mark)* **Medium**

iv The species will become eligible for inclusion in the endangered species list when the population falls below 200. When does the model predict that this will occur? *(2 marks)* **Hard**

(Q6b, **2006 HSC**)

51 A particle is moving in a straight line. Its displacement, x metres, from the origin, O, at time t seconds, where $t \geq 0$, is given by $x = 1 - \dfrac{7}{t+4}$.

i Find the initial displacement of the particle. *(1 mark)* **Easy**

ii Find the velocity of the particle as it passes through the origin. *(3 marks)* **Hard**

iii Show that the acceleration of the particle is always negative. *(1 mark)* **Medium**

iv Sketch the graph of the displacement of the particle as a function of time. *(2 marks)* **Medium**

(Q8a, **2006 HSC**)

52 A cone is inscribed in a sphere of radius a, centred at O. The height of the cone is x and the radius of the base is r, as shown in the diagram.

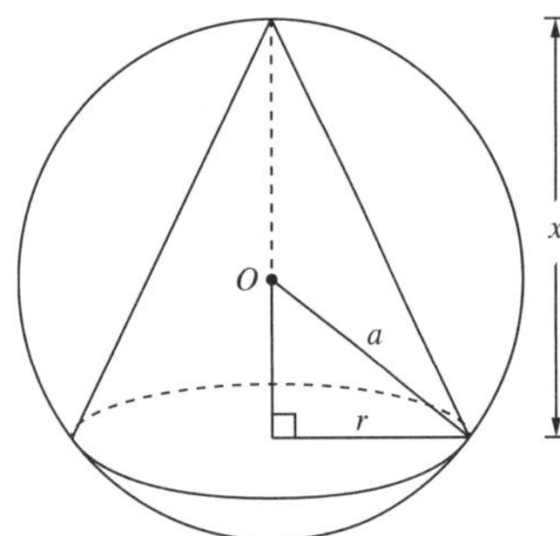

i Show that the volume, V, of the cone is given by $V = \frac{1}{3}\pi(2ax^2 - x^3)$. *(2 marks)* Medium

ii Find the value of x for which the volume of the cone is a maximum. You must give reasons why your value of x gives the maximum volume. *(3 marks)* Hard

(Q9c, **2006 HSC**)

53 Find the equation of the tangent to $y = \log_e x$ at the point $(e, 1)$. *(2 marks)*

(Q2d, **2005 HSC**) Easy

54 A function $f(x)$ is defined by $f(x) = (x + 3)(x^2 - 9)$.

i Find all solutions of $f(x) = 0$. *(2 marks)* Easy

ii Find the coordinates of the turning points of the graph of $y = f(x)$, and determine their nature. *(3 marks)* Medium

iii Hence sketch the graph of $y = f(x)$, showing the turning points and the points where the curve meets the x-axis. *(2 marks)* Easy

iv For what values of x is the graph of $y = f(x)$ concave down? *(1 mark)* Easy

(Q4b, **2005 HSC**)

55 Find the coordinates of the point P on the curve $y = 2e^x + 3x$ at which the tangent to the curve is parallel to the line $y = 5x - 3$. *(3 marks)*

(Q5c, **2005 HSC**) Hard

56 A tank initially holds 3600 litres of water. The water drains from the bottom of the tank. The tank takes 60 minutes to empty.

A mathematical model predicts that the volume, V litres, of water that will remain in the tank after t minutes is given by

$$V = 3600\left(1 - \frac{t}{60}\right)^2, \text{ where } 0 \le t \le 60.$$

i What volume does the model predict will remain after ten minutes? *(1 mark)* Easy

ii At what rate does the model predict that the water will drain from the tank after twenty minutes? *(2 marks)* Medium

iii At what time does the model predict that the water will drain from the tank at its fastest rate? *(2 marks)* Medium

(Q6b, **2005 HSC**)

57 A cylinder of radius x and height $2h$ is to be inscribed in a sphere of radius R centred at O as shown.

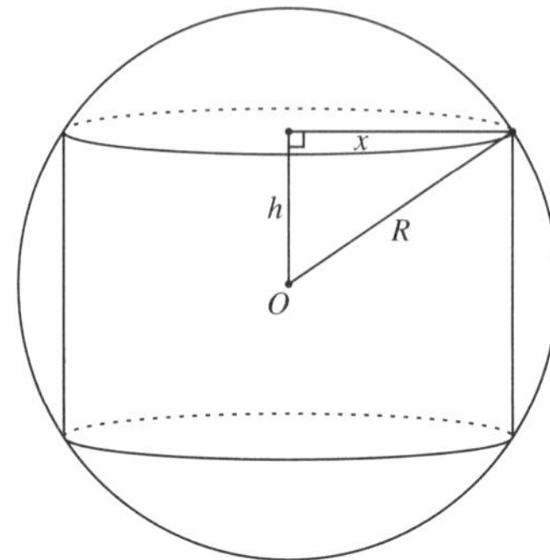

i Show that the volume of the cylinder is given by $V = 2\pi h(R^2 - h^2)$. *(1 mark)* Medium

ii Hence, or otherwise, show that the cylinder has a maximum volume when $h = \frac{R}{\sqrt{3}}$. *(3 marks)* Hard

(Q8a, **2005 HSC**)

58 Consider the function $f(x) = x^3 - 3x^2$.

i Find the coordinates of the stationary points of the curve $y = f(x)$ and determine their nature. *(3 marks)* Medium

ii Sketch the curve showing where it meets the axes. *(2 marks)* Easy

iii Find the values of x for which the curve $y = f(x)$ is concave up. *(2 marks)* Easy

(Q4b, **2004 HSC**)

59 A particle moves along a straight line so that its displacement, x metres, from a fixed point O is given by $x = 1 + 3\cos 2t$, where t is measured in seconds.

i What is the initial displacement of the particle? *(1 mark)* Medium

ii Sketch the graph of x as a function of t for $0 \le t \le \pi$. *(2 marks)* Medium

iii Hence, or otherwise, find when AND where the particle first comes to rest after $t = 0$. *(2 marks)* Medium

iv Find a time when the particle reaches its maximum speed. What is this speed? *(2 marks)* Hard

(Q5b, **2004 HSC**)

60 Consider the function $f(x) = \dfrac{\log_e x}{x}$, for $x > 0$.

- **i** Show that the graph of $y = f(x)$ has a stationary point at $x = e$. *(2 marks)* **Medium**
- **ii** By considering the gradient on either side of $x = e$, or otherwise, show that the stationary point is a maximum. *(1 mark)* **Medium**
- **iii** Use the fact that the maximum value of $f(x)$ occurs at $x = e$ to deduce that $e^x \geq x^e$ for all $x > 0$. *(2 marks)* **Hard**

(Q9c, **2004 HSC**)

---

61 The diagram shows a triangular piece of land ABC with dimensions $AB = c$ metres, $AC = b$ metres and $BC = a$ metres, where $a \leq b \leq c$.

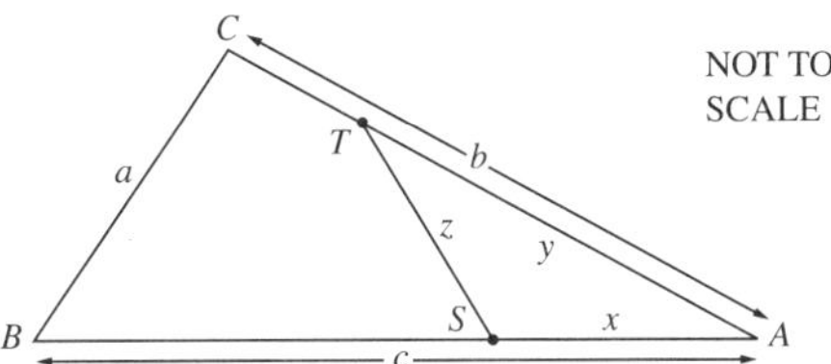

The owner of the land wants to build a straight fence to divide the land into two pieces of equal area. Let S and T be points on AB and AC respectively so that ST divides the land into two pieces of equal area.

Let $AS = x$ metres, $AT = y$ metres and $ST = z$ metres.

- **i** Show that $xy = \dfrac{1}{2}bc$. *(1 mark)* **Hard**
- **ii** Use the cosine rule in triangle AST to show that

$$z^2 = x^2 + \frac{b^2c^2}{4x^2} - bc\cos A.$$

(2 marks) **Hard**

- **iii** Show that the value of z^2 in the equation in part **ii** is a minimum when $x = \sqrt{\dfrac{bc}{2}}$. *(4 marks)* **Hard**
- **iv** Show that the minimum length of the fence is $\sqrt{\dfrac{(P-2b)(P-2c)}{2}}$ metres, where $P = a + b + c$. (You may assume that the value of x given in part **iii** is feasible.) *(2 marks)* **Hard**

(Q10b, **2004 HSC**)

---

62 Find the equation of the normal to the curve $y = 2\log_e x$ at the point $(e, 2)$. *(3 marks)*

(Q2a, **2003 HSC**) **Medium**

---

63 Consider the function $f(x) = x^4 - 4x^3$.

- **i** Show that $f'(x) = 4x^2(x - 3)$. *(1 mark)* **Easy**
- **ii** Find the coordinates of the stationary points of the curve $y = f(x)$, and determine their nature. *(3 marks)* **Medium**
- **iii** Sketch the graph of the curve $y = f(x)$, showing the stationary points. *(1 mark)* **Medium**
- **iv** Find the values of x for which the graph of $y = f(x)$ is concave down. *(2 marks)* **Medium**

(Q5a, **2003 HSC**)

---

64 The velocity of a particle is given by $v = 2 - 4\cos t$ for $0 \leq t \leq 2\pi$,

where v is measured in metres per second and t is measured in seconds.

- **i** At what times during this period is the particle at rest? *(2 marks)* **Medium**
- **ii** What is the maximum velocity of the particle during this period? *(2 marks)* **Medium**
- **iii** Sketch the graph of v as a function of t for $0 \leq t \leq 2\pi$. *(2 marks)* **Medium**

(Q7b, **2003 HSC**)

---

65 A fish is swimming at a constant speed against the current. The current is moving at a constant speed of u ms^{-1}. The speed of the fish relative to the water is v ms^{-1}, so that the actual speed of the fish is $(v - u)$ ms^{-1}.

The rate at which the fish uses energy is proportional to v^3, so the amount of energy used in t seconds is given by

$$E = av^3t,$$

where a is a constant.

- **i** Show that the energy used to swim L metres is given by $E = \dfrac{aLv^3}{v-u}$. *(1 mark)* **Hard**
- **ii** Migrating fish try to minimise the total energy used to swim a fixed distance. Find the value of v that minimises E. (You may assume $v > u > 0$.) *(4 marks)* **Hard**

(Q9c, **2003 HSC**)

---

66 A pulley P is attached to the ceiling at O by a piece of metal that can swing freely. One end of a rope is attached to the ceiling at A. The rope is passed through the pulley P and a weight is attached to the other end of the rope at M, as shown in the diagram.

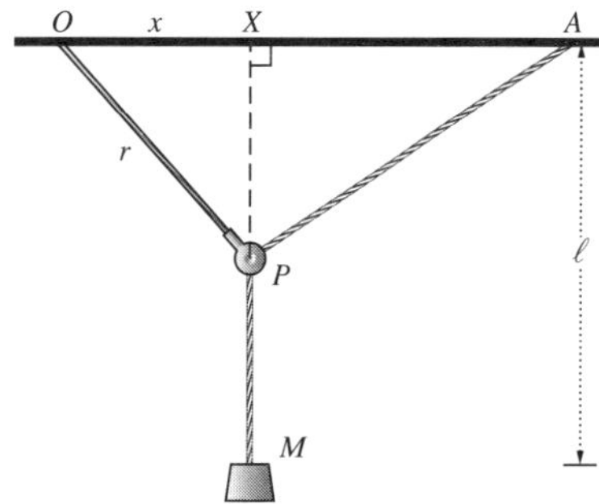

The distance OA is 1 m, the length of the rope is 2 m, and the length of the piece of metal $OP = r$ metres, where $0 < r < 1$. Let X be the point where the line MP produced meets OA. Let $OX = x$ metres and $XM = \ell$ metres.

i By considering triangles OXP and AXP, show that
$$\ell = 2 + \sqrt{r^2 - x^2} - \sqrt{1 - 2x + r^2}.$$
(1 mark) **Hard**

ii Show that
$$\frac{d\ell}{dx} = \frac{(r^2 - x^2) - x^2(1 - 2x + r^2)}{\sqrt{r^2 - x^2}\sqrt{1 - 2x + r^2}\left(\sqrt{r^2 - x^2} + x\sqrt{1 - 2x + r^2}\right)}$$
(2 marks) **Hard**

iii You are given the factorisation
$(r^2 - x^2) - x^2(1 - 2x + r^2)$
$= (x - 1)(2x^2 - r^2x - r^2)$
(Do NOT prove this.) Find the value of x for which M is closest to the floor. Justify your answer. *(2 marks)* **Hard**

(Q10b, **2003 HSC**)

67 Find the equation of the tangent to $y = e^{2x}$ at the point $(0, 1)$. *(2 marks)*

(Q2a, **2002 HSC**) **Easy**

68 A cooler, which is initially full, is drained so that at time t seconds the volume of water V, in litres, is given by

$$V = 25\left(1 - \frac{t}{60}\right)^2 \quad \text{for} \quad 0 \le t \le 60.$$

i How much water was initially in the cooler? *(1 mark)* **Easy**

ii After how many seconds was the cooler one-quarter full? *(2 marks)* **Medium**

iii At what rate was the water draining out when the cooler was one-quarter full? *(2 marks)* **Hard**

(Q7b, **2002 HSC**)

69 A particle moves in a straight line. At time t seconds, its distance x metres from a fixed point O on the line is given by

$$x = \sin 2t + 3.$$

i Sketch the graph of x as a function of t for $0 \le t \le 2\pi$. *(3 marks)* **Medium**

ii Using your graph, or otherwise, find the times when the particle is at rest, and the position of the particle at those times. *(2 marks)* **Medium**

iii Describe the motion completely. *(2 marks)* **Hard**

(Q8b, **2002 HSC**)

70 A particle moves in a straight line so that its displacement, in metres, is given by

$x = \dfrac{t-2}{t+2}$ where t is measured in seconds.

i What is the displacement when $t = 0$? *(1 mark)* **Easy**

ii Show that $x = 1 - \dfrac{4}{t+2}$.

Hence find expressions for the velocity and the acceleration in terms of t. *(3 marks)* **Medium**

iii Is the particle ever at rest? Give reasons for your answer. *(1 mark)* **Easy**

iv What is the limiting velocity of the particle as t increases indefinitely? *(1 mark)* **Easy**

(Q7c, **2001 HSC**)

71 The diagram shows a ten-centimetre high glass that is being filled with water at a constant rate (by volume). Let $y = f(t)$ be the depth of water in the glass as a function of time t.

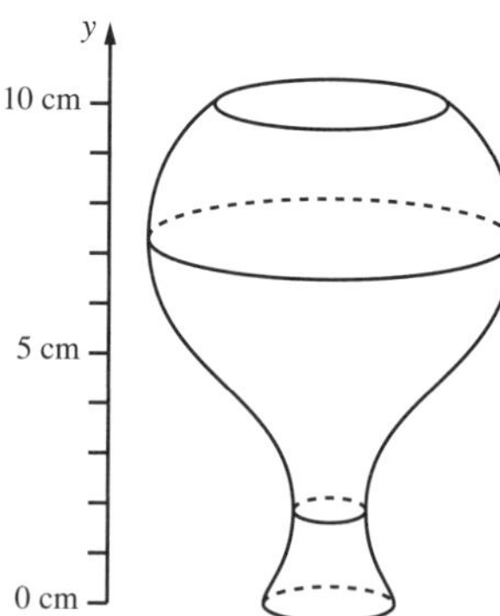

i Find the approximate depth y_1 at which $\dfrac{dy}{dt}$ is a maximum.

Find the approximate depth y_2 at which $\dfrac{dy}{dt}$ is a minimum. *(2 marks)* **Hard**

ii Assume that the glass takes 5 seconds to fill.

Graph $y = f(t)$ and identify any points on your graph where the concavity changes. *(2 marks)* **Hard**

(Q8c, **2001 HSC**)

72 The diagram shows a farmhouse F that is located 250 m from a straight section of road. The road begins at the bus depot D, which is situated 2000 m from the front gate G of the farmhouse. The school bus leaves the depot at 8 am and travels along the road at a speed of 15 ms^{-1}. Claire lives in the farmhouse, and she can run across the open paddock between the house and the road at a speed of 4 ms^{-1}. The bus will stop for Claire anywhere on the road, but will not wait for her.

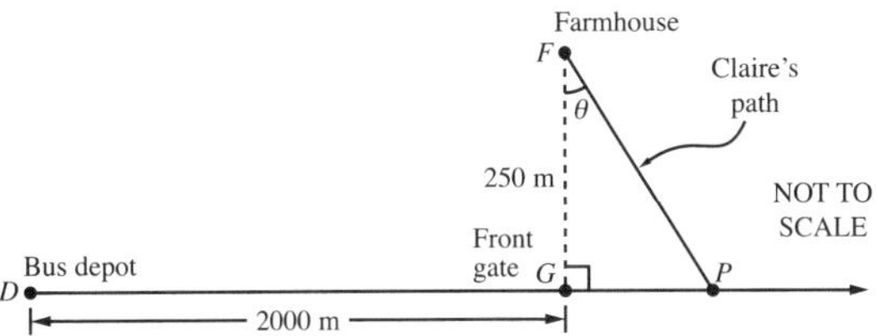

Assume that Claire catches the bus at the point P on the road where $\angle GFP = \theta$.

i Find two expressions in terms of θ, one expression for the time taken for the bus to travel from D to P and the other expression for the time taken by Claire to run from F to P. *(2 marks)* **Hard**

ii What is the latest time that Claire can leave home in order to catch the bus? *(4 marks)* **Hard**

(Q10b, **2001 HSC**)

73 Find the equation of the tangent to the curve $y = 2\log_e x$ at $(1, 0)$. *(3 marks)*

(Q3d, **2000 HSC**) **Medium**

74 The number N of students logged onto a website at any time over a five-hour period is approximated by the formula

$$N = 175 + 18t^2 - t^4, 0 \leq t \leq 5.$$

i What was the initial number of students logged onto the website? **Easy**

ii How many students were logged onto the website at the end of the five hours? **Easy**

iii What was the maximum number of students logged onto the website? **Medium**

iv When were the students logging onto the website most rapidly? **Medium**

v Sketch the curve $N = 175 + 18t^2 - t^4$ for $0 \leq t \leq 5$. **Medium**

(9 marks) (Q6b, **2000 HSC**)

75 A particle is moving in a straight line, starting from the origin. At time t seconds the particle has a displacement of x metres from the origin and a velocity v ms^{-1}. The displacement is given by $x = 2t - 3\log_e(t + 1)$.

i Find an expression for v. **Easy**

ii Find the initial velocity. **Easy**

iii Find when the particle comes to rest. **Medium**

(Q8a, **2000 HSC**)

76 An enclosure is to be built adjoining a barn, as in the diagram. The walls of the barn meet at 135°, and 117 metres of fencing is available for the enclosure, so that $x + y = 117$ where x and y are as shown in the diagram.

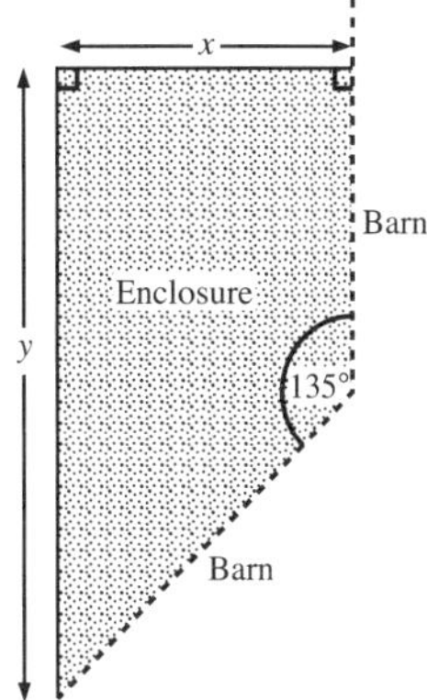

i Show that the shaded area of the enclosure in square metres is given by

$$A = 117x - \frac{3}{2}x^2.$$

ii Show that the largest area of the enclosure occurs when $y = 2x$.

(5 marks) (Q8b, **2000 HSC**) **Hard**

77 Consider the curve given by $y = x^3 - 6x^2 + 9x + 1$.

i Find $\frac{dy}{dx}$. **Easy**

ii Find the coordinates of the two stationary points. **Medium**

iii Determine the nature of the stationary points. **Easy**

iv Sketch the curve for $x \geq 0$. **Easy**

(8 marks) (Q5a, **1999 HSC**)

78 A particle P is moving along the x axis. Its position at time t seconds is given by

$$x = 2\sin t - t, t \geq 0.$$

i Find an expression for the velocity of the particle. **Easy**

ii In what direction is the particle moving at $t = 0$? **Easy**

iii Determine when the particle first comes to rest. **Easy**

iv When is the acceleration negative for $0 \leq t \leq 2\pi$? **Medium**

(Q7b, **1999 HSC**)

79 The function $f(x) = xe^{-2x} + 1$ has first derivative $f'(x) = e^{-2x} - 2xe^{-2x}$ and second derivative $f''(x) = 4xe^{-2x} - 4e^{-2x}$.

- **i** Find the value of x for which $y = f(x)$ has a stationary point. Medium
- **ii** Find the values of x for which $f(x)$ is increasing. Medium
- **iii** Find the value of x for which $y = f(x)$ has a point of inflection and determine where the graph of $y = f(x)$ is concave upwards. Medium
- **iv** Sketch the curve $y = f(x)$ for $-\frac{1}{2} \le x \le 4$. Medium
- **v** Describe the behaviour of the graph for very large positive values of x. Easy

(7 marks) (Q6b, **1998 HSC**)

80 Consider the function $y = 1 + \sqrt{3}\sin x + \cos x$.

- **i** Find the equation of the tangent to the graph of the function at $x = \frac{5\pi}{6}$. Medium
- **ii** Find the maximum and minimum values of $1 + \sqrt{3}\sin x + \cos x$ in the interval $0 \le x \le 2\pi$. Hard

(5 marks) (Q7b, **1998 HSC**)

81 The diagram shows a sector of a circle with radius r cm. The angle at the centre is θ radians and the perimeter of the sector is 8 cm.

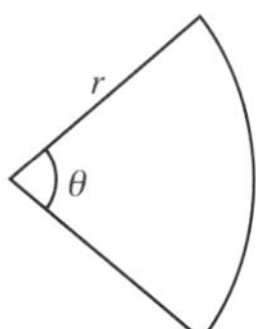

- **i** Find an expression for r in terms of θ. Medium
- **ii** Show that A, the area of the sector in cm^2, is given by $A = \dfrac{32\theta}{(\theta + 2)^2}$. Medium
- **iii** If $0 \le \theta \le \frac{\pi}{2}$, find the maximum area and the value of θ for which this occurs. Hard

(5 marks) (Q7c, **1998 HSC**)

82 A wire of length 5 metres is to be bent to form the hypotenuse and base of a right-angled triangle PQR, as shown in the diagram. Let the length of the base QR be x metres.

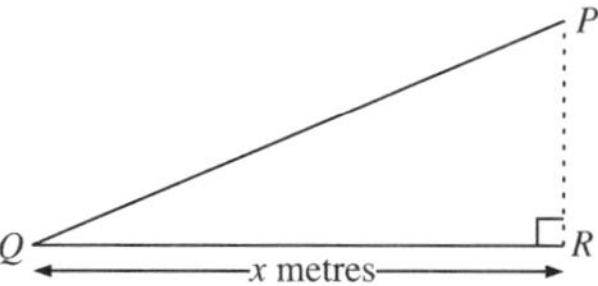

- **i** What is the length of the hypotenuse PQ in terms of x? Easy
- **ii** Show that the area of the triangle PQR is $\frac{1}{2}x\sqrt{25 - 10x}$ square metres. Medium
- **iii** What is the maximum possible area of the triangle? Hard

(6 marks) (Q6a, **1997 HSC**)

83 A particle is moving along the x axis. Its position at time t is given by

$$x = t + \sin t.$$

- **i** At what times during the period $0 < t < 3\pi$ is the particle stationary? Medium
- **ii** At what times during the period $0 < t < 3\pi$ is the acceleration equal to 0? Medium
- **iii** Carefully sketch the graph of $x = t + \sin t$ for $0 < t < 3\pi$. Clearly label any stationary points and any points of inflection. Hard

(8 marks) (Q8b, **1997 HSC**)

84 Lee takes some medicine. The amount, M, of medicine present in Lee's bloodstream t hours later is given by

$$M = 4t^2 - t^3, \text{ for } 0 \le t \le 3.$$

- **i** Sketch the curve $M = 4t^2 - t^3$ for $0 \le t \le 3$ showing any stationary points. Medium
- **ii** At what time is the amount of medicine in Lee's bloodstream a maximum? Easy
- **iii** When is the amount present increasing most rapidly? Medium

(7 marks) (Q6a, **1996 HSC**)

85 $PQRS$ is a rectangle with $PQ = 6$ cm and $QR = 4$ cm. T and U lie on the lines SP and SR respectively, so that T, Q, and U are collinear, as shown in the diagram. Let $PT = x$ cm and $RU = y$ cm.

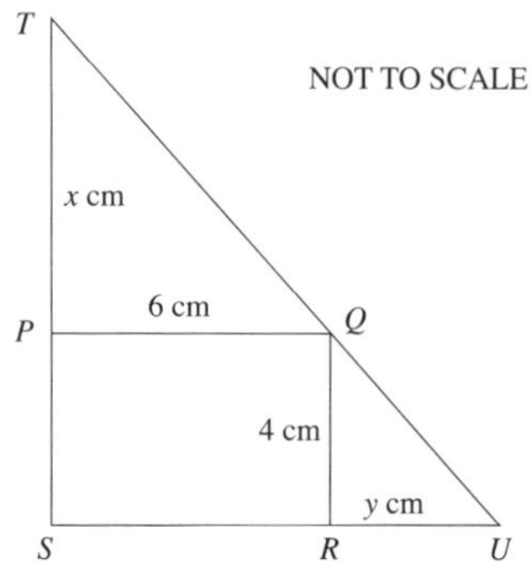

[i Show that triangles *TPQ* and *QRU* are similar.

ii Show that $xy = 24$.]

iii Show that the area, A, of triangle *TSU* is given by $A = 24 + 3x + \frac{48}{x}$. **Medium**

iv Find the height and base of the triangle *TSU* with minimum area. Justify your answer. **Medium**

(Q8b, **1996 HSC**)

86 i On the same set of axes, accurately draw the graphs of $y = \sin x$ and $y = \frac{2}{3}x$ for $0 \leq x \leq \pi$. **Medium**

ii Find the gradient of the tangent to $y = \sin x$ at the origin. **Easy**

iii For what values of m does the equation $\sin x = mx$ have a solution in the domain $0 < x < \pi$? **Medium**

(5 marks) (Q10a, **1996 HSC**)

87 Consider the curve given by $y = 7 + 4x^3 - 3x^4$.

i Find the coordinates of the two stationary points.

ii Find all values of x for which $\frac{d^2y}{dx^2} = 0$.

iii Determine the nature of the stationary points.

iv Sketch the curve for the domain $-1 \leq x \leq 2$. *(8 marks)*

(Q5b, **1995 HSC**) **Easy**

88 A rectangular beam of width w cm and depth d cm is cut from a cylindrical pine log as shown.

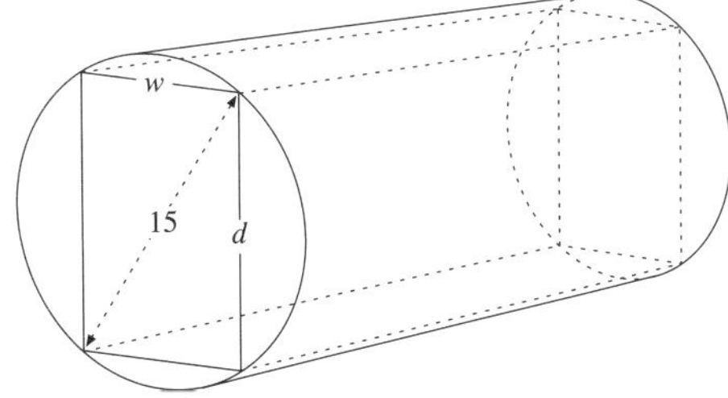

The diameter of the cross-section of the log (and hence the diagonal of the cross-section of the beam) is 15 cm.

The strength S of the beam is proportional to the product of its width and the square of its depth, so that

$$S = kd^2w.$$

i Show that $S = k(225w - w^3)$. **Easy**

ii What numerical dimensions will give a beam of maximum strength? Justify your answer. **Medium**

iii A square beam with diagonal 15 cm could have been cut from the log. Show that the rectangular beam of maximum strength is more than 8% stronger than this square beam. **Hard**

(7 marks) (Q9a, **1995 HSC**)

89 A can of baked beans is in the shape of a closed cylinder with height h cm and radius r cm, as shown in the diagram.

i The volume of the can is 500 cm^3. Find an expression for h in terms of r. **Easy**

ii Show that the surface area, S cm^2, of the can is given by

$S = 2\pi r^2 + \frac{1000}{r}$. **Easy**

iii If the area of metal used to make the can is to be minimized, find the radius of the can. **Medium**

(Q7a, **1994 HSC**)

90 Consider the curve given by $y = \frac{1}{4}x^4 - x^3$.

i Find any turning points and determine their nature.

ii Find any points of inflexion.

iii Sketch the curve for $-1.5 \leq x \leq 4.5$, indicating where the curve crosses the x axis.

iv For what values of x is the curve concave down?

(Q6a, **1993 HSC**) **Easy**

91 The diagram shows a graph of the parabola $y = x^2$ and the tangent to the parabola at $x = c$.

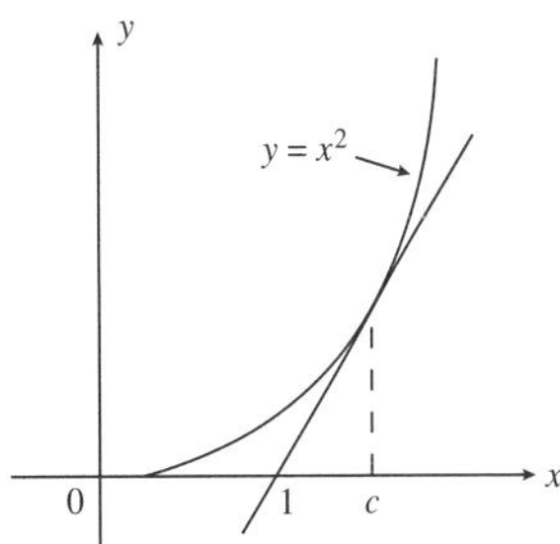

i Find the gradient of the tangent at $x = c$. **Easy**

ii Find the equation of the tangent at $x = c$. **Medium**

iii Find the value of c if the tangent intersects the x axis at $x = 1$. **Hard**

(Q6a, **1992 HSC**)

Year 12 Applications of the derivative—Worked answers

1 **a** $N(t) = 200e^{0.013t}$

$N(0) = 200e^{0.013(0)}$

$= 200$ ✓

(1 mark)

b $N(24) = 200e^{0.013(24)}$

$= 273.2309386...$

$= 273$ (nearest whole) ✓

(1 mark)

c $N'(t) = 0.013 \times 200e^{0.013t}$

$N'(24) = 0.013 \times 273.2309386...$ ✓ (from part **b**)

$= 3.552002202...$

$= 3.55$ (2 dec. pl.) ✓

$\therefore$ 3.55 bacteria/h

(2 marks)

2 $y = x^3 - 6x^2 + 8$

$\frac{dy}{dx} = 3x^2 - 12x = 0$ ✓

$3x(x-4) = 0$

$x = 0, 4$

$\therefore$ stationary points at $x = 0$ and $x = 4$. ✓

Compare y-values for $x = 0$, $x = 4$, $x = -1$ and $x = 7$:

$y(0) = 0^3 - 6(0)^2 + 8$

$= 8$

$y(4) = 4^3 - 6(4)^2 + 8$

$= -24$

$y(-1) = (-1)^3 - 6(-1)^2 + 8$ ✓

$= 1$

$y(7) = 7^3 - 6(7)^2 + 8$

$= 57$ ✓

$\therefore$ maximum value is 57, minimum value is -24. *(4 marks)*

3 **a** $f(x) = xe^{-2x}$

$f'(x) = e^{-2x} - 2xe^{-2x}$

$f''(x) = -2e^{-2x} - 2[e^{-2x} - 2xe^{-2x}]$ ✓

$= -2e^{-2x} - 2e^{-2x} + 4xe^{-2x}$

$= -4e^{-2x} + 4xe^{-2x}$

$= 4e^{-2x}(x-1)$

$= 4(x-1)e^{-2x}$ ✓

(2 marks)

b $f'(x) = e^{-2x} - 2xe^{-2x} = 0$

$e^{-2x}(1-2x) = 0$

$x = \frac{1}{2}$

(as $e^{-2x} \neq 0$)

Also, $f\left(\frac{1}{2}\right) = \frac{1}{2}e^{-2\left(\frac{1}{2}\right)}$

$= \frac{1}{2}e^{-1}$

$= \frac{1}{2e}$

$\therefore$ stationary point at $\left(\frac{1}{2}, \frac{1}{2e}\right)$ ✓

Now, $f''\left(\frac{1}{2}\right) = 4\left(\frac{1}{2} - 1\right)e^{-2\left(\frac{1}{2}\right)}$

$= -2e^{-1} < 0$

$\therefore$ maximum at $\left(\frac{1}{2}, \frac{1}{2e}\right)$ ✓ *(2 marks)*

c $f'(x) = 4(x-1)e^{-2x} = 0$

$x = 1$

(as $e^{-2x} \neq 0$)

Also, $f(1) = \frac{1}{2}e^{-2(1)}$

$= \frac{1}{2}e^{-2}$

$= \frac{1}{2e^2}$

$\therefore$ possible point of inflexion at $\left(1, \frac{1}{2e^2}\right)$ ✓

Check neighbourhood:

x	0	1	2
$f''(x)$	< 0	0	> 0

$\therefore$ as sign change, then a point of inflexion at $\left(1, \frac{1}{2e^2}\right)$

Also, when $y = 0$: $0 = xe^{-2x}$

$\therefore x = 0$ (as $e^{-2x} \neq 0$)

when $x = 0$: $y = 0e^{-2(0)}$

$\therefore y = 0$

$\therefore$ curve passes through $(0, 0)$, maximum at $\left(\frac{1}{2}, \frac{1}{2e}\right)$ and point of inflexion at $\left(1, \frac{1}{2e^2}\right)$. ✓

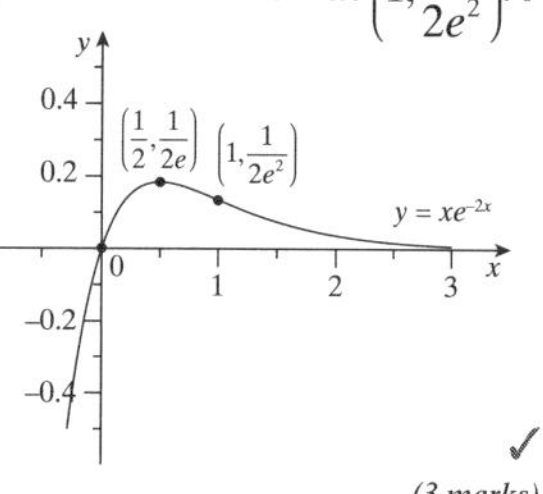

✓

(3 marks)

4 **a** Consider gradient PX = gradient XY:

$\frac{2-0}{1-x} = \frac{y}{x}$ ✓

$2x = y(1-x)$

$y = \frac{2x}{x-1}$ ✓

(2 marks)

b $A = \frac{1}{2} \times x \times y$

$= \frac{1}{2} \times x \times \frac{2x}{x-1}$

(from part **a**)

$\therefore A = \frac{x^2}{x-1}$ ✓

$\frac{dA}{dx} = \frac{(x-1).2x - x^2.1}{(x-1)^2}$

(using quotient rule)

$= \frac{2x^2 - 2x - x^2}{(x-1)^2}$

$= \frac{x^2 - 2x}{(x-1)^2} = 0$

$x^2 - 2x = 0$

$x(x-2) = 0$

$x = 0, 2$

But $x > 1$, hence $x = 2$. ✓

Determine the nature by checking neighbourhood:

x	1.5	2	2.5
$\frac{dA}{dx}$	< 0	0	> 0

$\therefore$ minimum value of $x = 2$. ✓

Subs. $x = 2$ into $A = \frac{x^2}{x-1}$:

$= \frac{2^2}{2-1}$

$= 4$

$\therefore$ minimum area of 4 units². ✓

(4 marks)

5 $P = 5000b^{-\frac{t}{10}}$

Substitute $P = 1250$, $t = 20$:

$1250 = 5000b^{-\frac{20}{10}}$

$1250 = 5000b^{-2}$

$1250 = \frac{5000}{b^2}$

$1250b^2 = 5000$

$b^2 = 4$

$b = 2$ (as $b > 0$) ✓

$\therefore P = 5000 \times 2^{-\frac{t}{10}}$

$\frac{dP}{dt} = 5000 \times (\ln 2) \times -\frac{1}{10} \times 2^{-\frac{t}{10}}$

$= -500 \ln 2 \times 2^{-\frac{t}{10}}$ ✓

Substitute $\frac{dP}{dt} = -30$:

$-30 = -500 \ln 2 \times 2^{-\frac{t}{10}}$

$2^{-\frac{t}{10}} = \frac{30}{500 \ln 2}$

$\ln 2^{-\frac{t}{10}} = \ln\left(\frac{3}{50 \ln 2}\right)$

$-\frac{t}{10} \ln 2 = \ln\left(\frac{3}{50 \ln 2}\right)$

$-\frac{t}{10} = \frac{\ln\left(\frac{3}{50 \ln 2}\right)}{\ln 2}$

$t = -10 \times \frac{\ln\left(\frac{3}{50 \ln 2}\right)}{\ln 2}$ ✓

$= 35.30127316...$

$= 35.3$ (1 dec. pl.)

$\therefore$ after 35.3 years. ✓

(4 marks)

6 **a** $y = -5t^2 + 70t + 100$

Maximum height when $y' = 0$:

$y' = -10t + 70 = 0$

$10t = 70$

$t = 7$ ✓

Maximum height after 7 seconds (as function is a concave down parabola)

Substitute $t = 7$:

$y = -5(7)^2 + 70(7) + 100$

$= 345$

$\therefore$ the maximum height is 345 m. ✓

(2 marks)

b Substitute $y = 0$:

$0 = -5t^2 + 70t + 100$

$5t^2 - 70t - 100 = 0$

$t^2 - 14t - 20 = 0$ ✓

$t = \dfrac{14 \pm \sqrt{(-14)^2 - 4(1)(-20)}}{2(1)}$

$= \dfrac{14 \pm \sqrt{276}}{2}$

$= \dfrac{14 \pm 2\sqrt{69}}{2}$

$= 7 \pm \sqrt{69}$

$= 7 + \sqrt{69}$ (as $t > 0$) ✓

Substitute into velocity equation:

$v = -10t + 70$

$v(7 + \sqrt{69})$

$= -10(7 + \sqrt{69}) + 70$

$= -70 - 10\sqrt{69} + 70$

$= -10\sqrt{69}$

$\therefore$ a velocity of $-10\sqrt{69}$ ms^{-1}. ✓

(3 marks)

7 **a** $T = 25 + 70(1.5)^{-0.4t}$

Substitute $t = 4$:

$T = 25 + 70(1.5)^{-0.4(4)}$

$= 61.589\,125\,15\ldots$

$= 61.6$ (1 dec. pl.)

$\therefore$ the temperature is 61.6°. ✓

(1 mark)

b $\dfrac{dT}{dt} = 70 \times \ln 1.5 \times -0.4(1.5)^{-0.4t}$

$= -28 \ln 1.5 \times (1.5)^{-0.4t}$ ✓

Substitute $t = 4$:

$\dfrac{dT}{dt}(4) = -28 \ln 1.5 \times (1.5)^{-0.4(4)}$

$= -5.934\,345\,433\ldots$

$= -5.9$ (1 dec. pl.)

$\therefore$ the temperature is dropping at 5.9°/min. ✓

(2 marks)

c Substitute $T = 55$:

$55 = 25 + 70(1.5)^{-0.4t}$

$70(1.5)^{-0.4t} = 30$

$1.5^{-0.4t} = \dfrac{3}{7}$ ✓

$\ln 1.5^{-0.4t} = \ln \dfrac{3}{7}$

$-0.4t \ln 1.5 = \ln \dfrac{3}{7}$

$-0.4t = \dfrac{\ln \frac{3}{7}}{\ln 1.5}$ ✓

$t = -\dfrac{1}{0.4} \dfrac{\ln \frac{3}{7}}{\ln 1.5}$

$t = 5.224\,234\,117\ldots$

$= 5$ min 13 s (nearest sec.)

$\therefore$ it will take 5 minutes 13 seconds. ✓

(3 marks)

8 **a**

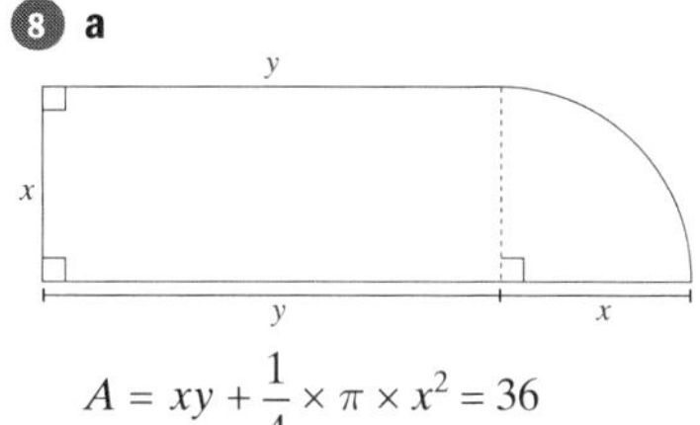

$A = xy + \dfrac{1}{4} \times \pi \times x^2 = 36$

$xy = 36 - \dfrac{\pi x^2}{4}$

$y = \dfrac{36}{x} - \dfrac{\pi x}{4}$ ✓

$P = 2x + 2y + \dfrac{1}{4} \times 2 \times \pi \times x$

$= 2y + 2x + \dfrac{\pi x}{2}$ ✓

Substitute $y = \dfrac{36}{x} - \dfrac{\pi x}{4}$:

$P = 2\left(\dfrac{36}{x} - \dfrac{\pi x}{4}\right) + 2x + \dfrac{\pi x}{2}$

$= \dfrac{72}{x} - \dfrac{\pi x}{2} + 2x + \dfrac{\pi x}{2}$

$= 2x + \dfrac{72}{x}$ ✓

(3 marks)

b $P = 2x + \dfrac{72}{x}$

$= 2x + 72x^{-1}$

$\dfrac{dP}{dx} = 2 - 72x^{-2}$ ✓

$= 2 - \dfrac{72}{x^2} = 0$

$2 = \dfrac{72}{x^2}$

$2x^2 = 72$

$x = 6$ ($x > 0$) ✓

$\dfrac{d^2P}{dx^2} = 144x^{-3}$

$= \dfrac{144}{x^3}$

$\dfrac{d^2P}{dx^2}(6) = \dfrac{144}{6^3} > 0$,

hence minimum perimeter when $x = 6$. ✓

$P(6) = 2(6) + \dfrac{72}{6}$

$= 24$

$\therefore$ the smallest possible perimeter is 24 m. ✓

(4 marks)

9 **a** $y = c \ln x$

$\dfrac{dy}{dx} = \dfrac{c}{x}$

$\dfrac{dy}{dx}(p) = \dfrac{c}{p}$ ✓

Also, substitute $x = p$ in y:

$y(p) = c \ln p$

Using $(p, c \ln p)$ and gradient $= \dfrac{c}{p}$:

$y - c \ln p = \dfrac{c}{p}(x - p)$

$y - c \ln p = \dfrac{c}{p}x - c$

$y = \dfrac{c}{p}x - c + c \ln p$ ✓

(2 marks)

b $y = \dfrac{c}{p}x - c + c \ln p$

As passes through origin, substitute $x = 0, y = 0$:

$0 = \dfrac{c}{p}(0) - c + c \ln p$

$c = c \ln p$

$1 = \ln p$

$p = e$ ✓

As $\dfrac{dy}{dx}(p) = \dfrac{c}{p}$,

then $\dfrac{dy}{dx}(e) = \dfrac{c}{e} = 1$

$\therefore c = e$ ✓

(2 marks)

10 **a** $m = a \sin\left(\dfrac{\pi}{26}t\right) + b$

Substitute $m = 35\,000, t = 13$:

$35\,000 = a \sin\left(\dfrac{\pi}{26}(13)\right) + b$

$a \sin \dfrac{\pi}{2} + b = 35\,000$

$a + b = 35\,000 \ldots$ ①

Also, substitute $m = 5000$ and $t = 39$:

$5000 = a \sin\left(\dfrac{\pi}{26}(39)\right) + b$

$a \sin \dfrac{3\pi}{2} + b = 5000$

$-a + b = 5000 \ldots$ ②

① + ②: $2b = 40\,000$

$b = 20\,000$ ✓

Substitute $b = 20\,000$ in ①:

$a + 20\,000 = 35\,000$

$a = 15\,000$ ✓

$\therefore a = 15\,000, b = 20\,000$

OR Average of maximum and minimum is $\dfrac{35\,000 + 5000}{2} = 20\,000$.

This means $b = 20\,000$.

Hence, $a = 35\,000 - 20\,000$

$= 15\,000$

$a = 15\,000, b = 20\,000$

(2 marks)

b From the graph, $m(t)$ is increasing when $0 < t < 13$ and $39 < t \leq 52$. ✓

As $f(x) = \cos x$ is decreasing from $0 < t < \pi$, etc, then $f(x) = -\cos x$ is increasing from $0 < t < \pi$, etc. This means $c(t)$ is increasing when

$0 < \left(\dfrac{\pi}{26}(t - 10)\right) < \pi$

$0 < t - 10 < \pi \div \frac{\pi}{26}$

$0 < t - 10 < 26$

$10 < t < 36$ ✓

Hence both populations are increasing when $10 < t < 13$. ✓ *(3 marks)*

c $c(t)$ reaches a maximum when $t = 36$.

$m(t) = 15000 \sin\left(\frac{\pi}{26}t\right) + 20000$

$m'(t) = \frac{15000\pi}{26} \cos\left(\frac{\pi}{26}t\right)$ ✓

$m'(36) = \frac{15000\pi}{26} \cos \frac{36\pi}{26}$

$= -642.7062162\ldots$

$= -643$ (nearest whole)

$\therefore$ the mice population is decreasing at the rate of 643 per week. ✓ *(2 marks)*

11 The particle is stationary at $t = 11$ and $t = 13$.

This means the particle reached a maximum speed between $t = 11$ and $t = 13$.

At that time,

$\frac{dv}{dt}$ (or acceleration) $= 0$.

Answer C

12 **i** $L(t) = 200000e^{-0.14t}$

Substitute $t = 31$:

$L(31) = 200000e^{-0.14(31)}$

$= 2607.305641\ldots$

$= 2607$ (nearest whole) ✓ *(1 mark)*

ii $\frac{dL}{dt} = (-0.14)200000e^{-0.14t}$

$= -28\,000e^{-0.14t}$ ✓

Substitute $t = 31$:

$\frac{dL}{dt}(31) = -28000e^{-0.14(31)}$

$= -365.0227897\ldots$

$= -365$ (nearest whole)

$\therefore$ decreasing by 365 leaves per day. ✓ *(2 marks)*

iii $L(t) = 200\,000e^{-0.14t}$

Substitute $L(t) = 100$:

$100 = 200000e^{-0.14t}$

$e^{-0.14t} = 0.0005$

$-0.14t = \ln 0.0005$ ✓

$t = \frac{\ln 0.0005}{-0.14}$

$= 54.29216043\ldots$

$= 54$ (nearest whole) ✓ *(2 marks)*

13 **i**

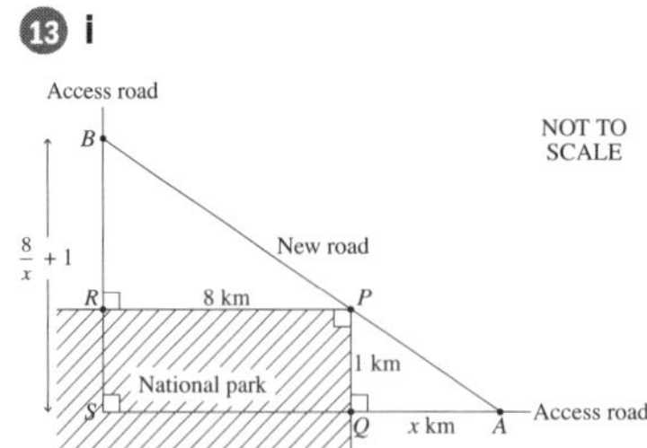

In triangles BPR and PAQ,

$\angle BRP = \angle PQA$ (given)

$\angle BPR = \angle PAQ$ (corresponding $\angle$s, $RP \| QA$)

$\therefore \Delta BPR \,|||\, \Delta PAQ$ (equiangular) ✓

$\therefore \frac{BR}{1} = \frac{8}{x}$ (matching sides of similar triangles in proportion)

$\therefore BR = \frac{8}{x}$ ✓

Consider BR and AQ both produced to S to form rectangle $RPQS$:

Using ΔBSA and Pythagoras' theorem:

$\therefore D^2 = (x + 8)^2 + \left(\frac{8}{x} + 1\right)^2$ ✓ *(3 marks)*

ii $D^2 = (x + 8)^2 + \left(\frac{8}{x} + 1\right)^2$

$= x^2 + 16x + 64 + \frac{64}{x^2} + \frac{16}{x} + 1$

$= x^2 + 16x + 65 + 64x^{-2} + 16x^{-1}$

$\frac{dD^2}{dx} = 2x + 16 - \frac{128}{x^3} - \frac{16}{x^2}$ ✓

Substitute $x = 2$:

$\frac{dD^2}{dx}(2) = 2(2) + 16 - \frac{128}{2^3} - \frac{16}{2^2}$

$= 4 + 16 - 16 - 4$

$= 0$

$\therefore$ stationary point at $x = 2$ ✓

As $\frac{dD^2}{dx} = 2x + 16 - 128x^{-3} - 16x^{-2}$

$\frac{d^2D^2}{dx^2} = 2x + 384x^{-4} + 32x^{-3}$

$= 2x + \frac{384}{x^4} + \frac{32}{x^3}$

$\frac{d^2D^2}{dx^2}(2) = 2(2) + \frac{384}{2^4} + \frac{32}{2^3} > 0$

$\therefore$ minimum value of D^2 when $x = 2$ ✓ *(3 marks)*

14 $y = \cos 2x$

$\frac{dy}{dx} = -2\sin 2x$ ✓

$\frac{dy}{dx}\left(\frac{\pi}{6}\right) = -2\sin 2\left(\frac{\pi}{6}\right)$

$= -2\sin \frac{\pi}{3}$

$= -\sqrt{3}$

Also, $y\left(\frac{\pi}{6}\right) = \cos 2\left(\frac{\pi}{6}\right)$

$= \cos \frac{\pi}{3}$

$= \frac{1}{2}$ ✓

Using $\left(\frac{\pi}{6}, \frac{1}{2}\right)$ and $m = -\sqrt{3}$,

$y - y_1 = m(x - x_1)$

$y - \frac{1}{2} = -\sqrt{3}\left(x - \frac{\pi}{6}\right)$

$y = \frac{1}{2} - \sqrt{3}x + \frac{\pi\sqrt{3}}{6}$

$6y = 3 - 6\sqrt{3}x + \pi\sqrt{3}$

$\therefore 6\sqrt{3}x + 6y - 3 - \pi\sqrt{3} = 0$ *(3 marks)* ✓

15 **i** $x = \frac{t^3}{3} - 2t^2 + 3t$

$\frac{dx}{dt} = t^2 - 4t + 3$

$\frac{dx}{dt}(0) = 0^2 - 4(0) + 3$

$= 3$

$\therefore$ initial velocity of 3 ms^{-1}. ✓ *(1 mark)*

ii $\frac{dx}{dt} = t^2 - 4t + 3 = 0$ ✓

$(t - 1)(t - 3) = 0$

$t = 1, 3$

$\therefore$ after 1 second and 3 seconds. ✓ *(2 marks)*

iii $\frac{d^2x}{dt^2} = 2t - 4 = 0$

$2t = 4$

$t = 2$ ✓

Substitute $t = 2$ into x:

$x(2) = \frac{2^3}{3} - 2(2)^2 + 3(2)$

$= \frac{8}{3} - 8 + 6$

$= \frac{2}{3}$

$\therefore$ acceleration zero when particle is $\frac{2}{3}$ m to the right of origin. ✓ *(2 marks)*

16 $y = 6x^2 - x^3$

i $y' = 12x - 3x^2 = 0$

$3x(4 - x) = 0$

$x = 0$ or 4 ✓

$y(0) = 6(0)^2 - 0^3$

$= 0$

$y(4) = 6(4)^2 - 4^3$

$= 32$

$\therefore$ stationary points at $(0, 0)$ and $(4, 32)$. ✓

$y'' = 12 - 6x$

$y''(0) = 12 - 6(0) > 0$

$\therefore$ minimum

$y''(4) = 12 - 6(4) < 0$

$\therefore$ maximum

$\therefore$ minimum at $(0, 0)$ and maximum at $(4, 32)$. ✓ *(3 marks)*

ii $y''(2) = 12 - 6(2) = 0$

$\therefore (2, 16)$ is a possible point of inflexion. ✓

Now, check the neighbourhood of $x = 2$:

x	1.9	2	2.1
y''	>0	0	<0

As sign change for y'', then a change in concavity.

$\therefore (2, 16)$ is a point of inflexion. ✓ *(2 marks)*

iii Let $x = 0, \therefore y = 0$.

Let $y = 0$, $\therefore 6x^2 - x^3 = 0$

$x^2(6 - x) = 0$

$x = 0, 6$

$\therefore$ intercepts at $(0, 0)$ and $(6, 0)$. ✓

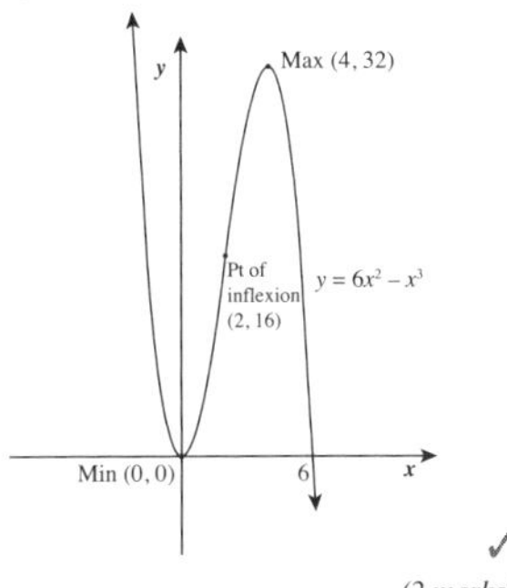

✓ *(2 marks)*

17 **i** Let h = perpendicular height of cone

$\therefore h = \sqrt{10^2 - x^2}$ (by Pythagoras)

$= \sqrt{100 - x^2}$

$\therefore V = \frac{1}{3} \times \pi \times x^2 \times \sqrt{100 - x^2}$

$= \frac{1}{3}\pi x^2\sqrt{100 - x^2}$ *(1 mark)*

ii $V = \frac{1}{3}\pi\, x^2(100 - x^2)^{\frac{1}{2}}$

$\frac{dV}{dx} = \frac{\pi}{3}[2x(100 - x^2)^{\frac{1}{2}} + x^2.\frac{1}{2}(100 - x^2)^{-\frac{1}{2}}.-2x]$ ✓

$= \frac{\pi x}{3}[2(100 - x^2)^{\frac{1}{2}} - x^2(100 - x^2)^{-\frac{1}{2}}]$

$= \frac{\pi x}{3}[(100 - x^2)^{-\frac{1}{2}}(200 - 2x^2 - x^2)]$

$= \frac{\pi x}{3}[(100 - x^2)^{-\frac{1}{2}}(200 - 3x^2)]$

$= \frac{\pi x(200 - 3x^2)}{3\sqrt{100 - x^2}}$ ✓ *(2 marks)*

iii $\frac{dV}{dx} = \frac{\pi x(200 - 3x^2)}{3\sqrt{100 - x^2}} = 0$

$x(200 - 3x^2) = 0$

$x = 0, \sqrt{\frac{200}{3}} \quad (x \geq 0)$ ✓

$\therefore x = 0, \frac{10\sqrt{2}}{\sqrt{3}}$

Consider the neighbourhood for $x = \frac{10\sqrt{2}}{\sqrt{3}}$:

x	8	$\frac{10\sqrt{2}}{\sqrt{3}}$	9
$\frac{dV}{dx}$	>0	0	<0

$\therefore$ maximum at $x = \frac{10\sqrt{2}}{\sqrt{3}}$. ✓

Now, arc length of sector = circumference of circular base of cone

$10\theta = 2\pi\frac{10\sqrt{2}}{\sqrt{3}}$

$\theta = \frac{2\pi\sqrt{2}}{\sqrt{3}}$

$\therefore \theta = \frac{2\pi\sqrt{6}}{3}$ ✓ *(3 marks)*

18 $y = 2x^3 + 3x^2 - 12x + 7$

i $y' = 6x^2 + 6x - 12 = 0$

$6(x^2 + x - 2) = 0$

$(x + 2)(x - 1) = 0$

$x = -2, 1$ ✓

$y(-2) = 2(-2)^3 + 3(-2)^2 - 12(-2) + 7$

$= 27$

$y(1) = 2(1)^3 + 3(1)^2 - 12(1) + 7$

$= 0$

$\therefore$ stationary points at $(-2, 27)$ and $(1, 0)$. ✓

$y'' = 12x + 6$

$y''(-2) = 12(-2) + 6 < 0$

$\therefore$ maximum ✓

$y''(1) = 12(1) + 6 > 0$

$\therefore$ minimum ✓

$\therefore$ maximum at $(-2, 27)$ and minimum at $(1, 0)$. *(4 marks)*

ii

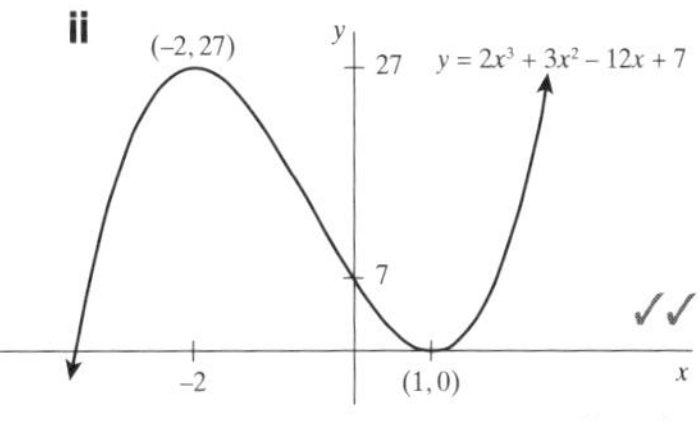

✓✓ *(2 marks)*

iii From the graph, $y' > 0$ when the curve is increasing: $x < -2$ and $x > 1$. ✓ *(1 mark)*

19

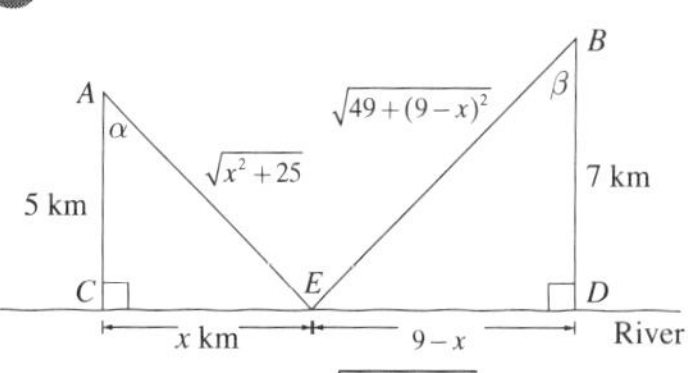

i $AE = \sqrt{x^2 + 5^2}$ (by Pythagoras)

$= \sqrt{x^2 + 25}$

As $DE = 9 - x$,

$EB = \sqrt{7^2 + (9 - x)^2}$ (by Pythagoras)

$= \sqrt{49 + (9 - x)^2}$

As $L = AE + EB$,

$\therefore L = \sqrt{x^2 + 25} + \sqrt{49 + (9 - x)^2}$ ✓ *(1 mark)*

ii $L = (x^2 + 25)^{\frac{1}{2}} + (49 + (9 - x)^2)^{\frac{1}{2}}$

$\frac{dL}{dx} = \frac{1}{2}(x^2 + 25)^{-\frac{1}{2}}.2x + \frac{1}{2}(49 + (9 - x)^2)^{-\frac{1}{2}}.(-18 + 2x) = 0$ ✓

$x(x^2 + 25)^{-\frac{1}{2}} - [49 + (9 - x)^2]^{-\frac{1}{2}}.(9 - x) = 0$

$\frac{x}{\sqrt{x^2 + 25}} - \frac{9 - x}{\sqrt{49 + (9 - x)^2}} = 0$ ✓

But, from ΔACE,

$\frac{x}{\sqrt{x^2 + 25}} = \sin\alpha$,

and from ΔBDE,

$\frac{9 - x}{\sqrt{49 + (9 - x)^2}} = \sin\beta$,

$\therefore \sin\alpha - \sin\beta = 0$

$\therefore \sin\alpha = \sin\beta$ ✓ *(3 marks)*

iii If $\sin\alpha = \sin\beta$, then $\alpha = \beta$ (for acute α, β) and hence ΔACE and ΔBDE are similar (2 $\angle$s equal).

$\therefore \frac{x}{5} = \frac{9 - x}{7}$ ✓

(matching sides of similar Δs)

$7x = 45 - 5x$

$12x = 45$

$x = 3.75$ ✓ *(2 marks)*

iv As $\frac{dL}{dx}(3.75) = 0$, then check nature of stationary point:

x	3	3.75	4
L'	<0	0	>0

As sign changes, then $x = 3.75$ gives a minimum value for L. ✓ *(1 mark)*

20

$y = \tan x$

$\frac{dy}{dx} = \sec^2 x$

$\frac{dy}{dx}\left(\frac{\pi}{8}\right) = \sec^2 \frac{\pi}{8}$ ✓

$= \frac{1}{\cos^2 \frac{\pi}{8}}$

$= 1.171\,572\,857\ldots$

$= 1.17$ (3 sig. figs.) ✓ *(2 marks)*

21 i

$y = 4x^3 - x^4$

$y' = 12x^2 - 4x^3 = 0$

$4x^2(3 - x) = 0$

$x = 0$ or 3 ✓

$y(0) = 4(0)^3 - (0)^4$

$= 0$

$y(3) = 4(3)^3 - (3)^4$

$= 27$

$\therefore$ stationary points at $(0, 0)$ and $(3, 27)$. ✓

$y'' = 24x - 12x^2$

$y''(0) = 24(0) - 12(0)^2 = 0$

$\therefore$ possible horizontal point of inflexion

Now, consider the slope of curve near $x = 0$:

x	-1	0	1
y'	>0	0	>0

$\therefore$ horizontal point of inflexion at $(0, 0)$ ✓

Also,

$y''(3) = 24(3) - 12(3)^2 < 0$

$\therefore$ maximum at $(3, 27)$ ✓

$\therefore$ horizontal point of inflexion at $(0, 0)$ and maximum at $(3, 27)$ *(4 marks)*

ii

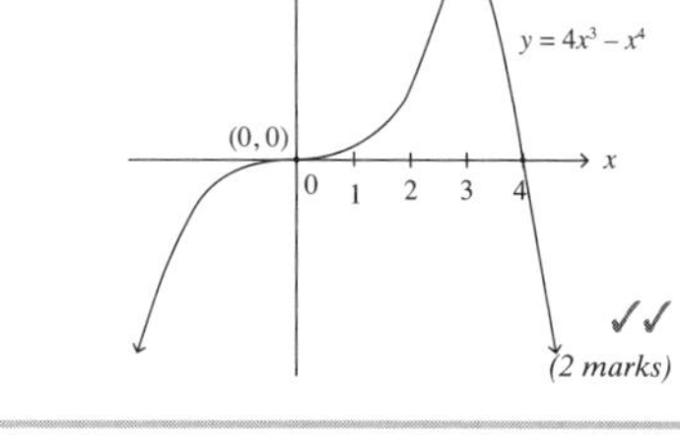

✓✓ *(2 marks)*

22 i Let the length of the enclosure be y m.

$\therefore$ Area $= xy = 720$

$\therefore y = \frac{720}{x}$

Now, $\ell = 5x + y$

$= 5x + \frac{720}{x}$ ✓

(1 mark)

ii $\ell = 5x + 720x^{-1}$

$\frac{d\ell}{dx} = 5 - 720x^{-2}$ ✓

$= 5 - \frac{720}{x^2} = 0$

$5x^2 = 720$

$x^2 = 144$

$x = 12$ (as $x > 0$) ✓

$\frac{d^2\ell}{dx^2} = 1440x^{-3}$

As $\frac{d^2\ell}{dx^2} > 0$, for all positive x, then minimum.

Subs. $x = 12$:

$\ell = 5(12) + \frac{720}{12}$

$= 120$

$\therefore$ the minimum length is 120 m. ✓ *(3 marks)*

23 i

$y = 200(1 + 19e^{-0.5t})^{-1}$

$\frac{dy}{dt} = -200(1 + 19e^{-0.5t})^{-2} \times -9.5e^{-0.5t}$ ✓

$= \frac{1900e^{-0.5t}}{(1+19e^{-0.5t})^2}$ ✓ *(2 marks)*

ii As $t \geq 0$, consider $t = 0$:

$y(0) = \frac{200}{1+19e^{-0.5(0)}}$

$= \frac{200}{1+19}$

$= 10$

As $t \to \infty$,

$y(\infty) = \frac{200}{1+19e^{-0.5(\infty)}}$

$= \frac{200}{1+0}$

$= 200$

As y is monotonic increasing, then range is $10 \leq y < 200$. ✓

(2 marks)

iii From $y = \frac{200}{1+19e^{-0.5t}}$

$\therefore 1 + 19e^{-0.5t} = \frac{200}{y}$

$19e^{-0.5t} = \frac{200}{y} - 1$

$19e^{-0.5t} = \frac{200 - y}{y}$

$\therefore 1900e^{-0.5t} = 100\left[\frac{200 - y}{y}\right]$

As $\frac{dy}{dt}$

$= \frac{1900e^{-0.5t}}{(1+19e^{-0.5t})^2}$

$= 100\left[\frac{200 - y}{y}\right] \div \left[\frac{200}{y}\right]^2$

$= 100\left[\frac{200 - y}{y}\right] \times \frac{y^2}{40\,000}$

$= \frac{y(200 - y)}{400}$

$= \frac{y}{400}(200 - y)$ ✓ *(1 mark)*

iv Let $R = \frac{y}{400}(200 - y)$

$\therefore R = \frac{y}{2} - \frac{y^2}{400}$

$\frac{dR}{dy} = \frac{1}{2} - \frac{y}{200} = 0$ ✓

$\therefore \frac{y}{200} = \frac{1}{2}$

$y = 100$

$\frac{d^2R}{dy^2} = -\frac{1}{200} < 0$

$\therefore$ maximum.

$\therefore$ fastest rate when 100 yabbies. ✓

(2 marks)

24 i

$y = x^3 - x^2 - x + 3$

$y' = 3x^2 - 2x - 1 = 0$

$(3x + 1)(x - 1) = 0$

$x = -\frac{1}{3}$ or 1 ✓

$y(-\frac{1}{3}) = (-\frac{1}{3})^3 - (-\frac{1}{3})^2 - (-\frac{1}{3}) + 3$

$= 3\frac{5}{27}$

$y(1) = 1^3 - 1^2 - 1 + 3$

$= 2$

$\therefore$ stationary points at $(-\frac{1}{3}, 3\frac{5}{27})$ and $(1, 2)$. ✓

$y'' = 6x - 2$

$y''(-\frac{1}{3}) = 6(-\frac{1}{3}) - 2 < 0$

$\therefore$ maximum at $(-\frac{1}{3}, 3\frac{5}{27})$. ✓

$y''(1) = 6(1) - 2 > 0$

$\therefore$ minimum at $(1, 2)$. ✓

(4 marks)

ii $y'' = 6x - 2 = 0$

$6x = 2$

$x = \frac{1}{3}$

$\therefore$ possible point of inflexion at $(\frac{1}{3}, \frac{70}{27})$. ✓

Consider neighbourhood:

x	$\frac{1}{4}$	$\frac{1}{3}$	$\frac{1}{2}$
y''	<0	0	>0

As change in concavity, then the point of inflexion is at $(\frac{1}{3}, \frac{70}{27})$. ✓

(2 marks)

iii Substitute $x = 0$ in $y = x^3 - x^2 - x + 3$:

$\therefore y = 3$.

$\therefore$ y-intercept of 3.

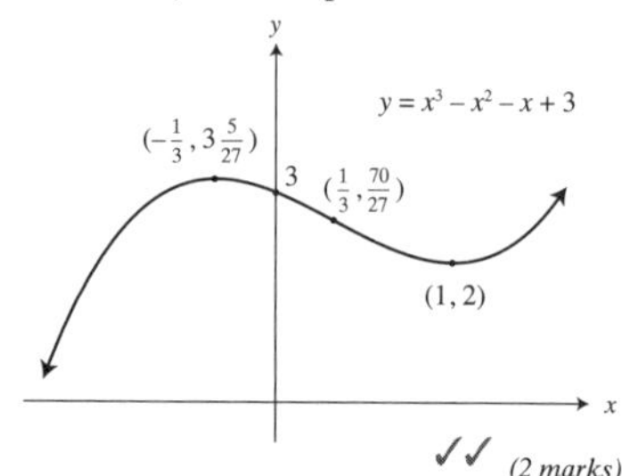

✓✓ *(2 marks)*

25 **i**

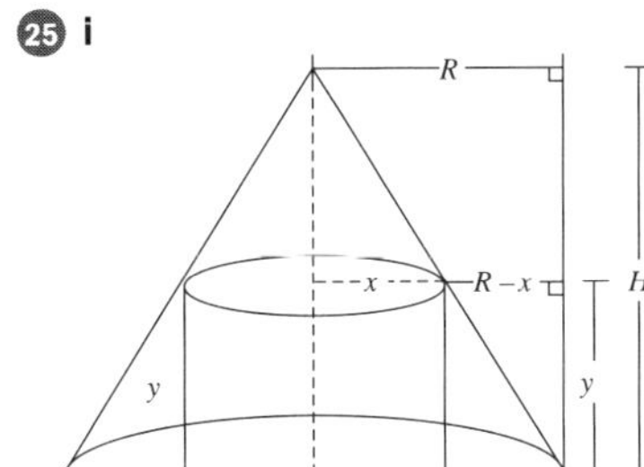

Consider sides of similar triangles:

$$\frac{y}{H} = \frac{R - x}{R}$$

$$y = \frac{H(R - x)}{R} \quad ✓$$

Now, volume of cylinder:

$$V = \pi x^2 y$$

$$= \pi x^2\left[\frac{H(R-x)}{R}\right] \quad ✓$$

$$= \frac{H\pi}{R}x^2(R - x) \quad ✓ \text{ (3 marks)}$$

ii

$$V = \frac{H\pi}{R}[x^2(R - x)]$$

$$\frac{dV}{dx} = \frac{H\pi}{R}[2x(R - x) + x^2(-1)]$$

$$= \frac{H\pi}{R}[2xR - 2x^2 - x^2]$$

$$= \frac{H\pi}{R}[2xR - 3x^2] \quad ✓$$

Maximum volume occurs when $\frac{dV}{dx} = 0$

$$\therefore 2xR - 3x^2 = 0$$

$$x(2R - 3x) = 0$$

$$x = 0, \frac{2R}{3}$$

$$x = \frac{2R}{3} \quad (\text{as } x > 0)$$

$$\frac{d^2V}{dx^2} = \frac{H\pi}{R}[2R - 6x]$$

$$\frac{d^2V}{dx^2}\left[\frac{2R}{3}\right] = \frac{H\pi}{R}\left[2R - 6\left(\frac{2R}{3}\right)\right] < 0$$

$$= \frac{H\pi}{R}[-2R] < 0$$

(as $H > 0, R > 0$)

$\therefore$ maximum volume when $x = \frac{2R}{3}$. ✓

Substitute $x = \frac{2R}{3}$ in $V = \frac{H\pi}{R}x^2(R - x)$:

$$V = \frac{H\pi}{R}\left(\frac{2R}{3}\right)^2\left(R - \frac{2R}{3}\right)$$

$$= \frac{H\pi}{R} \times \frac{4R^2}{9} \times \frac{R}{3}$$

$$\therefore V_{cylinder} = \frac{4H\pi R^2}{27} \quad ✓$$

Now, volume of cone with radius R and height H:

$$V_{cone} = \frac{1}{3}\pi R^2 H$$

But, $\frac{4}{9} \times \frac{1}{3}\pi R^2 H = \frac{4H\pi R^2}{27}$.

$\therefore$ volume of cylinder does not exceed $\frac{4}{9}$ of volume of cone. ✓ *(4 marks)*

26 $y = e^x - ex$

$$\frac{dy}{dx} = e^x - e = 0 \quad ✓$$

$$e^x = e$$

$$x = 1$$

$$y(1) = e^1 - e(1)$$

$$= 0 \therefore \text{stat. pt at } (1, 0) \quad ✓$$

$$\frac{d^2y}{dx^2} = e^x$$

$$\frac{d^2y}{dx^2}(1) = e^1 > 0 \therefore \text{minimum}$$

$\therefore$ minimum at $(1, 0)$. ✓ *(3 marks)*

27 **i**

$$2x + 2y + \frac{1}{2} \times \pi \times x = 10$$

$$4x + 4y + \pi x = 20 \quad ✓$$

$$4y = 20 - x(4 + \pi)$$

$$y = 5 - x\left(1 + \frac{\pi}{4}\right) \quad ✓$$

(2 marks)

ii

$$A = xy + \frac{1}{2} \times \pi \times \left(\frac{x}{2}\right)^2$$

$$L = 3xy + \frac{\pi x^2}{8}$$

$$= 3x\left(5 - x\left(1 + \frac{\pi}{4}\right)\right) + \frac{\pi x^2}{8} \quad ✓$$

$$= 3x\left(5 - x - \frac{\pi x}{4}\right) + \frac{\pi x^2}{8}$$

$$= 15x - 3x^2 - \frac{3\pi x^2}{4} + \frac{\pi x^2}{8}$$

$$= 15x - 3x^2 - \frac{5\pi x^2}{8}$$

$$= 15x - x^2\left(3 + \frac{5\pi}{8}\right) \quad ✓$$

(2 marks)

iii

$$L = 15x - x^2\left(3 + \frac{5\pi}{8}\right)$$

$$\frac{dL}{dx} = 15 - 2x\left(3 + \frac{5\pi}{8}\right) = 0$$

$$2x\left(3 + \frac{5\pi}{8}\right) = 15$$

$$x = \frac{15}{2\left(3 + \frac{5\pi}{8}\right)}$$

$$= \frac{15}{6 + \frac{5\pi}{4}}$$

$$= \frac{60}{24 + 5\pi} \quad ✓$$

$$\frac{d^2L}{dx^2} = -2\left(3 + \frac{5\pi}{8}\right) < 0$$

$\therefore$ maximum ✓

$$\therefore y\left(\frac{60}{24 + 5\pi}\right)$$

$$= 5 - \left(\frac{60}{24 + 5\pi}\right)\left(1 + \frac{\pi}{4}\right)$$

$$= 5 - \left(\frac{60 + 15\pi}{24 + 5\pi}\right)$$

$$= \frac{120 + 25\pi - 60 - 15\pi}{24 + 5\pi}$$

$$= \frac{60 + 10\pi}{24 + 5\pi} \quad ✓$$

$$\therefore x = \frac{60}{24 + 5\pi}, y = \frac{60 + 10\pi}{24 + 5\pi}$$

(3 marks)

28 $\dot{x} = -3$ means the particle is moving to the left; as $\ddot{x} = +2$, the particle is slowing down.

$\therefore$ particle is moving to the left with decreasing speed.

Answer D

29 **i** As $D = ST$, then:

Car A travels $80t$ km; car B travels $50t$ km.

$$\therefore AR = 80t$$

$$\text{and } BR = 100 - 50t \quad ✓$$

Using cosine rule:

$$r^2 = (100 - 50t)^2 + (80t)^2 - 2(100 - 50t)(80t) \times \cos 60°$$

$$= 10\,000 - 10\,000t + 2500t^2 + 6400t^2 - 8000t + 4000t^2$$

$$\therefore r^2 = 12\,900t^2 - 18\,000t + 10\,000$$

✓ *(2 marks)*

ii Min. when $\frac{d(r^2)}{dt} = 0$

$$\frac{d(r^2)}{dt} = 25\,800t - 18\,000 = 0$$

$$t = \frac{18\,000}{25\,800}$$

$$= \frac{30}{43} \quad ✓$$

Consider nature:

$$\frac{d^2(r^2)}{dt^2} = 25\,800 > 0$$

$\therefore$ Minimum ✓

Subs. in r^2:

$$r^2 = 12\,900\left(\frac{30}{43}\right)^2 - 18\,000\left(\frac{30}{43}\right) + 10\,000$$

$$r = 60.999\,428\,13\ldots$$

$$= 61 \text{ (nearest whole)}$$

$\therefore$ minimum distance is 61 km ✓ *(3 marks)*

30 $f(x) = 3x^4 + 4x^3 - 12x^2$

i $f'(x) = 12x^3 + 12x^2 - 24x = 0$

$12x(x^2 + x - 2) = 0$

$12x(x + 2)(x - 1) = 0$

$\therefore$ Stat. points at $x = 0, -2, 1$ ✓

$f(0) = 3(0)^4 + 4(0)^3 - 12(0)^2 = 0 \quad \therefore (0, 0)$

$f(-2) = 3(-2)^4 + 4(-2)^3 - 12(-2)^2 = -32 \quad \therefore (-2, -32)$

$f(1) = 3(1)^4 + 4(1)^3 - 12(1)^2 = -5 \quad \therefore (1, -5)$

$\therefore$ Stat. points at $(0, 0)$, $(-2, -32)$, $(1, -5)$ ✓

$f''(x) = 36x^2 + 24x - 24$

For $(0, 0)$:

$f''(0) = 36(0)^2 + 24(0) - 24 < 0 \quad \therefore$ Max $(0, 0)$

For $(-2, -32)$:

$f''(-2) = 36(-2)^2 + 24(-2) - 24 > 0 \quad \therefore$ Min $(-2, -32)$

For $(1, -5)$:

$f''(1) = 36(1)^2 + 24(1) - 24 > 0 \quad \therefore$ Min $(1, -5)$

$\therefore$ Max $(0, 0)$, Min $(-2, -32)$, Min $(1, -5)$ ✓

(3 marks)

ii

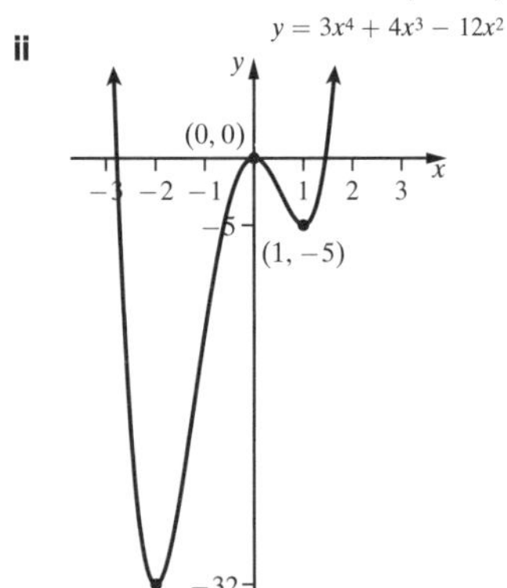

✓✓ *(2 marks)*

iii From **ii**, the function is increasing for $-2 < x < 0$ and $x > 1$ ✓ *(1 mark)*

iv $3x^4 + 4x^3 - 12x^2 + k = 0$

$3x^4 + 4x^3 - 12x^2 = -k$

Solution is point(s) of intersection of $y = 3x^4 + 4x^3 - 12x^2$ and $y = -k$.

Using part **ii**, no solutions if $-k < -32$ i.e. $k > 32$ ✓ *(1 mark)*

31

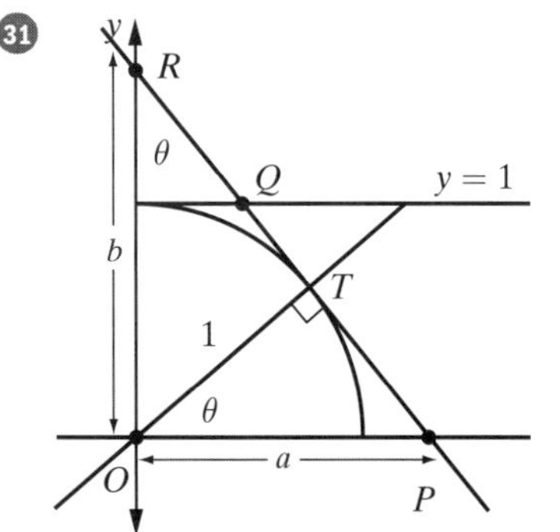

i Using $\frac{x}{a} + \frac{y}{b} = 1$:

In ΔOPT, is $\frac{OT}{OP} = \cos\theta$

$\frac{1}{a} = \cos\theta$

$a = \frac{1}{\cos\theta}$

In ΔORT, is $\frac{OT}{OR} = \sin\theta$

$\frac{1}{b} = \sin\theta$

$b = \frac{1}{\sin\theta}$ ✓

$\therefore$ equation is

$\frac{x}{\frac{1}{\cos\theta}} + \frac{y}{\frac{1}{\sin\theta}} = 1$

$x\cos\theta + y\sin\theta = 1$ ....**1** ✓ *(2 marks)*

ii Subs $y = 1$ in **1**:

$x\cos\theta + \sin\theta = 1$

$x\cos\theta = 1 - \sin\theta$

$x = \frac{1 - \sin\theta}{\cos\theta}$

$\therefore Q\left(\frac{1 - \sin\theta}{\cos\theta}, 1\right)$

As $B(0, 1)$, then

$BQ = \frac{1 - \sin\theta}{\cos\theta}$ ✓ *(1 mark)*

iii Area $= \frac{1}{2}h(a + b)$

$A = \frac{1}{2} \times 1 \times \left(\frac{1 - \sin\theta}{\cos\theta} + \frac{1}{\cos\theta}\right)$ ✓

$= \frac{1}{2} \times \left(\frac{2 - \sin\theta}{\cos\theta}\right)$

$= \frac{2 - \sin\theta}{2\cos\theta}$

$\therefore A = \frac{2 - \sin\theta}{2\cos\theta}$ units^2 ✓ *(2 marks)*

iv $A = \frac{2 - \sin\theta}{2\cos\theta}$

$\frac{dA}{d\theta} = \frac{2\cos\theta(-\cos\theta) - (2 - \sin\theta).(-2\sin\theta)}{4\cos^2\theta}$

$= \frac{-2\cos^2\theta + 4\sin\theta - 2\sin^2\theta}{4\cos^2\theta}$

$= \frac{4\sin\theta - 2(\sin^2\theta + \cos^2\theta)}{4\cos^2\theta}$

$= \frac{4\sin\theta - 2}{4\cos^2\theta}$

$= \frac{2(2\sin\theta - 1)}{4\cos^2\theta}$ ✓

$= \frac{(2\sin\theta - 1)}{2\cos^2\theta} = 0$

$2\sin\theta - 1 = 0$

$\sin\theta = \frac{1}{2}$

$\theta = \frac{\pi}{6}$ (as $0 < \theta < \frac{\pi}{2}$) ✓

Test for min:

θ	$\frac{\pi}{12}$	$\frac{\pi}{6}$	$\frac{\pi}{3}$
$\frac{dA}{d\theta}$	< 0	0	> 0

$\therefore \theta = \frac{\pi}{6}$ for minimum ✓ *(3 marks)*

32 i $f(x) = x^3 - 3x + 2$

$f'(x) = 3x^2 - 3 = 0$

$3x^2 = 3$

$x = \pm 1$ ✓

$f(1) = (1)^3 - 3(1) + 2 = 0$

$f(-1) = (-1)^3 - 3(-1) + 2 = 4$

Stat points at $(1, 0)$ and $(-1, 4)$

$f''(x) = 6x$

$f''(1) = 6(1) = 6 > 0$

$\therefore$ min $(1, 0)$ ✓

$f''(-1) = 6(-1) = -6 < 0$

$\therefore$ max $(-1, 4)$ ✓

(3 marks)

ii y-intercept of 2

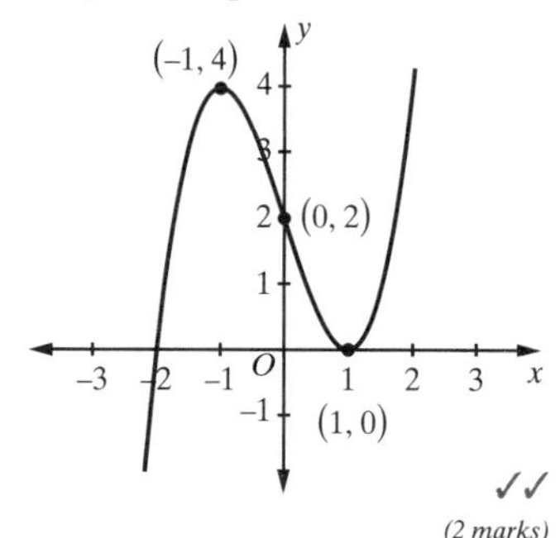

✓✓ *(2 marks)*

33 i $\dot{x} = 8 - 8e^{-2t}$

Subs $t = 0$,

$\dot{x} = 8 - 8e^{-2(0)}$

$= 8 - 8e^0$

$= 8 - 8$

$= 0$

$\therefore$ particle initially at rest. ✓ *(1 mark)*

ii $\ddot{x} = 16e^{-2t}$

As $e^{-2t} > 0$, for all values of t, then $\ddot{x}$ is always positive. ✓ *(1 mark)*

iii Since particle is initially at rest (from **i**) and always a positive acceleration applied (from **ii**), then the velocity must be positive for all $t > 0$. ✓ This means it is moving in the positive direction. ✓

(2 marks)

iv $\dot{x} = 8 - 8e^{-2t}$

As $t \to \infty, e^{-2t} \to 0$,

$\therefore \dot{x}$ approaches 8.

The constant is 8. ✓

(1 mark)

v

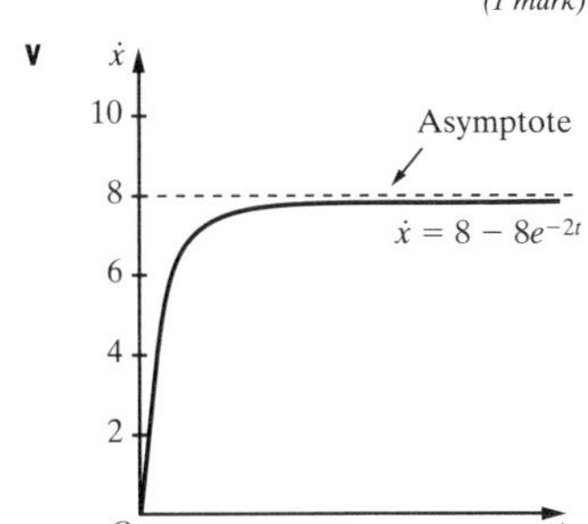

(2 marks)

34 $f(x) = e^{-x} - 2e^{-2x}$

i $f'(x) = -e^{-x} + 4e^{-2x}$ ✓

(1 mark)

ii At turning points $f'(x) = 0$.

i.e. $-e^{-x} + 4e^{-2x} = 0$

$e^{-x}(-1 + 4e^{-x}) = 0$

$e^{-x} = 0$ or $-1 + 4e^{-x} = 0$

No solution or

$$4e^{-x} = 1$$

$$e^{-x} = \frac{1}{4}$$

$$-x = \ln\left(\frac{1}{4}\right)$$

$$x = -\ln\left(\frac{1}{4}\right)$$

$$= \ln\left(\frac{1}{4}\right)^{-1}$$

$$x = \ln 4$$ ✓

$f''(x) = e^{-x} - 8e^{-2x}$

$f''(\ln 4) < 0$

So the maximum turning point is at $x = \ln 4$.

$$f(\ln 4) = e^{-\ln 4} - 2e^{-2(\ln 4)}$$

$$= \frac{1}{4} - \frac{2}{16}$$

$$= \frac{1}{8}$$

The maximum turning point is at $(\ln 4, \frac{1}{8})$. ✓

(2 marks)

iii $f(\ln 2) = e^{-\ln 2} - 2e^{-2(\ln 2)}$

$$= \frac{1}{2} - \frac{2}{4}$$

$$= 0$$ ✓

(1 mark)

iv As $x \to \infty, e^{-x} \to 0$

So $f(x) \to 0$ ✓

(1 mark)

v When $x = 0$,

$$e^{-x} - 2e^{-2x} = e^{-0} - 2e^{-2(0)}$$

$$= 1 - 2$$

$$= -1$$

The y-intercept is -1. ✓

(1 mark)

vi

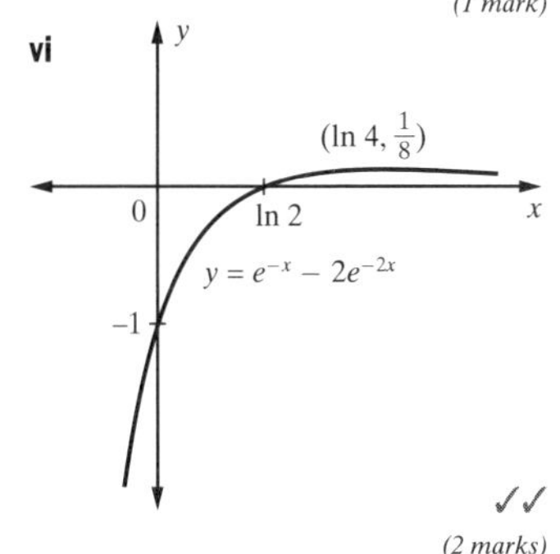

✓✓

(2 marks)

35 **i** Using arc length $= r\theta$

$$P = r + r + r\theta$$

$$= 2r + r\theta$$

$$= r(\theta + 2)$$ ✓

(1 mark)

ii $P = r(\theta + 2)$ from (i)

$$P = r\theta + 2r$$

$$r\theta = P - 2r$$

$$\theta = \frac{P - 2r}{r}$$

Using area of sector $= \frac{1}{2}r^2\theta$

$$A = \frac{1}{2}r^2 \times \frac{P - 2r}{r}$$

$$= \frac{1}{2}Pr - r^2$$

$$\therefore A = \frac{1}{2}Pr - r^2$$ ✓

(1 mark)

iii $A = \frac{1}{2}Pr - r^2$

$$\frac{dA}{dr} = \frac{1}{2}P - 2r = 0$$

$$2r = \frac{P}{2}$$

$$r = \frac{P}{4}$$ ✓

$$\frac{d^2A}{dr^2} = -2 < 0, \quad \therefore \text{maximum}$$

$\therefore$ max when $r = \frac{P}{4}$ ✓

(2 marks)

iv Subs $r = \frac{P}{4}$ in

$$P = r(\theta + 2)$$

$$P = \frac{P}{4}(\theta + 2)$$

$$4 = \theta + 2$$

$$\theta = 2$$

$\therefore$ angle is 2 radians. ✓ *(1 mark)*

v Using $0 < \theta < 2\pi$

From **ii**,

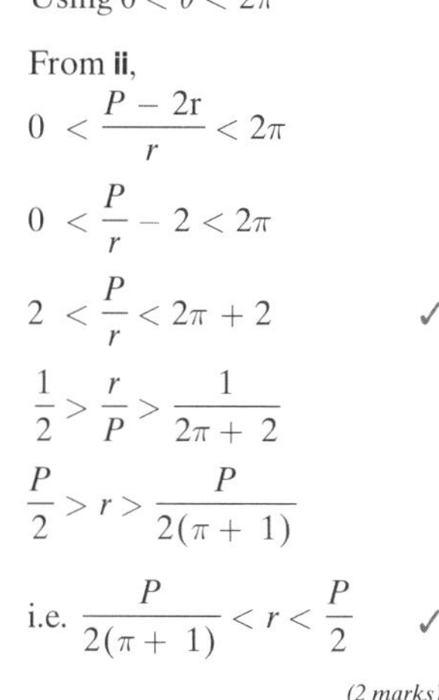

$$0 < \frac{P - 2r}{r} < 2\pi$$

$$0 < \frac{P}{r} - 2 < 2\pi$$

$$2 < \frac{P}{r} < 2\pi + 2$$ ✓

$$\frac{1}{2} > \frac{r}{P} > \frac{1}{2\pi + 2}$$

$$\frac{P}{2} > r > \frac{P}{2(\pi + 1)}$$

i.e. $\frac{P}{2(\pi + 1)} < r < \frac{P}{2}$ ✓

(2 marks)

Alternatively:

From part **i**, $P = r(\theta + 2)$

$\therefore P = r\theta + 2r > 2r$

$r < \frac{P}{2}$

Also from part **i**, since $\theta < 2\pi$,

$\therefore P = r(\theta + 2) < r(2\pi + 2)$

$r > \frac{P}{2(\pi+1)}$

$\therefore \frac{P}{2(\pi+1)} < r < \frac{P}{2}$

36 $y = \ln(3x)$

$$\frac{dy}{dx} = \frac{3}{3x}$$

$$= \frac{1}{x}$$ ✓

At $x = 2$, $\frac{dy}{dx} = \frac{1}{2}$

$\therefore$ The gradient is $\frac{1}{2}$. ✓

(2 marks)

37 **i** $V = \pi r^2 h$

$$10 = \pi r^2 h$$

$$\therefore h = \frac{10}{\pi r^2}$$ ✓

Substitute h in A:

$$A = 2\pi r^2 + 2\pi rh$$

$$= 2\pi r^2 + 2\pi r \times \frac{10}{\pi r^2}$$

$$= 2\pi r^2 + \frac{20}{r}$$ ✓

(2 marks)

ii

$$A = 2\pi r^2 + \frac{20}{r}$$
$$= 2\pi r^2 + 20r^{-1}$$
$$\frac{dA}{dr} = 4\pi r - 20r^{-2} \checkmark$$
$$\therefore 4\pi r - \frac{20}{r^2} = 0$$
$$4\pi r = \frac{20}{r^2}$$
$$4\pi r^3 = 20$$
$$r^3 = \frac{20}{4\pi}$$
$$r^3 = \frac{5}{\pi}$$
$$r = \sqrt[3]{\frac{5}{\pi}} \checkmark$$

As $\frac{dA}{dr} = 4\pi r - 20r^{-2}$,

then $\frac{d^2A}{dr^2} = 4\pi + 40r^{-3}$

But as $r > 0$, then $\frac{d^2A}{dr^2} > 0$,

so A is minimum when

$r = \sqrt[3]{\frac{5}{\pi}}$. ✓

[Or, use first derivative test:

r	1	$\sqrt[3]{\frac{5}{\pi}}$	2
$\frac{dA}{dr}$	< 0	0	> 0

$\therefore$ A is minimum when

$r = \sqrt[3]{\frac{5}{\pi}}$].

(3 marks)

38 $f(x) = (x + 2)(x^2 + 4)$

i Let $y = (x + 2)(x^2 + 4)$

$$= x^3 + 4x + 2x^2 + 8$$
$$= x^3 + 2x^2 + 4x + 8$$
$$\frac{dy}{dx} = 3x^2 + 4x + 4 \checkmark$$

For $3x^2 + 4x + 4$, check discriminant:

$$\Delta = b^2 - 4ac$$
$$= 4^2 - 4(3)(4)$$
$$= -32 < 0$$

$\therefore$ No real roots for $\frac{dy}{dx} = 0$

$\therefore$ No stationary points. ✓

(2 marks)

ii

$$\frac{dy}{dx} = 3x^2 + 4x + 4$$
$$\frac{d^2y}{dx^2} = 6x + 4$$

Now, concave down when

$\frac{d^2y}{dx^2} < 0$,

$6x + 4 < 0$ ✓

$6x < -4$

$x < -\frac{2}{3}$

$\therefore$ Concave down when

$x < -\frac{2}{3}$.

Similarly, concave up when

$x > -\frac{2}{3}$. ✓

(2 marks)

iii $y = (x + 2)(x^2 + 4)$

x-intercept: let $y = 0$, then $x = -2$

y-intercept: let $x = 0$, then $y = 8$

Also, at $x = \frac{-2}{3}$,

then

$$y = \left(\frac{-2}{3} + 2\right)\left(\left(\frac{-2}{3}\right)^2 + 4\right)$$
$$= 5.925925 \ldots.$$

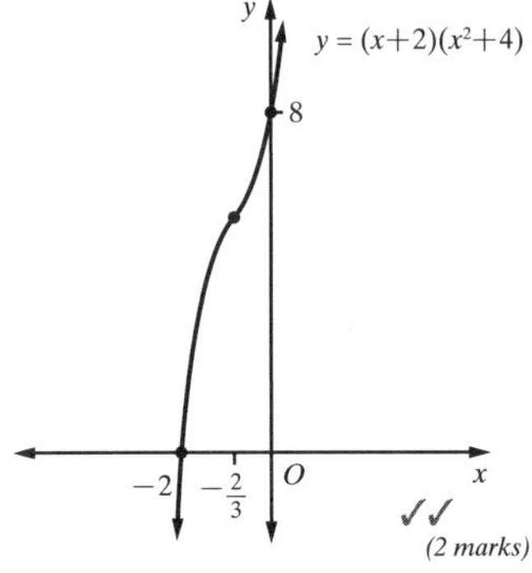

✓✓ *(2 marks)*

39

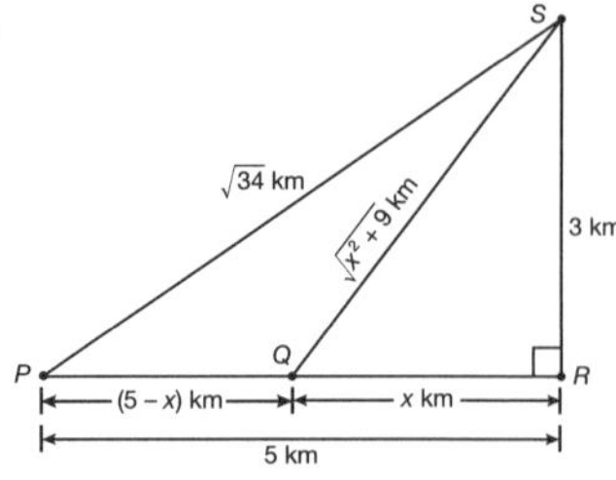

i Cost = \$1000 × 5 + \$2600 × 3
= \$5000 + \$7800
= \$12 800 ✓

(1 mark)

ii By Pythagoras' Theorem in ΔPRS,

$$PS^2 = PR^2 + RS^2$$
$$= 5^2 + 3^2$$
$$= 25 + 9$$
$$= 34$$
$$\therefore PS = \sqrt{34}$$

$\therefore$ Cost = \$2600 × $\sqrt{34}$
= \$15 160.474 …
= \$15 160
to the nearest dollar. ✓

(1 mark)

iii By Pythagoras' Theorem in ΔQRS,

$$QS^2 = QR^2 + RS^2$$
$$= x^2 + 3^2$$
$$= x^2 + 9$$
$$\therefore QS = \sqrt{x^2 + 9} \checkmark$$

and $PQ = 5 - x$

$$\therefore C = 1000 \times (5 - x) + 2600\sqrt{x^2 + 9}$$
$$= 1000\left(5 - x + 2.6\sqrt{x^2 + 9}\right) \checkmark$$

(2 marks)

iv $C = 1000(5 - x + 2.6\sqrt{x^2 + 9})$

$= 1000\left[5 - x + 2.6(x^2 + 9)^{\frac{1}{2}}\right]$

$\frac{dC}{dx} = 1000\left[-1 + 2.6\left(\frac{1}{2}\right)(x^2 + 9)^{-\frac{1}{2}} \times 2x\right]$

$= 1000\left[-1 + 2.6x(x^2 + 9)^{-\frac{1}{2}}\right]$

$= 1000\left(-1 + \frac{2.6x}{\sqrt{x^2 + 9}}\right)$ ✓

Stationary points occur when

$\frac{dC}{dx} = 0$

$1000\left(-1 + \frac{2.6x}{\sqrt{x^2 + 9}}\right) = 0$

$-1 + \frac{2.6x}{\sqrt{x^2 + 9}} = 0$

$\frac{2.6x}{\sqrt{x^2 + 9}} = 1$

$2.6x = \sqrt{x^2 + 9}$

$(2.6x)^2 = x^2 + 9$

$6.76x^2 = x^2 + 9$

$5.76x^2 = 9$

$x^2 = \frac{9}{5.76}$

$x = \frac{3}{2.4}$

$\therefore x = 1.25.$ ✓

To determine nature of stationary point:

x	1.25^-	1.25	1.25^+
$\frac{dC}{dx}$	–	0	+

$\therefore$ A minimum value occurs when $x = 1.25$. ✓

Now $C = 1000(5 - x + 2.6\sqrt{x^2 + 9})$

When $x = 1.25$,

$C = 1000(5 - 1.25 + 2.6\sqrt{1.25^2 + 9})$

$= 12\,200$

$\therefore$ The minimum cost of laying the cable is \$12 200.

✓ *(4 marks)*

v $C = 1000 \times (5 - x) + 1100\sqrt{x^2 + 9}$

$= 1000(5 - x + 1.1\sqrt{x^2 + 9})$

$= 1000\left[5 - x + 1.1(x^2 + 9)^{\frac{1}{2}}\right]$

$\frac{dC}{dx} = 1000\left[-1 + 1.1\left(\frac{1}{2}\right)(x^2 + 9)^{-\frac{1}{2}} \times 2x\right]$

$= 1000\left[-1 + 1.1x(x^2 + 9)^{-\frac{1}{2}}\right]$

$= 1000\left(-1 + \frac{1.1x}{\sqrt{x^2 + 9}}\right)$

Stationary points occur when

$\frac{dC}{dx} = 0$

$1000\left(-1 + \frac{1.1x}{\sqrt{x^2 + 9}}\right) = 0$

$\left(-1 + \frac{1.1x}{\sqrt{x^2 + 9}}\right) = 0$

$\frac{1.1x}{\sqrt{x^2 + 9}} = 1$

$1.1x = \sqrt{x^2 + 9}$

$(1.1x)^2 = x^2 + 9$

$1.21x^2 = x^2 + 9$

$0.21x^2 = 9$

$x^2 = \frac{9}{0.21}$

$\therefore x = \frac{3}{\sqrt{0.21}}$

$\doteqdot 6.5$ ✓

Since $x < 5$, this solution for x is invalid. (See diagram.)

Consider cost of laying cable from P to R, then from R to S:

Cost = \$1000 × 5 + \$1100 × 3 = \$8300

Consider cost of laying cable in a straight line underwater directly from P to S:

Cost = \$1100 × PS

= \$1100 × $\sqrt{34}$

= \$6414.047 …

= \$6414 to nearest dollar.

$\therefore$ Minimum cost of \$6414 can be achieved by laying cable underwater in a straight line directly from P to S. ✓ *(2 marks)*

40 **i** $I = Ae^{-ks}$

When $s = 0$, $I = 6000$

$\therefore 6000 = Ae^{-k \times 0}$

$6000 = Ae^0$

$6000 = A$

i.e. $A = 6000$. ✓ *(1 mark)*

ii $I = Ae^{-ks}$

When $s = 6$, $I = 1000$ and $A = 6000$

$\therefore 1000 = 6000 \times e^{-k \times 6}$

$\frac{1}{6} = e^{-6k}$

$\log_e \frac{1}{6} = -6k$

$k = -\frac{1}{6}\log_e \frac{1}{6}$ ✓

$= -\frac{1}{6}\log_e 6^{-1}$

$= \frac{1}{6}\log_e 6$

$= \frac{\log_e 6}{6}$

$= 0.2986 \ldots$

$\therefore k = 0.30$ to 2 decimal places. ✓ *(2 marks)*

iii $I = Ae^{-ks}$

$\therefore \dfrac{dI}{ds} = -kAe^{-ks}$

When $s = 6$, $A = 6000$

and $k = \dfrac{\log_e 6}{6}$ or $0.2986\ldots$,

$\therefore \dfrac{dI}{ds} = -\dfrac{\log_e 6}{6} \times 6000 \times e^{-\frac{\log_e 6}{6} \times 6}$ ✓

$= -\log_e 6 \times 1000 \times e^{-\log_e 6}$

$= -298.6265\ldots$

$\doteqdot -299.$

$\therefore$ At 6 metres, the light intensity is decreasing at 299 lux per metre. ✓

(2 marks)

41 **i** The initial velocity is when $t = 0$.

$\therefore$ Initial velocity is 20 m/s. ✓

(1 mark)

ii The velocity of the particle is equal to 0 when $t = 10$. ✓

(1 mark)

iii The acceleration of the particle is zero when $t = 6$. ✓

(1 mark)

42 **i** The curve crosses the x-axis when

$f(x) = 0$

$\therefore x^4 - 8x^2 = 0$

$x^2(x^2 - 8) = 0$ ✓

$x^2(x - 2\sqrt{2})(x + 2\sqrt{2}) = 0$

$\therefore x = 0$ or $2\sqrt{2}$ or $-2\sqrt{2}$

$\therefore$ The coordinates are $(0, 0)$, $(2\sqrt{2}, 0)$ and $(-2\sqrt{2}, 0)$. ✓

(2 marks)

ii For an even function,

$f(x) = f(-x)$

Now $f(x) = x^4 - 8x^2$

and $f(-x) = (-x)^4 - 8(-x)^2$

$= x^4 - 8x^2$

$\therefore f(x) = f(-x)$

$\therefore f(x)$ is an even function. ✓

(1 mark)

iii $f(x) = x^4 - 8x^2$

$f'(x) = 4x^3 - 16x$

$= 4x(x^2 - 4)$

$= 4x(x - 2)(x + 2)$ ✓

Stationary points occur when

$f'(x) = 0$

$\therefore 4x(x - 2)(x + 2) = 0$

$\therefore x = 0$ or 2 or -2 ✓

Now $f(0) = 0$

and $f(2) = (2)^4 - 8(2)^2$

$= -16$

and $f(-2) = (-2)^4 - 8(-2)^2$

$= -16$

$\therefore$ Coordinates of the stationary points are $(0, 0)$ $(2, -16)$ and $(-2, -16)$. ✓

Now $f''(x) = 12x^2 - 16$

$\therefore f''(0) = 12(0)^2 - 16$

$= -16 < 0$

$\therefore$ A maximum turning point occurs at $(0, 0)$.

and $f''(2) = 12(2)^2 - 16$

$= 32 > 0$

$\therefore$ A minimum turning point occurs at $(2, -16)$.

and $f''(-2) = 12(-2)^2 - 16$

$= 32 > 0$

$\therefore$ A minimum turning point occurs at $(-2, -16)$. ✓

(4 marks)

iv

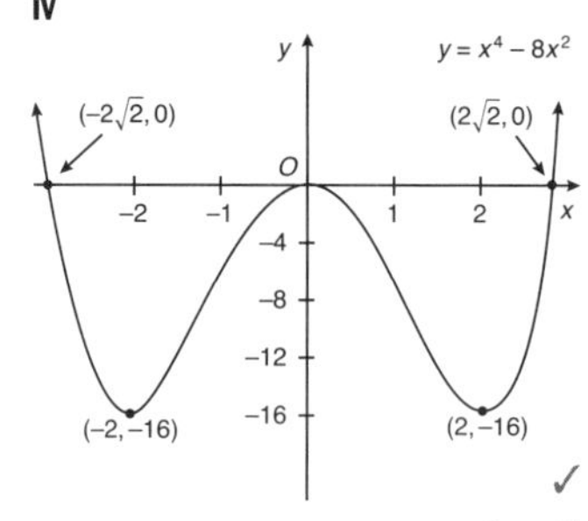

✓

(1 mark)

43

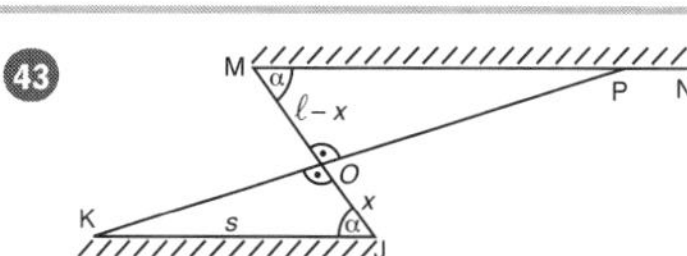

i In ΔOKJ and ΔOMP,

$\angle KOJ = \angle POM$ (vertically opposite angles)

$\angle KJO = \angle PMO$ (alternate angles, $KJ \parallel MP$)

$\therefore \Delta OKJ \,|||\, \Delta OMP$ (equiangular)

$\therefore \dfrac{MP}{s} = \dfrac{\ell - x}{x}$

$\therefore MP = \dfrac{(\ell - x)s}{x}$ ✓

Total area = Area ΔOKJ + Area ΔOMP

$A = \dfrac{1}{2}sx\sin\alpha + \dfrac{1}{2}MP(\ell - x)\sin\alpha$ ✓

$= \dfrac{1}{2}sx\sin\alpha + \dfrac{1}{2}\dfrac{(\ell - x)s}{x}(\ell - x)\sin\alpha$

$= \dfrac{1}{2}s\sin\alpha\left[x + \dfrac{(\ell - x)^2}{x}\right]$

$= \dfrac{1}{2}s\sin\alpha\left[x + \dfrac{\ell^2 - 2\ell x + x^2}{x}\right]$

$= \dfrac{1}{2}s\sin\alpha\left(x + \dfrac{\ell^2}{x} - 2\ell + x\right)$

$= \dfrac{1}{2}s\sin\alpha\left(2x - 2\ell + \dfrac{\ell^2}{x}\right)$

$= s\left(x - \ell + \dfrac{\ell^2}{2x}\right)\sin\alpha.$ ✓

(3 marks)

ii $A = s\left(x - \ell + \dfrac{\ell^2}{2x}\right)\sin\alpha$

$\dfrac{dA}{dx} = s\left(1 - \dfrac{\ell^2}{2x^2}\right)\sin\alpha$

Stationary points occur when

$\dfrac{dA}{dx} = 0$

$\therefore s\left(1 - \dfrac{\ell^2}{2x^2}\right)\sin\alpha = 0$ ✓

$\therefore 1 - \dfrac{\ell^2}{2x^2} = 0$

$2x^2 = \ell^2$

$x^2 = \dfrac{\ell^2}{2}$

$\therefore x = \dfrac{\ell}{\sqrt{2}},\ x > 0$ ✓

Now $\dfrac{dA}{dx} = s\left(1 - \dfrac{\ell^2}{2x^2}\right)\sin\alpha$

$\therefore \dfrac{d^2A}{dx^2} = s\sin\alpha\left(\dfrac{\ell^2}{x^3}\right)$

$= s\sin\alpha\left[\dfrac{\ell^2}{\left(\frac{\ell}{\sqrt{2}}\right)^3}\right]$, when $x = \dfrac{\ell}{\sqrt{2}}$

$= \dfrac{2\sqrt{2}\,s\sin\alpha}{\ell}$

> 0 as $0 < \alpha < 90°$

$\therefore$ A minimum value of A occurs when $x = \dfrac{\ell}{\sqrt{2}}$. ✓

(3 marks)

iii $MP = \dfrac{(\ell - x)s}{x}$ (from **i**)

When $x = \dfrac{\ell}{\sqrt{2}}$,

$$MP = \frac{\left(\ell - \dfrac{\ell}{\sqrt{2}}\right) \times s}{\dfrac{\ell}{\sqrt{2}}}$$

$$= \left(\ell - \frac{\ell}{\sqrt{2}}\right) \times s \times \frac{\sqrt{2}}{\ell}$$

$$= (\sqrt{2} - 1)s.$$

$\therefore$ The length of MP is $(\sqrt{2} - 1)s$ metres. ✓

(1 mark)

44 $y = x \sin x$

$\dfrac{dy}{dx} = x \,.\, \dfrac{d}{dx}(\sin x) + \sin x \,.\, \dfrac{d}{dx}(x)$

$= x \,.\, \cos x + \sin x \,.\, 1$

$= x \cos x + \sin x$ ✓

At $x = \pi$, $\dfrac{dy}{dx} = \pi \cos \pi + \sin \pi$

$= -\pi + 0$

$= -\pi$ ✓

$\therefore$ Equation of the tangent at $(\pi, 0)$ with gradient $= -\pi$ is

$y - 0 = -\pi(x - \pi)$

$\therefore y = -\pi x + \pi^2.$ ✓

(3 marks)

45 i The curve crosses the x-axis when $f(x) = 0$

$\therefore x^4 - 4x^3 = 0$

$x^3(x - 4) = 0$

$\therefore x = 0$ or 4. ✓

Now $f(0) = 0$

and $f(4) = (4)^4 - 4(4)^3$

$= 0$

$\therefore$ Coordinates are $(0, 0)$ and $(4, 0)$. ✓

(2 marks)

ii $f(x) = x^4 - 4x^3$

$f'(x) = 4x^3 - 12x^2$

$= 4x^2(x - 3)$

Stationary points occur when

$f'(x) = 0$

$\therefore 4x^2(x - 3) = 0$

$\therefore x = 0$ or 3 ✓

Now $f(0) = 0$

and $f(3) = (3)^4 - 4(3)^3$

$= 81 - 108$

$= -27$

$\therefore$ Coordinates of the stationary points are $(0, 0)$ and $(3, -27)$. ✓

Now $f''(x) = 12x^2 - 24x$

$= 12x(x - 2)$

$\therefore f''(0) = 12(0)(0 - 2)$

$= 0$

Checking the change in concavity:

x	-0.1	0	0.1
$f''(x)$	>0	0	<0

$\therefore$ There is a change in concavity.

$\therefore$ A horizontal point of inflexion occurs at $(0, 0)$. ✓

Now $f''(3) = 12(3)(3 - 2)$

$= 36 > 0$

$\therefore$ A minimum turning point occurs at $(3, -27)$. ✓

(4 marks)

iii Possible points of inflexion occur when

$f''(x) = 0$

$12x^2 - 24x = 0$

$12x(x - 2) = 0$

$\therefore x = 0$ or 2

From **ii**, a horizontal point of inflexion occurs at (0,0).

Now, at $x = 2$, checking the change in concavity:

x	1.9	2	2.1
$f''(x)$	<0	0	>0

$\therefore$ There is a change in concavity and a point of inflexion at $x = 2$.

Now, $f(2) = (2)^4 - 4(2)^3$

$= 16 - 32$

$= -16$

$\therefore$ A point of inflexion occurs at $(2, -16)$. ✓

(1 mark)

iv

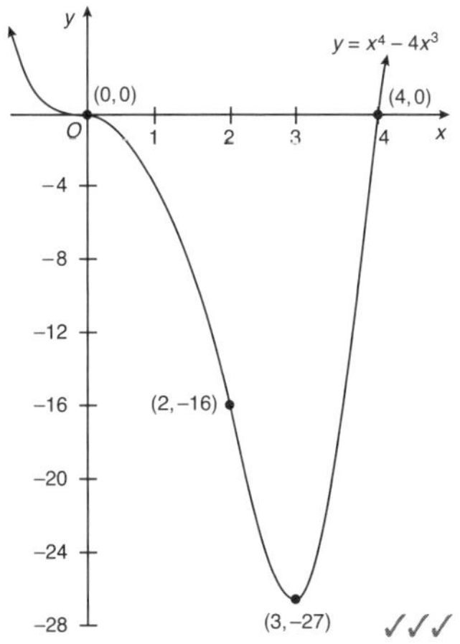

y
$y = x^4 - 4x^3$
(0,0)
(4,0)
O 1 2 3 4 x
-4 -8 -12 -16 -20 -24 -28
(2, –16)
(3, –27)

✓✓✓

(3 marks)

46 i $y = kx^n$

$\dfrac{dy}{dx} = knx^{n-1}$

$= kna^{n-1}$ at $x = a$.

$y = \log_e x$

$\dfrac{dy}{dx} = \dfrac{1}{x}$

$= \dfrac{1}{a}$ at $x = a$.

Since there is a common tangent, the two gradients are equal,

i.e. $kna^{n-1} = \dfrac{1}{a}$

$kna^n \,.\, \dfrac{1}{a} = \dfrac{1}{a}$

$kna^n = 1$

$a^n = \dfrac{1}{nk}.$ ✓

(1 mark)

ii Now $y = kx^n$

At $x = a$, $y = ka^n$

Also, $y = \log_e x$

At $x = a$, $y = \log_e a$

$\therefore ka^n = \log_e a$ . . . (1)

and from (i):

$a^n = \dfrac{1}{nk}$ ✓. . . (2) ✓

Substituting (2) into (1):

$k\left(\dfrac{1}{nk}\right) = \log_e a$

$\dfrac{1}{n} = \log_e a$

$e^{\frac{1}{n}} = a$

Substituting into (2):

$\left(e^{\frac{1}{n}}\right)^n = \dfrac{1}{nk}$

$e = \dfrac{1}{nk}$

$\therefore k = \dfrac{1}{ne}.$ ✓

(2 marks)

47

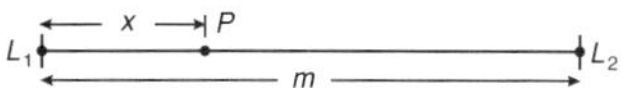

L_1 —— x —— P —— m —— L_2

i Distance of P from L_2 is $(m - x)$.

Now $N = \dfrac{L}{d^2}$

$\therefore N = \dfrac{L_1}{x^2} + \dfrac{L_2}{(m - x)^2}$ ✓

(1 mark)

ii $N = \dfrac{L_1}{x^2} + \dfrac{L_2}{(m-x)^2}$

$x \neq m,\ x \neq 0$

$= L_1 x^{-2} + L_2(m-x)^{-2}$

$\dfrac{dN}{dx} = -2L_1x^{-3} - 2L_2(m-x)^{-3} . -1$

$= -2L_1x^{-3} + 2L_2(m-x)^{-3}$

$= \dfrac{-2L_1}{x^3} + \dfrac{2L_2}{(m-x)^3}$ ✓

Stationary points occur when

$\dfrac{dN}{dx} = 0$

$\therefore \dfrac{-2L_1}{x^3} + \dfrac{2L_2}{(m-x)^3} = 0$

$\dfrac{2L_2}{(m-x)^3} = \dfrac{2L_1}{x^3}$

$2L_2x^3 = 2L_1(m-x)^3$ ✓

$L_2x^3 = L_1(m-x)^3$

$\sqrt[3]{L_2}\,x = \sqrt[3]{L_1}\,(m-x)$

$\sqrt[3]{L_2}\,x + \sqrt[3]{L_1}\,x = \sqrt[3]{L_1}\,m$

$x\left(\sqrt[3]{L_2} + \sqrt[3]{L_1}\right) = \sqrt[3]{L_1}\,m$

$x = \dfrac{\sqrt[3]{L_1}\,m}{\sqrt[3]{L_1} + \sqrt[3]{L_2}}$ ✓

Now $\dfrac{dN}{dx} = -2L_1x^{-3} + 2L_2(m-x)^{-3}$

$\therefore \dfrac{d^2N}{dx^2} = 6L_1x^{-4} - 6L_2(m-x)^{4} . -1$

$= \dfrac{6L_1}{x^4} + \dfrac{6L_2}{(m-x)^4}$

> 0

$\therefore$ A minimum noise level occurs at P

when $x = \dfrac{\sqrt[3]{L_1}\,m}{\sqrt[3]{L_1} + \sqrt[3]{L_2}}$. ✓

(4 marks)

48 $y = \cos 2x$

$\dfrac{dy}{dx} = -2\sin 2x$

At $x = \dfrac{\pi}{6}$,

$\dfrac{dy}{dx} = -2\sin 2\left(\dfrac{\pi}{6}\right)$

$= -2\sin\dfrac{\pi}{3}$

$= -2\left(\dfrac{\sqrt{3}}{2}\right)$

$= -\sqrt{3}$. ✓

At $x = \dfrac{\pi}{6}$,

$y = \cos 2x$

$= \cos 2\left(\dfrac{\pi}{6}\right)$

$= \cos\dfrac{\pi}{3}$

$= \dfrac{1}{2}$ ✓

$\therefore$ Equation of the tangent at $\left(\dfrac{\pi}{6}, \dfrac{1}{2}\right)$ with gradient of $-\sqrt{3}$ is:

$y - \dfrac{1}{2} = -\sqrt{3}\left(x - \dfrac{\pi}{6}\right)$

$2y - 1 = -2\sqrt{3}x + \dfrac{2\sqrt{3}\pi}{6}$

$12y - 6 = -12\sqrt{3}x + 2\sqrt{3}\pi$

$12\sqrt{3}x + 12y - 6 - 2\sqrt{3}\pi = 0$

$6\sqrt{3}x + 6y - 3 - \sqrt{3}\pi = 0$. ✓

(3 marks)

49 i $f(x) = 2x^2(3-x)$

$= 6x^2 - 2x^3$

$f'(x) = 12x - 6x^2$

Stationary points occur when

$f'(x) = 0$

$12x - 6x^2 = 0$

$6x(2-x) = 0$

$\therefore x = 0$ or 2

Now $f(0) = 2(0)^2[3-(0)]$

$= 0$

and $f(2) = 2(2)^2[3-(2)]$

$= 8$

$\therefore$ Coordinates of the stationary points are $(0, 0)$ ✓ and $(2, 8)$.

Now $f'(x) = 12x - 6x^2$

$\therefore f''(x) = 12 - 12x$

$\therefore f''(0) = 12 - 12(0)$

$= 12 - 0$

$= 12 > 0$

$\therefore$ A minimum turning point occurs at $(0, 0)$. ✓

and $f''(2) = 12 - 12(2)$

$= 12 - 24$

$= -12 < 0$

$\therefore$ A maximum turning point occurs at $(2, 8)$. ✓

(3 marks)

ii Possible points of inflexion occur when

$f''(x) = 0$

$12 - 12x = 0$

$12(1-x) = 0$

$x = 1$

Checking the change in concavity:

x	0.9	1	1.1
$f''(x)$	>0	0	<0

$\therefore$ There is change in concavity and a point of inflexion at $x = 1$.

Now, $f(1) = 2(1)^2[3-1]$

$= 4$

$\therefore (1, 4)$ are the coordinates of the point of inflexion.

✓ *(1 mark)*

iii Curve meets the x-axis where

$f(x) = 0$

$2x^2(3-x) = 0$

$\therefore x = 0$ or 3 ✓

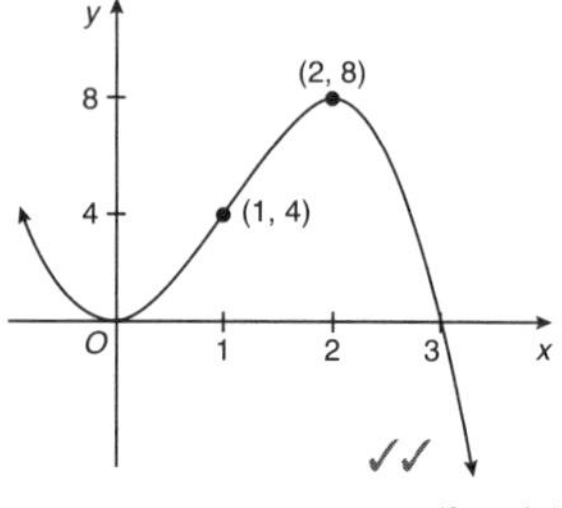

✓✓

(3 marks)

iv Minimum value of $f(x)$ for $-1 \le x \le 4$ occurs when $x = 4$,

$f(4) = 2(4)^2[3-(4)]$

$= -32$.

(1 mark)

50 i $P = 150 + 300e^{-0.05t}$

When $t = 0$,

$P = 150 + 300e^{-0.05 \times 0}$

$= 150 + 300$

$= 450$. ✓

(1 mark)

ii $P = 150 + 300e^{-0.05t}$

$\dfrac{dP}{dt} = -0.05 \times 300e^{-0.05t}$

$= -15e^{-0.05t}$ ✓

When $t = 10$,

$\dfrac{dP}{dt} = -15e^{-0.05 \times 10}$

$= -15e^{-0.5}$

$= -9.0979\ldots$

$\therefore$ The bird population is decreasing at a rate of $9.0979\ldots$ birds per year. ✓

(2 marks)

iii Now $\lim\limits_{t \to \infty} e^{-0.05t} = 0$

and $P = 150 + 300e^{-0.05t}$

$\therefore P = 150 + 300 \times 0$

$= 150$. ✓

(1 mark)

iv When $P = 200$,

$$200 = 150 + 300e^{-0.05t}$$
$$50 = 300e^{-0.05t}$$
$$\frac{50}{300} = e^{-0.05t} \quad ✓$$
$$\log_e \frac{50}{300} = -0.05t$$
$$t = \log_e \frac{50}{300} \div -0.05$$
$$= 35.8351\ldots$$

$\therefore$ The bird population will be eligible for inclusion in the endangered species list during the 36th year. ✓

(2 marks)

51 **i** $x = 1 - \dfrac{7}{t+4}$

When $t = 0$,

$$x = 1 - \frac{7}{4}$$
$$= -\frac{3}{4}$$

$\therefore$ The initial displacement of the particle is $\frac{3}{4}$ m to the left of the origin. ✓

(1 mark)

ii
$$v = \frac{dx}{dt}$$
$$= \frac{d}{dt}\left(1 - \frac{7}{t+4}\right)$$
$$= \frac{d}{dt}\left(1 - 7(t+4)^{-1}\right)$$
$$= 7(t+4)^{-2}$$
$$= \frac{7}{(t+4)^2} \quad ✓$$

Now $x = 1 - \dfrac{7}{t+4}$

When the particle is at the origin, $x = 0$.

$$\therefore 0 = 1 - \frac{7}{t+4}$$
$$-1 = -\frac{7}{t+4}$$
$$-1(t+4) = -7$$
$$-t - 4 = -7$$
$$\therefore t = 3$$

$\therefore$ The particle is at the origin after 3 seconds. ✓

Now $v = \dfrac{7}{(t+4)^2}$

When $t = 3$,

$$v = \frac{7}{(3+4)^2}$$
$$= \frac{1}{7}$$
$$\doteqdot 0.1428\ldots$$

$\therefore$ The velocity of the particle as it passes through the origin is $\frac{1}{7}$ ms^{-1}. ✓

(3 marks)

iii
$$a = \frac{dv}{dt}$$
$$= \frac{d}{dt}\left(\frac{7}{(t+4)^2}\right)$$
$$= \frac{d}{dt}\left(7(t+4)^{-2}\right)$$
$$= -14(t+4)^{-3}$$
$$= -\frac{14}{(t+4)^3}$$

Now for $t \geqslant 0$, $(t+4)^3 > 0$

$$\therefore -\frac{14}{(t+4)^3} < 0$$

$\therefore$ Acceleration of the particle is always negative. ✓ *(1 mark)*

iv

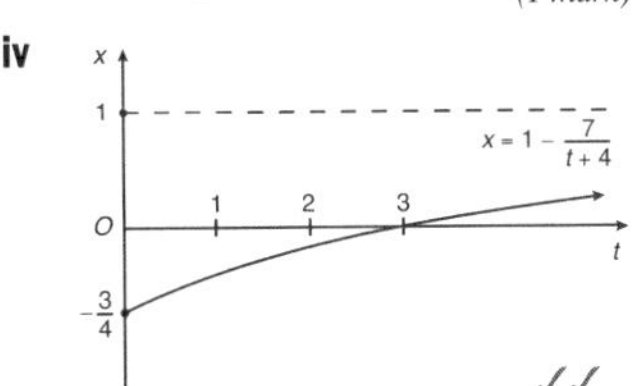

✓✓

N.B.

From **i**, when $t = 0, x = -\frac{3}{4}$

From **ii**, when $t = 3, x = 0$,
$v = \frac{1}{4}$

From **ii**, $a < 0$

Also as $t \to \infty, x \to 1$ *(2 marks)*

52 **i** $V = \frac{1}{3}\pi r^2 h$ (where $h = x$)

$$\therefore V = \frac{1}{3}\pi r^2 x$$

Now $a^2 = (x-a)^2 + r^2$ (Pythagoras' Theorem)

$$a^2 = x^2 - 2ax + a^2 + r^2$$
$$\therefore r^2 = 2ax - x^2 \quad ✓$$
$$\therefore V = \frac{1}{3}\pi(2ax - x^2)(x)$$
$$= \frac{1}{3}\pi(2ax^2 - x^3) \quad ✓$$

(2 marks)

ii
$$V = \frac{1}{3}\pi(2ax^2 - x^3)$$
$$= \frac{2}{3}\pi a x^2 - \frac{1}{3}\pi x^3$$
$$\frac{dV}{dx} = \frac{4}{3}\pi a x - \pi x^2$$
$$\frac{d^2V}{dx^2} = \frac{4}{3}\pi a - 2\pi x \quad ✓$$

A maximum volume occurs when $\dfrac{dV}{dx} = 0$

$$\frac{4}{3}\pi a x - \pi x^2 = 0$$
$$\pi x\left(\frac{4}{3}a - x\right) = 0$$
$$x = 0 \text{ or } \frac{4}{3}a \quad ✓$$
$$\therefore x = \frac{4}{3}a$$

When $x = \frac{4}{3}a$,

$$\frac{d^2V}{dx^2} = \frac{4}{3}\pi a - 2\pi\left(\frac{4}{3}a\right)$$
$$= \frac{4}{3}\pi a - \frac{8}{3}\pi a$$
$$= -\frac{4}{3}\pi a < 0$$

$\therefore$ Cone has a maximum value when $x = \frac{4}{3}a$. ✓

(3 marks)

53
$$y = \log_e x$$
$$\frac{dy}{dx} = \frac{1}{x}$$

At $x = e$, $\dfrac{dy}{dx} = \dfrac{1}{e}$ ✓

$\therefore$ Equation of tangent at $(e, 1)$ with gradient $= \frac{1}{e}$ is

$$y - 1 = \frac{1}{e}(x - e)$$
$$y - 1 = \frac{x}{e} - 1$$
$$\therefore y = \frac{x}{e}. \quad ✓$$

(2 marks)

54 **i**
$$f(x) = 0$$
$$\therefore (x+3)(x^2 - 9) = 0$$
$$(x+3)(x+3)(x-3) = 0$$
$$\therefore x = 3 \text{ or } -3.$$

✓✓ *(2 marks)*

ii
$$f(x) = (x+3)(x^2 - 9)$$
$$= x^3 - 9x + 3x^2 - 27$$
$$= x^3 + 3x^2 - 9x - 27$$
$$f'(x) = 3x^2 + 6x - 9$$

Stationary points occur when

$$f'(x) = 0$$
$$3x^2 + 6x - 9 = 0$$
$$3(x^2 + 2x - 3) = 0$$
$$3(x+3)(x-1) = 0$$
$$\therefore x = -3 \text{ or } 1. \quad ✓$$

Now
$f(-3) = (-3+3)[(-3)^2 - 9]$
$= 0$

and $f(1) = (1+3)(1^2 - 9)$
$= 4 \times -8$
$= -32$

$\therefore$ Coordinates of stationary points are $(-3, 0)$ and $(1, -32)$. ✓

Now $f'(x) = 3x^2 + 6x - 9$
$\therefore f''(x) = 6x + 6$

$\therefore f''(-3) = 6(-3) + 6$
$= -18 + 6$
$= -12 < 0.$

$\therefore$ A maximum turning point occurs at $(-3, 0)$.
and $f''(1) = 6(1) + 6$
$= 12 > 0.$

$\therefore$ A minimum turning point occurs at $(1, -32)$. ✓

(3 marks)

iii The curve meets the x-axis when

$f(x) = 0$
$(x+3)(x^2 - 9) = 0$
$(x+3)(x+3)(x-3) = 0$
$\therefore x = \pm 3.$

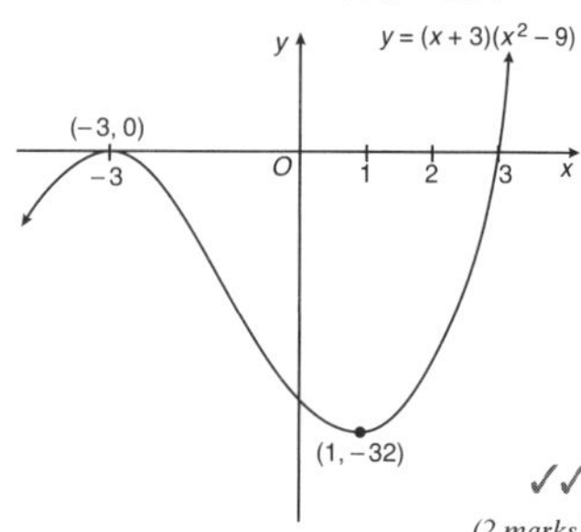

✓✓ *(2 marks)*

iv Graph of $y = f(x)$ is concave down when $f''(x) < 0$

$6x + 6 < 0$
$6x < -6$
$x < -1.$ ✓ *(1 mark)*

55 Gradient of $y = 5x - 3$ is 5.
$\therefore$ Gradient of tangent must be 5.

Now $y = 2e^x + 3x$

$\therefore \frac{dy}{dx} = 2e^x + 3$ ✓✓

If $\frac{dy}{dx} = 5$

$\therefore 2e^x + 3 = 5$
$2e^x = 2$
$e^x = 1$
$\therefore x = 0$ ✓

When $x = 0, y = 2e^0 + 3(0)$
$= 2$

$\therefore$ Coordinates of P are $(0, 2)$. ✓

(3 marks)

56 i $V = 3600\left(1 - \frac{t}{60}\right)^2,$
where $0 \leqslant t \leqslant 60$

When $t = 10$,

$V = 3600\left(1 - \frac{10}{60}\right)^2$

$= 2500$ litres ✓

(1 mark)

ii $V = 3600\left(1 - \frac{t}{60}\right)^2$

$\frac{dV}{dt} = 2 \times 3600\left(1 - \frac{t}{60}\right) \cdot \frac{d}{dt}\left(1 - \frac{t}{60}\right)$

$= 7200\left(1 - \frac{t}{60}\right) \cdot -\frac{1}{60}$

$= -120\left(1 - \frac{t}{60}\right)$ ✓

When $t = 20$,

$\frac{dv}{dt} = -120\left(1 - \frac{20}{60}\right)$

$= -120 \times \frac{40}{60}$

$= -80$

$\therefore$ After 20 minutes, water is draining at 80 L/min. ✓

(2 marks)

iii Now $\frac{dV}{dt} = -120\left(1 - \frac{t}{60}\right)$
$= -120 + 2t$

To find when water will drain fastest, consider $\frac{d^2V}{dt^2}$:

Now $\frac{d^2V}{dt^2} = 2$

Since, $\frac{d^2V}{dt^2}$ is a constant, check end-points: ✓

When $t = 0$, $\frac{dV}{dt} = -120$

When $t = 60$, $\frac{dV}{dt} = 0$

$\therefore$ The water will **drain** at the fastest rate when $t = 0$. ✓

(2 marks)

57 i $V = \pi r^2 h$
where $r = x$, $h = 2h$
$\therefore V = \pi x^2(2h)$

Now $R^2 = x^2 + h^2$
(Pythagoras' theorem)
$\therefore x^2 = R^2 - h^2$

$\therefore V = \pi(R^2 - h^2)(2h)$
$= 2\pi h(R^2 - h^2).$ ✓

(1 mark)

ii $V = 2\pi h(R^2 - h^2)$
$= 2\pi R^2 h - 2\pi h^3$

$\frac{dV}{dh} = 2\pi R^2 - 6\pi h^2$

$\frac{d^2V}{dh^2} = -12\pi h$ ✓

A maximum volume may occur when

$\frac{dV}{dh} = 0$

$2\pi R^2 - 6\pi h^2 = 0$
$6\pi h^2 = 2\pi R^2$

$h^2 = \frac{2\pi R^2}{6\pi}$

$= \frac{R^2}{3}$

$\therefore h = \frac{R}{\sqrt{3}}$ ✓

$\left(\text{since } h > 0, h \neq -\frac{R}{\sqrt{3}}\right)$

When $h = \frac{R}{\sqrt{3}}$,

$\frac{d^2V}{dh^2} = -12\pi h$

$= -12\pi\left(\frac{R}{\sqrt{3}}\right) < 0$

$\therefore$ Cylinder has a maximum volume when $h = \frac{R}{\sqrt{3}}$. ✓

(3 marks)

58 i $f(x) = x^3 - 3x^2$

$f'(x) = 3x^2 - 6x$
$= 3x(x - 2)$

Stationary points occur when
$f'(x) = 0$
$3x(x - 2) = 0$
$\therefore x = 0$ or $x = 2$ ✓

Now $f(0) = 0^3 - 3(0)^2$
$= 0$
and $f(2) = 2^3 - 3(2)^2$
$= 8 - 12$
$= -4$

$\therefore$ Coordinates of the stationary points are $(0, 0)$ and $(2, -4)$.
Now $f''(x) = 6x - 6$

$\therefore f''(0) = 6(0) - 6$
$= -6 < 0$

$\therefore$ A maximum turning point at $(0, 0)$ ✓

Now, $f''(2) = 6(2) - 6$
$= 6 > 0$

$\therefore$ A minimum turning point at $(2, -4)$. ✓ *(3 marks)*

ii When $y = f(x) = 0$

$x^3 - 3x^2 = 0$

$x^2(x - 3) = 0$

$\therefore x = 0$ or 3.

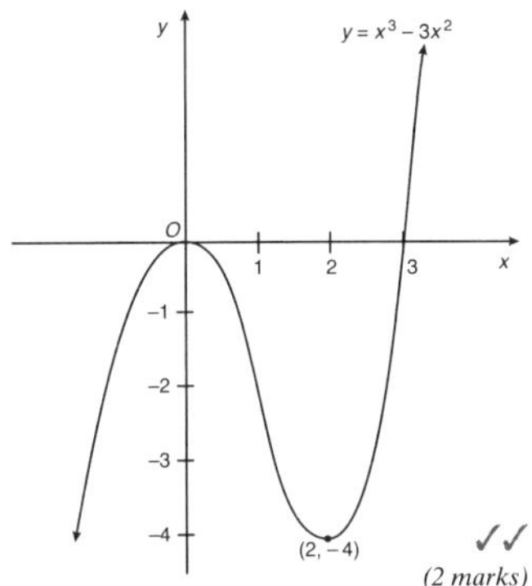

✓✓ *(2 marks)*

iii $f''(x) = 6x - 6$ ✓

The curve is concave up when

$f''(x) > 0$

$6x - 6 > 0$

$6x > 6$

$x > 1$. ✓ *(2 marks)*

59 i $x = 1 + 3\cos 2t$

When $t = 0$,

$x = 1 + 3\cos 0$

$= 1 + 3$

$= 4$

$\therefore$ Particle is initially 4 m to the right of O. ✓ *(1 mark)*

ii Period $= \dfrac{2\pi}{2} = \pi$

Range: $-2 \leqslant x \leqslant 4$ ✓

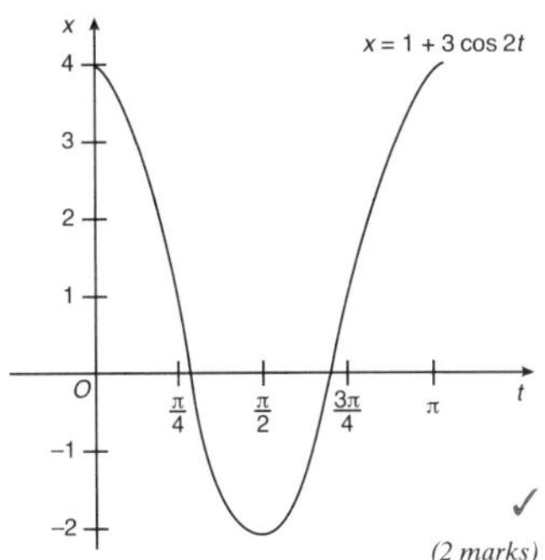

✓ *(2 marks)*

iii $x = 1 + 3\cos 2t$

$\therefore v = -6\sin 2t$

The particle comes to rest when

$v = 0$

i.e. $-6\sin 2t = 0$

$\sin 2t = 0$

$2t = 0, \pi, 2\pi \ldots$

$t = 0, \dfrac{\pi}{2}, \pi \ldots$

$\therefore$ Particle first comes to rest when $t = \dfrac{\pi}{2}$ seconds. ✓

When $t = \dfrac{\pi}{2}$,

$x = 1 + 3\cos 2\left(\dfrac{\pi}{2}\right)$

$= 1 + 3\cos\pi$

$= 1 + 3(-1)$

$= 1 - 3$

$= -2$

$\therefore$ Particle first comes to rest at 2 m to the left of O. ✓

OR

The particle is at rest at the turning points of the curve when $\dfrac{dx}{dt} = 0$.

i.e. at $t = 0, \dfrac{\pi}{2}, \pi \ldots$

$\therefore$ From the graph, the particle first comes to rest when $t = \dfrac{\pi}{2}$ seconds, 2 m to the left of O. *(2 marks)*

iv $x = 1 + 3\cos 2t$

$v = \dfrac{dx}{dt} = -6\sin 2t$

$a = \dfrac{dv}{dt} = -12\cos 2t$

The particle reaches its maximum speed when

$\dfrac{dv}{dt} = 0$

$-12\cos 2t = 0$

$\cos 2t = 0$

$2t = \dfrac{\pi}{2}, \dfrac{3\pi}{2}, \ldots$

$t = \dfrac{\pi}{4}, \dfrac{3\pi}{4}, \ldots$ ✓

$\therefore$ Maximum speed is reached at $\dfrac{\pi}{4}$ seconds, $\dfrac{3\pi}{4}$ seconds, etc.

When $t = \dfrac{\pi}{4}$ seconds,

$v = -6\sin 2\left(\dfrac{\pi}{4}\right)$

$= -6\sin\left(\dfrac{\pi}{2}\right)$

$= -6 \times 1$

$= -6$

$\therefore$ Maximum speed is 6 m s^{-1}. ✓ *(2 marks)*

60 i $f(x) = \dfrac{\log_e x}{x}$, for $x > 0$

$f'(x) = \dfrac{x \cdot \dfrac{d}{dx}(\log_e x) - \log_e x \cdot \dfrac{d}{dx}(x)}{x^2}$

$= \dfrac{x\left(\dfrac{1}{x}\right) - \log_e x \cdot 1}{x^2}$

$= \dfrac{1 - \log_e x}{x^2}$ ✓

Stationary points occur when

$f'(x) = 0$

$\therefore \dfrac{1 - \log_e x}{x^2} = 0$

$1 - \log_e x = 0$

$\log_e x = 1$

$\therefore x = e$.

$\therefore$ There is a stationary point at $x = e$. ✓ *(2 marks)*

ii Consider the graph of $y = \log_e x$:

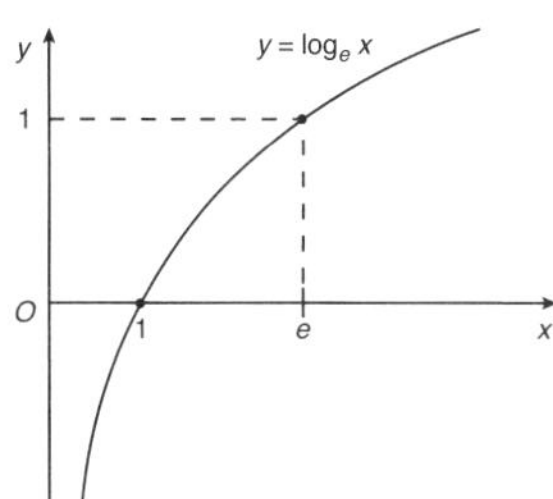

When $x < e$, $\log_e x < 1$

$\therefore 1 - \log_e x > 0$

$\therefore \dfrac{1 - \log_e x}{x^2} > 0$

i.e. $f'(x) > 0$

When $x > e$, $\log_e x > 1$

$\therefore 1 - \log_e x < 0$

$\therefore \dfrac{1 - \log_e x}{x^2} < 0$

i.e. $f'(x) < 0$

$\therefore$ The stationary point at $x = e$ is a maximum. ✓ *(1 mark)*

iii $f(x) = \dfrac{\log_e x}{x}$, for $x > 0$

$\therefore f(e) = \dfrac{\log_e e}{e}$

$= \dfrac{1}{e}$

Since the maximum value of $f(x) = \dfrac{\log_e x}{x}$ is $\dfrac{1}{e}$, then

$\dfrac{\log_e x}{x} \leqslant \dfrac{1}{e}$, $x > 0$ ✓

$\log_e x \leqslant \dfrac{x}{e}$

$e\log_e x \leqslant x$

$\log_e x^e \leqslant x$

$e^{\log_e x^e} \leqslant e^x$

$\therefore x^e \leqslant e^x$ (Note: $e^{\log a} = a$)

$\therefore e^x \geqslant x^e$, for all $x > 0$. ✓ *(2 marks)*

61 i Now, area of ΔABC

$= \dfrac{1}{2} bc\sin A$

and area of ΔAST

$= \dfrac{1}{2} xy\sin A$

Since area

$\Delta AST = \frac{1}{2}$ area ΔABC (given)

$\frac{1}{2}xy\sin A = \frac{1}{2}\times\frac{1}{2}bc\sin A$

$xy\sin A = \frac{1}{2}bc\sin A$

$xy = \frac{1}{2}bc.$ ✓

(1 mark)

ii By the cosine rule in ΔAST,

$z^2 = x^2 + y^2 - 2xy\cos A$

Now

$xy = \frac{1}{2}bc$

$\therefore y = \frac{bc}{2x}$

$\therefore z^2 = x^2 + \left(\frac{bc}{2x}\right)^2 - 2x\left(\frac{bc}{2x}\right)\cos A$ ✓

$\therefore z^2 = x^2 + \frac{b^2c^2}{4x^2} - bc\cos A$ ✓

(2 marks)

iii $z^2 = x^2 + \left(\frac{b^2c^2}{4}\right)x^{-2} - bc\cos A$

$\frac{d(z^2)}{dx} = 2x - \frac{2b^2c^2}{4}x^{-3}$

$= 2x - \frac{b^2c^2}{2x^3}$ ✓

Stationary points occur when

$\frac{d(z^2)}{dx} = 0$

$2x - \frac{b^2c^2}{2x^3} = 0$

$4x^4 - b^2c^2 = 0$

$4x^4 = b^2c^2$

$x^4 = \frac{b^2c^2}{4}$

$x^2 = \frac{bc}{2} \quad (x^2 \geqslant 0)$

$x = \sqrt{\frac{bc}{2}} \quad (x > 0)$ ✓

Now $\frac{d(z^2)}{dx} = 2x - \left(\frac{b^2c^2}{2}\right)x^{-3}$

$\therefore \frac{d^2(z^2)}{dx^2} = 2 + \frac{3b^2c^2}{2}x^{-4}$

$= 2 + \frac{3b^2c^2}{2x^4}$ ✓

> 0 for all x

$\therefore$ A minimum value of z^2 when $x = \sqrt{\frac{bc}{2}}$. ✓

(4 marks)

iv Now $z^2 = x^2 + \frac{b^2c^2}{4x^2} - bc\cos A$

and $\cos A = \frac{b^2 + c^2 - a^2}{2bc}$

$\therefore z^2 = x^2 + \frac{b^2c^2}{4x^2} - bc\left(\frac{b^2 + c^2 - a^2}{2bc}\right)$

When $x = \sqrt{\frac{bc}{2}}$,

$z^2 = \frac{bc}{2} + \frac{b^2c^2}{4\left(\frac{bc}{2}\right)} - \left(\frac{b^2+c^2-a^2}{2}\right)$

$= \frac{bc}{2} + \frac{bc}{2} - \left(\frac{b^2+c^2-a^2}{2}\right)$

$= \frac{a^2 - b^2 + 2bc - c^2}{2}$

$= \frac{a^2 - (b^2 - 2bc + c^2)}{2}$

$= \frac{1}{2}[a^2 - (b-c)^2]$ ✓

$= \frac{1}{2}\left[(a-(b-c))(a+(b-c))\right]$

$= \frac{1}{2}\left[(a-b+c)(a+b-c)\right]$

$= \frac{1}{2}\left[(a+b+c-2b)(a+b+c-2c)\right]$

$= \frac{(P-2b)(P-2c)}{2}$

where $P = a + b + c$

$\therefore z = \sqrt{\frac{(P-2b)(P-2c)}{2}}$ metres. ✓

(2 marks)

62 $y = 2\log_e x$

$\frac{dy}{dx} = 2\left(\frac{1}{x}\right)$

$= \frac{2}{x}$ ✓

At $x = e$, $\frac{dy}{dx} = \frac{2}{e}$

$\therefore$ Gradient of normal $= -\frac{1}{\frac{2}{e}}$

$= -\frac{e}{2}$ ✓

$\therefore$ Equation of normal through $(e, 2)$ with gradient $= -\frac{e}{2}$ is

$y - 2 = -\frac{e}{2}(x - e)$

$2y - 4 = -e(x - e)$

$2y - 4 = -ex + e^2$

$ex + 2y - 4 - e^2 = 0$ ✓

(3 marks)

63 i $f(x) = x^4 - 4x^3$

$f'(x) = 4x^3 - 12x^2$

$= 4x^2(x - 3)$

(1 mark)

ii Stationary points occur when

$f'(x) = 0$

$4x^2(x-3) = 0$

$\therefore x = 0$ or 3

When $x = 0$, $y = 0^4 - 4(0)^3$

$= 0$

When $x = 3$, $y = 3^4 - 4(3)^3$

$= 81 - 108$

$= -27$

$\therefore$ Coordinates of stationary points are $(0, 0)$ and $(3, -27)$. ✓

Now, $f''(x) = 12x^2 - 24x$

$= 12x(x-2)$

$\therefore f''(3) = 12(3)(3-2)$

$= 36 > 0$

$\therefore$ A minimum turning point at $(3, -27)$ ✓

Now, $f''(0) = 12(0)(-2)$

$= 0$

When $x < 0$, say $x = -0.1$

$f''(-0.1) = 12(-0.1)(-0.1-2)$

$= (+)(-)(-)$

> 0

When $x > 0$, say $x = 0.1$

$f''(0.1) = 12(0.1)(0.1-2)$

$= (+)(+)(-)$

< 0

$\therefore$ Concavity changes.

$\therefore$ A horizontal point of inflexion at $(0, 0)$. ✓ *(3 marks)*

iii

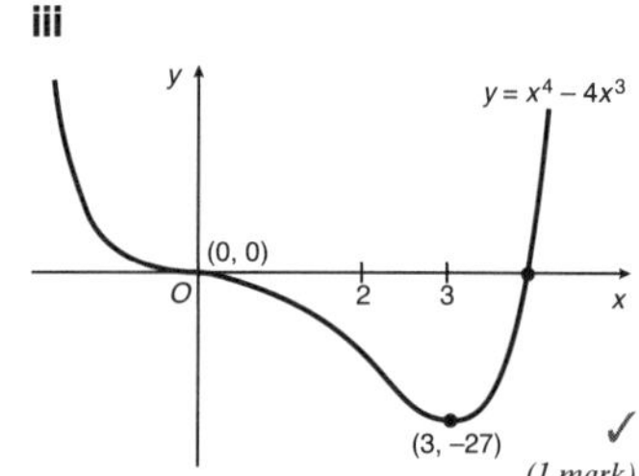

✓ *(1 mark)*

iv The curve is concave down when

$f''(x) < 0$

$\therefore 12x^2 - 24x < 0$

$\therefore 12x(x-2) < 0$ ✓

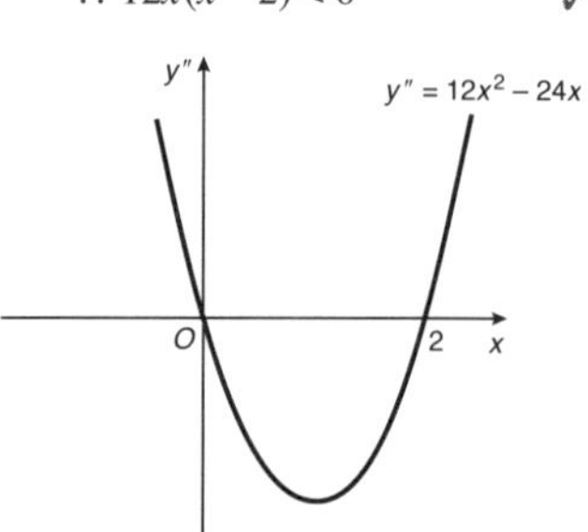

$\therefore 0 < x < 2$ (from graph) ✓

(2 marks)

64 **i** The particle is at rest when

$v = 0$

$\therefore 2 - 4\cos t = 0 \quad 0 \leqslant t \leqslant 2\pi$

$-4\cos t = -2$

$\cos t = \frac{1}{2}$ ✓

$\therefore t = \frac{\pi}{3}$ or $\left(2\pi - \frac{\pi}{3}\right)$

$= \frac{\pi}{3}$ or $\frac{5\pi}{3}$ seconds. ✓

(2 marks)

ii $v = 2 - 4\cos t$

$v' = -(-4)\sin t$

$= 4\sin t$

Maximum or minimum velocity occurs when

$v' = 0 \quad 0 \leqslant t \leqslant 2\pi$

$\therefore 4\sin t = 0$

$\sin t = 0$

$\therefore t = 0, \pi, 2\pi.$ ✓

When $t = 0$, $v = 2 - 4\cos 0$

$= 2 - 4$

$= -2$

When $t = \pi$, $v = 2 - 4\cos \pi$

$= 2 - 4(-1)$

$= 2 + 4$

$= 6$

When $t = 2\pi$, $v = 2 - 4\cos(2\pi)$

$= 2 - 4(1)$

$= 2 - 4$

$= -2$

$\therefore$ Maximum velocity is 6 m/s. ✓

(2 marks)

iii $v = 2 - 4\cos t$

Period $= 2\pi$

Maximum velocity $= 6$

Minimum velocity $= -2$

t intercepts: $\frac{\pi}{3}$ and $\frac{5\pi}{3}$.

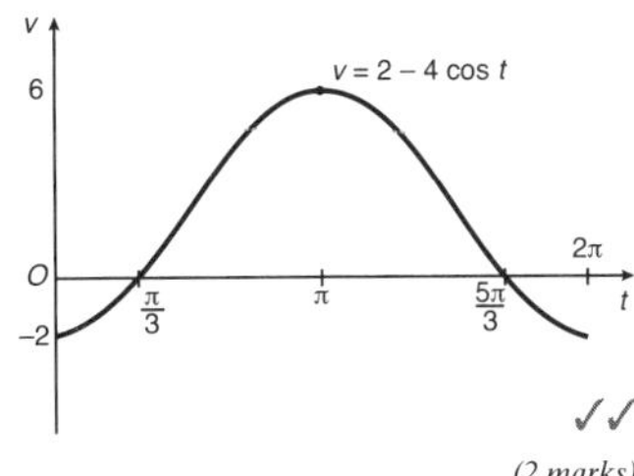

✓✓

(2 marks)

65 **i** distance = speed × time

i.e. $d = s \times t$

$\therefore t = \frac{d}{s}$

Now $s = v - u$

and $d = L$

$\therefore t = \frac{L}{v - u}$

and $E = av^3 t$

$\therefore E = av^3\left(\frac{L}{v - u}\right)$

$= \frac{aLv^3}{v - u}.$ ✓

(1 mark)

ii $\frac{dE}{dv} = \frac{(v-u)\cdot\frac{d}{dv}(aLv^3) - aLv^3\cdot\frac{d}{dv}(v-u)}{(v-u)^2}$

$= \frac{(v-u)3aLv^2 - aLv^3\cdot 1}{(v-u)^2}$

$= \frac{3aLv^3 - 3aLuv^2 - aLv^3}{(v-u)^2}$

$= \frac{2aLv^3 - 3aLuv^2}{(v-u)^2}$

$= \frac{aLv^2(2v - 3u)}{(v-u)^2}$ ✓

Possible maximum or minimum values occur when

$\frac{dE}{dv} = 0$

$\frac{aLv^2(2v-3u)}{(v-u)^2} = 0$

$\therefore v = 0$ or

$2v - 3u = 0$

$2v = 3u$

$v = \frac{3u}{2}$ ✓

Since $v > 0$, $v \neq 0 \quad \therefore v = \frac{3u}{2}$

To determine the nature of the stationary point:

When $v < \frac{3u}{2}\left(= \frac{6u}{4}\right)$, say $v = \frac{5u}{4}$,

$\frac{dE}{dv} = \frac{aL\left(\frac{5u}{4}\right)^2\left(2\left(\frac{5u}{4}\right) - 3u\right)}{\left(\frac{5u}{4} - u\right)^2}$

$= \frac{aL\left(\frac{5u}{4}\right)^2\left(\frac{5u}{2} - 3u\right)}{\left(\frac{u}{4}\right)^2}$

$= \frac{(+)(+)(-)}{(+)}$

< 0 ✓

When $v > \frac{3u}{2}$, say $v = \frac{4u}{2} = 2u$

$\frac{dE}{dv} = \frac{aL(2u)^2[2(2u) - 3u]}{(2u-u)^2}$

$= \frac{aL(2u)^2(4u - 3u)}{u^2}$

$= \frac{(+)(+)(+)}{(+)}$

> 0

$\therefore E$ is a minimum when $v = \frac{3u}{2}$. ✓

(4 marks)

66

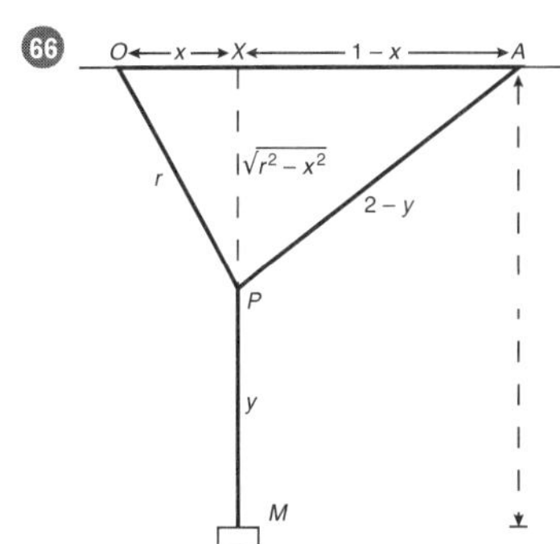

i Let $PM = y \quad \therefore AP = 2 - y$

Now $\ell = XP + PM$

In ΔOXP,

$OX^2 + XP^2 = OP^2$ (Pythagoras' theorem)

$\therefore x^2 + XP^2 = r^2$

$XP^2 = r^2 - x^2$

$\therefore XP = \sqrt{r^2 - x^2}$

In ΔAXP,

$AP^2 = AX^2 + XP^2$

$\therefore (2-y)^2 = (1-x)^2 + \left(\sqrt{r^2 - x^2}\right)^2$

$\therefore (2-y)^2 = 1 - 2x + x^2 + r^2 - x^2$

$(2-y)^2 = 1 - 2x + r^2$

$2 - y = \sqrt{1 - 2x + r^2}$

$y = 2 - \sqrt{1 - 2x + r^2}$

Now $\ell = XP + PM$

$= \sqrt{r^2 - x^2} + y$

$= \sqrt{r^2 - x^2} + 2 - \sqrt{1 - 2x + r^2}$

$= 2 + \sqrt{r^2 - x^2} - \sqrt{1 - 2x + r^2}.$ ✓

(1 mark)

ii $\ell = 2 + \left(r^2 - x^2\right)^{\frac{1}{2}} - \left(1 - 2x + r^2\right)^{\frac{1}{2}}$

$\frac{dl}{dx} = \frac{1}{2}\left(r^2 - x^2\right)^{-\frac{1}{2}}.-2x - \frac{1}{2}\left(1 - 2x + r^2\right)^{-\frac{1}{2}}.-2$

$= \frac{-x}{\sqrt{r^2 - x^2}} + \frac{1}{\sqrt{1 - 2x + r^2}}$ ✓

$$= \frac{-x\sqrt{1-2x+r^2}+\sqrt{r^2-x^2}}{\sqrt{r^2-x^2}\sqrt{1-2x+r^2}}$$

$$= \frac{\sqrt{r^2-x^2}-x\sqrt{1-2x+r^2}}{\sqrt{r^2-x^2}\sqrt{1-2x+r^2}}$$

$$\times \frac{\sqrt{r^2-x^2}+x\sqrt{1-2x+r^2}}{\sqrt{r^2-x^2}+x\sqrt{1-2x+r^2}}$$

(N.B. $(a-b)\times(a+b) = a^2-b^2$)

$$= \frac{(r^2-x^2)-x^2(1-2x+r^2)}{\sqrt{r^2-x^2}\sqrt{1-2x+r^2}\left(\sqrt{r^2-x^2}+x\sqrt{1-2x+r^2}\right)}$$

✓ *(2 marks)*

iii For M to be closest to the floor, l must be a maximum.

Now,

$$\frac{dl}{dx} = \frac{(x-1)(2x^2-r^2x-r^2)}{\sqrt{r^2-x^2}\sqrt{1-2x+r^2}\left(\sqrt{r^2-x^2}+x\sqrt{1-2x+r^2}\right)}$$

(using the factorisation given)

A possible maximum value occurs when

$$\frac{dl}{dx} = 0$$

$\therefore (x-1)(2x^2-r^2x-r^2) = 0$

$\therefore x = 1$ or $2x^2-r^2x-r^2 = 0$

Now if $x = 1$, $r = 1$, but $0 < r < 1$

$\therefore x \neq 1$.

Solving $2x^2-r^2x-r^2 = 0$:

$$x = \frac{-b \pm \sqrt{b^2-4ac}}{2a}$$

$$= \frac{-(-r^2) \pm \sqrt{(-r^2)^2-4(2)(-r^2)}}{2(2)}$$

$$= \frac{r^2 \pm \sqrt{r^4+8r^2}}{4}$$

$$\therefore x = \frac{r^2+\sqrt{r^4+8r^2}}{4} \text{ or } \frac{r^2-\sqrt{r^4+8r^2}}{4}$$

Since x is a length,

$$x = \frac{r^2+\sqrt{r^4+8r^2}}{4}$$ ✓

$$\left(\text{as } \frac{r^2-\sqrt{r^4+8r^2}}{4} < 0\right)$$

To confirm that this value of x gives a maximum value of l, we need to sketch the curve of $\frac{dl}{dx}$, with roots

$x_1 = 1$ and

$$x_2 = \frac{r^2-\sqrt{r^4+8r^2}}{4}$$

$$x_3 = \frac{r^2+\sqrt{r^4+8r^2}}{4}$$

Now, $x_2 < x_3$ and $x_2 < 0$ and $x_3 < 1$ (since $r < 1$)

Now, the numerator of $\frac{dl}{dx}$ is a cubic equation and the denominator is always positive.

Hence as $x \to \infty$, $\frac{dl}{dx} \to +\infty$.

Hence the curve has the following shape:

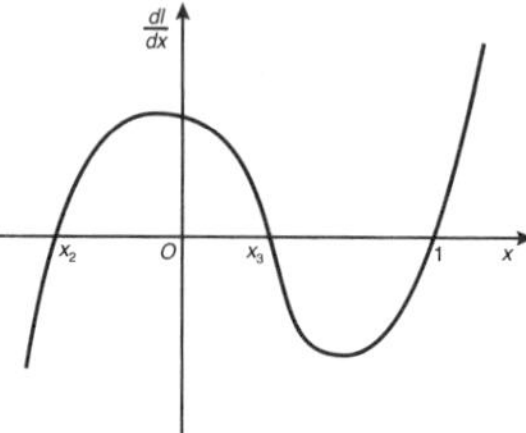

At $x_3 - \varepsilon$, $\frac{dl}{dx} > 0$

At $x_3 + \varepsilon$, $\frac{dl}{dx} < 0$

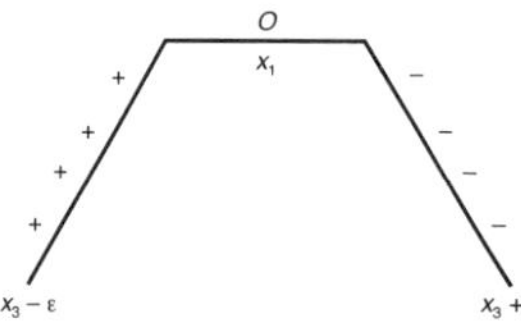

$\therefore$ At x_3, a maximum value of l occurs.

$\therefore$ The value of x for which M is closest to the floor is

$$x = \frac{r^2+\sqrt{r^4+8r^2}}{4}$$

$$= \frac{r^2+r\sqrt{r^2+8}}{4}$$

$$= r\left[\frac{r+\sqrt{r^2+8}}{4}\right].$$ ✓

(2 marks)

67

$$y = e^{2x}$$

$$\frac{dy}{dx} = 2e^{2x}$$

at $x = 0$, $\frac{dy}{dx} = 2e^0 = 2$ ✓

$\therefore$ Equation of tangent at $(0, 1)$ with gradient 2

is $y - 1 = 2(x - 0)$

$y - 1 = 2x$

$y = 2x + 1$. ✓

(2 marks)

68 **i** $V = 25\left(1-\frac{t}{60}\right)^2$ for $0 \le t \le 60$

At $t = 0$, $V = 25\left(1-\frac{0}{60}\right)^2$

$= 25 \times 1$

$= 25$ litres. ✓

(1 mark)

ii $V = 25\left(1-\frac{t}{60}\right)^2$

When $V = \frac{25}{4}$,

$$\frac{25}{4} = 25\left(1-\frac{t}{60}\right)^2$$

$$\frac{1}{4} = \left(1-\frac{t}{60}\right)^2$$

$$\pm\frac{1}{2} = 1-\frac{t}{60}$$

$$\frac{t}{60} = 1 \pm \frac{1}{2}$$ ✓

$$\frac{t}{60} = 1\frac{1}{2} \text{ or } \frac{t}{60} = \frac{1}{2}$$

$\therefore t = 90$ or $t = 30$

$\therefore t = 30$ seconds

$0 \le t \le 60$.

✓ *(2 marks)*

iii Now, $\frac{dV}{dt} = 2 \times 25\left(1-\frac{t}{60}\right)^1 \cdot \frac{d}{dt}\left(1-\frac{t}{60}\right)$

$$= 50\left(1-\frac{t}{60}\right) \cdot -\frac{1}{60}$$ ✓

When $t = 30$,

$$\frac{dV}{dt} = 50\left(1-\frac{30}{60}\right) \times -\frac{1}{60}$$

$$= 50\left(\frac{1}{2}\right)\left(-\frac{1}{60}\right)$$

$$= -\frac{50}{120}$$

$$= -\frac{5}{12} \text{ litres/second.}$$

So rate of draining out

$= +\frac{5}{12}$ litres/second ✓

(2 marks)

69 **i** Range: $-1+3 \leqslant x \leqslant 1+3$

$\therefore 2 \leqslant x \leqslant 4$ ✓

Period: $\frac{2\pi}{2} = \pi$. ✓

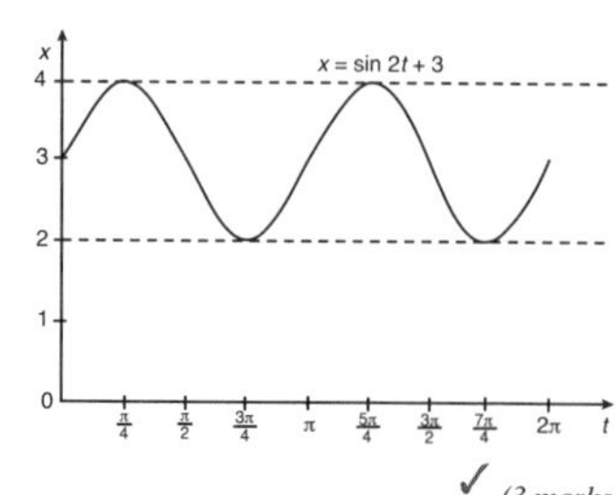

✓ *(3 marks)*

ii The particle is at rest when $\frac{dx}{dt} = 0$, i.e. the gradient of the tangent is 0.

Time	*Position*
$t = \frac{\pi}{4}$ s	$x = 4$ m
$t = \frac{3\pi}{4}$ s	$x = 2$ m
$t = \frac{5\pi}{4}$ s	$x = 4$ m
$t = \frac{7\pi}{4}$ s	$x = 2$ m ✓✓

(2 marks)

iii The particle oscillates between $x = 4$ and $x = 2$ with the centre of oscillation at $x = 3$. The particle starts at $x = 3$ and moves to the right to $x = 4$, where at $t = \frac{\pi}{4}$ s the particle stops, changes direction and moves to the left until at $x = 2$ and $t = \frac{3\pi}{4}$ s it again stops, changes direction and moves to the right again, returning to the centre of oscillation at $x = 3$ ($t = \pi$ s). This motion is then repeated from $t = \pi$ s to $t = 2\pi$ s. ✓✓

(2 marks)

70 i

$$x = \frac{t-2}{t+2}$$

When $t = 0$, $x = \frac{0-2}{0+2}$

$= \frac{-2}{2}$

$= -1$ m ✓

(1 mark)

ii $x = 1 - \frac{4}{t+2}$

RHS $= 1 - \frac{4}{t+2}$

$= \frac{t+2}{t+2} - \frac{4}{t+2}$

$= \frac{t+2-4}{t+2}$

$= \frac{t-2}{t+2}$

$= x$

$=$ LHS ✓

Now, $x = 1 - \frac{4}{t+2}$

$= 1 - 4(t+2)^{-1}$

$v = \frac{dx}{dt} = 0 + 4(t+2)^{-2}$

$= \frac{4}{(t+2)^2}$

$= 4(t+2)^{-2}$ ✓

$a = \frac{dv}{dt} = -8(t+2)^{-3}$

$= \frac{-8}{(t+2)^3}$ ✓

(3 marks)

iii The particle is at rest if

$v = 0$

$\frac{4}{(t+2)^2} = 0$

Since $t \geq 0$, $(t+2)^2 > 0$

$\therefore \frac{4}{(t+2)^2} > 0$

$\therefore$ Particle is never at rest. ✓

(1 mark)

iv $v = \frac{4}{(t+2)^2}$

As t increases indefinitely, $(t+2)^2$ increases indefinitely

$\therefore \frac{4}{(t+2)^2}$ approaches zero.

$\therefore$ Limiting velocity is zero. ✓

(1 mark)

71 i $\frac{dy}{dt}$ will be a maximum at the depth where the glass is narrowest, i.e. at $y_1 = 2$ cm. ✓

$\frac{dy}{dt}$ will be a minimum at the depth where the glass is widest, i.e. at $y_2 = 7$ cm. ✓

(2 marks)

ii

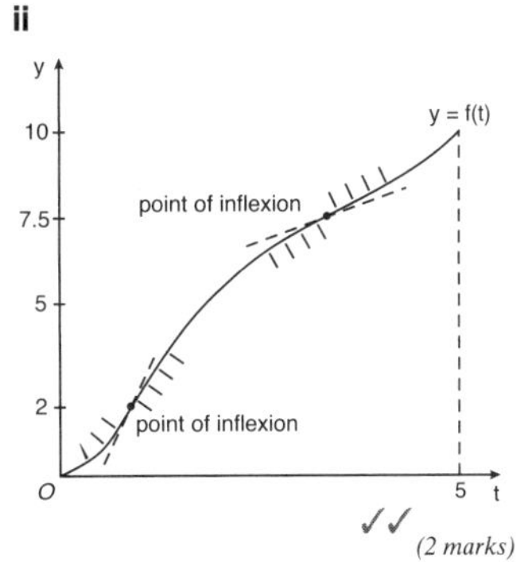

✓✓ *(2 marks)*

72 i Time taken by bus; T_{BUS}:

In ΔGFP,

$\tan\theta = \frac{GP}{250}$

$\therefore$ GP $= 250\tan\theta$

$\therefore$ DP $= 2000 + 250\tan\theta$

Now, time $= \frac{\text{distance}}{\text{speed}}$

$\therefore T_{BUS} = \frac{2000 + 250\tan\theta}{15}$ ✓

Time taken by Claire, T_{CLAIRE}:

In ΔGFP,

$\cos\theta = \frac{250}{FP}$

$\therefore FP = \frac{250}{\cos\theta}$

$= 250\sec\theta$

$\therefore T_{CLAIRE} = \frac{250\sec\theta}{4}$. ✓

(2 marks)

ii Let $T = T_{BUS} - T_{CLAIRE}$

i.e. $T = \frac{2000 + 250\tan\theta}{15} - \frac{250(\cos\theta)^{-1}}{4}$

We need to find the maximum value of this time difference (T).

$\frac{dT}{d\theta} = \frac{250\sec^2\theta}{15} + \frac{250(\cos\theta)^{-2}.(-\sin\theta)}{4}$

$= \frac{250\sec^2\theta}{15} - \frac{250\sin\theta\sec^2\theta}{4}$ ✓

Stationary points occur when

$\frac{dT}{d\theta} = 0$

$\frac{250\sec^2\theta}{15} - \frac{250\sin\theta\sec^2\theta}{4} = 0$

$250\sec^2\theta\left(\frac{1}{15} - \frac{\sin\theta}{4}\right) = 0$

Now, $\sec^2\theta > 0$,

$\therefore \frac{1}{15} - \frac{\sin\theta}{4} = 0$

$\frac{\sin\theta}{4} = \frac{1}{15}$

$\sin\theta = \frac{4}{15}$

$\therefore \theta = 0.2699$ radians ✓

Consider $\frac{dT}{d\theta}$:

θ	0.2	0.2699	0.3
$\frac{dT}{d\theta}$	+	0	–

$\therefore$ A maximum value at $\theta = 0.2699$ ✓

At $\theta = 0.2699$,

$T = \frac{2000 + 250\tan(0.2699)}{15} - \frac{250}{4\cos(0.2699)}$

$= 137.944 - 64.848$

$= 73.096$

$= 73$ secs ($= 1$ min 13 secs)

$\therefore$ The latest Claire can leave is 1 minute 13 seconds past 8 am.

✓ *(4 marks)*

73

$y = 2\log_e x$

$\frac{dy}{dx} = 2 \times \frac{1}{x}$

$= \frac{2}{x}$ ✓

At $x = 1$, $\frac{dy}{dx} = 2$ ✓

∴ Equation of tangent through $(1, 0)$ with gradient $= 2$ is:

$y - 0 = 2(x - 1)$

$y = 2x - 2.$ ✓

(3 marks)

74 **i** $N = 175 + 18t^2 - t^4, 0 \leq t \leq 5$

When $t = 0$,

$N = 175 + 0 - 0$

$= 175.$ ✓

ii When $t = 5$,

$N = 175 + 18(5)^2 - 5^4$

$= 175 + 450 - 625$

$= 0.$ ✓

iii $\frac{dN}{dt} = 36t - 4t^3$

$= 4t(9 - t^2)$

$= 4t(3 + t)(3 - t)$

$\frac{d^2N}{dt^2} = 36 - 12t^2$

$= 12(3 - t^2)$ ✓

Stationary points occur when

$\frac{dN}{dt} = 0$

$\therefore 4t(3 + t)(3 - t) = 0$

$\therefore t = 0$ or 3 since $0 \leqslant t \leqslant 5$.

When $t = 0$, $\frac{d^2N}{dt^2} = 36 > 0$.

∴ A minimum turning point at $t = 0$. ✓

When $t = 3$, $\frac{d^2N}{dt^2} = 12(3 - 9)$

$= 12 \times -6$

$= -72 < 0.$

∴ A maximum turning point at $t = 3$. ✓

When $t = 3$,

$N = 175 + 18(3)^2 - 3^4$

$= 175 + 162 - 81$

$= 256.$

∴ Maximum number of students was 256. ✓

iv Students were logging on most rapidly

when $\frac{d^2N}{dt^2} = 0$

$\therefore 12(3 - t^2) = 0$

$t^2 = 3$

$t = \pm\sqrt{3}$

$\therefore t = \sqrt{3}$ ✓

since $0 \leqslant t \leqslant 5$

v When $t = 0, N = 175.$

When $t = 5, N = 0.$

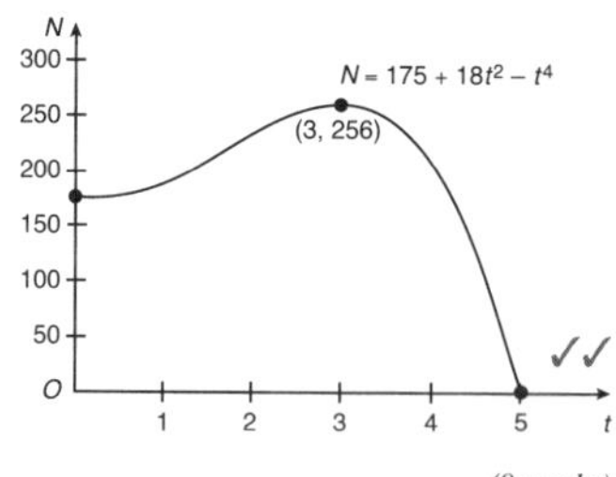

(9 marks)

(Total mark allocation only in exam)

75 **i** $x = 2t - 3\log_e(t + 1)$

$v = \frac{dx}{dt} = 2 - \frac{3}{t+1}.$

ii When $t = 0$, $v = 2 - \frac{3}{1}$

$= 2 - 3$

$= -1 \text{ ms}^{-1}.$

iii The particle comes to rest when $v = 0$

i.e. $2 - \frac{3}{t+1} = 0$

$2 = \frac{3}{t+1}$

$2t + 2 = 3$

$2t = 1$

$t = \frac{1}{2}.$

∴ Particle comes to rest when $t = \frac{1}{2}$ second.

(No mark allocation in exam)

76

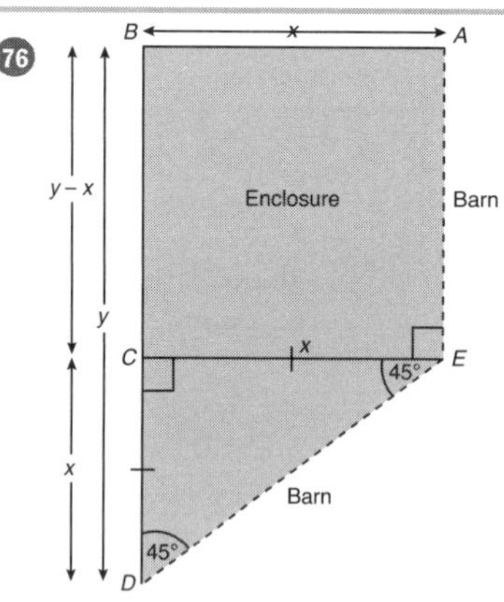

i Construct CE

Now $\angle DCE = 90°$

and $\angle CED = 45°$

$\therefore \angle CDE = 45°$

(angle sum ΔCDE)

$\therefore \Delta CDE$ is isosceles.

Since $AB = x$

$\therefore CE = x$

(opp. sides of rectangle)

$\therefore CD = x$

(ΔCDE isosceles)

Since $BD = y$

and $CD = x$

$\therefore BC = y - x$ ✓

Area of enclosure = Area of rectangle $ABCE$ + Area of ΔDCE

$\therefore A = AB \times BC + \frac{1}{2}(CE \times CD)$

$= x(y - x) + \frac{1}{2}x^2$

$= xy - x^2 + \frac{1}{2}x^2$

$= xy - \frac{1}{2}x^2 \quad \ldots (1)$ ✓

Now, $x + y = 117$ (given)

$\therefore y = 117 - x$

Substituting into (1):

$A = x(117 - x) - \frac{1}{2}x^2$

$= 117x - x^2 - \frac{1}{2}x^2$

$= 117x - \frac{3}{2}x^2$ ✓

ii $\frac{dA}{dx} = 117 - 3x$

$\frac{d^2A}{dx^2} = -3$

Stationary points occur when

$\frac{dA}{dx} = 0$

$\therefore 117 - 3x = 0$

$3x = 117$

$\therefore x = 39$ ✓

When $x = 39$, $\frac{d^2A}{dx^2} = -3 < 0$

∴ A maximum value occurs when $x = 39$.

Since $y = 117 - x$,

$\therefore y = 117 - 39$ at $x = 39$

$= 78.$

∴ The largest area occurs when $y = 78$ and $x = 39$,

i.e. when $y = 2x$. ✓ *(5 marks)*

(Total mark allocation only in exam)

77 **i** $y = x^3 - 6x^2 + 9x + 1$

$\frac{dy}{dx} = 3x^2 - 12x + 9$ ✓

ii Stationary points occur when

$\frac{dy}{dx} = 0$

$3x^2 - 12x + 9 = 0$

$3(x^2 - 4x + 3) = 0$

$3(x - 3)(x - 1) = 0$

$\therefore x = 3$ or 1 ✓

When $x = 3$,

$y = (3)^3 - 6(3)^2 + 9(3) + 1$

$= 27 - 54 + 27 + 1$

$= 1$

When $x = 1$,
$y = (1)^3 - 6(1)^2 + 9(1) + 1$
$= 1 - 6 + 9 + 1$
$= 5$

$\therefore$ Coordinates of stationary points are (3, 1) and (1, 5) ✓

iii $\dfrac{d^2y}{dx^2} = 6x - 12$

When $x = 3$, $\dfrac{d^2y}{dx^2} = 6(3) - 12$
$= 6 > 0$

$\therefore$ A minimum turning point at (3, 1) ✓

When $x = 1$, $\dfrac{d^2y}{dx^2} = 6(1) - 12$
$= -6 < 0$

$\therefore$ A maximum turning point at (1, 5) ✓

iv When $x = 0, y = 1$

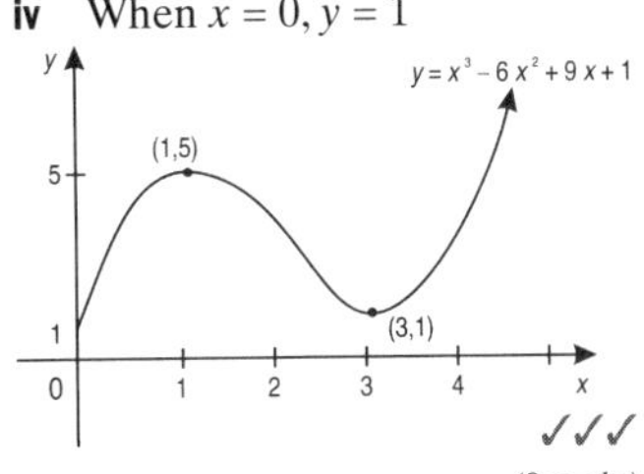

✓✓✓

(8 marks)
(Total mark allocation only in exam)

78 **i** $x = 2\sin t - t \quad t \geqslant 0$

$v = \dfrac{dx}{dt} = 2\cos t - 1$

ii When $t = 0$, $v = 2\cos(0) - 1$
$= 2(1) - 1$
$= +1$ units/s

Hence the particle is moving away from the origin to the right (positive x direction).

iii Particle comes to rest when
$v = 0$
$2\cos t - 1 = 0$
$2\cos t = 1$
$\cos t = \frac{1}{2}$

$\therefore t = \frac{\pi}{3}, 2\pi - \frac{\pi}{3}, \ldots$

$\therefore$ Particle first comes to rest at $t = \frac{\pi}{3}$ seconds

iv $a = \dfrac{dv}{dt} = -2\sin t$

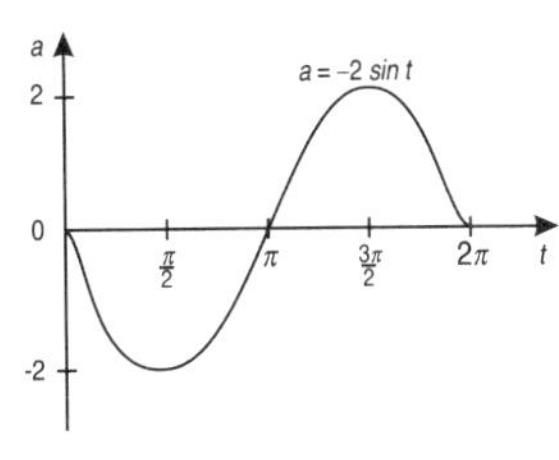

$a < 0$ when $0 < t < \pi$

(No mark allocation in exam)

79 **i** Stationary points occur when
$f'(x) = 0$
$\therefore e^{-2x} - 2xe^{-2x} = 0$
$e^{-2x}(1 - 2x) = 0$
$\therefore 1 - 2x = 0$
(since $e^{-2x} > 0$)
$2x = 1$
$x = \frac{1}{2}$. ✓

ii $f(x)$ is increasing when
$f'(x) > 0$
$\therefore e^{-2x}(1 - 2x) > 0$
Now, $e^{-2x} > 0$ for all x
but $1 - 2x > 0$
when $1 > 2x$
$\therefore 2x < 1$
$\therefore x < \frac{1}{2}$. ✓

$\therefore f(x)$ is increasing when $x < \frac{1}{2}$.

iii Possible points of inflection occur when
$f''(x) = 0$
$4xe^{-2x} - 4e^{-2x} = 0$
$4e^{-2x}(x - 1) = 0$
$e^{-2x} > 0 \quad \therefore x - 1 = 0$
$x = 1$

Checking for change of concavity:

x	$f''(x)$
0.9	$(+)(-) < 0$
1.0	0
1.1	$(+)(+) > 0$

$\therefore$ There is a change of concavity and a point of inflection at $x = 1$. ✓

The graph is concave up when $f''(x) > 0$
$4e^{-2x}(x - 1) > 0$
Now $e^{-2x} > 0$ and $x - 1 > 0$
when $x > 1$

$\therefore$ Graph is concave upwards when $x > 1$. ✓

iv Now $f(x) = xe^{-2x} + 1$
$f\left(-\frac{1}{2}\right) = -0.4$
$f(0) = 1.0$
$f(4) = 1.0013$

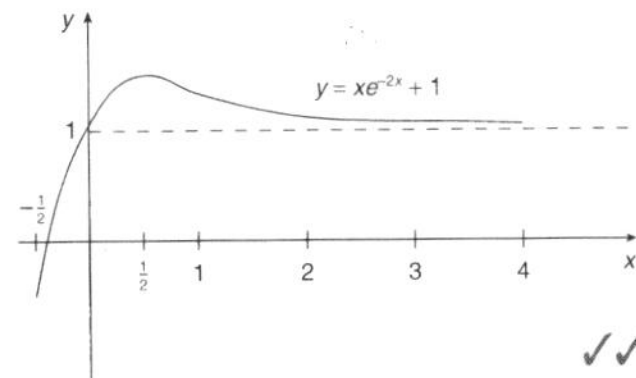

✓✓

v $f(x) = xe^{-2x} + 1$
$= \dfrac{x}{e^{2x}} + 1$

As x gets very large, e^{2x} gets infinitesimally larger and $\dfrac{x}{e^{2x}}$ becomes very small and approaches zero.
$\therefore f(x)$ will approach 1. ✓

(7 marks)
(Total mark allocation only in exam)

80 **i** $y = 1 + \sqrt{3}\sin x + \cos x$
$\dfrac{dy}{dx} = \sqrt{3}\cos x - \sin x$

At $x = \frac{5\pi}{6}$,
$\dfrac{dy}{dx} = \sqrt{3}\cos\left(\frac{5\pi}{6}\right) - \sin\left(\frac{5\pi}{6}\right)$
$= \sqrt{3}\left(-\frac{\sqrt{3}}{2}\right) - \left(\frac{1}{2}\right)$
$= -\frac{3}{2} - \frac{1}{2}$
$= -\frac{4}{2}$
$= -2$ ✓

and
$y = 1 + \sqrt{3}\sin\left(\frac{5\pi}{6}\right) + \cos\left(\frac{5\pi}{6}\right)$
$= 1 + \sqrt{3}\left(\frac{1}{2}\right) + \left(-\frac{\sqrt{3}}{2}\right)$
$= 1 + \frac{\sqrt{3}}{2} - \frac{\sqrt{3}}{2}$
$= 1$

Equation of tangent through $\left(\frac{5\pi}{6}, 1\right)$ with gradient $= -2$ is:
$y - 1 = -2\left(x - \frac{5\pi}{6}\right)$
$y - 1 = -2x + \frac{5\pi}{3}$
$y = -2x + \frac{5\pi}{3} + 1$ ✓

ii $y = 1 + \sqrt{3}\sin x + \cos x$
$\dfrac{dy}{dx} = \sqrt{3}\cos x - \sin x$
$\dfrac{d^2y}{dx^2} = -\sqrt{3}\sin x - \cos x$

Possible maximum and minimum values occur when
$\dfrac{dy}{dx} = 0$
$\sqrt{3}\cos x - \sin x = 0$
$\sqrt{3}\cos x = \sin x$
$\dfrac{\sqrt{3}\cos x}{\cos x} = \dfrac{\sin x}{\cos x}$
$\sqrt{3} = \tan x$
$\therefore x = \dfrac{\pi}{3}$ and $\dfrac{4\pi}{3}$, ✓
$0 \leqslant x \leqslant 2\pi$

and
$\dfrac{d^2y}{dx^2} = -\sqrt{3}\sin\frac{\pi}{3} - \cos\frac{\pi}{3}$
$= -\sqrt{3}\left(\frac{\sqrt{3}}{2}\right) - \frac{1}{2}$
$= -\frac{3}{2} - \frac{1}{2}$
$= -\frac{4}{2}$
$= -2$

$\therefore$ A maximum turning point at $\left(\frac{\pi}{3}, 3\right)$ ✓

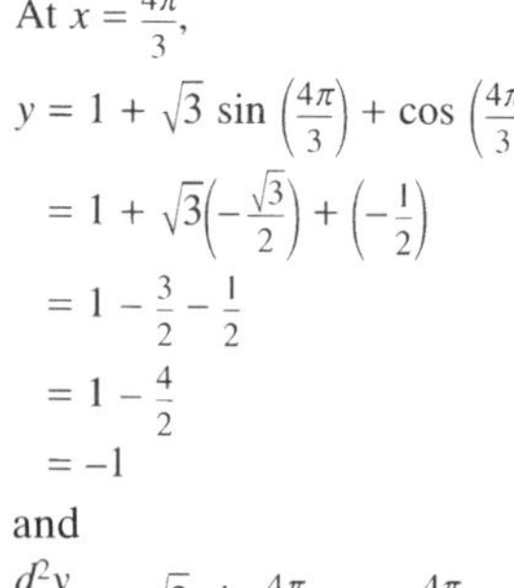

At $x = \frac{4\pi}{3}$,

$$y = 1 + \sqrt{3}\sin\left(\frac{4\pi}{3}\right) + \cos\left(\frac{4\pi}{3}\right)$$
$$= 1 + \sqrt{3}\left(-\frac{\sqrt{3}}{2}\right) + \left(-\frac{1}{2}\right)$$
$$= 1 - \frac{3}{2} - \frac{1}{2}$$
$$= 1 - \frac{4}{2}$$
$$= -1$$

and

$$\frac{d^2y}{dx^2} = -\sqrt{3}\sin\frac{4\pi}{3} - \cos\frac{4\pi}{3}$$
$$= -\sqrt{3}\left(-\frac{\sqrt{3}}{2}\right) - \left(-\frac{1}{2}\right)$$
$$= \frac{3}{2} + \frac{1}{2}$$
$$= 2$$

$\therefore$ A minimum turning point at $\left(\frac{4\pi}{3}, -1\right)$ ✓

$\therefore$ Maximum and minimum values of 3 and –1 respectively.

(5 marks)
(Total mark allocation only in exam)

81 **i**

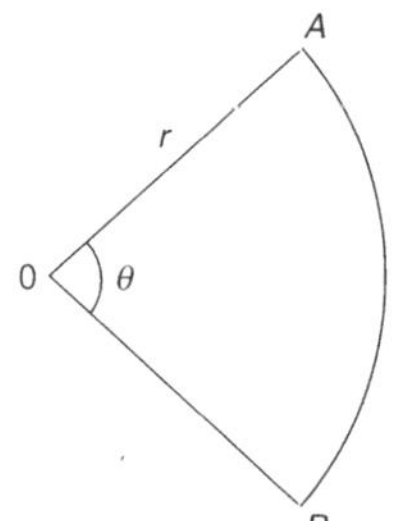

Perimeter = $OA + OB + \text{arc } AB$

$$8 = r + r + r\theta$$
$$8 = 2r + r\theta$$
$$8 = r(2 + \theta)$$
$$r = \frac{8}{2 + \theta}$$ ✓

ii Area of sector

$$= \frac{1}{2}r^2\theta$$
$$= \frac{1}{2}\left(\frac{8}{2 + \theta}\right)^2.\theta$$
$$= \frac{1}{2}.\frac{64}{(2 + \theta)^2}.\theta$$
$$= \frac{32\theta}{(2 + \theta)^2}$$ ✓

iii $A = \dfrac{32\theta}{(2 + \theta)^2}$ for $0 < \theta < \dfrac{\pi}{2}$

$$\frac{dA}{d\theta} = \frac{(2 + \theta)^2.\frac{d}{d\theta}(32\theta) - 32\theta.\frac{d}{d\theta}(2 + \theta)^2}{[(2 + \theta)^2]^2}$$
$$= \frac{(2 + \theta)^2.32 - 32\theta.2(2 + \theta)}{(2 + \theta)^4}$$
$$= \frac{32(2 + \theta)^2 - 64\theta(2 + \theta)}{(2 + \theta)^4}$$
$$= \frac{32(2 + \theta)[(2 + \theta) - 2\theta]}{(2 + \theta)^4}$$
$$= \frac{32(2 + \theta)(2 - \theta)}{(2 + \theta)^4}$$
$$= \frac{32(2 - \theta)}{(2 + \theta)^3}$$ ✓

Stationary points occur when

$$\frac{dA}{d\theta} = 0$$
$$\frac{32(2 - \theta)}{(2 + \theta)^3} = 0$$
$$\therefore \theta = 2$$

Since $0 \leqslant \theta \leqslant \frac{\pi}{2}$, this solution lies outside of the range of possible values of θ. ✓

$\therefore$ the maximum will lie on one of the endpoints of the interval.

Consider points on the graph of

$$A = \frac{32\theta}{(\theta + 2)^2} \text{ for } 0 < \theta < \frac{\pi}{2}:$$

At $\theta = 0$, $A = 0$

At $\theta = \frac{\pi}{2}$, $A = \dfrac{32\left(\frac{\pi}{2}\right)}{\left(\frac{\pi}{2} + 2\right)^2}$

$$= \frac{16\pi}{\left(\frac{\pi}{2} + 2\right)^2}$$
$$\doteqdot 3.94$$

There are no turning points in this interval and the curve must be rising in this interval. Therefore the maximum area will occur when $\theta = \frac{\pi}{2}$ and will be approximately

$$\frac{16\pi}{\left(\frac{\pi}{2} + 2\right)^2} \doteqdot 3.94 \text{ cm}^2.$$ ✓

(5 marks)
(Total mark allocation only in exam)

82 **i**

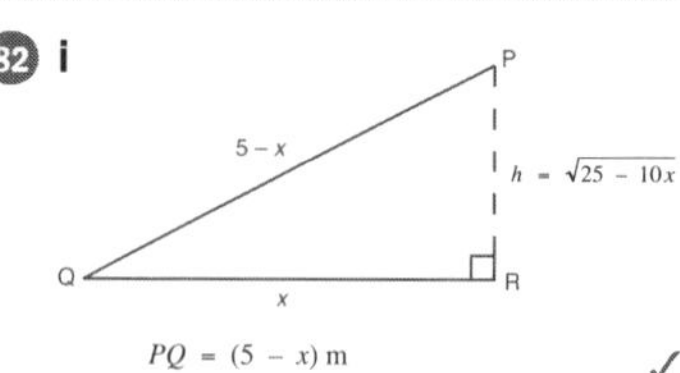

$PQ = (5 - x)$ m ✓

ii Now

$$(5 - x)^2 = x^2 + h^2$$
$$25 - 10x + x^2 = x^2 + h^2$$
$$h^2 = 25 - 10x$$
$$h = \sqrt{25 - 10x} \text{ m}$$

($h > 0$ as it is a height)

$\therefore$ Area of ΔPQR

$$= \tfrac{1}{2} \times b \times h$$
$$= \tfrac{1}{2} \times x \times \sqrt{25 - 10x}$$
$$= \tfrac{1}{2}x\sqrt{25 - 10x} \text{ m}^2$$ ✓

iii

$$A = \tfrac{1}{2}x(25 - 10x)^{\frac{1}{2}}$$
$$\frac{dA}{dx} = \tfrac{1}{2}\left[x.\tfrac{1}{2}(25 - 10x)^{-\frac{1}{2}}.-10 + (25 - 10x)^{\frac{1}{2}}.1\right]$$
$$= \frac{1}{2}\left[\frac{-10x(25 - 10x)^{-\frac{1}{2}}}{2} + (25 - 10x)^{\frac{1}{2}}\right]$$
$$= \frac{1}{2}\left[-\frac{5x}{\sqrt{25 - 10x}} + \sqrt{25 - 10x}\right]$$ ✓

Stationary points occur when

$$\frac{dA}{dx} = 0$$
$$\frac{1}{2}\left[-\frac{5x}{\sqrt{25 - 10x}} + \sqrt{25 - 10x}\right] = 0$$
$$-\frac{5x}{\sqrt{25 - 10x}} + \sqrt{25 - 10x} = 0$$
$$\sqrt{25 - 10x} = \frac{5x}{\sqrt{25 - 10x}}$$
$$\left(\sqrt{25 - 10x}\right)^2 = 5x$$
$$25 - 10x = 5x$$
$$15x = 25$$
$$x = \frac{25}{15}$$
$$= \frac{5}{3}$$ ✓

Now when $x < \frac{5}{3}$, say $x = 1$,

$$\frac{dA}{dx} = \frac{1}{2}\left[-\frac{5}{\sqrt{15}} + \sqrt{15}\right]$$
$$\doteqdot 1.29 > 0$$

When $x > \frac{5}{3}$, say $x = 2$,

$$\frac{dA}{dx} = \frac{1}{2}\left[-\frac{10}{\sqrt{5}} + \sqrt{5}\right]$$
$$\doteqdot -1.12 < 0$$

$\therefore$ A maximum value occurs at $x = \frac{5}{3}$. ✓

At $x = \frac{5}{3}$, $A = \frac{1}{2}\left(\frac{5}{3}\right)\sqrt{25 - 10\left(\frac{5}{3}\right)}$

$$= \frac{5}{6}\sqrt{25 - \frac{50}{3}}$$
$$= \frac{5}{6}\sqrt{\frac{25}{3}}$$
$$= \frac{5}{6} \times 5 \times \frac{1}{\sqrt{3}}$$
$$= \frac{25}{6} \times \frac{1}{\sqrt{3}} \times \frac{\sqrt{3}}{\sqrt{3}}$$
$$= \frac{25\sqrt{3}}{18} \text{ m}^2.$$ ✓

(6 marks)
(Total mark allocation only in exam)

83 **i** $x = t + \sin t$

$$v = \frac{dx}{dt} = 1 + \cos t$$ ✓

The particle is stationary when

$$v = 0$$
$$\therefore 1 + \cos t = 0$$
$$\cos t = -1$$
$$\therefore t = \pi$$

for $0 < t < 3\pi$. ✓

ii $a = \frac{dv}{dt} = 0 - \sin t$

$= -\sin t$ ✓

Now, $a = 0$ when

$0 = -\sin t$

$\therefore \sin t = 0$

$\therefore t = \pi, 2\pi$ for $0 < t < 3\pi$. ✓

iii Possible points of inflection occur when

$\frac{dv}{dt} = \frac{d^2x}{dt^2} = 0$

i.e. when $t = \pi, 2\pi$

When $t = \pi, x = \pi + \sin \pi = \pi$.

When $t = 2\pi, x = 2\pi + \sin 2\pi = 2\pi$.

Consider the point (π, π).

t	3	π	3.2
$\frac{d^2x}{dt^2} = -\sin t$	−0.14	0	0.06

$\therefore$ a change of concavity.

$\therefore$ There is a horizontal point of inflection at (π, π). (Since at $t = \pi$, $\frac{dx}{dt} = 0$ also.) ✓

Consider the point $(2\pi, 2\pi)$:

t	6.2	$2\pi \doteqdot (6.28)$	6.4
$\frac{dx}{dt} = -\sin t$	0.08	0	−0.12

$\therefore$ a change of concavity.

$\therefore$ There is a point of inflection at $(2\pi, 2\pi)$. ✓

End points:

When $t = 0, x = 0 + \sin 0 = 0$

When $t = 3\pi$,

$x = 3\pi + \sin 3\pi = 3\pi$

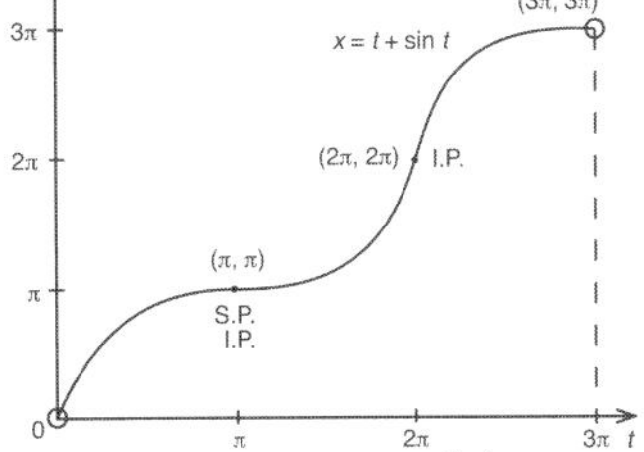

✓✓

(8 marks)

(Total mark allocation only in exam)

84 **i** $M = 4t^2 - t^3$

$\frac{dM}{dt} = 8t - 3t^2$

$= t(8 - 3t)$

$\frac{d^2M}{dt^2} = 8 - 6t$

Stationary points occur when

$\frac{dM}{dt} = 0$

$t(8 - 3t) = 0$

$t = 0$ or $\frac{8}{3}$ ✓

When $t = 0, M = 0$

and $\frac{d^2M}{dt^2} = 8 > 0$

$\therefore$ A minimum turning point at $(0, 0)$. ✓

When $t = \frac{8}{3}$,

$M = 4\left(\frac{8}{3}\right)^2 - \left(\frac{8}{3}\right)^3$

$= 4\left(\frac{64}{9}\right) - \frac{512}{27}$

$= \frac{4 \times 64 \times 3}{27} - \frac{512}{27}$

$= \frac{256}{27}$

$= 9\frac{13}{27}$

and $\frac{d^2M}{dt^2} = 8 - 6\left(\frac{8}{3}\right)$

$= 8 - 16$

$= -8$

< 0

$\therefore$ A maximum turning point at $\left(\frac{8}{3}, 9\frac{13}{27}\right)$ ✓

End point of domain:

When $t = 3, M = 4(3)^2 - (3)^3$

$= 36 - 27$

$= 9$

$\therefore$ End point is (3, 9).

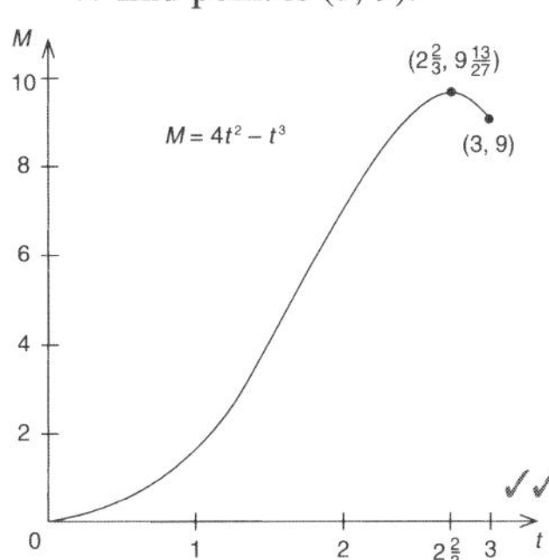

✓✓

ii Maximum occurs when

$t = \frac{8}{3}$

$= 2\frac{2}{3}$ hours. ✓

iii The amount is increasing most rapidly when the slope of the curve is the greatest,

i.e. when $\frac{d^2M}{dt^2} = 0$

$8 - 6t = 0$

$6t = 8$

$t = \frac{8}{6}$

$= \frac{4}{3}$

$= 1\frac{1}{3}$ hours. ✓

(7 marks)

(Total mark allocation only in exam)

85 [Parts **i** and **ii** not in topic.]

iii $\therefore A = \frac{1}{2} \times SU \times TS$

$= \frac{1}{2}(6 + y)(4 + x)$

but from (ii) $y = \frac{24}{x}$

$\therefore A = \frac{1}{2}\left(6 + \frac{24}{x}\right)(4 + x)$

$= \frac{1}{2}\left(24 + 6x + \frac{96}{x} + 24\right)$

$= \frac{1}{2}\left(48 + 6x + \frac{96}{x}\right)$

$= 24 + 3x + \frac{48}{x}$

iv $A = 24 + 3x + 48x^{-1}$

$\frac{dA}{dx} = 3 - 48x^{-2}$

$= 3 - \frac{48}{x^2}$

$\frac{d^2A}{dx^2} = 96x^{-3}$

$= \frac{96}{x^3}$

Stationary points occur when

$\frac{dA}{dx} = 0$

$3 - \frac{48}{x^2} = 0$

$3 = \frac{48}{x^2}$

$x^2 = 16$

$\therefore x = 4$

($x = -4$ is ignored as x is a positive length)

When $x = 4$

$\frac{d^2A}{dx^2} = \frac{96}{64} > 0$

$\therefore$ A minimum value occurs when $x = 4$

$\therefore$ Height $= x + 4$

$= 4 + 4$

$= 8$ cm

Base $= 6 + \frac{24}{x}$

$= 6 + \frac{24}{4}$

$= 12$ cm.

(No mark allocation in exam)

86 **i**

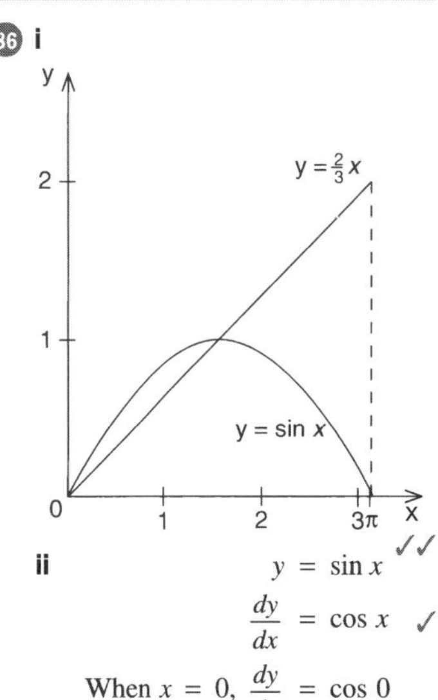

✓✓

ii $y = \sin x$

$\frac{dy}{dx} = \cos x$ ✓

When $x = 0$, $\frac{dy}{dx} = \cos 0$

$= 1$

$\therefore$ Gradient of tangent at the origin is 1. ✓

iii $0 < m < 1$ ✓

(5 marks)
(Total mark allocation only in exam)

87 **i** $y = 7 + 4x^3 - 3x^4$

$\frac{dy}{dx} = 12x^2 - 12x^3$

$= 12x^2(1 - x)$ ✓

Stationary points occur when

$\frac{dy}{dx} = 0$

$12x^2(1 - x) = 0$

$\therefore x = 0 \text{ or } 1$ ✓

When $x = 0$,

$y = 7 + 0 - 0$

$= 7$

When $x = 1$,

$y = 7 + 4 - 3$

$= 8$

$\therefore$ The coordinates of the stationary points are $(0, 7)$ and $(1, 8)$. ✓

ii $\frac{d^2y}{dx^2} = 24x - 36x^2$

$= 12x(2 - 3x)$

Now $\frac{d^2y}{dx^2} = 0$

when $12x(2 - 3x) = 0$

$\therefore x = 0 \text{ or } \frac{2}{3}$. ✓

iii When $x = 0$, $\frac{d^2y}{dx^2} = 0$

When $x < 0$, say $x = -0.1$

$\frac{d^2y}{dx^2} = 24(-0.1) - 36(-0.1)^2$

$= -2.76$

When $x > 0$, say $x = +0.1$

$\frac{d^2y}{dx^2} = 24(0.1) - 36(0.1)^2$

$= 2.04$

$\therefore$ A horizontal point of inflexion of $(0, 7)$. ✓

When $x = 1$,

$\frac{d^2y}{dx^2} = 12(1)(-1) < 0$

$\therefore$ A maximum turning point at $(1, 8)$. ✓

iv Values at end points of domain:

At $x = -1$,

$y = 7 + 4(-1)^3 - 3(-1)^4$

$= 7 - 4 - 3$

$= 0$

At $x = 2$,

$y = 7 + 4(2)^3 - 3(2)^4$

$= 7 + 32 - 48$

$= -9$ ✓

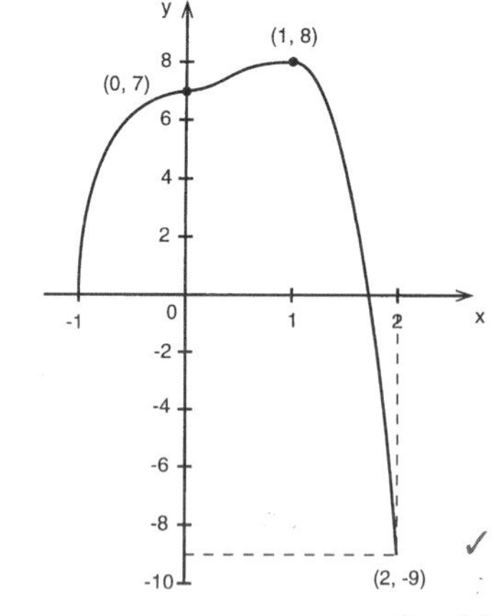

✓

(8 marks)
(Total mark allocation only in exam)

88 **i** $S = kd^2w$

Now $d^2 + w^2 = 15^2$
(Pythagoras' Theorem)

$\therefore d^2 = 225 - w^2$

$\therefore S = k(225 - w^2)w$

$= k(225w - w^3)$. ✓

ii $S = 225kw - kw^3$

$\frac{dS}{dw} = 225k - 3kw^2$

$= 3k(75 - w^2)$

$\frac{d^2S}{dw^2} = 3k(-2w)$

$= -6kw$

Stationary points occur when

$\frac{dS}{dw} = 0$

$3k(75 - w^2) = 0$

$\therefore 75 - w^2 = 0$

$75 = w^2$

$w = \sqrt{75}$

$= 5\sqrt{3}$ ✓

($-5\sqrt{3}$ is ignored as w is a positive length.)

When $w = 5\sqrt{3}$,

$\frac{d^2S}{dw^2} = -6k(5\sqrt{3})$

< 0 (since $k > 0$)

$\therefore$ A maximum value occurs when $w = 5\sqrt{3}$ ✓

Now $225 = d^2 + w^2$

$\therefore 225 = d^2 + (5\sqrt{3})^2$

$225 = d^2 + 75$

$d^2 = 225 - 75$

$= 150$

$\therefore d = \sqrt{150}$

$= 5\sqrt{6}$

($-5\sqrt{6}$ is ignored as before.)

$\therefore$ Numerical dimensions giving maximum strength are $d = 5\sqrt{6}$ and $w = 5\sqrt{3}$. ✓

iii If the beam was a square,

$d = w$

$\therefore 225 = w^2 + w^2$

$225 = 2w^2$

$w^2 = \frac{225}{2}$

$w = \frac{15}{\sqrt{2}}$

($w = -\frac{15}{\sqrt{2}}$ is ignored as before.)

and $S = kw^3$

$\therefore S = k\left(\frac{15}{\sqrt{2}}\right)^3$

$= \frac{3375k}{2\sqrt{2}}$

$= \frac{3375k}{2\sqrt{2}} \times \frac{\sqrt{2}}{\sqrt{2}}$

$= \frac{3375\sqrt{2}k}{4}$ ✓

For a beam of maximum strength, $d = 5\sqrt{6}$ and $w = 5\sqrt{3}$ (from **ii**)

$\therefore S = kd^2w$

$= k(5\sqrt{6})^2(5\sqrt{3})$

$= k(150)(5\sqrt{3})$

$= 750\sqrt{3}k$ ✓

Comparing the strength of the beam of maximum strength to that of the square beam as a percentage:

$\frac{750\sqrt{3}k}{\frac{3375\sqrt{2}k}{4}} \times 100$

$= \frac{750\sqrt{3}k \times 4}{3375\sqrt{2}k} \times 100$

$= \frac{3000\sqrt{3}}{3375\sqrt{2}} \times 100$

$\doteqdot 108.9\%$

$\therefore$ The rectangular beam of maximum strength is more than 8% stronger than the square beam. ✓ *(7 marks)*

(Total mark allocation only in exam)

89 **i** $V = \pi r^2 h$

$\therefore 500 = \pi r^2 h$

$\therefore h = \frac{500}{\pi r^2}$.

ii

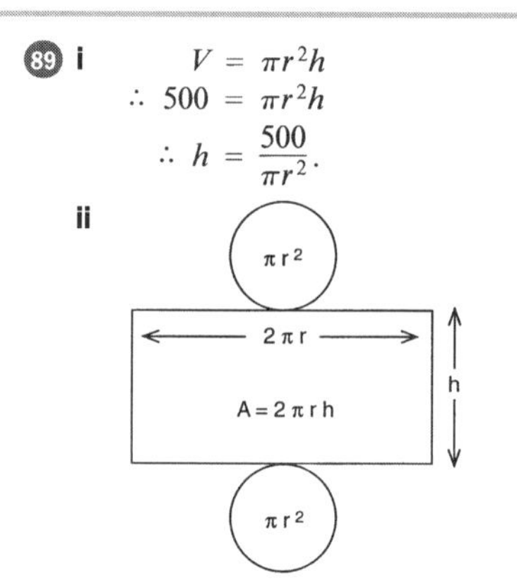

$$
\begin{aligned}
S &= 2\pi rh + 2\pi r^2 \\
&= 2\pi r\left(\frac{500}{\pi r^2}\right) + 2\pi r^2 \\
&= \frac{1000}{r} + 2\pi r^2 \\
&= 2\pi r^2 + \frac{1000}{r}.
\end{aligned}
$$

iii $S = 2\pi r^2 + 1000r^{-1}$

$$
\begin{aligned}
\frac{dS}{dr} &= 4\pi r - 1000r^{-2} \\
&= 4\pi r - \frac{1000}{r^2} \\
\frac{d^2S}{dr^2} &= 4\pi + 2000r^{-3} \\
&= 4\pi + \frac{2000}{r^3}
\end{aligned}
$$

Stationary points occur when

$$
\begin{aligned}
\frac{dS}{dr} &= 0 \\
4\pi r - \frac{1000}{r^2} &= 0 \\
4\pi r &= \frac{1000}{r^2} \\
4\pi r^3 &= 1000 \\
r^3 &= \frac{1000}{4\pi} \\
r &= \frac{10}{\sqrt[3]{4\pi}} \\
&\doteqdot 4.3
\end{aligned}
$$

When $r = 4.3$,

$$\frac{d^2S}{dr^2} = 4\pi + \frac{2000}{(4.3)^3} > 0$$

$\therefore$ A minimum value occurs when

$$r = \frac{10}{\sqrt[3]{4\pi}} \doteqdot 4.3$$

(No mark allocation in exam)

90 **i**

$$
\begin{aligned}
y &= \tfrac{1}{4}x^4 - x^3 \\
\frac{dy}{dx} &= x^3 - 3x^2 \\
&= x^2(x - 3) \\
\frac{d^2y}{d^2x} &= 3x^2 - 6x \\
&= 3x(x - 2)
\end{aligned}
$$

Stationary points occur when

$$
\begin{aligned}
\frac{dy}{dx} &= 0 \\
x^2(x - 3) &= 0 \\
\therefore x &= 0 \text{ or } 3
\end{aligned}
$$

When $x = 0, y = 0$

and $\dfrac{d^2y}{dx^2} = 0$

$\therefore$ Possible point of inflexion or turning point at $(0, 0)$.

When $x = 0 - \varepsilon$

$\dfrac{dy}{dx} = (+)(-) < 0$

When $x = 0 + \varepsilon$

$\dfrac{dy}{dx} = (+)(-) < 0$

$\therefore$ A horizontal point of inflexion at $(0, 0)$.

When $x = 3$,

$$
\begin{aligned}
y &= \tfrac{1}{4}(3)^4 - (3)^3 \\
&= \tfrac{81}{4} - 27 \\
&= \tfrac{81}{4} - \tfrac{108}{4} \\
&= -\tfrac{27}{4} \\
&= -6\tfrac{3}{4}
\end{aligned}
$$

and $\dfrac{d^2y}{dx^2} = 3(3)(1) > 0$

$\therefore$ A minimum turning point at $(3, -6\tfrac{3}{4})$.

ii Points of inflexion may occur when

$$
\begin{aligned}
\frac{d^2y}{dx^2} &= 0 \\
\text{i.e. } 3x(x - 2) &= 0 \\
\therefore x &= 0 \text{ or } 2
\end{aligned}
$$

From **i**, there is a horizontal point of inflexion at $(0, 0)$.

When $x = 2$,

$$
\begin{aligned}
y &= \tfrac{1}{4}(2)^4 - (2)^3 \\
&= \tfrac{16}{4} - 8 \\
&= 4 - 8 \\
&= -4
\end{aligned}
$$

When $x = 2 - \varepsilon$

$\dfrac{d^2y}{dx^2} = (+)(+)(-) < 0$

When $x = 2 + \varepsilon$

$\dfrac{d^2y}{dx^2} = (+)(+)(+) > 0$

$\therefore$ A point of inflexion at $(2, -4)$.

iii Values at end points:

When

$$
\begin{aligned}
x &= -1.5, \\
y &= \tfrac{1}{4}(-1.5)^4 - (-1.5)^3 \\
&= \tfrac{1}{4}(5.0625) - (-3.375) \\
&= 4.6406\ldots \\
&= 4.6 \text{ to 1 decimal place}
\end{aligned}
$$

When

$$
\begin{aligned}
x &= 4.5, \\
y &= \tfrac{1}{4}(4.5)^4 - (4.5)^3 \\
&= \tfrac{1}{4}(410.0625) - (91.125) \\
&= 11.3906\ldots \\
&= 11.4 \text{ to 1 decimal place.}
\end{aligned}
$$

The curve crosses the x-axis when

$$
\begin{aligned}
y &= 0 \\
\text{i.e. } x^3(\tfrac{1}{4}x - 1) &= 0 \\
\therefore x &= 0 \text{ or } 4
\end{aligned}
$$

iv The curve is concave down when

$$
\begin{aligned}
\frac{d^2y}{dx^2} &< 0 \\
\text{i.e. } 3x(x - 2) &< 0
\end{aligned}
$$

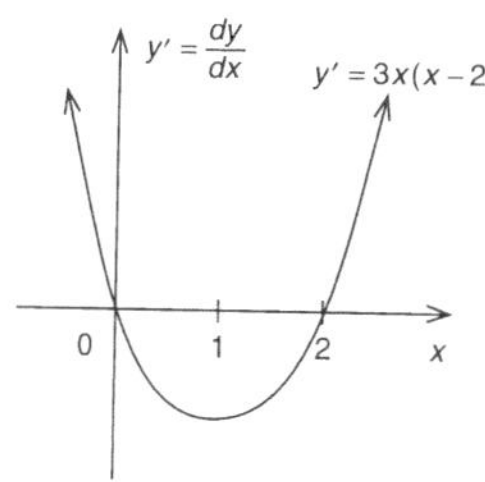

$\therefore 0 < x < 2$

(No mark allocation in exam)

91 **i**

$$
\begin{aligned}
y &= x^2 \\
\frac{dy}{dx} &= 2x \\
&= 2c \text{ at } x = c.
\end{aligned}
$$

ii Equation of tangent at (c, c^2) with gradient $= 2c$ is given by:

$$
\begin{aligned}
y - y_1 &= m(x - x_1) \\
y - c^2 &= 2c(x - c) \\
y - c^2 &= 2cx - 2c^2 \\
y &= 2cx - c^2
\end{aligned}
$$

iii At $x = 1, y = 0$

$$
\begin{aligned}
\therefore 0 &= 2c - c^2 \\
0 &= c(2 - c) \\
\therefore c &= 0 \text{ or } 2
\end{aligned}
$$

As the tangent intercepts the x-axis at $x = 1$, $c = 0$ is ignored and the answer is $c = 2$.

(No mark allocation in exam)

1 What is $\displaystyle\int \frac{1}{(2x+1)^2}\,dx$?

A $\dfrac{-2}{2x+1}+C$ **B** $\dfrac{-1}{2(2x+1)}+C$

C $2\ln(2x+1)+C$ **D** $\dfrac{1}{2}\ln(2x+1)+C$

(1 mark)

(Q6, **2022 HSC**) Easy

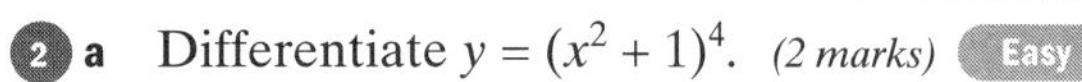

2 **a** Differentiate $y=(x^2+1)^4$. *(2 marks)* Easy

b Hence, or otherwise, find $\displaystyle\int x\left(x^2+1\right)^3 dx$. *(1 mark)* Easy

(Q18, **2022 HSC**)

3 What is $\displaystyle\int e+e^{3x}\,dx$?

A $ex+3e^{3x}+c$ **B** $ex+\dfrac{1}{3}e^{3x}+c$

C $e+3e^{3x}+c$ **D** $e+\dfrac{1}{3}e^{3x}+c$ *(1 mark)*

(Q4, **2020 HSC**) Medium

4 Find $\displaystyle\int \frac{x}{4+x^2}\,dx$. *(2 marks)*

(Q17, **2020 HSC**) Easy

5 **a** Differentiate $e^{2x}(2x+1)$. *(2 marks)* 

b Hence, find $\displaystyle\int (x+1)e^{2x}\,dx$. *(1 mark)* Medium

(Q18, **2020 HSC**)

6 Which of these is the antiderivative of $\dfrac{3x^3-2x^2}{x}$?

A x^4-x^3+C **B** x^3-x^2+C

C $\dfrac{x^4}{4}-\dfrac{x^3}{3}+C$ **D** $\dfrac{x^2-x}{2}+C$ *(1 mark)*

Bonus question (see page iv) Easy

7 Which of these is the anti-derivative of $2^x \ln 4$?

A $\dfrac{2^x}{2\ln 2}+c$ **B** $2^{2x}+c$

C $2^{x+1}+c$ **D** $\dfrac{2^x}{3\ln 2}+c$ *(1 mark)*

Bonus question Medium

8 Which expression is equal to $\displaystyle\int \tan^2 x\,dx$?

A $\tan x - x + C$ **B** $\tan x - 1 + C$

C $\dfrac{\tan^3 x^2}{6}+C$ **D** $\dfrac{\tan^3 x}{3}+C$ *(1 mark)*

(Q9, **2019 HSC**) Hard

9 **i** Differentiate $(\ln x)^2$. *(2 marks)* Easy

ii Hence, or otherwise, find

$\displaystyle\int \frac{\ln x}{x}\,dx$ *(1 mark)* Medium

(Q13c, **2019 HSC**)

10 A particle is moving along a straight line. The particle is initially at rest. The acceleration of the particle at time t seconds is given by $a=e^{2t}-4$, where $t \ge 0$.

Find an expression, in terms of t, for the velocity of the particle. *(2 marks)*

(Q14a, **2019 HSC**) Easy

11 The derivative of a function $y=f'(x)$ is given by $f'(x)=3x^2+2x-1$.

i Find the x-values of the two stationary points of $y=f(x)$, and determine the nature of the stationary points. *(2 marks)* Easy

ii The curve passes through the point $(0,4)$. Find an expression for $f(x)$. *(2 marks)* Easy

iii Hence sketch the curve, clearly indicating the stationary points. *(2 marks)* Easy

iv For what values of x is the curve concave down? *(1 mark)* Medium

(Q14b, **2019 HSC**)

12 The graph of $y=f'(x)$ is shown.

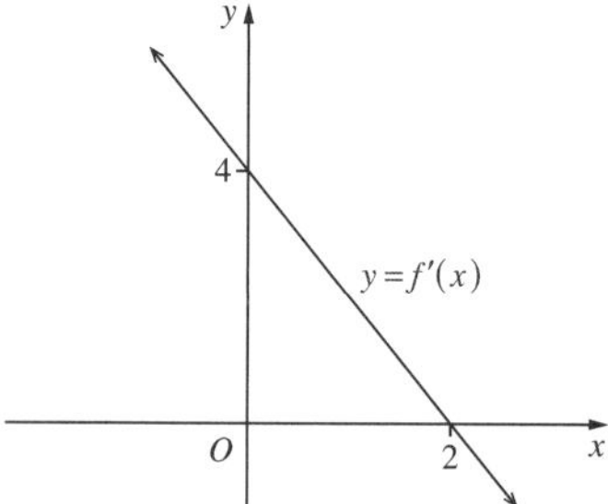

The curve $y=f(x)$ has a maximum value of 12.

What is the equation of the curve $y=f(x)$?

A $y=x^2-4x+12$ **B** $y=4+4x-x^2$

C $y=8+4x-x^2$ **D** $y=x^2-4x+16$

(1 mark)

(Q9, **2017 HSC**) Medium

13 Find $\displaystyle\int (2x+1)^4\,dx$. *(1 mark)*

(Q11b, **2017 HSC**) Easy

14 The rate at which water flows into a tank is given by

$$\frac{dV}{dt} = \frac{2t}{1+t^2},$$

where V is the volume of water in the tank in litres and t is the time in seconds.

Initially the tank is empty.

Find the exact amount of water in the tank after 10 seconds. *(3 marks)*

(Q13d, **2017 HSC**) 

---

15 Two particles move along the x-axis.

When $t = 0$, particle P_1 is at the origin and moving with velocity 3.

For $t \geq 0$, particle P_1 has acceleration given by $a_1 = 6t + e^{-t}$.

i Show that the velocity of particle P_1 is given by $v_1 = 3t^2 + 4 - e^{-t}$. *(2 marks)* Easy

When $t = 0$, particle P_2 is also at the origin.

For $t \geq 0$, particle P_2 has velocity given by $v_2 = 6t + 1 - e^{-t}$.

ii When do the two particles have the same velocity? *(2 marks)* Easy

iii Show that the two particles do not meet for $t > 0$. *(3 marks)* Hard

(Q15c, **2017 HSC**)

---

16 A particle is moving along the x-axis. The graph shows its velocity v metres per second at time t seconds.

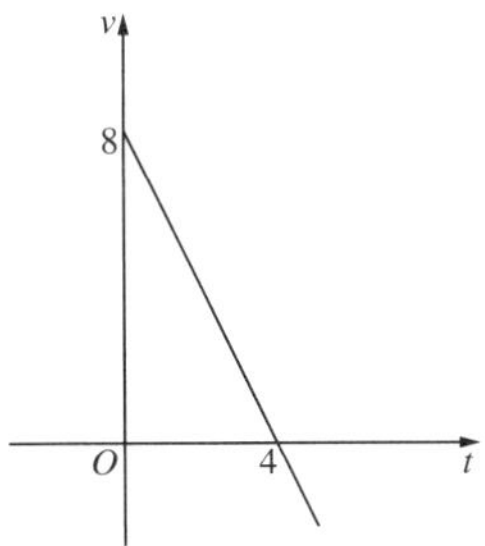

When $t = 0$ the displacement x is equal to 2 metres. What is the maximum value of the displacement x?

A 8 m
B 14 m
C 16 m
D 18 m *(1 mark)*

(Q9, **2015 HSC**) 

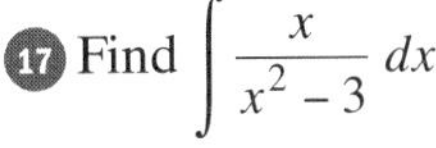

17 Find $\displaystyle\int \frac{x}{x^2-3}\,dx$. *(2 marks)*

(Q11h, **2015 HSC**) Easy

---

18 In a theme park ride, a chair is released from a height of 110 metres and falls vertically. Magnetic brakes are applied when the velocity of the chair reaches −37 metres per second.

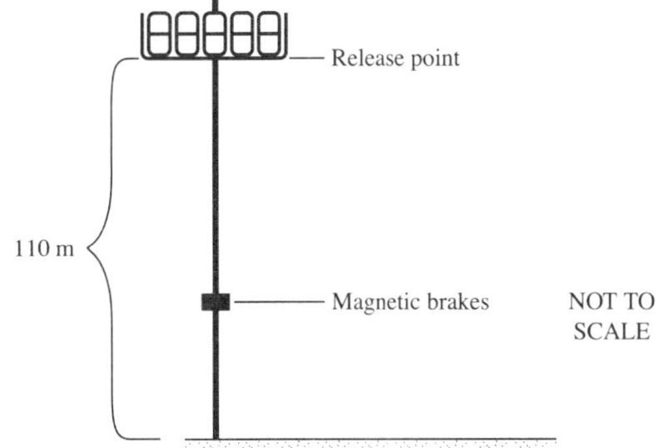

The height of the chair at time t seconds is x metres. The acceleration of the chair is given by $\ddot{x} = -10$. At the release point, $t = 0$, $x = 110$ and $\dot{x} = 0$.

i Using calculus, show that $x = -5t^2 + 110$. *(2 marks)* Easy

ii How far has the chair fallen when the magnetic brakes are applied? *(2 marks)* Medium

(Q14a, **2015 HSC**)

---

19 Water is flowing in and out of a rock pool. The volume of water in the pool at time t hours is V litres. The rate of change of the volume is given by

$$\frac{dV}{dt} = 80\sin(0.5t).$$

At time $t = 0$, the volume of water in the pool is 1200 litres and is increasing.

i After what time does the volume of water first start to decrease? *(2 marks)* Hard

ii Find the volume of water in the pool when $t = 3$. *(2 marks)* Medium

iii What is the greatest volume of water in the pool? *(1 mark)* Hard

(Q15c, **2015 HSC**)

---

20 Which expression is equal to $\displaystyle\int e^{2x}\,dx$?

A $e^{2x} + c$
B $2e^{2x} + c$
C $\dfrac{e^{2x}}{2} + c$
D $\dfrac{e^{2x+1}}{2x+1} + c$ *(1 mark)*

(Q4, **2014 HSC**) 

---

21 Find $\displaystyle\int \frac{1}{(x+3)^2}\,dx$. *(2 marks)*

(Q11d, **2014 HSC**) Easy

---

22 The gradient function of a curve $y = f(x)$ is given by $f'(x) = 4x - 5$. The curve passes through the point $(2, 3)$.

Find the equation of the curve. *(2 marks)*

(Q11f, **2014 HSC**) Easy

23 i Differentiate $3 + \sin 2x$. *(1 mark)* Easy

ii Hence, or otherwise, find $\int \frac{\cos 2x}{3 + \sin 2x}\, dx$. *(2 marks)* Medium

(Q13a, **2014 HSC**)

24 Find $\int e^{4x+1}\, dx$. *(2 marks)*

(Q11e, **2013 HSC**) Easy

25 The velocity of a particle moving along the x-axis is given by $\dot{x} = 10 - 2t$, where x is the displacement from the origin in metres and t is the time in seconds. Initially the particle is 5 metres to the right of the origin.

i Show that the acceleration of the particle is constant. *(1 mark)* Easy

ii Find the time when the particle is at rest. *(1 mark)* Easy

iii Show that the position of the particle after 7 seconds is 26 metres to the right of the origin. *(2 marks)* Easy

iv Find the distance travelled by the particle during the first 7 seconds. *(2 marks)* Easy

(Q14a, **2013 HSC**)

26 The derivative of a function $f(x)$ is $f'(x) = 4x - 3$. The line $y = 5x - 7$ is tangent to the graph of $f(x)$.

Find the function $f(x)$. *(3 marks)*

(Q16a, **2013 HSC**) Medium

27 Find $\int \frac{4x}{x^2 + 6}\, dx$. *(2 marks)*

(Q12b, **2012 HSC**) Easy

28 The velocity of a particle is given by

$$\dot{x} = 1 - 2\cos t,$$

where x is the displacement in metres and t is the time in seconds. Initially the particle is 3 m to the right of the origin.

i Find the initial velocity of the particle. *(1 mark)* Easy

ii Find the maximum velocity of the particle. *(1 mark)* Hard

iii Find the displacement, x, of the particle in terms of t. *(2 marks)* Medium

iv Find the position of the particle when it is at rest for the first time. *(2 marks)* Medium

(Q15b, **2012 HSC**)

29 Find $\int \frac{1}{3x^2}\, dx$. *(2 marks)*

(Q2e, **2011 HSC**) Medium

30 The gradient of a curve is given by $\frac{dy}{dx} = 6x - 2$. The curve passes through the point $(-1, 4)$.

What is the equation of the curve? *(2 marks)*

(Q4c, **2011 HSC**) Medium

31 i Differentiate $y = \sqrt{9 - x^2}$ with respect to x. *(2 marks)* Easy

ii Hence, or otherwise, find $\int \frac{6x}{\sqrt{9 - x^2}}\, dx$. *(2 marks)* Medium

(Q4d, **2011 HSC**)

32 A tap releases liquid A into a tank at the rate of $\left(2 + \frac{t^2}{t + 1}\right)$ litres per minute, where t is time in minutes. A second tap releases liquid B into the same tank at the rate of $\left(1 + \frac{1}{t + 1}\right)$ litres per minute. The taps are opened at the same time and release the liquids into an empty tank.

i Show that the rate of flow of liquid A is greater than the rate of flow of liquid B by t litres per minute. *(1 mark)* Hard

ii The taps are closed after 4 minutes. By how many litres is the volume of liquid A greater than the volume of liquid B in the tank when the taps are closed? *(2 marks)* Hard

(Q9b, **2011 HSC**)

33 i Find $\int \sqrt{5x + 1}\, dx$. *(2 marks)* Medium

ii Find $\int \frac{x}{4 + x^2}\, dx$. *(2 marks)* Medium

(Q2d, **2010 HSC**)

34 The acceleration of a particle is given by

$$\ddot{x} = 4\cos 2t$$

where x is displacement in metres and t is time in seconds.

Initially the particle is at the origin with a velocity of 1 ms^{-1}.

i Show that the velocity of the particle is given by $\dot{x} = 2\sin 2t + 1$. *(2 marks)* Easy

ii Find the time when the particle first comes to rest. *(2 marks)* Medium

iii Find the displacement, x, of the particle in terms of t. *(2 marks)* Medium

(Q7a, **2010 HSC**)

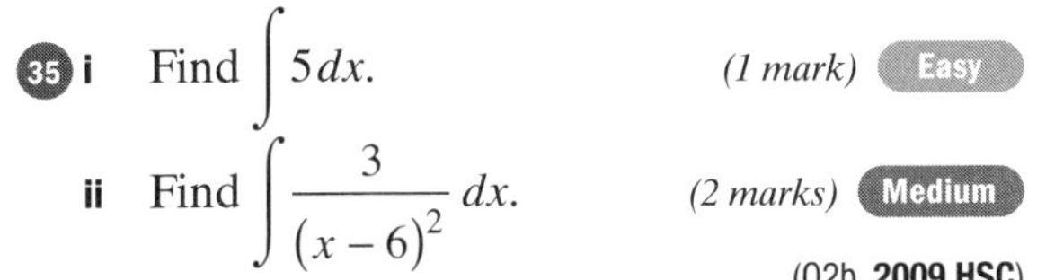

35 i Find $\int 5\,dx$. *(1 mark)* Easy

ii Find $\int \frac{3}{(x-6)^2}\,dx$. *(2 marks)* Medium

(Q2b, **2009 HSC**)

36 The acceleration of a particle is given by $\ddot{x} = 8e^{-2t} + 3e^{-t}$, where x is displacement in metres and t is time in seconds.

Initially its velocity is -6 ms^{-1} and its displacement is 5 m.

i Show that the displacement of the particle is given by $x = 2e^{-2t} + 3e^{-t} + t$. *(2 marks)* Medium

ii Find the time when the particle comes to rest. *(3 marks)* Hard

iii Find the displacement when the particle comes to rest. *(1 mark)* Hard

(Q7a, **2009 HSC**)

37 Find $\int \frac{dx}{x+5}$. *(1 mark)* Easy

(Q2c i, **2008 HSC**)

38 The gradient of a curve is given by $\frac{dy}{dx} = 1 - 6\sin 3x$. The curve passes through the point $(0, 7)$.

What is the equation of the curve? *(3 marks)*

(Q5a, **2008 HSC**) Medium

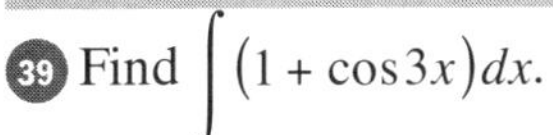

39 Find $\int (1 + \cos 3x)\,dx$. *(2 marks)*

(Q2b i, **2007 HSC**) Easy

40 A particle is moving on the x-axis and is initially at the origin. Its velocity, v metres per second, at time t seconds is given by

$$v = \frac{2t}{16 + t^2}.$$

i What is the initial velocity of the particle? *(1 mark)* Easy

ii Find an expression for the acceleration of the particle. *(2 marks)* Medium

iii Find the time when the acceleration of the particle is zero. *(1 mark)* Medium

iv Find the position of the particle when $t = 4$. *(3 marks)* Hard

(Q5b, **2007 HSC**)

41 Find $\int 1 + e^{7x}\,dx$. *(2 marks)*

(Q2b i, **2006 HSC**) Easy

42 During a storm, water flows into a 7000-litre tank at a rate of $\frac{dV}{dt}$ litres per minute,where $\frac{dV}{dt} = 120 + 26t - t^2$ and t is the time in minutes since the storm began.

i At what times is the tank filling at twice the initial rate? *(2 marks)* Hard

ii Find the volume of water that has flowed into the tank since the start of the storm as a function of t. *(1 mark)* Easy

iii Initially, the tank contains 1500 litres of water. When the storm finishes, 30 minutes after it began, the tank is overflowing. How many litres of water have been lost? *(2 marks)* Hard

(Q9b, **2006 HSC**)

43 Find a primitive of $4 + \sec^2 x$. *(2 marks)*

(Q1c, **2005 HSC**) Easy

44 Find $\int \frac{6x^2}{x^3 + 1}\,dx$. *(2 marks)*

(Q2c i, **2005 HSC**) Easy

45 Find $\int \frac{x}{x^2 - 3}\,dx$. *(2 marks)*

(Q3b ii, **2004 HSC**) Easy

46 Find a primitive of $3x^2 - 8$. *(2 marks)*

(Q1e, **2003 HSC**) Easy

47 Find $\int \frac{2x}{x^2 + 5}\,dx$. *(1 mark)*

(Q3d i, **2003 HSC**) Easy

48 Integrate $\frac{3}{x}$. *(1 mark)*

(Q1d, **2002 HSC**) Easy

49 Find $\int \cos 3x\,dx$. *(1 mark)*

(Q2d i, **2002 HSC**) Easy

50 The gradient function of a curve is given by

$$f'(x) = 3(x+1)(x-3)$$

and the curve $y = f(x)$ passes through the point $(0, 12)$.

i Find the equation of the curve $y = f(x)$. *(2 marks)* Easy

ii Sketch the curve $y = f(x)$, clearly labelling turning points and the y intercept. *(3 marks)* Hard

iii For what values of x is the curve concave up? *(1 mark)* Hard

(Q6b, **2002 HSC**)

51

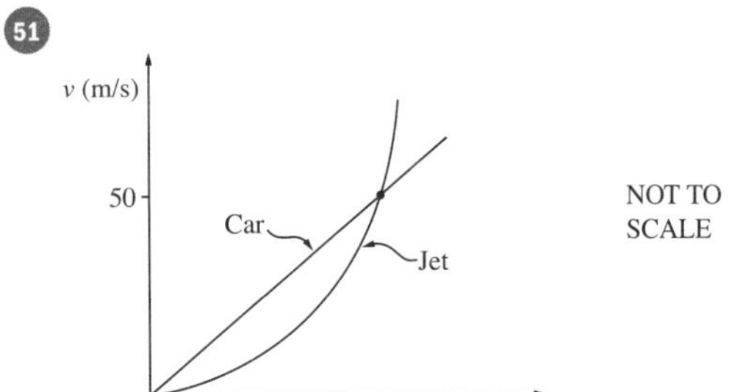

A car and a jet race one another from rest down a runway. The car increases its speed v_1 at a constant rate, while the speed of the jet is given by $v_2 = 2t^2$. After 5 seconds the car and the jet have the same speed of 50 m/s, as shown on the graph.

- **i** Find an equation for the speed v_1 of the car in terms of t. *(1 mark)* Easy
- **ii** How far behind the car is the jet after 5 seconds? *(2 marks)* Easy
- **iii** After how many seconds does the jet catch up with the car? *(1 mark)* Easy

(Q9c, **2002 HSC**) Easy

52 Find a primitive of $3 + \dfrac{1}{x}$. *(2 marks)*

(Q1d, **2001 HSC**) Easy

53 When a valve is released, a chemical flows into a large tank that is initially empty. The volume, V litres, of chemical in the tank increases at the rate

$$\frac{dV}{dt} = 2e^t + 2e^{-t}$$

where t is measured in hours from the time the valve is released.

- **i** At what rate does the chemical initially enter the tank? *(1 mark)* Easy
- **ii** Use integration to find an expression for V in terms of t. *(2 marks)* Medium
- **iii** Show that $2e^{2t} - 3e^t - 2 = 0$ when $V = 3$. *(1 mark)* Medium
- **iv** Find t, to the nearest minute, when $V = 3$. *(2 marks)* Hard

(Q9b, **2001 HSC**)

54 Find $\displaystyle\int \sec^2 5x\, dx$.

(Q3c i, **2000 HSC**) Easy

55

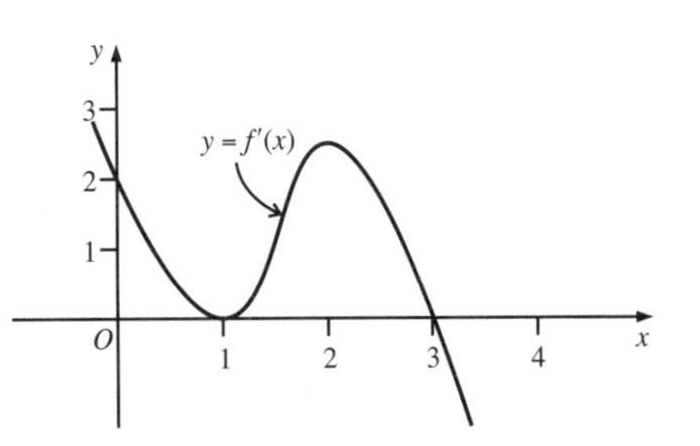

The above diagram shows a sketch of the gradient function of the curve $y = f(x)$.

In your Writing Booklet, draw a sketch of the function $y = f(x)$ given that $f(0) = 0$. *(2 marks)*

(Q9b, **2000 HSC**) Medium

56 The first snow of the season begins to fall during the night. The depth of the snow, h, increases at a constant rate through the night and the following day. At 6 am a snow plough begins to clear the road of snow. The speed, v km/h, of the snow plough is inversely proportional to the depth of snow. (This means $v = \dfrac{A}{h}$ where A is a constant.)

Let x km be the distance the snow plough has cleared and let t be the time in hours from the beginning of the snowfall. Let $t = T$ correspond to 6 am.

- **i** Explain carefully why, for $t \geq T$, $\dfrac{dx}{dt} = \dfrac{k}{t}$, where k is a constant.
- **ii** In the period from 6 am to 8 am the snow plough clears 1 km of road, but it takes a further 3·5 hours to clear the next kilometre.
  At what time did it begin snowing? *(6 marks)*

(Q10b, **2000 HSC**) Hard

57 Find:

- **i** $\displaystyle\int \left(\frac{1}{x^2} + \frac{1}{x}\right) dx$ Easy
- **ii** $\displaystyle\int \cos(2x + 1)\, dx.$ *(4 marks)* Easy

(Q2a, **1999 HSC**)

58 Find a primitive of $x^2 + 7$. *(2 marks)*

(Q1d, **1998 HSC**) Easy

59 Find $\displaystyle\int \frac{x}{x^2 + 3}\, dx.$ *(2 marks)*

(Q2c, **1998 HSC**) Easy

60 The graph of $y = f(x)$ passes through the point $(1, 4)$ and $f'(x) = 2x + 7$. Find $f(x)$. *(2 marks)*

(Q2b, **1997 HSC**) Easy

61 Find a primitive of $\dfrac{2x}{x^2 + 1}$. *(1 mark)*

(Q2c, **1997 HSC**) Easy

62 A pen moves along the x axis, ruling a line. The diagram shows the graph of the velocity of the tip of the pen as a function of time.

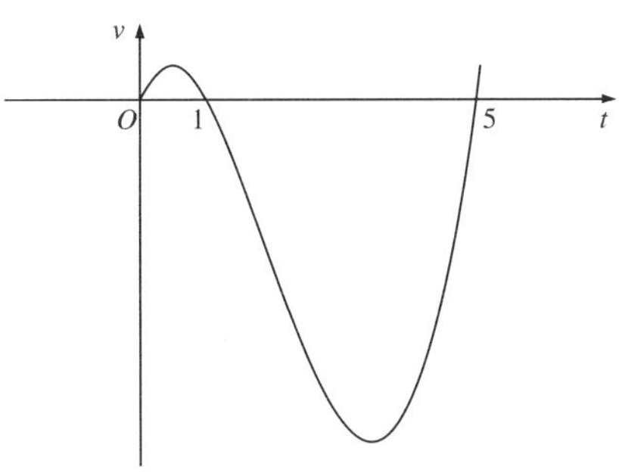

The velocity, in centimetres per second, is given by the equation

$$v = 4t^3 - 24t^2 + 20t,$$

where t is the time in seconds. When $t = 0$, the tip of the pen is at $x = 3$. That is, the tip is initially 3 centimetres to the right of the origin.

- **i** Find an expression for x, the position of the tip of the pen, as a function of time.
- **ii** What feature will the graph of x as a function of t have at $t = 1$?
- **iii** The pen uses 0·05 milligrams of ink per centimetre travelled. How much ink is used between $t = 0$ and $t = 2$?

(7 marks) (Q9b, **1997 HSC**) Medium

63 Find a primitive function of $6 - x^{-3}$. *(2 marks)*

(Q1e, **1996 HSC**) Easy

64 Find $\int \sec^2 6x \, dx$. *(1 mark)*

(Q3c, **1996 HSC**) Easy

65 The graph of $y = f(x)$ passes through the point $(1, 3)$, and $f'(x) = 3x^2 - 2$. Find $f(x)$. *(2 marks)*

(Q5a, **1996 HSC**) Easy

66 Find a primitive function of $2x + 11$. *(2 marks)*

(Q1e, **1995 HSC**) Easy

67 Find $\int e^{3x} \, dx$.

(Q3c i, **1995 HSC**) Easy

68 Find $\int e^{2x} \, dx$.

(Q3c i, **1994 HSC**) Easy

69 Find $\int (x - 1)^3 \, dx$.

(Q3c i, **1993 HSC**) Easy

70 Find:

- **i** $\int (2x + 3)^{10} \, dx$ Easy
- **ii** $\int \sin \frac{x}{2} \, dx$. Medium

(Q3b, **1992 HSC**)

Year 12 The antiderivative—Worked answers

1 $\int \frac{1}{(2x+1)^2}dx = \int (2x+1)^{-2}\,dx$

$= \frac{(2x+1)^{-1}}{-1\times 2} + C$

$= \frac{-1}{2(2x+1)} + C$

Answer B

2 **a** $y = (x^2+1)^4$

Using the chain rule:

$\frac{dy}{dx} = \frac{dy}{du}\times\frac{du}{dx}$

$\frac{dy}{dx} = 4(x^2+1)^3.2x$ ✓

$= 8x(x^2+1)^3$ ✓

(2 marks)

b From part **a**,

$\int 8x(x^2+1)^3\,dx$

$= (x^2+1)^4 + c$

$\therefore \int x(x^2+1)^3\,dx$

$= \frac{1}{8}(x^2+1)^4 + C$ ✓

(1 mark)

3 $\int (e+e^{3x})\,dx = ex + \frac{1}{3}e^{3x} + c$

Answer B

4 $\int \frac{x}{4+x^2}\,dx = \frac{1}{2}\int \frac{2x}{4+x^2}\,dx$ ✓

$= \frac{1}{2}\ln(4+x^2) + c$ ✓

(2 marks)

5 **a** Using the product rule,

Let $u = e^{2x}$ $\quad \frac{du}{dx} = 2e^{2x}$

Let $v = 2x+1$ $\quad \frac{dv}{dx} = 2$

$\frac{d}{dx}[e^{2x}(2x+1)]$

$= u\frac{dv}{dx} + v\frac{du}{dx}$

$= e^{2x}.2 + (2x+1).2e^{2x}$ ✓

$= 2e^{2x} + 2e^{2x}(2x+1)$

$= 2e^{2x} + 4xe^{2x} + 2e^{2x}$

$= 4xe^{2x} + 4e^{2x}$

$= 4e^{2x}(x+1)$ ✓

(2 marks)

b From part **a**,

$\int 4e^{2x}(x+1)\,dx = e^{2x}(2x+1)$

$\int (x+1)e^{2x}\,dx$

$= \frac{1}{4}e^{2x}(2x+1) + c$ ✓ *(1 mark)*

6 $\frac{3x^3-2x^2}{x} = 3x^2 - 2x$

$\int (3x-2x)dx = x^3 - x^2 + c$

Answer B

7 Using $\int a^x dx = \frac{a^x}{\ln a} + c,$

$\int 2^x \ln 4\,dx = \ln 4\int 2^x dx$

$= 2\ln 2 \times \frac{2^x}{\ln 2} + c$

$= 2\times 2^x + c$

$= 2^{x+1} + c$

Answer C

8 As $\sec^2 x = 1 + \tan^2 x$,

$\int \tan^2 x\,dx = \int(\sec^2 x - 1)\,dx$

$= \tan x - x + C$

Answer A

9 **i** $\frac{d}{dx}[(\ln x)^2] = 2\ln x \times \frac{1}{x}$ ✓

$= \frac{2\ln x}{x}$ ✓

(2 marks)

ii From **i**,

$\int \frac{2\ln x}{x}\,dx = (\ln x)^2 + c$

$\therefore \int \frac{\ln x}{x}\,dx = \frac{1}{2}(\ln x)^2 + c$ ✓

(1 mark)

10 $a = e^2t - 4$

$v = \int (e^{2t} - 4)\,dt$

$= \frac{1}{2}e^{2t} - 4t + c$ ✓

Substitute $v = 0, t = 0$:

$0 = \frac{1}{2}e^{2(0)} - 4(0) + c$

$0 = \frac{1}{2} + c$

$c = -\frac{1}{2}$

$\therefore v = \frac{1}{2}e^{2t} - 4t - \frac{1}{2}$ ✓

(2 marks)

11 **i** $f'(x) = \quad 3x^2 + 2x - 1 = 0$

$(3x-1)(x+1) = 0$

$x = \frac{1}{3}, -1$ ✓

$f''(x) = 6x + 2$

$f''\left(\frac{1}{3}\right) = 6\left(\frac{1}{3}\right) + 2 > 0,$

hence min.

$f''(-1) = 6(-1) + 2 < 0,$

hence max.

$\therefore$ minimum at $x = \frac{1}{3}$, and

maximum at $x = -1$ ✓ *(2 marks)*

ii As $f'(x) = 3x^2 + 2x - 1$,

$f(x) = x^3 + x^2 - x + C$ ✓

Substitute (0, 4):

$4 = (0)^3 + (0)^2 - 0 + C$

$C = 4$

$\therefore f(x) = x^3 + x^2 - x + 4$ ✓

(2 marks)

iii

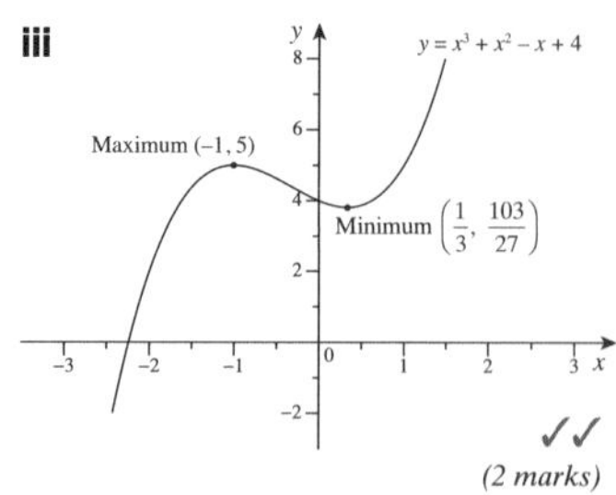

✓✓

(2 marks)

iv $f''(x) = 6x + 2 < 0$

$6x < -2$

$x < -\frac{1}{3}$ ✓

(1 mark)

12 Using gradient-intercept form, the line is $y' = 4 - 2x$.

$\therefore y = 4x - x^2 + c$

Now, maximum of $y = 12$ when $y' = 0$, i.e. when $x = 2$:

$\therefore 12 = 4(2) - (2)^2 + c$

$12 = 8 - 4 + c$

$c = 8$

$\therefore y = 4x - x^2 + 8$

i.e. $y = 8 + 4x - x^2$

Answer C

(1 mark)

13 $\int (2x+1)^4\,dx$

$= \frac{(2x+1)^5}{5\times 2} + c$

$= \frac{(2x+1)^5}{10} + c$ ✓

(1 mark)

14 $\frac{dV}{dt} = \frac{2t}{1+t^2}$

$V = \ln(1+t^2) + c$ ✓

Subs. $t = 0, V = 0$:

$0 = \ln(1+0^2) + c$

$c = 0$

$\therefore V = \ln(1+t^2)$ ✓

$V(10) = \ln(1+10^2)$

$= \ln(101)$

$\therefore$ the volume is ln (101) litres. ✓

(3 marks)

15 **i** $a_1 = 6t + e^{-t}$

$v_1 = 3t^2 - e^{-t} + c$ ✓

Subs. $t = 0, v_1 = 3$:

$3 = 3(0)^2 - e^0 + c$

$3 = 0 - 1 + c$

$c = 4$

$v_1 = 3t^2 + 4 - e^{-t}$ ✓

(2 marks)

ii Let $v_1 = v_2$

$3t^2 + 4 - e^{-t} = 6t + 1 - e^{-t}$

$3t^2 - 6t + 3 = 0$ ✓

$t^2 - 2t + 1 = 0$

$(t-1)^2 = 0$

$t = 1$

$\therefore$ after 1 second. ✓

(2 marks)

iii $v_1 = 3t^2 + 4 - e^{-t}$
$\therefore x_1 = t^3 + 4t + e^{-t} + k_1$
Subs. $t = 0, x_1 = 0$:
$0 = (0)^3 + 4(0) + e^0 + k_1$
$0 = 1 + k_1$
$k_1 = -1$
$x_1 = t^3 + 4t + e^{-t} - 1$ ✓
Also $v_2 = 6t + 1 - e^{-t}$
$\therefore x_2 = 3t^2 + t + e^{-t} + k_2$
Subs. $t = 0, x_2 = 0$:
$0 = 3(0)^2 + 0 + e^0 + k_2$
$0 = 1 + k_2$
$k_2 = -1$
$x_2 = 3t^2 + t + e^{-t} - 1$ ✓
Let $x_1 = x_2$:
$t^3 + 4t + e^{-t} - 1 = 3t^2 + t + e^{-t} - 1$
$t^3 - 3t^2 + 3t = 0$
$t(t^2 - 3t + 3) = 0$
For $t^2 - 3t + 3$,
$\Delta = 9 - 4(1)(3) < 0$, and
$a = 1 > 0$.
Hence $t^2 - 3t + 3 > 0$ for all t, and so particles do not meet for $t > 0$. ✓
(3 marks)

16 Maximum displacement when $v = 0$.
Distance travelled
= area of triangle
$= \frac{1}{2} \times 4 \times 8$
$= 16$
Displacement $= 2 + 16$
$= 18$
$\therefore$ the maximum displacement is 18 m.
Alternative method:
The graph has gradient of -2 and v-intercept of 8.
$\therefore v = -2t + 8$.
Now, $x = -t^2 + 8t + c$
Subs, $t = 0, x = 2$:
$2 = -(0)^2 + 8(0) + c$
$\therefore c = 2$
$x = -t^2 + 8t + 2$
Subs. $t = 4$:
$x = -(4)^2 + 8(4) + 2$
$= 18$
$\therefore$ the maximum displacement is 18 m.
Answer D

17 $\int \frac{x}{x^2 - 3}\,dx = \frac{1}{2}\int \frac{2x}{x^2 - 3}\,dx$ ✓
$= \frac{1}{2}\log_e(x^2 - 3) + c$ ✓
(2 marks)

18 **i** $\ddot{x} = -10$
$\dot{x} = -10t + c$
Substitute $\dot{x} = 0$ and $t = 0$:
$0 = -10(0) + c$
$c = 0$
$\therefore \dot{x} = -10t$ ✓
$x = -5t^2 + k$
Substitute $x = 110$ and $t = 0$:
$110 = -5(0)^2 + k$
$k = 110$
$\therefore x = -5t^2 + 110$ ✓
(2 marks)

ii Substitute $\dot{x} = -37$ in $\dot{x} = -10t$:
$-37 = -10t$
$\therefore t = 3.7$
Substitute $t = 3.7$ in $x = -5t^2 + 110$:
$x = -5(3.7)^2 + 110$
$= 41.55$ ✓
$\therefore$ the brakes are applied at a height of 41.55 m.
As $110 - 41.55 = 68.45$, the chair has fallen 68.45 m. ✓
(2 marks)

19 **i** 'First starts to decrease' when $\frac{dV}{dt} = 0$:
$80 \sin(0.5t) = 0$
$\sin(0.5t) = 0$
$0.5t = 0, \pi, 2\pi, \ldots$
$t = 0, 2\pi, 4\pi, \ldots$ ✓
As $t > 0$, then $t = 2\pi, 4\pi, \ldots$
$\frac{d^2V}{dt^2} = 40\cos(0.5t)$
$\frac{d^2V}{dt^2}(2\pi) = 40\cos(0.5(2\pi))$
< 0
$\therefore$ maximum volume when $t = 2\pi$.
$\therefore$ the volume starts to decrease after 2π hours. ✓
(2 marks)

ii $\frac{dV}{dt} = 80\sin(0.5t)$
$V = -160\cos(0.5t) + c$
Substitute $V = 1200$ and $t = 0$:
$1200 = -160\cos(0.5(0)) + c$
$1200 = -160 + c$
$c = 1360$
$\therefore V = 1360 - 160\cos(0.5t)$ ✓
Substitute $t = 3$:
$V = 1360 - 160\cos(0.5(3))$
$= 1348.682\,048\ldots$
$= 1349$ (nearest whole)
$\therefore$ the volume is 1349 L. ✓
(2 marks)

iii Greatest volume when $t = 2\pi$ (from part **i**).
$V = 1360 - 160\cos(0.5(2\pi))$
$= 1360 - 160\cos\pi$
$= 1360 + 160$
$= 1520$
$\therefore$ the greatest volume is 1520 L. ✓
(1 mark)

20 $\int e^{2x}\,dx = \frac{1}{2}e^{2x} + c$
$= \frac{e^{2x}}{2} + c$
Answer C

21 $\int \frac{1}{(x+3)^2}\,dx = \int (x+3)^{-2}\,dx$
$= \frac{(x+3)^{-1}}{-1} + c$ ✓
$= \frac{-1}{x+3} + c$ ✓
(2 marks)

22 As $f'(x) = 4x - 5$
$f(x) = 2x^2 - 5x + c$ ✓
Subs. (2, 3):
$3 = 2(2)^2 - 5(2) + c$
$3 = 8 - 10 + c$
$c = 5$
$\therefore f(x) = 2x^2 - 5x + 5$ ✓
(2 marks)

23 **i** $\frac{d}{dx}[3 + \sin 2x] = 2\cos 2x$ ✓
(1 mark)

ii $\int \frac{\cos 2x}{3 + \sin 2x}\,dx$
$= \frac{1}{2}\int \frac{2\cos 2x}{3 + \sin 2x}\,dx$ ✓
$= \frac{1}{2}\log_e(3 + \sin 2x) + c$ ✓
(using result in part **i**)
(2 marks)

24 $\int e^{4x+1}\,dx = \frac{1}{4}e^{4x+1} + c$ ✓✓
(2 marks)

25 $\dot{x} = 10 - 2t$
i $\ddot{x} = -2$, which is constant. ✓
(1 mark)
ii $\dot{x} = 10 - 2t = 0$
$2t = 10$
$t = 5$
$\therefore$ at rest after 5 seconds ✓
(1 mark)
iii $\dot{x} = 10 - 2t$
$x = 10t - t^2 + c$
When $t = 0, x = 5$: ✓
$5 = 10(0) - (0)^2 + c$
$c = 5$
$\therefore x = 10t - t^2 + 5$ ✓
When $t = 7$:
$x = 10(7) - (7)^2 + 5$
$= 26$
$\therefore$ after 7 seconds, 26 metres to the right. ✓
(2 marks)

iv $x = 10t - t^2 + 5$

Subs. $t = 0, x = 5$

Subs. $t = 5$,

$x = 10(5) - (5)^2 + 5$

$= 30$

Subs. $t = 7, x = 26$ ✓

From $x = 5$ to $x = 30$ is 25 metres, then back to $x = 26$ is 4 metres. ✓

$\therefore$ particle travels 29 metres.

(2 marks)

26

$f'(x) = 4x - 3$

$f(x) = 2x^2 - 3x + c$ ✓

$2x^2 - 3x + c = 5x - 7$

$2x^2 - 8x + c + 7 = 0$

If tangent, then $\Delta = 0$:

$\Delta = (-8)^2 - 4(2)(c + 7) = 0$ ✓

$64 - 8(c + 7) = 0$

$64 - 8c - 56 = 0$

$8 - 8c = 0$

$8c = 8$

$c = 1$

$\therefore f(x) = 2x^2 - 3x + 1$ ✓

OR:

Gradient of tangent = 5

$\therefore 4x - 3 = 5$

$4x = 8$

$x = 2$

Subs. in $y = 5x - 7$:

$y = 5(2) - 7$

$= 3$

$\therefore$ tangent is at the point (2, 3)

As $f'(x) = 4x - 3$

$f(x) = 2x^2 - 3x + c$

Subs. (2, 3):

$3 = 2(2)^2 - 3(2) + c$

$3 = 8 - 6 + c$

$c = 1$

$\therefore f(x) = 2x^2 - 3x + 1$

(3 marks)

27 $\int \frac{4x}{x^2 + 6}\, dx = 2\int \frac{2x}{x^2 + 6}\, dx$ ✓

$= 2\log_e(x^2 + 6) + c$ ✓

(2 marks)

28 i $\dot{x} = 1 - 2\cos t$

Subs $t = 0$,

$\dot{x}(0) = 1 - 2\cos(0)$

$= 1 - 2$

$= -1$

$\therefore$ initial velocity is -1 ms^{-1} ✓

(1 mark)

ii As $-1 \leq \cos t \leq 1$, then

$-2 \leq 2\cos t \leq 2$

$\therefore 2 \geq -2\cos t \geq -2$

$\therefore 3 \geq 1 - 2\cos t \geq -1$

$\therefore$ maximum velocity is 3 ms^{-1} ✓

OR: Maximum velocity when $\ddot{x} = 0$

$\ddot{x} = 2\sin t = 0$

$t = 0, \pi, 2\pi, \ldots$

$\dot{x}(0) = -1$ from (i)

$\dot{x}(\pi) = 1 - 2\cos\pi$

$= 1 - 2(-1)$

$= 3$

$\therefore$ maximum velocity is 3 ms^{-1}

(1 mark)

iii $\dot{x} = 1 - 2\cos t$

$x = t - 2\sin t + c$ ✓

When $t = 0, x = 3$,

$3 = 0 - 2\sin 0 + c$

$\therefore c = 3$

$\therefore x = t - 2\sin t + 3$ ✓

(2 marks)

iv $\dot{x} = 1 - 2\cos t = 0$

$\cos t = \frac{1}{2}$

$t = \frac{\pi}{3}$ ✓

$x\left(\frac{\pi}{3}\right) = \frac{\pi}{3} - 2\sin\frac{\pi}{3} + 3$

$= \frac{\pi}{3} - 2\left(\frac{\sqrt{3}}{2}\right) + 3$

$= \frac{\pi}{3} - \sqrt{3} + 3$ ✓

(2 marks)

29 $\int \frac{1}{3x^2}\, dx = \int \frac{1}{3}x^{-2}\, dx$

$= \frac{1}{-3}x^{-1} + c$ ✓

$= -\frac{1}{3x} + c$ ✓

(2 marks)

30 $\frac{dy}{dx} = 6x - 2$

$y = 3x^2 - 2x + c$ ✓

Subs in $(-1, 4)$:

$4 = 3 \times (-1)^2 - 2 \times (-1) + c$

$4 = 3 + 2 + c$

$c = 4 - 5$

$= -1$

$\therefore$ the equation is $y = 3x^2 - 2x - 1$ ✓

(2 marks)

31 i $y = \sqrt{9 - x^2}$

$= (9 - x^2)^{\frac{1}{2}}$

Using the function of a function rule:

$\frac{dy}{dx} = \frac{1}{2}(9 - x^2)^{-\frac{1}{2}}.(-2x)$ ✓

$= \frac{-x}{\sqrt{9 - x^2}}$ ✓

(2 marks)

ii From **i**,

$\int \frac{-x}{\sqrt{9 - x^2}}\, dx$

$= \sqrt{9 - x^2} + c$ ✓

$\therefore \int \frac{6x}{\sqrt{9 - x^2}}\, dx$

$= -6\sqrt{9 - x^2} + k$ ✓

(2 marks)

32 i Difference

$= 2 + \frac{t^2}{t + 1} - \left(1 + \frac{1}{t + 1}\right)$

$= 1 + \frac{t^2}{t + 1} - \frac{1}{t + 1}$

$= 1 + \frac{t^2 - 1}{t + 1}$

$= 1 + \frac{(t - 1)(t + 1)}{t + 1}$

$= 1 + t - 1$

$= t$

$\therefore A$ is greater by t litres per minute. ✓

(1 mark)

ii Volume difference

$= \int_0^4 \left(2 + \frac{t^2}{t + 1}\right) - \left(1 + \frac{1}{t + 1}\right) dt$

$= \int_0^4 t\, dt$ (from **i**)

$= \left[\frac{t^2}{2}\right]_0^4$ ✓

$= 8 - 0$

$= 8$

$\therefore A$ has 8 L more than B. ✓

(2 marks)

33 i $\int \sqrt{5x + 1}\, dx$

$= \int (5x + 1)^{\frac{1}{2}}\, dx$

$= \frac{(5x + 1)^{\frac{3}{2}}}{\frac{3}{2}.5} + c$ ✓

$= \frac{2(5x + 1)^{\frac{3}{2}}}{15} + c$

$= \frac{2\sqrt{(5x + 1)^3}}{15} + c$ ✓

(2 marks)

ii $\int \frac{x}{4+x^2}\,dx$

$= \frac{1}{2}\int \frac{2x}{4+x^2}\,dx$ ✓

$= \frac{1}{2}\log_e(4+x^2) + c$ ✓

(2 marks)

34 $\ddot{x} = 4\cos 2t$

i $\dot{x} = \int 4\cos 2t\, dt$

$\therefore \dot{x} = 2\sin 2t + c$ ✓

Subs $t = 0, v = \dot{x} = 1$:

$1 = 2\sin 0 + c$

$\therefore c = 1$

$\therefore \dot{x} = 2\sin 2t + 1$ ✓

(2 marks)

ii Subs $v = \dot{x} = 0$:

$0 = 2\sin 2t + 1$

$2\sin 2t = -1$

$\sin 2t = -\frac{1}{2}$ ✓

$2t = \frac{7\pi}{6}$

$t = \frac{7\pi}{12}$

$\therefore$ The particle comes to rest after $\frac{7\pi}{12}$ seconds. ✓ *(2 marks)*

iii $x = \int 2\sin 2t + 1\, dt$

$\therefore x = -\cos 2t + t + c$ ✓

Subs $t = 0, x = 0$:

$0 = -\cos 0 + 0 + c$

$0 = -1 + c$

$c = 1$

$\therefore x = -\cos 2t + t + 1$

$\therefore$ The displacement is $x = 1 - \cos 2t + t.$ ✓

(2 marks)

35 **i** $\int 5\, dx = 5x + C$ ✓

(1 mark)

ii $\int \frac{3}{(x-6)^2}\,dx$

$= 3\int (x-6)^{-2}\,dx$

$= \frac{3(x-6)^{-1}}{-1} + C$ ✓

$= \frac{-3}{x-6} + C.$ ✓

(2 marks)

36 **i** $\ddot{x} = 8e^{-2t} + 3e^{-t}$

$\therefore \dot{x} = \int \left(8e^{-2t} + 3e^{-t}\right) dt$

$= -4e^{-2t} - 3e^{-t} + C$

When $t = 0, \dot{x} = -6$

$\therefore -6 = -4e^{-2(0)} - 3e^{-(0)} + C$

$-6 = -4 - 3 + C$

$\therefore C = 1$

$\therefore \dot{x} = -4e^{-2t} - 3e^{-t} + 1$ ✓

$\therefore x = \int \left(-4e^{-2t} - 3e^{-t} + 1\right) dt$

$= 2e^{-2t} + 3e^{-t} + t + K$

When $t = 0, x = 5$

$\therefore 5 = 2e^{-2(0)} + 3e^{0} + 0 + K$

$5 = 2 + 3 + K$

$5 = 5 + K$

$\therefore K = 0$

$\therefore x = 2e^{-2t} + 3e^{-t} + t$ ✓

(2 marks)

ii The particle comes to rest when $\dot{x} = 0$

$\therefore -4e^{-2t} - 3e^{-t} + 1 = 0$

$-4(e^{-t})^2 - 3e^{-t} + 1 = 0$ ✓

Let $m = e^{-t}$,

$\therefore -4m^2 - 3m + 1 = 0$

$4m^2 + 3m - 1 = 0$

$(m + 1)(4m - 1) = 0$

$\therefore m + 1 = 0$ or $4m - 1 = 0$

$m = -1$ $\quad m = \frac{1}{4}$

$\therefore e^{-t} = -1$ or $e^{-t} = \frac{1}{4}$ ✓

no solution

$\log_e e^{-t} = \log_e \frac{1}{4}$

$-t = \log_e \frac{1}{4}$

$-t = -\log_e 4$

$t = \log_e 4$

$\therefore$ Particle comes to rest when $t = \log_e 4$ seconds. ✓

(3 marks)

iii $x = 2e^{-2t} + 3e^{-t} + t$

When $t = \log_e 4$,

$x = 2e^{-2\log_e 4} + 3e^{-\log_e 4} + \log_e 4$

$= 2e^{\log_e 4^{-2}} + 3e^{\log_e 4^{-1}} + \log_e 4$

$= 2e^{\log_e\left(\frac{1}{16}\right)} + 3e^{\log_e\left(\frac{1}{4}\right)} + \log_e 4$

$= 2\left(\frac{1}{16}\right) + 3\left(\frac{1}{4}\right) + \log_e 4$

$= \frac{7}{8} + \log_e 4.$ ✓

(1 mark)

37 $\int \frac{dx}{x+5} = \log_e(x+5) + C$ ✓

(1 mark)

38 $\frac{dy}{dx} = 1 - 6\sin 3x$

$\therefore y = \int (1 - 6\sin 3x)dx$

$= x + 2\cos 3x + C$ ✓

Now, the curve passes through $(0, 7)$,

$\therefore 7 = 0 + 2\cos 3(0) + C$

$7 = 0 + 2\cos 0 + C$ ✓

$7 = 2 + C$

$\therefore C = 5$

$\therefore y = x + 2\cos 3x + 5.$ ✓

(3 marks)

39 $\int (1 + \cos 3x)dx$

$= x + \frac{1}{3}\sin 3x + C$ ✓✓

(2 marks)

40 **i** $v = \frac{2t}{16 + t^2}$

The initial velocity of the particles occurs when $t = 0$.

$\therefore v = \frac{2 \times 0}{16 + 0^2}$

$= 0$

$\therefore$ The initial velocity of the particle is 0 m/s. The particle is stationary. ✓

(1 mark)

ii $a = \frac{dv}{dt}$

$= \frac{d}{dt}\left(\frac{2t}{16 + t^2}\right)$

$= \frac{(16 + t^2)\cdot\frac{d}{dt}(2t) - (2t)\cdot\frac{d}{dt}(16 + t^2)}{(16 + t^2)^2}$ ✓

$= \frac{(16 + t^2)(2) - (2t)(2t)}{(16 + t^2)^2}$

$= \frac{32 + 2t^2 - 4t^2}{(16 + t^2)^2}$

$\therefore a = \frac{32 - 2t^2}{(16 + t^2)^2}.$ ✓

(2 marks)

iii $a = \frac{32 - 2t^2}{(16 + t^2)^2}$

When $a = 0$, $0 = \frac{32 - 2t^2}{(16 + t^2)^2}$

$0 = 32 - 2t^2$

$0 = 2(16 - t^2)$

$16 - t^2 = 0$

$(4 + t)(4 - t) = 0$

$\therefore 4 + t = 0$ or $4 - t = 0$

$t = -4$ $\quad t = 4$

Since $t = -4$ is not a possible solution, $t = 4$.

$\therefore$ The acceleration of the particle is zero after 4 seconds. ✓

(1 mark)

iv $v = \dfrac{2t}{16+t^2}$

$\therefore x = \displaystyle\int \frac{2t}{16+t^2}\,dt$

i.e. $x = \log_e\left(16+t^2\right) + C$ ✓

When $t = 0, x = 0$

$\therefore 0 = \log_e\left(16+(0)^2\right) + C$

$= \log_e 16 + C$

$\therefore C = -\log_e 16$

$\therefore x = \log_e\left(16+t^2\right) - \log_e 16$ ✓

When $t = 4$,

$x = \log_e\left(16+(4)^2\right) - \log_e 16$

$= \log_e 32 - \log_e 16$

$= \log_e\left(\dfrac{32}{16}\right)$

$= \log_e 2$

$= 0.6931\ldots$

$= 0.693$ to 3 decimal places

$\therefore$ The position of the particle when $t = 4$ is $\log_e 2$ metres. ✓

(3 marks)

41 $\displaystyle\int 1 + e^{7x}\,dx = x + \frac{1}{7}e^{7x} + C.$ ✓✓

(2 marks)

42 i $\dfrac{dV}{dt} = 120 + 26t - t^2$

At $t = 0$, $\dfrac{dV}{dt} = 120 + 26(0) - (0)^2$

$= 120$ ✓

When $\dfrac{dV}{dt} = 240$,

$240 = 120 + 26t - t^2$

$t^2 - 26t + 120 = 0$

$(t-20)(t-6) = 0$

$\therefore t = 6$ or 20

$\therefore$ The tank is filling at twice the initial rate at 6 min and 20 min. ✓

(2 marks)

ii $\dfrac{dV}{dt} = 120 + 26t - t^2$

$\therefore V = \displaystyle\int\left(120 + 26t - t^2\right)dt$

$= 120t + \dfrac{26}{2}t^2 - \dfrac{t^3}{3} + C$

$= 120t + 13t^2 - \dfrac{1}{3}t^3 + C$

If $t = 0, V = 0$ $\therefore C = 0$

$\therefore V = 120t + 13t^2 - \dfrac{1}{3}t^3.$ ✓

(1 mark)

iii When $t = 30$,

$V = 120(30) + 13(30)^2 - \dfrac{1}{3}(30)^3$

$= 3600 + 11\,700 - 9000$

$= 6300.$ ✓

$\therefore$ Total water in the tank and overflowing

$= 6300 + 1500$

$= 7800$ litres.

$\therefore$ Amount of water lost

$= 7800 - 7000$

$= 800$ litres. ✓

(2 marks)

43 $\displaystyle\int(4 + \sec^2 x)\,dx = 4x + \tan x + C$ ✓✓

(2 marks)

44 $\displaystyle\int\frac{6x^2}{x^3+1}\,dx = 2\int\frac{3x^2}{x^3+1}\,dx$ ✓

$= 2\ln(x^3+1) + C.$ ✓

(2 marks)

45 $\displaystyle\int\frac{x}{x^2-3}\,dx = \frac{1}{2}\int\frac{2x}{x^2-3}\,dx$ ✓

$= \dfrac{1}{2}\ln(x^2-3) + C.$ ✓

(2 marks)

46 $\displaystyle\int(3x^2-8)dx = \frac{3x^3}{3} - 8x + C$ ✓

$= x^3 - 8x + C.$ ✓

(2 marks)

47 $\displaystyle\int\frac{2x}{x^2+5}\,dx = \ln(x^2+5) + C$ ✓

(1 mark)

48 $\displaystyle\int\frac{3}{x}\,dx = 3\log_e x + C.$ ✓

(1 mark)

49 $\displaystyle\int\cos 3x\,dx = \frac{1}{3}\sin 3x + C.$ ✓

(1 mark)

50 i $f'(x) = 3(x+1)(x-3)$

$= 3(x^2 - 2x - 3)$

$= 3x^2 - 6x - 9$

$\therefore f(x) = \dfrac{3x^3}{3} - \dfrac{6x^2}{2} - 9x + C$

$= x^3 - 3x^2 - 9x + C$ ✓

The curve passes through $(0, 12)$,

$\therefore 12 = 0 - 0 - 0 + C$

$\therefore C = 12$

$\therefore f(x) = x^3 - 3x^2 - 9x + 12$ ✓

(2 marks)

ii Stationary points occur when

$f'(x) = 0$

$3(x+1)(x-3) = 0$

$\therefore x = 3$ or $-1.$ ✓

When $x = 3$,

$y = 3^3 - 3(3)^2 - 9(3) + 12$

$= 27 - 27 - 27 + 12$

$= -15$

When $x = -1$,

$y = (-1)^3 - 3(-1)^2 - 9(-1) + 12$

$= -1 - 3 + 9 + 12$

$= 17$

Stationary points are $(3, -15)$ and $(-1, 17)$

Now $f'(x) = 3x^2 - 6x - 9$

$\therefore f''(x) = 6x - 6$

Now $f''(3) = 6(3) - 6$

$= 12 > 0$

$\therefore$ A minimum turning point at $(3, -15)$

and $f''(-1) = 6(-1) - 6$

$= -12 < 0$

$\therefore$ A maximum turning point at $(-1, 17)$ ✓

y intercept occurs when $x = 0$, i.e.

$y = f(0) = 0^3 - 3(0)^2 - 9(0) + 12$

$= 12$

$\therefore$ y intercept is 12.

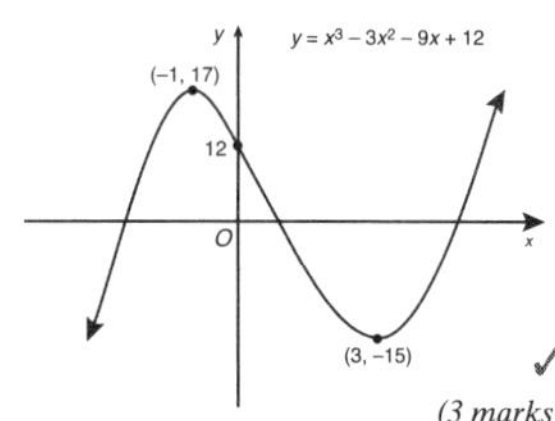

✓

(3 marks)

iii Curve is concave up when

$f''(x) > 0$

$6x - 6 > 0$

$6x > 6$

$x > 1.$ ✓

(1 mark)

51 i The equation of the line for the car is given by $v_1 = mt$ where m is the slope of the line.

Now $m = \dfrac{50}{5} = 10$

$\therefore v_1 = 10t$ ✓

(1 mark)

ii **Car:**

$v_1 = 10t$

i.e. $\dfrac{dx_1}{dt} = 10t$

$\therefore x_1 = 5t^2 + C$

At $t = 0$, $x_1 = 0$ $\therefore C = 0$

$\therefore x_1 = 5t^2$

At $t = 5,\ x_1 = 5 \times 5^2$
$= 125$ m ✓

Jet:

$v_2 = 2t^2$

i.e. $\dfrac{dx_2}{dt} = 2t^2$

$\therefore x_2 = \dfrac{2t^3}{3} + C_1$

At $t = 0,\ x_2 = 0 \quad \therefore C_1 = 0$

$\therefore x_2 = \dfrac{2t^3}{3}$

At $t = 5,\ x_2 = \dfrac{2(5)^3}{3}$

$= \dfrac{250}{3}$

$= 83\frac{1}{3}$ m.

At $t = 5,\ x_1 - x_2 = 125 - 83\frac{1}{3}$

$= 41\frac{2}{3}$ m ✓

The jet is $41\frac{2}{3}$ m behind the car.

(2 marks)

iii The jet catches up with the car when $x_1 = x_2$

$\therefore 5t^2 = \dfrac{2}{3}t^3$

$5t^2 - \dfrac{2}{3}t^3 = 0$

$t^2\left(5 - \dfrac{2}{3}t\right) = 0$

$\therefore t = 0$

or $5 - \dfrac{2}{3}t = 0$

$-\dfrac{2}{3}t = -5$

$t = -5 \times -\dfrac{3}{2}$

$= \dfrac{15}{2}$

$\therefore$ The jet catches up with car after $7\frac{1}{2}$ s. ✓

(1 mark)

52 $\displaystyle\int\left(3 + \frac{1}{x}\right)dx = 3x + \log_e x + C$ ✓✓

(2 marks)

53 i $\dfrac{dV}{dt} = 2e^t + 2e^{-t}$

When $t = 0,\ \dfrac{dV}{dt} = 2e^0 + 2e^0$

$= 2 + 2$

$= 4$

i.e. Rate = 4 litres/hour. ✓

(1 mark)

ii $V = 2e^t - 2e^{-t} + C$ ✓

When $t = 0, V = 0$

$\therefore 0 = 2e^0 - 2e^0 + C$

$0 = 2 - 2 + C$

$\therefore C = 0$

$\therefore V = 2e^t - 2e^{-t}$ ✓

(2 marks)

iii $3 = 2e^t - 2e^{-t}$

$3 = 2e^t - \dfrac{2}{e^t}$

$\therefore 3e^t = 2e^{2t} - 2$

$\therefore 2e^{2t} - 3e^t - 2 = 0$ ✓

(1 mark)

iv Let $m = e^t$

$\therefore 2m^2 - 3m - 2 = 0$

$(2m + 1)(m - 2) = 0$

$\therefore m = -\dfrac{1}{2}$ or 2

but $m = e^t$ and $e^t > 0$ for $t > 0$

$\therefore m = 2$

$\therefore e^t = 2$ ✓

$\therefore \ln e^t = \ln 2$

$t \ln e = \ln 2$

$\therefore t = \ln 2$

$= 0.6931\ldots$ hours

$= 41.58\ldots$ min

$= 42$ minutes ✓ to nearest minute

(2 marks)

54 $\displaystyle\int \sec^2 5x\,dx = \frac{1}{5}\tan 5x + C$

(No mark allocation in exam)

55

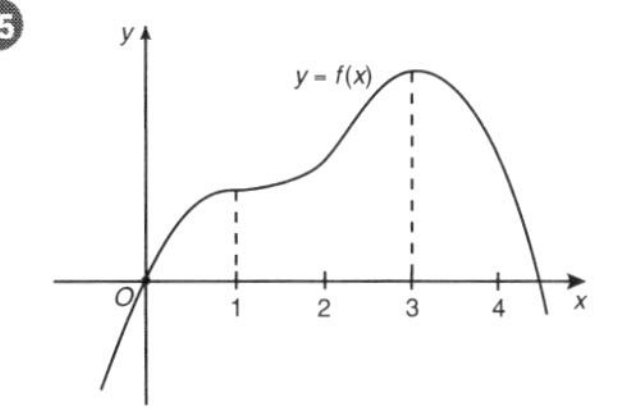

✓

Explanation:

$f'(1) = f'(3) = 0$

$\therefore$ Stationary points occur at $x = 1$ and $x = 3$.

Now $f'(1 - \varepsilon) > 0$
and $f'(1 + \varepsilon) > 0$

$\therefore$ a horizontal point of inflexion at $x = 1$.

Also, $f'(3 - \varepsilon) > 0$
and $f'(3 + \varepsilon) < 0$

$\therefore$ a maximum turning point at $x = 3$.

For $x < 3$, the gradient of $y = f(x)$ is always positive except at the above stationary points. ✓

(2 marks)

56 i If the depth of snow increases at a constant rate b, then after t hours,

$h = bt$

Now $v = \dfrac{A}{h}$ (given)

i.e. $\dfrac{dx}{dt} = \dfrac{A}{h}$ ✓

$\therefore \dfrac{dx}{dt} = \dfrac{A}{bt}$

$= \dfrac{A}{b} \times \dfrac{1}{t}$

$= \dfrac{k}{t}$ ✓

(since $\dfrac{A}{b}$ is a constant).

ii $\dfrac{dx}{dt} = \dfrac{k}{t}$

$\therefore x = k \ln t + C$ ✓

At 6 am, $t = T,\ x = 0$

$\therefore 0 = k \ln T + C$

$\therefore C = -k \ln T$

$\therefore x = k \ln t - k \ln T$

$= k(\ln t - \ln T)$

$= k \ln\left(\dfrac{t}{T}\right)$ ✓

At 8 am, $t = T + 2,\ x = 1$

$\therefore 1 = k \ln\left(\dfrac{T+2}{T}\right) \quad \ldots(1)$

At 11.30 am, $t = T + 5.5,\ x = 2$

$\therefore 2 = k \ln\left(\dfrac{T+5.5}{T}\right) \quad \ldots(2)$

Multiplying (1) × 2:

$\therefore 2 = 2k \ln\left(\dfrac{T+2}{T}\right) \quad \ldots(3)$

Equating (2) and (3):

$k \ln\left(\dfrac{T+5.5}{T}\right) = 2k \ln\left(\dfrac{T+2}{T}\right)$ ✓

$\ln\left(\dfrac{T+5.5}{T}\right) = 2 \ln\left(\dfrac{T+2}{T}\right)$

$\ln\left(\dfrac{T+5.5}{T}\right) = \ln\left(\dfrac{T+2}{T}\right)^2$

$\therefore \dfrac{T+5.5}{T} = \dfrac{T^2+4T+4}{T^2}$

$(n.b.\ T \neq 0)$

$\therefore T + 5.5 = \dfrac{T^2+4T+4}{T}$

$T^2 + 5.5T = T^2 + 4T + 4$

$1.5T = 4$

$T = 2.\dot{6}$ hours

$= 2$ h 40 min.

Since T is the time from the beginning of the snowfall and corresponds to 6 am, it started snowing at 3.20 am. ✓

(6 marks)

(Total mark allocation only in exam)

57 i $\int\left(\frac{1}{x^2}+\frac{1}{x}\right)dx$

$= \int\left(x^{-2}+\frac{1}{x}\right)dx$

$= \frac{x^{-1}}{-1} + \ln x + C$ ✓

$= -\frac{1}{x} + \ln x + C.$ ✓

ii $\int \cos(2x+1)\,dx$

$= \frac{\sin(2x+1)}{2} + C$ ✓✓

(4 marks)
(Total mark allocation only in exam)

58 $\int (x^2+7)\,dx = \frac{x^3}{3} + 7x + c.$ ✓✓

(2 marks)

59 $\int \frac{x}{x^2+3}\,dx = \frac{1}{2}\int \frac{2x}{x^2+3}\,dx$ ✓

$= \frac{1}{2}\ln(x^2+3) + c.$ ✓

(2 marks)

60 $f'(x) = 2x + 7$

$f(x) = \frac{2x^2}{2} + 7x + C$

$= x^2 + 7x + C$ ✓

Now $f(1) = 4$

$\therefore 4 = (1)^2 + 7(1) + C$

$4 = 1 + 7 + C$

$\therefore c = -4$

$\therefore f(x) = x^2 + 7x - 4$ ✓

(2 marks)

61 $\int \frac{2x}{x^2+1}\,dx = \ln(x^2+1) + C$ ✓

(1 mark)

62 i $v = \frac{dx}{dt} = 4t^3 - 24t^2 + 20t$ ✓

$\therefore x = \frac{4t^4}{4} - \frac{24t^3}{3} + \frac{20t^2}{2} + C$

$= t^4 - 8t^3 + 10t^2 + C$

When $t = 0, x = 3 \therefore C = 3$

$\therefore x = t^4 - 8t^3 + 10t^2 + 3$ ✓

ii When $t = 1$,

$v = \frac{dx}{dt} = 4 - 24 + 20$

$= 0$

$\therefore$ There is a stationary point at $t = 1$. ✓

Now $\frac{d^2x}{dt^2} = 12t^2 - 48t + 20$ ✓

When $t = 1$,

$\frac{d^2x}{dt^2} = 12 - 48 + 20$

$= -16 < 0$

$\therefore$ At $t = 1$, there will be a maximum turning point. ✓

iii $x = t^4 - 8t^3 + 10t^2 + 3$

t	0	1	2
x	3	6	–5

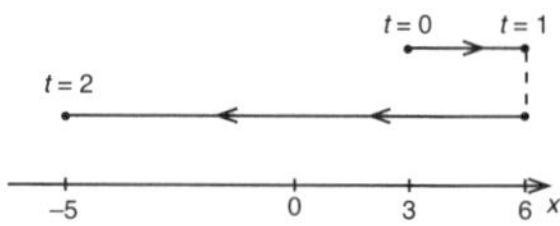

The pen changes direction at $t = 1$,

$\therefore$ Total distance travelled

$= 3 + 11$

$= 14$ cm ✓

The amount of ink used

$= 14 \times 0.05$

$= 0.7$ milligrams of ink. ✓

(7 marks)
(Total mark allocation only in exam)

63 $\int (6 - x^{-3})dx = 6x - \frac{x^{-2}}{-2} + C$

$= 6x + \frac{x^{-2}}{2} + C$ ✓

$= 6x + \frac{1}{2x^2} + C.$ ✓

(2 marks)

64 $\int \sec^2 6x\,dx = \frac{1}{6}\tan 6x + C$ ✓

(1 mark)

65 $f'(x) = 3x^2 - 2$

$f(x) = \int (3x^2 - 2)\,dx$

$= \frac{3x^3}{3} - 2x + C$

$\therefore f(x) = x^3 - 2x + C$ ✓

Now, $y = f(x)$ passes through $(1, 3)$

$\therefore 3 = 1^3 - 2(1) + C$

$3 = 1 - 2 + C$

$\therefore C = 4$

$\therefore f(x) = x^3 - 2x + 4$ ✓

(2 marks)

66 $\int (2x + 11)\,dx = x^2 + 11x + C$ ✓✓

(2 marks)

67 $\int e^{3x}\,dx = \frac{e^{3x}}{3} + C.$

(No mark allocation in exam)

68 $\int e^{2x}\,dx = \frac{1}{2}e^{2x} + C$

(No mark allocation in exam)

69 $\int (x-1)^3\,dx$

$= \frac{(x-1)^4}{4 \times 1} + C$

$= \frac{(x-1)^4}{4} + C.$

(No mark allocation in exam)

70 i $\int (2x+3)^{10}\,dx$

$= \frac{(2x+3)^{11}}{11 \times 2} + C$

$= \frac{(2x+3)^{11}}{22} + C$

ii $\int \sin\frac{x}{2}\,dx = -2\cos\frac{x}{2} + C$

(No mark allocation in exam)

1 The graph of the even function $y = f(x)$ is shown.

The area of the shaded region A is $\frac{1}{2}$ and the area of the shaded region B is $\frac{3}{2}$.

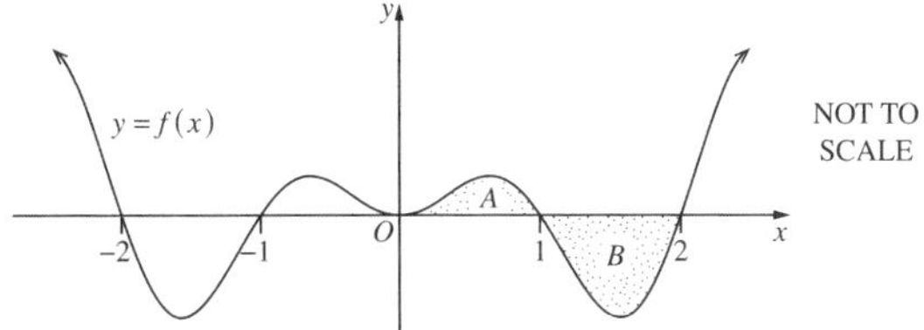

What is the value of $\int_{-2}^{2} f(x)\,dx$?

A 4 **B** 2
C −2 **D** −4 *(1 mark)*

(Q8, **2022 HSC**) Easy

---

2 Use two applications of the trapezoidal rule to find an approximate value of $\int_{0}^{2} \sqrt{1 + x^2}\,dx$.

Give your answer correct to 2 decimal places. *(2 marks)*

(Q13, **2022 HSC**) Easy

---

3 The parabola $y = x^2$ meets the line $y = 2x + 3$ at the points $(-1, 1)$ and $(3, 9)$ as shown in the diagram.

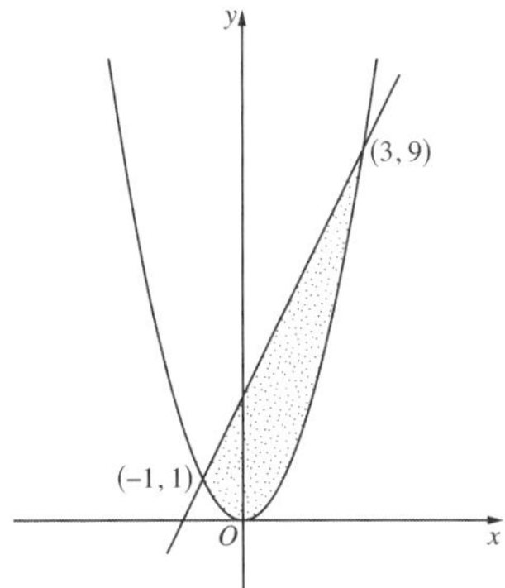

Find the area enclosed by the parabola and the line. *(3 marks)*

(Q16, **2022 HSC**) Easy

---

4 The graph of the circle $x^2 + y^2 = 2$ is shown.

The interval connecting the origin, O, and the point $(1, 1)$ makes an angle θ with the positive x-axis.

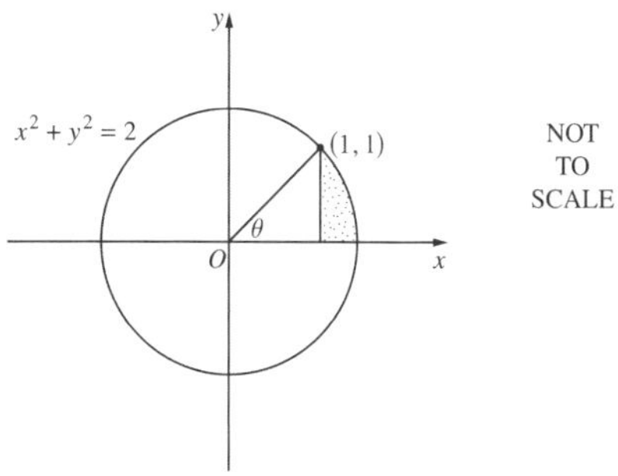

a By considering the value of θ, find the exact area of the shaded region, as shown on the diagram. *(2 marks)* Medium

Part of the hyperbola $y = \frac{a}{b - x} - 1$ which passes through the points $(0, 0)$ and $(1, 1)$ is drawn with the circle $x^2 + y^2 = 2$ as shown.

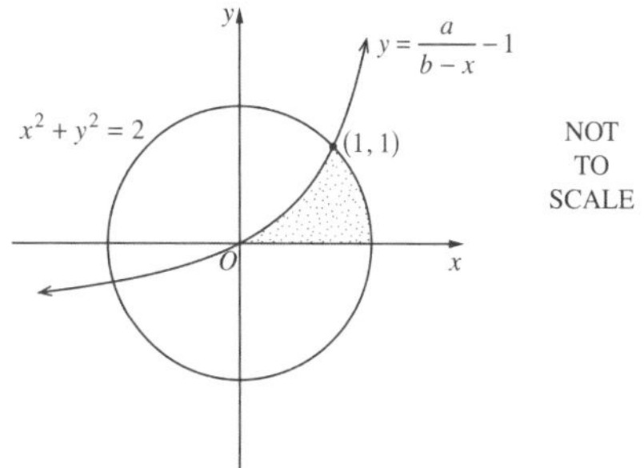

b Show that $a = b = 2$. *(2 marks)* Easy

c Using parts **a** and **b**, find the exact area of the region bounded by the hyperbola, the positive x-axis and the circle as shown on the diagram. *(3 marks)* Hard

(Q28, **2022 HSC**)

---

5 Evaluate $\int_{-2}^{0} \sqrt{2x + 4}\,dx$. *(2 marks)*

(Q15, **2021 HSC**) Easy

---

6 The curve $y = \frac{3}{x - 1}$ intersects the line $y = \frac{3}{2}x$ at the point $(2, 3)$.

The region bounded by the curve $y = \frac{3}{x - 1}$, the line $y = \frac{3}{2}x$, the x-axis and the line $x = 4$ is shaded in the diagram.

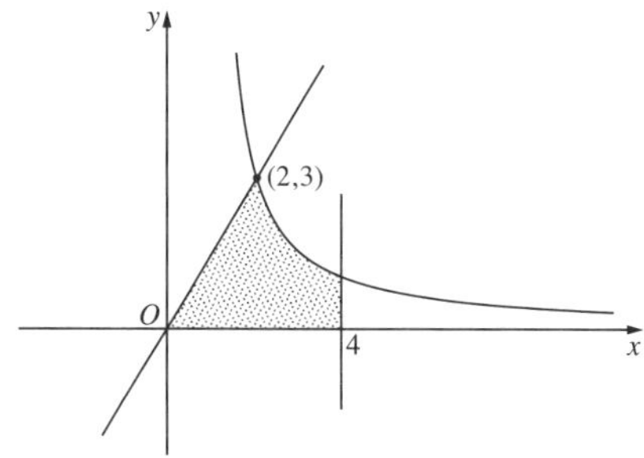

Find the exact area of the shaded region.

(3 marks)

(Q24, **2021 HSC**) Medium

7 Kenzo has a solar powered phone charger. Its power, P, can be modelled by the function

$$P(t) = 400\sin\left(\frac{\pi}{12}t\right), 0 \le t \le 12,$$

where t is the number of hours after sunrise.

a Sketch the graph of P for $0 \le t \le 12$.

(2 marks) Medium

Power is the rate of change of energy. Hence the amount of energy, E units, generated by the solar powered phone charger from $t = a$ to $t = b$, where $0 \le a \le b \le 12$ is given by

$$E = \int_a^b P(t)\,dt.$$

b Show that $E = \frac{4800}{\pi}\left(\cos\frac{a\pi}{12} - \cos\frac{b\pi}{12}\right)$.

(2 marks) Medium

c To make a phone call, a phone battery needs at least 300 units of energy. Kenzo woke up 3 hours after sunrise and found that his phone battery had no units of energy. He immediately began to use his solar powered charger to charge his phone battery.
Find the least amount of time he needed to wait before he could make a phone call. Give your answer correct to the nearest minute. *(3 marks)* Hard

d The next day, Kenzo woke up 6 hours after sunrise and again found that his phone battery had no units of energy. He immediately began to use his solar powered charger to charge his phone battery.
Would it take more time or less time or the same amount of time, compared to the answer in part **c**, to charge his phone battery in order to make a phone call? Explain your answer by referring to the graph drawn in part **a**. *(1 mark)* Hard

(Q27, **2021 HSC**)

8 The region bounded by the graph of the function $f(x) = 8 - 2^x$ and the coordinate axes is shown.

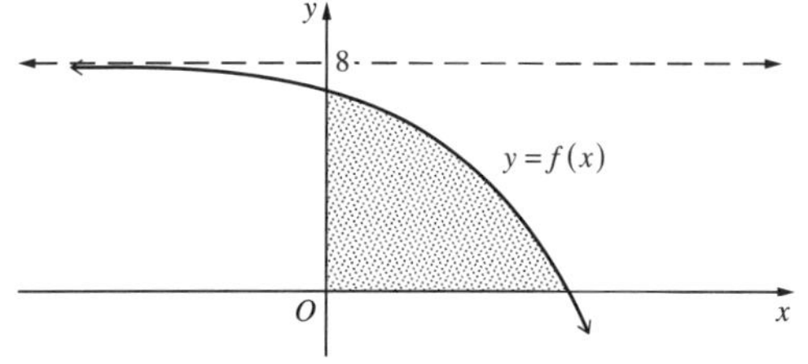

a Show that the exact area of the shaded region is given by $24 - \frac{7}{\ln 2}$.

(3 marks) Medium

b A new function $g(x)$ is found by taking the graph of $y = -f(-x)$ and translating it by 5 units to the right.
Sketch the graph of $y = g(x)$ showing the x-intercept and the asymptote.

(2 marks) Medium

c Hence, find the exact value of $\int_2^5 g(x)\,dx$.

(1 mark) Hard

(Q28, **2021 HSC**)

9 The diagram shows the graph $y = f(x)$, which is made up of line segments and a semicircle.

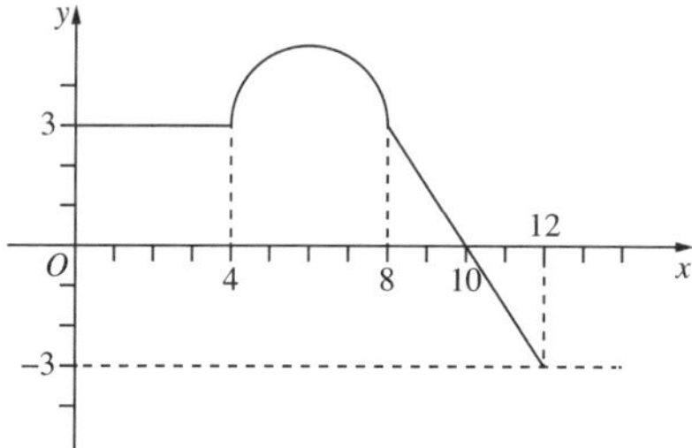

What is the value of $\int_0^{12} f(x)\,dx$?

A $24 + 2\pi$ **B** $24 + 4\pi$
C $30 + 2\pi$ **D** $30 + 4\pi$ *(1 mark)*

(Q7, **2020 HSC**) Medium

10 Evaluate $\int_0^{\frac{\pi}{4}} \sec^2 x\,dx$. *(2 marks)*

(Q13, **2020 HSC**) Easy

11 Kenzo is driving his car along a road while his friend records the velocity of the car, $v(t)$, in km/h every minute over a 5-minute period. The table gives the velocity $v(t)$ at time t hours.

t	0	$\frac{1}{60}$	$\frac{2}{60}$	$\frac{3}{60}$	$\frac{4}{60}$	$\frac{5}{60}$
$v(t)$	60	55	65	68	70	67

The distance covered by the car over the 5-minute period is given by

$$\int_0^{\frac{5}{60}} v(t)\,dt.$$

Use the trapezoidal rule and the velocity at each of the six time values to find the approximate distance in kilometres the car has travelled in the 5-minute period. Give your answer correct to one decimal place. *(2 marks)*

(Q20, **2020 HSC**) Medium

12 The diagram shows two parabolas $y = 4x - x^2$ and $y = ax^2$, where $a > 0$. The two parabolas intersect at the origin, O, and at A.

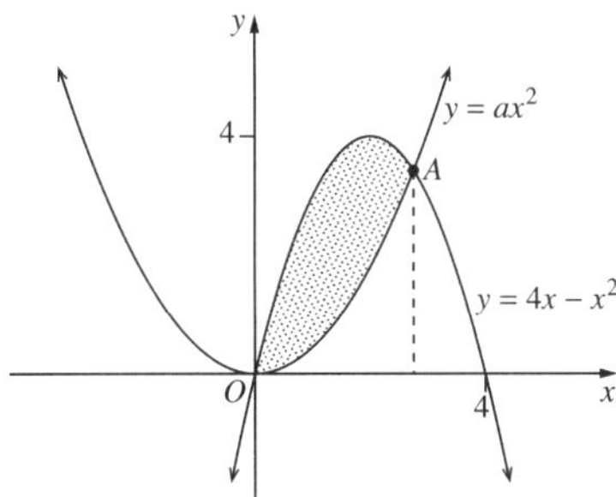

a Show that the x-coordinate of A is $\frac{4}{a+1}$. *(2 marks)* Medium

b Find the value of a such that the shaded area is $\frac{16}{3}$. *(4 marks)* Medium

(Q30, **2020 HSC**)

13 **i** Find the value of $\int_0^{\pi} \sin x \, dx$. *(1 mark)* Easy

ii Using the trapezoidal rule with seven function values, find an approximation to the integral $\int_0^{\pi} \sin x \, dx$, leaving your answer in terms of π and $\sqrt{3}$. *(2 marks)* Medium

iii Using parts **i** and **ii**, explain why $\frac{12}{2+\sqrt{3}}$ is a poor approximation for π. *(1 mark)* Medium

Bonus question (see page iv)

14 The diagram shows a paddock bounded by a river. One section of the fence along the river needs to be moved to alleviate erosion.

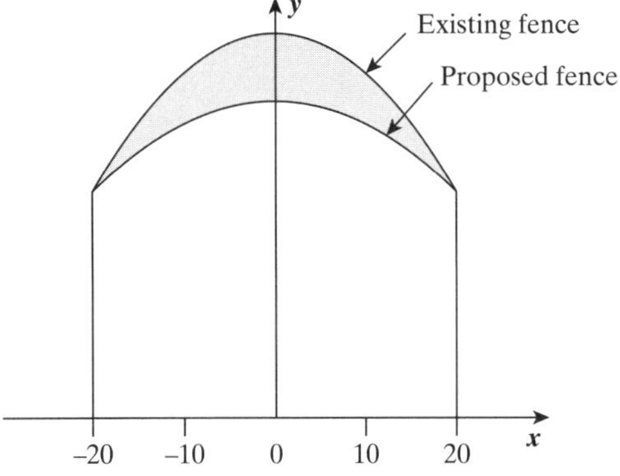

The distances, in metres, at 10-m intervals are shown in the tables below.

With existing fence:

x	–20	–10	0	10	20
y	104	148	170	148	104

With proposed fence:

x	–20	–10	0	10	20
y	104	127	136	127	104

Use the trapezoidal rule with the measurements given to calculate the approximate decrease in area. *(3 marks)*

Bonus question Medium

15 Use the trapezoidal rule with five function values to show that

$$\int_{-\frac{\pi}{3}}^{\frac{\pi}{3}} \sec x \, dx \approx \frac{\pi}{6}\left[\frac{4+3\sqrt{3}}{\sqrt{3}}\right].$$ *(3 marks)*

Bonus question Medium

16 A river is 10 m wide at a certain location and the depth of the river, in metres, has been recorded at 2.5-m intervals. The cross-section, including those measurements, is shown below.

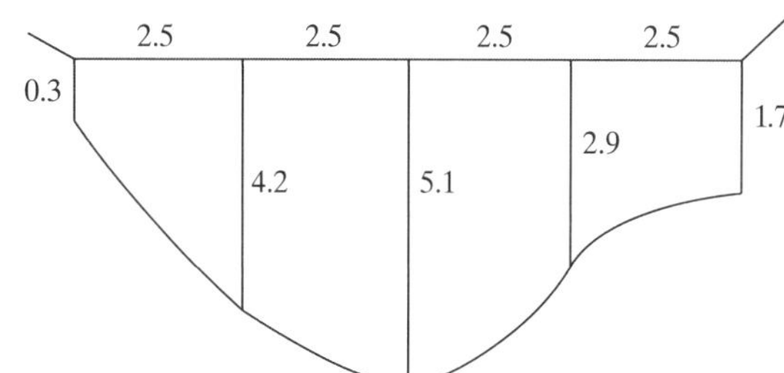

i Use the trapezoidal rule with the measurements to calculate the approximate area of the cross-section. *(3 marks)* Medium

ii If the flow of the river is 0.3 metres per second, find the approximate volume of water flowing through the cross-section each minute. *(1 mark)* Medium

Bonus question

17 A runner is training for an important race. His speed, v, in metres per minute has been recorded at four-minute intervals over a 20-minute period. The table shows the speed at time t minutes.

t	0	4	8	12	16	20
v	184	167	142	131	138	172

The distance covered by the runner over this period is given by $\int_0^{20} v \, dt$.

Use the trapezoidal rule and the values in the table to find the approximate distance. *(3 marks)*

Bonus question Medium

18 Use the trapezoidal rule with three subintervals to find an approximation for $\int_0^{\pi} x\sin^2 x\,dx$, using three subintervals. *(3 marks)*

Bonus question **Hard**

19 Evaluate $\displaystyle\int_0^1 \frac{1}{(3x+2)^2}\,dx$. *(3 marks)*

(Q11e, **2019 HSC**) **Medium**

20 The diagram shows the graph of $y = \dfrac{3x}{x^2+1}$.

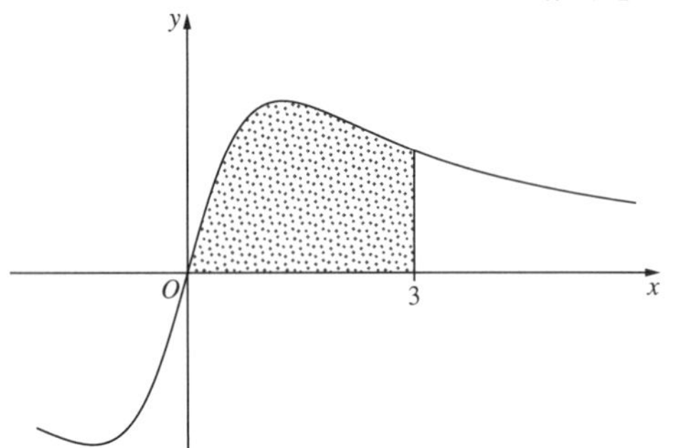

The region enclosed by the graph, the x-axis and the line $x = 3$ is shaded.

Calculate the exact value of the area of the shaded region. *(3 marks)*

(Q12d, **2019 HSC**) **Easy**

21 The diagram shows the region R, bounded by the curve $y = x^r$, where $r \geq 1$, the x-axis and the tangent to the curve at the point $(1, 1)$.

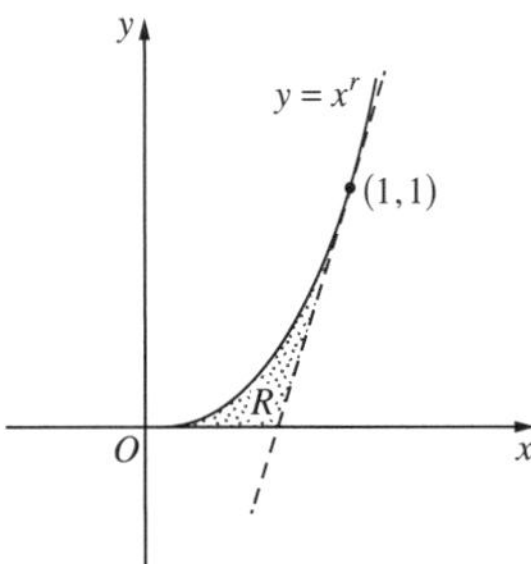

i Show that the tangent to the curve at $(1, 1)$ meets the x-axis at $\left(\dfrac{r-1}{r}, 0\right)$. *(2 marks)* **Easy**

ii Using the result of part **i**, or otherwise, show that the area of the region R is $\dfrac{r-1}{2r(r+1)}$. *(2 marks)* **Medium**

iii Find the exact value of r for which the area of R is a maximum. *(3 marks)* **Hard**

(Q16c, **2019 HSC**)

22 The diagram shows the graph of $y = f(x)$ with intercepts at $x = -1, 0, 3$ and 4.

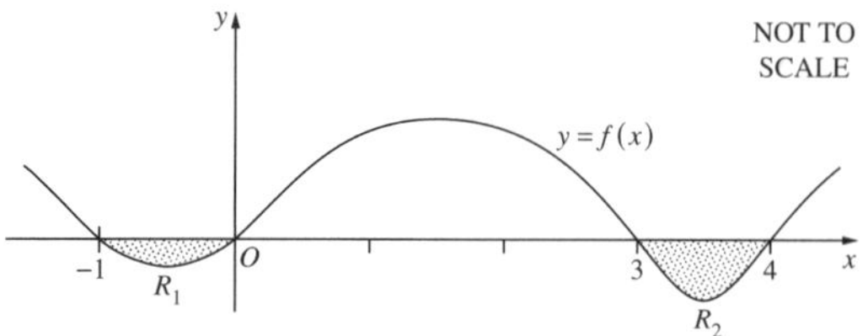

The area of shaded region R_1 is 2.

The area of shaded region R_2 is 3.

It is given that $\displaystyle\int_0^4 f(x)\,dx = 10$

What is the value of $\displaystyle\int_{-1}^{3} f(x)\,dx$?

A 5 **B** 9

C 11 **D** 15 *(1 mark)*

(Q7, **2018 HSC**) **Hard**

23 A trigonometric function $f(x)$ satisfies the condition

$$\int_0^{\pi} f(x)\,dx \neq \int_{\pi}^{2\pi} f(x)\,dx.$$

Which function could be $f(x)$?

A $f(x) = \sin(2x)$ **B** $f(x) = \cos(2x)$

C $f(x) = \sin\left(\dfrac{x}{2}\right)$ **D** $f(x) = \cos\left(\dfrac{x}{2}\right)$

(1 mark)

(Q10, **2018 HSC**) **Medium**

24 Evaluate $\displaystyle\int_0^3 e^{5x}\,dx$. *(2 marks)*

(Q11e, **2018 HSC**) **Easy**

25 The diagram shows the region bounded by the curve $y = \dfrac{1}{x+3}$ and the lines $x = 0, x = 45$ and $y = 0$. The region is divided into two parts of equal area by the line $x = k$, where k is a positive integer.

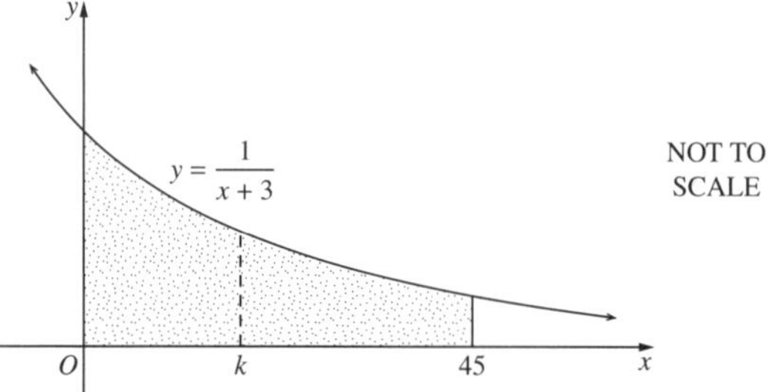

What is the value of the integer k, given that the two parts have equal areas? *(3 marks)*

(Q15b, **2018 HSC**) **Medium**

26 The shaded region is enclosed by the curve $y = x^3 - 7x$ and the line $y = 2x$, as shown in the diagram. The line $y = 2x$ meets the curve $y = x^3 - 7x$ at $O(0, 0)$ and $A(3, 6)$. Do NOT prove this.

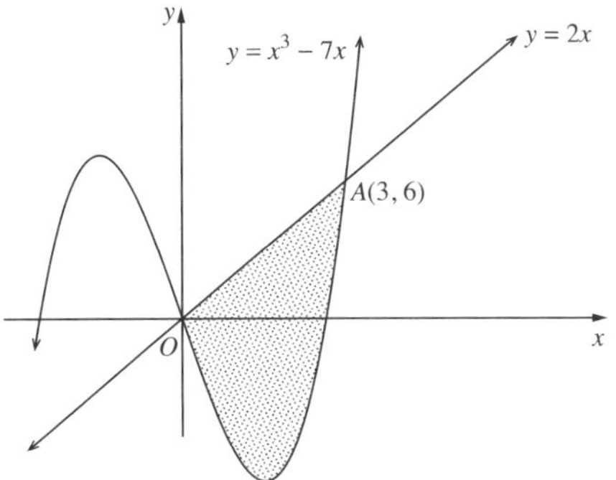

NOT TO SCALE

i Use integration to find the area of the shaded region. *(2 marks)* **Medium**

[Part **ii** not in topic]

The point P is chosen on the curve $y = x^3 - 7x$ so that the tangent at P is parallel to the line $y = 2x$ and the x-coordinate of P is positive.

iii Show that the coordinates of P are $(\sqrt{3}, -4\sqrt{3})$. *(2 marks)* **Medium**

(Q15c, **2018 HSC**)

27 The shaded region shown is enclosed by two parabolas, each with x-intercepts at $x = -1$ and $x = 1$.

The parabolas have equations $y = 2k(x^2 - 1)$ and $y = k(1 - x^2)$, where $k > 0$.

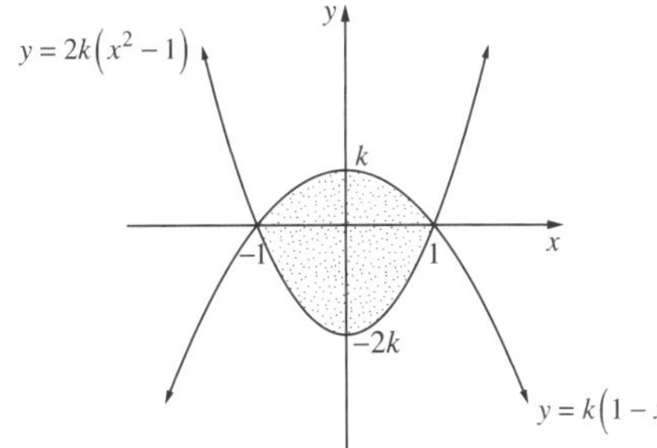

Given that the area of the shaded region is 8, find the value of k. *(3 marks)*

(Q14d, **2017 HSC**) **Medium**

28 What is the value of $\int_{-3}^{2} |x + 1|\,dx$?

A $\frac{5}{2}$ **B** $\frac{11}{2}$

C $\frac{13}{2}$ **D** $\frac{17}{2}$ *(1 mark)*

(Q9, **2016 HSC**) **Hard**

29 Evaluate $\int_{0}^{1} (2x + 1)^3\,dx$. *(2 marks)*

(Q11d, **2016 HSC**) **Easy**

30 **i** Differentiate $y = xe^{3x}$. *(1 mark)* **Easy**

ii Hence find the exact value of

$$\int_{0}^{2} e^{3x}(3 + 9x)\,dx.$$ *(2 marks)* **Medium**

(Q12d, **2016 HSC**)

31 The curve $y = \sqrt{2}\cos\left(\frac{\pi}{4}x\right)$ meets the line $y = x$ at $P(1, 1)$, as shown in the diagram.

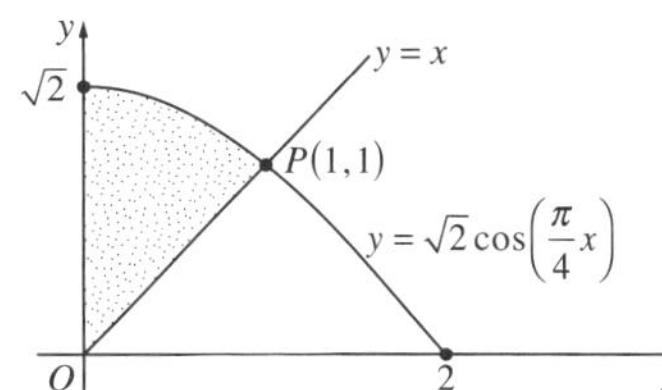

Find the exact value of the shaded area. *(3 marks)*

(Q13d, **2016 HSC**) **Medium**

32 A particle moves in a straight line. Its velocity v ms^{-1} at time t seconds is given by

$$v = 2 - \frac{4}{t + 1}.$$

i Find the initial velocity. *(1 mark)* **Easy**

ii Find the acceleration of the particle when the particle is stationary. *(2 marks)* **Medium**

iii By considering the behaviour of v for large t, sketch a graph of v against t for $t \geq 0$, showing any intercepts. *(2 marks)* **Medium**

iv Find the exact distance travelled by the particle in the first 7 seconds. *(3 marks)* **Medium**

(Q16a, **2016 HSC**)

33 Using the trapezoidal rule with 4 subintervals, which expression gives the approximate area under the curve $y = xe^x$ between $x = 1$ and $x = 3$?

A $\frac{1}{4}(e^1 + 6e^{1.5} + 4e^2 + 10e^{2.5} + 3e^3)$

B $\frac{1}{4}(e^1 + 3e^{1.5} + 4e^2 + 5e^{2.5} + 3e^3)$

C $\frac{1}{2}(e^1 + 6e^{1.5} + 4e^2 + 10e^{2.5} + 3e^3)$

D $\frac{1}{2}(e^1 + 3e^{1.5} + 4e^2 + 5e^{2.5} + 3e^3)$ *(1 mark)*

(Q5, **2015 HSC**) **Easy**

34 The diagram shows the parabola $y = 4x - x^2$ meeting the line $y = 2x$ at $(0, 0)$ and $(2, 4)$.

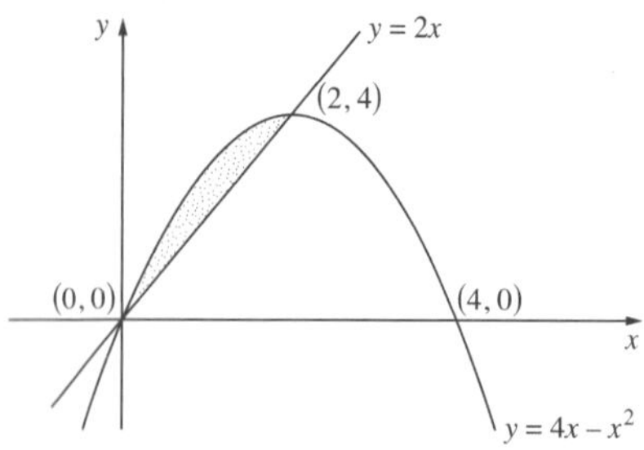

Which expression gives the area of the shaded region bounded by the parabola and the line?

A $\int_0^2 x^2 - 2x\,dx$ **B** $\int_0^2 2x - x^2\,dx$

C $\int_0^4 x^2 - 2x\,dx$ **D** $\int_0^4 2x - x^2\,dx$ *(1 mark)*

(Q7, **2015 HSC**) Easy

35 The diagram shows the area under the curve $y = \dfrac{2}{x}$ from $x = 1$ to $x = d$.

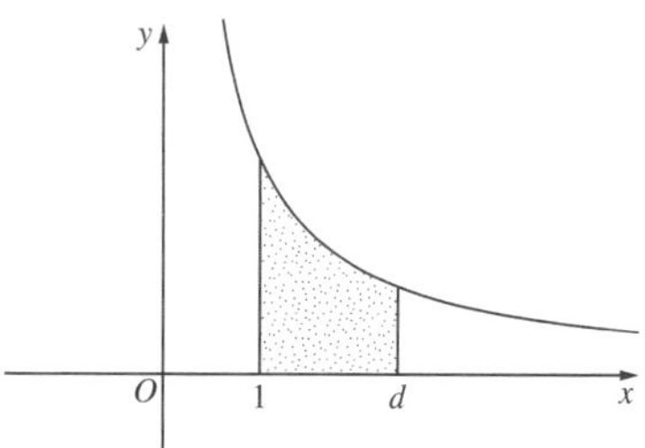

What value of d makes the shaded area equal to 2?

A e **B** $e + 1$

C $2e$ **D** e^2 *(1 mark)*

(Q10, **2015 HSC**) Medium

36 Evaluate $\displaystyle\int_0^{\frac{\pi}{4}} \cos 2x\,dx$. *(2 marks)*

(Q11g, **2015 HSC**) Easy

37 The diagram shows the curve with equation $y = x^2 - 7x + 10$. The curve intersects the x-axis at points A and B. The point C on the curve has the same y-coordinate as the y-intercept of the curve.

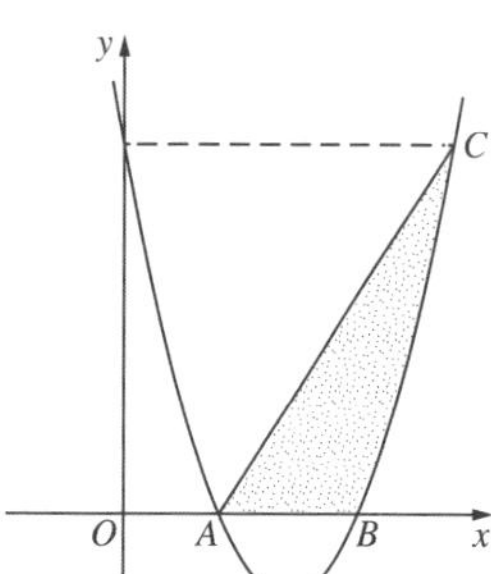

i Find the x-coordinates of points A and B. *(1 mark)* Easy

ii Write down the coordinates of C. *(1 mark)* Easy

iii Evaluate $\displaystyle\int_0^2 \left(x^2 - 7x + 10\right)dx$. *(1 mark)* Easy

iv Hence, or otherwise, find the area of the shaded region. *(2 marks)* Medium

(Q16a, **2015 HSC**)

38 Evaluate $\displaystyle\int_0^{\frac{\pi}{2}} \sin\frac{x}{2}\,dx$. *(3 marks)*

(Q11e, **2014 HSC**) Easy

39 The parabola $y = -2x^2 + 8x$ and the line $y = 2x$ intersect at the origin and at the point A.

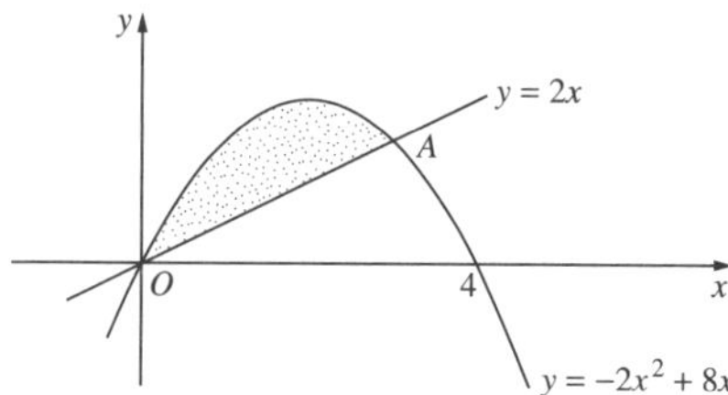

i Find the x-coordinate of the point A. *(1 mark)* Easy

ii Calculate the area enclosed by the parabola and the line. *(3 marks)* Easy

(Q12d, **2014 HSC**)

40 Evaluate $\displaystyle\int_0^1 \frac{x^2}{x^3 + 1}\,dx$. *(3 marks)*

(Q11f, **2013 HSC**) Medium

41 The diagram shows the graphs of the functions $f(x) = 4x^3 - 4x^2 + 3x$ and $g(x) = 2x$. The graphs meet at O and at T.

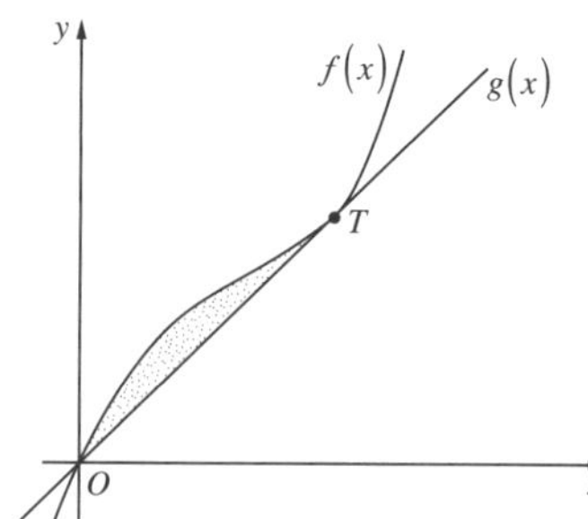

i Find the x-coordinate of T. *(1 mark)* Medium

ii Find the area of the shaded region between the graphs of the functions $f(x)$ and $g(x)$. *(3 marks)* Medium

(Q13b, **2013 HSC**)

42 The diagram shows the graph $y = f(x)$.

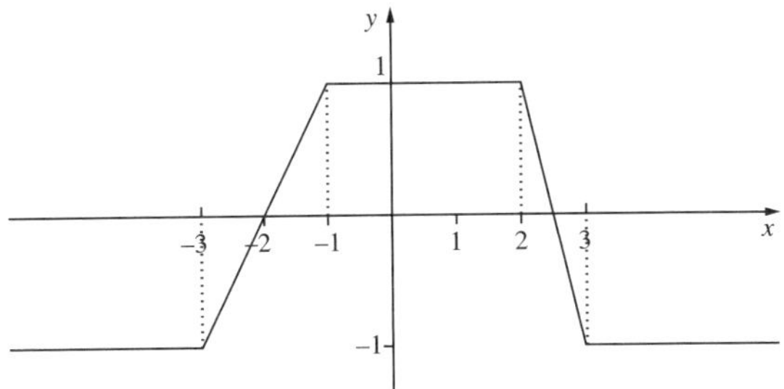

What is the value of a, where $a > 0$,

so that $\int_{-a}^{a} f(x)\,dx = 0$? *(1 mark)*

(Q14d, **2013 HSC**) **Hard**

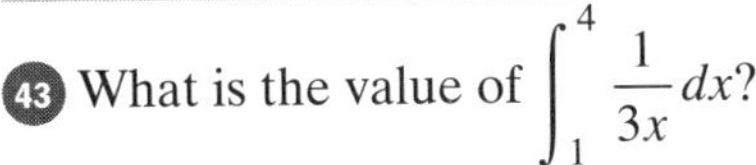

43 What is the value of $\int_{1}^{4} \frac{1}{3x}\,dx$?

A $\frac{1}{3}\ln 3$ **B** $\frac{1}{3}\ln 4$

C $\ln 9$ **D** $\ln 12$ *(1 mark)*

(Q9, **2012 HSC**) **Easy**

---

44 The graph of $y = f(x)$ has been drawn to scale for $0 \le x \le 8$.

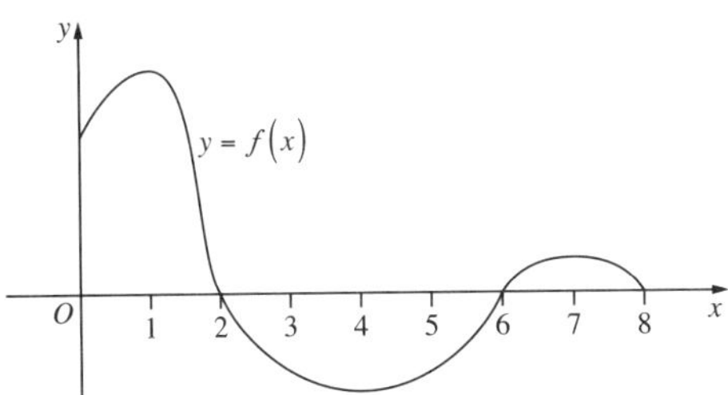

Which of the following integrals has the greatest value?

A $\int_{0}^{1} f(x)\,dx$ **B** $\int_{0}^{2} f(x)\,dx$

C $\int_{0}^{7} f(x)\,dx$ **D** $\int_{0}^{8} f(x)\,dx$ *(1 mark)*

(Q10, **2012 HSC**) **Easy**

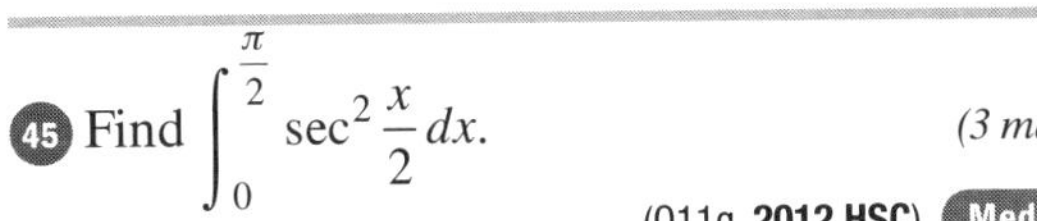

45 Find $\int_{0}^{\frac{\pi}{2}} \sec^2 \frac{x}{2}\,dx$. *(3 marks)*

(Q11g, **2012 HSC**) **Medium**

---

46 The diagram shows the parabolas $y = 5x - x^2$ and $y = x^2 - 3x$. The parabolas intersect at the origin O and the point A. The region between the two parabolas is shaded.

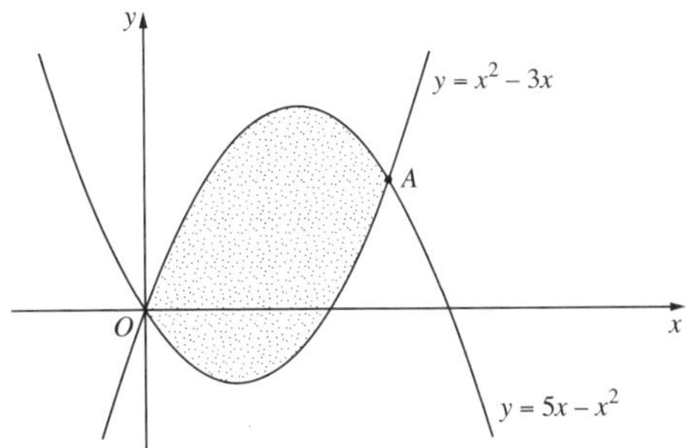

i Find the x-coordinate of the point A. *(1 mark)* **Easy**

ii Find the area of the shaded region. *(3 marks)* **Easy**

(Q13b, **2012 HSC**)

---

47 Evaluate $\int_{e}^{e^3} \frac{5}{x}\,dx$. *(2 marks)*

(Q4b, **2011 HSC**) **Easy**

---

48 The diagram shows the graph $y = 2\cos x$.

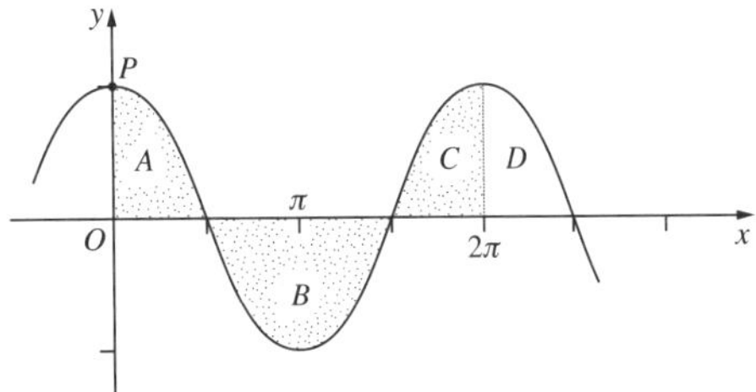

i State the coordinates of P. *(1 mark)* **Easy**

ii Evaluate the integral $\int_{0}^{\frac{\pi}{2}} 2\cos x\,dx$. *(2 marks)* **Easy**

iii Indicate which area in the diagram, A, B, C or D, is represented by the integral

$\int_{\frac{3\pi}{2}}^{2\pi} 2\cos x\,dx$. *(1 mark)* **Easy**

iv Using parts **ii** and **iii**, or otherwise, find the area of the region bounded by the curve $y = 2\cos x$ and the x-axis, between $x = 0$ and $x = 2\pi$. *(1 mark)* **Easy**

v Using the parts above, write down the value of $\int_{\frac{\pi}{2}}^{2\pi} 2\cos x\,dx$. *(1 mark)* **Easy**

(Q6c, **2011 HSC**)

---

49 Given that $\int_{0}^{6} (x + k)\,dx = 30$, and k is a constant, find the value of k. *(2 marks)*

(Q2e, **2010 HSC**) **Easy**

50 i Sketch the curve $y = \ln x$. *(1 mark)* **Medium**

ii Use the trapezoidal rule with three function values to find an approximation to $\int_1^3 \ln x \, dx$. *(2 marks)* **Medium**

iii State whether the approximation found in part ii is greater than or less than the exact value of $\int_1^3 \ln x \, dx$. Justify your answer. *(1 mark)* **Hard**

(Q3b, **2010 HSC**)

51 The curves $y = e^{2x}$ and $y = e^{-x}$ intersect at the point $(0, 1)$ as shown in the diagram.

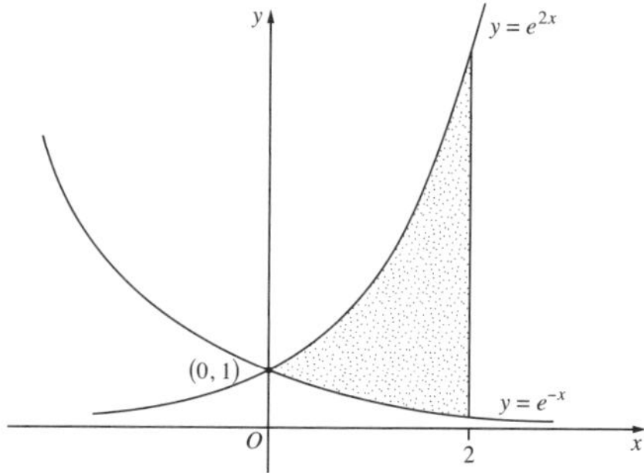

Find the exact area enclosed by the curves and the line $x = 2$. *(3 marks)*

(Q4b, **2010 HSC**) **Medium**

52 The diagram shows the curve $\frac{1}{x}$, for $x > 0$.

The area under the curve between $x = a$ and $x = 1$ is A_1. The area under the curve between $x = 1$ and $x = b$ is A_2.

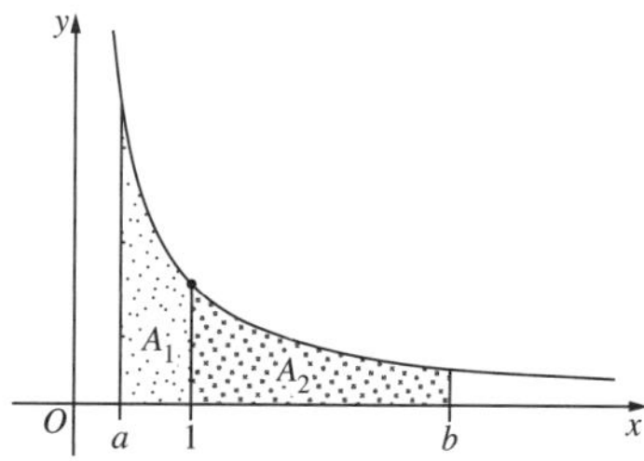

The areas A_1 and A_2 are each equal to 1 square unit.

Find the values of a and b. *(3 marks)*

(Q5c, **2010 HSC**) **Medium**

53 Let $y = f(x)$ be a function defined for $0 \leq x \leq 6$, with $f(0) = 0$.

The diagram shows the graph of the derivative of f, $y = f'(x)$.

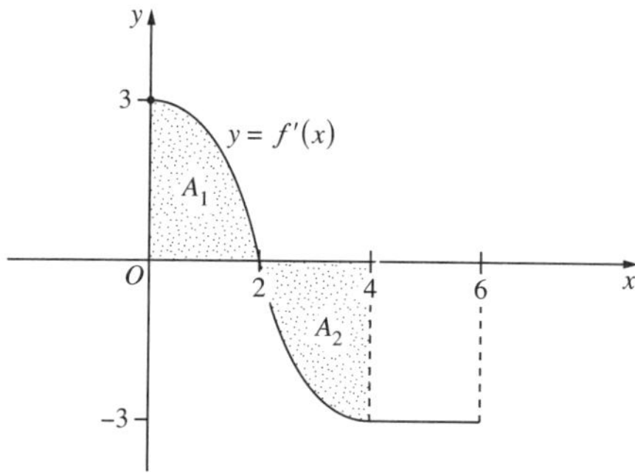

The shaded region A_1 has area 4 square units.
The shaded region A_2 has area 4 square units.

i For which values of x is $f(x)$ increasing? *(1 mark)* **Medium**

ii What is the maximum value of $f(x)$? *(1 mark)* **Hard**

iii Find the value of $f(6)$. *(1 mark)* **Hard**

iv Draw a graph of $y = f(x)$ for $0 \leq x \leq 6$. *(2 marks)* **Hard**

(Q9b, **2010 HSC**)

54 Evaluate $\int_1^4 x^2 + \sqrt{x} \, dx$. *(3 marks)*

(Q2b iii, **2009 HSC**) **Easy**

55 Evaluate $\int_0^{\frac{\pi}{12}} \sec^2 3x \, dx$. *(3 marks)*

(Q2c ii, **2008 HSC**) **Easy**

56 i Differentiate $\log_e (\cos x)$ with respect to x. *(2 marks)*

ii Hence, or otherwise, evaluate

$$\int_0^{\frac{\pi}{4}} \tan x \, dx.$$ *(2 marks)*

(Q3b, **2008 HSC**) **Medium**

57 A beam is supported at $(-b, 0)$ and $(b, 0)$ as shown in the diagram.

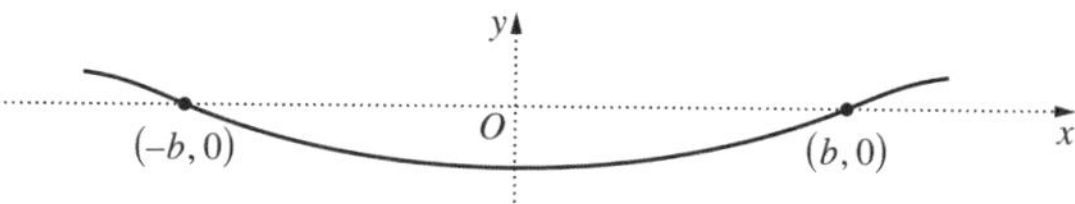

It is known that the shape formed by the beam has equation $y = f(x)$, where $f(x)$ satisfies

$$f''(x) = k\,(b^2 - x^2) \text{ (} k \text{ is a positive constant)}$$

and $f'(-b) = -f'(b)$.

i Show that

$$f'(x) = k\left(b^2 x - \frac{x^3}{3}\right).$$ *(2 marks)*

ii How far is the beam below the x-axis at $x = 0$? *(2 marks)*

(Q9c, **2008 HSC**) **Hard**

58 In the diagram, the shaded region is bounded by $y = \log_e(x - 2)$, the x-axis and the line $x = 7$.

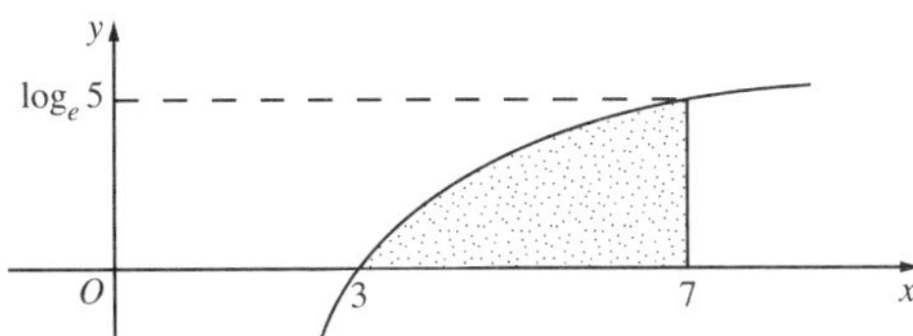

Find the exact value of the area of the shaded region. *(5 marks)*

(Q10a, **2008 HSC**) **Hard**

59 Evaluate 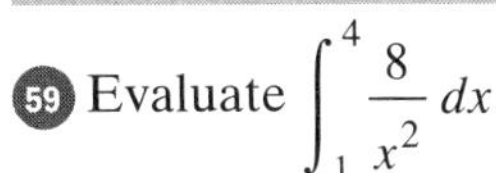 $\displaystyle\int_1^4 \frac{8}{x^2}\,dx$. *(3 marks)*

(Q2b ii, **2007 HSC**) **Medium**

60 The diagram shows the graphs of $y = \sqrt{3}\cos x$ and $y = \sin x$. The first two points of intersection to the right of the y-axis are labelled A and B.

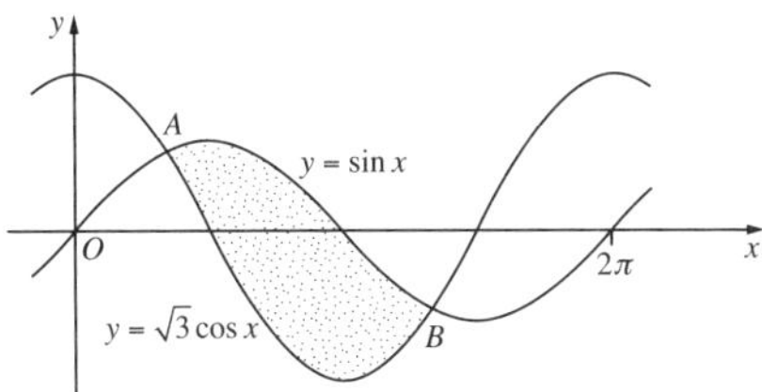

i Solve the equation $\sqrt{3}\cos x = \sin x$ to find the x-coordinates of A and B. *(2 marks)*

ii Find the area of the shaded region in the diagram. *(3 marks)*

(Q7b, **2007 HSC**) **Medium**

61 Evaluate $\displaystyle\int_0^3 \frac{8x}{1+x^2}\,dx$. *(3 marks)*

(Q2b ii, **2006 HSC**) **Medium**

62 i Show that $\dfrac{d}{dx}\log_e(\cos x) = -\tan x$. *(1 mark)* **Easy**

ii

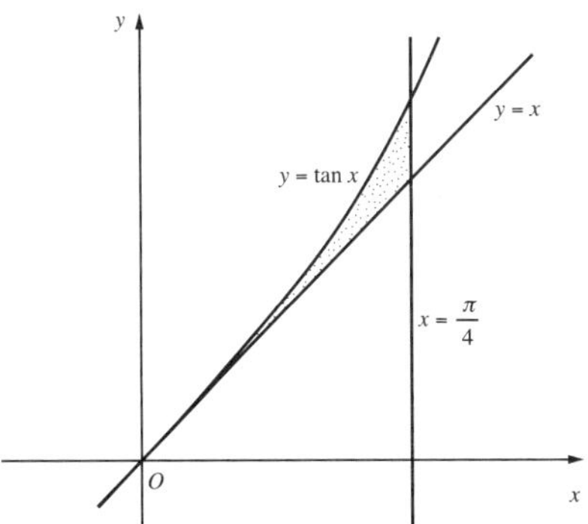

The shaded region in the diagram is bounded by the curve $y = \tan x$ and the lines $y = x$ and $x = \dfrac{\pi}{4}$.

Using the result of part **i**, or otherwise, find the area of the shaded region.

(3 marks) **Medium**

(Q5b, **2006 HSC**)

63 A function $f(x)$ is defined by $f(x) = 1 + 2\cos x$.

i Show that the graph of $y = f(x)$ cuts the x-axis at $x = \dfrac{2\pi}{3}$. *(1 mark)* **Medium**

ii Sketch the graph of $y = f(x)$ for $-\pi \le x \le \pi$ showing where the graph cuts each of the axes. *(3 marks)* **Medium**

iii Find the area under the curve $y = f(x)$ between $x = -\dfrac{\pi}{2}$ and $x = \dfrac{2\pi}{3}$. *(3 marks)* **Hard**

(Q7b, **2006 HSC**)

64 Evaluate $\displaystyle\int_0^{\frac{\pi}{6}} \cos 3x\,dx$. *(2 marks)*

(Q2c ii, **2005 HSC**) **Easy**

65 The graph shows the velocity, $\dfrac{dx}{dt}$, of a particle as a function of time. Initially the particle is at the origin.

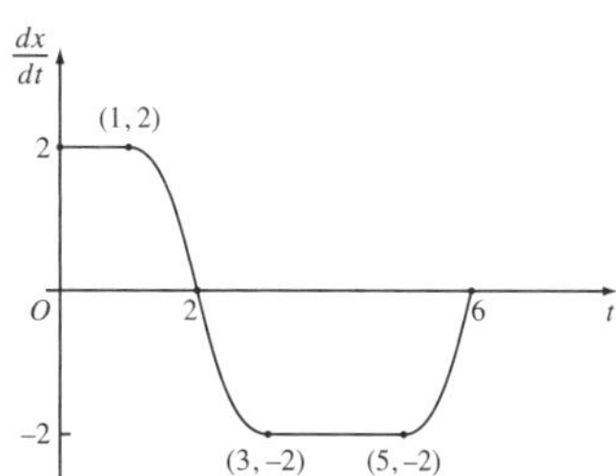

i At what time is the displacement, x, from the origin a maximum? *(1 mark)* **Medium**

ii At what time does the particle return to the origin? Justify your answer. *(2 marks)* **Medium**

iii Draw a sketch of the acceleration, $\dfrac{d^2x}{dt^2}$, as a function of time for $0 \le t \le 6$. *(2 marks)* **Hard**

(Q7b, **2005 HSC**)

66 The shaded region in the diagram is bounded by the circle of radius 2 centred at the origin, the parabola $y = x^2 - 3x + 2$, and the x-axis.

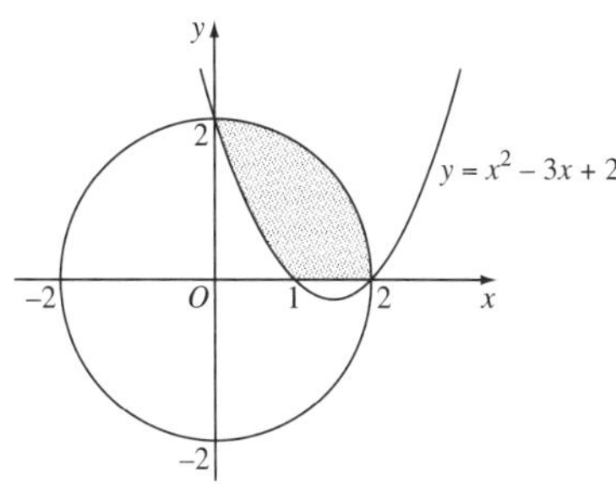

By considering the difference of two areas, find the area of the shaded region. *(3 marks)*

(Q8b, **2005 HSC**) **Medium**

67 A particle is initially at rest at the origin. Its acceleration as a function of time, t, is given by

$$\ddot{x} = 4\sin 2t.$$

i Show that the velocity of the particle is given by $\dot{x} = 2 - 2\cos 2t$. *(2 marks)* **Medium**

ii Sketch the graph of the velocity for $0 \le t \le 2\pi$ AND determine the time at which the particle first comes to rest after $t = 0$. *(3 marks)* **Medium**

iii Find the distance travelled by the particle between $t = 0$ and the time at which the particle first comes to rest after $t = 0$. *(2 marks)* **Hard**

(Q9a, **2005 HSC**)

68 Evaluate $\int_1^2 e^{3x}\,dx$. *(2 marks)*

(Q3b i, **2004 HSC**) **Easy**

69 A particle moves along the x-axis. Initially it is at rest at the origin. The graph shows the acceleration, a, of the particle as a function of time t for $0 \le t \le 5$.

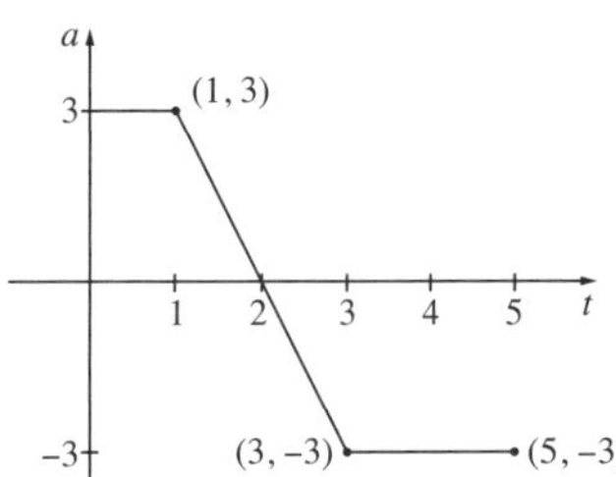

i Write down the time at which the velocity of the particle is a maximum. *(1 mark)* **Medium**

ii At what time during the interval $0 \le t \le 5$ is the particle furthest from the origin? Give brief reasons for your answer. *(2 marks)* **Hard**

(Q9b, **2004 HSC**)

70 Evaluate $\int_{\frac{\pi}{4}}^{\frac{\pi}{3}} \sec^2 x\,dx$. *(2 marks)*

(Q3d ii, **2003 HSC**) **Easy**

71 The graphs of $y = x - 4$ and $y = x^2 - 4x$ intersect at the points $(4, 0)$ and A, as shown in the diagram.

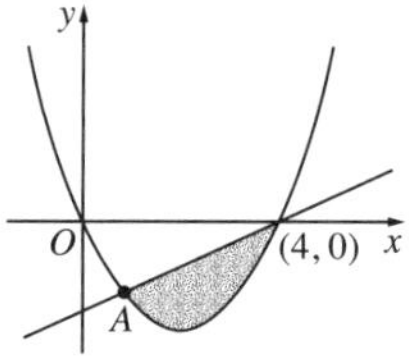

i Find the coordinates of A. *(2 marks)* **Easy**

ii Find the area of the shaded region bounded by $y = x^2 - 4x$ and $y = x - 4$. *(3 marks)* **Medium**

(Q4c, **2003 HSC**)

72 $\int_0^1 \left(e^{5x} - 1\right)dx$. *(2 marks)*

(Q2d ii, **2002 HSC**) **Easy**

73 The graphs of $y = 2x$ and $y = 6x - x^2$ intersect at the origin and point B.

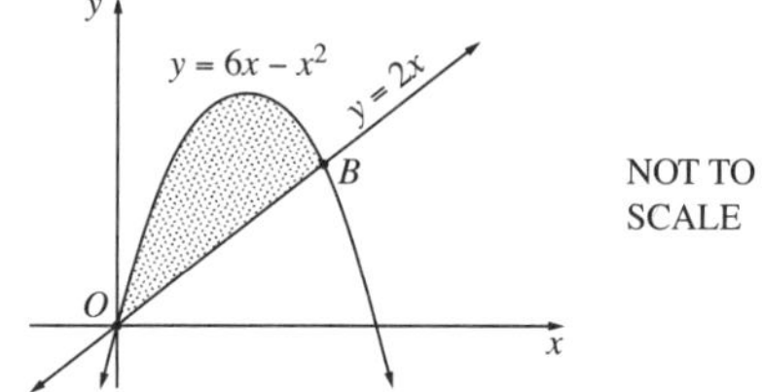

i Show that the coordinates of B are $(4, 8)$. *(1 mark)* **Easy**

ii Find the shaded area bounded by $y = 6x - x^2$ and $y = 2x$. *(3 marks)* **Medium**

(Q4d, **2002 HSC**)

74 Evaluate $\int_0^1 \frac{dx}{x+4}$. *(2 marks)*

(Q3a, **2001 HSC**) **Easy**

75 **i** Sketch the curve $y = 3\sin 2x$ for $-\frac{\pi}{2} \le x \le \frac{\pi}{2}$. *(2 marks)* **Medium**

ii On your diagram for part **i**, sketch the line $y = \frac{1}{4}x$, and shade the region represented by $\int_0^{\frac{\pi}{4}} \left(3\sin 2x - \frac{1}{4}x\right)dx$. *(2 marks)* **Easy**

iii Find the exact value of the integral in part **ii**. *(2 marks)* **Medium**

(Q4c, **2001 HSC**)

76 The diagram shows the cross-section of a creek, with the depths of the creek shown in metres, at 4 metre intervals. The creek is 12 metres in width.

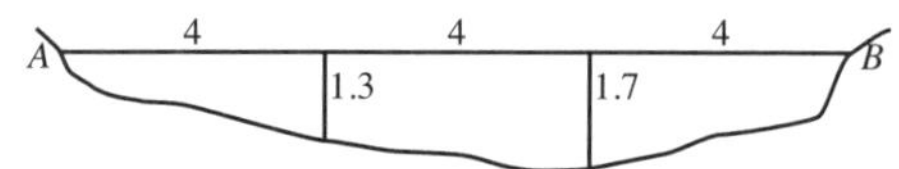

i Use the trapezoidal rule to find an approximate value for the area of the cross-section. *(2 marks)* **Easy**

ii Water flows through this section of the creek at a speed of 0.5 ms^{-1}. Calculate the approximate volume of water that flows past this section in one hour.

(2 marks) **Medium**

(Q5d, **2001 HSC**)

77 Find:

$$\int_{-2}^{1} \frac{2}{x+3}\,dx$$

(Q3c ii, **2000 HSC**) **Easy**

78 The diagram shows the graphs of $y = x^2 - 2$ and $y = x$.

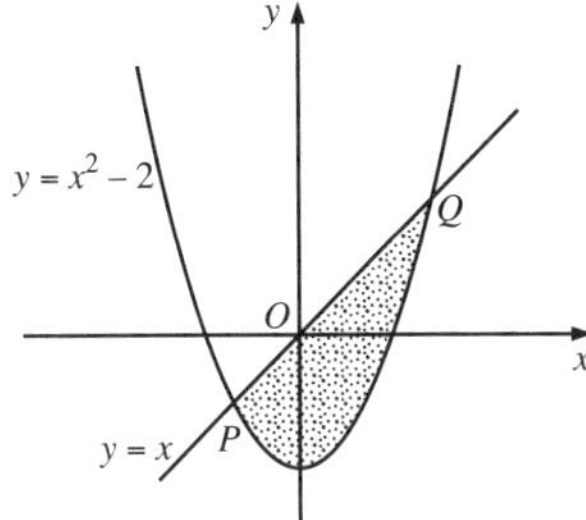

i Find the x values of the points of intersection, P and Q. **Easy**

ii Calculate the area of the shaded region.

(5 marks) **Hard**

(Q7c, **2000 HSC**)

79 The diagram shows the graph of the function $y = 4x - x^2$.

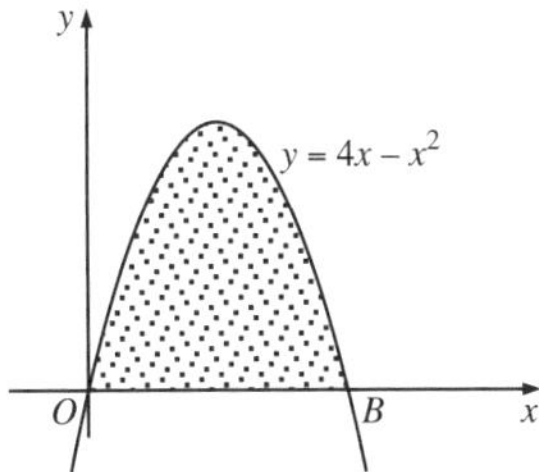

i Find the x coordinate of the point B where the curve crosses the positive x axis.

Easy

ii Find the area of the shaded region contained by the curve $y = 4x - x^2$ and the x axis. **Medium**

(Q4c, **1999 HSC**)

80 The graph of the function f is shown in the diagram. The shaded areas are bounded by $y = f(x)$ and the x axis. The shaded area A is 8 square units, the shaded area B is 3 square units and the shaded area C is 1 square unit.

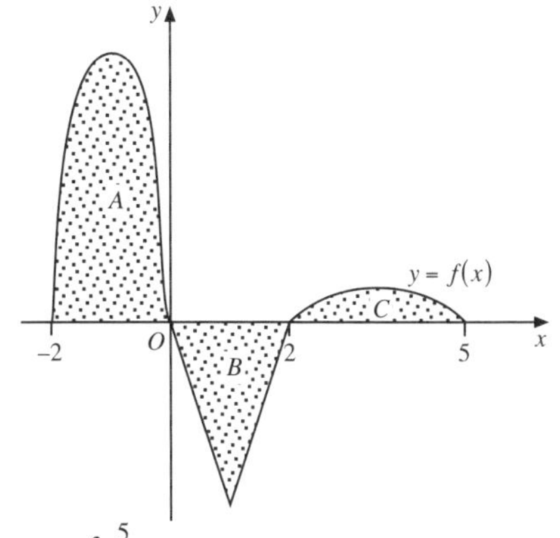

Evaluate $\int_{-2}^{5} f(x)\,dx$. *(2 marks)*

(Q5c, **1999 HSC**) **Medium**

81 The diagram shows the graphs of the functions $y = \sec^2 x$ and $y = x$ between $x = 0$ and $\frac{\pi}{4}$.

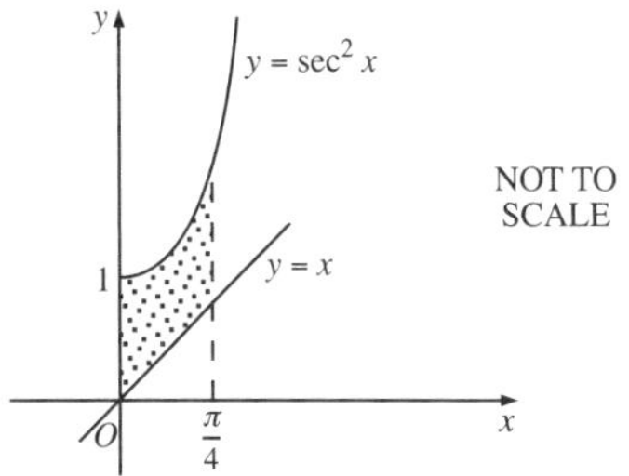

Calculate the area of the shaded region.

(3 marks)

(Q9a, **1999 HSC**) **Medium**

82 Evaluate the following integrals:

i $\int_{1}^{2} \frac{1}{x^2}\,dx$ **Medium**

ii $\int_{0}^{3} e^{4x}\,dx$ *(4 marks)* **Easy**

(Q2b, **1998 HSC**)

83 The following table lists the values of a function for three values of x.

x	1·0	2·0	3·0
$f(x)$	1·7	9·0	4·3

Use these three function values to estimate $\int_{1}^{3} f(x)\,dx$ by the trapezoidal rule.

(Q4a ii, **1998 HSC**) **Easy**

84 Sand is tipped from a truck onto a pile. The rate, R kg/s, at which the sand is flowing is given by the expression $R = 100t - t^3$, for $0 \le t \le T$, where t is the time in seconds after the sand begins to flow.

i Find the rate of flow at time $t = 8$. **Easy**

ii What is the largest value of T for which the expression for R is physically reasonable? **Medium**

iii Find the maximum rate of flow of sand. **Medium**

iv When the sand starts to flow, the pile already contains 300 kg of sand. Find an expression for the amount of sand in the pile at time t. **Medium**

v Calculate the total weight of sand that was tipped from the truck in the first 8 seconds. **Medium**

(7 marks) (Q8a, **1998 HSC**)

85 The diagram shows the graphs of the functions $y = \cos 2x$ and $y = \sin x$ between $x = -\frac{\pi}{2}$ and $x = \frac{\pi}{2}$. The two graphs intersect at $x = \frac{\pi}{6}$ and $x = -\frac{\pi}{2}$. Calculate the area of the shaded region.

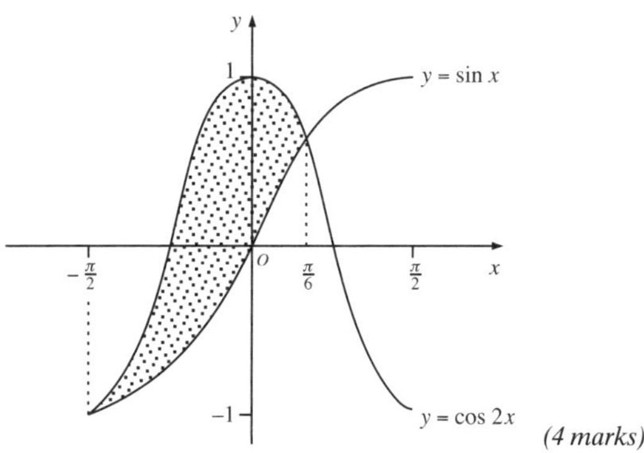

(4 marks)

(Q9b, **1998 HSC**) **Hard**

86 The diagram shows the graphs of the functions $y = \cos x$ and $y = \sin 2x$ between $x = 0$ and $x = \frac{\pi}{2}$. The two graphs intersect at $x = \frac{\pi}{6}$ and $x = \frac{\pi}{2}$.

Calculate the area of the shaded region.

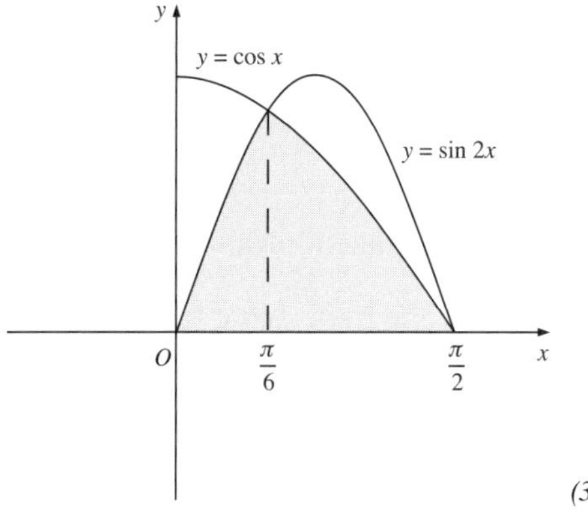

(3 marks)

(Q4c, **1997 HSC**) **Medium**

87 Evaluate $\int_0^{\ln 7} e^{-x}\,dx$. *(2 marks)*

(Q5a, **1997 HSC**) **Medium**

88 Evaluate $\int_1^{e^3} \frac{5}{x}\,dx$. *(3 marks)*

(Q3d, **1996 HSC**) **Medium**

89 Let $f(x) = \sqrt{25 - x^2}$.

i Copy the following table of values into your Writing Booklet and supply the missing values. **Easy**

x	0	1	2	3	4	5
$f(x)$	5·000		4·583			0·000

ii Use these six values of the function and the trapezoidal rule to find the approximate value of $\int_0^5 \sqrt{25 - x^2}\,dx$. **Medium**

iii Draw the graph of $x^2 + y^2 = 25$ and shade the region whose area is represented by the integral $\int_0^5 \sqrt{25 - x^2}\,dx$. **Medium**

iv Use your answer to part **iii** to explain why the exact value of the integral is $\frac{25\pi}{4}$. **Medium**

v Use your answers to part **ii** and part **iv** to find an approximate value for π. **Hard**

(9 marks) (Q4b, **1996 HSC**)

90 The diagram shows the graphs of $y = e^x - 1$ and $y = e^{-x}$.

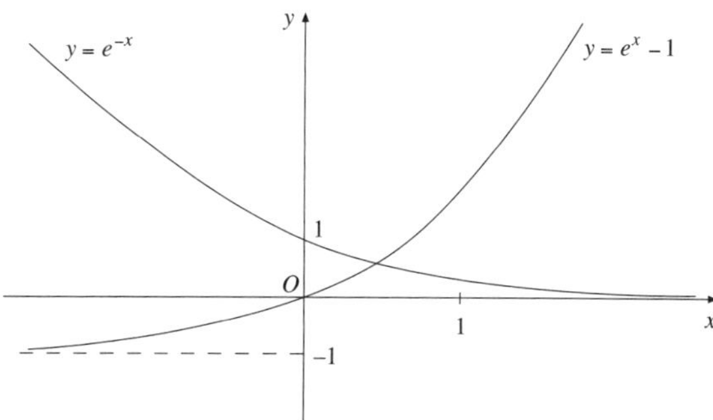

i Find the area between the curves from $x = 1$ to $x = 2$. Leave your answer in terms of e.

ii Show that the curves intersect when $e^{2x} - e^x - 1 = 0$.

(Q5c, **1996 HSC**) **Medium**

91 Two particles P and Q start moving along the x axis at time $t = 0$ and never meet. Particle P is initially at $x = 4$ and its velocity v at time t is given by $v = 2t + 4$.

The position of particle Q is given by $x = 1 + 3\log_e(t + 1)$. The diagram shows the graph of $x = 1 + 3\log_e(t + 1)$.

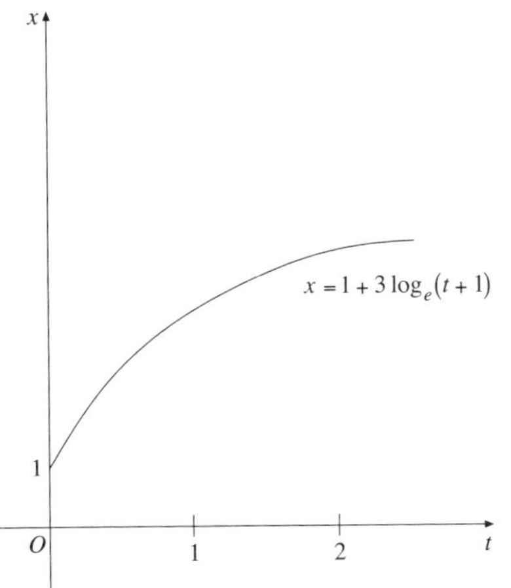

a Find an expression for the position of P at time t. *(2 marks)* Medium

b Copy the diagram into your Writing Booklet and, on the same set of axes, draw the graph of the function found in part **a**. *(2 marks)* Easy

c P and Q are joined by an elastic string and M is the midpoint of the string. Show that the position of M at time t is given by $x = \frac{1}{2}\left[t^2 + 4t + 3\log_e(t+1) + 5\right]$. *(1 mark)* Easy

d Find the time at which the acceleration of M is zero. *(3 marks)* Hard

e Find the minimum distance between P and Q. *(4 marks)* Hard

(Q9, **1996 HSC**)

92 Find:

$$\int_0^{\frac{\pi}{2}} \sin 2x \, dx.$$

(Q3c ii, **1995 HSC**) Easy

93 The diagram shows the graphs of $y = \dfrac{4}{x}$ and $y = 5 - x$.

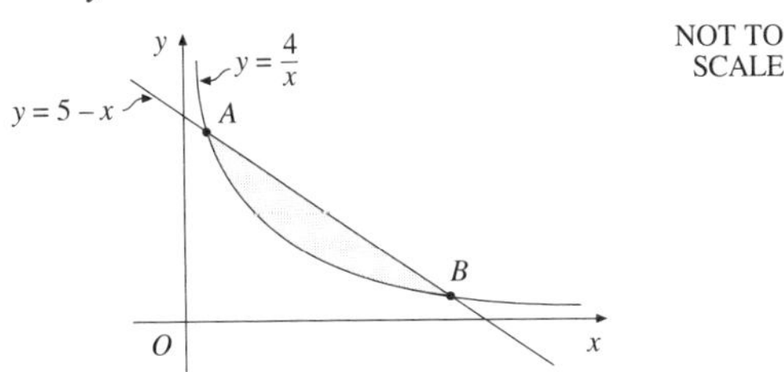

The graphs intersect at the points A and B as shown.

i Find the x coordinates of the points A and B.

ii Find the area of the shaded region between $y = \dfrac{4}{x}$ and $y = 5 - x$.

(5 marks) (Q6b, **1995 HSC**) Medium

94

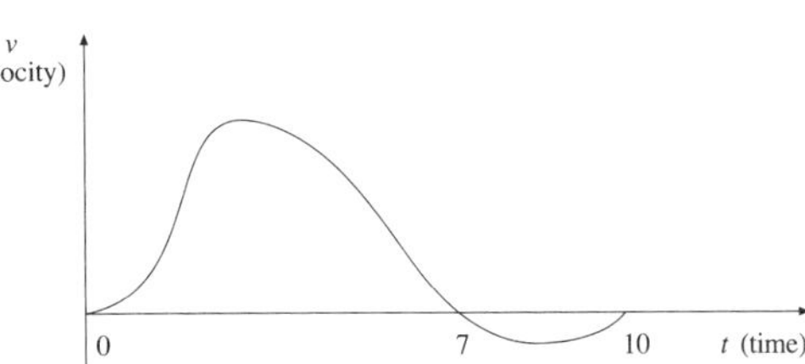

A particle is observed as it moves in a straight line in the period between $t = 0$ and $t = 10$. Its velocity v at time t is shown on the graph above. Copy or trace this graph into your Writing Booklet.

i *On the time axis*, mark and clearly label with the letter Z the times when the acceleration of the particle is zero. Medium

ii *On the time axis*, mark and clearly label with the letter G the time when the acceleration is greatest. Medium

iii There are three occasions when the particle is at rest, i.e. $t = 0$, $t = 7$, and $t = 10$. The particle is furthest from its initial position on one of these occasions. Indicate which occasion, giving reasons for your answer. Hard

(5 marks) (Q9b, **1995 HSC**)

95 Find $\displaystyle\int_1^9 \left(1 + \sqrt{x}\right) dx$.

(Q3c ii, **1994 HSC**) Medium

96 Given that $\dfrac{d}{dx}\left(e^{x^2}\right) = 2xe^{x^2}$, evaluate $\displaystyle\int_0^1 xe^{x^2} \, dx$.

(Q4c, **1994 HSC**) Easy

97 **i** Prove that the line $y = x + 2$ is a tangent to the parabola $y = x^2 - 5x + 11$. Medium

ii Let Q be the point where the line $y = x + 2$ touches the parabola $y = x^2 - 5x + 11$. Show that the normal to the parabola at Q is $y = -x + 8$. Medium

iii Find the area of the region enclosed between the parabola and the line $y = -x + 8$. Hard

(Q9a, **1994 HSC**)

98 Find $\displaystyle\int_0^{\frac{\pi}{2}} \sin 2x \, dx$.

(Q3c ii, **1993 HSC**) Easy

Year 12 Areas and the definite integral—Worked answers

1 $f(x)$ is an even function, which means it is symmetrical about the y-axis.

$$\int_{-2}^{-1} f(x)dx = \int_{1}^{2} f(x)dx = -\frac{3}{2}$$

$$\int_{-1}^{0} f(x)dx = \int_{0}^{1} f(x)dx = \frac{1}{2}$$

$$\int_{-2}^{2} f(x)dx = 2\left[\int_{0}^{1} f(x)dx + \int_{1}^{2} f(x)dx\right] = 2\left[\frac{1}{2} - \frac{3}{2}\right] = -2$$

Answer C

2 $f(x) = \sqrt{1+x^2}$

Two applications requires 3 function values.

x	0	1	2
$f(x)$	1	$\sqrt{2}$	$\sqrt{5}$

✓

$$\int_{0}^{2} \sqrt{1+x^2} \doteqdot \frac{1}{2}\left[1 + \sqrt{5} + 2(\sqrt{2})\right]$$

$= 3.032247551 \ldots$

$= 3.03$ (2 dec. pl.) ✓

(2 marks)

3 $A = \int_{-1}^{3} (2x+3-x^2)dx$

$= \left[x^2 + 3x - \frac{x^3}{3}\right]_{-1}^{3}$ ✓

$= 3^2 + 3(3) - \frac{3^3}{3} - \left[(-1)^2 + 3(-1) - \frac{(-1)^3}{3}\right]$ ✓

$= 9 + 9 - 9 - \left(1 - 3 + \frac{1}{3}\right)$

$= 9 + 1\frac{2}{3}$

$= 10\frac{2}{3}$ ✓

$\therefore 10\frac{2}{3}$ unit^2

(3 marks)

4 a Radius $= \sqrt{2}$ and $\theta = \frac{\pi}{4}$

Shaded area

= area of sector – area of triangle

$= \frac{1}{2} \times (\sqrt{2})^2 \times \frac{\pi}{4} - \frac{1}{2} \times 1 \times 1$ ✓

$= \frac{\pi}{4} - \frac{1}{2}$

$\therefore$ area is $\left(\frac{\pi}{4} - \frac{1}{2}\right)$ units^2. ✓

(2 marks)

b Substitute $(0, 0)$ into $y = \frac{a}{b-x} - 1$:

$0 = \frac{a}{b-0} - 1$

$\frac{a}{b} = 1$

$a = b$ ✓

Substitute $(1, 1)$ and $a = b$ into $y = \frac{a}{b-x} - 1$:

$1 = \frac{b}{b-1} - 1$

$\frac{b}{b-1} = 2$

$b = 2b - 2$

$b = 2$ ✓

$\therefore a = b = 2$

(2 marks)

c $y = \frac{2}{2-x} - 1$

Area $= \int_{0}^{1}\left(\frac{2}{2-x} - 1\right)dx + \left(\frac{\pi}{4} - \frac{1}{2}\right)$ (from **a**) ✓

$= \left[-2\ln|2-x| - x\right]_0^1 + \frac{\pi}{4} - \frac{1}{2}$ ✓

$= -2\ln(2-1) - 1 + 2\ln(2-0) - 0 + \frac{\pi}{4} - \frac{1}{2}$

$= 0 - 1 + 2\ln 2 + \frac{\pi}{4} - \frac{1}{2}$

$= 2\ln 2 + \frac{\pi}{4} - 1\frac{1}{2}$

$\therefore$ the area is $\left(2\ln 2 + \frac{\pi}{4} - 1\frac{1}{2}\right)$ units^2 ✓

(3 marks)

5 $\int_{-2}^{0} \sqrt{2x+4}\, dx$

$= \int_{-2}^{0} (2x+4)^{\frac{1}{2}} dx$

$= \left[\frac{2(2x+4)^{\frac{3}{2}}}{3.2}\right]_{-2}^{0}$ ✓

$= \frac{1}{3}\left[(2(0)+4)^{\frac{3}{2}} - (2(-2)+4)^{\frac{3}{2}}\right]$

$= \frac{8}{3}$ ✓

(2 marks)

6 Area of triangle $+ \int_{2}^{4} \frac{3}{x-1}dx$ ✓

$= \frac{1}{2} \times 2 \times 3 + \left[3\ln|x-1|\right]_2^4$ ✓

$= 3 + 3(\ln(4-1) - \ln(2-1))$

$= 3 + 3(\ln 3 - \ln 1)$

$= 3 + 3\ln 3$

$= 3(1 + \ln 3)$

$\therefore$ the area is $3(1 + \ln 3)$ units^2. ✓

(3 marks)

7 a $P = 400\sin\left(\frac{\pi}{12}t\right)$

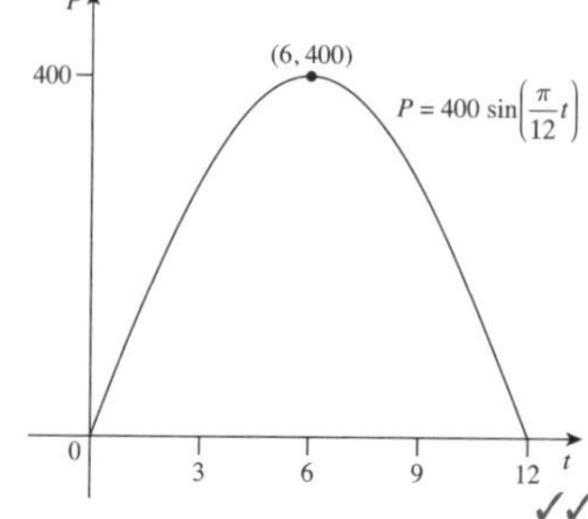

✓✓

(2 marks)

b $E = \int_a^b 400\sin\left(\frac{\pi}{12}t\right)dt$

$= \left[-\frac{4800}{\pi}\cos\left(\frac{\pi}{12}t\right)\right]_a^b$ ✓

$= -\frac{4800}{\pi}\left[\cos\left(\frac{b\pi}{12}\right) - \cos\left(\frac{a\pi}{12}\right)\right]$

$= \frac{4800}{\pi}\left(\cos\left(\frac{a\pi}{12}\right) - \cos\left(\frac{b\pi}{12}\right)\right)$ ✓

(2 marks)

c Substitute $E = 300$ and $a = 3$:

$300 = \frac{4800}{\pi}\left(\cos\frac{3\pi}{12} - \cos\frac{b\pi}{12}\right)$

$300 = \frac{4800}{\pi}\left(\frac{1}{\sqrt{2}} - \cos\frac{b\pi}{12}\right)$

$\frac{300\pi}{4800} = \frac{1}{\sqrt{2}} - \cos\frac{b\pi}{12}$

$\cos\frac{b\pi}{12} = \frac{1}{\sqrt{2}} - \frac{\pi}{16}$ ✓

$\cos\frac{b\pi}{12} = 0.51075724\ldots$

$\frac{b\pi}{12} = 1.034730973\ldots$ radians

$b = 3.952381179\ldots$

$= 3$ h 57 min ✓

As the charging time is $b - a$, then 3 h 57 min minus 3 h is 57 minutes.

∴ Kenzo needs 57 minutes. ✓

(3 marks)

d The graph has a maximum at $t = 6$. From the graph in **a** if the areas represent a phone charge, less time is required after 6 hours. ✓

(1 mark)

8 a For $y = 8 - 2^x$, substitute $y = 0$:

$0 = 8 - 2^x$

$2^x = 8$

$x = 3$ ✓

Area $= \int_0^3 (8 - 2^x)\,dx$

$= \left[8x - \frac{2^x}{\ln 2}\right]_0^3$ ✓

$= 8(3) - \frac{2^3}{\ln 2} - \left(8(0) - \frac{2^0}{\ln 2}\right)$

$= 24 - \frac{8}{\ln 2} - \left(-\frac{1}{\ln 2}\right)$

$= 24 - \frac{7}{\ln 2}$

∴ the area is $\left(24 - \frac{7}{\ln 2}\right)$ units^2. ✓

(3 marks)

b As $f(x) = 8 - 2^x$, then the

- reflection about the y-axis: $f(-x) = 8 - 2^{-x}$
- reflection about the x-axis: $-f(-x) = 2^{-x} - 8$
- translating 5 units to right means replace x with $x - 5$:

$g(x) = 2^{-(x-5)} - 8$

$g(x) = 2^{5-x} - 8$

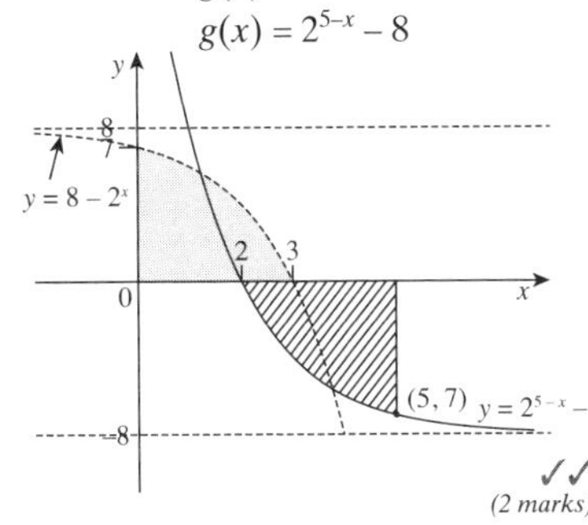

✓✓

(2 marks)

c From the diagram in **b**, the areas are identical in size but for $y = g(x)$ it is below the x-axis.

This means $\int_2^5 g(x)\,dx$

$= -\left(24 - \frac{7}{\ln 2}\right)$.

Alternatively,

$\int_2^5 g(x)\,dx = \int_{-3}^0 -f(-x)\,dx$

$= \int_0^3 -f(x)\,dx$

$= -\int_0^3 f(x)\,dx$

$= -\left(24 - \frac{7}{\ln 2}\right)$ ✓

(1 mark)

9 Consider the areas:

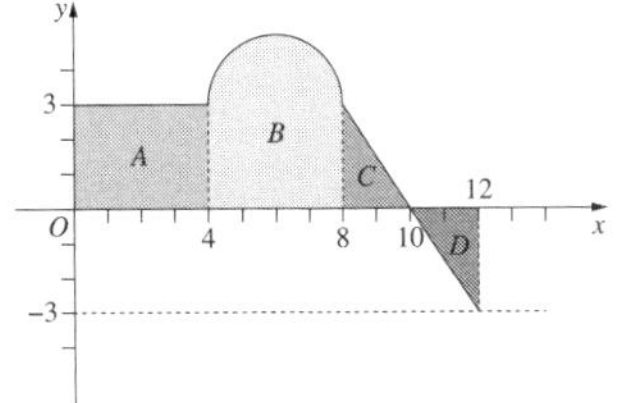

Area $A = 4 \times 3 = 12$

Area $B = 4 \times 3 + \frac{1}{2} \times \pi \times 2^2$

$= 12 + 2\pi$

Area $C = \frac{1}{2} \times 2 \times 3 = 3$

Area $D = \frac{1}{2} \times 2 \times 3 = 3$

$\int_0^{12} f(x)\,dx = 12 + 12 + 2\pi + 3 - 3$

$= 24 + 2\pi$

Answer A

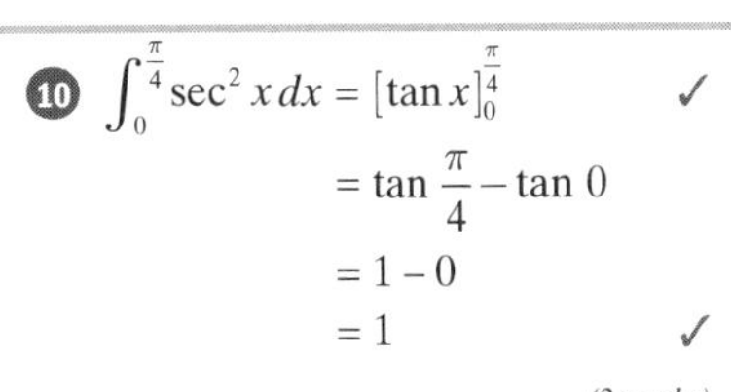

10 $\int_0^{\frac{\pi}{4}} \sec^2 x\,dx = [\tan x]_0^{\frac{\pi}{4}}$ ✓

$= \tan\frac{\pi}{4} - \tan 0$

$= 1 - 0$

$= 1$ ✓

(2 marks)

11 $\int_a^b f(x)dx$

$\approx \frac{b-a}{2n}\{f(a) + f(b) + 2[f(x_1) + \ldots + f(x_{n-1})]\}$

$\int_0^{\frac{5}{60}} v(t)\,dt$

$\approx \frac{\frac{5}{60} - 0}{2(5)}\{60 + 67 + 2(55 + 65 + 68 + 70)\}$ ✓

$= 5.35833333\ldots$

$= 5.4$ (1 dec. pl.)

∴ Kenzo travelled approximately 5.4 km. ✓

(2 marks)

12 a $ax^2 = 4x - x^2$

$ax^2 + x^2 - 4x = 0$

$x^2(a + 1) - 4x = 0$

$x[x(a + 1) - 4] = 0$

$x = 0$ or $x(a + 1) = 4$ ✓

$x = \frac{4}{a+1}$

∴ A has x-coordinate of $\frac{4}{a+1}$. ✓

(2 marks)

b Area =

$\int_0^{\frac{4}{a+1}} (4x - x^2 - ax^2)\,dx = \frac{16}{3}$

$\int_0^{\frac{4}{a+1}} (4x - (1+a)x^2)dx = \frac{16}{3}$ ✓

$\left[2x^2 - (a+1)\frac{x^3}{3}\right]_0^{\frac{4}{a+1}} = \frac{16}{3}$

$2\left(\frac{4}{a+1}\right)^2 - (a+1)\frac{\left(\frac{4}{a+1}\right)^3}{3} - 0$

$= \frac{16}{3}$

$\frac{32}{(a+1)^2} - \frac{64(a+1)}{3(a+1)^3} = \frac{16}{3}$ ✓

$\frac{32}{(a+1)^2} - \frac{64}{3(a+1)^2} = \frac{16}{3}$

$96 - 64 = 16(a + 1)^2$

(by multiplying by $3(a + 1)^2$)

$16(a + 1)^2 = 32$

$(a + 1)^2 = 2$ ✓

$a + 1 = \pm\sqrt{2}$

$a = -1 + \sqrt{2}$ ✓

(as $a > 0$)

(4 marks)

13 i $\int_0^{\pi} \sin x\,dx = [-\cos x]_0^{\pi}$

$= -(\cos\pi - \cos 0)$

$= -(-1 - 1)$

$= 2$ ✓

(1 mark)

ii

x	0	$\frac{\pi}{6}$	$\frac{\pi}{3}$	$\frac{\pi}{2}$	$\frac{2\pi}{3}$	$\frac{5\pi}{6}$	π
$f(x)$	0	$\frac{1}{2}$	$\frac{\sqrt{3}}{2}$	1	$\frac{\sqrt{3}}{2}$	$\frac{1}{2}$	0
	$f(a)$	$f(x_1)$	$f(x_2)$	$f(x_3)$	$f(x_4)$	$f(x_5)$	$f(b)$

Using

$$\approx \frac{b-a}{2n}\{f(a)+f(b)+2[f(x_1)+...+f(x_{n-1})]\}$$

$$= \frac{\pi-0}{2(6)}\left\{f(0)+f(\pi)+2\left[f\left(\frac{\pi}{6}\right)+f\left(\frac{\pi}{3}\right)+f\left(\frac{\pi}{2}\right)+f\left(\frac{2\pi}{3}\right)+f\left(\frac{5\pi}{6}\right)\right]\right\}$$ ✓

$$= \frac{\pi}{12}\left\{0+0+2\left[\frac{1}{2}+\frac{\sqrt{3}}{2}+1+\frac{\sqrt{3}}{2}+\frac{1}{2}\right]\right\}$$

$$= \frac{\pi}{12}(4+2\sqrt{3})$$

$$= \frac{\pi}{6}(2+\sqrt{3})$$ ✓ *(2 marks)*

iii Let $\frac{\pi}{6}(2+\sqrt{3}) = 2$ (from **i** and **iii**)

$$\pi = \frac{12}{2+\sqrt{3}}$$

As $\frac{12}{2+\sqrt{3}} = 3.215\,390\ldots$, hence it is a poor approximation. The graph of $y = \sin x$ is concave down over the given domain and so the trapezoidal rule will underestimate the area. ✓

(1 mark)

14 Form a table using the differences in the y-values.

x	-20	-10	0	10	20
y	0	21	34	21	0

✓

Using the trapezoidal rule:

Decrease

$$\approx \frac{b-a}{2n}\{f(a)+f(b)+2[f(x_1)+...+f(x_{n-1})]\}$$

$$= \frac{20-(-20)}{2\times 4}\{0+0+2[21+34+21]\}$$ ✓

$$= 760$$

The decrease is 760 m^2. ✓ *(3 marks)*

15

x	$-\frac{\pi}{3}$	$-\frac{\pi}{6}$	0	$\frac{\pi}{6}$	$\frac{\pi}{3}$
$f(x)$	2	$\frac{2}{\sqrt{3}}$	1	$\frac{2}{\sqrt{3}}$	2
	$f(a)$	$f(x_1)$	$f(x_2)$	$f(x_3)$	$f(b)$

✓

$$\int_{-\frac{\pi}{3}}^{\frac{\pi}{3}} \sec x\,dx$$

$$\approx \frac{b-a}{2n}\{f(a)+f(b)+2[f(x_1)+...+f(x_{n-1})]\}$$

$$= \frac{\frac{\pi}{3}-\left(-\frac{\pi}{3}\right)}{2(4)}\left\{2+2+2\left[\frac{2}{\sqrt{3}}+1+\frac{2}{\sqrt{3}}\right]\right\}$$ ✓

$$= \frac{\pi}{12}\left\{4+2\left[\frac{4+\sqrt{3}}{\sqrt{3}}\right]\right\}$$

$$= \frac{\pi}{6}\left[2+\frac{4+\sqrt{3}}{\sqrt{3}}\right]$$

$$= \frac{\pi}{6}\left[\frac{2\sqrt{3}+4+\sqrt{3}}{\sqrt{3}}\right]$$

$$= \frac{\pi}{6}\left[\frac{4+3\sqrt{3}}{\sqrt{3}}\right]$$ ✓ *(3 marks)*

16 **i** $A \approx \frac{b-a}{2n}\{f(a)+f(b)+2[f(x_1)+...+f(x_{n-1})]\}$ ✓

$$= \frac{2.5}{2}\{0.3+1.7+2[4.2+5.1+2.9]\}$$ ✓

$$= 33$$

The area of the cross-section is approximately 33 m^2. ✓

(3 marks)

ii Volume

$\approx$ 33 m^2 × 0.3 m per second

= 9.9 m^3 per second

= 594 m^3 per minute ✓ *(1 mark)*

17 $\int_0^{20} v\,dt$

$$\approx \frac{b-a}{2n}\{f(a)+f(b)+2[f(x_1)+...+f(x_{n-1})]\}$$

$$= \frac{20-0}{2\times 5}\{f(0)+f(20)+2[f(4)+f(8)+f(12)+f(16)]\}$$ ✓

$$= 2\{184 + 172 + 2[167 + 142 + 131 + 138]\}$$ ✓

$$= 3024$$

The runner covered approximately 3024 m, or a little over 3 km, in that time. ✓

(3 marks)

18 Complete the table for $f(x) = x\sin^2 x$:

x	0	$\frac{\pi}{3}$	$\frac{2\pi}{3}$	π
$f(x)$	0	$\frac{\pi}{4}$	$\frac{\pi}{2}$	0
	$f(a)$	$f(x_1)$	$f(x_2)$	$f(b)$

Area

$$\approx \frac{\pi-0}{2(3)}[f(0) + f(\pi) + 2\{f(\tfrac{\pi}{3}) + f(\tfrac{2\pi}{3})\}]$$ ✓

$$= \frac{\pi}{6}[0 + 0 + 2(\tfrac{\pi}{4} + \tfrac{\pi}{2})]$$ ✓

$$= \frac{\pi^2}{4}$$

$$\therefore \int_0^{\pi} x\sin^2 x\,dx \approx \frac{\pi^2}{4}$$ ✓ *(3 marks)*

19 $\int_0^1 \frac{1}{(3x+2)^2}\,dx$

$$= \int_0^1 (3x+2)^{-2}\,dx$$ ✓

$$= \left[\frac{(3x+2)^{-1}}{-1\times 3}\right]_0^1$$

$$= -\frac{1}{3}\left[\frac{1}{3x+2}\right]_0^1$$

$$= -\frac{1}{3}\left[\frac{1}{3(1)+2} - \frac{1}{3(0)+2}\right]$$ ✓

$$= -\frac{1}{3}\left[\frac{1}{5}-\frac{1}{2}\right]$$

$$= \frac{1}{10}$$ ✓

(3 marks)

20 Area $= \int_0^3 \frac{3x}{x^2+1}\,dx$

$$= \frac{3}{2}\int_0^3 \frac{2x}{x^2+1}\,dx$$

$$= \frac{3}{2}[\ln(x^2+1)]_0^3$$ ✓

$$= \frac{3}{2}[\ln(3^2+1) - \ln(0^2+1)]$$

$$= \frac{3}{2}[\ln 10 - \ln 1]$$ ✓

$$= \frac{3}{2}\ln 10$$

$\therefore$ area is $\frac{3}{2}\ln 10$ units^2. ✓

(3 marks)

21 **i** $y = x^r$

$$\frac{dy}{dx} = rx^{r-1}$$

Substitute $x = 1$:

$$\frac{dy}{dx}(1) = r(1)^{r-1} = r$$ ✓

Equation of tangent:

$$y - 1 = r(x - 1)$$

$$y = rx - r + 1$$

Substitute $y = 0$:

$$0 = rx - r + 1$$

$$rx = r - 1$$

$$x = \frac{r-1}{r}$$

$\therefore$ tangent meets the x-axis at $\left(\frac{r-1}{r}, 0\right)$. ✓

(2 marks)

ii Area

$$= \int_0^1 x^r\,dx - \frac{1}{2}\times\left(1-\frac{r-1}{r}\right)\times 1$$

$$= \left[\frac{x^{r+1}}{r+1}\right]_0^1 - \frac{1}{2}\times\frac{r-r+1}{r}\times 1$$

$$= \left[\frac{1^{r+1}}{r+1} - \frac{0^{r+1}}{r+1}\right] - \frac{1}{2r}$$ ✓

$$= \frac{1}{r+1} - \frac{1}{2r}$$

$= \frac{2r-(r+1)}{2r(r+1)}$

$= \frac{r-1}{2r(r+1)}$ ✓ *(2 marks)*

iii Consider $A = \frac{r-1}{2r(r+1)}$

$A = \frac{r-1}{2r^2+2r}$

$\frac{dA}{dr} = \frac{(2r^2+2r).1-(r-1)(4r+2)}{(2r^2+2r)^2}$

$= \frac{2r^2+2r-(4r^2-2r-2)}{(2r^2+2r)^2}$ ✓

$= \frac{-2r^2+4r+2}{(2r^2+2r)^2} = 0$

$-2r^2+4r+2=0$

$r^2-2r-1=0$

$r = \frac{2 \pm \sqrt{(-2)^2-4(1)(-1)}}{2(1)}$

$= \frac{2 \pm \sqrt{8}}{2}$

$= \frac{2 \pm 2\sqrt{2}}{2}$

$= 1+\sqrt{2}$ (as $r > 0$) ✓

Consider neighbourhood of $r = 1+\sqrt{2}$:

r	2.4	$1+\sqrt{2}$	2.5
A'	>0	0	<0

$\therefore$ maximum area when $r = 1+\sqrt{2}$. ✓ *(3 marks)*

22 Given

$\int_{-1}^{0} f(x)\,dx = -2$,

$\int_{3}^{4} f(x)\,dx = -3$ and

$\int_{0}^{4} f(x)\,dx = 10$:

First, $\int_{0}^{4} f(x)\,dx$

$= \int_{0}^{3} f(x)\,dx + \int_{3}^{4} f(x)\,dx$

$10 = \int_{0}^{3} f(x)\,dx - 3$

$\therefore \int_{0}^{3} f(x)\,dx = 13$

Also $\int_{-1}^{3} f(x)\,dx$

$= \int_{-1}^{0} f(x)\,dx + \int_{0}^{3} f(x)\,dx$

$= -2 + 13$

$= 11$

Answer C

23 Consider the graphs of each of the four options.

This is the graph of $y = \cos\frac{x}{2}$, between 0 and 2π:

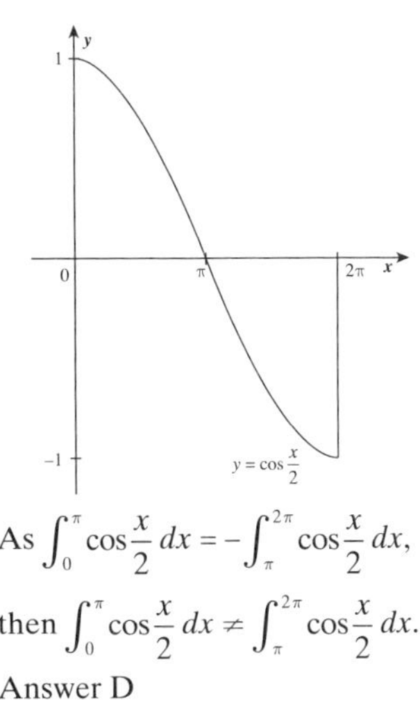

As $\int_0^{\pi} \cos\frac{x}{2}\,dx = -\int_{\pi}^{2\pi} \cos\frac{x}{2}\,dx$,

then $\int_0^{\pi} \cos\frac{x}{2}\,dx \neq \int_{\pi}^{2\pi} \cos\frac{x}{2}\,dx$.

Answer D

24 $\int_0^3 e^{5x}dx = \left[\frac{1}{5}e^{5x}\right]_0^3$ ✓

$= \frac{1}{5}[e^{15} - e^0]$

$= \frac{e^{15}-1}{5}$ ✓ *(2 marks)*

25 $\int_0^k \frac{1}{x+3}\,dx = \int_k^{45} \frac{1}{x+3}\,dx$

$[\ln(x+3)]_0^k = [\ln(x+3)]_k^{45}$ ✓

$\ln(k+3) - \ln(0+3)$

$= \ln(45+3) - \ln(k+3)$

$\ln(k+3) - \ln 3$

$= \ln 48 - \ln(k+3)$

$2\ln(k+3) = \ln 48 + \ln 3$

$\ln(k+3)^2 = \ln 144$ ✓

$(k+3)^2 = 144$

$k+3 = 12$

$k = 9$ ✓ *(3 marks)*

26 **i** Area $= \int_0^3 (2x-(x^3-7x))\,dx$

$= \int_0^3 (9x-x^3)\,dx$

$= \left[\frac{9x^2}{2} - \frac{x^4}{4}\right]_0^3$ ✓

$= \frac{9(3)^2}{2} - \frac{3^4}{4} - 0$

$= 20.25$

$\therefore$ the area is 20.25 u^2. ✓ *(2 marks)*

[Part **ii** not in topic]

iii $y = x^3 - 7x$

$y' = 3x^2 - 7 = 2$

$3x^2 = 9$

$x^2 = 3$

$x = \sqrt{3}$ (as $x > 0$) ✓

$y(\sqrt{3}) = (\sqrt{3})^3 - 7(\sqrt{3})$

$= 3\sqrt{3} - 7\sqrt{3}$

$= -4\sqrt{3}$

$\therefore P(\sqrt{3}, -4\sqrt{3})$ ✓ *(2 marks)*

27 $A = 2\int_0^1 k(1-x^2) - 2k(x^2-1)\,dx$

$= 8$

$2\int_0^1 k - kx^2 - 2kx^2 + 2k\,dx = 8$

$\int_0^1 3k - 3kx^2\,dx = 4$ ✓

$\left[3kx - kx^3\right]_0^1 = 4$ ✓

$3k(1) - k(1)^3 - 0 = 4$

$3k - k = 4$

$2k = 4$

$k = 2$ ✓ *(3 marks)*

28

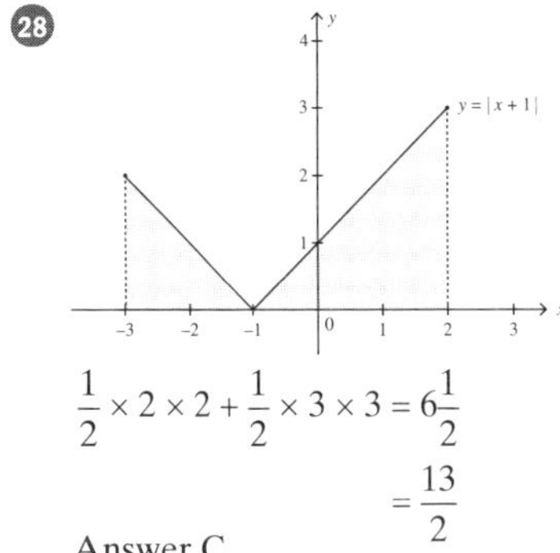

$\frac{1}{2} \times 2 \times 2 + \frac{1}{2} \times 3 \times 3 = 6\frac{1}{2}$

$= \frac{13}{2}$

Answer C

29 $\int_0^1 (2x+1)^3\,dx$

$= \left[\frac{(2x+1)^4}{4\times 2}\right]_0^1$ ✓

$= \left[\frac{(2x+1)^4}{8}\right]_0^1$

$= \frac{1}{8}[(2(1)+1)^4 - (2(0)+1)^4]$

$= \frac{1}{8}[81-1]$

$= 10$ ✓ *(2 marks)*

30 **i** $y = xe^{3x}$

Using the product rule,

Let $u = x$ $\quad \frac{du}{dx} = 1$

Let $v = e^{3x}$ $\quad \frac{dv}{dx} = 3e^{3x}$

$\frac{dy}{dx} = u\frac{dv}{dx} + v\frac{du}{dx}$

$= x.3e^{3x} + e^{3x}.1$

$= e^{3x} + 3xe^{3x}$

$= e^x(1+3x)$ ✓ *(1 mark)*

ii $\int_0^2 e^{3x}(3+9x)dx$

$= 3\int_0^2 e^{3x}(1+3x)dx$

$= 3\left[xe^{3x}\right]_0^2 dx$ (from part **i**)✓

$= 3[2e^6 - 0]$

$= 6e^6$ ✓ *(2 marks)*

31 Shaded area

$= \int_0^1 \sqrt{2}\cos\left(\frac{\pi}{4}x\right) dx - \frac{1}{2} \times 1 \times 1$

$= \left[\frac{4\sqrt{2}}{\pi}\sin\frac{\pi}{4}x\right]_0^1 - \frac{1}{2}$ ✓

$= \frac{4\sqrt{2}}{\pi}\left[\sin\frac{\pi}{4} - \sin 0\right] - \frac{1}{2}$

$= \frac{4\sqrt{2}}{\pi}\left[\frac{1}{\sqrt{2}} - 0\right] - \frac{1}{2}$ ✓

$= \frac{4}{\pi} - \frac{1}{2}$

$= \frac{8-\pi}{2\pi}$

$\therefore$ the area is $\frac{8-\pi}{2\pi}$ units^2 ✓

(3 marks)

32 **i** Let $t = 0$:

$\therefore v(0) = 2 - \frac{4}{1}$

$= -2$

$\therefore$ the initial velocity is -2 ms^{-1}. ✓

(1 mark)

ii Let $v = 0$:

$\therefore 2 - \frac{4}{t+1} = 0$

$\frac{4}{t+1} = 2$

$2t + 2 = 4$

$2t = 2$

$t = 1$ ✓

As $v = 2 - 4(t+1)^{-1}$

$\therefore a = 4(t+1)^{-2}.1$

$= \frac{4}{(t+1)^2}$

Subs. $t = 1$:

$a(1) = \frac{4}{(1+1)^2}$

$= 1$

$\therefore$ the acceleration is 1 ms^{-2}. ✓

(2 marks)

iii [graph: v against t, with $v = 2 - \frac{4}{t+1}$, passing through $(0, -2)$ and $(1, 0)$, asymptote $v = 2$]

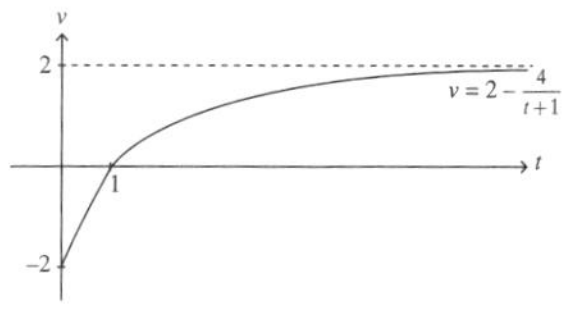

(2 marks)

iv $d = \left|\int_0^1 \left(2 - \frac{4}{t+1}\right) dt\right| + \int_1^7 \left(2 - \frac{4}{t+1}\right) dt$

$= \left|[2t - 4\ln(t+1)]_0^1\right| + [2t - 4\ln(t+1)]_1^7$ ✓

$= |2 - 4\ln 2 - 0| + 2(7) - 4\ln(7+1) - [2(1) - 4\ln(1+1)]$

$= 4\ln 2 - 2 + 14 - 4\ln 8 - 2 + 4\ln 2$ ✓

(as $|2 - 4\ln 2| = 4\ln 2 - 2$)

$= 8\ln 2 + 10 - 4\ln 8$

$= 8\ln 2 + 10 - 4\ln 2^3$

$= 8\ln 2 + 10 - 12\ln 2$

$= 10 - 4\ln 2$

$\therefore$ the distance is $(10 - 4\ln 2)$ m. ✓

(3 marks)

33

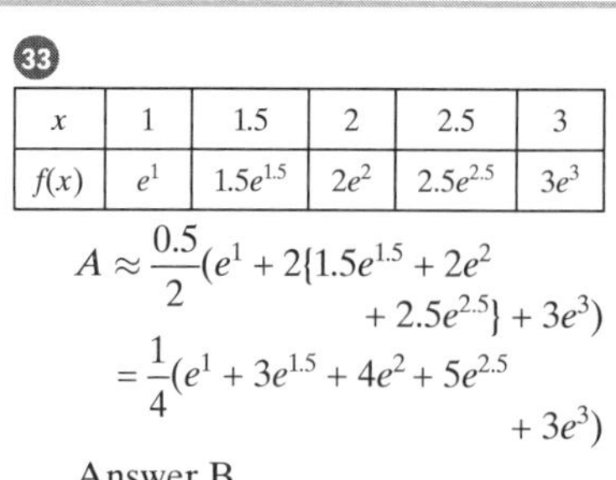

x	1	1.5	2	2.5	3
$f(x)$	e^1	$1.5e^{1.5}$	$2e^2$	$2.5e^{2.5}$	$3e^3$

$A \approx \frac{0.5}{2}(e^1 + 2\{1.5e^{1.5} + 2e^2 + 2.5e^{2.5}\} + 3e^3)$

$= \frac{1}{4}(e^1 + 3e^{1.5} + 4e^2 + 5e^{2.5} + 3e^3)$

Answer B

34 Area $= \int_0^2 4x - x^2 - 2x\ dx$

$= \int_0^2 2x - x^2\ dx$

Answer B

35 $\int_1^d \frac{2}{x}\, dx = 2$

$2\int_1^d \frac{1}{x}\, dx = 2$

$\int_1^d \frac{1}{x}\, dx = 1$

$\log_e d - \log_e 1 = 1$

$\log_e d = 1$

$d = e$

Answer A

36 $\int_0^{\frac{\pi}{4}} \cos 2x\, dx = \left[\frac{1}{2}\sin 2x\right]_0^{\frac{\pi}{4}}$ ✓

$= \frac{1}{2}\left[\sin\frac{\pi}{2} - \sin 0\right]$

$= \frac{1}{2}(1 - 0)$

$= \frac{1}{2}$ ✓

(2 marks)

37 **i** $x^2 - 7x + 10 = 0$

$(x-2)(x-5) = 0$

$x = 2, 5$

$\therefore$ the x-coordinates are 2 and 5. ✓

(1 mark)

ii y-intercept of $x^2 - 7x + 10$ is 10.

Substitute $y = 10$ in $y = x^2 - 7x + 10$:

$10 = x^2 - 7x + 10$

$x^2 - 7x = 0$

$x(x-7) = 0$

$x = 0, 7$

$\therefore C(7, 10)$ ✓

(1 mark)

iii $\int_0^2 (x^2 - 7x + 10)\ dx$

$= \left[\frac{x^3}{3} - \frac{7x^2}{2} + 10x\right]_0^2$

$= \frac{2^3}{3} - \frac{7(2)^2}{2} + 10(2) - 0$

$= 8\frac{2}{3}$ ✓

(1 mark)

iv

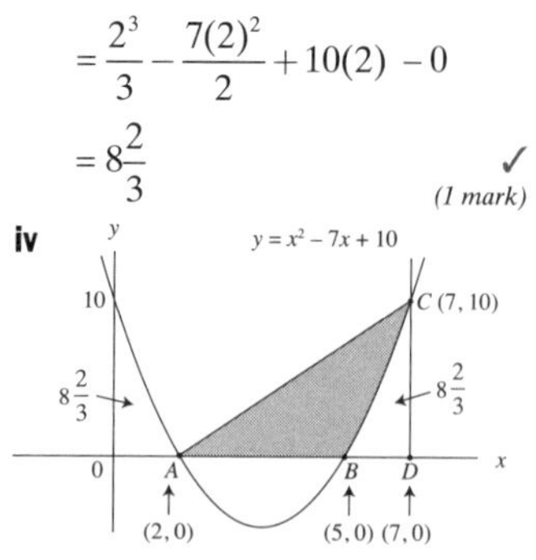

Parabola is symmetrical:

Area = Area of $\Delta ACD - \int_0^2 (x^2 - 7x + 10)\ dx$ ✓

$= \frac{1}{2} \times 5 \times 10 - 8\frac{2}{3}$

$= 16\frac{1}{3}$

$\therefore$ shaded area is $16\frac{1}{3}$ units^2. ✓

(2 marks)

38 $\int_0^{\frac{\pi}{2}} \sin\frac{x}{2}\, dx = \left[-2\cos\frac{x}{2}\right]_0^{\frac{\pi}{2}}$ ✓

$= -2(\cos\frac{\pi}{4} - \cos 0)$

$= -2\left[\frac{1}{\sqrt{2}} - 1\right]$ ✓

$= 2 - \frac{2}{\sqrt{2}}$

$= 2 - \frac{2}{\sqrt{2}} \times \frac{\sqrt{2}}{\sqrt{2}}$

$= 2 - \sqrt{2}$ ✓

(3 marks)

39 **i** $2x = -2x^2 + 8x$

$2x^2 + 2x - 8x = 0$

$2x^2 - 6x = 0$

$2x(x-3) = 0$

$x = 0, 3$

$\therefore$ At A, $x = 3$. ✓

(1 mark)

ii Area $= \int_0^3 -2x^2 + 8x - 2x\ dx$

$= \int_0^3 -2x^2 + 6x\ dx$ ✓

$= \left[\frac{-2x^3}{3} + 3x^2\right]_0^3$ ✓

$= \frac{-2(3)^3}{3} + 3(3)^2 - 0$

$= -18 + 27$

$= 9$

$\therefore$ area is 9 units^2. ✓

(3 marks)

40

$$\int_0^1 \frac{x^2}{x^3+1}\,dx$$

$$= \frac{1}{3}\int_0^1 \frac{3x^2}{x^3+1}\,dx$$

$$= \frac{1}{3}[\log_e(x^3+1)]_0^1 \quad ✓$$

$$= \frac{1}{3}[\log_e(1+1) - \log_e(0+1)]$$

$$= \frac{1}{3}[\log_e 2 - \log_e 1] \quad ✓$$

$$= \frac{1}{3}\log_e 2 \quad ✓$$

(3 marks)

41 **i**

$$4x^3 - 4x^2 + 3x = 2x$$
$$4x^3 - 4x^2 + x = 0$$
$$x(4x^2 - 4x + 1) = 0$$
$$x(2x-1)^2 = 0$$
$$x = 0, \frac{1}{2}$$

$\therefore T$ has x-coordinate of $\frac{1}{2}$ ✓

(1 mark)

ii Area

$$= \int_0^{\frac{1}{2}} 4x^3 - 4x^2 + 3x - 2x\,dx$$

$$= \int_0^{\frac{1}{2}} 4x^3 - 4x^2 + x\,dx$$

$$= \left[x^4 - \frac{4x^3}{3} + \frac{x^2}{2}\right]_0^{\frac{1}{2}} \quad ✓$$

$$= \left(\frac{1}{2}\right)^4 - \frac{4\left(\frac{1}{2}\right)^3}{3} + \frac{\left(\frac{1}{2}\right)^2}{2} - 0 \quad ✓$$

$$= \frac{1}{16} - \frac{1}{6} + \frac{1}{8}$$

$$= \frac{1}{48}$$

$\therefore$ area is $\frac{1}{48}$ units^2 ✓

(3 marks)

42

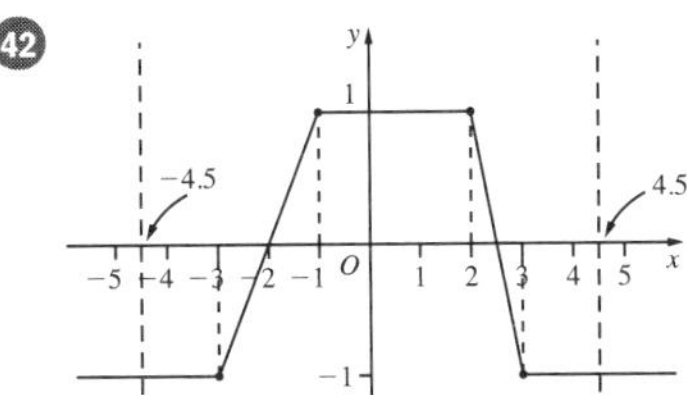

Consider the area between the x-axis and the graph $y = f(x)$. You need to find a value of a, such that the area above the x-axis is identical to the area below the x-axis.

When $a = 4.5$, $\int_{-4.5}^{4.5} f(x)\,dx = 0$.

$\therefore a = 4.5$

OR:

Firstly, $\int_{-3}^{-1} f(x)\,dx = 0$ and

$\int_2^3 f(x)\,dx = 0$.

Now $\int_{-1}^{2} f(x)\,dx = 3$, so

$\int_{-4.5}^{-3} f(x)\,dx + \int_3^{4.5} f(x)\,dx = 3$.

$\therefore a = 4.5$ ✓

(1 mark)

43

$$\int_1^4 \frac{1}{3x}\,dx = \frac{1}{3}\int_1^4 \frac{1}{x}\,dx$$

$$= \frac{1}{3}[\ln x]_1^4$$

$$= \frac{1}{3}[\ln 4 - \ln 1]$$

$$= \frac{1}{3}\ln 4$$

Answer B *(1 mark)*

44 As $\int_2^6 f(x)\,dx < 0$ and

$\int_6^8 f(x)\,dx < \left|\int_2^6 f(x)\right|dx$,

then the greatest value is $\int_0^2 f(x)\,dx$.

Answer B *(1 mark)*

45

$$\int_0^{\frac{\pi}{2}} \sec^2\frac{x}{2}\,dx$$

$$= \left[2\tan\frac{x}{2}\right]_0^{\frac{\pi}{2}} \quad ✓$$

$$= 2\left[\tan\frac{\pi}{4} - \tan 0\right] \quad ✓$$

$$= 2[1-0]$$
$$= 2$$

✓ *(3 marks)*

46 **i** Solve simultaneously:

$$x^2 - 3x = 5x - x^2$$
$$2x^2 - 8x = 0$$
$$2x(x-4) = 0$$
$$x = 0, 4$$

$\therefore A$ has x-coordinate of 4 ✓

(1 mark)

ii Area

$$= \int_0^4 5x - x^2 - (x^2 - 3x)\,dx$$

$$= \int_0^4 8x - 2x^2\,dx \quad ✓$$

$$= \left[4x^2 - \frac{2x^3}{3}\right]_0^4 \quad ✓$$

$$= 4 \times 4^2 - \frac{2(4)^3}{3} - 0$$

$$= 64 - \frac{128}{3}$$

$$= \frac{64}{3}$$

$\therefore$ area is $21\frac{1}{3}$ units^2 ✓

(3 marks)

47

$$\int_e^{e^3} \frac{5}{x}\,dx = 5\int_e^{e^3} \frac{1}{x}\,dx$$

$$= 5[\log_e e^3 - \log_e e] \quad ✓$$
$$= 5[3-1]$$
$$= 10 \quad ✓$$

(2 marks)

48 **i** Subs $x = 0$ in

$$y = 2\cos x$$
$$y = 2\cos 0$$
$$= 2 \times 1$$
$$= 2 \quad \therefore P(0, 2)$$

✓ *(1 mark)*

ii $\int_0^{\frac{\pi}{2}} 2\cos x\,dx = [2\sin x]_0^{\frac{\pi}{2}}$ ✓

$$= 2\left(\sin\frac{\pi}{2} - \sin 0\right)$$
$$= 2(1-0)$$
$$= 2 \quad ✓$$

(2 marks)

iii C ✓ *(1 mark)*

iv Area A = Area C

Area B is twice Area A

$\therefore$ Total area $= 4 \times$ Area A

$= 4 \times 2$

$= 8$

The area is 8 units^2 ✓

(1 mark)

v Area B is 4 units^2 below x-axis,
Area C is 2 units^2.

$\therefore$ value of integral $= -4 + 2$
$= -2$ ✓

(1 mark)

49

$$\int_0^6 (x + k)\,dx = 30$$

$$\left[\frac{x^2}{2} + kx\right]_0^6 = 30 \quad ✓$$

$$\left[\frac{6^2}{2} + 6k\right] - \left[\frac{0^2}{2} + 0\right] = 30$$

$$18 + 6k = 30$$

$$6k = 30 - 18$$

$$6k = 12$$

$$k = 2 \quad ✓$$

(2 marks)

50 i

✓

(1 mark)

ii

x	1	2	3
y	0	ln 2	ln 3

Trapezoidal rule:

$$\int_1^3 \ln x\,dx$$

$$\approx \frac{1}{2}[0 + \ln 3 + 2\ln 2] \quad ✓$$

$$\approx \frac{1}{2}[\ln 3 + \ln 4]$$

$$\approx \frac{1}{2}\ln 12$$

$\therefore$ The area is approximately $\frac{1}{2}\ln 12$ units^2. ✓

(2 marks)

iii The shaded region on the diagram in **i** is the area using the trapezoidal rule. It is less than the area under the curve. ✓ *(1 mark)*

51 Area

$$= \int_0^2 e^{2x} - e^{-x}\,dx$$

$$= \left[\frac{1}{2}e^{2x} + e^{-x}\right]_0^2 \quad ✓$$

$$= \left[\frac{1}{2}e^4 + e^{-2}\right] - \left[\frac{1}{2}e^0 + e^0\right] \quad ✓$$

$$= \left[\frac{1}{2}e^4 + e^{-2}\right] - 1\frac{1}{2}$$

$$= \frac{1}{2}\left[e^4 + \frac{2}{e^2} - 3\right]$$

$\therefore$ The area is

$\frac{1}{2}\left[e^4 + \frac{2}{e^2} - 3\right]$ units^2. ✓

(3 marks)

52

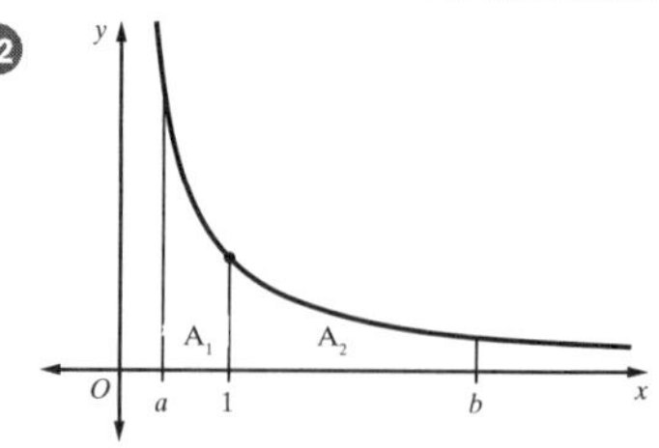

$$A_1 = \int_a^1 \frac{1}{x}\,dx = 1$$

$$\therefore [\log_e x]_a^1 = 1$$

$$\log_e 1 - \log_e a = 1$$

$$0 - \log_e a = 1$$

$$\log_e a = -1$$

$$e^{\log a} = e^{-1}$$

$$a = \frac{1}{e} \quad ✓$$

$$A_2 = \int_1^b \frac{1}{x}\,dx = 1$$

$$\therefore [\log_e x]_1^b = 1$$

$$\log_e b - \log_e 1 = 1$$

$$\log_e b - 0 = 1$$

$$\log_e b = 1 \quad ✓$$

$$e^{\log b} = e^1$$

$$b = e$$

$\therefore a = \frac{1}{e}$ and $b = e$ ✓

(3 marks)

53 i $f(x)$ is increasing when $f'(x) > 0$.
From the graph, $0 < x < 2$. ✓

(1 mark)

ii Maximum value of $f(x)$ when $f'(x) = 0$.
$\therefore$ when $x = 2$.

Now, $\int_0^2 f'(x)\,dx = 4$

$$[f(x)]_0^2 = 4$$

$$f(2) - f(0) = 4$$

But $f(0) = 0$, (given)

$$\therefore f(2) - 0 = 4$$

$$f(2) = 4$$

$\therefore$ The maximum value of $f(x)$ is 4. ✓

(1 mark)

iii Method 1:
As $f(0) = 0$ and $A_1 = A_2$, then $f(4) = 0$.
From $x = 4$ to $x = 6$, the gradient is -3.
As $m = -3$, then $f(6) = -6$.

Method 2:

Now, $\int_2^4 f'(x)\,dx = -4$

$$[f(x)]_2^4 = -4$$

$$f(4) - f(2) = -4$$

But $f(2) = 4$, (from ii)

$$\therefore f(4) - 4 = -4$$

$$f(4) = 0$$

From $x = 4$ to $x = 6$, the gradient is -3.
As $m = -3$, then $f(6) = -6$. ✓

(1 mark)

iv

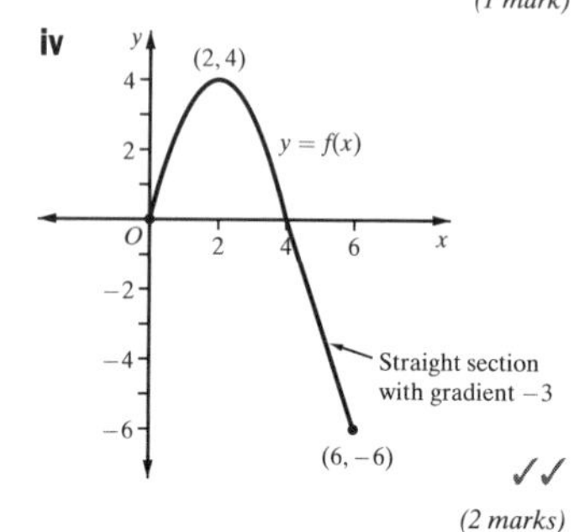

✓✓

(2 marks)

54

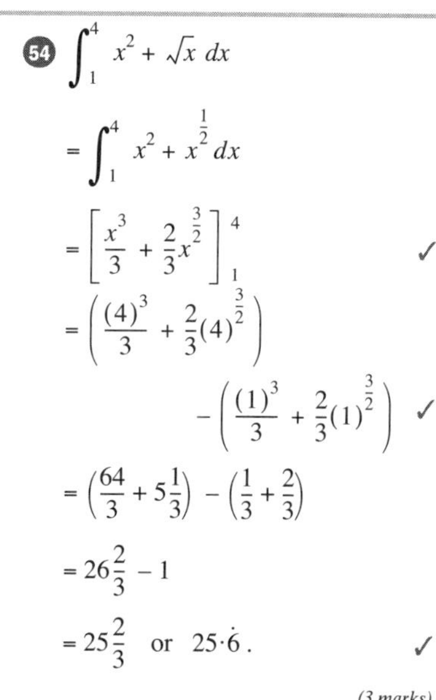

$$\int_1^4 x^2 + \sqrt{x}\,dx$$

$$= \int_1^4 x^2 + x^{\frac{1}{2}}\,dx$$

$$= \left[\frac{x^3}{3} + \frac{2}{3}x^{\frac{3}{2}}\right]_1^4 \quad ✓$$

$$= \left(\frac{(4)^3}{3} + \frac{2}{3}(4)^{\frac{3}{2}}\right) - \left(\frac{(1)^3}{3} + \frac{2}{3}(1)^{\frac{3}{2}}\right) \quad ✓$$

$$= \left(\frac{64}{3} + 5\frac{1}{3}\right) - \left(\frac{1}{3} + \frac{2}{3}\right)$$

$$= 26\frac{2}{3} - 1$$

$$= 25\frac{2}{3} \text{ or } 25{\cdot}\dot{6}. \quad ✓$$

(3 marks)

55

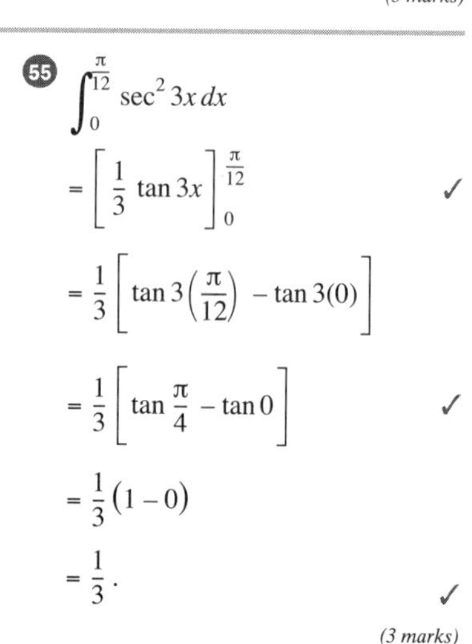

$$\int_0^{\frac{\pi}{12}} \sec^2 3x\,dx$$

$$= \left[\frac{1}{3}\tan 3x\right]_0^{\frac{\pi}{12}} \quad ✓$$

$$= \frac{1}{3}\left[\tan 3\left(\frac{\pi}{12}\right) - \tan 3(0)\right]$$

$$= \frac{1}{3}\left[\tan\frac{\pi}{4} - \tan 0\right] \quad ✓$$

$$= \frac{1}{3}(1 - 0)$$

$$= \frac{1}{3}. \quad ✓$$

(3 marks)

56 i

$y = \log_e(\cos x)$

$\frac{dy}{dx} = \frac{1}{\cos x} \times \frac{d}{dx}(\cos x)$

$= \frac{-\sin x}{\cos x}$ ✓

$= -\tan x.$ ✓

(2 marks)

ii $\int_0^{\frac{\pi}{4}} \tan x \, dx$

$= -\left[\log_e(\cos x)\right]_0^{\frac{\pi}{4}}$ ✓

$= -\left[\log_e\left(\cos\frac{\pi}{4}\right) - \log_e(\cos 0)\right]$

$= -\left[\log_e\frac{1}{\sqrt{2}} - \log_e 1\right]$

$= -\left[\log_e\frac{1}{\sqrt{2}} - 0\right]$

$= -\log_e\frac{1}{\sqrt{2}}$

$= \log_e\sqrt{2}$

$= 0.3465\ldots$

$= 0.35$ to 2 decimal places. ✓

(2 marks)

57 i

$f''(x) = k(b^2 - x^2)$

$\therefore f'(x) = \int k(b^2 - x^2)dx$

$= kb^2x - \frac{kx^3}{3} + C$ ✓

Now $f'(b) = kb^2(b) - k\frac{(b)^3}{3} + C$

$= kb^3 - \frac{kb^3}{3} + C$

$\therefore -f'(b) = -\left(kb^3 - \frac{kb^3}{3} + C\right)$

$= -kb^3 + \frac{kb^3}{3} - C$

and

$f'(-b) = kb^2(-b) - \frac{k(-b)^3}{3} + C$

$= -kb^3 + \frac{kb^3}{3} + C$

Now $f'(-b) = -f'(b)$

$\therefore -kb^3 + \frac{kb^3}{3} + C$

$= -kb^3 + \frac{kb^3}{3} - C$

$\therefore C = -C$

$2C = 0$

$\therefore C = 0$

$\therefore f'(x) = kb^2x - \frac{kx^3}{3}$

$= k\left(b^2x - \frac{x^3}{3}\right).$ ✓

(2 marks)

ii

$f'(x) = k\left(b^2x - \frac{x^3}{3}\right)$

$\therefore f(x) = \int k\left(b^2x - \frac{x^3}{3}\right)dx$

$= \frac{kb^2x^2}{2} - \frac{kx^4}{12} + C_2$

Now $f(x)$ passes through $(b, 0)$,

$\therefore f(b) = \frac{kb^2(b^2)}{2} - \frac{k(b^4)}{12} + C_2$

$= \frac{kb^4}{2} - \frac{kb^4}{12} + C_2$

$= \frac{6kb^4 - kb^4}{12} + C_2$

$= \frac{5kb^4}{12} + C_2$ ✓

$= 0$

$\therefore \frac{5kb^4}{12} + C_2 = 0$

$C_2 = -\frac{5kb^4}{12}$

$\therefore f(x) = \frac{kb^2x^2}{2} - \frac{kx^4}{12} - \frac{5kb^4}{12}$

$\therefore f(0) = \frac{kb^2(0)^2}{2} - \frac{k(0)^4}{12} - \frac{5kb^4}{12}$

$= -\frac{5kb^4}{12}$

$\therefore$ The beam is $\frac{5kb^4}{12}$ units below the x-axis, at $x = 0$. ✓

(2 marks)

58 $y = \log_e(x - 2)$

(N.B. We cannot integrate $y = \log_e(x - 2)$)

$\therefore (x - 2) = e^y$

$\therefore x = e^y + 2$ ✓

Required area = Area of rectangle – Area under the curve bounded by the y-axis

$A = L \times B - \int_0^{\log_e 5} e^y + 2\, dy$ ✓

$= 7 \times \log_e 5 - \left[e^y + 2y\right]_0^{\log_e 5}$ ✓

$= 7\log_e 5 - \left[\left(e^{\log_e 5} + 2\log_e 5\right) - \left(e^0 - 2(0)\right)\right]$

$= 7\log_e 5 - \left[5 + 2\log_e 5 - (1 - 0)\right]$ ✓

$= 7\log_e 5 - \left[4 + 2\log_e 5\right]$

$= 7\log_e 5 - 4 - 2\log_e 5$

$= \left(5\log_e 5 - 4\right)$ units^2. ✓

(5 marks)

59 $\int_1^4 \frac{8}{x^2}dx$

$= 8\int_1^4 x^{-2}dx$

$= 8\left[\frac{x^{-1}}{-1}\right]_1^4$ ✓

$= 8\left[-\frac{1}{x}\right]_1^4$

$= 8\left[\left(-\frac{1}{4}\right) - (-1)\right]$ ✓

$= 8\left[-\frac{1}{4} + 1\right]$

$= 8 \times \frac{3}{4}$

$= 6.$ ✓ *(3 marks)*

60 i $\sqrt{3}\cos x = \sin x$

$\sqrt{3} = \frac{\sin x}{\cos x}$

$\tan x = \sqrt{3}$ ✓

$\therefore x = \frac{\pi}{3}, \pi + \frac{\pi}{3}$

$= \frac{\pi}{3}, \frac{4\pi}{3}$

$\therefore$ x-coordinates of A and B are $\frac{\pi}{3}$ and $\frac{4\pi}{3}$ respectively. ✓

(2 marks)

ii $A = \int_{\frac{\pi}{3}}^{\frac{4\pi}{3}} (\sin x - \sqrt{3}\cos x)\,dx$

$= \left[-\cos x - \sqrt{3}\sin x\right]_{\frac{\pi}{3}}^{\frac{4\pi}{3}}$ ✓

$= -\left[\cos x + \sqrt{3}\sin x\right]_{\frac{\pi}{3}}^{\frac{4\pi}{3}}$

$= -\left\{\left(\cos\frac{4\pi}{3} + \sqrt{3}\sin\frac{4\pi}{3}\right) - \left(\cos\frac{\pi}{3} + \sqrt{3}\sin\frac{\pi}{3}\right)\right\}$

$= -\left\{\left(-\cos\frac{\pi}{3} - \sqrt{3}\sin\frac{\pi}{3}\right) - \left(\cos\frac{\pi}{3} + \sqrt{3}\sin\frac{\pi}{3}\right)\right\}$

$= -\left(-\cos\frac{\pi}{3} - \sqrt{3}\sin\frac{\pi}{3} - \cos\frac{\pi}{3} - \sqrt{3}\sin\frac{\pi}{3}\right)$

$= -\left(-2\cos\frac{\pi}{3} - 2\sqrt{3}\sin\frac{\pi}{3}\right)$ ✓

$= 2\left(\frac{1}{2}\right) + 2\sqrt{3}\left(\frac{\sqrt{3}}{2}\right)$

$= 1 + 3$

$= 4$ units^2. ✓

(3 marks)

61 $\int_0^3 \frac{8x}{1+x^2}\,dx$

$= 4\int_0^3 \frac{2x}{1+x^2}\,dx$ ✓

$= 4\Big[\log_e(1+x^2)\Big]_0^3$ ✓

$= 4\log_e 10 - 4\log_e 1$

$= 4\log_e 10 - 0$

$= 4\log_e 10.$ ✓

(3 marks)

62 **i** $\frac{d}{dx}\log_e(\cos x)$

$= \frac{1}{\cos x}\cdot\frac{d}{dx}(\cos x)$

$= \frac{1}{\cos x}\cdot -\sin x$

$= -\frac{\sin x}{\cos x}$

$= -\tan x.$ ✓

(1 mark)

ii Required area = area under curve – area of triangle

$= \int_0^{\frac{\pi}{4}} \tan x\,dx - \frac{1}{2}\times\frac{\pi}{4}\times\frac{\pi}{4}$ ✓

$= -\int_0^{\frac{\pi}{4}} (-\tan x)\,dx - \frac{\pi^2}{32}$

$= -\Big[\log_e(\cos x)\Big]_0^{\frac{\pi}{4}} - \frac{\pi^2}{32}$ ✓

$= -\Big[\log_e\left(\cos\frac{\pi}{4}\right) - \log_e(\cos 0)\Big] - \frac{\pi^2}{32}$

$= -\Big[\log_e\left(\frac{1}{\sqrt{2}}\right) - \log_e 1\Big] - \frac{\pi^2}{32}$

$= -\Big[-\log_e\sqrt{2}\Big] - \frac{\pi^2}{32}$

$= -\Big[-\log_e 2^{\frac{1}{2}}\Big] - \frac{\pi^2}{32}$

$= -\Big[-\frac{1}{2}\log_e 2\Big] - \frac{\pi^2}{32}$

$= \frac{1}{2}\log_e 2 - \frac{\pi^2}{32}$ ✓

$\therefore$ Area is $(\frac{1}{2}\log_e 2 - \frac{\pi^2}{32})$ units^2.

(3 marks)

63 **i** The curve cuts the x-axis when

$f(x) = 0$

$1 + 2\cos x = 0$

$2\cos x = -1$

$\cos x = -\frac{1}{2}$

$x = -\frac{2\pi}{3}$ and $\frac{2\pi}{3}$.

(for $-\pi \leqslant x \leqslant \pi$)

✓ *(1 mark)*

ii Period $= 2\pi$ ✓

Range: $-1 \leqslant \cos x \leqslant 1$

$\therefore -2 \leqslant 2\cos x \leqslant 2$

$\therefore -1 \leqslant 1 + 2\cos x \leqslant 3$ ✓

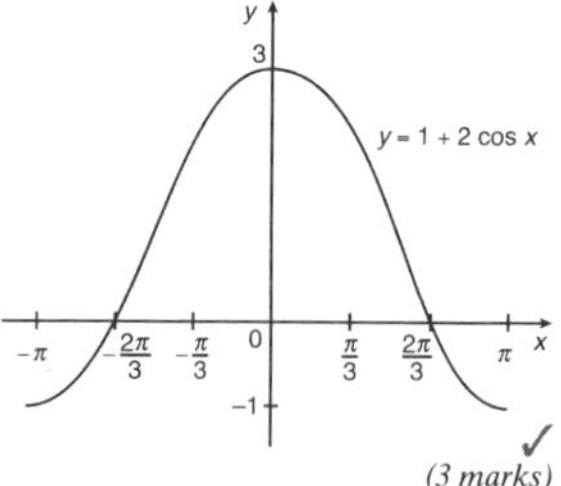

✓

(3 marks)

iii Area $= \int_{-\frac{\pi}{2}}^{\frac{2\pi}{3}} (1 + 2\cos x)\,dx$

$= \Big[x + 2\sin x\Big]_{-\frac{\pi}{2}}^{\frac{2\pi}{3}}$ ✓

$= \left[\frac{2\pi}{3} + 2\sin\frac{2\pi}{3}\right] - \left[-\frac{\pi}{2} + 2\sin\left(-\frac{\pi}{2}\right)\right]$ ✓

$= \left[\frac{2\pi}{3} + 2\left(\frac{\sqrt{3}}{2}\right)\right] - \left[-\frac{\pi}{2} + 2(-1)\right]$

$= \frac{2\pi}{3} + \sqrt{3} + \frac{\pi}{2} + 2$

$= \left(\frac{7\pi}{6} + 2 + \sqrt{3}\right)$ units^2. ✓

(3 marks)

64 $\int_0^{\frac{\pi}{6}} \cos 3x\,dx$

$= \left[\frac{1}{3}\sin 3x\right]_0^{\frac{\pi}{6}}$ ✓

$= \frac{1}{3}\left[\sin 3\left(\frac{\pi}{6}\right) - \sin 3(0)\right]$

$= \frac{1}{3}\left(\sin\frac{\pi}{2} - \sin 0\right)$

$= \frac{1}{3}(1 - 0)$

$= \frac{1}{3}.$ ✓

(2 marks)

65 **i** The displacement x from the origin is a maximum when $t = 2$.
(N.B. $v > 0$ when $t < 2$
$v = 0$ when $t = 2$
$v < 0$ when $t > 2$) ✓

(1 mark)

ii The particle returns to the origin at $t = 4$ by symmetry. (N.B. Area under curve from $t = 0$ to $t = 2$ is of the same magnitude but opposite in sign as area under the curve between $t = 2$ and $t = 4$.)
The particle is initially at the origin. It moves in a positive direction away from the origin for 2 seconds. It then comes to rest at $t = 2$. It then moves back towards the origin for the next 2 seconds, returning to the origin when $t = 4$. (Note the symmetry.)

✓✓ *(2 marks)*

iii

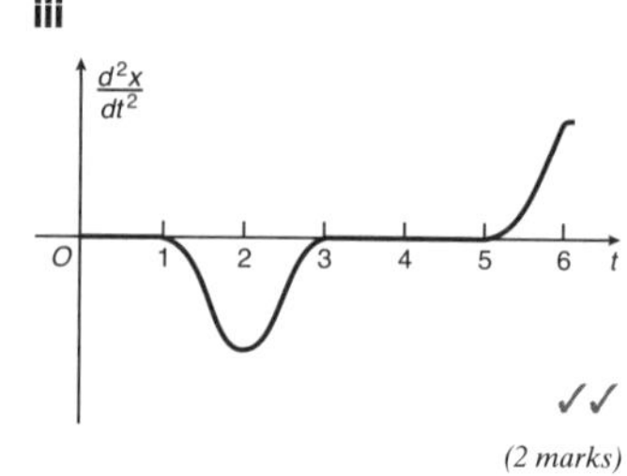

✓✓

(2 marks)

66 Area of shaded region

= Area of quadrant – $\int_0^1 (x^2 - 3x + 2)\,dx$

Now area of circle $= \pi r^2$

$= \pi(2)^2$

$= 4\pi$

$\therefore$ Area of quadrant $= \frac{1}{4}\times 4\pi$

$= \pi$

$\therefore$ Area of shaded region

$= \pi - \int_0^1 (x^2 - 3x + 2)\,dx$ ✓

$= \pi - \left[\frac{x^3}{3} - \frac{3x^2}{2} + 2x\right]_0^1$ ✓

$= \pi - \left[\left(\frac{1}{3} - \frac{3}{2} + 2\right) - 0\right]$

$= \pi - \frac{5}{6}$

$\therefore$ Area $= \left(\pi - \frac{5}{6}\right)$ units^2. ✓

(3 marks)

67 **i** $\ddot{x} = 4\sin 2t$

$\dot{x} = \int 4\sin 2t\,dt$

$= -\frac{1}{2}(4\cos 2t) + C$

$= -2\cos 2t + C$ ✓

When $t = 0, \dot{x} = 0$

$\therefore 0 = -2\cos 0 + C$

$\therefore 0 = -2 + C$

$\therefore C = 2$

$\therefore \dot{x} = -2\cos 2t + 2$

$= 2 - 2\cos 2t.$ ✓

(2 marks)

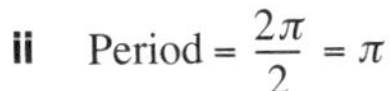

ii Period $= \frac{2\pi}{2} = \pi$ ✓

Range: $-1 \leqslant -\cos t \leqslant 1$

$\therefore -2 \leqslant -2\cos t \leqslant 2$

$\therefore 0 \leqslant 2 - 2\cos t \leqslant 4$ ✓

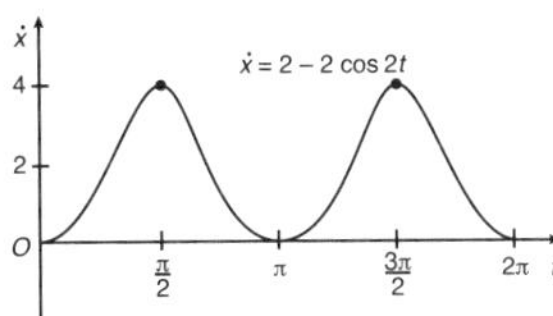

The particle is at rest when $\dot{x} = 0$.

The first time this occurs after $t = 0$ is $t = \pi$. ✓

(3 marks)

iii Distance travelled

$= \int_0^{\pi} (2 - 2\cos 2t)\, dt$

$= \left[2t - \frac{2\sin 2t}{2}\right]_0^{\pi}$ ✓

$= \left[2t - \sin 2t\right]_0^{\pi}$

$= (2\pi - 0) - (0 - 0)$

$= 2\pi$.

$\therefore$ Distance travelled is 2π units. ✓ *(2 marks)*

68 $\int_1^2 e^{3x}\, dx = \left[\frac{1}{3}e^{3x}\right]_1^2$ ✓

$= \frac{1}{3}e^6 - \frac{1}{3}e^3$

$= \frac{1}{3}(e^6 - e^3)$. ✓

(2 marks)

69 i The velocity of the particle is a maximum when $t = 2$.

Note: When $t = 2$, $a = 0$,

when $t < 2$, $a > 0$

and when $t > 2$, $a < 0$

$\therefore$ There is a maximum velocity at $t = 2$. ✓ *(1 mark)*

ii From $t = 0$ to $t = 2$, the acceleration of the particle is positive and velocity reaches a maximum at $t = 2$.

From $t = 2$ to $t = 4$, the acceleration is of the same magnitude as the period $t = 0$ to $t = 2$, but opposite in direction (i.e. negative).

Hence, since the velocity was zero at $t = 0$, the velocity has returned to zero by $t = 4$.

From $t = 4$, the acceleration of the particle continues to be negative and the velocity becomes negative.

$\therefore$ The particle is furthest from the origin at $t = 4$. ✓✓

(2 marks)

70 $\int_{\frac{\pi}{4}}^{\frac{\pi}{3}} \sec^2 x\, dx = \left[\tan x\right]_{\frac{\pi}{4}}^{\frac{\pi}{3}}$ ✓

$= \tan\frac{\pi}{3} - \tan\frac{\pi}{4}$

$= \sqrt{3} - 1$. ✓

(2 marks)

71 i Solving simultaneously:

$y = x - 4$ ...(1)

$y = x^2 - 4x$ ...(2)

Equating (1) and (2):

$x^2 - 4x = x - 4$

$x^2 - 5x + 4 = 0$

$(x - 4)(x - 1) = 0$

$\therefore x = 4$ or 1. ✓

For A, $x = 1$ $\therefore y = 1 - 4$

$= -3$

$\therefore$ Coordinates of A are $(1, -3)$.

✓ *(2 marks)*

ii Required area

$= \int_1^4 (x - 4)\, dx - \int_1^4 (x^2 - 4x)\, dx$ ✓

$= \int_1^4 (x - 4 - x^2 + 4x)\, dx$

$= \int_1^4 (5x - x^2 - 4)\, dx$

$= \left[\frac{5x^2}{2} - \frac{x^3}{3} - 4x\right]_1^4$ ✓

$= \left[\frac{5(4)^2}{2} - \frac{(4)^3}{3} - 4(4)\right] - \left[\frac{5(1)^2}{2} - \frac{(1)^3}{3} - 4(1)\right]$

$= \left[40 - \frac{64}{3} - 16\right] - \left[\frac{5}{2} - \frac{1}{3} - 4\right]$

$= \left[\frac{120}{3} - \frac{64}{3} - \frac{48}{3}\right] - \left[\frac{15}{6} - \frac{2}{6} - \frac{24}{6}\right]$

$= \frac{8}{3} - \left(-\frac{11}{6}\right)$

$= \frac{16}{6} + \frac{11}{6}$

$= \frac{27}{6}$

$= \frac{9}{2}$

$= 4\frac{1}{2}$ units^2. ✓

(3 marks)

72 $\int_0^1 (e^{5x} - 1) = \left[\frac{1}{5}e^{5x} - x\right]_0^1$ ✓

$= \left(\frac{1}{5}e^5 - 1\right) - \left(\frac{1}{5}e^0 - 0\right)$

$= \frac{1}{5}e^5 - 1 - \frac{1}{5}$

$= \frac{1}{5}e^5 - \frac{6}{5}$

$= \frac{1}{5}(e^5 - 6)$. ✓

(2 marks)

73 i Solving simultaneously:

$y = 2x$ ...(1)

$y = 6x - x^2$ ...(2)

Equating (1) and (2):

$6x - x^2 = 2x$

$4x - x^2 = 0$

$x(4 - x) = 0$

$\therefore x = 0$ or 4

At $x = 4$, $y = 2(4)$

$= 8$

$\therefore$ Coordinates of B are $(4, 8)$. ✓

(1 mark)

ii Required area

$= \int_0^4 (6x - x^2)dx - \int_0^4 2x\, dx$

$= \int_0^4 (4x - x^2)dx$ ✓

$= \left[2x^2 - \frac{x^3}{3}\right]_0^4$ ✓

$= \left(2(4)^2 - \frac{(4)^3}{3}\right) - (0)$

$= 32 - \frac{64}{3}$

$= \frac{96 - 64}{3}$

$= \frac{32}{3}$

$= 10\frac{2}{3}$ units^2. ✓

(3 marks)

74 $\int_0^1 \frac{dx}{x + 4} = \left[\ln(x + 4)\right]_0^1$ ✓

$= \ln 5 - \ln 4$

$= \ln\frac{5}{4}$

$= \ln 1.25$

$\doteqdot 0.22$. ✓

(2 marks)

75 i and ii

Range: $-3 \le y \le 3$ ✓

Period: $\frac{2\pi}{2} = \pi$ ✓

✓

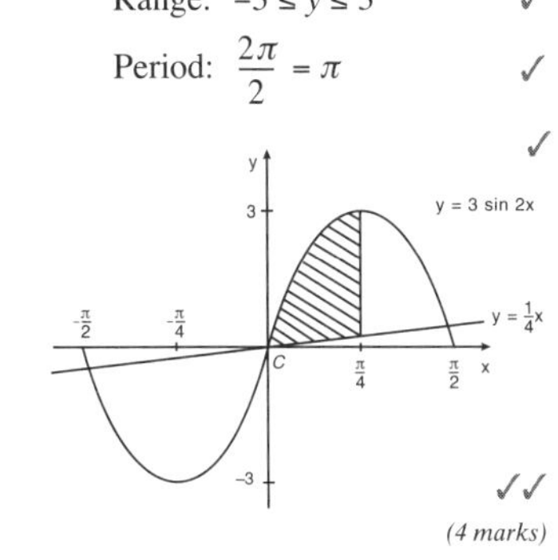

✓✓

(4 marks)

iii $\int_0^{\frac{\pi}{4}} \left(3 \sin 2x - \frac{1}{4}x\right) dx$

$= \left[-\frac{3}{2} \cos 2x - \frac{x^2}{8}\right]_0^{\frac{\pi}{4}}$ ✓

$= \left[\left(-\frac{3}{2} \cos \frac{2\pi}{4} - \frac{(\frac{\pi}{4})^2}{8}\right) - \left(-\frac{3}{2} \cos 0 - 0\right)\right]$

$= \left[\left(-\frac{3}{2} \cos \frac{\pi}{2} - \frac{\pi^2}{128}\right) - \left(-\frac{3}{2}\right)\right]$

$= 0 - \frac{\pi^2}{128} + \frac{3}{2}$

$= \frac{3}{2} - \frac{\pi^2}{128}$ ✓

(2 marks)

76 i

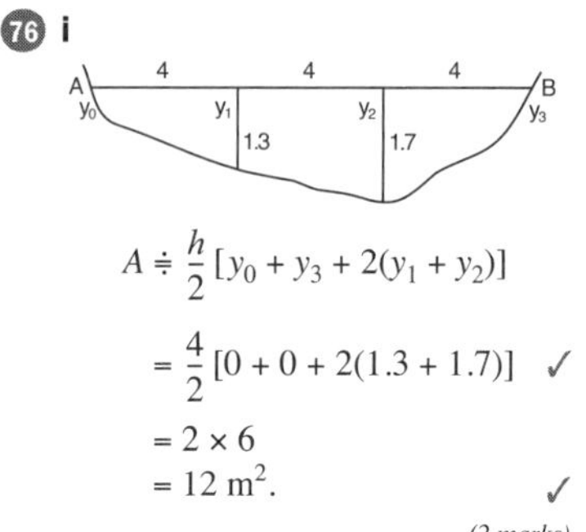

$A \doteqdot \frac{h}{2}[y_0 + y_3 + 2(y_1 + y_2)]$

$= \frac{4}{2}[0 + 0 + 2(1.3 + 1.7)]$ ✓

$= 2 \times 6$

$= 12 \text{ m}^2.$ ✓

(2 marks)

ii $V = A \times \text{length}$

length = speed × time
$= 0.5 \times (60 \times 60)$
$= 1800$ m ✓

$V = 12 \times 1800$
$= 21\,600 \text{ m}^3.$ ✓

(2 marks)

77 $\int_{-2}^{1} \frac{2}{x+3}\, dx = 2\int_{-2}^{1} \frac{1}{x+3}\, dx$

$= 2\left[\ln (x+3)\right]_{-2}^{1}$

$= 2(\ln 4 - \ln 1)$

$= 2 \ln 4.$

(No mark allocation in exam)

78 i Solving $y = x^2 - 2$ and $y = x$ simultaneously:

$x^2 - 2 = x$

$x^2 - x - 2 = 0$

$(x-2)(x+1) = 0$ ✓

$\therefore x = 2$ or -1

∴ The x values are 2 and −1. ✓

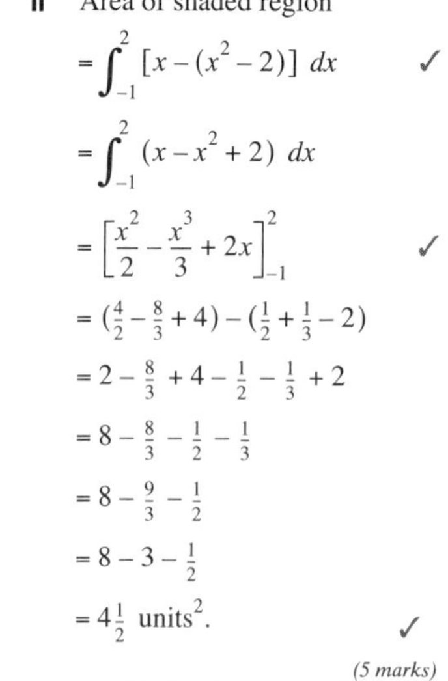

ii Area of shaded region

$= \int_{-1}^{2} [x - (x^2 - 2)]\, dx$ ✓

$= \int_{-1}^{2} (x - x^2 + 2)\, dx$

$= \left[\frac{x^2}{2} - \frac{x^3}{3} + 2x\right]_{-1}^{2}$ ✓

$= (\frac{4}{2} - \frac{8}{3} + 4) - (\frac{1}{2} + \frac{1}{3} - 2)$

$= 2 - \frac{8}{3} + 4 - \frac{1}{2} - \frac{1}{3} + 2$

$= 8 - \frac{8}{3} - \frac{1}{2} - \frac{1}{3}$

$= 8 - \frac{9}{3} - \frac{1}{2}$

$= 8 - 3 - \frac{1}{2}$

$= 4\frac{1}{2} \text{ units}^2.$ ✓

(5 marks)
(Total mark allocation only in exam)

79 i At B, $y = 0$

$\therefore 4x - x^2 = 0$

$x(4 - x) = 0$

$\therefore x = 0$ or 4

∴ At B, $x = 4$.

ii $A = \int_0^4 (4x - x^2)dx$

$= \left[2x^2 - \frac{x^3}{3}\right]_0^4$

$= \left(2(4)^2 - \frac{(4)^3}{3}\right) - 0$

$= 32 - \frac{64}{3}$

$= \frac{96}{3} - \frac{64}{3}$

$= \frac{32}{3}$

$= 10\frac{2}{3} \text{ units}^2.$

(No mark allocation in exam)

80 $\int_{-2}^{5} f(x)dx$ = Area A – Area B + Area C

$= 8 - 3 + 1$ ✓

$= 6$ ✓

(2 marks)

81 $A = \int_0^{\frac{\pi}{4}} (\sec^2 x - x)\, dx$

$= \left[\tan x - \frac{x^2}{2}\right]_0^{\frac{\pi}{4}}$ ✓

$= \left(\tan (\frac{\pi}{4}) - \frac{(\frac{\pi}{4})^2}{2}\right) - (\tan 0 - 0)$ ✓

$= 1 - \frac{\pi^2}{32} - 0$

$= (1 - \frac{\pi^2}{32}) \text{ units}^2.$ ✓

(3 marks)

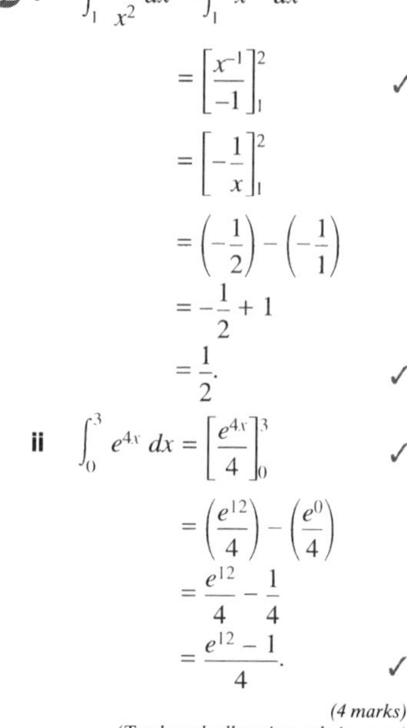

82 i $\int_1^2 \frac{1}{x^2}\, dx = \int_1^2 x^{-2}\, dx$

$= \left[\frac{x^{-1}}{-1}\right]_1^2$ ✓

$= \left[-\frac{1}{x}\right]_1^2$

$= \left(-\frac{1}{2}\right) - \left(-\frac{1}{1}\right)$

$= -\frac{1}{2} + 1$

$= \frac{1}{2}.$ ✓

ii $\int_0^3 e^{4x}\, dx = \left[\frac{e^{4x}}{4}\right]_0^3$ ✓

$= \left(\frac{e^{12}}{4}\right) - \left(\frac{e^0}{4}\right)$

$= \frac{e^{12}}{4} - \frac{1}{4}$

$= \frac{e^{12} - 1}{4}.$ ✓

(4 marks)
(Total mark allocation only in exam)

83

x	1.0	2.0	3.0
$f(x)$	1.7	9.0	4.3
	y_0	y_1	y_2

$\int_1^3 f(x)\, dx$

$\doteqdot \frac{h}{2}[y_0 + y_2 + 2(y_1)]$

$= \frac{1}{2}[1.7 + 4.3 + 2(9.0)]$

$= \frac{1}{2}[24]$

$= 12.$

(No mark allocation in exam)

84 i $R = 100t - t^3$

At $t = 8$, $R = 100(8) - (8)^3$
$= 800 - 512$
$= 288$

∴ Flow rate is 288 kg/s. ✓

ii The expression is reasonable until $R = 0$

$\therefore 0 = 100t - t^3$
$= t(100 - t^2)$

$\therefore t = 0, 10$, or -10

(not possible)

∴ The largest value of T is 10. ✓

iii $R = 100t - t^3$

$\frac{dR}{dt} = 100 - 3t^2$ $\frac{d^2R}{dt^2} = -6t$

Stationary points occur when $\frac{dR}{dt} = 0$

$$100 - 3t^2 = 0$$
$$3t^2 = 100$$
$$t^2 = \frac{100}{3}$$
$$t = \pm\sqrt{\frac{100}{3}}$$
$$= \frac{10}{\sqrt{3}} \quad (\text{since } t > 0)$$

When $t = \frac{10}{\sqrt{3}}$, $\frac{d^2R}{dt^2} = -6\left(\frac{10}{\sqrt{3}}\right)$

$$= -\frac{60}{\sqrt{3}} < 0$$

∴ A maximum value occurs at $t = \frac{10}{\sqrt{3}}$. ✓

When $t = \frac{10}{\sqrt{3}}$,

$$R = 100\left(\frac{10}{\sqrt{3}}\right) - \left(\frac{10}{\sqrt{3}}\right)^3$$
$$= \frac{1000}{\sqrt{3}} - \frac{1000}{3\sqrt{3}}$$
$$= \frac{3000 - 1000}{3\sqrt{3}}$$
$$= \frac{2000}{3\sqrt{3}}$$
$$= \frac{2000\sqrt{3}}{9}$$
$$\doteqdot 384.9$$

∴ The maximum flow rate is approximately 384.9 kg/s. ✓

iv $R = 100t - t^3$

If M is the amount in the pile at time t:

$$\frac{dM}{dt} = 100t - t^3$$
$$\therefore M = 50t^2 - \frac{t^4}{4} + C$$

When $t = 0$, $M = 300$ ∴ $C = 300$

$$\therefore M = 50t - \frac{t^4}{4} + 300. \quad ✓$$

v $\int_0^8 (100t - t^3)\,dt$

$$= \left[\frac{100t^2}{2} - \frac{t^4}{4}\right]_0^8$$
$$= \left[50t^2 - \frac{t^4}{4}\right]_0^8 \quad ✓$$
$$= \left(50(64) - \frac{4096}{4}\right) - 0$$
$$= 3200 - 1024$$
$$= 2176$$

∴ Total weight of sand tipped from truck in first 8 seconds was 2176 kg. ✓

(7 marks)
(Total mark allocation only in exam)

85 The area of the shaded region is given by the integral of the upper function less the lower function as the curves are continuous and do not intersect in the interval $\left[-\frac{\pi}{2}, \frac{\pi}{6}\right]$.

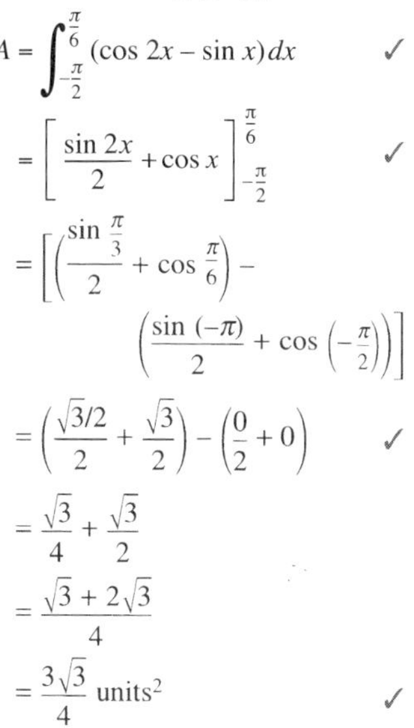

$$A = \int_{-\frac{\pi}{2}}^{\frac{\pi}{6}} (\cos 2x - \sin x)\,dx \quad ✓$$
$$= \left[\frac{\sin 2x}{2} + \cos x\right]_{-\frac{\pi}{2}}^{\frac{\pi}{6}} \quad ✓$$
$$= \left[\left(\frac{\sin\frac{\pi}{3}}{2} + \cos\frac{\pi}{6}\right) - \left(\frac{\sin(-\pi)}{2} + \cos\left(-\frac{\pi}{2}\right)\right)\right]$$
$$= \left(\frac{\sqrt{3}/2}{2} + \frac{\sqrt{3}}{2}\right) - \left(\frac{0}{2} + 0\right) \quad ✓$$
$$= \frac{\sqrt{3}}{4} + \frac{\sqrt{3}}{2}$$
$$= \frac{\sqrt{3} + 2\sqrt{3}}{4}$$
$$= \frac{3\sqrt{3}}{4} \text{ units}^2 \quad ✓$$

(4 marks)

86 Area of shaded region

$$= \int_0^{\frac{\pi}{6}} \sin 2x\,dx + \int_{\frac{\pi}{6}}^{\frac{\pi}{2}} \cos x\,dx \quad ✓$$
$$= \left[-\frac{\cos 2x}{2}\right]_0^{\frac{\pi}{6}} + \left[\sin x\right]_{\frac{\pi}{6}}^{\frac{\pi}{2}} \quad ✓$$
$$= \left[-\frac{\cos\frac{\pi}{3}}{2} - \left(-\frac{\cos 0}{2}\right)\right] + \left(\sin\frac{\pi}{2} - \sin\frac{\pi}{6}\right)$$
$$= -\tfrac{1}{4} - \left(-\tfrac{1}{2}\right) + 1 - \tfrac{1}{2}$$
$$= -\tfrac{1}{4} + \tfrac{1}{2} + 1 - \tfrac{1}{2}$$
$$= \tfrac{3}{4} \text{ units}^2. \quad ✓$$

(3 marks)

87 $\int_0^{\ln 7} e^{-x}\,dx$

$$= \left[-e^{-x}\right]_0^{\ln 7} \quad ✓$$
$$= -e^{-\ln 7} - (-e^0)$$
$$= -\frac{1}{e^{\ln 7}} - (-1)$$
$$= -\frac{1}{7} + 1$$
$$= \frac{6}{7} \quad ✓$$

(2 marks)

88 $\int_1^{e^3} \frac{5}{x}\,dx = \left[5 \ln x\right]_1^{e^3}$ ✓

$$= 5(\ln e^3 - \ln 1)$$
$$= 5(3 \ln e - 0) \quad ✓$$
$$= 5 \times 3$$
$$= 15. \quad ✓$$

(3 marks)

89 i

x	0	1	2	3	4	5
$f(x)$	5.000	4.899	4.583	4.000	3.000	0.000
	y_0	y_1	y_2	y_3	y_4	y_5

✓✓

ii $\int_0^5 \sqrt{25 - x^2}\,dx$

$$\doteqdot \frac{b-a}{2n}[y_0 + 2(y_1 + y_2 + y_3 + y_4) + y_5] \quad ✓$$
$$= \frac{5-0}{2 \times 5}[5.000 + 2(4.899 + 4.583 + 4.000 + 3.000) + 0.000]$$
$$= \frac{5}{10}[37.964]$$
$$= 18.982 \quad ✓$$

iii

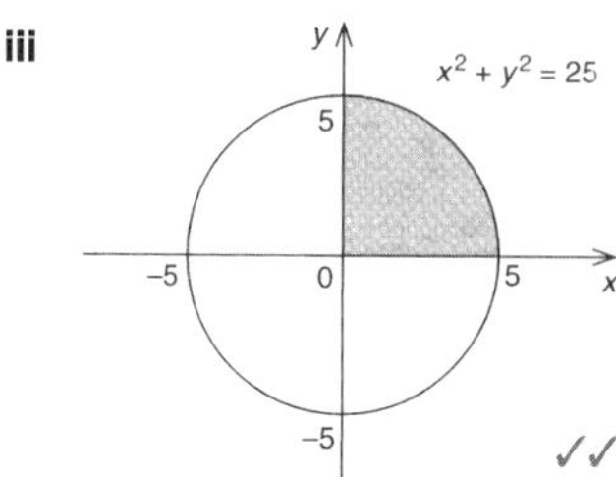

✓✓

iv Area of circle $= \pi r^2$

$$\therefore \int_0^5 \sqrt{25 - x^2}\,dx = \frac{\pi r^2}{4}$$

(quarter of circle)

$$= \frac{\pi(5)^2}{4}$$
$$= \frac{25\pi}{4}. \quad ✓$$

v From parts **ii** and **iv**:

$$\frac{25\pi}{4} \doteqdot 18.982 \quad ✓$$
$$\pi \doteqdot 18.982 \times \frac{4}{25}$$
$$\pi \doteqdot 3.037\ldots$$
$$\pi \doteqdot 3.04 \quad ✓$$

(9 marks)
(Total mark allocation only in exam)

90 i $A = \int_1^2 (e^x - 1)\,dx - \int_1^2 e^{-x}\,dx$

$$= \int_1^2 (e^x - 1 - e^{-x})\,dx$$
$$= \left[e^x - x + e^{-x}\right]_1^2$$
$$= (e^2 - 2 + e^{-2}) - (e^1 - 1 + e^{-1})$$
$$= e^2 - 2 + \frac{1}{e^2} - e + 1 - \frac{1}{e}$$
$$= \left(e^2 + \frac{1}{e^2} - e - \frac{1}{e} - 1\right) \text{ units}^2$$

ii The curves intersect when

$$e^x - 1 = e^{-x}$$
$$\therefore e^x - 1 = \frac{1}{e^x}$$
$$\therefore e^{2x} - e^x = 1$$
$$\therefore e^{2x} - e^x - 1 = 0.$$

(No mark allocation in exam)

91 **a**

$$v = 2t + 4$$

i.e. $\frac{dx}{dt} = 2t + 4$

$\therefore x = t^2 + 4t + C$ ✓

When $t = 0, x = 4 \therefore C = 4$

$\therefore x = t^2 + 4t + 4$ ✓

(2 marks)

b

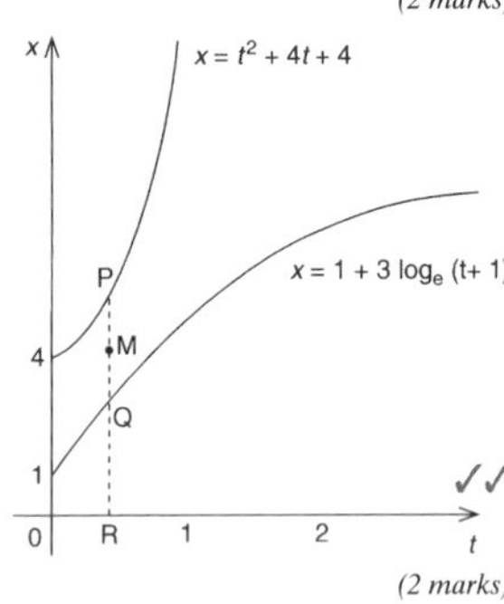

✓✓

(2 marks)

c The position of M is given by

$$\frac{PR + QR}{2}$$

i.e. $x = \frac{1}{2}[t^2 + 4t + 4 + 1 + 3\log_e(t + 1)]$

$= \frac{1}{2}[t^2 + 4t + 3\log_e(t + 1) + 5]$

✓ *(1 mark)*

d $v = \frac{dx}{dt}$

$= \frac{1}{2}\left[2t + 4 + \frac{3}{t + 1}\right]$

$= \frac{1}{2}[2t + 4 + 3(t + 1)^{-1}]$ ✓

$a = \frac{d^2x}{dt^2}$

$= \frac{1}{2}[2 - 3(t + 1)^{-2}]$

$= \frac{1}{2}\left[2 - \frac{3}{(t + 1)^2}\right]$ ✓

When $a = 0$:

$\therefore \frac{1}{2}\left[2 - \frac{3}{(t + 1)^2}\right] = 0$

$2 - \frac{3}{(t + 1)^2} = 0$

$2(t + 1)^2 - 3 = 0$

$2(t + 1)^2 = 3$

$(t + 1)^2 = \frac{3}{2}$

$t + 1 = \pm\sqrt{\frac{3}{2}}$

$\therefore t = \sqrt{\frac{3}{2}} - 1$ ✓

($-\sqrt{\frac{3}{2}} - 1$ is ignored as it is negative)

(3 marks)

e

$D = (t^2 + 4t + 4) - (1 + 3\log_e(t + 1))$

$= t^2 + 4t - 3\log_e(t + 1) + 3$

$\frac{dD}{dt} = 2t + 4 - \frac{3}{t + 1}$

$= 2t + 4 - 3(t + 1)^{-1}$ ✓

$\frac{d^2D}{dt^2} = 2 + 3(t + 1)^{-2}$

$= 2 + \frac{3}{(t + 1)^2}$ ✓

Stationary points occur when

$\frac{dD}{dt} = 0$

$2t + 4 - \frac{3}{t + 1} = 0$

$(2t + 4)(t + 1) - 3 = 0$

$2t^2 + 6t + 4 - 3 = 0$

$2t^2 + 6t + 1 = 0$

$t = \frac{-6 \pm \sqrt{(6)^2 - 4(2)(1)}}{2(2)}$

$= \frac{-6 \pm \sqrt{28}}{4}$ ✓

These are both negative times which do not exist.

∴ The minimum distance occurs when $t = 0$ (as the curves are diverging

i.e. $D = 4 - 1$

$= 3$ ✓ *(4 marks)*

92 $\int_0^{\frac{\pi}{2}} \sin 2x \, dx$

$= \left[-\frac{1}{2}\cos 2x\right]_0^{\frac{\pi}{2}}$

$= -\frac{1}{2}(\cos \pi - \cos 0)$

$= -\frac{1}{2}(-1 - 1)$

$= -\frac{1}{2} \times -2$

$= 1.$

(No mark allocation in exam)

93 **i** Solving simultaneously:

$\frac{4}{x} = 5 - x$

$4 = 5x - x^2$

$x^2 - 5x + 4 = 0$ ✓

$(x - 4)(x - 1) = 0$

$\therefore x = 4 \text{ or } 1$ ✓

ii Required area

$= \int_1^4 (5 - x)\, dx - \int_1^4 \frac{4}{x}\, dx$ ✓

$= \int_1^4 \left(5 - x - \frac{4}{x}\right) dx$

$= \left[5x - \frac{x^2}{2} - 4\ln x\right]_1^4$ ✓

$= (20 - 8 - 4\ln 4) - (5 - \frac{1}{2} - 4\ln 1)$

$= (12 - 4\ln 4) - (4\frac{1}{2})$

$= (7\frac{1}{2} - 4\ln 4) \text{ units}^2.$ ✓

(5 marks)
(Total mark allocation only in exam)

94 **i** and **ii**

✓✓

iii Between $t = 0$ and $t = 7$, the particle is moving away from its initial position because its velocity is positive. ✓

Between $t = 0$ and $t = 10$, the particle is moving towards its initial position because its velocity is negative. ✓

The particle is furthest from its initial position at $t = 7$. ✓

(5 marks)
(Total mark allocation only in exam)

95 $\int_1^9 (1 + \sqrt{x})\, dx$

$= \int_1^9 (1 + x^{\frac{1}{2}})\, dx$

$= \left[x + \frac{2}{3}x^{\frac{3}{2}}\right]_1^9$

$= [9 + \frac{2}{3}(9)^{\frac{3}{2}}] - [1 + \frac{2}{3}(1)]$

$= [9 + \frac{2}{3}(27)] - [1 + \frac{2}{3}]$

$= [9 + 18] - [1\frac{2}{3}]$

$= 27 - 1\frac{2}{3}$

$= 25\frac{1}{3}.$

(No mark allocation in exam)

96 $\frac{d}{dx}(e^{x^2}) = 2xe^{x^2}$

$\therefore \frac{1}{2} \cdot \frac{d}{dx}(e^{x^2}) = xe^{x^2}$

$\therefore \int_0^1 xe^{x^2}\, dx = \frac{1}{2}\left[e^{x^2}\right]_0^1$

$= \frac{1}{2}(e^1 - e^0)$

$= \frac{1}{2}(e - 1).$

(No mark allocation in exam)

97 **i** Solving simultaneously:

$y = x + 2$ ...(1)

$y = x^2 - 5x + 11$ ...(2)

Substituting (2) into (1):

$x^2 - 5x + 11 = x + 2$

$x^2 - 6x + 9 = 0$

$(x - 3)^2 = 0$

∴ There is only 1 point of intersection which is $x = 3$.

∴ The line is a tangent to the parabola.

ii At Q, $x = 3$

$\therefore y = 3 + 2 = 5$

∴ Q is the point (3, 5)

The gradient of $y = x + 2$ is 1.

∴ The gradient of the normal is –1.

The equation of the normal through the point $Q(3, 5)$ with gradient = –1 is

$$y - 5 = -1(x - 3)$$
$$y - 5 = -x + 3$$
$$y = -x + 8$$

iii Solving simultaneously to find the upper and lower limits:

$$y = -x + 8 \quad \ldots (1)$$
$$y = x^2 - 5x + 11 \quad \ldots (2)$$

Substituting (2) into (1):

$$x^2 - 5x + 11 = -x + 8$$
$$x^2 - 4x + 3 = 0$$
$$(x - 1)(x - 3) = 0$$
$$\therefore x = 1 \text{ or } 3$$

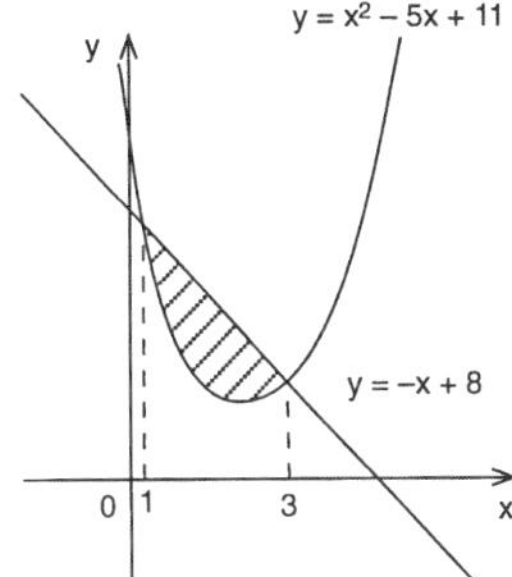

Required area

$$= \int_1^3 (-x + 8)\,dx - \int_1^3 (x^2 - 5x + 11)\,dx$$
$$= \int_1^3 (-x + 8 - x^2 + 5x - 11)\,dx$$
$$= \int_1^3 (-x^2 + 4x - 3)\,dx$$
$$= \left[-\frac{x^3}{3} + 2x^2 - 3x\right]_1^3$$
$$= (-\tfrac{27}{3} + 18 - 9) - (-\tfrac{1}{3} + 2 - 3)$$
$$= (-9 + 18 - 9) - (-\tfrac{1}{3} - 1)$$
$$= 0 - (-1\tfrac{1}{3})$$
$$= 1\tfrac{1}{3} \text{ units}^2.$$

(No mark allocation in exam)

98 $\displaystyle\int_0^{\frac{\pi}{2}} \sin 2x\,dx = \left[-\frac{1}{2}\cos 2x\right]_0^{\frac{\pi}{2}}$

$$= -\frac{1}{2}\left[\cos 2\left(\frac{\pi}{2}\right) - \cos 2(0)\right]$$
$$= -\tfrac{1}{2}[\cos \pi - \cos 0]$$
$$= -\tfrac{1}{2}[-1 - 1]$$
$$= -\tfrac{1}{2} \times -2$$
$$= 1$$

(No mark allocation in exam)

1 Eli is choosing between two investment options.

Option 1: Depositing a single amount of \$40000 today, earning interest of 1.2% per annum, compounded monthly.

Option 2: Depositing \$1000 at the end of each quarter, earning interest of 2.4% per annum, compounded quarterly.

A table of future value interest factors for an annuity of \$1 is shown.

N \ r	0.002	0.006	0.020	0.024	0.060	0.240
	Interest rate per period as a decimal					
10	10.09048	10.27437	10.94972	11.15211	13.18079	31.64344
20	20.38460	21.18211	24.29737	25.28909	36.78559	303.60062
30	30.88646	32.76227	40.56808	43.20983	79.05819	2640.91639
40	41.60026	45.05630	60.40198	65.92708	154.76197	22 728.80260

a What is the value of Eli's investment after 10 years using Option 1? *(2 marks)* Easy

b What is the difference between the future values after 10 years using Option 1 and Option 2? *(2 marks)* Easy

CQ (Q21, **2022 HSC**)

2 A table of future value interest factors for an annuity of \$1 is shown.

Table of future value interest factors

Number of periods	Interest rate per period				
	0.25%	0.5%	0.75%	1%	1.25%
2	2.0025	2.0050	2.0075	2.0100	2.0125
4	4.0150	4.0301	4.0452	4.0604	4.0756
6	6.0376	6.0755	6.1136	6.1520	6.1907
8	8.0704	8.1414	8.2132	8.2857	8.3589
10	10.1133	10.2280	10.3443	10.4622	10.5817

Simone deposits \$1000 into a savings account at the end of each year for 8 years.

The interest rate for these 8 years is 0.75% per annum, compounded annually.

After the 8th deposit, Simone stops making deposits but leaves the money in the savings account. The money in her savings account then earns interest at 1.25% per annum, compounded annually, for a further two years.

Find the amount of money in Simone's savings account at the end of ten years. *(3 marks)*

CQ (Q25, **2021 HSC**) Medium

3 A table of future values for an annuity of \$1 is shown.

Ian is paid an annual salary of \$108000 and he contributes 9.5% of his salary into a superannuation account at the end of each year.

At the contribution stage his fund manager charges \$2120 each year for fees. If the remainder of the money earns interest of 6% p.a. what will be the balance in his fund after 7 years? Give your answer to the nearest dollar. *(2 marks)*

Table of future value interest factors

Period	Interest rate per period							
	0.50%	**1.00%**	**1.50%**	**2.00%**	**3.00%**	**4.00%**	**5.00%**	**6.00%**
1	1.0000	1.0000	1.0000	1.0000	1.0000	1.0000	1.0000	1.0000
2	2.0050	2.0100	2.0150	2.0200	2.0300	2.0400	2.0500	2.0600
3	3.0150	3.0301	3.0452	3.0604	3.0909	3.1216	3.1525	3.1836
4	4.0301	4.0604	4.0909	4.1216	4.1836	4.2465	4.3101	4.3746
5	5.0503	5.1010	5.1523	5.2040	5.3091	5.4163	5.5256	5.6371
6	6.0755	6.1520	6.2296	6.3081	6.4684	6.6330	6.8019	6.9753
7	7.1059	7.2135	7.3230	7.4343	7.6625	7.8983	8.1420	8.3938
8	8.1414	8.2857	8.4328	8.5830	8.8923	9.2142	9.5491	9.8975

Bonus question (see page iv) Medium

4 A table of future values for an annuity of \$1 is shown.

i Lincoln contributes \$6000 at the end of each quarter into his superannuation account which gains interest at 6% p.a. compounded quarterly for two years.
What is the total amount of interest earned? *(2 marks)* Medium

Table of future value interest factors

Period	Interest rate per period							
	0.50%	**1.00%**	**1.50%**	**2.00%**	**3.00%**	**4.00%**	**5.00%**	**6.00%**
1	1.0000	1.0000	1.0000	1.0000	1.0000	1.0000	1.0000	1.0000
2	2.0050	2.0100	2.0150	2.0200	2.0300	2.0400	2.0500	2.0600
3	3.0150	3.0301	3.0452	3.0604	3.0909	3.1216	3.1525	3.1836
4	4.0301	4.0604	4.0909	4.1216	4.1836	4.2465	4.3101	4.3746
5	5.0503	5.1010	5.1523	5.2040	5.3091	5.4163	5.5256	5.6371
6	6.0755	6.1520	6.2296	6.3081	6.4684	6.6330	6.8019	6.9753
7	7.1059	7.2135	7.3230	7.4343	7.6625	7.8983	8.1420	8.3938
8	8.1414	8.2857	8.4328	8.5830	8.8923	9.2142	9.5491	9.8975

ii Mia makes six-monthly contributions into her superannuation account.
What contribution would provide a future value of $67 581.48 after 4 years at 6% p.a. with interest compounded six-monthly? *(1 mark)* **Medium**

Bonus question

5 A table of future value interest factors is used to compare annuities of $1.

Table of future value interest factors						
	Interest rate per period					
Period	**0.50%**	**1.50%**	**2.00%**	**3.00%**	**4.00%**	**6.00%**
4	4.0301	4.0909	4.1216	4.1836	4.2465	4.3746
8	8.1414	8.4328	8.5830	8.8923	9.2142	9.8975
16	16.6142	17.9324	18.6393	20.1569	21.8245	25.6725
32	34.6086	40.6883	44.2270	52.5028	62.7015	90.8898
48	54.0978	69.5652	79.3535	104.4084	139.2632	256.5645
96	122.8285	211.7202	284.6467	535.8502	1054.2960	4462.6505

Lucy and Archie were offered interest of 6% p.a. for 4 years. Lucy invested $1500 each quarter compounded quarterly. Archie invested $500 every month compounded monthly. Which person earned the greater amount of interest? Support your answer with calculations. *(3 marks)*

Bonus question **Medium**

6 The table shows the present value interest factors for various interest rates and periods of time.

Table of present value interest factors							
Period	Interest rate per period						
	0.50%	**1.00%**	**1.25%**	**1.50%**	**2.00%**	**3.00%**	**4.00%**
20	18.9874	18.0456	17.5993	17.1686	16.3514	14.8775	13.5903
30	27.7941	25.8077	24.8889	24.0158	22.3965	19.6004	17.2920
40	36.1722	32.8347	31.3269	29.9158	27.3555	23.1148	19.7928
50	44.1428	39.1961	37.0129	34.9997	31.4236	25.7298	21.4822
120	90.0735	69.7005	61.9828	55.4985	45.3554	32.3730	24.7741
180	118.5035	83.3217	71.4496	62.0956	48.5844	33.1703	24.9785
240	139.5808	90.8194	75.9423	64.7957	49.5686	33.3057	24.9980

Mel is 40 years old today and plans to retire when she turns 60 years old. In retirement, she would like to receive an annual annuity of $60 000 until she turns 90. What single amount should she deposit today in a savings account in order to have an investment large enough to receive her annual annuity in retirement, assuming an interest rate of 4% per annum for the entire 50 years? *(4 marks)*

Bonus question **Hard**

7 The market value of Kate's house has been increasing at a constant value of 2.5% per annum since she bought it at the start of 2007.

At the start of 2019 the house was valued at $435 000.

i What was the value of the house at the start of 2007 when Kate bought it?
Express your answer to the nearest dollar. *(2 marks)* **Medium**

ii Daniel also bought a house at the start of 2007 for $305 000.
The value of his house has been increasing at the rate of 3.2% per annum.
In what year did the two houses have the same value? *(2 marks)* **Hard**

Bonus question

8 Layla wants to invest $30 000 for a total of 4 years. She has three investment options.

Option *A*—simple interest is paid at the rate of 5% per annum

Option *B*—compound interest is paid at a rate of 4.5% per annum, compounded annually

Option *C*—compound interest is paid at a rate of 4.2% per annum, compounded quarterly

Determine Layla's best investment option. Support your answer with calculations. *(3 marks)* **Medium**

Bonus question

9 The table shows the future value of an annuity of $1 for varying interest rates and time periods. The contribution is made at the end of each period.

Table of future value interest factors						
	Interest rate per period					
Period	**0.50%**	**1.50%**	**2.00%**	**3.00%**	**4.00%**	**6.00%**
1	1.0000	1.0000	1.0000	1.0000	1.0000	1.0000
2	2.0050	2.0150	2.0200	2.0300	2.0400	2.0600
3	3.0150	3.0452	3.0604	3.0909	3.1216	3.1836
4	4.0301	4.0909	4.1216	4.1836	4.2465	4.3746
5	5.0503	5.1523	5.2040	5.3091	5.4163	5.6371
6	6.0755	6.2296	6.3081	6.4684	6.6330	6.9753
7	7.1059	7.3230	7.4343	7.6625	7.8983	8.3938
8	8.1414	8.4328	8.5830	8.8923	9.2142	9.8975
9	9.1821	9.5593	9.7546	10.1591	10.5828	11.4913
10	10.2280	10.7027	10.9497	11.4639	12.0061	13.1808
11	11.2792	11.8633	12.1687	12.8078	13.4864	14.9716
12	12.3356	13.0412	13.4121	14.1920	15.0258	16.8699

Erica is planning to take a holiday in 3 years.

She needs $8000 for this holiday.

She is going to make regular quarterly payments into an account that earns interest at the rate of 2% per annum compounded quarterly.

What is the minimum investment, to the nearest $10, Erica needs to make each quarter in order to take this holiday? Support your answer with calculations. *(2 marks)*

Bonus question **Medium**

10 The graphs show the future values over time of \$$P$, invested at three different rates of compound interest.

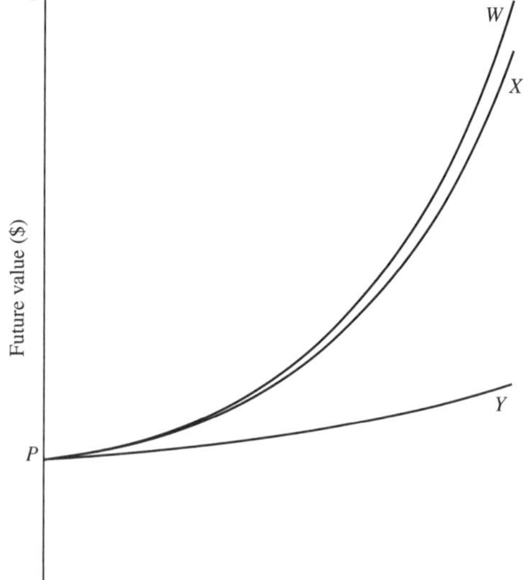

Which of the following correctly identifies each graph?

A

W	5% pa, compounding annually
X	10% pa, compounding annually
Y	10% pa, compounding quarterly

B

W	5% pa, compounding annually
X	10% pa, compounding quarterly
Y	10% pa, compounding annually

C

W	10% pa, compounding quarterly
X	10% pa, compounding annually
Y	5% pa, compounding annually

D

W	10% pa, compounding annually
X	10% pa, compounding quarterly
Y	5% pa, compounding annually

(1 mark)

(Q13, **2019 MS2 HSC**) Easy

11 The table shows the future values of an annuity of \$1 for different interest rates for 4, 5 and 6 years. The contributions are made at the end of each year.

Future value of an annuity of \$1

Years	Interest rate per annum			
	1%	2%	3%	4%
4	4.060	4.122	4.184	4.246
5	5.101	5.204	5.309	5.416
6	6.152	6.308	6.468	6.633

An annuity account is opened and contributions of \$2000 are made at the end of each year for 7 years.

For the first 6 years, the interest rate is 4% per annum, compounding annually.

For the 7th year, the interest rate increases to 5% per annum, compounding annually.

Calculate the amount in the account immediately after the 7th contribution is made. *(3 marks)*

(Q42, **2019 MS2 HSC**) 

12 A table of future value interest factors for an annuity of \$1 is shown.

Table of future value interest factors

Period	Interest rate per period				
	1%	2%	3%	4%	5%
3	3.0301	3.0604	3.0909	3.1216	3.1525
4	4.0604	4.1216	4.1836	4.2465	4.3101
5	5.1010	5.2040	5.3091	5.4163	5.5256
6	6.1520	6.3081	6.4684	6.6330	6.8019

An annuity involves contributions of \$12 000 per annum for 5 years. The interest rate is 4% per annum, compounded annually.

i Calculate the future value of this annuity. *(1 mark)* Easy

ii Calculate the interest earned on this annuity. *(1 mark)* Medium

(Q27c, **2017 MS2 HSC**)

13 The table gives the contribution per period for an annuity with a future value of \$1 at different interest rates and different periods of time.

Contribution per period for an annuity with a future value of \$1

Number of periods	Interest rate (% per period)					
	0.25%	0.5%	0.75%	1%	1.25%	1.5%
6	0.1656	0.1646	0.1636	0.1625	0.1615	0.1605
12	0.0822	0.0811	0.0800	0.0788	0.0778	0.0767
18	0.0544	0.0532	0.0521	0.0510	0.0499	0.0488
24	0.0405	0.0393	0.0382	0.0371	0.0360	0.0349
30	0.0321	0.0310	0.0298	0.0287	0.0277	0.0266
36	0.0266	0.0254	0.0243	0.0232	0.0222	0.0212

Margaret needs to save \$75 000 over 6 years for a deposit on a new apartment. She makes regular quarterly contributions into an investment account which pays interest at 3% pa.

How much will Margaret need to contribute each quarter to reach her savings goal? *(2 marks)*

(Q28d, **2016 MS2 HSC**) Easy

14 The table gives the present value interest factors for an annuity of \$1 per period, for various interest rates (r) and numbers of periods (N).

Table of present value interest factors

r / N	Interest rate per period (as a decimal)				
	0.0075	0.0080	0.0085	0.0090	0.0095
70	54.30462	53.43960	52.59397	51.76724	50.95891
71	54.89293	54.00754	53.14226	52.29657	51.46995
72	55.47685	54.57097	53.68593	52.82118	51.97618
73	56.05643	55.12993	54.22502	53.34111	52.47764
74	56.63169	55.68446	54.75957	53.85641	52.97438

i Oscar plans to invest \$200 each month for 74 months. His investment will earn interest at the rate of 0.0080 (as a decimal) per month. Use the information in the table to calculate the present value of this annuity. *(1 mark)* Easy

ii Lucy is using the same table to calculate the loan repayment for her car loan. Her loan is $21 500 and will be repaid in equal monthly repayments over 6 years. The interest rate on her loan is 10.8% per annum. Calculate the amount of each monthly repayment, correct to the nearest dollar. *(2 marks)* Medium

(Q30c, **2015 MS2 HSC**)

15 A table of future value interest factors is shown.

Table of future value interest factors

Period	*Interest rate per period*				
	1%	*2%*	*3%*	*4%*	*5%*
1	1.0000	1.0000	1.0000	1.0000	1.0000
2	2.0100	2.0200	2.0300	2.0400	2.0500
3	3.0301	3.0604	3.0909	3.1216	3.1525
4	4.0604	4.1216	4.1836	4.2465	4.3101

A certain annuity involves making equal contributions of $25 000 into an account every 6 months for 2 years at an interest rate of 4% per annum.

Based on the information provided, what is the future value of this annuity?

A $50 500 **B** $51 000
C $103 040 **D** $106 162 *(1 mark)*

(Q21, **2014 MS2 HSC**) Easy

16 An amount of $5000 is invested at 10% per annum, compounded six-monthly.

Compounded values of $1

Period	Interest rate per period				
	1%	*5%*	*10%*	*15%*	*20%*
1	1.010	1.050	1.100	1.150	1.200
2	1.020	1.103	1.210	1.323	1.440
3	1.030	1.158	1.331	1.521	1.728
4	1.041	1.216	1.464	1.750	2.074
5	1.051	1.276	1.611	2.011	2.488
6	1.062	1.340	1.772	2.313	2.986

Use the table to find the value of this investment at the end of three years. *(2 marks)*

(Q23c, **2011 MS2 HSC**) Medium

17 Lou bought a plasma TV which was priced at $3499. He paid $1000 deposit and got a loan for the balance that was paid off by 24 monthly instalments of $135.36.

What simple interest rate per annum, to the nearest percent, was charged on his loan?

A 11% **B** 15%
C 30% **D** 46% *(1 mark)*

(Q20, **2009 MS2 HSC**) Easy

18 The table shows the future value of a $1 annuity at different interest rates over different numbers of time periods.

Future values of a $1 annuity

Time Period	*Interest rate*				
	1%	*2%*	*3%*	*4%*	*5%*
1	1.0000	1.0000	1.0000	1.0000	1.0000
2	2.0100	2.0200	2.0300	2.0400	2.0500
3	3.0301	3.0604	3.0909	3.1216	3.1525
4	4.0604	4.1216	4.1836	4.2465	4.3101
5	5.1010	5.2040	5.3091	5.4163	5.5256
6	6.1520	6.3081	6.4684	6.6330	6.8019
7	7.2135	7.4343	7.6625	7.8983	8.1420
8	8.2857	8.5830	8.8923	9.2142	9.5491

i What would be the future value of a $5000 per year annuity at 3% per annum for 6 years, with interest compounding yearly? *(1 mark)* Easy

ii What is the value of an annuity that would provide a future value of $407 100 after 7 years at 5% per annum compound interest? *(1 mark)* Easy

iii An annuity of $1000 per quarter is invested at 4% per annum, compounded quarterly for 2 years. What will be the amount of interest earned? *(3 marks)* Medium

(Q27a, **2009 MS2 HSC**)

19 Heidi's funds in a superannuation scheme have a future value of $740 000 in 20 years time. The interest rate is 4% per annum and earnings are calculated six-monthly.

What single amount could be invested now to produce the same result over the same period of time at the same interest rate? *(3 marks)*

(Q24c, **2008 MS2 HSC**) Medium

20 A plasma TV depreciated in value by 15% per annum. Two years after it was purchased it had depreciated to a value of $2023, using the declining balance method.

What was the purchase price of the plasma TV? *(2 marks)*

(Q27c, **2008 MS2 HSC**) Easy

21 Bill borrows $420 000 to buy a house. Interest is charged at 7.2% per annum, compounded monthly.

How much does he owe at the end of the first month, after he has made a $4000 repayment?

A $418 496 **B** $418 520
C $445 952 **D** $446 240 *(1 mark)*

(Q21, **2006 MS2 HSC**) Easy

22 Liliana wants to borrow money to buy a house. The bank sent her an email with the following table attached.

Monthly repayments

Amount borrowed	Term of loan				
	10 years 120 months	15 years 180 months	20 years 240 months	25 years 300 months	30 years 360 months
\$80 000	\$970.62	\$764.52	\$669.15	\$617.45	\$587.01
\$90 000	\$1091.95	\$860.09	\$752.80	\$694.63	\$660.39
\$100 000	\$1213.28	\$955.65	\$836.44	\$771.82	\$733.76
\$110 000	\$1334.60	\$1051.22	\$920.08	\$849.00	\$807.14
\$120 000	\$1455.93	\$1146.78	\$1003.73	\$926.18	\$880.52
\$130 000	\$1577.26	\$1242.35	\$1087.37	\$1003.36	\$953.89
\$140 000	\$1698.59	\$1337.91	\$1171.02	\$1080.54	\$1027.27
\$150 000	\$1819.91	\$1433.48	\$1254.66	\$1157.72	\$1100.65
\$160 000	\$1941.24	\$1529.04	\$1338.30	\$1234.91	\$1174.02

i Liliana decides that she can afford \$1000 per month on repayments.
What is the maximum amount she can borrow, and how many years will she have to repay the loan? *(1 mark)* Easy

ii Zali intends to borrow \$160000 over 15 years from the same bank.
If she chooses to borrow \$160000 over 20 years instead, how much more interest will she pay? *(2 marks)* Medium

(Q27a, **2006 MS2 HSC**)

23 Rod is saving for a holiday. He deposits \$3600 into an account at the end of every year for four years. The account pays 5% per annum interest, compounding annually.

The table shows future values of an annuity of \$1.

Future values of an annuity of \$1

End of year	Interest rate				
	1%	2%	3%	4%	5%
1	1.0000	1.0000	1.0000	1.0000	1.0000
2	2.0100	2.0200	2.0300	2.0400	2.0500
3	3.0301	3.0604	3.0909	3.1216	3.1525
4	4.0604	4.1216	4.1836	4.2465	4.3101
5	5.1010	5.2040	5.3091	5.4163	5.5256
6	6.1520	6.3081	6.4684	6.6330	6.8019
7	7.2135	7.4343	7.6625	7.8983	8.1420
8	8.2857	8.5830	8.8923	9.2142	9.5491

i Use the table to find the value of Rod's investment at the end of four years. *(2 marks)* Easy

ii How much interest does Rod earn on his investment over the four years? *(2 marks)* Easy

(Q26b, **2005 MS2 HSC**)

24 Josh invests \$1000 in a term deposit that earns 3.5% per annum compounded annually.

What is the value of the investment at the end of 20 years? *(2 marks)*

(Q3a, **2002 HSC**) Easy

25 At the end of 2000, Zara purchased a new computer for \$4999. Use the declining balance method to determine the value of the computer at the end of 2002, assuming a depreciation rate of 40% per annum. (Answer to the nearest dollar.)

A \$800 **B** \$1000
C \$1800 **D** \$2000 *(1 mark)*

(Q11, **2002 MS2 HSC**) Easy

26 **i** Katherine invests \$50000 with Standard Credit Union for a term of 5 years. Her investment earns interest at 3.1% per annum compounded annually.

How much will Katherine's investment be worth at the end of 5 years?

Give your answer to the nearest dollar. *(2 marks)* Easy

ii Katherine's sister Liz also has \$50 000 to invest for a term of 5 years.
She invests with General Bank. Her investment earns interest at 3% per annum, compounded monthly.
Which sister makes the better investment? Justify your answer with appropriate calculations. *(2 marks)* Medium

(Q23b, **2002 MS2 HSC**)

27 A car is purchased for \$42000.

Use the declining balance method to calculate the salvage value of the car after 4 years at a depreciation rate of 15% per annum. *(2 marks)*

(Q27b, **2001 MS2 HSC**) Easy

28 Kim invests \$1000 at 8% per year compound interest, compounded quarterly. Calculate the value of the investment after 5 years.

(Q1f, **1994 HSC**) Easy

29 A sum of \$10000 is placed in a bank account and earns 12% interest per annum, compounded annually.

How much money is in the account at the end of 6 years, just after the final interest has been paid?

(Q1f, **1992 HSC**) Easy

Year 12 Modelling investments and loans—Worked Answers

1 **a** 1.2% per annum = 0.1% per month 10 years = 120 months

Using $A = P(1 + r)^n$

$= 40000(1.001)^{120}$ ✓

$= 45097.16995\ldots$

$= 45097.17$

(2 dec. pl.)

$\therefore$ \$45097.17 ✓

(2 marks)

b 2.4% per annum

= 0.6% per quarter

10 years = 40 quarters

The table gives a factor of 45.05630

Value = 1000 × 45.05630

= 45056.30

$\therefore$ \$45056.30 ✓

Difference

= \$45097.17 − \$45056.30

= \$40.87 ✓

(2 marks)

2 From the table, 0.75% for 8 years has an interest factor of 8.2132. ✓

Amount after 8 years

= \$1000 × 8.2132

= \$8213.20 ✓

Now, $A = P(1 + r)^n$

$= 8213.20 \times 1.0125^2$

$= 8419.813313\ldots$

$= 8419.81$ (2 dec. pl.)

$\therefore$ Simone has \$8419.81 in her account. ✓

(3 marks)

3 Contribution

= 0.095 × \$108000 − \$2120

= \$8140 ✓

Balance = \$8140 × 8.3938

= \$68325.532

= \$68326

(nearest whole)

The balance in his fund is \$68326. ✓

(2 marks)

4 **i** 1.5% per quarter, for 8 quarters:

Total contribution

= \$6000 × 8

= \$48000

Future value

= \$6000 × 8.4328

= \$50596.80 ✓

Amount of interest

= \$50596.80 − \$48000

= \$2596.80

Lincoln has earned interest of \$2596.80. ✓

(2 marks)

ii 3% per half-year, for 8 half-years:

Contribution

= \$67581.48 ÷ 8.8923

= \$7600

Mia's contribution is \$7600 every six months. ✓

(1 mark)

5 Lucy: 6% p.a. = 1.5% per quarter, 4 years = 16 quarters

Future value = 1500 × 17.9324

= 26898.60

Interest = 26898.60 − 1500 × 16

= 2898.60 ✓

Archie: 6% p.a. = 0.5% per month, 4 years = 48 months

Future value = 500 × 54.0978

= 27048.90

Interest = 27048.90 − 500 × 48

= 3048.90 ✓

Difference = 3048.90 − 2898.60

= 150.30

Archie earned \$150.30 more interest than Lucy. ✓

(3 marks)

6 From 60 years old to 90 years old is 30 years. Also, using 4% from the table:

$PV = 60000 \times 17.2920$ ✓

$= 1037520$

$\therefore$ Mel needs \$1037520 in her savings account when she turns 60. ✓

Using $A = 1037520$, $r = 0.04$ and $n = 20$:

$A = P(1 + r)^n$

$1037520 = P(1.04)^{20}$ ✓

$$P = \frac{1037\ 520}{1.04^{20}}$$

$= 473510.5844\ldots$

$= 473511$ (nearest whole)

$\therefore$ Mel needs to deposit \$473511 in her savings account today. ✓

(4 marks)

7 **i** Using $P = 435000$, $r = 0.025$, $n = 12$:

$P = A(1 + r)^n$

$435000 = A \times 1.025^{12}$

$$A = \frac{435\ 000}{1.025^{12}}$$ ✓

$= 323446.81$

$= 323447$ (nearest whole)

Kate paid \$323447 for the house. ✓

(2 marks)

ii Let $323447(1.025)^n$

$= 305000(1.032)^n$

$$\frac{1.025^n}{1.032^n} = \frac{305\ 000}{323\ 447}$$

$$\left(\frac{1.025}{1.032}\right)^n = \frac{305\ 000}{323\ 447}$$ ✓

$$n \ln\left(\frac{1.025}{1.032}\right) = \ln\left(\frac{305\ 000}{323\ 447}\right)$$

$$n = \ln\left(\frac{305\ 000}{323\ 447}\right) \div \ln\left(\frac{1.025}{1.032}\right)$$

$= 8.628125312\ldots$

As 2007 + 8.628… = 2015.628…, the two houses had the same value in 2015. ✓

(2 marks)

8 Option A:

Interest = 30000 × 0.05 × 4

= 6000 ✓

Option B:

$A = 30000(1.045)^4$

$= 35775.55802\ldots$

$= 35775.56$ (nearest cent)

Interest = 35775.56 − 30000

= 5775.56 ✓

Option C: 4.2% p.a. = 1.05% per quarter, 4 years = 16 quarters

$A = 30000(1.0105)^{16}$

$= 35457.02882\ldots$

$= 35457.03$ (2 dec. pl.)

Interest = 35457.03 − 30000

= 5457.03

The best strategy is Option A with interest of \$6000. ✓

(3 marks)

9 3 years = 12 quarters, 2% p.a. = 0.5% per quarter

Let investment = A

$8000 = A \times 12.3356$ ✓

$$A = \frac{8000}{12.3356}$$

$= 648.5294595\ldots$

$= 650$ (nearest ten)

Erica needs to invest \$650 each quarter. ✓ *(2 marks)*

10 The future value will be least at 5% p.a. compounding annually. The future value will be greatest at 10% p.a. compounding quarterly. So Y will be 5% p.a. compounding annually, W will be 10% p.a. compounding quarterly and X 10% p.a. compounding annually.

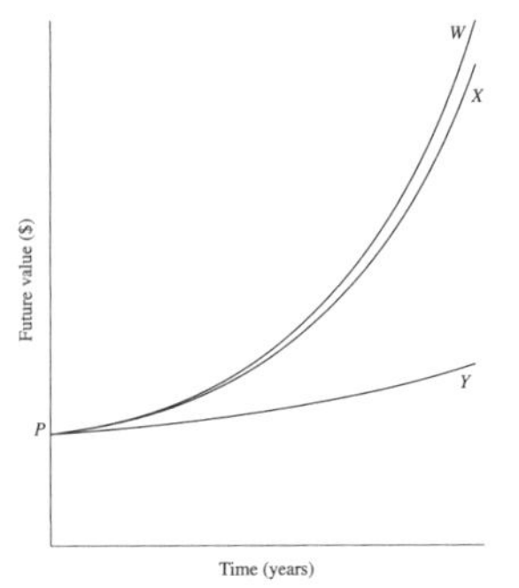

Answer C

11 Future value at end of 6 years
= 6.633 × $2000
= $13266 ✓
Interest for 7th year
= 0.05 × $13266
= $663.30 ✓
Total at end of 7 years
= $13266 + $663.30 + $2000
= $15929.30 ✓
(3 marks)

12 **i** Period 5 years, interest rate
= 4%
Interest rate factor is 5.4163.
Future value
= 5.4163 × $12000
= $64995.60 ✓
(1 mark)

ii Total contributions
= 5 × $12000
= $60000
Interest
= $64995.60 – $60000
= $4995.60 ✓
(1 mark)

13 Number of quarters = 6 × 4 = 24
Interest rate per quarter
= 3% ÷ 4 = 0.75% ✓
Contribution required
= 0.0382 × $75000
= $2865 ✓
(2 marks)

14 **i** For 74 months at 0.0080 per month the interest rate factor is 55.68446
Present value
= $200 × 55.684 46
= $11 136.892
= $11 137 (nearest dollar)
The present value of the annuity is $11 137 to the nearest dollar. ✓
(1 mark)

ii rate = 10.8% pa
= 0.9% per month
So $r = 0.0090$

6 years = 6 × 12 months
= 72 months
So $N = 72$
The required interest factor is 52.821 18 ✓
Amount of each repayment
= $21 500 ÷ 52.821 18
= $407.033 693....
= $407 (nearest dollar) ✓
(2 marks)

15 Number of periods = 2 × 2 = 4
Interest rate per period
= 4% ÷ 2 = 2%
Interest rate factor = 4.1216
FV = $25000 × 4.1216
= $103040
Answer C

16 Rate = 10% pa
= 5% per six-months
Period = 3 years
= 6 six-months ✓
Value of investment
= $5000 × 1.340
= $6700 ✓
(2 marks)

17 Price paid = $1000 + 24 × $135.36
= $4284.64
Interest = $4284.64 – $3499
= $749.64

$I = Prn$ (formulae sheet)
where n = 24 months = 2 years

$749.64 = 2499 \times r \times 2$
$749.64 = 4998r$
$r = \dfrac{749.64}{4998}$
= 0.1499 . . .
= 0.1499 . . . × 100
= 15% to nearest percent

Answer B

18 **i** Future value of $1 = $6.4684
∴ Future value of $5000
= $6.4684 × 5000
= $32 342. ✓
(1 mark)

ii Future value of $1 = $8.1420
Annuity which yields $407 100
= $407 100 ÷ 8.1420
= $50 000. ✓
(1 mark)

iii $1000 invested at 4% per annum compounded quarterly for 2 years is equal to $1000 invested for 8 periods at 1% per period. ✓
∴ Future value
= 1000 × $8.2857
= $8285.70 ✓

∴ Interest
= $8285.70 – 8 × $1000
= $285.70. ✓ *(3 marks)*

19 $N = \dfrac{A}{(1+r)^n}$

A = $740 000
r = 4% per annum
= 2% per 6 months
= 0.02 per 6 months ✓
n = 20 years = 40 (six-monthly) ✓

$\therefore N = \dfrac{740\,000}{(1+0.02)^{40}}$
= $335 138.91 ✓
(3 marks)

20 $S = V_0(1-r)^n$

r = 15% = 0.15
n = 2 years
S = $2023

$\therefore 2023 = V_0(1-0.15)^2$ ✓
$2023 = V_0 \times 0.7225$
$V_0 = 2023 \div 0.7225$
= 2800

∴ The purchase price is $2800. ✓
(2 marks)

21 r = 7.2% per annum
= (7.2 ÷ 12)% per month
= 0.6%

Interest for 1 month
$= \dfrac{0.6}{100} \times \$420\,000$
= $2520
∴ Amount owing after 1 month
= $420 000 + $2520 – $4000
= $418 520

Answer B

22 **i** From the table:
Over 10 years, Liliana can borrow $80000.
Over 15 years, Liliana can borrow $100000.
Continuing to look at the table, the maximum she can borrow is $130000 over 30 years. ✓
(1 mark)

ii Interest over 15 years
= (180 months)
= ($1529.04 × 180) – $160 000
= $275 227.20 – $160 000
= $115 227.20 ✓

Interest over 20 years
= (240 months)
= ($1338.30 × 240) – $160 000
= $321 192 – $160 000
= $161 192

∴ Additional interest
= $161 192 – $115 227.20
= $45 964.80. ✓

(2 marks)

23 i $1 invested at the end of every year for 4 years at 5% p.a. compounded annually becomes $4.3101. ✓
∴ Rod's investment will become $4.3101 × 3600
= $15 516.36 ✓

(2 marks)

ii Interest
= $15 516.36 – 4 × $3600 ✓
= $15 516.36 – $14 400
= $1116.36 ✓

(2 marks)

24 $A = P(1 + r)^n$

$= 1000\left(1 + \frac{3.5}{100}\right)^{20}$

$= 1000(1 + 0.035)^{20}$ ✓

$= 1000(1.035)^{20}$

= $1989.79 to nearest cent. ✓

(2 marks)

25 $S = V_0(1 - r^n)$

$= \$4999(1 - 0.4)^2$

= $1799.64

≑ $1800.

Answer C

26 i $A = P(1 + r)^n$

$= 50\,000(1 + 0.031)^5$ ✓

= $58 245.63

= $58 246 to nearest dollar. ✓

(2 marks)

ii $A = P(1 + r)^n$

$P = \$50\,000$

$r = 3 \div 12 = 0.25\%$ per month

$n = 5 \times 12 = 60$ months ✓

$\therefore A = 50\,000(1 + 0.0025)^{60}$

= $58 080.84

= $58 081 to nearest dollar.

∴ Katherine's investment is marginally better by $165. ✓

(2 marks)

27 $S = V_0(1 - r)^n$

$= 42\,000(1 - 0.15)^4$ ✓

= $21 924.26 ✓

(2 marks)

28 $A = P\left(1 + \frac{r}{100}\right)^n$

$P = \$1000$

$r = 8\%$ pa = 2% per quarter

$n = 5$ years = 20 quarters

$\therefore A = 1000\left(1 + \frac{2}{100}\right)^{20}$

$= 1000(1.02)^{20}$

= $1485.9473 . . .

= $1485.95 to nearest cent.

(No mark allocation in exam)

29 $A = P\left(1 + \frac{r}{100}\right)^n$

$= 10\,000\left(1 + \frac{12}{100}\right)^6$

$= 10\,000(1.12)^6$

= 19 738.2268 . . .

= $19 738.23 to the nearest cent.

(No mark allocation in exam)

1 Cards are stacked to build a 'house of cards'. A house of cards with 3 rows is shown.

A house of cards requires 3 cards in the top row, 6 cards in the next row, and each successive row has 3 more cards than the previous row.

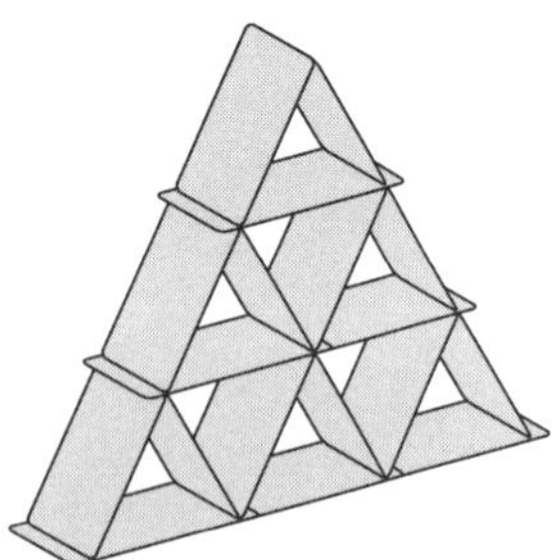

a Show that a house of cards with 12 rows has a total of 234 cards. *(2 marks)* Easy

b Another house of cards has a total of 828 cards.
How many rows are in this house of cards? *(3 marks)* Medium

(Q17, **2022 HSC**)

2 The first term of an arithmetic sequence is 5. The sum of the first 43 terms is 2021.

What is the common difference of the sequence? *(2 marks)*

(Q14, **2021 HSC**) Medium

3 Calculate the sum of the arithmetic series $4 + 10 + 16 + \ldots + 1354$. *(3 marks)*

(Q12, **2020 HSC**) Easy

4 A sequence is generated by the recurrence relation $T_n = T_{n-1} + 7, n \geq 2$. The first term of the sequence is 3. What is the twelfth term?

A 73 **B** 80
C 83 **D** 87 *(1 mark)*

Bonus question (see page iv) Easy

5 An employee is hired at an initial salary of \$37 500 per annum. At the start of each of the next seven years she receives an increase of \$1500.

i What will she be paid in her eighth year? *(2 marks)*

ii What will her total earnings be for the first eight years? *(2 marks)*

Bonus question Easy

6 Kim has two job offers. Job A has an initial salary of \$48 000 which will increase by \$1800 each year. Job B has an initial salary of \$45 000 which will increase by \$2400 each year. Which job will pay more in total over 10 years and by how much? *(3 marks)*

Bonus question Easy

7 In an arithmetic series, the fourth term is 6 and the sum of the first 16 terms is 120.

Find the common difference. *(3 marks)*

(Q12b, **2019 HSC**) Easy

8 In an arithmetic series, the third term is 8 and the twentieth term is 59.

i Find the common difference. *(1 mark)* Easy

ii Find the 50th term. *(2 marks)* Easy

(Q11d, **2018 HSC**)

9 In an arithmetic series, the fifth term is 200 and the sum of the first four terms is 1200.

Find the value of the tenth term. *(3 marks)*

(Q12c, **2017 HSC**) Easy

10 Write $\log 2 + \log 4 + \log 8 + \ldots + \log 512$ in the form $a \log b$ where a and b are integers greater than 1. *(2 marks)*

(Q14e, **2016 HSC**) Hard

11 The first three terms of an arithmetic series are 3, 7 and 11.

What is the 15th term of this series?

A 59 **B** 63
C 465 **D** 495 *(1 mark)*

(Q3, **2015 HSC**) Easy

12 Evaluate the arithmetic series $2 + 5 + 8 + 11 + \ldots + 1094$. *(2 marks)*

(Q12a, **2014 HSC**) Medium

13 A skyscraper of 110 floors is to be built. The first floor to be built will cost \$3 million. The cost of building each subsequent floor will be \$0.5 million more than the floor immediately below.

i What will be the cost of building the 25th floor? *(2 marks)* Easy

ii What will be the cost of building all 110 floors of the skyscraper? *(2 marks)* Easy

(Q3a, **2011 HSC**)

14 An arithmetic series has 21 terms. The first term is 3 and the last term is 53.

Find the sum of the series. *(2 marks)*

(Q3a, **2009 HSC**) Easy

15 Find the sum of the first 21 terms of the arithmetic series $3 + 7 + 11 + \dots$. *(2 marks)*

(Q1f, **2008 HSC**)

16 The first three terms of an arithmetic series are $-1 + 4 + 9 + \dots$

i Find the 60th term. *(2 marks)* Easy

ii Hence, or otherwise, find the sum of the first 60 terms of the series. *(2 marks)* Easy

(Q6a, **2001 HSC**)

17 In the construction of a 5 km expressway a truck delivers materials from a base. After depositing each load, the truck returns to the base to collect the next load. The first load is deposited 200 m from the base, the second 350 m from the base, the third 500 m from the base. Each subsequent load is deposited 150 m from the previous one.

i How far is the fifteenth load deposited from the base? Easy

ii How many loads are deposited along the total length of the 5 km expressway? (The last load is deposited at the end of the expressway.) Easy

iii How many kilometres has the truck travelled in order to make all the deposits and then return to the base? Medium

(6 marks) (Q4b, **2000 HSC**)

18 The third term of an arithmetic series is 32 and the sixth term is 17.

i Find the common difference. Easy

ii Find the sum of the first ten terms. Easy

(3 marks) (Q4b, **1998 HSC**)

19 The positive multiples of 7 are 7, 14, 21, ...

i What is the largest multiple of 7 less than 1000? Medium

ii What is the sum of the positive multiples of 7 which are less than 1000? Medium

(Q4d, **1994 HSC**)

Year 12 Arithmetic sequences and series—Worked Answers

1 a Arithmetic series 3, 6, 9 …
$a = 3, d = 3, n = 12$: ✓
$S_n = \frac{n}{2}[2a + (n-1)d]$
$S_{12} = \frac{12}{2}[2(3) + (12-1)3]$
$= 6 \times 39$
$= 234$ ✓
(2 marks)

b $S_n = \frac{n}{2}[2a + (n-1)d]$
$\frac{n}{2}[2(3) + (n-1)3] = 828$
$n(6 + 3n - 3) = 1656$
$n(3n + 3) = 1656$
$3n^2 + 3n - 1656 = 0$ ✓
$n^2 + n - 552 = 0$
$(n-23)(n+24) = 0$ ✓
$n = 23$ ✓
(as $n > 0$)
$\therefore$ 23 rows *(3 marks)*

2 As $S_n = \frac{n}{2}[2a + (n-1)d]$, and
using $a = 5, n = 43$ and $S_{43} = 2021$:
$2021 = \frac{43}{2}[2 \times 5 + (43-1)d]$ ✓
$4042 = 43[10 + 42d]$
$4042 = 430 + 1806d$
$1806d = 3612$
$d = 2$
$\therefore$ the common difference is 2. ✓
(2 marks)

3 $4 + 10 + 16 + \ldots + 1354$.
Arithmetic series with $a = 4$,
$d = 6$ and $\ell = T_n = 1354$. ✓
First, find the number of terms:
$T_n = a + (n-1)d$
$1354 = 4 + (n-1)6$
$1354 = 4 + 6n - 6$
$1354 = 6n - 2$
$6n = 1356$
$n = 226$ ✓
Now, $S_n = \frac{n}{2}[a + \ell]$
$S_{226} = \frac{226}{2}[4 + 1354]$
$= 153\,454$ ✓ *(3 marks)*

4 $a = 3, d = 7, n = 12$
$T_n = a + (n-1)d$
$= 3 + (12-1) \times 7$
$= 80$
The twelfth term is 80.
Answer B

5 i $a = 37\,500, d = 1500, n = 8$ ✓
$T_n = a + (n-1)d$
$= 37\,500 + (8-1) \times 1500$
$= 48\,000$
She will be paid $48 000 in her eighth year. ✓ *(2 marks)*

ii $a = 37\,500,\ n = 8,\ l = 48\,000$ ✓
$S_n = \frac{n}{2}(a + l)$
$S_8 = \frac{8}{2}(37\,500 + 48\,000)$
$= 342\,000$
Her total earnings for the first eight years will be $342 000. ✓ *(2 marks)*

6 Job A: $a = 48\,000,\ d = 1800,$
$n = 10$
$S_n = \frac{n}{2}[2a + (n-1)d]$
$S_{10} = \frac{10}{2}[2 \times 48\,000 + (10-1) \times 1800]$
$= 561\,000$ ✓
Job A will pay a total of $561 000.
Job B: $a = 45\,000,\ d = 2400,$
$n = 10$
$S_n = \frac{n}{2}[2a + (n-1)d]$
$S_{10} = \frac{10}{2}[2 \times 45\,000 + (10-1) \times 2400]$
$= 558\,000$
Job B will pay a total of $558 000. ✓
Job A will pay $3000 more over 10 years. ✓ *(3 marks)*

7 $T_4 = 6, S_{16} = 120$:
$T_4 = a + 3d = 6 \ldots ①$ ✓
$S_{16} = \frac{16}{2}[2a + 15d] = 120$
$8(2a + 15d) = 120$
$2a + 15d = 15 \ldots ②$ ✓
$2 \times ① : 2a + 6d = 12 \ldots ③$
$② - ③ : 9d = 3$
$d = \frac{1}{3}$ ✓ *(3 marks)*

8 i Using $T_n = a + (n-1)d$:
$T_3 = a + 2d = 8 \ldots ①$
$T_{20} = a + 19d = 59 \ldots ②$
$② - ① : 17d = 51$
$d = 3$
$\therefore$ common difference is 3. ✓
(1 mark)

ii Substitute into ①:
$a + 6 = 8$ ✓
$a = 2$
$T_{50} = 2 + 49(3)$
$= 149$
$\therefore$ 50th term is 149. ✓ *(2 marks)*

9 $n = 5, T_5 = 200$:
$T_n = a + (n-1)d$
$200 = a + (5-1)d$
$a + 4d = 200 \ldots ①$ ✓
$n = 4, S_4 = 1200$:
$S_n = \frac{n}{2}(2a + (n-1)d)$
$1200 = \frac{4}{2}(2a + (4-1)d)$
$2(2a + 3d) = 1200$
$2a + 3d = 600 \ldots ②$
$2 \times ① : 2a + 8d = 400 \ldots ③$
$③ - ② : 5d = -200$
$d = -40$ ✓
Subs. in ②:
$2a + 3(-40) = 600$
$2a - 120 = 600$
$2a = 720$
$a = 360$
$\therefore a = 360, d = -40, n = 10$:
$T_{10} = 360 + (10-1)(-40)$
$= 0$ ✓ *(3 marks)*

10 Consider $\log 2 + \log 4 + \log 8 + \ldots + \log 512$
$= \log 2^1 + \log 2^2 + \log 2^3 + \ldots + \log 2^9$
$= \log 2 + 2\log 2 + 3\log 2 + \ldots + 9\log 2$
$= \log 2(1 + 2 + 3 + \ldots + 9)$ ✓
As $1 + 2 + 3 + \ldots + 9$ is arithmetic series with $a = 1$,
$d = 1, n = 9, S_n = \frac{n}{2}[2a + (n-1)d]$:
$\text{Sum} = \log 2 \times \frac{9}{2}[2(1) + (9-1)1]$
$= \log 2 \times 45$
$= 45 \log 2$
$\therefore \log 2 + \log 4 + \log 8 + \ldots + \log 512 = 45 \log 2$
✓ *(2 marks)*

11 3, 7, 11 …
$a = 3, d = 4, n = 15$:
$T_n = a + (n-1)d$
$T_{15} = 3 + (15-1) \times 4$
$= 59$
$\therefore$ the 15th term is 59.
Answer A

12 Arithmetic series:
$a = 2, d = 3, T_n = 1094$
To find n: $T_n = a + (n-1)d$
$1094 = 2 + (n-1)3$
$1094 = 3n - 1$
$3n = 1095$
$n = 365$ ✓

To find sum: $S_n = \frac{n}{2}(a + l)$
$= \frac{365}{2}(2 + 1094)$
$= 200\,020$ ✓
(2 marks)

13 Using 3 to represent \$3 million, … the series is 3, 3.5, 4, …

Arithmetic series with $a = 3$, $d = 0.5$

i $n = 25$, ✓

$T_n = a + (n - 1)d$

$T_{25} = 3 + (25 - 1) \times 0.5$

$= 3 + 24 \times 0.5$

$= 15$

$\therefore$ the cost is \$15 million. ✓

(2 marks)

ii $n = 110$,

$S_n = \frac{n}{2}[2a + (n - 1)d]$

$S_{110} = \frac{110}{2}[2(3) + (110 - 1) \times 0.5]$ ✓

$= 55[6 + 109 \times 0.5]$

$= 3327.5$

$\therefore$ the cost is \$3327.5 million. ✓ *(2 marks)*

14 $S_n = \frac{n}{2}(a + L)$

$n = 21,\ a = 3,\ L = 53$ ✓

$S_{21} = \frac{21}{2}(3 + 53)$

$= \frac{21}{2} \times 56$

$= 588.$ ✓

(2 marks)

15 $3 + 7 + 11 + \ldots$

$S_n = \frac{n}{2}\left[2a + (n - 1)d\right]$

$a = 3, d = 4, n = 21$ ✓

$\therefore S_{21} = \frac{21}{2}\left[2 \times 3 + (21 - 1) \times 4\right]$

$= 10\frac{1}{2} \times (6 + 20 \times 4)$

$= 903$ ✓

(2 marks)

16 i $U_n = a + (n - 1)d$

$a = -1, n = 60, d = 5$ ✓

$U_{60} = -1 + (60 - 1)5$

$= -1 + 59(5)$

$= -1 + 295$

$= 294.$ ✓

(2 marks)

ii $S_n = \frac{n}{2}(a + l)$

$n = 60,\ a = -1,\ l = 294$ ✓

$S_{60} = \frac{60}{2}(-1 + 294)$

$= 30 \times 293$

$= 8790.$ ✓

(2 marks)

17

i 200, 350, 500, is an arithmetic sequence with

$a = 200$ and $d = 150$.

$U_n = a + (n - 1)d$

$U_{15} = 200 + 14(150)$

$= 200 + 2100$

$= 2300$ m. ✓

ii $a = 200,\ d = 150,\ U_n = 5000$

$\therefore 5000 = 200 + (n - 1)150$

$5000 = 200 + 150n - 150$

$5000 = 50 + 150n$ ✓

$150n = 4950$

$n = \frac{4950}{150}$

$= 33$

$\therefore$ 33 loads are deposited along the total length. ✓

iii Outward distance travelled from base is given by:

$S_n = \frac{n}{2}(2a + (n - 1)d)$

$n = 33,\ a = 200,\ d = 150$ ✓

$\therefore S_{33} = \frac{33}{2}(2(200) + 32 \times 150)$

$= \frac{33}{2}(400 + 4800)$

$= \frac{33}{2} \times 5200$

$= 85\,800$ m ✓

$\therefore$ Backward distance travelled = 85 800 m

$\therefore$ Total distance travelled

$= 85\,800 \times 2$

$= 171\,600$ m

$= 171.6$ km ✓

(6 marks)

(Total mark allocation only in exam)

18 i $u_n = a + (n - 1)d$

$u_3 = a + 2d = 32 \quad \ldots (1)$

$u_6 = a + 5d = 17 \quad \ldots (2)$

$(1) - (2)$: $-3d = 15$

$d = -5$ ✓

(Substituting $d = -5$ into (1):

$a + 2(-5) = 32$

$a - 10 = 32$

$\therefore a = 42.$) ✓

ii $S_n = \frac{n}{2}[2a + (n - 1)d]$

$S_{10} = \frac{10}{2}[2(42) + 9(-5)]$

$= 5\ [84 - 45]$

$= 5\ [39]$

$= 195.$ ✓

(3 marks)

(Total mark allocation only in exam)

19 i $\frac{1000}{7} = 142\frac{6}{7}$

$\therefore$ Largest multiple

$= 142 \times 7$

$= 994.$

ii $S_n = \frac{n}{2}(a + l)$

$n = 142, a = 7, l = 994$

$\therefore S_{142} = \frac{142}{2}(7 + 994)$

$= 71 \times 1001$

$= 71\,071.$

(No mark allocation in exam)

1 **a** The diagram shows the graph of $y = 2^{-x}$. Also shown on the diagram are the first 5 of an infinite number of rectangular strips of width 1 unit and height $y = 2^{-x}$ for non-negative integer values of x. For example, the second rectangle shown has width 1 and height $\frac{1}{2}$.

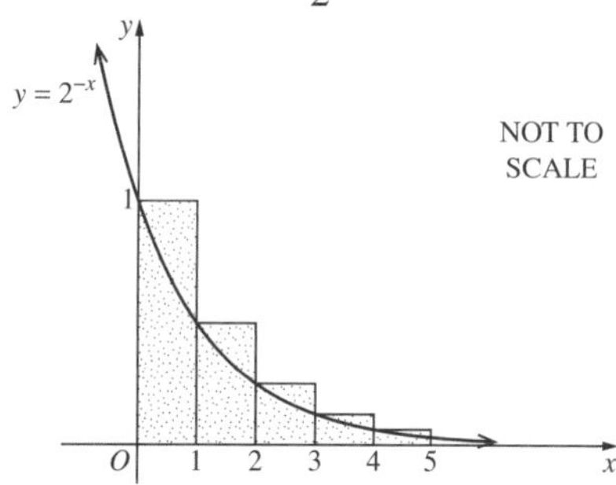

The sum of the areas of the rectangles forms a geometric series. Show that the limiting sum of this series is 2. *(1 mark)* Easy

b Show that $\int_0^4 2^{-x}\,dx = \frac{15}{16\ln 2}$. *(2 marks)* Medium

c Use parts **a** and **b** to show that $e^{15} < 2^{32}$. *(2 marks)* Medium

(Q29, **2022 HSC**)

2 A discrete random variable has probability distribution as shown in the table where n is a finite positive integer.

x	r	r^2	r^3	...	r^k	...	r^n
$P(X = x)$	r^n	r^{n-1}	r^{n-2}	...	r^{n-k+1}	...	r

Show that $E(X) = n(2r - 1)$. *(3 marks)*

(Q34, **2021 HSC**) Hard

3 What is the limiting sum of the following geometric series?

$2000 - 1200 + 720 - 432 \ldots$ *(2 marks)*

(Q11d, **2019 HSC**) Easy

4 An artist posted a song online. Each day there were $2^n + n$ downloads, where n is the number of days after the song was posted.

i Find the number of downloads on each of the first 3 days after the song was posted. *(1 mark)* Easy

ii What is the total number of times the song was downloaded in the first 20 days after it was posted? *(2 marks)* Hard

(Q14d, **2018 HSC**)

5 A geometric series has first term a and limiting sum 2.

Find all possible values for a. *(3 marks)* Hard

(Q16b, **2017 HSC**)

6 By summing the geometric series $1 + x + x^2 + x^3 + x^4$, or otherwise, find $\lim_{x \to 1} \frac{x^5 - 1}{x - 1}$. *(2 marks)*

(Q14d, **2016 HSC**) Medium

7 An eight- sided die is marked with numbers $1, 2, \ldots, 8$. A game is played by rolling the die until an 8 appears on the uppermost face. At this point the game ends.

i Using a tree diagram, or otherwise, explain why the probability of the game ending before the fourth roll is

$\frac{1}{8} + \frac{7}{8} \times \frac{1}{8} + \left(\frac{7}{8}\right)^2 \times \frac{1}{8}$. *(2 marks)* Medium

ii What is the smallest value of n for which the probability of the game ending before the nth roll is more than $\frac{3}{4}$? *(3 marks)* Hard

(Q15b, **2016 HSC**)

8 Find the limiting sum of the geometric series $1 - \frac{1}{4} + \frac{1}{16} - \frac{1}{64} + \ldots$. *(2 marks)*

(Q11d, **2015 HSC**) Easy

9 Which expression is a term of the geometric series $3x - 6x^2 + 12x^3 - \ldots$?

A $3072x^{10}$ **B** $-3072x^{10}$
C $3072x^{11}$ **D** $-3072x^{11}$ *(1 mark)*

(Q8, **2014 HSC**) Medium

10 At the beginning of every 8-hour period, a patient is given 10 mL of a particular drug.

During each of these 8-hour periods, the patient's body partially breaks down the drug. Only $\frac{1}{3}$ of the total amount of the drug present in the patient's body at the beginning of each 8-hour period remains at the end of that period.

i How much of the drug is in the patient's body immediately after the second dose is given? *(1 mark)* Easy

ii Show that the total amount of the drug in the patient's body never exceeds 15 mL. *(2 marks)* Medium

(Q14d, **2014 HSC**)

11 Kim and Alex start jobs at the beginning of the same year. Kim's annual salary in the first year is \$30000, and increases by 5% at the beginning of each subsequent year. Alex's annual salary in the first year is \$33000, and increases by \$1500 at the beginning of each subsequent year.

- i Show that in the 10th year Kim's annual salary is higher than Alex's annual salary. *(2 marks)* **Easy**
- ii In the first 10 years how much, in total, does Kim earn? *(2 marks)* **Medium**
- iii Every year, Alex saves $\frac{1}{3}$ of her annual salary. How many years does it take her to save \$87500? *(3 marks)* **Medium**

(Q12c, **2013 HSC**)

12 Pat and Chandra are playing a game. They take turns throwing two dice. The game is won by the first player to throw a double six. Pat starts the game.

- i Find the probability that Pat wins the game on the first throw. *(1 mark)* **Easy**
- ii What is the probability that Pat wins the game on the first or on the second throw? *(2 marks)* **Hard**
- iii Find the probability that Pat eventually wins the game. *(2 marks)* **Hard**

(Q15d, **2013 HSC**)

13 Rectangles of the same height are cut from a strip and arranged in a row. The first rectangle has width 10 cm. The width of each subsequent rectangle is 96% of the width of the previous rectangle.

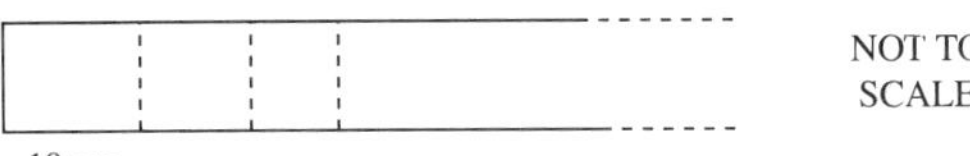

- i Find the length of the strip required to make the first ten rectangles. *(2 marks)* **Easy**
- ii Explain why a strip of length 3 m is sufficient to make any number of rectangles. *(1 mark)* **Medium**

(Q15a, **2012 HSC**)

14 The number of members of a new social networking site doubles every day. On Day 1 there were 27 members and on Day 2 there were 54 members.

- i How many members were there on Day 12? *(1 mark)* **Easy**
- ii On which day was the number of members first greater than 10 million? *(2 marks)* **Medium**
- iii The site earns 0.5 cents per member per day. How much money did the site earn in the first 12 days? Give your answer to the nearest dollar. *(2 marks)* **Medium**

(Q5a, **2011 HSC**)

15 Find the limiting sum of the geometric series

$1 - \frac{1}{3} + \frac{1}{9} - \frac{1}{27} + \cdots.$ *(2 marks)*

(Q1f, **2010 HSC**) **Easy**

16 A tree grows from ground level to a height of 1.2 metres in one year. In each subsequent year, it grows $\frac{9}{10}$ as much as it did in the previous year.

Find the limiting height of the tree. *(2 marks)*

(Q4a, **2009 HSC**) **Easy**

17 The zoom function in a software package multiplies the dimensions of an image by 1.2. In an image, the height of a building is 50 mm. After the zoom function is applied once, the height of the building in the image is 60 mm. After a second application, its height is 72 mm.

- i Calculate the height of the building in the image after the zoom function has been applied eight times. Give your answer to the nearest mm. *(2 marks)* **Easy**
- ii The height of the building in the image is required to be more than 400 mm. Starting from the original image, what is the least number of times the zoom function must be applied? *(2 marks)* **Hard**

(Q4b, **2008 HSC**)

18 Consider the geometric series

$$5 + 10x + 20x^2 + 40x^3 + \dots.$$

- i For what values of x does this series have a limiting sum? *(2 marks)* **Medium**
- ii The limiting sum of this series is 100. Find the value of x. *(2 marks)* **Medium**

(Q5b, **2008 HSC**)

19 Find the limiting sum of the geometric series

$\frac{3}{4} + \frac{3}{16} + \frac{3}{64} + \cdots.$ *(2 marks)*

(Q1d, **2007 HSC**) **Easy**

20 Find the limiting sum of the geometric series

$\frac{13}{5} + \frac{13}{25} + \frac{13}{125} + \cdots.$ *(2 marks)*

(Q1f, **2006 HSC**) **Easy**

21 Anne and Kay are employed by an accounting firm.

Anne accepts employment with an initial annual salary of \$50 000. In each of the following years her annual salary is increased by \$2500.

Kay accepts employment with an initial annual salary of \$50 000. In each of the following years her annual salary is increased by 4%.

i What is Anne's annual salary in her thirteenth year? *(2 marks)* Easy

ii What is Kay's annual salary in her thirteenth year? *(2 marks)* Easy

iii By what amount does the total amount paid to Kay in her first twenty years exceed that paid to Anne in her first twenty years? *(3 marks)* Medium

(Q7a, **2005 HSC**)

22 The triangle ABC has a right angle at B, $\angle BAC = \theta$ and $AB = 6$. The line BD is drawn perpendicular to AC. The line DE is then drawn perpendicular to BC. This process continues indefinitely as shown in the diagram.

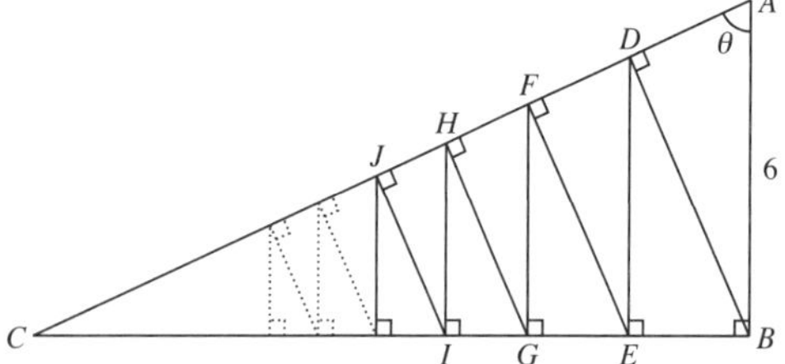

i Find the length of the interval BD, and hence show that the length of the interval EF is $6\sin^3\theta$. *(2 marks)* Hard

ii Show that the limiting sum $BD + EF + GH + \ldots$ is given by $6\sec\theta\tan\theta$. *(3 marks)* Hard

(Q9b, **2005 HSC**)

23 Consider the geometric series $1 - \tan^2\theta + \tan^4\theta - \ldots$

i When the limiting sum exists, find its value in simplest form. *(2 marks)* Medium

ii For what values of θ in the interval $-\dfrac{\pi}{2} < \theta < \dfrac{\pi}{2}$ does the limiting sum of the series exist? *(2 marks)* Hard

(Q9a, **2004 HSC**)

24 i Find the limiting sum of the geometric series

$$2 + \frac{2}{\sqrt{2}+1} + \frac{2}{(\sqrt{2}+1)^2} + \ldots$$ *(2 marks)* Medium

ii Explain why the geometric series

$$2 + \frac{2}{\sqrt{2}-1} + \frac{2}{(\sqrt{2}-1)^2} + \ldots$$

does NOT have a limiting sum. *(1 mark)* Medium

(Q7a, **2003 HSC**)

25 Consider the geometric series

$$1 + (\sqrt{5}-2) + (\sqrt{5}-2)^2 + \ldots$$

i Explain why the geometric series has a limiting sum. *(1 mark)* Easy

ii Find the exact value of the limiting sum. Write your answer with a rational denominator. *(2 marks)* Medium

(Q7a, **2002 HSC**)

26 An infinite geometric series has a first term of 8 and a limiting sum of 12. Calculate the common ratio. *(2 marks)*

(Q4a, **1999 HSC**) Easy

27 The first term of a geometric series is 16 and the fourth term is $\frac{1}{4}$.

i Find the common ratio. Easy

ii Find the limiting sum of the series. *(2 marks)* Easy

(Q4c, **1998 HSC**)

28 A ball is dropped from a height of 2 metres onto a hard floor and bounces. After each bounce, the maximum height reached by the ball is 75% of the previous maximum height. Thus, after it first hits the floor, it reaches a height of 1·5 metres before falling again, and after the second bounce, it reaches a height of 1·125 metres before falling again.

i What is the maximum height reached after the third bounce? Easy

ii What kind of sequence is formed by the successive maximum heights? Easy

iii What is the total distance travelled by the ball from the time it was first dropped until it eventually comes to rest on the floor? *(4 marks)* Medium

(Q5d, **1997 HSC**)

29 David has invented a game for one person. He throws two ordinary dice repeatedly until the sum of the two numbers shown is either 7 or 9. If the sum is 9, David wins. If the sum is 7, David loses. If the sum is any other number, he continues to throw until it is 7 or 9.

i Show that the probability that David wins on his first throw of the dice is $\frac{1}{9}$. Easy

ii Calculate the probability that a second throw is needed. Easy

iii What is the probability that David wins on his first, second, or third throw? Hard
Leave your answer in unsimplified form.

iv Calculate the probability that David wins the game. Hard

(Q10a, **1994 HSC**)

30 i Sketch the graph of $y = \cos x$ for $0 \leq x \leq 2\pi$. Easy

ii By using **i**, or otherwise, find those values of x satisfying $0 \leq x \leq 2\pi$ for which the geometrical series
$1 + 2\cos x + 4\cos^2 x + 8\cos^3 x + \ldots$
has a limiting sum. Hard

(Q4c, **1993 ME1 HSC**)

31 i For what values of r does the geometric series $a + ar + ar^2 + \ldots$ have a limiting sum?
For these values of r write down the limiting sum. Easy

ii Find a geometric series with common ratio $\frac{1}{w}$ that has limiting sum $\frac{1}{1-w}$. Hard

(Q10a, **1992 HSC**)

Year 12 Geometric sequences and series—Worked answers

1 a Areas are $1 \times 1, \frac{1}{2} \times 1, \frac{1}{4} \times 1,$

which is the series $1, \frac{1}{2}, \frac{1}{4} \ldots$

$a = 1, r = \frac{1}{2},$

$$\text{limiting sum} = \frac{a}{1-r} = \frac{1}{1-\frac{1}{2}} = 2 \quad ✓$$

$\therefore$ the limiting sum is 2. *(1 mark)*

b $\int_0^4 2^{-x}\,dx = \left[-\frac{1}{\ln 2}.2^{-x}\right]_0^4$ ✓

[using $\int a^{-x}\,dx = \frac{1}{f'(x)}.\frac{1}{\ln a}.a^{f(x)} + c$]

$$= -\frac{1}{\ln 2}(2^{-4} - 2^0)$$

$$= -\frac{1}{\ln 2}\left(\frac{1}{16} - 1\right)$$

$$= -\frac{1}{\ln 2}\left(-\frac{15}{16}\right) \quad ✓$$

$$= \frac{15}{16\ln 2}$$ *(2 marks)*

c From the graph in part **a**, area under curve < sum of areas of rectangles

$$\frac{15}{16\ln 2} < 2 \quad ✓$$

$15 < 32\ln 2$

$15 < \ln 2^{32}$

$e^{15} < 2^{32}$ ✓ *(2 marks)*

2 $E(X)$

$= r(r^n) + r^2(r^{n-1}) + r^3(r^{n-2}) + \ldots r^n(r)$

$= r^{n+1} + r^{n+1} + r^{n+1} + \ldots + r^{n+1}$

$\therefore E(X) = n(r^{n+1}) \ldots *$ ✓

Now, as sum of probabilities = 1:

$r^n + r^{n-1} + r^{n-2} + \ldots + r = 1$

$r + r^2 + r^3 + \ldots + r^n = 1$

This is a geometric series where $a = r, r = r,$ $n = n$ and $S_n = \frac{a(r^n - 1)}{r-1}$:

$$\frac{r(r^n - 1)}{r-1} = 1 \quad ✓$$

$r(r^n - 1) = r - 1$

$r^{n+1} - r = r - 1$

$r^{n+1} = 2r - 1$

Substituting into *:

$E(X) = n(2r - 1)$ ✓ *(3 marks)*

3 $a = 2000$

$$r = \frac{-1200}{2000} = -0.6 \quad ✓$$

$$\text{limiting sum} = \frac{a}{1-r} = \frac{2000}{1+0.6} = 1250 \quad ✓$$ *(2 marks)*

4 i $n = 1$: $\therefore 2^1 + 1 = 3$

$n = 2$: $\therefore 2^2 + 2 = 6$

$n = 3$: $\therefore 2^3 + 3 = 11$ ✓ *(1 mark)*

ii Need to calculate the sum of 2 series:

Geometric: 2, 4, 8, 16, …

$a = 2, r = 2, n = 20$

$$Sn = \frac{a(r^n - 1)}{r-1}$$

$$S_{20} = \frac{2(2^{20} - 1)}{2-1} = 2\,097\,150 \quad ✓$$

Arithmetic: 1, 2, 3, 4, …

$a = 1, d = 1, n = 20$

$$Sn = \frac{n}{2}[2a + (n-1)d]$$

$$S_{20} = \frac{20}{2}[2(1) + (20-1)1] = 210$$

Total sum = 210 + 2 097 150 = 2 097 360 ✓ *(2 marks)*

5 Using $S = \frac{a}{1-r}$:

$$2 = \frac{a}{1-r}$$

$2 - 2r = a$

$2r = 2 - a$

$$r = \frac{2-a}{2} \quad ✓$$

Now, limiting sum when $-1 < r < 1$:

$$-1 < \frac{2-a}{2} < 1 \quad ✓$$

$-2 < 2 - a < 2$

$-4 < -a < 0$

$0 < a < 4$ ✓ *(3 marks)*

6 For the geometric series, $1 + x + x^2 + x^3 + x^4$,

$a = 1, r = x, n = 5, S_n = \frac{a(r^n - 1)}{r-1}$:

$$S_5 = \frac{1(x^5 - 1)}{x-1} \quad ✓$$

$$\therefore \lim_{x\to 1}\frac{x^5 - 1}{x-1} = \lim_{x\to 1}(1 + x + x^2 + x^3 + x^4) = 1 + 1 + 1 + 1 + 1 = 5 \quad ✓$$

[OR: $\lim_{x\to 1}\frac{x^5 - 1}{x-1}$

$$= \lim_{x\to 1}\frac{(x-1)(x^4 + x^3 + x^2 + x + 1)}{x-1}$$

$$= \lim_{x\to 1}(x^4 + x^3 + x^2 + x + 1)$$

= 5] *(2 marks)*

7 i Let A = appear and N = not appear:

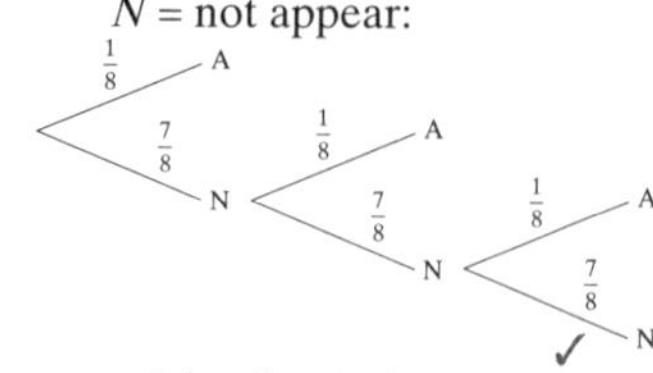

✓

$\therefore$ P(ending before 4th roll)

= P(ending after 1st roll or after 2nd roll or after 3rd roll)

= P(ending after 1st roll) + P(ending after 2nd roll) + P(ending after 3rd roll)

$$= \frac{1}{8} + \frac{7}{8} \times \frac{1}{8} + \frac{7}{8} \times \frac{7}{8} \times \frac{1}{8}$$

$$= \frac{1}{8} + \frac{7}{8} \times \frac{1}{8} + \left(\frac{7}{8}\right)^2 \times \frac{1}{8} \quad ✓$$ *(2 marks)*

ii P(ending before 4th roll)

$$= \frac{1}{8} + \frac{7}{8} \times \frac{1}{8} + \left(\frac{7}{8}\right)^2 \times \frac{1}{8}$$

$\therefore$ P(ending before nth roll)

$$= \frac{1}{8} + \frac{7}{8} \times \frac{1}{8} + \left(\frac{7}{8}\right)^2 \times \frac{1}{8} + \ldots + \left(\frac{7}{8}\right)^{n-2} \times \frac{1}{8}$$

which is geometric sum with $a = \frac{1}{8}, r = \frac{7}{8}, n = n - 1,$

$$S_n = \frac{a(1 - r^n)}{1-r}:$$

$$\frac{\frac{1}{8}\left[1 - \left(\frac{7}{8}\right)^{n-1}\right]}{1 - \frac{7}{8}} > \frac{3}{4} \quad ✓$$

$$1 - \left(\frac{7}{8}\right)^{n-1} > \frac{3}{4}$$

$$\left(\frac{7}{8}\right)^{n-1} < \frac{1}{4} \quad ✓$$

$$(n-1)\ln\left(\frac{7}{8}\right) < \ln\left(\frac{1}{4}\right)$$

$$n - 1 > \ln\left(\frac{1}{4}\right) \div \ln\left(\frac{7}{8}\right)$$

(as $\ln\left(\frac{7}{8}\right) < 0$)

$n - 1 > 10.381\,786\,14 \ldots$

$n > 11.381\,786\,14 \ldots$

$\therefore$ before the 12th roll. ✓ *(3 marks)*

8 $a = 1, r = -\frac{1}{4}$ ✓

$$S_\infty = \frac{a}{1-r} = \frac{1}{1+\frac{1}{4}}$$

$= 1 \div \frac{5}{4}$

$= \frac{4}{5}$ ✓ *(2 marks)*

9 Geometric series, where $a = 3x$, $r = -2x$,
Consider $n = 11$:

$T_n = ar^{n-1}$

$T_{11} = 3x \times (-2x)^{10}$

$= 3x \times 1024x^{10}$

$= 3072x^{11}$

Answer C

10 **i** Let A_n = amount of drug in body after n doses

$A_2 = 10 + 10 \times \frac{1}{3}$

$= 10 \times 1\frac{1}{3}$

$= 13\frac{1}{3}$

$\therefore 13\frac{1}{3}$ mL ✓ *(1 mark)*

ii $A_3 = 10 + 10 \times \frac{1}{3} + 10 \times \left(\frac{1}{3}\right)^2$

$\therefore A_n = 10 + 10 \times \frac{1}{3} + \ldots + 10 \times \left(\frac{1}{3}\right)^{n-1}$

Total amount is limiting sum:

Using $a = 10, r = \frac{1}{3}$, ✓

limiting sum $= \frac{a}{1-r}$:

$A = \frac{10}{1 - \frac{1}{3}}$

$= 10 \div \frac{2}{3}$

$= 15$

The maximum amount of the dose is 15 mL. ✓ *(2 marks)*

11 **i** Kim: 30 000 + 31 500 + …

$a = 30\,000, r = 1.05, n = 10$

$T_n = ar^{n-1}$

$T_{10} = 30\,000 \times (1.05)^{10-1}$

$= 30\,000 \times (1.05)^9$

$= 46\,539.846\,48\ldots$

$= 46\,540$ (nearest whole) ✓

Alex: 33 000 + 34 500 + …

$a = 33\,000, d = 1500, n = 10$

$T_n = a + (n-1)d$

$T_{10} = 33\,000 + (10-1) \times 1500$

$= 46\,500$

∴ Kim's salary is $40 more than Alex. ✓ *(2 marks)*

ii $S_n = \frac{a(r^n - 1)}{r - 1}$

$S_{10} = \frac{30\,000(1.05^{10} - 1)}{1.05 - 1}$ ✓

$= 377\,336.7761\ldots$

$= 377\,336.78$ (2 dec. pl.)

∴ Kim earns $377 336.78 ✓ *(2 marks)*

iii Total salary $= \$87\,500 \times 3$
$= \$262\,500$

$S_n = \frac{n}{2}[2a + (n-1)d]$

$= \frac{n}{2}[2(33\,000) + (n-1)1500]$

$= 262\,500$ ✓

$n[33\,000 + (n-1)750] = 262\,500$

$33\,000n + 750n^2 - 750n - 262\,500 = 0$

$750n^2 + 32\,250n - 262\,500 = 0$

$n^2 + 43n - 350 = 0$ ✓

$(n-7)(n+50) = 0$

$= 7, -50$ (but, $n > 0$)

∴ takes 7 years ✓ *(3 marks)*

12 **i** $P(\text{Pat: double 6}) = \frac{1}{36}$ ✓ *(1 mark)*

ii Pat wins on first or second
$= P(\text{Pat: double 6}) + P(\text{Pat: no double 6, Chandra: no double 6, Pat: double 6})$

$= \frac{1}{36} + \frac{35}{36} \times \frac{35}{36} \times \frac{1}{36}$ ✓

$= \frac{2521}{46\,656}$ ✓ *(2 marks)*

iii $P(\text{Pat eventually wins})$

$= \frac{1}{36} + \left(\frac{35}{36}\right)^2 \times \frac{1}{36} + \left(\frac{35}{36}\right)^4 \times \frac{1}{36} + \ldots$ ✓

$= \frac{1}{36}\left[1 + \left(\frac{35}{36}\right)^2 + \left(\frac{35}{36}\right)^4 + \ldots\right]$

Geometric series, $a = 1$, $r = \left(\frac{35}{36}\right)^2$,

limiting sum

$= \frac{1}{1 - r}$

$= \frac{1}{36}\left[\frac{1}{1 - \left(\frac{35}{36}\right)^2}\right]$

$= \frac{36}{71}$ ✓ *(2 marks)*

13 10, 9.6, …

i $a = 10, r = 0.96, n = 10$

$S_n = \frac{a(1 - r^n)}{1 - r}$

$= \frac{10(1 - 0.96^{10})}{1 - 0.96}$ ✓

$= 83.791\,841\ldots$

$= 83.79$ (2 dec. pl.)

∴ length of the strip is 83.79 cm ✓ *(2 marks)*

ii lim sum $= \frac{a}{1 - r}$

$= \frac{10}{1 - 0.96}$

$= 250$

As limiting sum is 250, then the strip has a maximum length of 250 cm, or 2.5 m, which means that 3 m is sufficient. ✓ *(1 mark)*

14 A geometric series: 27, 54, 108, …
$a = 27, r = 2$

i $n = 12$, $T_n = ar^{n-1}$

$T_{12} = 27 \times 2^{11}$

$= 55\,296$

∴ 55 296 members ✓ *(1 mark)*

ii Let $T_n = 10\,000\,000$

$10\,000\,000 = 27 \times 2^{n-1}$

$2^{n-1} = \frac{10000000}{27}$ ✓

$\log_e 2^{n-1} = \log_e \frac{10000000}{27}$

$(n-1)\log_e 2 = \log_e \frac{10000000}{27}$

$n - 1 = \frac{\log_e \frac{10000000}{27}}{\log_e 2}$

$n - 1 = 18.498\ldots$

$n = 19.498\ldots$

∴ on the 20th day the number exceeds 10 million. *(2 marks)*

iii $n = 12, \quad S_n = \dfrac{a(r^n - 1)}{r - 1}$

$S_{12} = \dfrac{27(2^{12} - 1)}{2 - 1}$

$= 110\,565$ ✓

Total money

$= 110\,565 \times \$0.005$

$= \$552.825$

$= \$553$ (nearest \$) ✓

(2 marks)

15 $a = 1, r = -\dfrac{1}{3},$

$S_\infty = \dfrac{a}{1 - r}$

$= \dfrac{1}{1 - \left(-\frac{1}{3}\right)}$ ✓

$= \dfrac{1}{1\frac{1}{3}}$

$= 1 \div 1\frac{1}{3}$

$= \dfrac{3}{4}$

$\therefore$ The limiting sum is $\dfrac{3}{4}$. ✓

(2 marks)

16 $S_\infty = \dfrac{a}{1-r}$

$a = 1.2, \ r = \dfrac{9}{10}$ ✓

$\therefore S_\infty = \dfrac{1.2}{1 - \frac{9}{10}}$

$= 12$

The limiting height of the tree is 12 m. ✓

(2 marks)

17 i The heights of the building are a geometric series:

50, 60, 72 …

$T_n = ar^{n-1}$

$r = 1.2, a = 50$

Let $n = 9$

(zoom has been applied 8 times, *or* the 9th image) ✓

$\therefore T_9 = 50(1.2)^{9-1}$

$= 50(1.2)^8$

$= 214.9908 \ldots$

$= 215$ mm to the nearest mm.

$\therefore$ The height of the building is 215 mm after 8 zoom function applications. ✓

(2 marks)

ii $T_n = ar^{n-1}$

$r = 1.2, \ a = 50, \ n =$ image no.

$T_n > 400$

$\therefore ar^{n-1} > 400$

$50(1.2)^{n-1} > 400$

$(1.2)^{n-1} > \dfrac{400}{50}$

$(1.2)^{n-1} > 8$ ✓

$\ln (1.2)^{n-1} > \ln 8$

$(n-1)\ln 1.2 > \ln 8$

$n - 1 > \dfrac{\ln 8}{\ln 1.2}$

$n > \dfrac{\ln 8}{\ln 1.2} + 1$

$> 12.4053 \ldots$

$\therefore n = 13$

The 13th image will be more than 400 mm in height. Therefore, the zoom function must be applied at least $13 - 1 = 12$ times. ✓

(2 marks)

18 $5 + 10x + 20x^2 + 40x^3 + \ldots$

i This is an infinite series with $a = 5, r = 2x$. ✓

For a limiting sum:

$-1 < r < 1$

$\therefore -1 < 2x < 1$

$\therefore -\dfrac{1}{2} < x < \dfrac{1}{2}$ ✓

(2 marks)

ii $S_\infty = \dfrac{a}{1-r}$

If $S_\infty = 100, a = 5$ and $r = 2x$,

$\therefore 100 = \dfrac{5}{1 - 2x}$ ✓

$100 \times (1 - 2x) = 5$

$1 - 2x = \dfrac{5}{100}$

$-2x = -\dfrac{19}{20}$

$\therefore x = \dfrac{19}{40}$ ✓

(2 marks)

19 $\dfrac{3}{4} + \dfrac{3}{16} + \dfrac{3}{64} + \ldots$

$S_\infty = \dfrac{a}{1-r}$, where $a = \dfrac{3}{4}, r = \dfrac{1}{4}$ ✓

$\therefore S_\infty = \dfrac{\frac{3}{4}}{1 - \frac{1}{4}}$

$= 1.$ ✓

(2 marks)

20 $S = \dfrac{a}{1-r}$

$a = \dfrac{13}{5}, r = \dfrac{1}{5}$ ✓

$S = \dfrac{\frac{13}{5}}{1 - \frac{1}{5}}$

$= 3\frac{1}{4}$ ✓

(2 marks)

21 i Anne's salary represents an arithmetic sequence:

50 000, 52 500, 55 000 …

with $a = 50\,000, d = 2500$, $n = 13$. ✓

$T_n = a + (n-1)d$

$\therefore T_{13} = 50\,000 + 12 \times 2500$

$= 50\,000 + 30\,000$

$= 80\,000$

$\therefore$ Anne's salary in her thirteenth year is \$80 000. ✓

(2 marks)

ii Kay's salary represents a geometric sequence with $a = 50\,000, r = 1.04, n = 13$. ✓

$T_n = ar^{n-1}$

$\therefore T_{13} = 50\,000(1.04)^{12}$

$= 80\,051.6109 \ldots$

$\therefore$ Kay's salary in her thirteenth year is \$80 051.61. ✓

(2 marks)

iii Total amount for Anne in first 20 years:

$S_n = \dfrac{n}{2}\left(2a + (n-1)d\right)$

$n = 20, \ a = 50\,000, \ d = 2500$

$\therefore S_{20} = \dfrac{20}{2}\left[2(50\,000) + (20 - 1)2500\right]$

$= 10(100\,000 + 19 \times 2500)$

$= 1\,475\,000$ ✓

$\therefore$ Total amount paid to Anne is \$1 475 000.

Total amount paid to Kay in first 20 years is given by

$S_n = a\dfrac{(r^n - 1)}{r - 1}$

with $n = 20, a = 50\,000$, $r = 1.04$

$\therefore S_{20} = \dfrac{50\,000(1.04^{20} - 1)}{1.04 - 1}$

$= 1\,488\,903.929$ ✓

Total amount paid to Kay is \$1 488 903.93.

Kay's total exceeds Anne's total by

\$1 488 903.93 – \$1 475 000

= \$13 903.93 ✓ *(3 marks)*

22 **i**

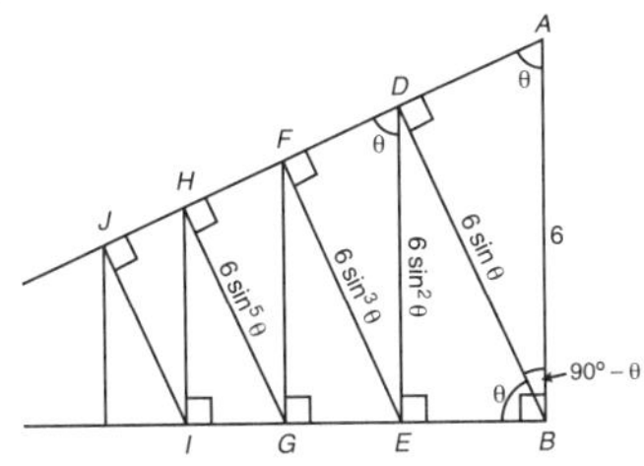

In ΔABD,

$$\sin\theta = \frac{BD}{6}$$

$\therefore BD = 6\sin\theta$

Now $\angle ABD = 90° - \theta$ (angle sum of triangle = 180°)

$\therefore \angle DBE = \theta$

($\angle ABE = 90°$, and $\angle ABE = \angle ABD + \angle DBE$)

(N.B. Each triangle has angles of θ, $90° - \theta$ and $90°$. Alternate and corresponding angles are equal; $AB \parallel DE \parallel FG$, and so on. Therefore all triangles are similar because they are equiangular.)

In ΔBDE,

$$\sin\theta = \frac{ED}{BD}$$

$ED = BD\sin\theta$

$= 6\sin\theta\sin\theta$

$= 6\sin^2\theta$ ✓

In ΔDEF,

$\angle FDE = \angle DAB = \theta$ (corresponding angles; $DE \parallel AB$)

$$\therefore \sin\theta = \frac{EF}{ED}$$

$EF = ED\sin\theta$

$= 6\sin^2\theta \,.\, \sin\theta$

$= 6\sin^3\theta$ ✓

(2 marks)

ii Now $BD = 6\sin\theta$

and $EF = 6\sin^3\theta$

$\therefore GH = 6\sin^5\theta$

(on continuing the pattern)

$\therefore BD + EF + GH + \ldots$

$= 6\sin\theta + 6\sin^3\theta + 6\sin^5\theta + \ldots$

Which is a geometric series with $a = 6\sin\theta$ and $r = \sin^2\theta$. ✓

For a limiting sum to exist:

$|r| < 1$

i.e. $|\sin^2\theta| < 1$

i.e. $-1 < \sin^2\theta < 1$

Since $0 < \theta < 90°$, $-1 < \sin^2\theta < 1$, therefore a limiting sum exists. ✓

$$S_\infty = \frac{a}{1-r}$$

$$= \frac{6\sin\theta}{1-\sin^2\theta}$$

$$= \frac{6\sin\theta}{\cos^2\theta} \quad (\text{since } \sin^2\theta + \cos^2\theta = 1)$$

$$= \frac{6\sin\theta}{\cos\theta \,.\, \cos\theta}$$

$$= \frac{6\sin\theta}{\cos\theta} \times \frac{1}{\cos\theta}$$

$= 6\tan\theta \times \sec\theta$

$= 6\sec\theta\tan\theta$ ✓

(3 marks)

23 **i** $1 - \tan^2\theta + \tan^4\theta - \ldots$

$$S = \frac{a}{1-r}$$

$$\text{Now } r = \frac{-\tan^2\theta}{1}$$

$= -\tan^2\theta$

$$\therefore S = \frac{1}{1-(-\tan^2\theta)} \quad ✓$$

$$= \frac{1}{1+\tan^2\theta}$$

$$= \frac{1}{\sec^2\theta}$$

$= \cos^2\theta$ ✓ *(2 marks)*

ii For the limiting sum to exist,

$|r| < 1$

i.e. $|-\tan^2\theta| < 1$

$\therefore \tan^2\theta < 1$

$\therefore -1 < \tan\theta < 1$ ✓

$\therefore -\frac{\pi}{4} < \theta < \frac{\pi}{4}$ ✓

(2 marks)

24 **i**

$$S = \frac{a}{1-r}$$

Now $a = 2$,

$$\text{and } r = \frac{\dfrac{2}{\sqrt{2}+1}}{2}$$

$$= \frac{1}{\sqrt{2}+1} \times \frac{\sqrt{2}-1}{\sqrt{2}-1}$$

$$= \frac{\sqrt{2}-1}{(\sqrt{2})^2 - 1^2}$$

$$= \frac{\sqrt{2}-1}{2-1}$$

$= \sqrt{2} - 1.$ ✓

$$\therefore S = \frac{2}{1-(\sqrt{2}-1)}$$

$$= \frac{2}{2-\sqrt{2}} \times \frac{2+\sqrt{2}}{2+\sqrt{2}}$$

$$= \frac{2(2+\sqrt{2})}{2^2 - (\sqrt{2})^2}$$

$$= \frac{2(2+\sqrt{2})}{2}$$

$= 2 + \sqrt{2}$ ✓

(2 marks)

ii

$$r = \frac{\dfrac{2}{\sqrt{2}-1}}{2}$$

$$= \frac{1}{\sqrt{2}-1} \times \frac{\sqrt{2}+1}{\sqrt{2}+1}$$

$$= \frac{\sqrt{2}+1}{(\sqrt{2})^2 - 1^2}$$

$$= \frac{\sqrt{2}+1}{2-1}$$

$$= \frac{\sqrt{2}+1}{1}$$

$= \sqrt{2} + 1$

$\doteqdot 2.4$

For a limiting sum to exist, $-1 < r < 1$ and $\sqrt{2} + 1$ does not satisfy this inequality. ✓

(1 mark)

25 **i** A limiting sum exists if $|r| < 1$, i.e. $-1 < r < 1$

Now $r = \sqrt{5} - 2$

$\doteqdot 0.236$

and $-1 < 0.236 < 1$

$\therefore$ A limiting sum exists. ✓

(1 mark)

ii

$$S = \frac{a}{1-r}$$

$$= \frac{1}{1-(\sqrt{5}-2)}$$

$$= \frac{1}{1-\sqrt{5}+2}$$

$$= \frac{1}{3-\sqrt{5}} \quad ✓$$

$$= \frac{1}{3-\sqrt{5}} \times \frac{3+\sqrt{5}}{3+\sqrt{5}}$$

$$= \frac{3+\sqrt{5}}{9-5}$$

$$= \frac{3+\sqrt{5}}{4} \quad ✓$$

(2 marks)

26

$$S = \frac{a}{1-r}$$

$$\therefore 12 = \frac{8}{1-r} \quad ✓$$

$12 - 12r = 8$

$-12r = -4$

$$r = \frac{-4}{-12}$$

$$= \frac{1}{3} \quad ✓$$

(2 marks)

27 **i** $U_n = ar^{n-1}$

$U_4 = ar^3$

$$\frac{1}{4} = 16r^3$$

$$r^3 = \frac{1}{64}$$

$$\therefore r = \frac{1}{4}. \quad ✓$$

ii $S = \dfrac{a}{1-r}$

$= \dfrac{16}{1-\frac{1}{4}}$

$= \dfrac{16}{\frac{3}{4}}$

$= \dfrac{64}{3}$ ✓

(2 marks)

(Total mark allocation only in exam)

28 **i**

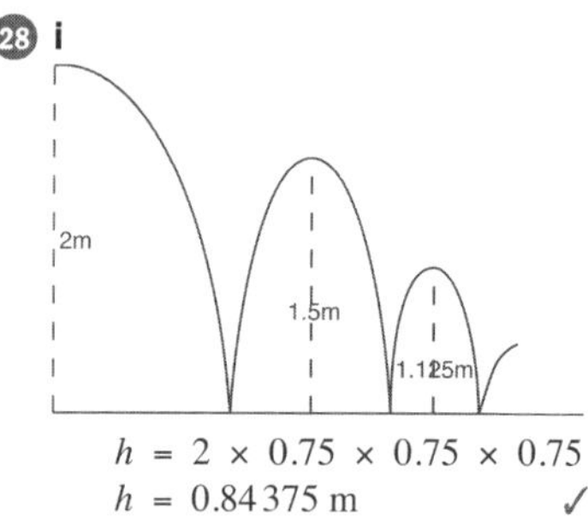

$h = 2 \times 0.75 \times 0.75 \times 0.75$

$h = 0.84375$ m ✓

ii Geometric sequence or progression (infinite). ✓

iii The falls constitute one series:

$2 + 1.5 + 1.25 + \ldots$

$S_\infty = \dfrac{a}{1-r}$

$a = 2, r = \frac{3}{4}$

$\therefore S_\infty = \dfrac{2}{1-\frac{3}{4}}$

$= \dfrac{2}{\frac{1}{4}}$

$= 8$ m ✓

The rises constitute another series:

$1.5 + 1.25 + 0.84375 + \ldots$

$a = 1.5, r = \frac{3}{4}$

$\therefore S_\infty = \dfrac{1.5}{1-\frac{3}{4}}$

$= \dfrac{1.5}{\frac{1}{4}}$

$= 6$ m

$\therefore$ Total distance travelled $= 8 + 6 = 14$ m. ✓

(4 marks)

(Total mark allocation only in exam)

29

		Die 1					
		1	2	3	4	5	6
	1	2	3	4	5	6	7*
	2	3	4	5	6	7*	8
Die 2	3	4	5	6	7*	8	(9)
	4	5	6	7*	8	(9)	10
	5	6	7*	8	(9)	10	11
	6	7*	8	(9)	10	11	12

i $P(9) = \dfrac{4}{36}$

$= \dfrac{1}{9}.$

ii $P(7 \text{ or } 9) = \dfrac{10}{36}$

$= \dfrac{5}{18}$

Probability that 2nd throw needed

$= 1 - P(7 \text{ or } 9)$

$= 1 - \dfrac{5}{18}$

$= \dfrac{13}{18}.$

iii Probability of win on throw 1

$= \dfrac{1}{9}$

Probability of win on throw 2

$= \dfrac{13}{18} \times \dfrac{1}{9}$

Probability of win on throw 3

$= \dfrac{13}{18} \times \dfrac{13}{18} \times \dfrac{1}{9}$

$= \left(\dfrac{13}{18}\right)^2 \times \dfrac{1}{9}$

Probability of win on throw 1, 2 or 3

$= \dfrac{1}{9}\left(1 + \dfrac{13}{18} + \left(\dfrac{13}{18}\right)^2\right)$

iv Probability that David wins game

$= \dfrac{1}{9} + \dfrac{13}{18} \times \dfrac{1}{9} + \left(\dfrac{13}{18}\right)^2 \times \dfrac{1}{9} + \ldots$

Limiting sum, $S_\infty = \dfrac{a}{1-r}$

$a = \dfrac{1}{9}, r = \dfrac{13}{18}$

$\therefore S_\infty = \dfrac{\frac{1}{9}}{1 - \frac{13}{18}}$

$= \dfrac{\frac{1}{9}}{\frac{5}{18}}$

$= \dfrac{1}{9} \times \dfrac{18}{5}$

$= \dfrac{2}{5}.$

(No mark allocation in exam)

30 **i**

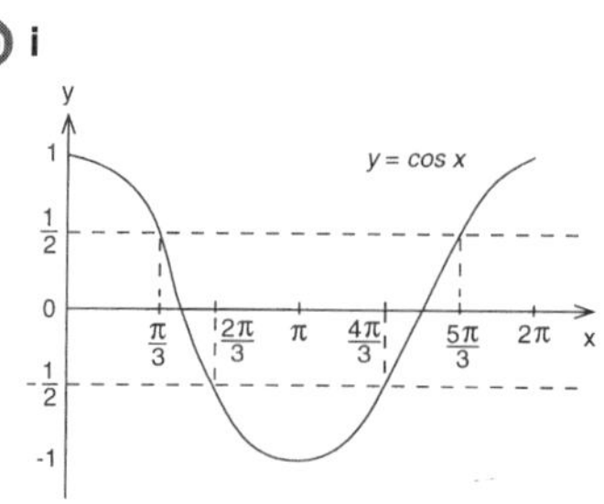

ii $1 + 2\cos x + 4\cos^2 x + 8\cos^3 x + \ldots$

is a geometric series with common ratio, $r = 2\cos x$.

A limiting sum exists if

$-1 < 2\cos x < 1$

$\therefore -\dfrac{1}{2} < \cos x < \dfrac{1}{2}$

Now $\cos x = \dfrac{1}{2}$

$(0 \leqslant x \leqslant 2\pi)$

when $x = \dfrac{\pi}{3},$

$2\pi - \dfrac{\pi}{3} = \dfrac{5\pi}{3}$

and $\cos x = -\dfrac{1}{2}$

$(0 \leqslant x \leqslant 2\pi)$

when $x = \pi - \dfrac{\pi}{3} = \dfrac{2\pi}{3},$

$\pi + \dfrac{\pi}{3} = \dfrac{4\pi}{3}$

$\therefore$ The series has a limiting sum for

$\dfrac{\pi}{3} < x < \dfrac{2\pi}{3}$ and

$\dfrac{4\pi}{3} < x < \dfrac{5\pi}{3}$ (from graph).

(No mark allocation in exam)

31 **i** $|r| < 1$ or $-1 < r < 1$

$S = \dfrac{a}{1-r}$

ii $S = \dfrac{a}{1-r}$

$S = \dfrac{1}{1-w}, \ r = \dfrac{1}{w}$

$\therefore \dfrac{1}{1-w} = \dfrac{a}{1-\frac{1}{w}}$

$\therefore a = \dfrac{1-\frac{1}{w}}{1-w}$

$= \dfrac{\frac{w-1}{w}}{1-w}$

$= \dfrac{w-1}{w} \times \dfrac{1}{1-w}$

$= \dfrac{w-1}{w(1-w)}$

$= \dfrac{w-1}{-w(w-1)}$

$= -\dfrac{1}{w}$

As $a = -\dfrac{1}{w}$ and $r = \dfrac{1}{w},$

the geometric series is:

$-\dfrac{1}{w} - \dfrac{1}{w^2} - \dfrac{1}{w^3} - \ldots$

(No mark allocation in exam)

1 In a reducing-balance loan, an amount $\$P$ is borrowed for a period of n months at an interest rate of 0.25% per month, compounded monthly. At the end of each month, a repayment of $\$M$ is made. After the nth repayment has been made, the amount owing, $\$A_n$, is given by

$$A_n = P(1.0025)^n - M(1 + (1.0025)^1 + (1.0025)^2 + \ldots + (1.0025)^{n-1}).$$

(Do NOT prove this.)

a Jane borrows \$200 000 in a reducing-balance loan as described.
The loan is to be repaid in 180 monthly repayments.
Show that $M = 1381.16$, when rounded to the nearest cent. *(2 marks)* **Medium**

b After 100 repayments of \$1381.16 have been made, the interest rate changes to 0.35% per month.
At this stage, the amount owing to the nearest dollar is \$100 032.
(Do NOT prove this.)
Jane continues to make the same monthly repayments.
For how many more months will Jane need to make full monthly payments of \$1381.16? *(3 marks)* **Hard**

c The final repayment will be less than \$1381.16.
How much will Jane need to pay in the final payment in order to pay off the loan? *(2 marks)* **Hard**

(Q32, **2022 HSC**)

2 **a** On the day that Megan was born, her grandfather deposited \$5000 into an account earning 3% per annum compounded annually. On each birthday after this, her grandfather deposited \$1000 into the same account, making his final deposit on Megan's 17th birthday. That is, a total of 18 deposits were made.
Let A_n be the amount in the account on Megan's nth birthday, after the deposit is made.
Show that $A_3 = \$8554.54$. *(2 marks)* **Medium**

b On her 17th birthday, just after the final deposit is made, Megan has \$30 025.83 in her account. You are NOT required to show this.
Megan then decides to leave all the money in the same account continuing to earn interest at 3% per annum compounded annually. On her 18th birthday, and on each birthday after this, Megan withdraws \$2000 from the account.
How many withdrawals of \$2000 will Megan be able to make? *(3 marks)* **Medium**

(Q29, **2021 HSC**)

3 Tina inherits \$60 000 and invests it in an account earning interest at a rate of 0.5% per month. Each month, immediately after the interest has been paid, Tina withdraws \$800.

The amount in the account immediately after the nth withdrawal can be determined using the recurrence relation

$$A_n = A_{n-1}(1.005) - 800,$$

where $n = 1, 2, 3, \ldots$ and $A_0 = 60000$.

a Use the recurrence relation to find the amount of money in the account immediately after the third withdrawal. *(2 marks)* **Easy**

b Calculate the amount of interest earned in the first three months. *(2 marks)* **Medium**

c Calculate the amount of money in the account immediately after the 94th withdrawal. *(3 marks)* **Medium**

CQ (Q26, **2020 HSC**)

4 Elliot deposited \$5000 at the end of each year for 10 years in an annuity earning 2.15% per annum, compounded annually.

He completed a table to show the amount in the account at the end of each year.

Part of the table has been shaded.

Year	Balance at beginning of year	Interest on balance	Annual deposit	Total at end of year
1	0.00	0.00	5000.00	5000.00
2	5000.00	107.50	5000.00	10 107.50
3	10 107.50	217.31	5000.00	15 324.81
4	15 324.81	329.48	5000.00	20 654.29
5				A
6				
7				
8				
9				
10				B

i Find the value of A. *(1 mark)* **Easy**

ii Use the sum of a geometric series to find the value of B. *(2 marks)* **Medium**

Bonus question (see page iv)

5 A person wins \$1 000 000 in a competition and decides to invest this money in an account that earns interest at 6% per annum compounded quarterly. The person decides to withdraw \$80 000 from this account at the end of every fourth quarter. Let A_n be the amount remaining in the account after the nth withdrawal.

i Show that the amount remaining in the account after the withdrawal at the end of the eighth quarter is $A_2 = 1000000 \times 1.015^8 - 80000(1 + 1.015^4)$. *(2 marks)* **Medium**

ii For how many years can the full amount of \$80000 be withdrawn? *(3 marks)* **Hard**

(Q16a, **2019 HSC**)

---

6 Kara deposits an amount of \$300000 into an account which pays compound interest of 4% per annum, added to the account at the end of each year. Immediately after the interest is added, Kara makes a withdrawal for expenses for the coming year. The first withdrawal is \$$P$. Each subsequent withdrawal is 5% greater than the previous one.
Let \$$A_n$ be the amount in the account after the nth withdrawal.

i Show that $A_2 = 300000(1.04)^2 - P[(1.04)+(1.05)]$. *(1 mark)* **Medium**

ii Show that $A_3 = 300000(1.04)^3 - P[(1.04)^2 + (1.04)(1.05) + (1.05)^2]$. *(1 mark)* **Hard**

iii Show that there will be money in the account when

$$\left(\frac{105}{104}\right)^n < 1 + \frac{3000}{P}$$

(3 marks) **Hard**

(Q16c, **2018 HSC**)

---

7 Anita opens a savings account. At the start of each month she deposits \$$X$ into the savings account. At the end of each month, after interest is added into the savings account, the bank withdraws \$2500 from the savings account as a loan repayment. Let M_n be the amount in the savings account after the nth withdrawal.

The savings account pays interest of 4.2% per annum compounded monthly.

i Show that after the second withdrawal the amount in the savings account is given by $M_2 = X(1.0035^2 + 1.0035) - 2500(1.0035 + 1)$. *(2 marks)* **Medium**

ii Find the value of X so that the amount in the savings account is \$80000 after the last withdrawal of the fourth year. *(3 marks)* **Medium**

(Q15b, **2017 HSC**)

---

8 A gardener develops an eco-friendly spray that will kill harmful insects on fruit trees without contaminating the fruit. A trial is to be conducted with 100000 insects. The gardener expects the spray to kill 35% of the insects each day and that exactly 5000 new insects will be produced each day.

The number of insects expected at the end of the nth day of the trial is A_n.

i Show that $A_2 = 0.65(0.65 \times 100000 + 5000) + 5000$. *(2 marks)* **Medium**

ii Show that $A_n = 0.65^n \times 100000 + 5000\dfrac{(1 - 0.65^n)}{0.35}$. *(1 mark)* **Hard**

(Q14b, **2016 HSC**)

---

9 Sam borrows \$100000 to be repaid at a reducible interest rate of 0.6% per month. Let \$$A_n$ be the amount owing at the end of n months and \$$M$ be the monthly repayment.

i Show that $A_2 = 100000(1.006)^2 - M(1+1.006)$. *(1 mark)* **Easy**

ii Show that $A_n = 100000(1.006)^n - M\left(\dfrac{(1.006)^n - 1}{0.006}\right)$. *(2 marks)* **Easy**

iii Sam makes monthly repayments of \$780. Show that after making 120 monthly repayments the amount owing is \$68500 to the nearest \$100. *(1 mark)* **Easy**

iv Immediately after making the 120th repayment, Sam makes a one-off payment, reducing the amount owing to \$48500. The interest rate and monthly repayment remain unchanged. After how many more months will the amount owing be completely repaid? *(3 marks)* **Hard**

(Q14c, **2015 HSC**)

---

10 At the start of a month, Jo opens a bank account and makes a deposit of \$500. At the start of each subsequent month, Jo makes a deposit which is 1% more than the previous deposit.

At the end of every month, the bank pays interest of 0.3% (per month) on the balance of the account.

i Explain why the balance of the account at the end of the second month is \$$500(1.003)^2$ + \$$500(1.01)(1.003)$. *(2 marks)* **Easy** ii Find the balance of the account at the end of the 60th month, correct to the nearest dollar. *(3 marks)* Medium

(Q16b, **2014 HSC**)

11 A family borrows \$500 000 to buy a house. The loan is to be repaid in equal monthly instalments. The interest, which is charged at 6% per annum, is reducible and calculated monthly. The amount owing after n months, $\$A_n$, is given by

$$A_n = Pr^n - M(1 + r + r^2 + \ldots + r^{n-1}),$$

(Do NOT prove this)

where $\$P$ is the amount borrowed, $r = 1.005$ and $\$M$ is the monthly repayment.

i The loan is to be repaid over 30 years. Show that the monthly repayment is \$2998 to the nearest dollar. *(2 marks)* Easy

ii Show that the balance owing after 20 years is \$270 000 to the nearest thousand dollars. *(1 mark)* Medium

iii After 20 years the family borrows an extra amount, so that the family then owes a total of \$370 000. The monthly repayment remains \$2998, and the interest rate remains the same. How long will it take to repay the \$370 000? *(2 marks)* Easy

(Q13d, **2013 HSC**)

12 Zina opened an account to save for a new car. Six months after opening the account, she made her first deposit of \$1200 and continued depositing \$1200 at the end of each six month period. Interest was paid at 3% per annum, compounded half-yearly.

How much was in Zina's account two years after first opening it?

A \$4909.08 **B** \$4982.72
C \$5018.16 **D** \$5094.55 *(1 mark)*

(Q23, **2013 MS2 HSC**) Easy

13 Ari takes out a loan of \$360 000. The loan is to be repaid in equal monthly repayments, $\$M$, at the end of each month, over 25 years (300 months). Reducible interest is charged at 6% per annum, calculated monthly.

Let $\$A_n$ be the amount owing after the nth repayment.

i Write down an expression for the amount owing after two months, $\$A_2$. *(1 mark)* Medium

ii Show that the monthly repayment is approximately \$2319.50. *(2 marks)* Medium

iii After how many months will the amount owing, $\$A_n$, become less than \$180 000? *(3 marks)* Hard

(Q15c, **2012 HSC**)

14 When Jules started working she began paying \$100 at the beginning of each month into a superannuation fund.

The contributions are compounded monthly at an interest rate of 6% per annum.

She intends to retire after having worked for 35 years.

i Let $\$P$ be the final value of Jules's superannuation when she retires after 35 years (420 months). Show that $\$P = \$143\,183$ to the nearest dollar. *(2 marks)* Medium

ii Fifteen years after she started working Jules read a magazine article about retirement, and realised that she would need \$800 000 in her fund when she retires. At the time of reading the magazine article she had \$29 227 in her fund. For the remaining 20 years she intends to work, she decides to pay a total of $\$M$ into her fund at the beginning of each month. The contributions continue to attract the same interest rate of 6% per annum, compounded monthly. At the end of n months after starting the new contributions, the amount in the fund is $\$A_n$.

1 Show that $A_2 = 29\,227 \times 1.005^2 + M(1.005 + 1.005^2)$. *(1 mark)* Hard

2 Find the value of M so that Jules will have \$800 000 in her fund after the remaining 20 years (240 months). *(3 marks)* Medium

(Q8c, **2011 HSC**)

15 i When Chris started a new job, \$500 was deposited into his superannuation fund at the beginning of each month. The money was invested at 0.5% per month, compounded monthly. Let $\$P$ be the value of the investment after 240 months, when Chris retires. Show that $P = 232\,175.55$. *(2 marks)* Medium

ii After retirement, Chris withdraws \$2000 from the account at the end of each month, without making any further deposits. The account continues to earn interest at 0.5% per month. Let $\$A_n$ be the amount left in the account n months after Chris's retirement.

1 Show that
$A_n = (P - 400000) \times 1.005^n + 400000$.
(3 marks) **Hard**

2 For how many months after retirement will there be money left in the account?
(2 marks) **Medium**

(Q9a, **2010 HSC**)

16 One year ago Daniel borrowed \$350 000 to buy a house. The interest rate was 9% per annum, compounded monthly. He agreed to repay the loan in 25 years with equal monthly repayments of \$2937.

i Calculate how much Daniel owed after his first monthly repayment. *(1 mark)* **Easy**

ii Daniel has just made his 12th monthly repayment. He now owes \$346 095. The interest rate now decreases to 6% per annum, compounded monthly.
The amount, $\$A_n$, owing on the loan after the nth monthly repayment is now calculated using the formula
$A_n = 346095 \times 1.005^n - 1.005^{n-1}M - \dots - 1.005M - M$
where $\$M$ is the monthly repayment, and $n = 1, 2, \dots, 288$. (Do NOT prove this formula.)
Calculate the monthly repayment if the loan is to be repaid over the remaining 24 years (288 months). *(3 marks)* **Medium**

iii Daniel chooses to keep his monthly repayments at \$2937. Use the formula in part **ii** to calculate how long it will take him to repay the \$346 095. *(3 marks)* **Hard**

iv How much will Daniel save over the term of the loan by keeping his monthly repayments at \$2937, rather than reducing his repayments to the amount calculated in part **ii**? *(1 mark)* **Medium**

(Q8b, **2009 HSC**)

17 Peter retires with a lump sum of \$100 000. The money is invested in a fund which pays interest each month at a rate of 6% per annum, and Peter receives a fixed monthly payment of $\$M$ from the fund. Thus, the amount left in the fund after the first monthly payment is $\$(100500 - M)$.

i Find a formula for the amount, $\$A_n$, left in the fund after n monthly payments.
(2 marks) **Easy**

ii Peter chooses the value of M so that there will be nothing left in the fund at the end of the 12th year (after 144 payments). Find the value of M. *(3 marks)* **Hard**

(Q9b, **2008 HSC**)

18 Mr and Mrs Caine each decide to invest some money each year to help pay for their son's university education. The parents choose different investment strategies.

i Mr Caine makes 18 yearly contributions of \$1000 into an investment fund. He makes his first contribution on the day his son is born, and his final contribution on his son's seventeenth birthday. His investment earns 6% compound interest per annum.
Find the total value of Mr Caine's investment on his son's eighteenth birthday. *(3 marks)* **Medium**

ii Mrs Caine makes her contributions into another fund. She contributes \$1000 on the day of her son's birth, and increases her annual contribution by 6% each year. Her investment also earns 6% compound interest per annum.
Find the total value of Mrs Caine's investment on her son's third birthday (just before she makes her fourth contribution).
(2 marks) **Hard**

iii Mrs Caine also makes her final contribution on her son's seventeenth birthday.
Find the total value of Mrs Caine's investment on her son's eighteenth birthday. *(1 mark)* **Medium**

(Q9c, **2007 HSC**)

19 Joe borrows \$200 000 which is to be repaid in equal monthly instalments. The interest rate is 7.2% per annum reducible, calculated monthly.

It can be shown that the amount, $\$A_n$, owing after the nth repayment is given by the formula:

$A_n = 200000r^n - M(1 + r + r^2 + \dots + r^{n-1})$,

where $r = 1.006$ and $\$M$ is the monthly repayment. (Do NOT show this.)

i The minimum monthly repayment is the amount required to repay the loan in 300 instalments. Find the minimum monthly repayment. *(3 marks)* **Medium**

ii Joe decides to make repayments of \$2800 each month from the start of the loan.
How many months will it take for Joe to repay the loan? *(2 marks)* **Hard**

(Q8b, **2006 HSC**)

20 Weelabarrabak Shire Council borrowed \$3 000 000 at the beginning of 2005. The annual interest rate is 12%. Each year, interest is calculated on the balance at the beginning of the year and added to the balance owing. The debt is to be repaid by equal annual repayments of \$480 000, with the first repayment being made at the end of 2005.

Let A_n be the balance owing after the n-th repayment.

i Show that $A_2 = (3 \times 10^6)(1.12)^2 - (4.8 \times 10^5)(1 + 1.12)$. *(1 mark)* Medium

ii Show that $A_n = 10^6[4 - (1.12)^n]$. *(2 marks)* Hard

iii In which year will Weelabarrabak Shire Council make the final repayment? *(2 marks)* Hard

(Q8c, **2005 HSC**)

21 Betty decides to set up a trust fund for her grandson, Luis. She invests \$80 at the beginning of each month. The money is invested at 6% per annum, compounded monthly.

The trust fund matures at the end of the month of her final investment, 25 years after her first investment. This means that Betty makes 300 monthly investments.

i After 25 years, what will be the value of the first \$80 invested? *(2 marks)* Easy

ii By writing a geometric series for the value of all Betty's investments, calculate the final value of Luis' trust fund. *(3 marks)* Medium

(Q7c, **2004 HSC**)

22 Barbara borrows \$120 000 to be repaid over a period of 25 years at 6% per annum reducible interest. Each year there are k regular repayments of \$$F$. Interest is calculated and charged just before each repayment.

i Write down an expression for the amount owing after two repayments. *(1 mark)* Medium

ii Show that the amount owing after n repayments is

$$A_n = 120\,000\,\alpha^n - \frac{kF(\alpha^n - 1)}{0.06},$$

where $\alpha = 1 + \frac{0.06}{k}$ *(2 marks)* Hard

iii Calculate the amount of each repayment if the repayments are made quarterly (ie. $k = 4$). *(2 marks)* Hard

iv How much would Barbara have saved over the term of the loan if she had chosen to make monthly rather than quarterly repayments? *(2 marks)* Medium

(Q10a, **2003 HSC**)

23 A superannuation fund pays an interest rate of 8.75% per annum which compounds annually. Stephanie decides to invest \$5000 in the fund at the beginning of each year, commencing on 1 January 2003.

What will be the value of Stephanie's superannuation when she retires on 31 December 2023? *(4 marks)*

(Q9b, **2002 HSC**) Medium

24 Helen sets up a prize fund with a single investment of \$1000 to provide her school with an annual prize valued at \$72. The fund accrues interest at a rate of 6% per annum, compounded annually. The first prize is awarded one year after the investment is set up.

i Calculate the balance in the fund at the beginning of the second year. *(1 mark)* Easy

ii Let \$$B_n$ be the balance in the fund at the end of n years (and after the nth prize has been awarded). Show that $B_n = 1200 - 200 \times (1.06)^n$. *(2 marks)* Medium

iii At the end of the tenth year (and after the tenth prize has been awarded) it is decided to increase the prize value to \$90. For how many more years can the prize fund be used to award the prize? *(3 marks)* Hard

(Q10a, **2001 HSC**)

25 A store offers a loan of \$5000 on a computer for which it charges interest at the rate of 1% per month. As a special deal, the store does not charge interest for the first three months however, the first repayment is due at the end of the first month.

A customer takes out the loan and agrees to repay the loan over three years by making 36 equal monthly repayments of \$$M$.

Let \$$A_n$ be the amount owing at the end of the nth repayment.

i Find an expression for A_3. Easy

ii Show that $A_5 = (5000 - 3M)1{\cdot}01^2 - M(1 + 1{\cdot}01)$ Easy

iii Find an expression for A_{36}. Easy

iv Find the value of M. *(6 marks)* Hard

(Q10a, **2000 HSC**)

26 Isabella invests \$$P$ at 8% per annum compounded annually. She intends to withdraw \$3000 at the end of each of the next six years to cover school fees.

i Write down an expression for the amount \$$A_1$ remaining in the account following the withdrawal of the first \$3000. Easy

ii Find an expression for the amount $$A_2$ remaining in the account after the second withdrawal. **Easy**

iii Calculate the amount $$P$ that Isabella needs to invest if the account balance is to be $0 at the end of six years.

(5 marks) **Medium**

(Q7a, **1999 HSC**)

---

27 A fish farmer began business on 1 January 1998 with a stock of 100 000 fish. He had a contract to supply 15 400 fish at a price of $10 per fish to a retailer in December each year. In the period between January and the harvest in December each year, the number of fish increases by 10%.

i Find the number of fish just after the second harvest in December 1999. **Easy**

ii Show that F_n, the number of fish just after the nth harvest, is given by $F_n = 154000 - 54000(1{\cdot}1)^n$. **Medium**

iii When will the farmer have sold all his fish, and what will his total income be? **Hard**

iv Each December the retailer offers to buy the farmer's business by paying $15 per fish for his entire stock. When should the farmer sell to maximise his total income?

(7 marks) **Hard**

(Q10b, **1998 HSC**)

---

28 On 1 July 1985, Anna invested $10 000 in a bank account that paid interest at a fixed rate of 8% per annum, compounded annually.

i How much would be in the account after the payment of interest on 1 July 1995 if no additional deposits were made? **Easy**

ii In fact, Anna added $1000 to her account on 1 July each year, beginning on 1 July 1986.
How much was in her account on 1 July 1995 after the payment of interest and her deposit? **Medium**

iii Anna's friend, Jennifer, invested $10 000 in an account at another bank on 1 July 1985 and made no further deposits. On 1 July 1995, the balance of Jennifer's account was $35 478.
What was the annual rate of compound interest paid on Jennifer's account?

(8 marks) **Hard**

(Q8b, **1995 HSC**)

---

29 Ezzat invests $50 000 in an account which earns 8% interest, compounded annually. He intends to withdraw $$M$ at the end of each year, immediately after the interest has been paid. He wishes to be able to do this for exactly 20 years, so that the account will then be empty.

i How much money does he have in the account immediately after he has made his first withdrawal? **Easy**

ii Write an expression in terms of M for the amount of money in the account, immediately after his 20th withdrawal. **Medium**

iii Calculate the value of M which leaves his account empty after the 20th withdrawal. **Medium**

iv Suppose Ezzat wished to be able to withdraw $8000 per year for the 20 years. By using your calculator alone, estimate, to the nearest per cent, the interest rate he would then need to earn on his account. **Hard**

(Q8b, **1993 HSC**)

1 **a** Using $P = 20000, a = 1, r = 1.0025, n = 180$ and $S_n = \dfrac{a(r^n - 1)}{r - 1}$:

$$A_n = 200000 \times 1.0025^{180} - M\left(\frac{1(1.0025^{180} - 1)}{1.0025 - 1}\right) = 0 \quad ✓$$

$$M\left(\frac{1.0025^{180} - 1}{0.0025}\right) = 200000 \times 1.0025^{180}$$

$$M = 200000 \times 1.0025^{180} \div \frac{(1.0025^{180} - 1)}{0.0025}$$

$$= 1381.163281\ldots$$

$$= 1381.16 \text{ (2 dec. pl.)} \quad ✓$$

(2 marks)

b

$$A_n = 100\,032 \times 1.0035^n - 1381.16\left(\frac{1(1.0035^n - 1)}{1.0035 - 1}\right) > 1381.16 \quad ✓$$

$$100\,032 \times 1.0035^n - 394\,617.14(1.0035^n - 1) > 1381.16$$

$$100032 \times 1.0035^n - 394617.14 \times 1.0035^n + 394\,617.14 > 1381.16$$

$$294\,585.14 \times 1.0035^n > 393\,235.96$$

$$1.0035^n > \frac{393235.96}{294585.14} \quad ✓$$

$$1.0035^n > 1.33488$$

$$\ln 1.0035^n > \ln 1.33488$$

$$n > \frac{\ln 1.339\,56}{\ln 1.0035}$$

$$n > 82.67 \qquad \therefore 83 \text{ months} \quad ✓$$

(3 marks)

c

$$A_{83} = 100\,032 \times 1.0035^{83} - 1381.16\left(\frac{(1.0035^{83} - 1)}{0.0035}\right) \quad ✓$$

$$= 928.29 \text{ (2 dec. pl.)}$$

Now, this amount owing after 83 repayments will be charged interest of 0.35%:

Amount = 928.29×1.0035

= 931.54 (2 dec. pl.)

∴ \$931.54 will be paid. ✓

(2 marks)

2 **a**

$$A_1 = 5000 \times 1.03 + 1000$$

$$A_2 = A_1 \times 1.03 + 1000$$

$$= 5000 \times 1.03^2 + 1000 \times 1.03 + 1000 \quad ✓$$

$$A_3 = A_2 \times 1.03 + 1000$$

$$= 5000 \times 1.03^3 + 1000 \times 1.03^2 + 1000 \times 1.03 + 1000$$

$$= 8554.535$$

$$= 8554.54 \text{ (2 dec. pl.)}$$

∴ A_3 is \$8554.54. ✓

(2 marks)

b Let the amount in Megan's account on her 18th birthday be M_1.

$$M_1 = 30\,025.83 \times 1.03 - 2000$$

$$M_2 = M_1 \times 1.03 - 2000$$

$$= 30\,025.83 \times 1.03^2 - 2000 \times 1.03 - 2000$$

$$= 30\,025.83 \times 1.03^2 - 2000(1 + 1.03) \quad ✓$$

Let n = number of possible withdrawals.

$$M_n = 30\,025.83 \times 1.03^n - 2000(1 + 1.03 + 1.032 + \ldots + 1.03^{n-1}) = 0$$

Using $a = 1, r = 1.03, n = n$ and $S_n = \dfrac{a(r^n - 1)}{r - 1}$:

$$30025.83 \times 1.03^n = 2000\left(\frac{1(1.03^n - 1)}{1.03 - 1}\right) \quad ✓$$

$$30025.83 \times 1.03^n = \frac{2000 \times 1.03^n}{0.03} - \frac{2000}{0.03}$$

$$\frac{2000 \times 1.03^n}{0.03} - 30025.83 \times 1.03^n = \frac{2000}{0.03}$$

$$1.03^n\left(\frac{2000}{0.03} - 30025.83\right) = \frac{2000}{0.03}$$

$$1.03^n = \frac{2000}{0.03} \div 36640.837$$

$$1.03^n = 1.819 \text{ (3 dec. pl.)}$$

$$n \ln 1.03 = \ln 1.819$$

$$n = \frac{\ln 1.819}{\ln 1.03}$$

$$= 20.249\ldots$$

∴ Megan can make 20 withdrawals. ✓

(3 marks)

3 **a**

$$A_1 = A_0(1.005) - 800$$

$$= 60000(1.005) - 800$$

$$= 59500$$

$$A_2 = A_1(1.005) - 800$$

$$= 59500(1.005) - 800$$

$$= 58997.50 \quad ✓$$

$$A_3 = A_2(1.005) - 800$$

$$= 58997.50(1.005) - 800$$

$$= 58492.49 \text{ (2 dec. pl.)}$$

∴ the amount of money in the account was \$58492.49. ✓

(2 marks)

b Interest = $60000 \times 0.005 + 59500 \times 0.005 + 58977.5 \times 0.005$ ✓

= 892.4875

= 892.49 (2 dec. pl.)

∴ the interest is \$892.49. ✓

(2 marks)

c $A_1 = 60000(1.005) - 800$

$A_2 = [60000(1.005) - 800] \times 1.005 - 800$

$= 60000(1.005)^2 - 800(1 + 1.005)$

Hence $A_{94} = 60000(1.005)^{94} - 800(1 + 1.005 + 1.005^2 + \ldots + 1.005^{93})$ ✓

$= 60000(1.005)^{94} - 800 \times \dfrac{1(1.005^{94} - 1)}{1.005 - 1}$ ✓

(Using a sum of a geometric series with $a = 1, r = 1.005, n = 94$)

$= 187.8459979\ldots$

$= 187.85$

$\therefore$ there will be \$187.85 in the account. ✓

(3 marks)

---

4 i $A = 20654.29 \times 1.0215 + 5000$

$= 26098.36$ (2 dec. pl.) ✓

(1 mark)

ii $B = 5000 + 5000 \times 1.0215 + 5000 \times 1.0215^2 + \ldots + 5000 \times 1.0215^9$

$= 5000(1 + 1.0215 + 1.0215^2 + \ldots + 1.0215^9)$ ✓

Using $a = 1, r = 1.0215, n = 10, S_n = \dfrac{a(r^n - 1)}{r - 1}$:

$B = 5000 \times \dfrac{1(1.0215^{10} - 1)}{1.0215 - 1}$

$= 55125.55941\ldots$

$= 55125.56$ (2 dec. pl.) ✓

(2 marks)

---

5 i 6% per annum = 1.5% per quarter

$A_1 = 1000000 \times 1.015^4 - 80000$

$A_2 = (1000000 \times 1.015^4 - 80000) \times 1.015^4 - 80000$ ✓

$= 1000000 \times 1.015^8 - 80000 \times 1.015^4 - 80000$

$= 1000000 \times 1.015^8 - 80000(1 + 1.015^4)$ ✓

(2 marks)

ii $A_n = 1000000 \times 1.015^{4n} - 80000(1 + 1.015^4 + 1.015^8 + \ldots + 1.015^{4(n-1)})$

Consider the series with $a = 1, r = 1.015^4, n = n$:

$A_n = 1000000 \times 1.015^{4n} - 80000\left[\dfrac{1((1.015^4)^n - 1)}{1.015^4 - 1}\right]$ ✓

$= 1000000 \times 1.015^{4n} - 1303706(1.015^{4n} - 1) \le 80000$

$1000000 \times 1.015^{4n} - 1303706 \times 1.015^{4n} + 1303706 \le 80000$

$303706 \times 1.015^{4n} \ge 1223706$

$1.015^{4n} \ge \dfrac{1\ 223\ 706}{303\ 706}$ ✓

$4n \ln 1.015 \ge \ln\dfrac{1\ 223\ 706}{303\ 706}$

$n \ge \dfrac{\ln\dfrac{1\ 223\ 706}{303\ 706}}{4\ln 1.015}$

$\ge 23.4000\ldots$

$\therefore$ \$80000 can be withdrawn annually for 24 years. ✓

(3 marks)

---

6 i $A_1 = 300000 \times 1.04 - P$

$A_2 = (300000 \times 1.04 - P) \times 1.04 - P \times 1.05$

$= 300000 \times 1.04^2 - P(1.04 + 1.05)$ ✓

(1 mark)

ii $A_3 = [300000 \times 1.04^2 - P(1.04 + 1.05)] \times 1.04 - P \times 1.05^2$

$= 300000 \times 1.04^3 - P(1.04^2 + 1.04 \times 1.05 + 1.05^2)$ ✓

(1 mark)

iii $A_n = 300000 \times 1.04^n - P(1.04^{n-1} + 1.04^{n-2} \times 1.05 + \ldots + 1.04 \times 1.05^{n-2} + 1.05^{n-1})$

A geometric series with $a = 1.04^{n-1}, r = \dfrac{1.05}{1.04} = \dfrac{105}{104}, n = n$ ✓

$$A_n = 300000 \times 1.04^n - P\left[\frac{1.04^{n-1}\left(\left(\frac{105}{104}\right)^n - 1\right)}{\frac{105}{104} - 1}\right] > 0$$

$$300000 \times 1.04^n - P\left[\frac{1.04^{n-1}\left(\left(\frac{105}{104}\right)^n - 1\right)}{\frac{1}{104}}\right] > 0 \quad ✓$$

$$300000 \times 1.04^n - 104P\left[1.04^{n-1}\left(\left(\frac{105}{104}\right)^n - 1\right)\right] > 0$$

$$3000 \times 1.04^n - 1.04P\left[1.04^{n-1}\left(\left(\frac{105}{104}\right)^n - 1\right)\right] > 0$$

$$3000 \times 1.04^n - P\left[1.04^n\left(\left(\frac{105}{104}\right)^n - 1\right)\right] > 0$$

$$3000 - P\left(\left(\frac{105}{104}\right)^n - 1\right) > 0$$

$$\frac{3000}{P} - \left(\frac{105}{104}\right)^n + 1 > 0$$

$$\left(\frac{105}{104}\right)^n < 1 + \frac{3000}{P} \quad ✓$$

(3 marks)

---

7 i 4.2% p.a. = 0.35% per month = 0.0035 per month,

4 years = 48 months

$M_1 = X \times 1.0035 - 2500$

$M_2 = M_1 \times 1.0035 + X \times 1.0035 - 2500$ ✓

$= (X \times 1.0035 - 2500) \times 1.0035 + X \times 1.0035 - 2500$

$= X \times 1.0035^2 - 2500 \times 1.0035 + X \times 1.0035 - 2500$

$= X(1.0035^2 + 1.0035) - 2500(1.0035 + 1)$ ✓

(2 marks)

ii $M_{48} = X(1.0035^{48} + 1.0035^{47} + \ldots + 1.0035) - 2500(1.0035^{47} + 1.0035^{46} + \ldots + 1)$

$= X(1.0035 + 1.0035^2 + \ldots + 1.0035^{48}) - 2500(1 + 1.0035^2 + \ldots + 1.0035^{47})$ ✓

Using geometric series and $S_n = \dfrac{a(r^n - 1)}{r - 1}$:

$M_{48} = X\left[\dfrac{1.0035(1.0035^{48} - 1)}{1.0035 - 1}\right] - 2500\left[\dfrac{1(1.0035^{48} - 1)}{1.0035 - 1}\right] = 80000$ ✓

$52.351X - 130421 = 80000$

$X = 4019$ (nearest whole) ✓

(3 marks)

---

8 i As 35% of insects die, then 65% of insects survive.

$\therefore A_1 = 0.65 \times 100000 + 5000$ ✓

$A_2 = 0.65 \times A_1 + 5000$

$= 0.65(0.65 \times 100000 + 5000) + 5000$ ✓

(2 marks)

ii $A_2 = 0.65^2 \times 100000 + 0.65 \times 5000 + 5000$
$= 0.65^2 \times 100000 + 5000(1 + 0.65)$
$\therefore A_n = 0.65^n \times 100000 + 5000(1 + 0.65 + 0.65^2 + \ldots + 0.65^{n-1})$

Using a geometric sum with $a = 1, r = 0.65, n = n$, $S_n = \frac{a(1-r^n)}{1-r}$:

$A_n = 0.65^n \times 100000 + 5000\left[\frac{1(1-0.65^n)}{1-0.65}\right]$

$= 0.65^n \times 100000 + 5000\frac{(1-0.65^n)}{0.35}$ ✓ *(1 mark)*

9 i $A_1 = 100\,000 \times 1.006 - M$
$A_2 = [100\,000 \times 1.006 - M] \times 1.006 - M$
$= 100\,000 \times 1.006^2 - 1.006M - M$
$= 100\,000 \times 1.006^2 - M(1 + 1.006)$ ✓ *(1 mark)*

ii $A_n = 100\,000 \times 1.006^n - M(1 + 1.006 + 1.006^2 + \ldots + 1.006^{n-1})$ ✓

Considering the geometric series with $a = 1$, $r = 1.006, n = n$, and using $S_n = \frac{a(r^n-1)}{r-1}$:

$A_n = 100\,000 \times 1.006^n - M\left[\frac{1(1.006^n-1)}{1.006-1}\right]$

$= 100\,000 \times 1.006^n - M\left[\frac{1.006^n-1}{0.006}\right]$ ✓ *(2 marks)*

iii Let $M = 780, n = 120$:
$A_{120} = 100\,000 \times 1.006^{120} - 780\left[\frac{1.006^{120}-1}{0.006}\right]$
$= 68\,499.4583\ldots$
$= 68\,500$ (nearest 500)
$\therefore$ the amount owing is \$68 500. ✓ *(1 mark)*

iv Let amount be 48 500 and $A_n = 0$:
$0 = 48\,500 \times 1.006^n - 780\left[\frac{1.006^n-1}{0.006}\right]$
$48\,500 \times 1.006^n = 130\,000(1.006^n - 1)$ ✓
$48\,500 \times 1.006^n = 130\,000 \times 1.006^n - 130\,000$
$81\,500 \times 1.006^n = 130\,000$
$1.006^n = \frac{1300}{815}$
$1.006^n = \frac{260}{163}$ ✓
$\log_e 1.006^n = \log_e \frac{260}{163}$
$n = \frac{\log_e \frac{260}{163}}{\log_e 1.006}$
$= 78.055\,137\,98\ldots$
$\therefore$ Sam will need 79 repayments. ✓ *(3 marks)*

10 i Deposits: $500, 500(1.01), 500(1.01^2), \ldots$
Let A_n = balance of account after n months
$A_1 = 500(1.003)$ ✓
$A_2 = 500(1.003)(1.003) + 500(1.01)(1.003)$
$= 500(1.003)^2 + 500(1.01)(1.003)$ ✓ *(2 marks)*

ii $A_3 = A_2(1.003) + 500(1.01)^2(1.003)$
$= 500(1.003^3) + 500(1.01)(1.003)^2 + 500(1.01)^2(1.003)$
$= 500[1.003^3 + (1.01)(1.003)^2 + (1.01)^2(1.003)]$
$A_{60} = 500[1.003^{60} + (1.01)(1.003)^{59} + \ldots + (1.01)^{59}(1.003)]$

Geom. series, $a = 1.003^{60}, r = \frac{1.01}{1.003}$, $n = 60$ ✓

$A_{60} = 500\left[\frac{1.003^{60}\left[\left(\frac{1.01}{1.003}\right)^{60} - 1\right]}{\frac{1.01}{1.003} - 1}\right]$ ✓

$= 44\,404.378\,66\ldots$
$= 44\,404$ (nearest whole)
$\therefore$ the balance is \$44 404. ✓ *(3 marks)*

11 i $A_n = Pr^n - M(1 + r + r^2 + \ldots + r^{n-1})$
$n = 30 \times 12 = 360$
$A_{360} = 500\,000 \times 1.005^{360} - M(1 + 1.005 + \ldots + 1.005^{359})$

Now, $1 + 1.005 + \ldots + 1.005^{359}$ is Geom. series, with $a = 1, r = 1.005, n = 360$.

Using $S_n = \frac{a(r^n - 1)}{r - 1}$,
$A_{360} = 500\,000 \times 1.005^{360} - M \times \frac{1(1.005^{360} - 1)}{1.005 - 1}$ ✓

Let $A_{360} = 0$:
$M = 500\,000 \times 1.005^{360} \div \frac{1(1.005^{360} - 1)}{1.005 - 1}$
$= 2997.752\,626\ldots$
$= 2998$ (nearest whole)
$\therefore$ \$2998 ✓ *(2 marks)*

ii Let $n = 240, M = 2998$:
$A_{240} = 500\,000 \times 1.005^{240} - 2998 \times \frac{1(1.005^{240} - 1)}{1.005 - 1}$
$= 269\,903.6342\ldots$
$= 270\,000$ (nearest thousand)
$\therefore$ \$270 000 ✓ *(1 mark)*

iii Let $P = 370\,000, M = 2998$:
$A_n = 370\,000 \times 1.005^n - 2998 \times \frac{1(1.005^n - 1)}{1.005 - 1}$
$A_n = 370\,000 \times 1.005^n - 599\,600(1.005^n - 1)$
$= 0$

$370\,000 \times 1.005^n - 599\,600 \times 1.005^n + 599\,600 = 0$
$229\,600 \times 1.005^n = 599\,600$
$1.005^n = \frac{599\,600}{229\,600}$ ✓
$n \log_e 1.005 = \log_e \frac{5996}{2296}$
$n = \log_e \frac{5996}{2296} \div \log_e 1.005$
$= 192.464\,3834\ldots$
$\therefore$ will take 193 repayments or 16 years 1 month ✓ *(2 marks)*

12 $a = 1200$

3% pa = 1.5% per half year

So $r = 1.015$

2 years = 4 half years

So $n = 4$

$$S_n = \frac{a(r^n - 1)}{r - 1}$$
$$= \frac{1200(1.015^4 - 1)}{1.015 - 1}$$
$$= 4909.08405$$
$$= 4909.08 \quad (2\text{ d.p.})$$

The amount in the account was \$4909.08

Answer A

13 **i** 6% pa = 0.5% per month

As 0.5% = 0.005,

$A_1 = 360000 \times 1.005 - M$

$A_2 = A_1 \times 1.005 - M$

$= (360000 \times 1.005 - M) \times 1.005 - M$

$= 360000 \times 1.005^2 - M(1 + 1.005)$ ✓ *(1 mark)*

ii $A_{300} = 360000 \times 1.005^{300}$

$- M(1 + 1.005 + 1.005^2 + \ldots + 1.005^{299})$

Now, $A_{300} = 0$

Using geometric series with

$a = 1, r = 1.005, n = 300$ ✓

$$M = 360000 \times 1.005^{300} \div \frac{1(1.005^{300} - 1)}{1.005 - 1}$$
$$= 2319.485045$$

$\therefore$ approximately \$2319.50 ✓ *(2 marks)*

iii Let $A_n = 180000$

360000×1.005^n

$- 2319.50\,(1 + 1.005 + \ldots + 1.005^{n-1})$

$= 180000$

360000×1.005^n

$$- 2319.50 \times \frac{1(1.005^n - 1)}{0.005} = 180000$$ ✓

1800×1.005^n

$- 2319.5 \times (1.005^n - 1) = 900$

1800×1.005^n

$- 2319.5 \times 1.005^n + 2319.5 = 900$

$519.5 \times 1.005^n = 1419.5$

$$1.005^n = \frac{1419.5}{519.5}$$ ✓

$$\log_e 1.005^n = \log_e \left[\frac{1419.5}{519.5}\right]$$

$$n \log_e 1.005 = \log_e \left[\frac{1419.5}{519.5}\right]$$

$$n = \frac{\log_e \left[\frac{1419.5}{519.5}\right]}{\log_e 1.005}$$

$$= 201.5408119\ldots$$

$\therefore$ after 202 months the amount will be less than \$180000. ✓ *(3 marks)*

14 **i** 6% pa = 0.5% per month

$P = 100 \times 1.005^{420} + 100 \times 1.005^{419} + \ldots + 100 \times 1.005$

$= 100[1.005 + 1.005^2 + \ldots + 1.005^{420}]$ ✓

Using $a = 1.005, r = 1.005, n = 420$,

$$P = 100 \times \frac{1.005(1.005^{420} - 1)}{1.005 - 1}$$
$$= 143\,183.385$$

$\therefore$ \$143 183 to nearest dollar. ✓

(2 marks)

ii **1** $A_1 = (29\,227 + M) \times 1.005$

$= 29\,227 \times 1.005 + 1.005M$

$A_2 = [(29\,227 \times 1.005 + 1.005M) + M] \times 1.005$

$= 29\,227 \times 1.005^2 + 1.005^2M + 1.005M$

$= 29\,227 \times 1.005^2 + M(1.005 + 1.005^2)$ ✓

(1 mark)

2 $A_{240} = 29\,227 \times 1.005^{240} + M(1.005 + 1.005^2 + \ldots + 1.005^{240})$ ✓

Using $a = 1.005, r = 1.005, n = 240$,

$$800\,000 = 29\,227 \times 1.005^{240} + M \times \frac{1.005(1.005^{240} - 1)}{1.005 - 1}$$ ✓

$800\,000 = 96\,747.34621 + 464.3510996M$

$464.3510996M = 703\,252.6538$

$M = 1514.48$ (2 dec. places)

$\therefore$ she needs to contribute \$1514.48 each month. ✓ *(3 marks)*

15 **i** $P = 500 \times 1.005^{240} + 500 \times 1.005^{239} + 500 \times 1.005^{238} + \ldots + 500 \times 1.005$

$= 500[1.005 + 1.005^2 + \ldots + 1.005^{240}]$ ✓

Using $a = 1.005, r = 1.005, n = 240$,

$$S_n = \frac{a(r^n - 1)}{r - 1}$$

$$\therefore P = 500 \times \frac{1.005(1.005^{240} - 1)}{1.005 - 1}$$
$$= 232\,175.5498\ldots$$
$$= 232\,175.55 \text{ (to 2 decimal places)}$$ ✓

(2 marks)

ii **1** $A_1 = P \times 1.005 - 2000$

$A_2 = A_1 \times 1.005 - 2000$

$= (P \times 1.005 - 2000) \times 1.005 - 2000$

$= P \times 1.005^2 - 2000(1 + 1.005)$

Hence,

$A_n = P \times 1.005^n - 2000(1 + 1.005 + \ldots + 1.005^{n-1})$ ✓

Using $a = 1, r = 1.005, n = n$,

$$S_n = \frac{a(r^n - 1)}{r - 1}$$

$$A_n = P \times 1.005^n - 2000 \times \frac{1(1.005^n - 1)}{1.005 - 1}$$ ✓

$= P \times 1.005^n - 400\,000 \times [1.005^n - 1]$

$= (P - 400\,000) \times 1.005^n + 400\,000$ ✓

(3 marks)

2 Let $A_n = 0$

$(P - 400000) \times 1.005^n + 400000 = 0$

$(P - 400000) \times 1.005^n = -400000$

$1.005^n = \frac{-400000}{232175.55 - 400000}$, using $P = 232175.55$ ✓

$= 2.383442937 \ldots.$

$\log_e(1.005)^n = \log_e 2.383\ldots.$

$n = \frac{\log_e 2.383\ldots}{\log_e 1.005}$

$= 174.14312\ldots$

$\therefore$ There will be money left for 175 months. ✓

(2 marks)

16 **i** $A_n = P(1 + r)^n - M$

$P = \$350\,000$, $n = 1$, $M = \$2937$,
$r = 0.09$ p.a. $= 0.0075$ per month

$\therefore A_1 = 350\,000 \times (1 + 0.0075)^1 - 2937$
$= 349\,688$

$\therefore$ Daniel owed \$349 688 after his first monthly repayment. ✓ *(1 mark)*

ii $A_n = 346\,095 \times 1.005^n - 1.005^{n-1}M - \ldots - 1.005M - M$

After 288 months, $A_n = 0$.

$\therefore 0 = 346\,095 \times 1.005^{288} - 1.005^{287}M - \ldots - 1.005M - M$

$0 = 346\,095 \times 1.005^{288} - M \times (1.005^{287} + \ldots + 1.005 + 1)$

$0 = 346\,095 \times 1.005^{288} - M \times (1 + 1.005 + \ldots + 1.005^{287})$ ✓

Now, $1 + 0.005 + \ldots + 0.005^{287}$ is a geometric series, where $a = 1, r = 1.005, n = 288$

$S_n = \frac{a(r^n - 1)}{r - 1}$

$\therefore S_{288} = \frac{1 \times (1.005^{288} - 1)}{1.005 - 1}$

$= \left[\frac{1.005^{288} - 1}{0.005}\right]$

$\therefore 0 = 346\,095 \times 1.005^{288} - M \times \left[\frac{1.005^{288} - 1}{0.005}\right]$ ✓

$M \times \left[\frac{1.005^{288} - 1}{0.005}\right] = 346\,095 \times 1.005^{288}$

$M = 346\,095 \times 1.005^{288} \div \left[\frac{1.005^{288} - 1}{0.005}\right]$

$M = 1\,455\,529.832\ldots \div 641.115\ldots$
$= 2270.310\ldots$

$\therefore$ The monthly repayment of the loan to be repaid over the remaining 24 years is \$2270.31 to the nearest cent. ✓

(3 marks)

iii $A_n = 346\,095 \times 1.005^n - 1.005^{n-1}M - \ldots - 1.005M - M$

$M = \$2937$, $r = 0.06$ p.a. $= 0.005$ per month, $A_n = 0$

$\therefore 0 = 346\,095 \times 1.005^n - 1.005^{n-1}(2937) - \ldots - 1.005(2937) - 2937$

$= 346\,095 \times 1.005^n - 2937(1.005^{n-1} + \ldots + 1.005 + 1)$

$= 346\,095 \times 1.005^n - 2937(1 + 1.005 + \ldots + 1.005^{n-1})$ ✓

Now $1 + 1.005 + \ldots + 1.005^{n-1}$ is a geometric series. where $a = 1, r = 1.005, n = n.$

$S_n = \frac{a(r^n - 1)}{r - 1}$

$\therefore S_n = \frac{1 \times (1.005^n - 1)}{1.005 - 1}$

$= \frac{1.005^n - 1}{0.005}$

$\therefore 0 = 346\,095 \times 1.005^n - 2937\left(\frac{1.005^n - 1}{0.005}\right)$ ✓

$= 346\,095 \times 1.005^n(0.005) - 2937(1.005^n - 1)$

$= 1730.475 \times 1.005^n - 2937(1.005^n) + 2937$

$= 1.005^n(1730.475 - 2937) + 2937$

$= -1206.525(1.005)^n + 2937$

$1206.525(1.005)^n = 2937$

$1.005^n = \frac{2937}{1206.525}$

$n \ln 1.005 = \ln \frac{2937}{1206.525}$

$n = \ln \frac{2937}{1206.525} \div \ln 1.005$

$= 178.373\ldots$

$\therefore$ Daniel will repay the loan in 179 months or 15 years. ✓ *(3 marks)*

iv Amount saved = \$2270.31 × 288 months – \$2937 × 179 months
= \$653 844.28 – \$525 723
= \$128 126.28

Daniel will save \$128 126 over the term of the loan to the nearest dollar. ✓ *(1 mark)*

17 **i** Let each monthly repayment be \$$M$

Interest = 6% per annum
= 0.06 per annum
= 0.005 per month.

$A_1 = 100\,000 \times 1.005 - M$

$A_2 = A_1 \times 1.005 - M$
$= (100\,000 \times 1.005 - M) \times 1.005 - M$
$= 100\,000 \times 1.005^2 - 1.005M - M$

$A_3 = A_2 \times 1.005 - M$
$= (100\,000 \times 1.005^2 - 1.005M - M) \times 1.005 - M$
$= 100\,000 \times 1.005^3 - 1.005^2M - 1.005M - M$
$= 100\,000 \times 1.005^3 - M(1.005^2 + 1.005 + 1)$
$= 100\,000 \times 1.005^3 - M(1 + 1.005 + 1.005^2)$

$\therefore A_n = 100\,000 \times 1.005^n - M(1 + 1.005 + 1.005^2 + \ldots + 1.005^{n-1})$ ✓

Now $(1 + 1.005 + 1.005^2 + \ldots + 1.005^{n-1})$ is a geometric series with $a = 1, r = 1.005$ and n terms.

$\therefore A_n = 100\,000 \times 1.005^n - M\left[\frac{a(r^n - 1)}{r - 1}\right]$

$= 100\,000 \times 1.005^n - M\left[\dfrac{1(1.005^n - 1)}{1.005 - 1}\right]$

$= 100\,000 \times 1.005^n - \dfrac{M(1.005^n - 1)}{0.005}$. ✓

(2 marks)

ii $A_n = 100\,000 \times 1.005^n - \dfrac{M(1.005^n - 1)}{0.005}$

After 144 payments, $A_n = 0$

$\therefore 0 = 100\,000 \times 1.005^{144} - \dfrac{M(1.005^{144} - 1)}{0.005}$ ✓

$\dfrac{M(1.005^{144} - 1)}{0.005} = 100\,000 \times 1.005^{144}$

$M(1.005^{144} - 1) = 100\,000 \times 1.005^{144} \times 0.005$

$M = \dfrac{100\,000 \times 1.005^{144} \times 0.005}{1.005^{144} - 1}$ ✓

$= \dfrac{1025.375408}{1.0507508}$

$= 975.8502136$

$= 975.85$

$\therefore$ The fixed monthly payment, \$$M$, is \$975.85. ✓

(3 marks)

18 i $A_1 = 1000\left(1 + \dfrac{6}{100}\right)^{18}$

$A_2 = 1000\left(1 + \dfrac{6}{100}\right)^{17}$

$A_3 = 1000\left(1 + \dfrac{6}{100}\right)^{16}$

$A_{18} = 1000\left(1 + \dfrac{6}{100}\right)^{1}$

The total value of the investment fund,

$A = 1000\left(1 + \dfrac{6}{100}\right)^{18} + 1000\left(1 + \dfrac{6}{100}\right)^{17} + 1000\left(1 + \dfrac{6}{100}\right)^{16} \ldots + 1000\left(1 + \dfrac{6}{100}\right)^{1}$

$= 1000 \times \left[1.06^{18} + 1.06^{17} + 1.06^{16} + \ldots + 1.06^{1}\right]$

$= 1000 \times \left[1.06^{1} + 1.06^{2} + 1.06^{3} \ldots + 1.06^{18}\right]$ ✓

Now $1.06^1 + 1.06^2 + 1.06^3 \ldots + 1.06^{18}$ is a geometric series with $a = 1.06$, $r = 1.06$, $n = 18$.

$S_n = \dfrac{a(r^n - 1)}{r - 1}$

$\therefore S_{18} = \dfrac{1.06(1.06^{18} - 1)}{1.06 - 1}$ ✓

$= \dfrac{1.06(1.06^{18} - 1)}{0.06}$

$= 32.7599\ldots$

$\therefore A = 1000 \times 32.7599\ldots$

$= 32\,759.991\ldots$

$=$ \$32 759.99 to the nearest cent

$\therefore$ The total value of Mr Caine's investment is \$32 759.99. ✓

(3 marks)

ii $A_1 = 1000 \times (1.06)^3$

$A_2 = 1000 \times 1.06 \times 1.06^2$

$= 1000 \times (1.06)^3$

$A_3 = 1000 \times 1.06^2 \times 1.06$

$= 1000 \times (1.06)^3$ ✓

$\therefore A = 3 \times 1000 \times (1.06)^3$

$=$ \$3573.048 . . .

$=$ \$3573.05 to the nearest cent.

$\therefore$ The total value of Mrs Caine's investment on her son's third birthday is \$3573.05. ✓

(2 marks)

iii $A_{18} = 18 \times 1000 \times (1.06)^{18}$

$=$ \$51 378.104 . . .

$=$ \$51 378.10 to the nearest cent

$\therefore$ The total value of Mrs Caine's investment is \$51 378.10. ✓

(1 mark)

19 i $A_n = 200\,000r^n - M\left(1 + r + r^2 + \ldots r^{n-1}\right)$

where $r = 1.006$, $n = 300$, $M =$ monthly repayment

$\therefore A_{300} = 200\,000(1.006)^{300} - M\left(1 + 1.006 + 1.006^2 + \ldots + 1.006^{299}\right)$

Now $\left(1 + 1.006 + 1.006^2 + \ldots + 1.006^{299}\right)$ is a geometric series,

where $a = 1$, $r = 1.006$, $n = 300$.

Now $S_n = \dfrac{a(r^n - 1)}{r - 1}$

$\therefore S_{300} = \dfrac{1 \times (1.006^{300} - 1)}{1.006 - 1}$

$= \left[\dfrac{1.006^{300} - 1}{0.006}\right]$

$\therefore A_{300} = 200\,000(1.006)^{300} - M\left[\dfrac{1.006^{300} - 1}{0.006}\right]$ ✓

The loan will be repaid when $A_{300} = 0$.

$\therefore 0 = 200\,000(1.006)^{300} - M\left[\dfrac{1.006^{300} - 1}{0.006}\right]$

$M\left[\dfrac{1.006^{300} - 1}{0.006}\right] = 200\,000(1.006)^{300}$ ✓

$M = 200\,000(1.006)^{300} \div \left[\dfrac{1.006^{300} - 1}{0.006}\right]$

$M = 1\,203\,439.36 \div 836.1994\ldots$

$= 1439.18$

$\therefore$ The minimum monthly repayment is \$1439 to the nearest dollar. ✓ *(3 marks)*

ii $A_n = 200\,000(1.006)^n - M\left[\dfrac{1.006^n - 1}{0.006}\right]$

When $M = 2800$,

$\therefore A_n = 200\,000(1.006)^n - 2800\left[\dfrac{1.006^n - 1}{0.006}\right]$

The loan will be repaid when $A_n = 0$.

$$\therefore 0 = 200\,000(1.006)^n - 2800\left(\frac{1.006^n - 1}{0.006}\right) \quad ✓$$

$$0 = 200\,000(1.006)^n - \left(\frac{2800 \times 1.006^n - 2800 \times 1}{0.006}\right)$$

$$0 = 200\,000(1.006)^n \times 0.006 - (2800 \times 1.006^n - 2800)$$
$$0 = 1200(1.006)^n - 2800(1.006)^n + 2800$$
$$0 = 1.006^n \times (1200 - 2800) + 2800$$

$$0 = -1600(1.006)^n + 2800$$
$$1600(1.006)^n = 2800$$
$$1.006^n = \frac{2800}{1600} = 1.75$$
$$n \log_e 1.006 = \log_e 1.75$$
$$n = \log_e 1.75 \div \log_e 1.006$$
$$\doteqdot 93.5488\ldots$$

∴ The loan will be repaid in 94 months. ✓

(2 marks)

20 **i** $A_1 = 3\,000\,000 \times 1.12 - 480\,000$
$= (3 \times 10^6)(1.12) - (4.8 \times 10^5)$

$A_2 = A_1 \times 1.12 - (4.8 \times 10^5)$
$= [(3 \times 10^6)(1.12) - (4.8 \times 10^5)] \times 1.12 - (4.8 \times 10^5)$
$= (3 \times 10^6)(1.12)^2 - (4.8 \times 10^5)(1.12) - (4.8 \times 10^5)$
$= (3 \times 10^6)(1.12)^2 - (4.8 \times 10^5)(1.12 + 1)$
$= (3 \times 10^6)(1.12)^2 - (4.8 \times 10^5)(1 + 1.12)$ ✓

(1 mark)

ii $A_3 = A_2 \times 1.12 - 4.8 \times 10^5$
$= [(3 \times 10^6)(1.12)^2 - (4.8 \times 10^5)(1 + 1.12)] \times 1.12 - (4.8 \times 10^5)$
$= (3 \times 10^6)(1.12)^3 - (4.8 \times 10^5)(1.12 + 1.12^2) - (4.8 \times 10^5)$
$= (3 \times 10^6)(1.12)^3 - (4.8 \times 10^5)(1 + 1.12 + 1.12^2)$

Continuing the pattern:

$\therefore A_n = (3 \times 10^6)(1.12)^n - (4.8 \times 10^5)(1 + 1.12 + \ldots + 1.12^{n-1})$ ✓

Now $(1 + 1.12 + 1.12^2 + \ldots + 1.12^{n-1})$ is a geometric series with $a = 1, r = 1.12$ and $n = n$.

$$\therefore A_n = (3 \times 10^6)(1.12)^n - (4.8 \times 10^5)\left(\frac{a(r^n - 1)}{r - 1}\right)$$

$$= (3 \times 10^6)(1.12)^n - (4.8 \times 10^5)\left(\frac{1(1.12^n - 1)}{1.12 - 1}\right)$$

$$= (3 \times 10^6)(1.12)^n - (4.8 \times 10^5)\left(\frac{1.12^n - 1}{0.12}\right)$$
$$= (3 \times 10^6)(1.12)^n - (4 \times 10^6)(1.12^n - 1)$$
$$= (3 \times 10^6)(1.12)^n - (4 \times 10^6)(1.12^n) + (4 \times 10^6)$$
$$= (1.12)^n[(3 \times 10^6) - (4 \times 10^6)] + (4 \times 10^6)$$
$$= (4 \times 10^6) - (1 \times 10^6)(1.12)^n$$
$$= 10^6[4 - (1.12)^n] \quad ✓$$

(2 marks)

iii The loan will be repaid when

$$A_n = 0$$
$$10^6[4 - (1.12^n)] = 0$$
$$4 - 1.12^n = 0$$
$$1.12^n = 4 \quad ✓$$
$$\ln (1.12)^n = \ln 4$$
$$n \ln (1.12) = \ln 4$$
$$n = \frac{\ln 4}{\ln 1.12}$$
$$= 12.2325\ldots$$

∴ The loan will be repaid in the 13th year, 2017. ✓

(2 marks)

21 **i** $A = P\left(1 + \frac{r}{100}\right)^n$

$P = \$80$
$n = 300$ months
$r = 6\%$ per annum
$= \frac{6}{12}\%$ per month
$= 0.5\%$ per month

$\therefore A = 80\left(1 + \frac{0.5}{100}\right)^{300}$ ✓
$= 80(1 + 0.005)^{300}$
$= 80(1.005)^{300}$
$= 357.1975\ldots$
$= \$357.20$ to the nearest cent. ✓ *(2 marks)*

ii $A_1 = 80(1.005)^{300}$
$A_2 = 80(1.005)^{299}$
$A_3 = 80(1.005)^{298}$
$\vdots$
$A_{300} = 80(1.005)^1$

The final value of the trust fund
$= 80(1.005)^{300} + 80(1.005)^{299} + \ldots + 80(1.005)^1$ ✓
$= 80[1.005^{300} + 1.005^{299} + \ldots + 1.005^1]$
$= 80[1.005^1 + 1.005^2 + \ldots + 1.005^{300}]$
(on reversing the order of terms)

$= 80\left[\frac{a(r^n - 1)}{r - 1}\right]$
where $a = 1.005, r = 1.005, n = 300$ ✓

$= 80\left[\frac{1.005(1.005^{300} - 1)}{1.005 - 1}\right]$
$= 55\,716.714\ldots$
$= \$55\,716.71$ to the nearest cent. ✓

(3 marks)

22 **i** $A_1 = 120\,000\left(1 + \frac{0.06}{k}\right)^1 - F$

$$A_2 = \left[120\,000\left(1 + \frac{0.06}{k}\right)^1 - F\right]\left(1 + \frac{0.06}{k}\right) - F$$

$$= 120\,000\left(1 + \frac{0.06}{k}\right)^2 - \left(1 + \frac{0.06}{k}\right)F - F \quad ✓$$

(1 mark)

ii $A_1 = 120\,000\,\alpha - F$
$A_2 = 120\,000\,\alpha^2 - \alpha F - F$
$A_3 = (120\,000\,\alpha^2 - \alpha F - F)(\alpha) - F$
$= 120\,000\,\alpha^3 - \alpha^2 F - \alpha F - F$

$\therefore A_n = 120\,000\alpha^n - \alpha^{n-1}F - \alpha^{n-2}F - \ldots - \alpha F - F$
$= 120\,000\alpha^n - F(\alpha^{n-1} + \alpha^{n-2} + \ldots + \alpha + 1)$
$= 120\,000\alpha^n - F(1 + \alpha + \alpha^2 + \ldots + \alpha^{n-1})$ ✓
$= 120\,000\alpha^n - F\left(\dfrac{a(r^n - 1)}{r-1}\right)$
where $a = 1, r = \alpha$ and $n = n$
$= 120\,000\alpha^n - F\left(\dfrac{1(\alpha^n - 1)}{\alpha - 1}\right)$
$= 120\,000\alpha^n - F\left(\dfrac{\alpha^n - 1}{1 + \frac{0.06}{k} - 1}\right)$
$= 120\,000\alpha^n - F\left(\dfrac{\alpha^n - 1}{\frac{0.06}{k}}\right)$
$= 120\,000\alpha^n - \dfrac{kF(\alpha^n - 1)}{0.06}$. ✓

(2 marks)

iii $n = 25 \times 4,\ k = 4,\ A_n = 0$
$= 100$
$\therefore \alpha = 1 + \dfrac{0.06}{4}$
$= 1.015$
$\therefore 0 = 120\,000\,(1.015)^{100} - \dfrac{4F(1.015^{100} - 1)}{0.06}$
$= 7200\,(1.015)^{100} - 4F\,(1.015^{100} - 1)$ ✓
$4F\,(1.015^{100} - 1) = 7200\,(1.015)^{100}$
$F = \dfrac{7200\,(1.015)^{100}}{4(1.015^{100} - 1)}$
$= \$2324.47$ ✓

(2 marks)

iv $n = 25 \times 12,\ k = 12,\ A_n = 0$
$= 300$
$\therefore \alpha = 1 + \dfrac{0.06}{12}$
$= 1.005$
$\therefore 0 = 120\,000(1.005)^{300} - \dfrac{12F(1.005^{300} - 1)}{0.06}$
$= 7200\,(1.005)^{300} - 12F\,(1.005^{300} - 1)$
$12F\,(1.005^{300} - 1) = 7200\,(1.005)^{300}$
$\therefore F = \dfrac{7200\,(1.005)^{300}}{12\,(1.005^{300} - 1)}$
$= \$773.16$ ✓

Amount repaid if quarterly repayments:
$= \$2324.47 \times 100$
$= \$232\,447$

Amount repaid if monthly repayments:
$= \$773.16 \times 300$
$= \$231\,948$

Saving $= \$232\,447 - \$231\,948$
$= \$499$ ✓

(2 marks)

23 $A = P\left(1 + \dfrac{r}{100}\right)^n$
$P = 5000, r = 8.75$
$\therefore A = 5000\left(1 + \dfrac{8.75}{100}\right)^n$
$= 5000\,(1.0875)^n$

$A_1 = 5000\,(1.0875)^{21}$
$A_2 = 5000\,(1.0875)^{20}$
$\vdots$
$A_{21} = 5000\,(1.0875)^1$ ✓

Total value on 31 December 2023
$= 5000\,(1.0875)^{21} + 5000\,(1.0875)^{20} + \ldots + 5000\,(1.0875)^1$
$= 5000\,[(1.0875)^{21} + (1.0875)^{20} + \ldots + (1.0875)^1]$
$= 5000\,[(1.0875)^1 + (1.0875)^2 + \ldots + (1.0875)^{21}]$ ✓
$= 5000\left[\dfrac{a(r^n - 1)}{r - 1}\right]$
where $a = 1.0875,\ r = 1.0875,\ n = 21$
$= 5000\left[\dfrac{1.0875(1.0875^{21} - 1)}{1.0875 - 1}\right]$ ✓
$= \$299\,605$ to the nearest dollar. ✓

(4 marks)

24 **i** $B_1 = 1000(1.06) - 72$
$= 988$
Balance at beginning of year 2 is \$988 ✓

(1 mark)

ii $B_2 = [1000(1.06) - 72]1.06 - 72$
$= 1000(1.06)^2 - 72(1.06) - 72$
$= 1000(1.06)^2 - 72(1.06 + 1)$

$B_3 = [1000(1.06)^2 - 72(1.06 + 1)]1.06 - 72$
$= 1000(1.06)^3 - 72(1.06)^2 - 72(1.06) - 72$
$= 1000(1.06)^3 - 72(1.06^2 + 1.06 + 1)$
$= 1000(1.06)^3 - 72(1 + 1.06 + 1.06^2)$
$\therefore B_n = 1000(1.06)^n - 72[1 + (1.06) + \ldots + (1.06)^{n-1}]$ ✓
$= 1000(1.06)^n - 72\left[\dfrac{a(r^n - 1)}{r - 1}\right]$
where $a = 1, r = 1.06$
$= 1000(1.06)^n - 72\left[\dfrac{1(1.06^n - 1)}{1.06 - 1}\right]$
$= 1000(1.06)^n - 72\left[\dfrac{1.06^n - 1}{0.06}\right]$
$= 1000(1.06)^n - 1200(1.06^n - 1)$
$= 1000(1.06)^n - 1200(1.06)^n + 1200$
$\therefore B_n = 1200 - 200(1.06)^n$ ✓

(2 marks)

iii Let \$$R$ be the balance in the fund at the end of n years where n is the number of years after the end of the tenth year (i.e. the number of years remaining for the fund).

After one more year,
$R_1 = [1200 - 200(1.06)^{10}]1.06 - 90$
$= 1200(1.06) - 200(1.06)^{11} - 90$

After two more years,
$R_2 = [1200(1.06) - 200(1.06)^{11} - 90]1.06 - 90$
$= 1200(1.06)^2 - 200(1.06)^{12} - 90(1.06) - 90$

After n more years,

$R_n = 1200(1.06)^n - 200(1.06)^{n+10} - 90(1.06)^{n-1} - \ldots - 90$ ✓

$= 1200(1.06)^n - 200(1.06)^{n+10} - 90[(1.06)^{n-1} + (1.06)^{n-2} + \ldots + 1]$

$= 1200(1.06)^n - 200(1.06)^{n+10} - 90[1 + \ldots + (1.06)^{n-2} + (1.06)^{n-1}]$

$= 1200(1.06)^n - 200(1.06)^{n+10} - 90\left[\frac{a(r^n - 1)}{r - 1}\right]$

where $a = 1, r = 1.06$

$= 1200(1.06)^n - 200(1.06)^{n+10} - 90\left[\frac{1[1.06^n - 1]}{1.06 - 1}\right]$

$= 1200(1.06)^n - 200(1.06)^{n+10} - 90\left[\frac{(1.06)^n - 1}{0.06}\right]$

$= 1200(1.06)^n - 200(1.06)^{n+10} - 1500[(1.06)^n - 1]$

$= 1200(1.06)^n - 200(1.06)^{n+10} - 1500(1.06)^n + 1500$

$= 1500 - 200(1.06)^{n+10} - 300(1.06)^n$

When $R_n = 0$, ✓

$0 = 1500 - 200(1.06)^{n+10} - 300(1.06)^n$

$\therefore 1500 = 200(1.06)^{n+10} + 300(1.06)^n$

$15 = 2(1.06)^n(1.06)^{10} + 3(1.06)^n$

$15 = 3.582(1.06)^n + 3(1.06)^n$

$15 = 6.582(1.06)^n$

$\therefore (1.06)^n = \frac{15}{6.582}$

$= 2.2789$

$\ln (1.06)^n = \ln 2.2789$

$n \ln (1.06) = \ln 2.2789$

$n = \frac{\ln 2.2789}{\ln 1.06}$

$\doteqdot 14.13$

$\doteqdot 14$

$\therefore$ The prize fund can be used for 14 more years. ✓

(3 marks)

25 **i** $A_1 = 5000 - M$

$A_2 = 5000 - M - M$

$A_3 = 5000 - M - M - M$

$= 5000 - 3M.$ ✓

ii $A_4 = (5000 - 3M)1.01 - M$

$A_5 = [(5000 - 3M)1.01 - M]1.01 - M$

$= (5000 - 3M)1.01^2 - M(1.01) - M$

$= (5000 - 3M)1.01^2 - M(1.01 + 1)$

$= (5000 - 3M)1.01^2 - M(1 + 1.01).$ ✓

iii $A_{36} = (5000 - 3M)1.01^{33} - M(1 + 1.01 + \ldots + (1.01)^{32}).$ ✓

iv After the 36th repayment, the loan is repaid. $\therefore A_{36} = 0$

i.e. $(5000 - 3M)1.01^{33} - M(1 + 1.01 + \ldots + (1.01)^{32}) = 0$

$\therefore (5000 - 3M)1.01^{33} - M\left[\frac{a(r^n - 1)}{r - 1}\right] = 0$

Where $a = 1, r = 1.01, n = 33$

$\therefore (5000 - 3M)1.01^{33} - M\frac{1(1.01^{33} - 1)}{1.01 - 1} = 0$ ✓

$\therefore (5000 - 3M)1.01^{33} - \frac{M(1.01^{33} - 1)}{0.01} = 0$

$\therefore (5000 - 3M)1.01^{33} - 100M(1.01^{33} - 1) = 0$

$\therefore 5000(1.01)^{33} - 3M(1.01)^{33} - 100M(1.01^{33} - 1) = 0$

$\therefore 5000(1.01)^{33} = 3M(1.01)^{33} + 100M(1.01^{33} - 1)$

$\therefore 5000(1.01)^{33} = M[3(1.01)^{33} + 100(1.01^{33} - 1)]$ ✓

$\therefore M = \frac{5000(1.01)^{33}}{103(1.01)^{33} - 100}$

$= \$161.34$ ✓

(6 marks)
(Total mark allocation only in exam)

26 **i** $\$A_1 = \$P(1.08) - \$3000$ ✓

ii $\$A_2 = \$A_1(1.08) - \$3000$

$= [\$P(1.08) - \$3000]1.08 - \$3000$

$= \$P(1.08)^2 - \$3000(1.08) - \$3000$ ✓

iii $\$A_6 = \$P(1.08)^6 - \$3000(1.08)^5 - \$3000(1.08)^4 - \ldots - \$3000$

After 6 years, $\$A_6 = 0$

$\therefore \$P(1.08)^6 - \$3000(1.08)^5 - \$3000(1.08)^4 - \ldots - \$3000 = 0$ ✓

$\therefore \$P(1.08)^6 = \$3000 + \$3000(1.08)^1 + \ldots + \$3000(1.08)^5$

$= \$3000[1 + (1.08)^1 + \ldots + (1.08)^5]$

$= \$3000\left[\frac{a(r^n - 1)}{r - 1}\right]$

where $a = 1, r = 1.08, n = 6$

$= \$3000\left[\frac{1((1.08)^6 - 1)}{1.08 - 1}\right]$ ✓

$\therefore \$P = \$3000\left[\frac{1.08^6 - 1}{0.08}\right] \div (1.08)^6$

$= \$13\,868.64$

$= \$13\,869$ to the nearest dollar. ✓

(5 marks)
(Total mark allocation only in exam)

27 **i** No. of fish after 1st harvest

$= 100\,000(1.1) - 15\,400$

No. of fish after 2nd harvest

$= [100\,000(1.1) - 15\,400]1.1 - 15\,400$

$= 100\,000(1.1)^2 - 15\,400(1.1) - 15\,400$

$= 88\,660$ fish ✓

ii No. of fish after nth harvest, F_n

$= 100\,000(1.1)^n - 15\,400(1.1)^{n-1} - 15\,400(1.1)^{n-2} \ldots - 15\,400(1.1) - 15\,400$

$= 100\,000(1.1)^n - [15\,400 + 15\,400(1.1) + \ldots + 15\,400(1.1)^{n-1}]$

$= 100\,000(1.1)^n - \left[\frac{a(r^n - 1)}{r - 1}\right]$

where $a = 15\,400$, $r = 1.1$ and $n = n$

$= 100\,000(1.1)^n - \left[\frac{15\,400[(1.1)^n - 1]}{1.1 - 1}\right]$ ✓

$= 100\,000(1.1)^n - \frac{15\,400[(1.1)^n - 1]}{0.1}$

$= 100\,000(1.1)^n - 154\,000[(1.1)^n - 1]$

$= 100\,000(1.1)^n - 154\,000(1.1)^n + 154\,000$

$= 154\,000 - 54\,000(1.1)^n$ ✓

iii When all fish are sold, $F_n = 0$

i.e. $154000 - 54000(1.1)^n = 0$

$$54000(1.1)^n = 154000$$
$$(1.1)^n = \frac{154000}{54000}$$
$$(1.1)^n = 2.851\ldots$$
$$\ln(1.1)^n = \ln 2.851\ldots$$
$$n\ln(1.1) = \ln 2.851\ldots$$
$$n = \frac{\ln 2.851\ldots}{\ln 1.1}$$
$$= 10.995\ldots$$
$$\doteqdot 11$$

∴ The farmer will have sold all his fish after the 11th harvest. ✓

His total income will be

$11 \times 15400 \times \$10 = \$1694000$ ✓

iv If I is the total income received after n harvests:

$I = 15 \times F_n + 15\,400 \times 10 \times n$
$= 15[154000 - 54000(1.1)^n] + 154000n$
$= 2310000 - 810000(1.1)^n + 154000n$ ✓

By trial and error:

When $n = 6$, $I = \$1799036$.
When $n = 7$, $I = \$1809539$.
When $n = 8$, $I = \$1805693$.

∴ The farmer should sell after the 7th harvest. ✓

(A more advanced approach:

$$\frac{dI}{dn} = -810000\ln(1.1).(1.1)^n + 154000$$

$$\left(\text{NB } \frac{d}{dn}(1.1)^n = \ln(1.1).(1.1)^n\right)$$

Stationary points occur when

$$\frac{dI}{dn} = 0$$
$$\therefore 0 = -810000\ln(1.1).(1.1)^n + 154000$$
$$154000 = 810000\ln(1.1).(1.1)^n$$
$$\frac{154000}{810000\ln(1.1)} = (1.1)^n$$
$$1.9948 = (1.1)^n$$
$$\ln 1.9948 = \ln(1.1)^n$$
$$\ln 1.9948 = n\ln(1.1)$$
$$n = \frac{\ln 1.9948}{\ln 1.1}$$
$$\doteqdot 7.25$$

Now when $n = 7$, $I = \$1809539$.
and when $n = 8$, $I = \$1805693$.

∴ The farmer should sell after the 7th harvest.

(7 marks)
(Total mark allocation only in exam)

28 **i** $A = P\left(1 + \frac{r}{100}\right)^n$

$P = 10000,\ r = 8,\ n = 10$

$A = 10000\left(1 + \frac{8}{100}\right)^{10}$
$= 10000(1.08)^{10}$
$= \$21589.25$ ✓

ii Amount in account on 1 July 1986
$= 10000(1.08)^1 + 1000$ ✓

Amount in account on 1 July 1987
$= [10000(1.08)^1 + 1000]1.08 + 1000$
$= 10000(1.08)^2 + 1000(1.08) + 1000$ ✓

Amount in account on 1 July 1988
$= [10000(1.08)^2 + 1000(1.08) + 1000]1.08 + 1000$
$= 10000(1.08)^3 + 1000(1.08)^2 + 1000(1.08) + 1000$

Amount in account on 1 July 1995 ✓
$= 10000(1.08)^{10} + 1000(1.08)^9 + 1000(1.08)^8 + \ldots + 1000$
$= 10000(1.08)^{10} + 1000[1.08^9 + 1.08^8 + \ldots + 1]$
$= 10000(1.08)^{10} + 1000[1 + 1.08^1 + \ldots + 1.08^9]$
$= 10000(1.08)^{10} + 1000\left[\frac{1(1.08^{10} - 1)}{1.08 - 1}\right]$ ✓

(using the formula for the sum of a geometric series)

$= 10000(1.08)^{10} + 1000\left[\frac{1.08^{10} - 1}{0.08}\right]$
$= 21589.25 + 14486.56$
$= \$36075.81$ ✓

iii $35478 = 10000\left(1 + \frac{r}{100}\right)^{10}$

$3.5478 = \left(1 + \frac{r}{100}\right)^{10}$ ✓

$1 + \frac{r}{100} = \sqrt[10]{3.5478}$

$1 + \frac{r}{100} = 1.1350\ldots$

$\frac{r}{100} = 0.1350\ldots$

$r \doteqdot 13.5\%$ ✓

(8 marks)
(Total mark allocation only in exam)

29 **i** $A = P\left(1 + \frac{r}{100}\right)^n$
$= 50000(1 + \frac{8}{100})^1$
$= 50000(1.08)$

Amount after 1st withdrawal,
$A_1 = 50000(1.08) - M$

ii Amount after 2nd withdrawal,
$A_2 = A_1(1.08) - M$
$= [50000(1.08) - M]1.08 - M$
$= 50000(1.08)^2 - 1.08M - M$

Amount after 3rd withdrawal,
$A_3 = A_2(1.08) - M$
$= [50000(1.08)^2 - 1.08M - M]1.08 - M$
$= 50000(1.08)^3 - 1.08^2M - 1.08M - M$

Amount after 20th withdrawal,
$A_{20} = 50000(1.08)^{20} - 1.08^{19}M - 1.08^{18}M - \ldots - M$
$= 50000(1.08)^{20} - M(1.08^{19} + 1.08^{18} + \ldots + 1)$
$= 50000(1.08)^{20} - M(1 + 1.08^1 + \ldots + 1.08^{19})$

$$= 50\,000(1.08)^{20} - M\left[\frac{1(1.08^{20} - 1)}{1.08 - 1}\right]$$

(using the formula for the sum of a geometric series)

$$= 50\,000(1.08)^{20} - M\left[\frac{1.08^{20} - 1}{0.08}\right].$$

iii $0 = 50\,000(1.08)^{20} - M\left[\frac{1.08^{20} - 1}{0.08}\right].$

$$M\left[\frac{1.08^{20} - 1}{0.08}\right] = 50\,000(1.08)^{20}$$

$$M = \frac{50\,000(1.08)^{20} \times 0.08}{1.08^{20} - 1}$$

$= 5092.61$ to the nearest cent.

iv

$$M = \frac{50\,000(1 + x)^{20} \times x}{(1 + x)^{20} - 1}$$

(from **iii**)

where $x = \frac{r}{100}$

If $M = 8000$,

$$8000 = \frac{50\,000(1 + x)^{20} \times x}{(1 + x)^{20} - 1}$$

Say $x = 0.14$, RHS $\doteqdot 7549$
Say $x = 0.15$, RHS $\doteqdot 7988$
Say $x = 0.16$, RHS $\doteqdot 8433$
$\therefore x = 0.15$ will give the required rate.

$\therefore \frac{r}{100} = 0.15$

$\therefore \quad r = 15$

The required interest rate would be 15%.

(No mark allocation in exam)

1 Consider the following dataset.

13 16 17 17 21 24

Which row of the table shows how the median and mean are affected when a score of 5 is added to the dataset?

	Median	*Mean*
A	Changes	Changes
B	Stays the same	Stays the same
C	Changes	Stays the same
D	Stays the same	Changes

(1 mark)

 (Q2, **2022 HSC**) 

2 The table shows the types of customer complaints received by an online business in a month.

Type of complaint	*Frequency*	*Cumulative frequency*	*Cumulative percentage*
Stock shortage	98	98	49
Delivery fee	62	*A*	80
Delivery time	24	184	92
Damaged item	8	192	*B*
Returns policy	6	198	99
Product information	2	200	100
Total	200		

a What are the values of ***A*** and ***B***?

(2 marks) 

b The data from the table are shown in the following Pareto chart.

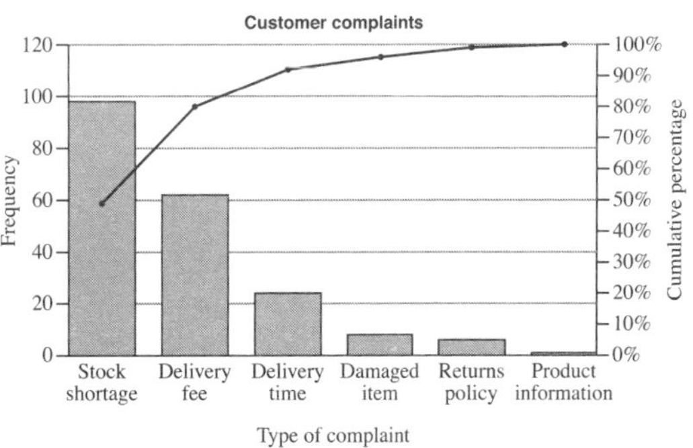

The manager will address 80% of the complaints.
Which types of complaints will the manager address? *(1 mark)*

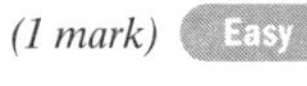 (Q11, **2022 HSC**)

3 The number of downloads of a song on each of twenty consecutive days is shown in the following graph.

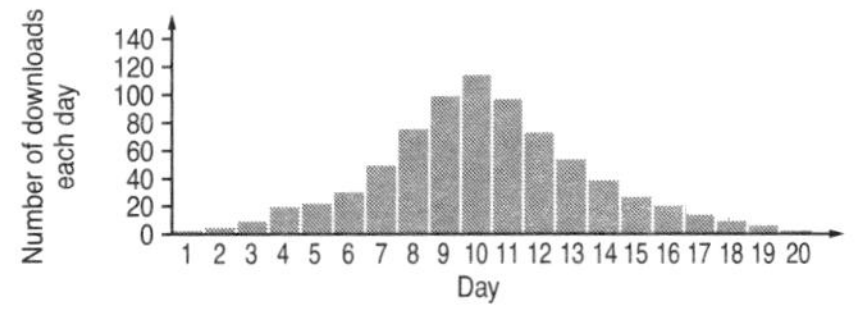

Which of the following graphs best shows the cumulative number of downloads up to and including each day?

A

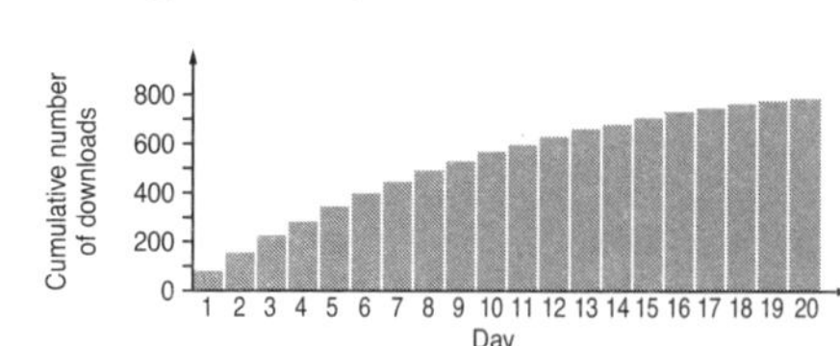

B

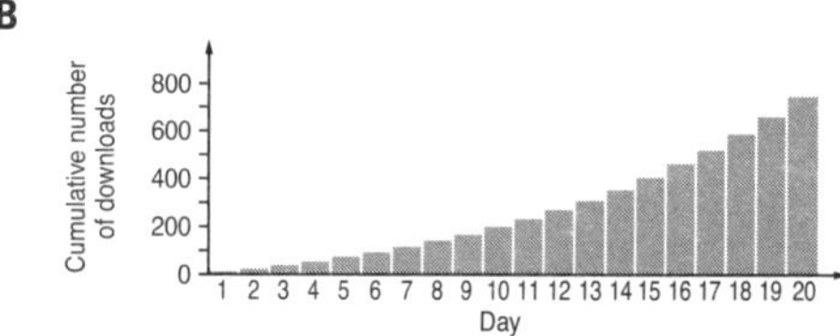

C

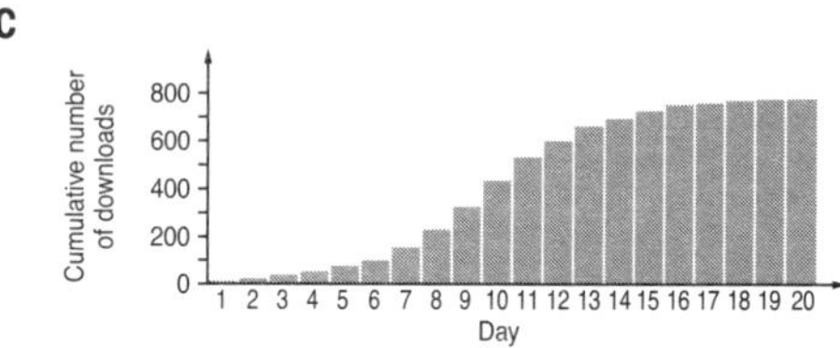

D

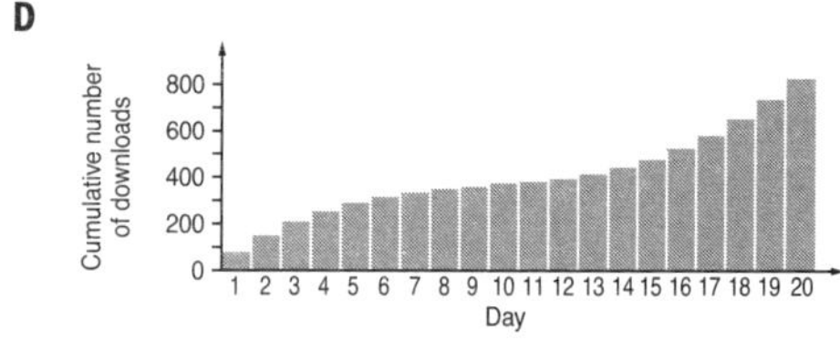

(1 mark)

 (Q4, **2021 HSC**) 

4 The five-number summary of a dataset is given.

Lowest score = 1
Lower quartile (Q_1) = 4
Median (Q_2) = 7
Upper quartile (Q_3) = 10
Highest score = 20

Is 20 an outlier? Justify your answer with calculations. *(2 marks)*

(Q17, **2021 MS2 HSC**) 

5 Consider the following dataset.

1 5 9 10 15

Suppose a new value, x, is added to this dataset, giving the following.

1 5 9 10 15 x

It is known that x is greater than 15. It is also known that the difference between the means of the two datasets is equal to ten times the difference between the medians of the two datasets.

Calculate the value of x. *(4 marks)*

(Q28, **2020 MS2 HSC**) **Medium**

---

6 Milo and Aurora each sat twelve class tests. Milo's results on the tests are displayed in the box-and-whisker plot.

i Aurora's 5-number summary for the tests is 65, 68, 70, 72, 75.
Draw a box-and-whisker plot to display Aurora's results below that of Milo's results. *(1 mark)* **Easy**

Milo

Aurora

64 65 66 67 68 69 70 71 72 73 74 75 76 77

ii What percentage of Milo's results were below 68? *(1 mark)* **Easy**

iii Aurora claims that her results were better than Milo's. Is she correct?
Justify your answer by referring to the summary statistics and the skewness of the distributions. *(3 marks)* **Easy**

Bonus question (see page iv)

---

7 A set of data has a lower quartile (Q_L) of 24 and an upper quartile (Q_U) of 42.

What is the maximum possible range for this set of data if there are no outliers? *(2 marks)*

Bonus question **Medium**

---

8 A school collected data related to the reasons given by students for arriving late. The Pareto chart shows the data collected.

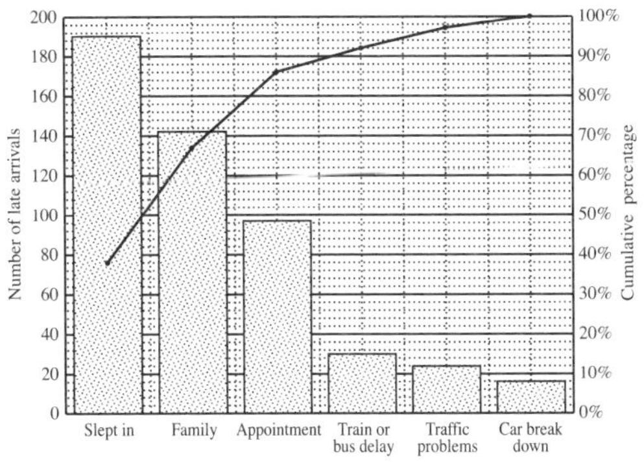

What percentage of students gave the reason 'Train or bus delay'?

A 6% **B** 15%
C 30% **D** 92% *(1 mark)*

(Q10, **2019 MS2 HSC**) **Medium**

---

9 The heights, in centimetres, of 10 players on a basketball team are shown.

170, 180, 185, 188, 192, 193, 193, 194, 196, 202

Is the height of the shortest player on the team considered an outlier? Justify your answer with calculations. *(3 marks)*

(Q19, **2019 MS2 HSC**) **Easy**

---

10 Two netball teams, Team A and Team B, each played 15 games in a tournament. For each team, the number of goals scored in each game was recorded.

The frequency table shows the data for Team A.

Number of goals	*Frequency*
19	1
20	0
21	1
22	1
23	1
24	3
25	0
26	4
27	3
28	1

The data for Team B was analysed to create the box-plot shown.

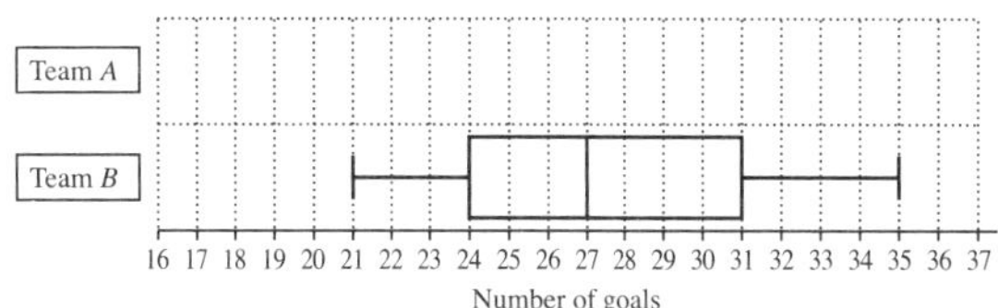

Compare the distributions of the number of goals scored by the two teams. Support your answer with the construction of a box-plot for the data for Team A. *(5 marks)*

(Q39, **2019 MS2 HSC**) **Medium**

---

11 A set of data is displayed in this dot plot.

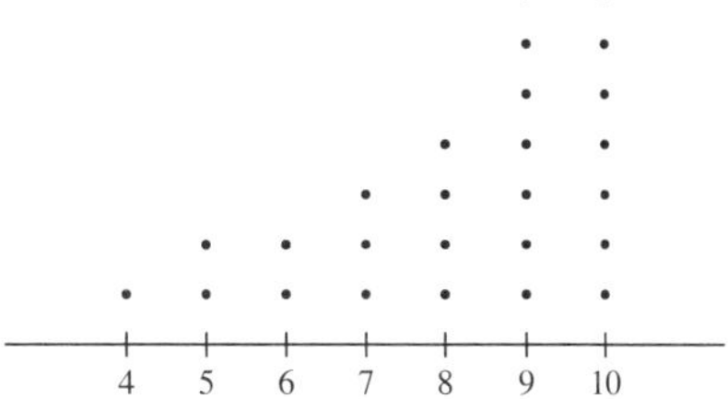

Which of the following best describes this set of data?

A Symmetrical **B** Positively skewed
C Negatively skewed **D** Normally distributed

(1 mark)

(Q6, **2018 MS2 HSC**) **Easy**

12 The graph displays the mean monthly rainfall in Sydney and Perth.

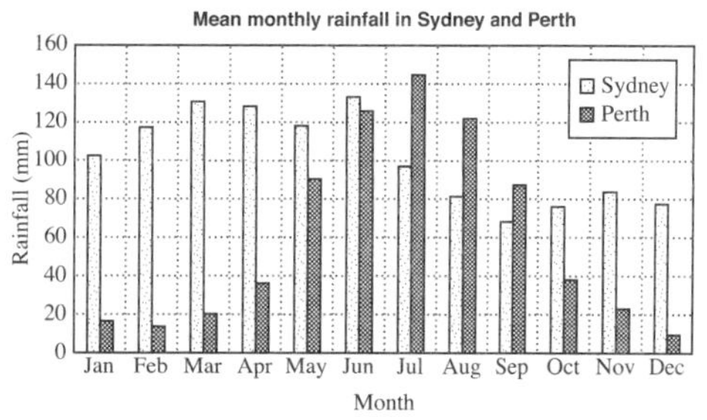

i For how many months is the mean monthly rainfall higher in Perth than in Sydney? *(1 mark)* Easy

ii For which of the two cities is the standard deviation of the mean monthly rainfall smaller? Justify your answer WITHOUT calculations. *(1 mark)* Easy

(Q26d, **2018 MS2 HSC**)

13 A cumulative frequency table for a data set is shown.

Score	*Cumulative frequency*
1	5
2	9
3	16
4	20
5	34
6	42

What is the interquartile range of this data set? *(2 marks)*

(Q26e, **2018 MS2 HSC**) 

14 A set of data has a lower quartile (Q_L) of 10 and an upper quartile (Q_U) of 16.

What is the maximum possible range for this set of data if there are no outliers? *(2 marks)*

(Q30a, **2017 MS2 HSC**) 

15 A soccer referee wrote down the number of goals scored in 9 different games during the season.

2, 3, 3, 3, 5, 5, 8, 9, ☐

The last number has been omitted. The range of the data is 10.

What is the five-number summary for this data set?

A 2, 3, 5, 8.5, 12 **B** 2, 3, 5, 8.5, 10
C 2, 3, 5, 8, 12 **D** 2, 3, 5, 8, 10 *(1 mark)*

(Q19, **2016 MS2 HSC**) Easy

16 A grouped data frequency table is shown.

Class interval	*Frequency*
1–5	3
6–10	6
11–15	8
16–20	9

What is the mean for this set of data?

A 6.5 **B** 10.5
C 11.9 **D** 12.4 *(1 mark)*

(Q21, **2016 MS2 HSC**) 

17 The box-and-whisker plots show the results of a History test and a Geography test.

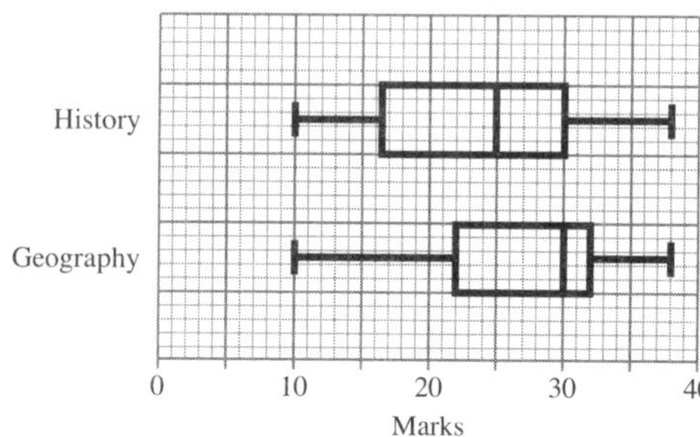

In History, 112 students completed the test. The number of students who scored above 30 marks was the same for the History test and the Geography test.

How many students completed the Geography test?

A 8 **B** 50
C 56 **D** 112 *(1 mark)*

(Q22, **2016 MS2 HSC**) 

18 The heights of 400 students were measured. The results are displayed in this cumulative frequency polygon.

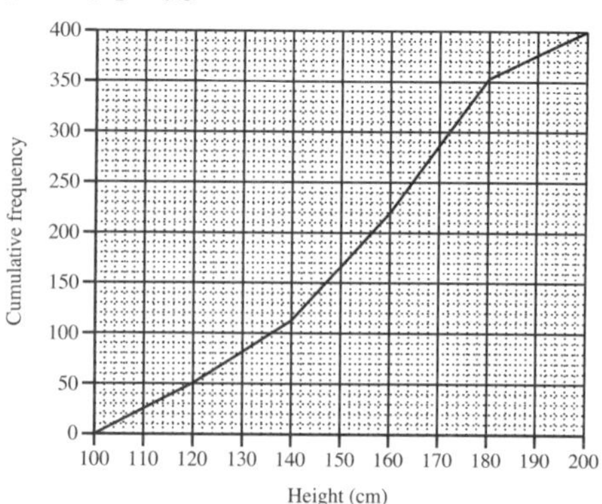

Use the polygon to estimate the interquartile range. *(2 marks)*

(Q27c, **2016 MS2 HSC**) 

19 A small population consists of three students of heights 153 cm, 168 cm and 174 cm. Samples of varying sizes can be taken from this population.

What is the mean of the mean heights of all the possible samples? Justify your answer. *(2 marks)*

(Q27b, **2016 MS2 HSC**) **Medium**

20 The ages of members of a dance class are shown in the back-to-back stem-and-leaf plot.

Women		*Men*
2	3	4 6
4 2	4	2 2 5 6 8
8 8 5 4 0 0	5	3
9 4 3 3	6	3

Pat claims that the women who attend the dance class are generally older than the men.

Is Pat correct? Justify your answer by referring to the median and skewness of the two sets of data. *(3 marks)*

(Q29c, **2016 MS2 HSC**) **Easy**

21 In a small business, the seven employees earn the following wages per week:

$300, $490, $520, $590, $660, $680, $970.

i Is the wage of $970 an outlier for this set of data? Justify your answer with calculations. *(3 marks)* **Easy**

ii Each employee receives a $20 pay increase. What effect will this have on the standard deviation? *(1 mark)* **Easy**

(Q27d, **2015 MS2 HSC**)

22 Data from 200 recent house sales are grouped into class intervals and a cumulative frequency histogram is drawn.

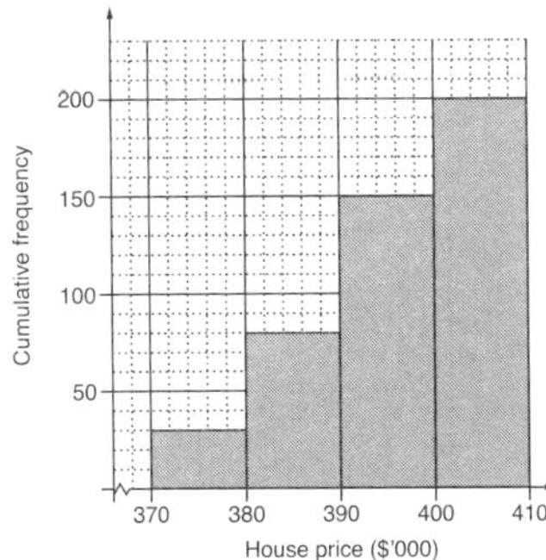

i Use the graph to estimate the median house price. *(1 mark)* **Medium**

Class Centre ($'000)	*Frequency*

ii By completing the table, calculate the mean house price. *(3 marks)* **Medium**

(Q29d, **2015 MS2 HSC**)

23 Twenty Year 12 students were surveyed. These students were asked how many hours of sport they play per week, to the nearest hour.
The results are shown in the frequency table.

Hours per week	*Frequency*
0–2	5
3–5	10
6–8	3
9–11	2

What is the mean number of hours of sport played by the students per week?

A 3.3 **B** 4.3

C 5.0 **D** 5.3 *(1 mark)*

(Q14, **2014 MS2 HSC**) **Easy**

24 Terry and Kim each sat twenty class tests. Terry's results on the tests are displayed in the box-and-whisker plot shown in part **i**.

i Kim's 5-number summary for the tests is 67, 69, 71, 73, 75.

Draw a box-and-whisker plot to display Kim's results below that of Terry's results.

Terry

Kim

65 66 67 68 69 70 71 72 73 74 75 76 77

(1 mark) **Easy**

ii What percentage of Terry's results were below 69? *(1 mark)* **Medium**

iii Terry claims that his results were better than Kim's. Is he correct? Justify your answer by referring to the summary statistics and the skewness of the distributions. *(4 marks)* **Medium**

(Q29c, **2014 MS2 HSC**)

25 A survey was conducted where people were asked which of two brands of smartphones they preferred. The results were:

- 48% preferred Brand X
- 52% preferred Brand Y.

A graph displaying the data is to be included in a magazine article. The editor of the magazine wishes to ensure that the graph is not misleading in any way.

Which graph should the editor choose to include in the article?

A

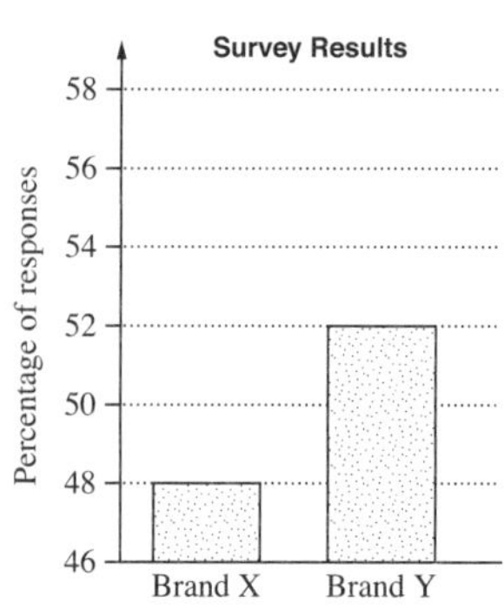

B

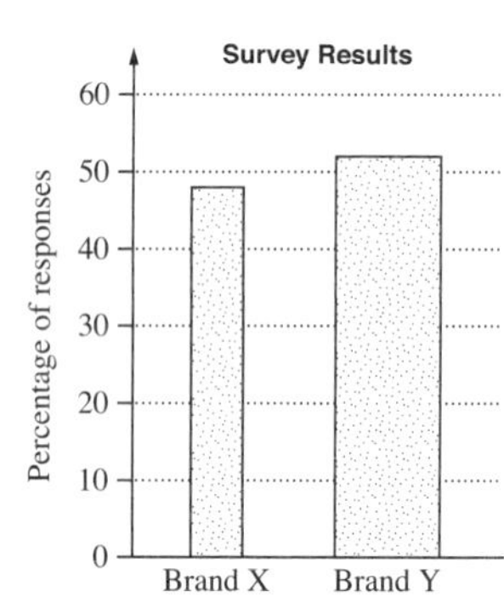

C

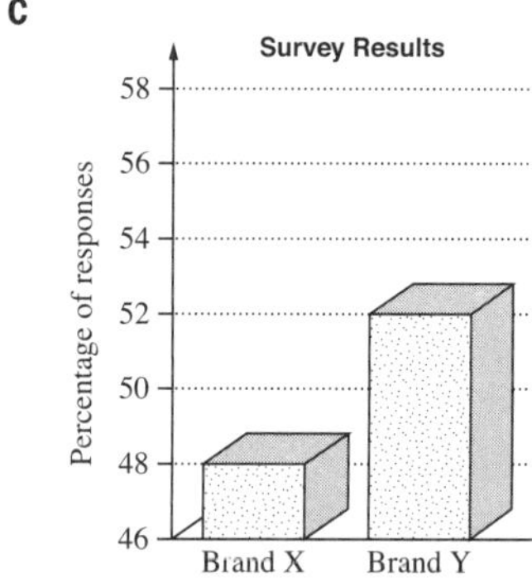

D

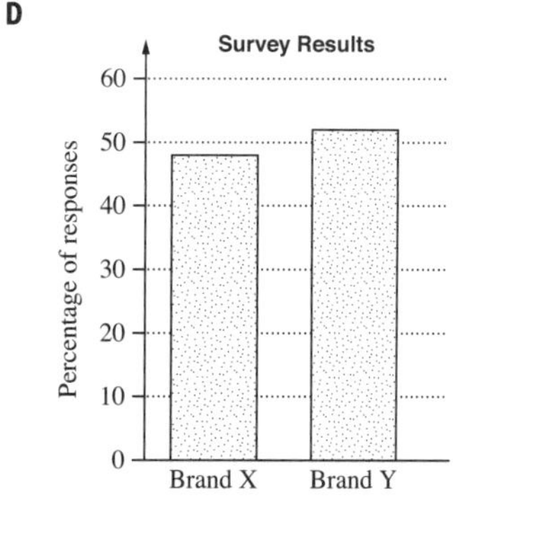

(1 mark)

(Q6, **2013 MS2 HSC**) Easy

26 The July sale prices for properties in a suburb were:

\$552 000, \$595 000, \$607 000, \$607 000, \$682 000 and \$685 000.

On 1 August, another property in the same suburb was sold for over one million dollars. If this property had been sold in July, what effect would it have had on the mean and median sale prices for July?

A Both the mean and the median would have changed.

B Neither the mean nor the median would have changed.

C The mean would have changed and the median would have stayed the same.

D The mean would have stayed the same and the median would have changed *(1 mark)*

(Q14, **2013 MS2 HSC**) Easy

27 Jason travels to work by car on all five days of his working week, leaving home at 7 am each day. He compares his travel times using roads without tolls and roads with tolls over a period of 12 working weeks.

He records his travel times (in minutes) in a back-to-back stem-and-leaf plot.

Travel time (minutes)

Without tolls		*With tolls*
9	3	5 8 9 9
9 9 8 7 7 6 5 5 4 4 3 2 0	4	0 1 2 6 7 7 8 8 8 9
9 8 7 5 4 3 3 3 2 2 2 2 1 1 0	5	2 4 4 5 6 8 9
1	6	1 3 5 7
	7	0 2 8
	8	2
	9	0

i What is the modal travel time when he uses roads without tolls? *(1 mark)* Easy

ii What is the median travel time when he uses roads without tolls? *(1 mark)* Medium

iii Describe how the two data sets differ in terms of the spread and skewness of their distributions. *(2 marks)* Medium

(Q26f, **2013 MS2 HSC**)

28 A retailer has collected data on the number of televisions that he sold each week in 2012.

He grouped the data into classes and displayed the data using a cumulative frequency histogram and polygon (ogive).

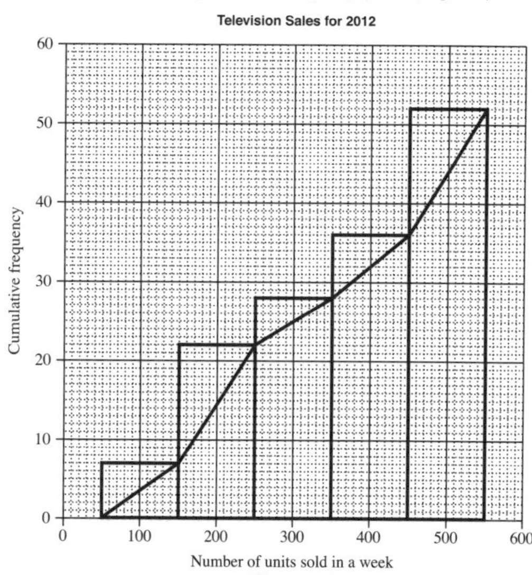

i Use the cumulative frequency polygon to determine the interquartile range. *(2 marks)* **Medium**

ii Oscar said that the retailer sold 300 televisions in 6 of the weeks in 2012. Is he correct? Give a reason for your answer. *(1 mark)* **Medium**

(Q27c, **2013 MS2 HSC**)

29 The test results in English and Mathematics for a class were recorded and displayed in the box-and-whisker plots.

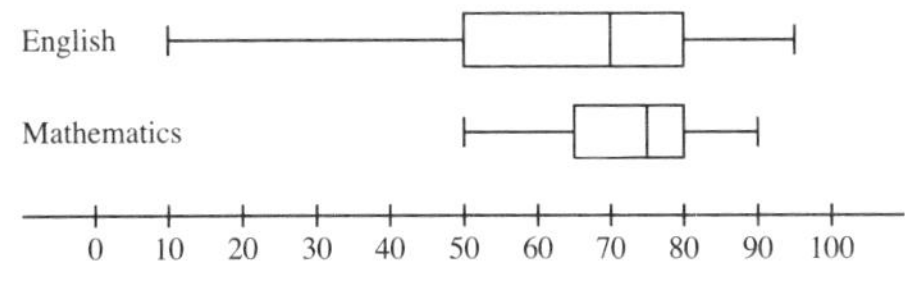

i What is the interquartile range for English? *(1 mark)* **Easy**

ii Compare and contrast the two data sets by referring to the skewness of the distributions and the measures of location and spread. *(3 marks)* **Medium**

(Q28d, **2012 MS2 HSC**)

30 A data set of nine scores has a median of 7.

The scores 6, 6, 12 and 17 are added to this data set.

What is the median of the data set now?

A 6 **B** 7
C 8 **D** 9 *(1 mark)*

(Q14, **2011 MS2 HSC**) 

31 The mean of a set of ten scores is 14. Another two scores are included and the new mean is 16.

What is the mean of the two additional scores?

A 4 **B** 16
C 18 **D** 26 *(1 mark)*

(Q21, **2009 HSC**) 

32 The marks for a Science test and a Mathematics test are presented in box-and-whisker plots.

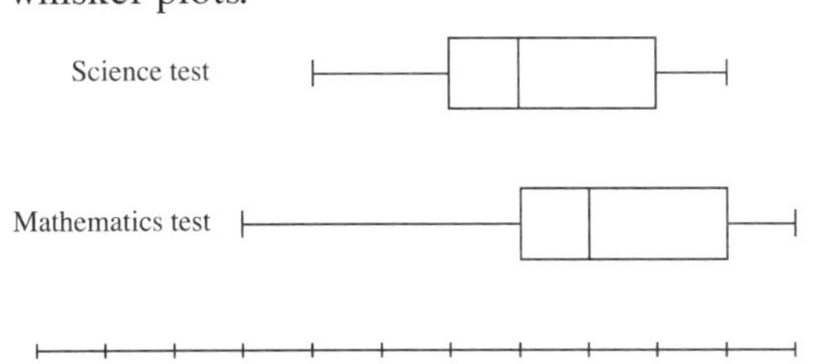

Which measure must be the same for both tests?

A Mean **B** Range
C Median **D** Interquartile range *(1 mark)*

(Q10, **2008 MS2 HSC**) **Easy**

33 Christina has completed three Mathematics tests. Her mean mark is 72%.

What mark (out of 100) does she have to get in her next test to increase her mean mark to 73%? *(2 marks)*

(Q23f, **2008 MS2 HSC**) **Medium**

34 Leanne copied a two-way table into her book.

	Male	*Female*	*Totals*
Full-time work	279	356	635
Part-time work	187	439	716
Totals	466	885	1351

Leanne made an error in copying one of the values in the shaded section of the table.

Which value has been incorrectly copied?

A The number of males in full-time work
B The number of males in part-time work
C The number of females in full-time work
D The number of females in part-time work *(1 mark)*

(Q16, **2007 MS2 HSC**) **Medium**

35 This set of data is arranged in order from smallest to largest.

$$5, 6, 11, x, 13, 18, 25$$

The range is six less than twice the value of x.

Which one of the following is true?

A The median is 12 and the interquartile range is 7.
B The median is 12 and the interquartile range is 12.
C The median is 13 and the interquartile range is 7.
D The median is 13 and the interquartile range is 12. *(1 mark)*

(Q21, **2007 MS2 HSC**) **Medium**

36 A set of examination results is displayed in a cumulative frequency histogram and polygon (ogive).

Sanath knows that his examination mark is in the 4th decile.

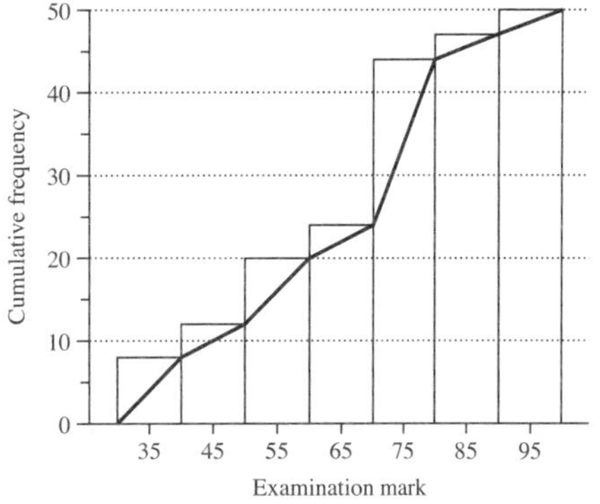

Which of the following could have been Sanath's examination mark?

A 37 **B** 57
C 67 **D** 77 *(1 mark)*

(Q22, **2007 MS2 HSC**) Medium

---

37 Consider the following set of scores:

$$3, 5, 5, 6, 8, 8, 9, 10, 10, 50.$$

i Calculate the mean of the set of scores. *(1 mark)* Easy

ii What is the effect on the mean and on the median of removing the outlier? *(2 marks)* Easy

(Q24a, **2007 MS2 HSC**)

---

38 Barry constructed a back-to-back stem-and-leaf plot to compare the ages of his students.

Ages of students attending Barry's Ballroom Dancing Studio

Females		*Males*
9	1	1 2 3
7	2	0 2 2 2 4 5
5	3	0 0 1 7
5 2	4	6 7
3 2 0	5	2
4 4 2 1	6	4 4

i Write a brief statement that compares the distribution of the ages of males and females from this set of data. *(1 mark)* 

ii What is the mode of this set of data? *(1 mark)* Easy

iii Liam decided to use a grouped frequency distribution table to calculate the mean age of the students at Barry's Ballroom Dancing Studio.
For the age group 30–39 years, what is the value of the product of the class centre and the frequency? *(2 marks)* Easy

iv Liam correctly calculated the mean from the grouped frequency distribution table to be 39.5.
Caitlyn correctly used the original data in the back-to-back stem-and-leaf plot and calculated the mean to be 38.2.
What is the reason for the difference in the two answers? *(1 mark)* Easy

(Q24d, **2007 MS2 HSC**)

---

39 A set of data is represented by the cumulative frequency histogram and ogive.

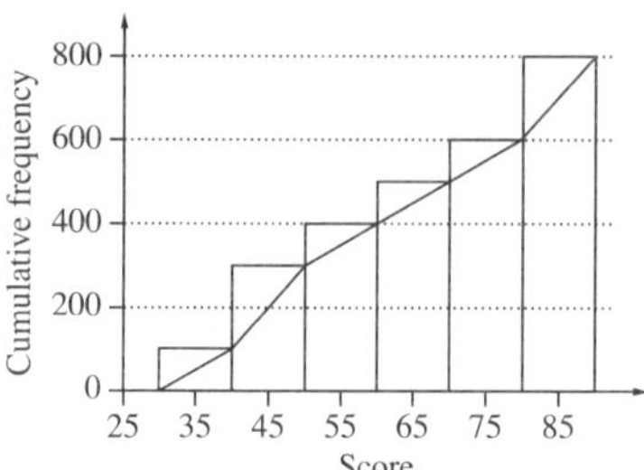

What is the best approximation for the interquartile range for this set of data?

A 25 **B** 30
C 35 **D** 40 *(1 mark)*

(Q9, **2005 MS2 HSC**) 

---

40 Two groups of people were surveyed about their weekly wages. The results are shown in the box-and-whisker plots.

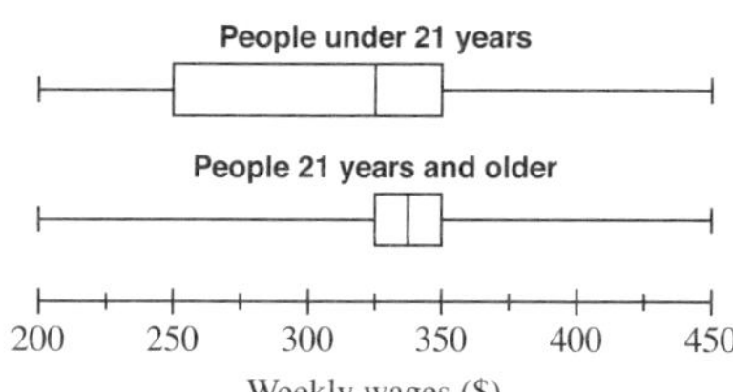

Which of the following statements is true for the people surveyed?

A The same percentage of people in each group earned more than \$325 per week.

B Approximately 75% of people under 21 years earned less than \$350 per week.

C Approximately 75% of people 21 years and older earned more than \$350 per week.

D Approximately 50% of people in each group earned between \$325 and \$350 per week. *(1 mark)*

(Q22, **2005 MS2 HSC**) 

---

41 The sector graph shows the proportion of people, as a percentage, living in each region of Sumcity. There are 24 000 people living in the Eastern Suburbs.

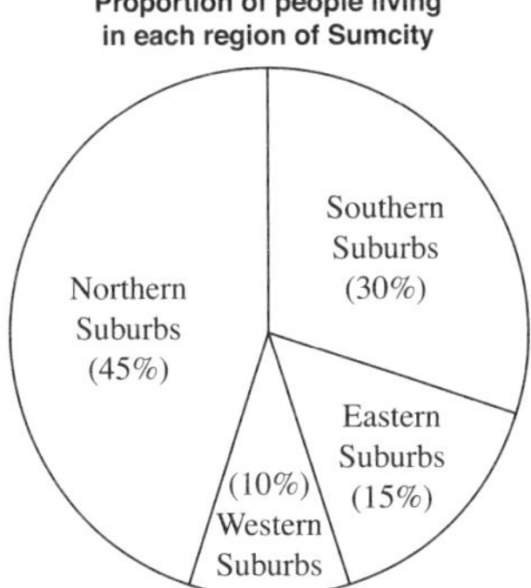

i Show that the total number of people living in Sumcity is 160000. *(1 mark)* Easy

Jake used the information above to draw a column graph.

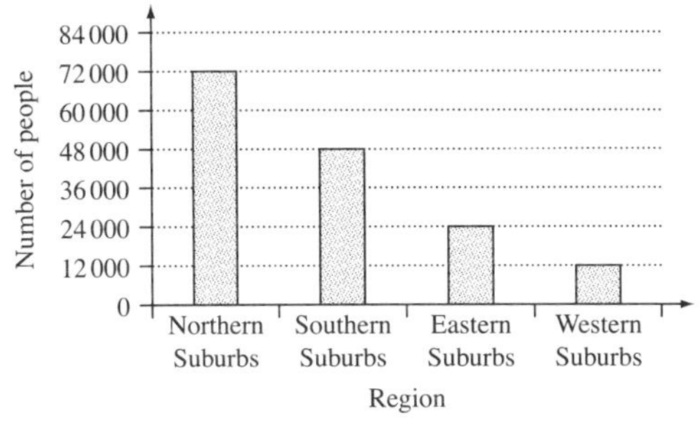

ii The column graph height is incorrect for one region.
Identify this region and justify your answer. *(2 marks)* Easy

(Q24d, **2005 MS2 HSC**)

42 A census was conducted of the 33171 households in Sunnytown. Each household was asked to indicate the number of cars registered to that household. The results are summarised in the following table.

Number of cars	*Frequency*
0	2 735
1	12 305
2	13 918
3	3 980
4	233
Total	33 171

i **1** Determine the mode number of cars in a household. *(1 mark)* Easy

2 Explain what is meant by *the mode number of cars in a household.* *(1 mark)* Easy

ii Sunnytown Council issued a 'free parking' sticker for each car registered to a household in Sunnytown.
How many parking stickers were issued? *(2 marks)* Easy

iii The council represented the results of the census in a sector graph.
What is the angle in the sector representing the households with no cars?
Give your answer to the nearest degree. *(1 mark)* Easy

iv Visitors to Sunnytown Airport have to pay for parking. The following step graph shows the cost of parking for t hours.

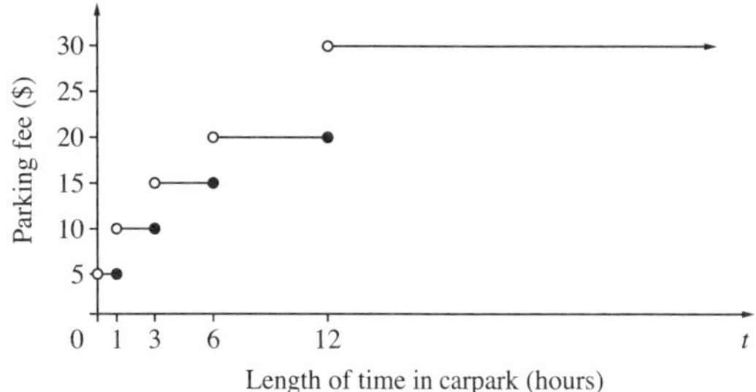

What is the cost for a car that is parked one evening from 6 pm to 8:30 pm? *(1 mark)* Easy

(Q25a, **2003 MS2 HSC**)

43 Students were surveyed about the number of movies they had watched in the last week. The results are shown in this *cumulative* frequency histogram.

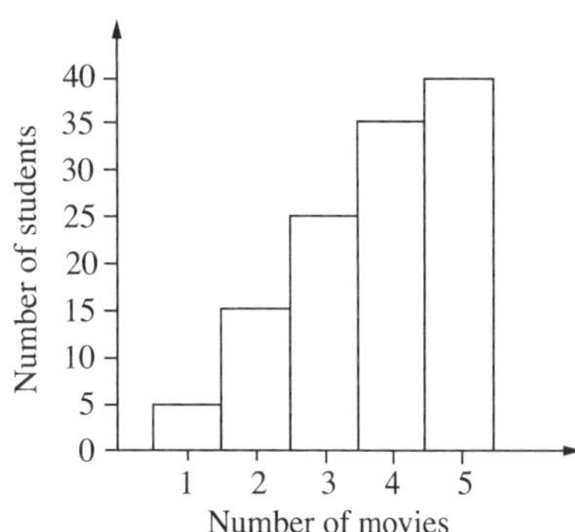

How many students said they watched four movies last week?

A 5 **B** 10
C 25 **D** 35 *(1 mark)*

(Q17, **2002 MS2 HSC**) Easy

Year 12 Data and summary statistics—Worked answers

1 Existing median = 17 and mean

$= \dfrac{13+16+17+17+21+24}{6}$

= 18.

When 5 is added the scores are 5, 13, 16, 17, 17, 21, 24.

Median = 17, which means median stays the same.

As 5 < 18, the mean changes.

Answer D

2 **a** $A = 98 + 62$

$= 160$ ✓

$B = \dfrac{192}{200} \times 100\%$

$= 96\%$ ✓

$\therefore A = 160$ and $B = 96$

(2 marks)

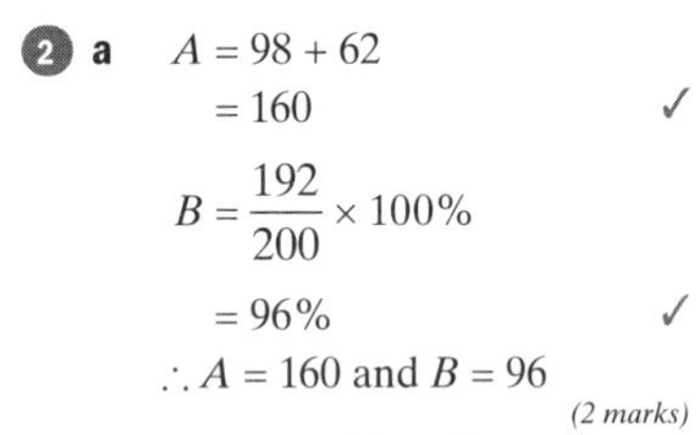

b

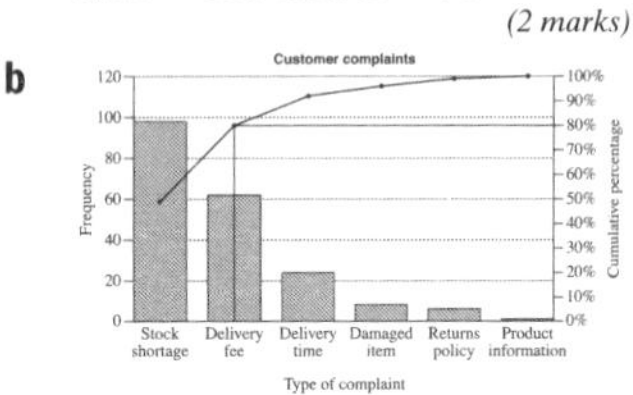

∴ Stock shortage and Delivery fee. ✓

(1 mark)

3 In the graph the highest number of downloads occurred on day 10. This means the cumulative frequency graph will show the greatest rise from day 9 to day 10. Also there is a small number of downloads on day 20. This means the cumulative frequency graph will show a very small rise from day 19 to day 20.

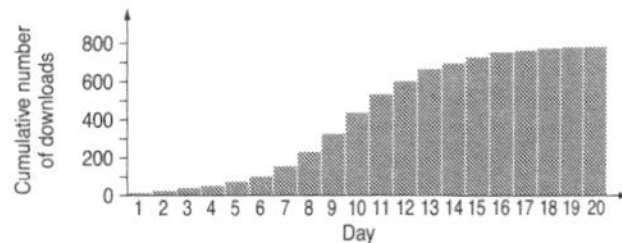

Answer C

4 $IQR = Q_3 - Q_1$

$= 10 - 4$

$= 6$

Now $Q_3 + 1.5 \times IQR$

$= 10 + 1.5 \times 6$

$= 19$ ✓

So 20 is an outlier because it is more than $Q_3 + 1.5 \times IQR$. ✓

(2 marks)

5 1 5 9 10 15

Mean $= \dfrac{1+5+9+10+15}{5}$

$= 8$

Median = 9 ✓

1 5 9 10 15 x

Median = 9.5

The difference between the medians is 0.5. ✓

So the difference between the means is 10 × 0.5 or 5.

As $x > 15$ the mean will have increased by 5.

New mean = 13

So $\dfrac{1+5+9+10+15+x}{6} = 13$ ✓

$40 + x = 13 \times 6$

$40 + x = 78$

$x = 38$ ✓

(4 marks)

6 **i**

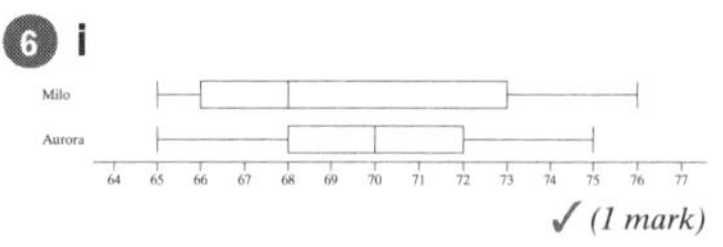

✓ *(1 mark)*

ii 50% of Milo's scores are below 68. ✓

(1 mark)

iii Yes, Aurora's scores were better:

- Aurora's median (70) is higher than Milo's median (68) ✓
- 75% of Aurora's scores are above 68 while only 50% of Milo's scores are above 68. ✓
- Aurora's scores are symmetrical whereas Milo's scores are positively skewed. ✓

(3 marks)

7 $IQR = 42 - 24$

$= 18$

$Q_L - 1.5 \times IQR = 24 - 1.5(18)$

$= -3$ ✓

$Q_U + 1.5 \times IQR = 42 + 1.5(18)$

$= 69$

∴ the maximum range of scores is –3 to 69, or a range of 72. ✓

(2 marks)

8 Cumulative percentage for train or bus delay = 92%

Previous cumulative percentage = 86%

So percentage for train or bus delay

= (92 – 86)%

= 6%

Answer A

9 170, 180, 185, 188, 192, 193, 193, 194, 196, 202

Median = 192.5

$Q_1 = 185$

$Q_3 = 194$

$IQR = 194 - 185$

$= 9$ ✓

Now, $Q_1 - 1.5 \times IQR$

$= 185 - 1.5 \times 9$

$= 171.5$ ✓

Now, 170 < 171.5

So the height of the shortest player could be considered an outlier. ✓

(3 marks)

10 Team A: lower extreme is 19 and upper extreme 28.

Median = 26

$Q_1 = 23$ and $Q_3 = 27$. ✓

Team A / Team B box plots on scale 16–37, Number of goals ✓

The spread is much greater for Team B than for Team A. Both the range and interquartile range are greater for Team B. ✓

In general, Team B scored more goals per game than Team A. The upper extreme for Team B is 7 goals higher than for Team A and the lower extreme, median and both quartiles are all higher for Team B than for Team A. In 50% of games Team B scored 27 goals or more whereas in only 25% of games Team A had 27 or more goals. ✓

The data for Team B is fairly symmetrical, perhaps slightly positively skewed, while that for Team A is negatively skewed. ✓

(5 marks)

11 The data is negatively skewed.

Answer C

12 **i** Rainfall is higher in Perth in July, August and September. So the mean monthly rainfall is higher in Perth than in Sydney for 3 months. ✓

(1 mark)

ii The standard deviation of the mean monthly rainfall will be smaller in Sydney because the monthly rainfall is more consistent in Sydney and there is a greater spread in Perth. ✓

(1 mark)

13 The median is the average of the 21st and 22nd scores.

Lower quartile is the 11th score.

So $Q_L = 3$

Upper quartile is the 32nd score.

So $Q_U = 5$ ✓

$IQR = Q_U - Q_L$
$= 5 - 3$
$= 2$ ✓

(2 marks)

14 $Q_L = 10, \quad Q_U = 16$

$IQR = Q_U - Q_L$
$= 16 - 10$
$= 6$

$Q_L - 1.5 \times IQR = 10 - 1.5 \times 6$
$= 1$

$Q_U + 1.5 \times IQR = 16 + 1.5 \times 6$
$= 25$ ✓

If there are no outliers, there are no scores less than 1 or greater than 25.

Maximum range = 25 – 1
= 24 ✓ *(2 marks)*

15 First number = 2 and range = 10
So last number = 2 + 10 = 12
2, 3, 3, 3, 5, 5, 8, 9, 12
Median = 5
Lower quartile = 3
Upper quartile = $\frac{8+9}{2} = 8.5$
So five-number summary is
2, 3, 5, 8.5, 12.
Answer A

16

Class interval	*Class centre*	*Frequency*
1–5	3	3
6–10	8	6
11–15	13	8
16–20	18	9

$\bar{x} = 12.423076\ldots$
$= 12.4$ (1 d.p.)
Answer D

17 For the History test, 30 is the upper quartile. So $\frac{1}{4}$ of 112 or 28 scored above 30.

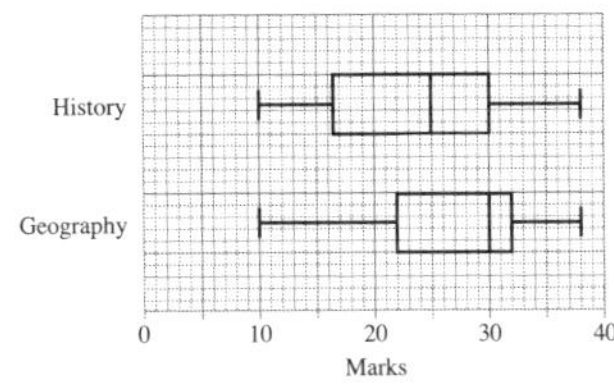

For the Geography test, 30 is the median. So 28 students is half the number of those who did the test.

Number completing the Geography test
$= 2 \times 28 = 56$
Answer C

18 Upper quartile ≈ 172
Lower quartile ≈ 136

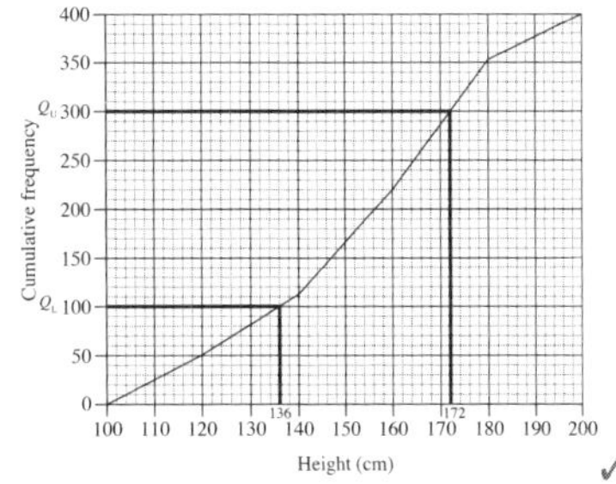

✓

Interquartile range = 172 – 136
= 36 ✓

(2 marks)

19 153 cm, 168 cm, 174 cm

$\bar{x} = \frac{153 + 168 + 174}{3}$ cm = 165 cm ✓

So the mean of the mean heights of all the possible samples will be 165 cm because the mean of all possible sample means is equal to the mean of the population. ✓

(2 marks)

20 Median for women = 55
Median for men = 45 ✓

Yes, Pat is correct. The median age of the women is 10 years older than the median age of the men. Only one man is older than the median age of the women. ✓

The data is also skewed. The data for the women is negatively skewed, meaning there are more ages towards the top end of the range, whereas the data for the men is positively skewed.

✓ *(3 marks)*

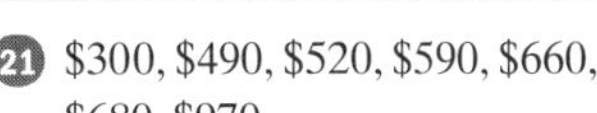

21 \$300, \$490, \$520, \$590, \$660, \$680, \$970

i Median = \$590
Q_U = \$680
Q_L = \$490
$IQR = Q_U - Q_L$
= \$680 – \$490
= \$190 ✓
$Q_U + 1.5 \times IQR$
= \$680 + 1.5 × \$190
= \$965 ✓
Now, \$970 > \$965
So \$970 is an outlier for this set of data. ✓

(3 marks)

ii The standard deviation will not change because the spread of marks from the mean will not change. ✓

(1 mark)

22 **i**

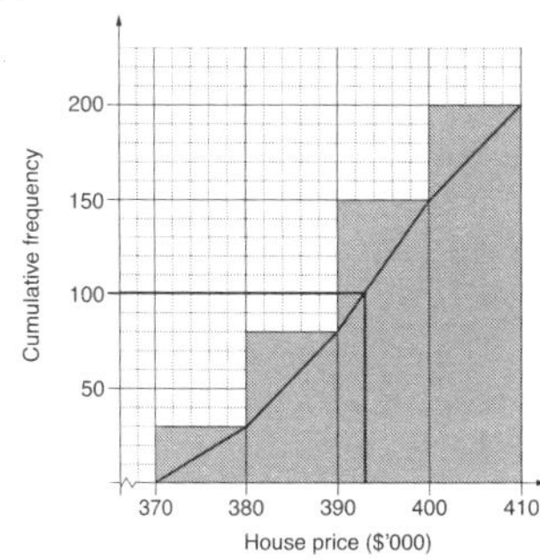

Using the graph the median house price is about \$393 000. ✓

(1 mark)

ii

Class Centre (\$'000)	Frequency
375	30
385	50
395	70
405	50

✓✓

By calculator the mean price is \$392 000. ✓ *(3 marks)*

23

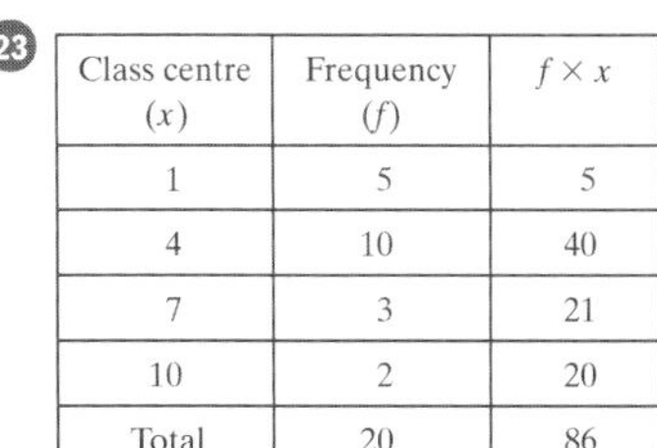

Class centre (x)	Frequency (f)	$f \times x$
1	5	5
4	10	40
7	3	21
10	2	20
Total	20	86

Mean = 86 ÷ 20
= 4.3

Answer B

24 **i**

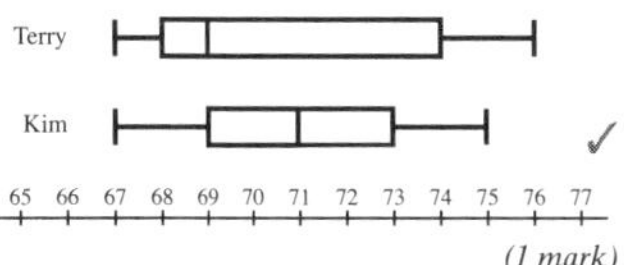

✓

(1 mark)

ii 69 is the median for Terry. So 50% of Terry's results were below 69. ✓ *(1 mark)*

iii No, Terry is not correct. ✓ Although he has the highest score, his box plot is positively skewed and only 50% of his scores are above 69 whereas 75% of Kim's scores are above that mark. ✓

Kim's box plot is symmetrical and both the range and interquartile range are smaller than those of Terry, meaning Kim's scores are more consistent. ✓

The upper quartile for Terry was 1 mark higher than that of Kim, as is the upper extreme, so Terry did perform well with the top 25% of scores.

Terry and Kim had the same lowest score. ✓ *(4 marks)*

25 The graph that should be included is D.

[The graph in option A doesn't begin at 0 on the vertical axis. The columns in graph B are not the same width and graph C uses 3D columns.]

Answer D

26 The mean would have increased because the amount being added is greater than any of the July prices.

The median would still be $607 000 (because the two middle prices were the same.)

So the mean would have changed but the median would have stayed the same.

Answer C

27

Travel time (minutes)

Without tolls		With tolls
9	3	5 8 9 9
9 9 8 7 7 6 5 5 4 4 3 2 0	4	0 1 2 6 7 7 8 8 8 9
9 8 7 5 4 3 3 3 2 2 2 2 1 1 0	5	2 4 4 5 6 8 9
1	6	1 3 5 7
	7	0 2 8
	8	2
	9	0

i Without tolls, there are 4 travel times of 52 minutes. No other travel time occurs as often.

So the modal travel time on roads without tolls is 52 minutes. ✓ *(1 mark)*

ii There are 30 travel times shown for roads without tolls. So the median is the average of the 15th and 16th times,

$$\text{Median} = \frac{50 + 51}{2} = 50.5$$

The median travel time on roads without tolls is 50.5 minutes. ✓ *(1 mark)*

iii The data set for travel times without tolls is fairly symmetrical while that for the times with tolls is positively skewed. The ✓ spread of times is much greater for the travel times with tolls. The range for the travel times with tolls is $2\frac{1}{2}$ times larger than for those without tolls. ✓ *(2 marks)*

28

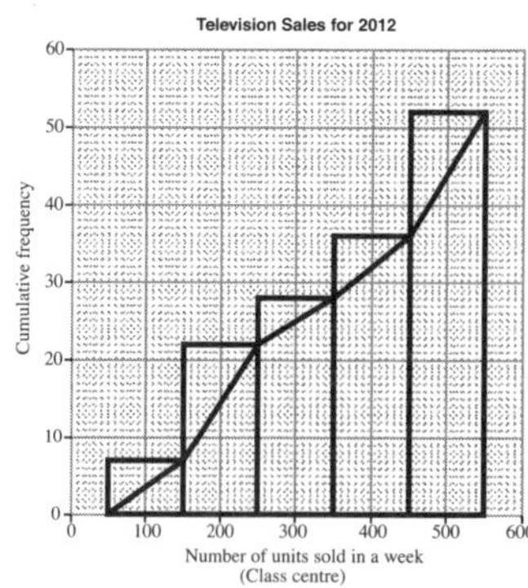

i Maximum cumulative frequency is 52.

Now $52 \div 4 = 13$

and $13 \times 3 = 39$

So the lower quartile will be found at a cumulative frequency of 13 and the upper quartile at 39. ✓

From the graph the lower quartile is 190 and the upper quartile is 470.

Interquartile range

$= 470 - 190$

$= 280$ ✓ *(2 marks)*

ii No, Oscar is not correct. The retailer has grouped the data, so Oscar should have said he sold between 250 and 350 televisions in 6 of the weeks. ✓ *(1 mark)*

29

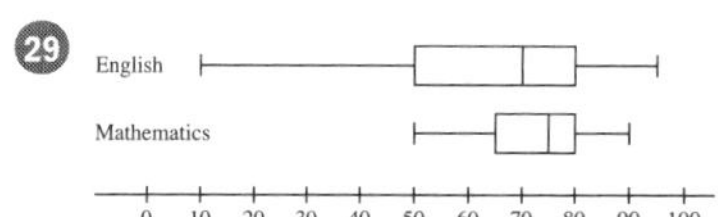

i English: lower quartile = 50

upper quartile = 80

interquartile range = 80 – 50

= 30 ✓ *(1 mark)*

ii [Different answers are possible.]

Both data sets are negatively skewed. ✓

The spread of marks for English is much greater than for Mathematics. It is more than double. (The range for English is 85 while the range for Maths is 40.) The interquartile range for English (30) is also double that for Maths (15). ✓

The highest mark in English (95) was higher than the highest mark in Maths (90), but the upper quartiles are the same (80). The median (70 for English, 75 for Maths), lower quartile (50 for English, 65 for Maths) and lower extreme (10 for English, 50 for Maths) are all higher for Maths than English.

25% of students scored marks in English that were less than any marks scored in Maths. ✓ *(3 marks)*

30 Two of the new scores are less than the original median and two are greater than the original median. So the median will not change.

The median will be 7.

Answer B

31 Total of ten scores $= 14 \times 10 = 140$

$$\therefore \frac{140 + \text{two new scores}}{12} = 16$$

$$140 + \text{two new scores} = 192$$

$$\text{two new scores} = 52$$

$$\therefore \text{Mean of two new scores} = \frac{52}{2} = 26.$$

Answer D

32 Interquartile range is the same, as the size of the rectangles is the same.

∴ The upper quartile – the lower quartile will have the same value.

Answer D

33 Total marks from 3 tests

$= 72 \times 3$

$= 216$

$$\text{New mean} = \frac{216 + \text{new mark}}{4} \quad ✓$$

$$73 = \frac{216 + \text{new mark}}{4}$$

$$292 = 216 + \text{new mark}$$

$$\text{new mark} = 292 - 216 = 76.$$ ✓ *(2 marks)*

34 In line 1,

$279 + 356 = 635$

∴ 279 and 356 are correct.

In column 1,

$279 + 187 = 466$

∴ 279 and 187 are correct

∴ 439 is incorrect which is females in part-time work.

Answer D

35 Range = 25 – 5
= 20

∴ 20 = 2x – 6
2x = 26
x = 13

∴ The set of data is
5, 6, 11, 13, 13, 18, 25

∴ The median is 13.

∴ The interquartile range
= upper quartile – lower quartile
= 18 – 6
= 12.

Answer D

36 The 4th decile is the 40th percentile.
40% of 50 = 20
∴ Scores between 55 and 60.
Answer B

37 **i** Mean

$= \frac{\text{sum of scores}}{\text{number of scores}}$

$= \frac{3 + 2(5) + 6 + 2(8) + 9 + 2(10) + 50}{10}$

$= \frac{114}{10}$

= 11.4. ✓

(1 mark)

ii Removing the outlier of 50 does not affect the median which will still be 8. ✓
However, the mean will now be $\frac{64}{9} = 7.1$,
which is significantly lower. ✓

(2 marks)

38 **i** There are more males attending the studio and the distribution of their ages is positively skewed. There are more younger males than females. The distribution of ages of females is negatively skewed. ✓ *(1 mark)*

ii Mode for the whole group is 64. ✓

(1 mark)

iii For the group 30–39 years,

the class centre $= \frac{30 + 39}{2}$
= 34.5 ✓

∴ Class centre × frequency
= 34.5 × 5
= 172.5. ✓

(2 marks)

iv The grouped frequency distribution table uses the class centres to calculate the mean, while Caitlyn used the raw data making her mean more accurate. ✓

(1 mark)

39

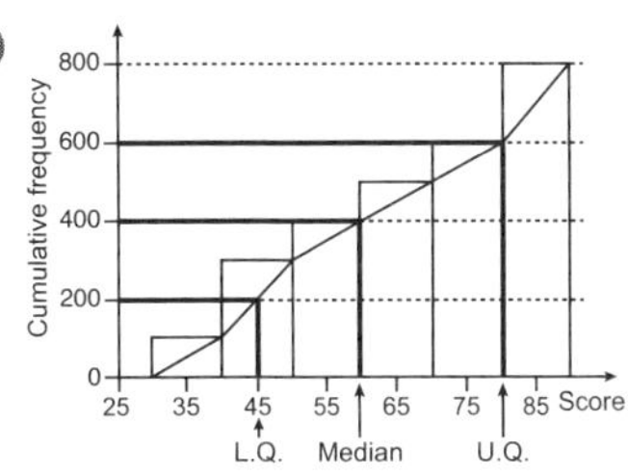

From graph,
Median = 60
Upper quartile = 80
Lower quartile = 45
∴ Interquartile range
= 80 – 45
= 35.
Answer C

40 A is not true because 50% of people under 21 earned more than \$325 but 75% of the others earned more than \$325. (∴ C is not true either.)
D is not true because only 25% of people under 21 earned between \$325 and \$350 per week.
B is true. 75% of people under 21 earned less than \$350 per week.
Answer B

41 **i** Eastern Suburbs population:
15% of total pop. = 24 000

∴ 1% of total pop. $= \frac{24\,000}{15}$

∴ 100% of total pop.

$= \frac{24\,000}{15} \times 100$

= 160 000 (by calc.)

∴ Total number of people in Sumcity is 160 000. ✓

(1 mark)

ii Western Suburbs is incorrect.
Western Suburbs population:
= 10% × 160 000

$= \frac{10}{100} \times 160\,000$

= 16 000 ✓

The graph shows 12 000. ✓

(Note also that on the column graph, Western Suburbs has $\frac{1}{2}$ the population of Eastern Suburbs but 10% is not half of 15%.)

(2 marks)

42 **i** **1** Most households have 2 cars. ✓

(1 mark)

2 *The mode number of cars in a household* refers to the number of cars with the greatest frequency registered per household; i.e. of the households surveyed, the greatest number of households had 2 cars. ✓

(1 mark)

ii Number of stickers issued
= 12 305 + 2 × 13 918 + 3 × 3980 + 4 × 233 ✓
= 53 013 ✓

(2 marks)

iii Required angle

$= \frac{2735}{33\,171} \times 360°$

= 29.682. . .°
= 30° to nearest degree. ✓

(1 mark)

iv This car is parked for $2\frac{1}{2}$ hours.
Parking for 1 to 3 hours costs \$10.
∴ Cost to park this car is \$10.
✓ *(1 mark)*

43 No. of students who watched 4 or less movies = 35
No. of students who watched 3 or less movies = 25
∴ 10 students watched 4 movies.
Answer B.

1 Jo is researching the relationship between the ages of teenage characters in television series and the ages of actors playing these characters.

After collecting the data, Jo finds that the correlation coefficient is 0.4564.

A scatterplot showing the data is drawn. The line of best fit with equation $y = -7.51 + 1.85x$, is also drawn.

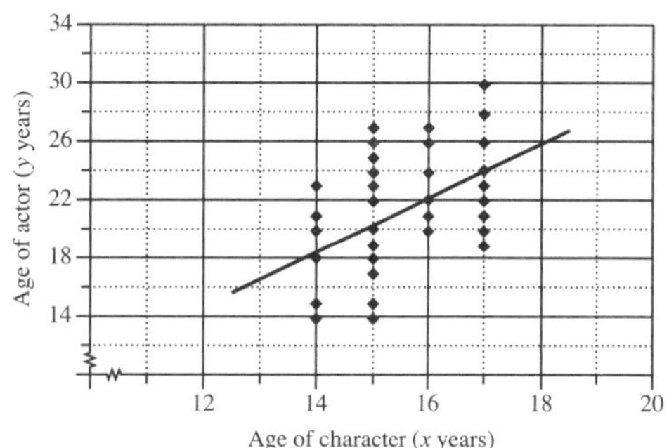

Describe and interpret the data and other information provided, with reference to the context given. *(4 marks)*

CQ (Q24, **2022 HSC**) Medium

2 For a sample of 17 inland towns in Australia, the height above sea level, x (metres), and the average maximum daily temperature, y (°C), were recorded.

The graph shows the data as well as a regression line.

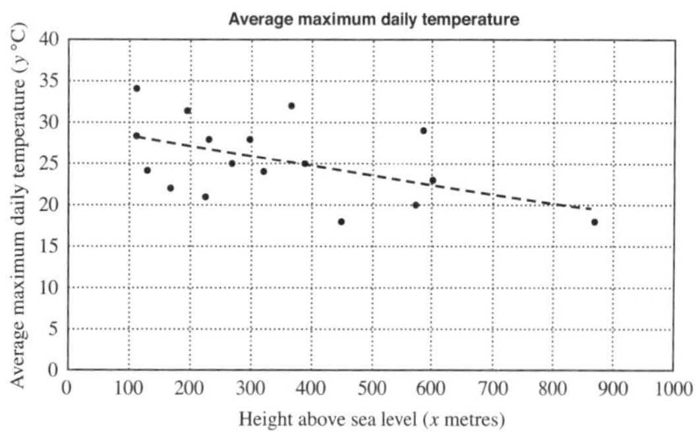

The equation of the regression line is $y = 29.2 - 0.011x$.

The correlation coefficient is $r = -0.494$.

a **i** By using the equation of the regression line, predict the average maximum daily temperature, in degrees Celsius, for a town that is 540 m above sea level. Give your answer correct to one decimal place. *(1 mark)* Easy

ii The gradient of the regression line is -0.011. Interpret the value of this gradient in the given context. *(2 marks)* Medium

b The graph below shows the relationship between the latitude, x (degrees south), and the average maximum daily temperature, y (°C), for the same 17 towns, as well as a regression line.

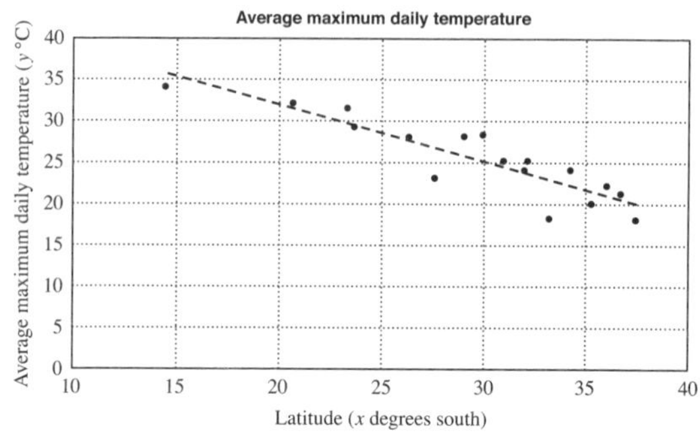

The equation of the regression line is $y = 45.6 - 0.683x$.

The correlation coefficient is $r = -0.897$.

Another inland town in Australia is 540 m above sea level. Its latitude is 28 degrees south.

Which measurement, height above sea level or latitude, would be better to use to predict this town's average maximum daily temperature? Give a reason for your answer. *(1 mark)* Medium

CQ (Q17, **2021 HSC**)

3 For a set of bivariate data, Pearson's correlation coefficient is –1.

Which graph could best represent this set of bivariate data?

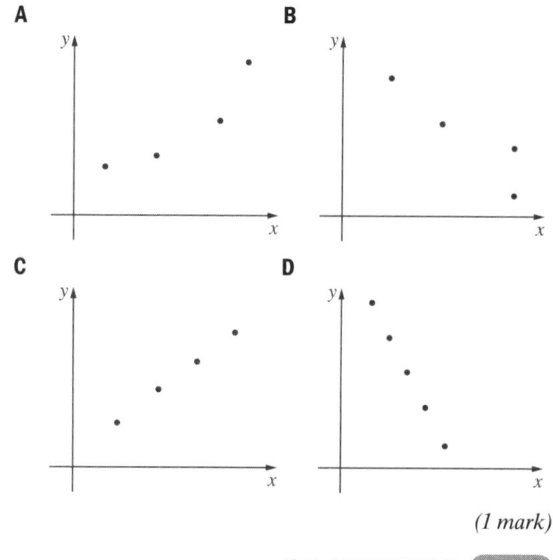

(1 mark)

(Q12, **2020 MS2 HSC**) Easy

4 A cricket is an insect. The male cricket produces a chirping sound.

A scientist wants to explore the relationship between the temperature in degrees Celsius and the number of cricket chirps heard in a 15-second time interval.

Once a day for 20 days, the scientist collects data. Based on the 20 data points, the scientist provides the information below.

- A box-plot of the temperature data is shown.

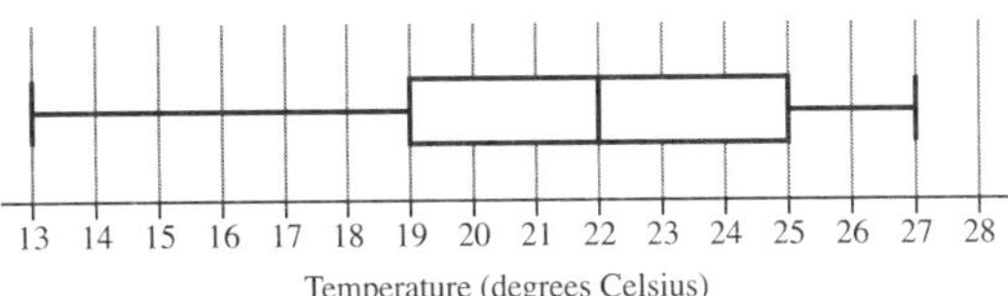

- The mean temperature in the dataset is 0.525°C below the median temperature in the dataset.
- A total of 684 chirps was counted when collecting the 20 data points.

The scientist fits a least-squares regression line using the data (x, y), where x is the temperature in degrees Celsius and y is the number of chirps heard in a 15-second time interval. The equation of this line is

$$y = -10.6063 + bx,$$

where b is the slope of the regression line.

The least-squares regression line passes through the point $(\bar{x}, \bar{y})$ where $\bar{x}$ is the sample mean of the temperature data and $\bar{y}$ is the sample mean of the chirp data.

Calculate the number of chirps expected in a 15-second interval when the temperature is 19° Celsius. Give your answer correct to the nearest whole number. *(5 marks)*

(Q27, **2020 HSC**) **Medium**

5 The height (h) and mass (M) of Baby Logan are measured and recorded over his first two years.

Height (cm), h	45	50	55	60	65	70	75	80	85
Mass (kg), M	2.6	3.5	5.3	6.4	7.6	8.7	10.7	12.0	13.3

This information is displayed in a scatter graph and a line of best fit by eye has been drawn.

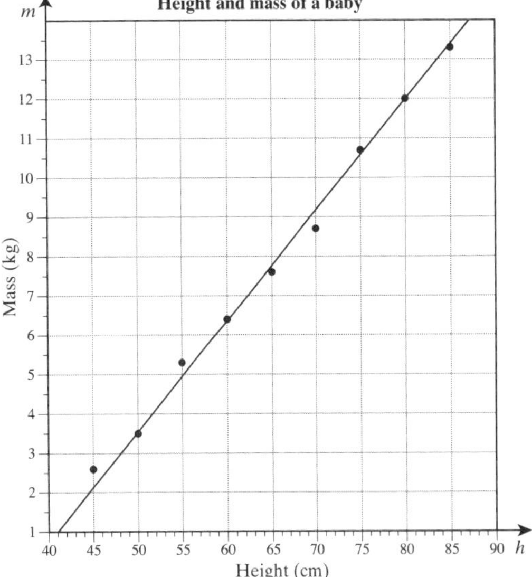

i Calculate Pearson's coefficient (r), correct to 3 decimal places. *(1 mark)* **Easy**

ii Hence, comment on the correlation between Logan's height and mass. *(1 mark)* **Easy**

iii Explain why the equation of this line of best fit is approximately $M = 0.28h - 10.4$. *(2 marks)* **Easy**

iv Based on the line of best fit, what is Logan's height when he has a mass of 12.5 kg? Leave your answer correct to 1 decimal place. *(1 mark)* **Easy**

v Logan's cousin Ella is born five weeks later.
In her first two years of life her height and mass were also plotted on a scatterplot.
A line of best fit for her data was drawn and its equation was $M = 0.3h - 11.5$.
What was the difference in the babies' mass when they both had a height of 75 cm? *(1 mark)* **Medium**

Bonus question (see page iv)

6 Jim is a plumber and wants to buy a second-hand work truck.

The following table shows the price (y) in dollars of eight trucks, and the distance (x) in kilometres they had travelled.

Distance travelled (x km)	25 000	60 000	85 000	40 000	50 000	75 000	30 000	45 000
Price (y dollars)	43 000	26 000	22 000	32 000	38 000	28 000	40 000	34 000

i Comment on the association between x and y with reference to the value of Pearson's correlation coefficient. *(2 marks)* **Easy**

ii The data is recorded on a scatterplot and Jim has drawn a regression line by eye.

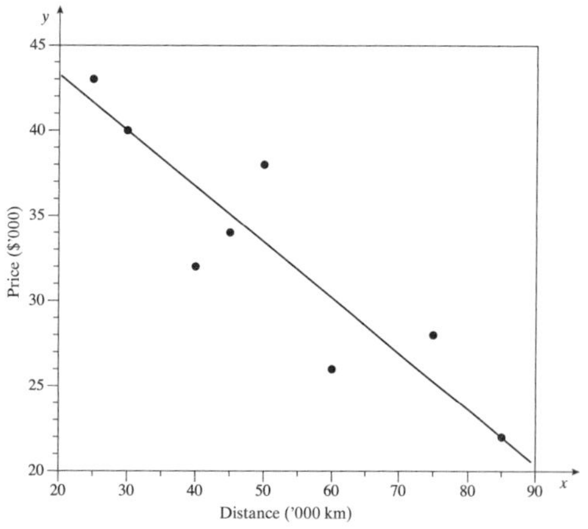

The relationship between x and y can be modelled by the regression equation $y = mx + c$.
Find the values of m and c for Jim's regression line. Give your answers correct to 2 decimal places. *(2 marks)* **Medium**

iii Use Jim's regression equation to estimate the price of a vehicle that has travelled 55 000 kilometres. *(1 mark)* **Easy**

iv At the start of 2020, Jim decides to buy the truck that had travelled 55 000 kilometres. The value of the truck decreases by 25% each year. During which year will the truck be valued at less than \$5000? *(2 marks)* **Medium**

Bonus question

7 The maximum temperature, T, in °C, on eight randomly selected days is shown in the table below. The table also shows the number of drinks, D, purchased from a vending machine.

Maximum temperature, T	28	25	36	34	37	31	32	30
Number of drinks sold, D	20	16	48	34	46	24	40	32

i Find the value of r, the correlation coefficient, correct to 4 decimal places. *(1 mark)* **Easy**

The relationship between the variables can be modelled by the regression line $D = mT + c$.
The data is displayed on a scatterplot and a regression line, by eye, is drawn.

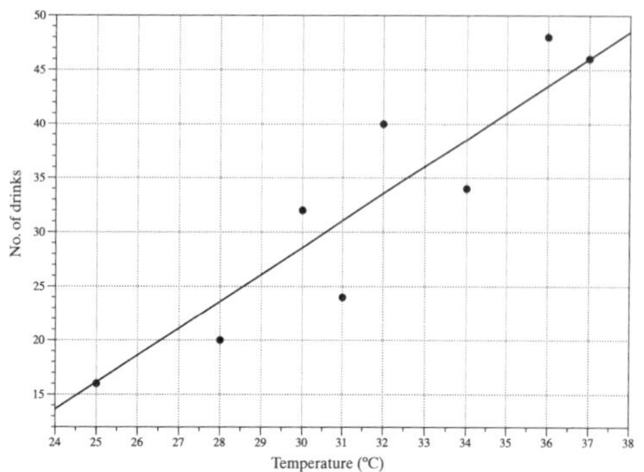

ii Find the equation of this regression line, in the form $D = mT + c$. *(2 marks)* **Easy**

iii Explain why the model is not appropriate to determine sales when the temperature is 16 °C. *(1 mark)* **Easy**

Bonus question

8 The heights (cm) of 10 mothers (m) and their daughters (d) were recorded in the table below.

Mother's height in cm (m)	170	163	160	172	164	158	164	166	167	165
Daughter's height in cm (d)	175	170	168	176	170	166	171	170	173	174

The data was used to create a scatter plot and Melia constructed a regression line by eye, as shown below.

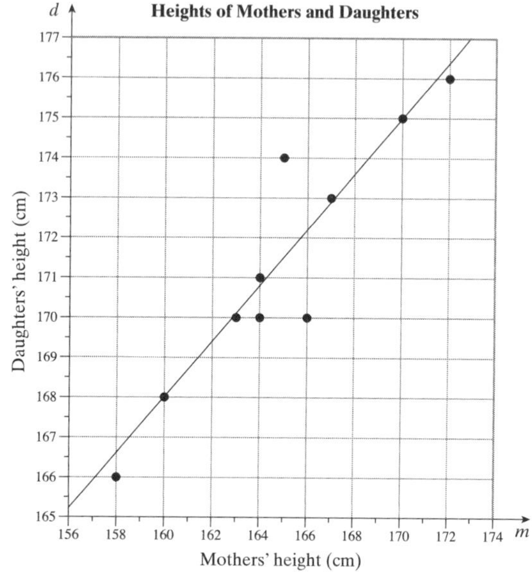

i Determine the equation of Melia's line of best fit as shown. *(2 marks)* **Easy**

ii Why would this model NOT be useful for predicting the height of the daughter of a woman of height 182 cm? *(1 mark)* **Easy**

Bonus question

9 A set of bivariate data is collected by measuring the height and arm span of seven children. The graph shows a scatterplot of these measurements.

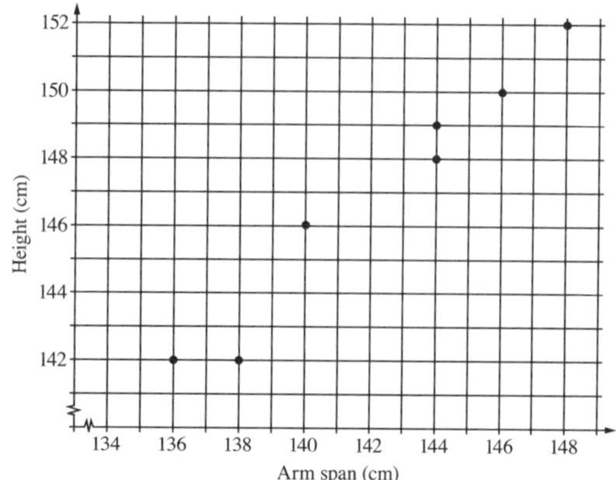

a Calculate Pearson's correlation coefficient for the data, correct to two decimal places. *(1 mark)* Easy

b Identify the direction and the strength of the linear association between height and arm span. *(1 mark)* Easy

c The equation of the least-squares regression line is shown.
Height = 0.866 × (arm span) + 23.7
A child has an arm span of 143 cm.
Calculate the predicted height for this child using the equation of the least-squares regression line. *(1 mark)* Easy

(Q23, **2019 MS2 HSC**)

10 Which of the data sets graphed below has the largest positive correlation coefficient value?

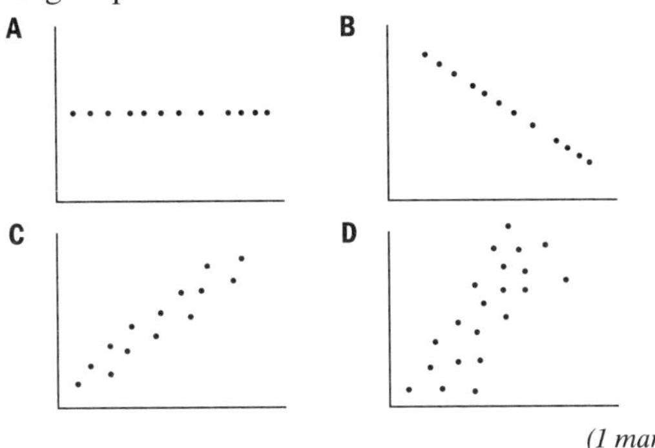

(1 mark)

(Q12, **2017 MS2 HSC**) Easy

11 The graph shows a scatterplot for a set of data.

Which of the following is the best approximation for the correlation coefficient of this set of data?

A −1
B − 0.3
C 0.3
D 1

(1 mark)

(Q3, **2016 MS2 HSC**) Easy

12 The shoe size and height of ten students were recorded.

Shoe size	6	7	7	8	8.5	9.5	10	11	12	12
Height	155	150	165	175	170	170	190	185	200	195

i Complete the scatter plot AND draw a line of fit by eye.

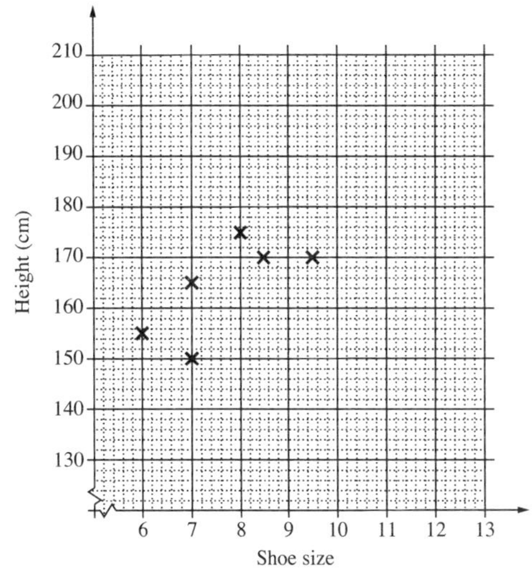

(2 marks) Easy

ii Use the line of fit to estimate the height difference between a student who wears a size 7.5 shoe and one who wears a size 9 shoe. *(1 mark)* Easy

iii A student calculated the correlation coefficient to be 1 for this set of data. Explain why this cannot be correct. *(1 mark)* Easy

(Q28e, **2015 MS2 HSC**)

13 Which graph best shows data with a correlation closest to 0.3?

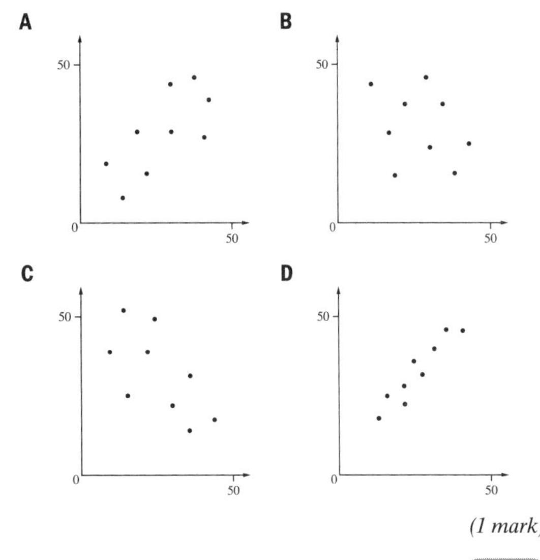

(1 mark)

(Q2, **2013 MS2 HSC**) Easy

14 Ahmed collected data on the age (a) and height (h) of males aged 11 to 16 years. He created a scatterplot of the data and constructed a line of best fit to model the relationship between the age and height of males.

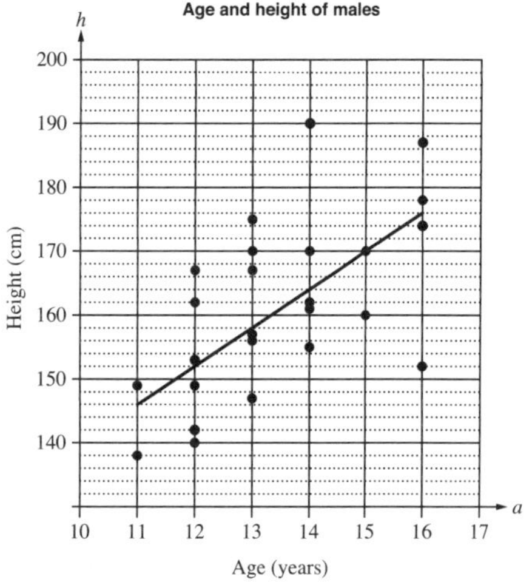

i Determine the gradient of the line of best fit shown on the graph. *(1 mark)* Easy

ii Explain the meaning of the gradient in the context of the data. *(1 mark)* Easy

iii Determine the equation of the line of best fit shown on the graph. *(2 marks)* Easy

(Q28b, **2013 MS2 HSC**)

15 Which of the following relationships would most likely show a negative correlation?

A The population of a town and the number of hospitals in that town.

B The hours spent training for a race and the time taken to complete the race.

C The price per litre of petrol and the number of people riding bicycles to work.

D The number of pets per household and the number of computers per household.

(1 mark)

(Q11, **2012 MS2 HSC**) Medium

16 Tourists visit a park where steam erupts from a particular geyser. The brochure for the park has a graph of the data collected for this geyser over a period of time.

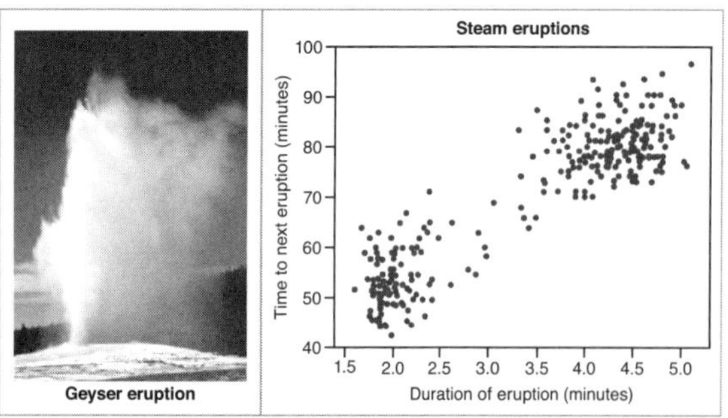

The graph shows the duration of an eruption and the time until the next eruption, timed from the end of one eruption to the beginning of the next.

i Tony sees an eruption that lasts 4 minutes. Based on the data in the graph, what is the minimum time that he can expect to wait for the next eruption? *(1 mark)* Easy

ii Julia saw two consecutive eruptions, one hour apart. Based on the data in the graph, what was the longest possible duration of the first eruption that she saw? *(1 mark)* Easy

iii What does the graph suggest about the relationship between the duration of an eruption and the time to the next eruption? *(1 mark)* Easy

(Q29a, **2012 MS2 HSC**)

17 In which graph would the data have a correlation coefficient closest to –0.9?

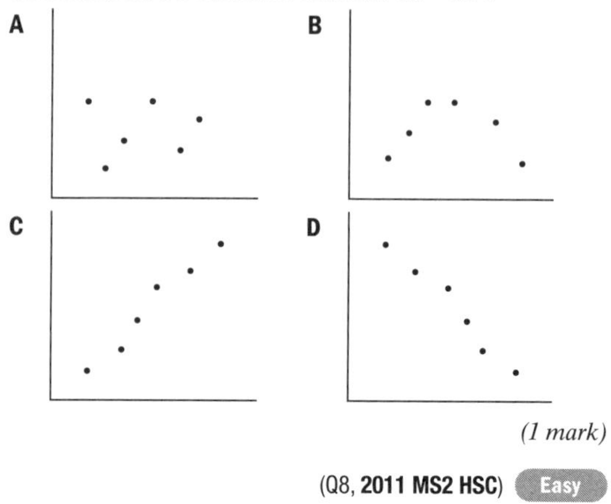

(1 mark)

(Q8, **2011 MS2 HSC**) Easy

18 The height and mass of a child are measured and recorded over its first two years.

Height (cm), H	45	50	55	60	65	70	75	80
Mass (kg), M	2.3	3.8	4.7	6.2	7.1	7.8	8.8	10.2

This information is displayed in a scatter graph.

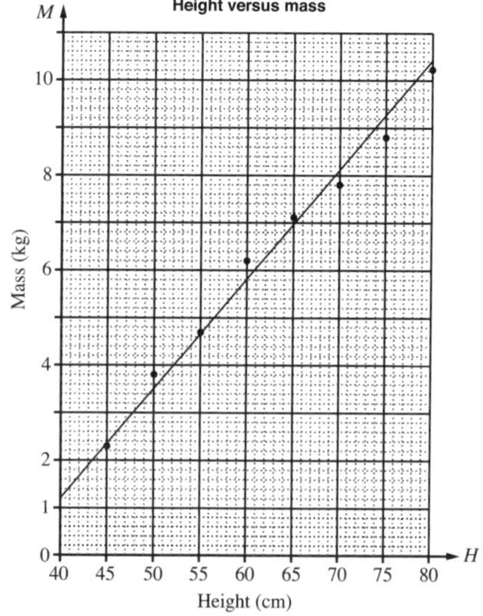

i Describe the correlation between the height and mass of this child, as shown in the graph. *(1 mark)* Easy

ii A line of best fit has been drawn on the graph. Find the equation of this line. *(2 marks)* Easy

(Q28b, **2009 MS2 HSC**)

Year 12 Bivariate data analysis—Worked answers

1 Possible answers include:

- As $r = 0.4564$, there is a moderate positive association between age of character and age of actor. ✓
- The line of best fit indicates there is a linear relationship between the age of character and age of actor. ✓
- The equation has a gradient of 1.85. This suggests that for every 1 additional year of a character, the actor is 1.85 years older. ✓
- Only once is the actor younger than the character (14 year old playing a 15 year old). ✓
- A total of 33 teenage characters was studied.
- As $30 - 14 = 16$, the range of age of the actors was 16.
- The median age of the characters is 15.
- There were two actors playing characters of the same age; however, generally speaking, the actor is older than the character's age.

(4 marks)

2 **a** **i** Substitute $x = 540$:
$y = 29.2 - 0.011 \times 540$
$= 23.26$
$= 23.3$ (1 dec. pl.)
$\therefore$ the temperature is 23.3°. ✓
(1 mark)

ii The average maximum daily temperature decreases 0.011 °C for every 1 m increase in elevation (or a decrease of 1.1 °C for every 100 m rise in elevation). ✓✓
(2 marks)

b $r = -0.897$ shows a stronger negative correlation than $r = -0.494$ so it would be better to use latitude for this prediction. ✓
(1 mark)

3 A correlation coefficient of -1 means perfect negative correlation.

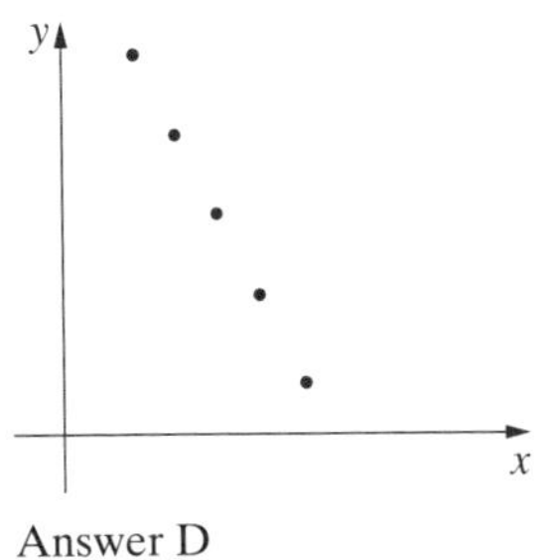

Answer D

4 From the box plot, median = 22. ✓
Temperature sample mean
$(\bar{x}) = 22 - 0.525$
$= 21.475$ ✓
Chirp data sample mean
$(\bar{y}) = \dfrac{684}{20}$
$= 34.2$ ✓
$\bar{y} = -10.6063 + b\bar{x}$
$34.2 = -10.6063 + b(21.475)$
$21.475b = 44.8063$
$b = 44.8063 \div 21.375$
$= 2.086440047\ldots$
$= 2.0864$ (4 dec. pl.)
$\therefore y = -10.6063 + 2.0864x$ ✓
Substitute $x = 19$:
$y = -10.6063 + 2.0864(19)$
$= 29.0353$
$= 29$ (nearest whole)
$\therefore$ 29 chirps ✓ *(5 marks)*

5 **i** $r = 0.997940765\ldots$
$= 0.998$ (3 dec. pl.) ✓
(1 mark)

ii As 0.998 is close to 1, there is a very strong positive correlation between the height and mass of this baby.
✓ *(1 mark)*

iii The line passes through the points (50, 3.5) and (80, 12).
Gradient $= \dfrac{12 - 3.5}{80 - 50}$
$= 0.28333\ldots$
$= 0.28$ (2 dec. pl.) ✓
The line has equation
$M = 0.28h + c$
As the line passes through (80, 12), then
$12 = 0.28(80) + c$
$12 = 22.4 + c$
$c = -10.4$
The equation is
$M = 0.28h - 10.4$ ✓
(2 marks)

iv Let $M = 12.5$:
$12.5 = 0.28h - 10.4$
$0.28h = 22.9$
$h = 81.78571429\ldots$
$= 81.8$ cm
Logan's height is about 81.8 cm. ✓
(1 mark)

v Logan: $M = 0.28(75) - 10.4$
$= 10.6$
Ella: $M = 0.3(75) - 11.5$
$= 11$
Difference $= 11 - 10.6$
$= 0.4$
Ella's mass is 0.4 kg more than Logan at a height of 75 cm. ✓ *(1 mark)*

6 **i** Pearson's coefficient
$r = -0.900631491\ldots$
$= -0.9006$ (4 dec. pl.) ✓
The association between distance and price is strong and negative. ✓
(2 marks)

ii The line passes through the points (30, 40) and (85, 22).
Gradient $m = \dfrac{22 - 40}{85 - 30}$
$= -0.32727272\ldots$
$= -0.33$ ✓
(2 dec. pl.)
The line has equation
$y = -0.33x + c$
Substitute (30, 40):
$40 = -0.33(30) + c$
$c = 49.8181818\ldots$
$= 49.82$ (2 dec. pl.)
$\therefore m = -0.33$ and $c = 49.82$ ✓
(2 marks)

iii $y = -0.33x + 49.82$
Let $x = 55$:
$y = -0.33(55) + 49.82$
$= 31.67$
The price would be \$31 670. ✓
(1 mark)

iv $5000 = 31\,670(0.75)^n$
$0.75^n = \dfrac{5000}{31\,670}$
$n\ln 0.75 = \ln\dfrac{5000}{31\,670}$ ✓
$n = \dfrac{\ln\dfrac{5000}{31\,670}}{\ln 0.75}$
$n = 6.416569279\ldots$
During 2026 the truck will be worth less than \$5000. ✓
(2 marks)

7 **i** $r = 0.919549645\ldots$
$= 0.9195$ (4 dec. pl.) ✓
(1 mark)

ii The line passes through the points (25, 16) and (37, 46).
Gradient $= \dfrac{46 - 16}{37 - 25}$
$= \dfrac{5}{2}$ ✓
The line has equation
$D = 2.5T + c$
As the line passes through (25, 16), then
$16 = 2.5(25) + c$
$16 = 62.5 + c$
$c = -46.5$
The equation is
$D = 2.5T - 46.5$ ✓
(2 marks)

iii The line is based on temperatures from 25 to 37 and 16 is beyond the bounds of the data points to which it is fitted. So it is not valid to draw conclusions from it. ✓ *(1 mark)*

8 i The line passes through the points (160, 168) and (170, 175).

$\text{Gradient} = \frac{175-168}{170-160}$

$= 0.7$ ✓

The line has equation $d = 0.7m + c$

As the line passes through (160, 168), then

$168 = 0.7(160) + c$

$c = 56$

The equation is $d = 0.7m + 56$ ✓ *(2 marks)*

ii The model was only designed for mothers with a height between 158 cm and 172 cm. ✓ *(1 mark)*

9 a $r = 0.98$ (2 decimal places; by calculator) ✓ *(1 mark)*

b Very strong positive association ✓ *(1 mark)*

c Arm span = 143 cm

$\text{Height} = 0.866 \times 143 + 23.7$

$= 147.538$

So the predicted height is 148 cm to the nearest centimetre. ✓ *(1 mark)*

10 Only graphs C and D show positive correlation.

The dots in C more closely form a straight line, so the graph has the largest positive correlation coefficient.

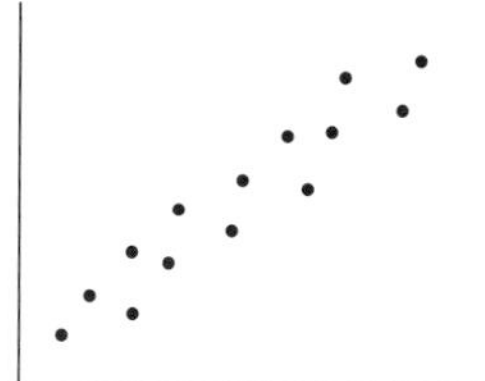

Answer C

11 [The correlation is negative, but weak, so the best approximation is –0.3.]

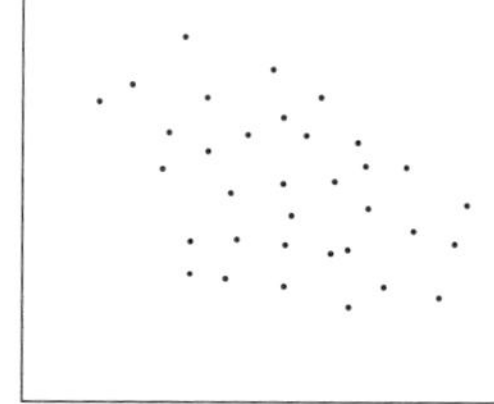

Answer B

12 i [One possible line of fit is shown.]

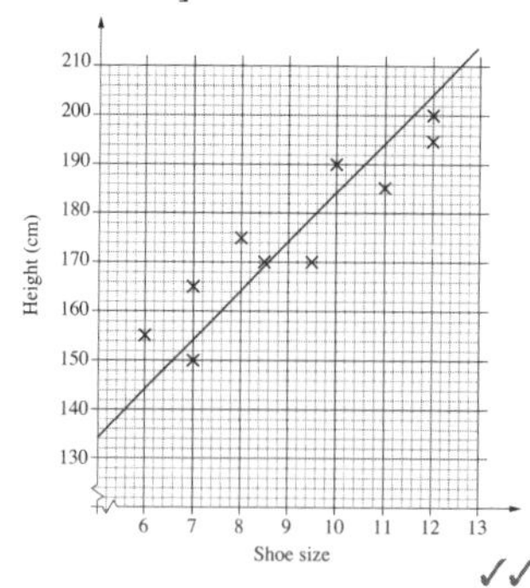

✓✓ *(2 marks)*

ii Using the line of fit, a student who wears size 7.5 shoes would be 159 cm tall and one who wears size 9 shoes would be 174 cm tall.

Difference in height

$= (174 - 159)$ cm

$= 15$ cm ✓ *(1 mark)*

iii A correlation coefficient of 1 means perfect correlation. The points would all need to form a straight line. As they do not do this, the coefficient cannot be correct. ✓ *(1 mark)*

13 Correlation of 0.3 means that there is a weak positive relationship. This is shown in option A.

Answer A

14 i $m = \frac{\text{vertical change in position}}{\text{horizontal change in position}}$

$= \frac{6}{1}$

$= 6$ ✓ *(1 mark)*

ii The gradient gives the number of centimetres that a male grows per year according to the data. ✓ *(1 mark)*

iii The equation is of the form

$h = ma + b$

But, $m = 6$

So $h = 6a + b$

When $a = 15, h = 170$

$170 = 6 \times 15 + b$ ✓

$= 90 + b$

$b = 170 - 90$

$= 80$

So the equation is

$h = 6a + 80$ ✓ *(2 marks)*

15 For a relationship to show negative correlation, one part should increase as the other decreases. This happens in option B.

As the hours spent training increase, the time to complete the race should decrease. This relationship is most likely to show negative correlation.

Answer B

16 i Using the graph, for an eruption of 4 minutes the minimum time to wait for the next eruption is 70 minutes. ✓ *(1 mark)*

ii For eruptions one hour apart, the longest possible duration of the first eruption, based on the data, is 3 minutes. ✓ *(1 mark)*

iii The graph suggests a strong positive relationship between the duration of an eruption and the length of time until the next eruption. The longer the duration of the first eruption, the longer the wait until the next eruption. ✓ *(1 mark)*

17 A correlation coefficient of –0.9 means there is a very strong negative association between the two variables. This is shown in [D].

Answer D

18 i As the height increases, the mass increases. There is a strong positive correlation between the height and the mass. ✓ *(1 mark)*

ii $y = mx + b$

m is the gradient

Using the points (60, 5.8) and (40, 1.2),

$m = \frac{5.8-1.2}{60-40}$

$= \frac{4.6}{20}$

$= 0.23$ ✓

$\therefore y = 0.23x + b$

$5.8 = 0.23(60) + b$

$b = -8$

i.e. $M = 0.23H - 8$ ✓

(2 marks)

1 Consider the following graph of a probability density function $f(x)$.

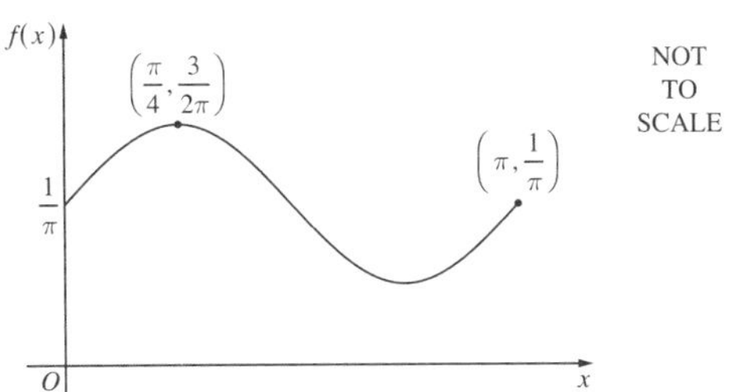

What is the value of the mode?

A $\frac{1}{\pi}$ **B** $\frac{3}{2\pi}$

C $\frac{\pi}{4}$ **D** π *(1 mark)*

(Q7, **2022 HSC**) Easy

2 A continuous random variable X has cumulative distribution function given by

$$F(x) = \begin{cases} 1 & x > e^3 \\ \frac{1}{k}\ln x & 1 \leq x \leq e^3 \\ 0 & x < 1 \end{cases}.$$

a Show that $k = 3$. *(1 mark)* Easy

b Given that $P(X < c) = 2P(X > c)$, find the exact value of c. *(2 marks)* Hard

(Q30, **2022 HSC**)

3 The number of hours for which light bulbs will work before failing can be modelled by the random variable X with cumulative distribution function

$$F(x) = \begin{cases} 1 - e^{-0.01x}, & x \geq 0 \\ 0, & x < 0 \end{cases}.$$

Jane sells light bulbs and promises that they will work for longer than exactly 99% of all light bulbs.

Find how long, according to Jane's promise, a light bulb bought from her should work. Give your answer in hours, rounded to two decimal places. *(2 marks)*

(Q30, **2021 HSC**) Hard

4 People are given a maximum of six hours to complete a puzzle. The time spent on the puzzle, in hours, can be modelled using the continuous random variable X which has probability density function

$$f(x) = \begin{cases} \frac{Ax}{x^2 + 4}, & \text{for } 0 \leq x \leq 6, \text{ (where } A > 0) \\ 0, & \text{for all other values of } x \end{cases}.$$

The graph of the probability density function is shown below. The graph has a local maximum.

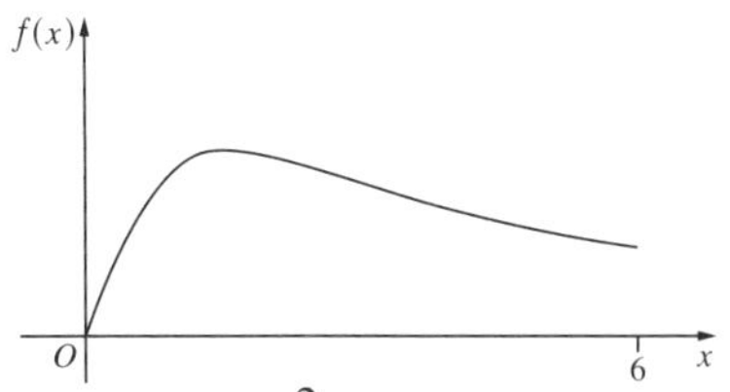

a Show that $A = \frac{2}{\ln 10}$. *(2 marks)* Medium

b Show that the mode of X is two hours. *(2 marks)* Hard

c Show that $P(X < 2) = \log_{10}2$. *(2 marks)* Hard

d The Intelligence Quotient (IQ) scores of people are normally distributed with a mean of 100 and standard deviation of 15. It has been observed that the puzzle is generally completed more quickly by people with a high IQ.
It is known that 80% of people with an IQ greater than 130 can complete the puzzle in less than two hours.
A person chosen at random can complete the puzzle in less than two hours.
What is the probability that this person has an IQ greater than 130? Give your answer correct to three decimal places. *(2 marks)* Hard

(Q33, **2021 HSC**)

5 A continuous random variable, X, has the following probability density function.

$$f(x) = \begin{cases} \sin x & \text{for } 0 \leq x \leq k \\ 0 & \text{for all other values of } x \end{cases}$$

a Find the value of k. *(2 marks)* Medium

b Find $P(X \leq 1)$. Give your answer correct to four decimal places. *(2 marks)* Hard

(Q23, **2020 HSC**)

6 The probability density function, f, of a random variable X is given by

$$f(x) = \begin{cases} \frac{3}{76}(x^2 + 1) & 0 \leq x \leq k \\ 0 & \textit{elsewhere} \end{cases}$$

i Explain why $k = 4$. *(2 marks)* Medium

ii Find $P(X \geq 2)$ correct to 2 decimal places. *(1 mark)* Medium

Bonus question (see page iv)

7 Suppose a train arrives shortly after 2 pm each day, and that the number of minutes after 2 pm that the train arrives can be modelled as a continuous random variable with probability density function, f, given by

$$f(x) = \begin{cases} k(2-x) & 0 \leq x \leq 2 \\ 0 & \text{elsewhere} \end{cases}$$

Show that $k = \frac{1}{2}$, and hence find the probability that the train is between 60 and 90 seconds late. *(4 marks)* Hard

Bonus question

8 The probability density function, f, of a random variable X is given by

$$f(x) = \begin{cases} \frac{3}{7}(x^2+2) & 0 \leq x \leq 1 \\ 0 & \text{elsewhere} \end{cases}$$

Find, correct to 2 decimal places:

i $P(X \leq 0.8)$. *(2 marks)* Medium

ii The expected value, or mean, of a continuous random variable X is given by the formula $E(X) = \int_{-\infty}^{\infty} x f(x)dx$. Find the mean. *(2 marks)* Medium

Bonus question

9 The probability density function, f, of a random variable X is given by

$$f(x) = \begin{cases} \frac{3}{32}x(k-x) & 0 \leq x \leq k \\ 0 & \text{elsewhere} \end{cases}$$

i Show that $k = 4$. *(2 marks)* Easy

ii Write down the value of $E(X)$. *(1 mark)* Easy

iii Using the formula $E(X^2) = \int_{-\infty}^{\infty} x^2 f(x)dx$, calculate $Var(X)$. *(2 marks)* Medium

Bonus question

10 A pizza shop estimates that the time X, in hours, to deliver a pizza from when it is ordered is a continuous random variable with probability density function given by

$$f(x) = \begin{cases} \frac{2}{5}(3-x) & 0 \leq x \leq 1 \\ 0 & \text{elsewhere} \end{cases}$$

i What is the probability of the pizza being delivered within 30 minutes of being ordered? *(1 mark)* Easy

ii The expected value, or mean, of a continuous random variable X is given by the formula $E(X) = \int_{-\infty}^{\infty} x f(x)dx$. Calculate the mean delivery time. *(2 marks)* Medium

iii Using the formula $E(X^2) = \int_{-\infty}^{\infty} x^2 f(x)dx$, find the standard deviation of the delivery time to the nearest minute. *(3 marks)* Medium

Bonus question

11 The probability density function, f, of a random variable X is given by

$$f(x) = \begin{cases} \frac{x+2}{k} & 0 \leq x \leq 4 \\ 0 & \text{elsewhere} \end{cases}$$

i Show that $k = 16$. *(2 marks)* Medium

ii Hence, find the value of a such that $P(X \leq a) = \frac{3}{8}$. *(2 marks)* Medium

Bonus question

12 The probability density function, f, of a random variable X is given by

$$f(x) = \begin{cases} \frac{1}{k}(x^2+5) & 0 \leq x \leq 3 \\ 0 & \text{elsewhere} \end{cases}$$

i Show that $k = 24$. *(2 marks)* Medium

ii Hence, find $P(0.5 \leq X \leq 2)$. *(1 mark)* Easy

Bonus question

13 The probability density function, f, of a random variable X is given by

$$f(x) = \begin{cases} kx^2(2-x) & 0 \leq x \leq 2 \\ 0 & \text{elsewhere} \end{cases}$$

i Show that $k = \frac{3}{4}$. *(2 marks)* Medium

ii Find X if $P(X \geq 1.5)$, correct to two decimal places. *(2 marks)* Medium

iii Find the mode. *(2 marks)* Hard

Bonus question

14 The probability density function, f, of a random variable X is given by

$$f(x) = \begin{cases} \frac{x}{8}(x^2+2) & 0 \leq x \leq k \\ 0 & \text{elsewhere} \end{cases}$$

i Explain why $k = 2$. *(2 marks)* Medium

ii Find $P(0.5 \leq X \leq 1.5)$. *(2 marks)* Medium

iii Show $P(X \leq 1) = \frac{5}{32}$. *(1 mark)* Medium

Bonus question

Year 12 Continuous random variables—Worked answers

1 The mode occurs at the maximum point on a probability density function.

As maximum at $\left(\frac{\pi}{4}, \frac{3}{2\pi}\right)$, the mode is $\frac{\pi}{4}$.

Answer C

2 a Substitute $(e^3, 1)$:

$\therefore \frac{1}{k}\ln e^3 = 1$

$\frac{3}{k} = 1$

$k = 3$ ✓

(1 mark)

b $P(X < c) = 2P(X > c)$

$\frac{1}{3}[\ln x]_1^c = 2 \times \frac{1}{3}[\ln x]_c^{e^3}$ ✓

$[\ln x]_1^c = 2[\ln x]_c^{e^3}$

$\ln c - \ln 1 = 2(\ln e^3 - \ln c)$

$\ln c = 2(3 - \ln c)$

$\ln c = 6 - 2\ln c$

$3\ln c = 6$

$\ln c = 2$

$c = e^2$ ✓

(2 marks)

3 $F(x) = 1 - e^{-0.01x}$

Substitute $F(x) = 0.99$:

$0.99 = 1 - e^{-0.01x}$ ✓

$e^{-0.01x} = 0.01$

$-0.01x = \ln 0.01$

$x = \frac{\ln 0.01}{-0.01}$

$= 460.5170186\ldots$

$= 460.52$ (2 dec. pl.)

$\therefore$ the light bulb should work for 460.52 hours. ✓

(2 marks)

4 a $\int_0^6 \frac{Ax}{x^2+4}\,dx = 1$

$\frac{A}{2}\int_0^6 \frac{2x}{x^2+4}\,dx = 1$ ✓

$\left[\ln|x^2+4|\right]_0^6 = \frac{2}{A}$ ✓

$\ln(6^2+4) - \ln(0^2+4) = \frac{2}{A}$

$\ln\frac{40}{4} = \frac{2}{A}$

$\ln 10 = \frac{2}{A}$

$A = \frac{2}{\ln 10}$ ✓

(2 marks)

b $f(x) = \frac{2}{\ln 10}\cdot\frac{x}{x^2+4}$

For $\frac{d}{dx}\left[\frac{x}{x^2+4}\right]$,

$u = x \qquad \frac{du}{dx} = 1$

$v = x^2 + 4 \qquad \frac{dv}{dx} = 2x$

Using quotient rule:

$\frac{d}{dx}\left[\frac{x}{x^2+4}\right] = \frac{(x^2+4).1 - x.2x}{(x^2+4)^2}$

$= \frac{x^2+4-2x^2}{(x^2+4)^2}$

$= \frac{4-x^2}{(x^2+4)^2}$

$\therefore f'(x) = \frac{2}{\ln 10}\cdot\frac{4-x^2}{(x^2+4)^2}$

Let $f'(x) = 0$

$2(4 - x^2) = 0$

$2(2 - x)(2 + x) = 0$

$\therefore x = 2$

(as $x > 0$)

$\therefore$ the mode is 2 hours. ✓

(2 marks)

c $P(X < 2)$

$= \frac{2}{\ln 10}\int_0^2 \frac{x}{x^2+4}\,dx$ ✓

$= \frac{1}{\ln 10}\int_0^2 \frac{2x}{x^2+4}\,dx$

$= \frac{1}{\ln 10}\left[\ln|x^2+4|\right]_0^2$

$= \frac{1}{\ln 10}[\ln(2^2+4) - \ln(0^2+4)]$

$= \frac{1}{\ln 10}[\ln 8 - \ln 4)]$

$= \frac{1}{\ln 10}[\ln\frac{8}{4}]$

$= \frac{\ln 2}{\ln 10}$

$= \log_{10} 2$ ✓

(using change of base rule)

(2 marks)

d $\mu = 100$ and $\sigma = 15$.

IQ = 130 is a z-score = 2

$\therefore P(\text{IQ} > 130) = 2.5\%$

$= 0.025$

$P(X < 2) = \log_{10} 2$ (from **c**)

$P(\text{IQ} > 130|X < 2)$

$= \frac{P(IQ > 130 \cap X < 2)}{P(X < 2)}$ ✓

$= \frac{0.025 \times 0.8}{\log_{10} 2}$

$= 0.066438561\ldots$

$= 0.066$ (3 dec. pl.) ✓

(2 marks)

5 a $\int_0^k \sin x\,dx = 1$

$[-\cos x]_0^k = 1$ ✓

$[-\cos k - (-\cos 0)] = 1$

$-\cos k + 1 = 1$

$\cos k = 0$

$k = \frac{\pi}{2}$ ✓

(2 marks)

b $P(X \le 1) = \int_0^1 \sin x\,dx$

$= [-\cos x]_0^1$ ✓

$= -\cos 1 - (-\cos 0)$

$= -\cos 1 + 1$

$= 0.459697694\ldots$

$= 0.4597$ ✓

(4 dec. pl.)

(2 marks)

6 i $\int_0^k \frac{3}{76}(x^2+1)\,dx = 1$

(probabilities sum to 1)

$\frac{3}{76}\int_0^k (x^2+1)\,dx = 1$

$\left[\frac{x^3}{3} + x\right]_0^k = \frac{76}{3}$ ✓

$\frac{k^3}{3} + k - 0 = \frac{76}{3}$

$k^3 + 3k - 76 = 0$

Substitute $k = 4$:

$4^3 + 3(4) - 76 = 64 + 12 - 76$

$= 0$

$\therefore k = 4$ is a solution. ✓

(2 marks)

ii $\frac{3}{76}\int_2^4 (x^2+x)\,dx$

$= \frac{3}{76}\left[\frac{x^3}{3} + x\right]_2^4$

$= \frac{3}{76}\left[\frac{4^3}{3} + 4 - \left(\frac{2^3}{3} + 2\right)\right]$

$= 0.815789473\ldots$

$= 0.82$ (2 dec. pl.) ✓

(1 mark)

7 $\int_0^2 k(2-x)\,dx = 1$

$k\int_0^2 (2-x)\,dx = 1$

$\left[2x - \frac{x^2}{2}\right]_0^2 = \frac{1}{k}$ ✓

$4 - \frac{2^2}{2} - 0 = \frac{1}{k}$

$2 = \frac{1}{k}$

$k = \frac{1}{2}$ ✓

$P(1 \le X \le 1.5)$

$= \int_1^{1.5} \frac{1}{2}(2-x)\,dx$

$= \frac{1}{2}\left[2x - \frac{x^2}{2}\right]_1^{1.5}$ ✓

$= \frac{1}{2}\left[2(1.5) - \frac{1.5^2}{2} - [2(1) - \frac{1^2}{2}]\right]$

$= \frac{1}{2}[1.875 - 1.5]$
$= 0.1875$
The probability is 0.1875. ✓
(4 marks)

8 **i** $\int_0^{0.8} \frac{3}{7}(x^2+2)\,dx$
$= \frac{3}{7}\int_0^{0.8}(x^2+2)\,dx$
$= \frac{3}{7}\left[\frac{x^3}{3}+2x\right]_0^{0.8}$ ✓
$= \frac{3}{7}\left(\frac{0.8^3}{3}+2(0.8)-0\right)$
$= 0.758\,857\,142\ldots$
$= 0.76$ (2 dec. pl.) ✓
(2 marks)

ii $\mu = \int_0^1 x \times \frac{3}{7}(x^2+2)dx$
$= \frac{3}{7}\int_0^1 (x^3+2x)dx$
$= \frac{3}{7}\left[\frac{x^4}{4}+x^2\right]_0^1$ ✓
$= \frac{3}{7}\left[\frac{1^4}{4}+1^2-0\right]$
$= 0.535\,714\,285\ldots$
$= 0.54$ (2 dec. pl.) ✓ *(2 marks)*

9 **i** $\int_0^k \frac{3}{32}\left(kx-x^2\right)dx = 1$
$\frac{3}{32}\left[\frac{kx^2}{2}-\frac{x^3}{3}\right]_0^k = 1$ ✓
$\frac{k(k)^2}{2}-\frac{k^3}{3} = \frac{32}{3}$
$3k^3 - 2k^3 = 64$
$k^3 = 64$
$k = 4$ ✓
(2 marks)

ii As $f(x) = \frac{3}{32}\left(4x-x^2\right)$ is a parabola, then due to symmetry, $E(X) = \frac{0+4}{2} = 2.$ ✓
(1 mark)

iii $E(X^2)$
$= \int_0^4 \left(x^2 \times \frac{3}{32}\left(4x-x^2\right)\right)dx$
$= \frac{3}{32}\int_0^4 (4x^3-x^4)\,dx$
$= \frac{3}{32}\left[x^4-\frac{x^5}{5}\right]_0^4$ ✓
$= \frac{3}{32}\left(4^4-\frac{4^5}{5}-0\right)$
$= 4.8$
$\text{Var}(X) = 4.8 - 2^2$
$= 0.8$ ✓ *(2 marks)*

10 **i** $\int_0^{0.5}\left(\frac{2}{5}(3-x)\right)dx$
$= \frac{2}{5}\int_0^{0.5}(3-x)\,dx$
$= \frac{2}{5}\left[3x-\frac{x^2}{2}\right]_0^{0.5}$
$= \frac{2}{5}\left(3(0.5)-\frac{(0.5)^2}{2}-0\right)$
$= 0.55$ ✓
(1 mark)

ii $E(X) = \frac{2}{5}\int_0^1 x(3-x)dx$
$= \frac{2}{5}\int_0^1 (3x-x^2)dx$
$= \frac{2}{5}\left[\frac{3x^2}{2}-\frac{x^3}{3}\right]_0^1$ ✓
$= \frac{2}{5}\left(\frac{3(1)^2}{2}-\frac{1^3}{3}-0\right)$
$= 0.466\,666\,66\ldots$
$= 0.467$ (3 dec. pl.)
$\therefore$ the mean time is 0.467 hours or 28 minutes. ✓
(2 marks)

iii $E(X^2) = \frac{2}{5}\int_0^1 x^2(3-x)\,dx$
$= \frac{2}{5}\int_0^1 (3x^2-x^3)\,dx$
$= \frac{2}{5}\left[x^3-\frac{x^4}{4}\right]_0^1$ ✓
$= \frac{2}{5}\left(1^3-\frac{1^4}{4}-0\right)$
$= 0.3$ ✓
$Var(X) = E(X^2) - [E(X)]^2$
$= 0.3 - 0.467^2$
$= 0.082\,222\,222\ldots$
$\sigma = 0.286\,744\,176\ldots$
$= 0.29$ (2 dec. pl.)
$\therefore$ the standard deviation is 0.29 hours or 17 minutes to the nearest minute. ✓
(3 marks)

11 **i** $\int_0^4 \frac{x+2}{k}\,dx = 1$
$\frac{1}{k}\int_0^4 (x+2)\,dx = 1$
$\left[\frac{x^2}{2}+2x\right]_0^4 = k$ ✓
$\left[\frac{4^2}{2}+2(4)-0\right] = k$
$8+8 = k$
$k = 16$ ✓
(2 marks)

ii $\int_0^a \frac{1}{16}(x+2)\,dx = \frac{3}{8}$
$\frac{1}{16}\left[\frac{x^2}{2}+2x\right]_0^a = \frac{3}{8}$ ✓
$\frac{a^2}{2}+2a-0 = 6$
$a^2+4a-12 = 0$
$(a-2)(a+6) = 0$
$a = 2$
(as $a > 0$) ✓
(2 marks)

12 **i** As probability sums to 1, you need to solve
$\int_0^3 \frac{1}{k}(x^2+5)\,dx = 1.$
$\frac{1}{k}\int_0^3 (x^2+5)\,dx = 1$
$\int_0^3 (x^2+5)\,dx = k$
$k = \left[\frac{x^3}{3}+5x\right]_0^3$ ✓
$= \frac{3^3}{3}+5(3)-0$
$= 9+15$
$= 24$ ✓
(2 marks)

ii $P(0.5 \le X \le 2)$
$= \int_{0.5}^2 \frac{1}{24}(x^2+5)\,dx$
$= \frac{1}{24}\left[\frac{x^3}{3}+5x\right]_{0.5}^2$
$= \frac{1}{24}\left[\frac{2^3}{3}+5(2)-\left(\frac{0.5^3}{3}+5(0.5)\right)\right]$
$= 0.421\,875$ ✓
(1 mark)

13 **i** $\int_0^2 (kx^2(2-x))\,dx = 1$
(probability sums to 1)
$k\int_0^2 (2x^2-x^3)\,dx = 1$
$k\left[\frac{2x^3}{3}-\frac{x^4}{4}\right]_0^2 = 1$ ✓
$k\left(\frac{2(2)^3}{3}-\frac{2^4}{4}-0\right) = 1$
$\frac{4k}{3} = 1$
$k = \frac{3}{4}$ ✓

(2 marks)

ii $\frac{3}{4}\int_{1.5}^{2}(x^2(2-x))\,dx$

$=\frac{3}{4}\int_{1.5}^{2}(2x^2-x^3)\,dx$

$=\frac{3}{4}\left[\frac{2x^3}{3}-\frac{x^4}{4}\right]_{1.5}^{2}$ ✓

$=\frac{3}{4}\left[\frac{2(2)^3}{3}-\frac{2^4}{4}-\left(\frac{2(1.5)^3}{3}-\frac{1.5^4}{4}\right)\right]$

$=0.261\,718\,75\ldots$

$=0.26$ (2 dec. pl.) ✓

(2 marks)

iii Consider $f(x)=\frac{3}{4}x^2(2-x)$.

$f(x)=\frac{3}{2}x^2-\frac{3}{4}x^3$

$f'(x)=3x-\frac{9}{4}x^2=0$

$3x\left(1-\frac{3}{4}x\right)=0$

$x=0,\frac{4}{3}$ ✓

$f''(x)=3-\frac{9}{2}x$

$f''(0)=3-\frac{9}{2}(0)>0\quad\therefore$ minimum

$f''\left(\frac{4}{3}\right)=3-\frac{9}{2}\left(\frac{4}{3}\right)<0\quad\therefore$ maximum

$\therefore$ the mode is $\frac{4}{3}$. ✓

(2 marks)

14 i $\int_0^k\frac{x}{8}(x^2+2)\,dx$

$=\frac{1}{8}\int_0^k(x^3+2x)\,dx$

$=\frac{1}{8}\left[\frac{x^4}{4}+x^2\right]_0^k$ ✓

$=\frac{1}{8}\left(\frac{k^4}{4}+k^2-0\right)$

$=\frac{1}{8}\left(\frac{k^4}{4}+k^2\right)$

Substitute $k=2$:

$=\frac{1}{8}\left(\frac{2^4}{4}+2^2\right)$

$=1$

PDF has an area of 1 when $k=2$. ✓

(2 marks)

ii $P(0.5\le X\le 1.5)$

$=\int_{0.5}^{1.5}\frac{x}{8}(x^2+2)\,dx$

$=\frac{1}{8}\int_{0.5}^{1.5}(x^3+2x)\,dx$

$=\frac{1}{8}\left[\frac{x^4}{4}+x^2\right]_{0.5}^{1.5}$ ✓

$=\frac{1}{8}\left[\frac{1.5^4}{4}+1.5^2-\left(\frac{0.5^4}{4}+0.5^2\right)\right]$

$=0.406\,25$ ✓

(2 marks)

iii $P(X\le 1)=\int_0^1\frac{x}{8}(x^2+2)dx$

$=\frac{1}{8}\int_0^1(x^3+2x)\,dx$

$=\frac{1}{8}\left[\frac{x^4}{4}+x^2\right]_0^1$

$=\frac{1}{8}\left[\frac{1^4}{4}+1^2-0\right]$

$=\frac{1}{8}\left(\frac{5}{4}\right)$

$=\frac{5}{32}$ ✓

(1 mark)

1 The life span of batteries from a particular factory is normally distributed with a mean of 840 hours and a standard deviation of 80 hours.

It is known from statistical tables that for this distribution approximately 60% of the batteries have a life span of less than 860 hours.

What is the approximate percentage of batteries with a life span between 820 and 920 hours? *(3 marks)*

(Q26, **2022 HSC**) Medium

2 A random variable is normally distributed with mean 0 and standard deviation 1. The table gives the probability that this random variable lies between 0 and z for different values of z.

z	0.1	0.2	0.3	0.4	0.5	0.6
Probability	0.0398	0.0793	0.1179	0.1554	0.1915	0.2257

The probability values given in the table for different values of z are represented by the shaded area in the following diagram.

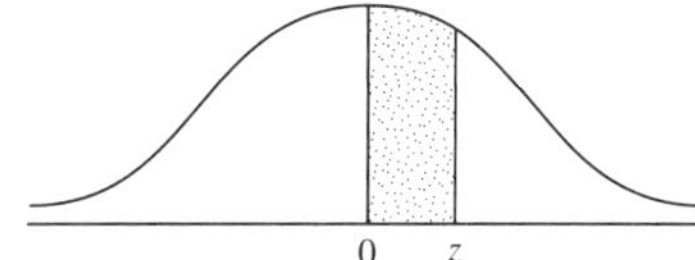

a Using the table, find the probability that a value from a random variable that is normally distributed with mean 0 and standard deviation 1 lies between 0.1 and 0.5. *(1 mark)* Hard

b Birth weights are normally distributed with a mean of 3300 grams and a standard deviation of 570 grams. By first calculating a z-score, find how many babies, out of 1000 born, are expected to have a birth weight greater than 3528 grams. *(3 marks)* Medium

(Q22, **2021 HSC**)

3 In a particular city, the heights of adult females and the heights of adult males are each normally distributed.

Information relating to two females from that city is given in Table 1.

Table 1

Height	*Gender*	*Percentage of females in this city shorter than this person*
175 cm	Female	97.5%
160.6 cm	Female	16%

The means and standard deviations of adult females and males, in centimetres, are given in Table 2.

Table 2

	Mean	*Standard deviation*
Females	μ	σ
Males	1.05μ	1.1σ

A selected male is taller than 84% of the population of adult males in this city.

By first labelling the normal distribution curve below with the heights of the two females given in Table 1, calculate the height of the selected male, in centimetres, correct to two decimal places.

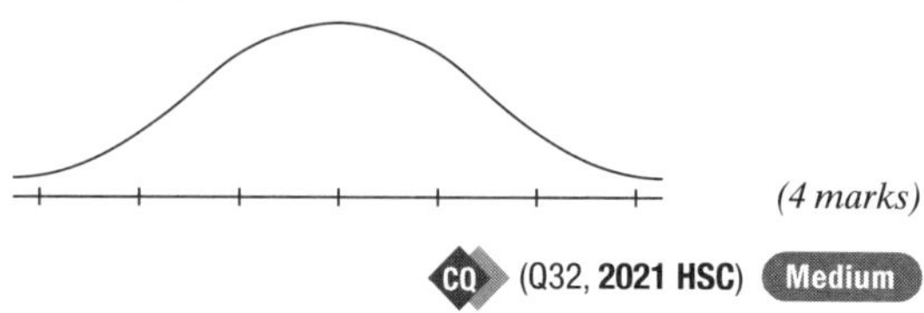

(4 marks)

(Q32, **2021 HSC**) Medium

4 John recently did a class test in each of three subjects. The class scores on each test were normally distributed.

The table shows the subjects and John's scores as well as the mean and standard deviation of the class scores on each test.

Subject	*John's score*	*Mean*	*Standard deviation*
French	82	70	8
Commerce	80	65	5
Music	74	50	12

Relative to the rest of the class, which row of the table below shows John's strongest subject and his weakest subject?

	Strongest subject	*Weakest subject*
A	Commerce	French
B	French	Music
C	Music	French
D	Commerce	Music

(1 mark)

(Q3, **2020 HSC**) Easy

5 Suppose the weight of melons is normally distributed with a mean of μ and a standard deviation of σ.

A melon has a weight below the lower quartile of the distribution but NOT in the bottom 10% of the distribution.

Which of the following most accurately represents the region in which the weight of this melon lies?

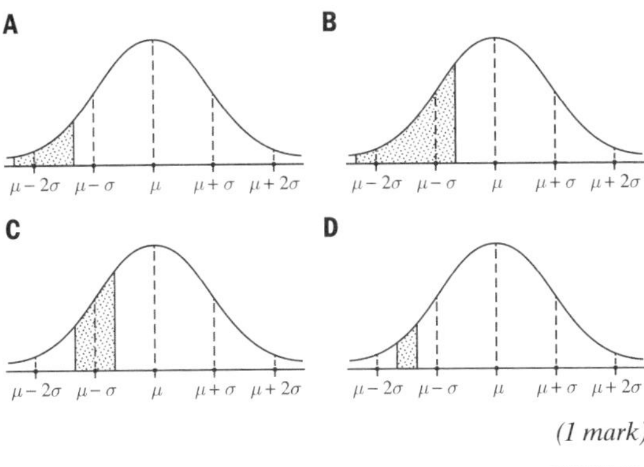

(1 mark)

(Q9, **2020 HSC**) **Medium**

---

6 In a particular country, the hourly rate of pay for adults who work is normally distributed with a mean of \$25 and a standard deviation of \$5.

a Two adults who both work are chosen at random. Find the probability that at least one of them earns between \$15 and \$30 per hour. *(3 marks)* **Hard**

b The number of adults who work is equal to three times the number of adults who do not work. One adult is chosen at random. Find the probability that the chosen adult works and earns more than \$25 per hour. *(2 marks)* **Medium**

(Q28, **2020 HSC**)

---

7 The Intelligence Quotient (IQ) scores for adults in City *A* are normally distributed with a mean of 108 and a standard deviation of 10.

The IQ scores for adults in City *B* are normally distributed with a mean of 112 and a standard deviation of 16.

a Yin is an adult who lives in City *A* and has an IQ score of 128.
What percentage of the adults in this city have an IQ score higher than Yin's? *(2 marks)* **Easy**

b There are 1 000 000 adults living in City *B*. Calculate the number of adults in City *B* that would be expected to have an IQ score lower than Yin's. *(2 marks)* **Easy**

c Simon, an adult who lives in City *A*, moves to City *B*. The z-score corresponding to his IQ score in City *A* is the same as the z-score corresponding to his IQ score in City *B*.
By first forming an equation, calculate Simon's IQ score. Give your answer correct to one decimal place. *(3 marks)* **Medium**

(Q35, **2020 MS2 HSC**)

---

8 The table below provides some values of the probabilities for the cumulative normal distribution.

i Explain why $P(z \leq -1.86) = 0.0314$. *(1 mark)* **Medium**

ii Find $P(-1.86 \leq z \leq 2.07)$. *(1 mark)* **Medium**

iii In a state-wide mathematics competition, students' results are normally distributed. The mean is 78 and the standard deviation is 6.9. Students who score in the top 10% of the results are awarded a voucher. Estimate the minimum score required to be awarded a voucher. *(3 marks)* **Hard**

z	.00	.01	.02	.03	.04	.05	.06	.07	.08	.09
1.2	.8849	.8869	.8888	.8907	.8925	.8944	.8962	.8980	.8997	.9015
1.3	.9032	.9049	.9066	.9082	.9099	.9115	.9131	.9147	.9162	.9177
1.4	.9192	.9207	.9222	.9236	.9251	.9265	.9279	.9292	.9306	.9319
1.5	.9332	.9345	.9357	.9370	.9382	.9394	.9406	.9418	.9430	.9441
1.6	.9452	.9463	.9474	.9484	.9495	.9505	.9515	.9525	.9535	.9545
1.7	.9554	.9564	.9573	.9582	.9591	.9599	.9608	.9616	.9625	.9633
1.8	.9641	.9649	.9656	.9664	.9671	.9678	.9686	.9693	.9700	.9706
1.9	.9713	.9719	.9726	.9732	.9738	.9744	.9750	.9756	.9762	.9767
2.0	.9772	.9778	.9783	.9788	.9793	.9798	.9803	.9808	.9812	.9817

Bonus question (see page iv)

---

9 Mae's class sits two Ancient History tests.

In the first test, the mean for the class was 74.2 and the standard deviation was 4.5.

In this test Mae scored 85.

i Find Mae's z-score in the first test. *(1 mark)* **Easy**

ii In the second Ancient History test, the mean for the class was 78.6 and the standard deviation was 3.5. Calculate the mark that Mae needed to obtain in the second test to ensure that her performance relative to the class was maintained. *(2 marks)* **Medium**

Bonus question

---

10 A group of 20 000 students sat a state-wide examination.

The results of the examination were normally distributed.

In the examination, Freya achieved a score of 81, which corresponded to a z-score of 1.4.

In the same examination, Cooper achieved a score of 63, which corresponded to a z-score of –2.2.

i Using this information, form two equations and solve them simultaneously to find the values of the mean and the standard deviation. *(3 marks)* **Medium**

ii Hence, find the number of students who scored at least 69 in the examination. *(2 marks)* **Medium**

Bonus question

11 The table below provides some values of the probabilities for the cumulative normal distribution.

i Find $P(-0.93 \leq z \leq 1.79)$. *(1 mark)*  Easy

ii The heights of males in a large population are normally distributed. The mean height is 175.6 cm and the standard deviation is 7.2 cm. Jack has calculated he is taller than 80% of the males in the population. Estimate Jack's height, to one decimal place. *(3 marks)* **Hard**

z	.00	.01	.02	.03	.04	.05	.06	.07	.08	.09
0.8	.7881	.7910	.7939	.7967	.7995	.8023	.8051	.8078	.8106	.8133
0.9	.8159	.8186	.8212	.8238	.8264	.8289	.8315	.8340	.8365	.8389
1.0	.8413	.8438	.8461	.8485	.8508	.8531	.8554	.8577	.8599	.8621
1.1	.8643	.8665	.8686	.8708	.8729	.8749	.8770	.8790	.8810	.8830
1.2	.8849	.8869	.8888	.8907	.8925	.8944	.8962	.8980	.8997	.9015
1.3	.9032	.9049	.9066	.9082	.9099	.9115	.9131	.9147	.9162	.9177
1.4	.9192	.9207	.9222	.9236	.9251	.9265	.9279	.9292	.9306	.9319
1.5	.9332	.9345	.9357	.9370	.9382	.9394	.9406	.9418	.9430	.9441
1.6	.9452	.9463	.9474	.9484	.9495	.9505	.9515	.9525	.9535	.9545
1.7	.9554	.9564	.9573	.9582	.9591	.9599	.9608	.9616	.9625	.9633

Bonus question

12 The lifetime of a particular make of lightbulb is normally distributed with mean 940 hours and standard deviation 65 hours.

Using the Cumulative Normal Distribution Table below, find the probability that a lightbulb of the same make chosen at random from a batch of 5000 lightbulbs has a lifetime of at least 1031 hours.

z	.0	.1	.2	.3	.4	.5	.6	.7	.8	.9
0.	0.5000	0.5398	0.5793	0.6179	0.6554	0.6915	0.7257	0.7580	0.788	0.8159
1.	0.8413	0.8643	0.8849	0.9032	0.9192	0.9332	0.9452	0.9554	0.9641	0.9731
2.	0.9772	0.9821	0.9861	0.9893	0.9918	0.9938	0.9953	0.9965	0.9974	0.9981
3.	0.9987	0.9990	0.9993	0.9995	0.9997	0.9998	0.9998	0.9999	0.9999	1.000

A 8.08% **B** 8.86%
C 8.92% **D** 9.16% *(1 mark)*

Bonus question **Medium**

13 In Jerry's Plains, the maximum temperature for each day is recorded. The mean of these maximum temperatures during September over the past 150 years is 26 °C, and the standard deviation is 2.4 °C. You may assume that these maximum temperatures are normally distributed.

i What percentage of September days have a maximum temperature less than 23.6 °C? *(1 mark)* Easy

ii Calculate the z-score for a maximum temperature of 22.4 °C. *(1 mark)* Easy

iii The probability that a September day chosen at random has a maximum temperature between 22.4 °C and 23.6 °C can be found by evaluating $\int_a^b f(x)\,dx$ where a and b are constants and where

$$f(x) = \frac{1}{\sqrt{2\pi}} e^{-\frac{x^2}{2}}$$

is the normal probability density function with mean 0 and standard deviation 1.
By first finding the values of a and b, calculate an approximate value for this probability by using the trapezoidal rule with 3 function values. *(3 marks)* **Hard**

iv Using the results above, find the approximate probability that a September day chosen at random has a maximum temperature of 22.4 °C or less. *(1 mark)* **Medium**

Bonus question

14 The results (as percentages) of a large group of students in a Mathematics competition are represented by a random variable X and are normally distributed with a mean of 70 and a variance of 100. All results are expressed as whole percentages.

i Determine the z-score for a student who scored 66. *(1 mark)* Easy

ii Students who scored in the top 2.5% of the group receive a monetary prize.
What is the minimum mark required to receive the prize? *(1 mark)* Easy

iii Students who gained a mark with a z-score range of $0 \leq z \leq 1$ received a credit certificate.
Those students with a mark with a z-score range of $1 \leq z \leq 2$ received a distinction certificate.
If 816 credit certificates were awarded, how many distinction certificates were issued? *(2 marks)* **Medium**

Bonus question

15 The heights of Year 12 girls are normally distributed with a mean of 163 cm and a standard deviation of 6 cm.

i What percentage of Year 12 girls have a height of more than 169 cm? *(1 mark)* Easy

ii Calculate the z-score for a girl who has a height of 160 cm. *(1 mark)* Easy

iii The probability that a Year 12 girl chosen at random has a height between 160 cm and 169 cm can be found by evaluating

$\int_a^b f(x)\,dx$ where a and b are constants and

where $f(x) = \frac{1}{\sqrt{2\pi}} e^{-\frac{x^2}{2}}$

is the normal probability density function with mean 0 and standard deviation 1.
By first finding the values of a and b, calculate an approximate value for this probability by using the trapezoidal rule with 4 function values. *(3 marks)* **Hard**

iv In a particular year there were 54 600 students enrolled in Year 12 in NSW. 52% of the students were girls. Using the results above, approximately how many girls had a height of less than 160 cm? *(2 marks)* **Hard**

Bonus question

16 The scores on an examination are normally distributed with a mean of 70 and a standard deviation of 6. Michael received a score on the examination between the lower quartile and the upper quartile of the scores.

Which shaded region most accurately represents where Michael's score lies?

A

58 64 70 76 82

B

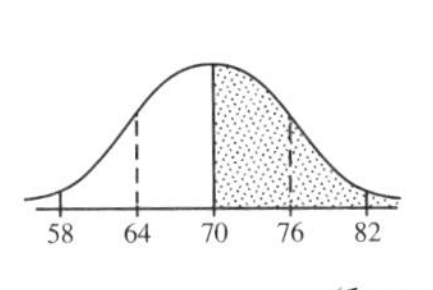

C

58 64 70 76 82

D

58 64 70 76 82

(1 mark)

(Q15, **2019 MS2 HSC**) **Medium**

17 In a particular country, the birth weight of babies is normally distributed with a mean of 3000 grams. It is known that 95% of these babies have a birth weight between 1600 grams and 4400 grams.

One of these babies has a birth weight of 3497 grams. What is the z-score of this baby's birth weight? *(2 marks)*

(Q38, **2019 MS2 HSC**) **Easy**

18 A set of data is normally distributed with a mean of 48 and a standard deviation of 3.

Approximately what percentage of the scores lies between 39 and 45?

A 15.85% **B** 31.7%
C 47.5% **D** 49.85% *(1 mark)*

(Q23, **2018 MS2 HSC**) **Easy**

19 The heights of Year 12 girls are normally distributed with a mean of 165 cm and a standard deviation of 5.5 cm.

What is the z-score for a height of 154 cm?

A -2 **B** -0.5
C 0.5 **D** 2 *(1 mark)*

(Q13, **2017 MS2 HSC**) **Easy**

20 All the students in a class of 30 did a test.

The marks, out of 10, are shown in the dot plot.

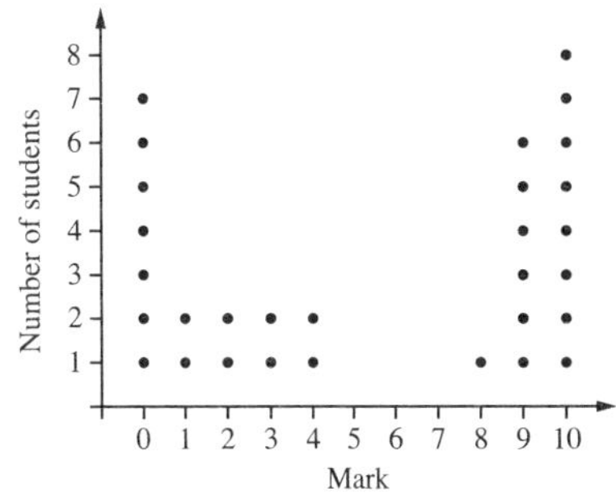

i Find the median test mark. *(1 mark)* **Easy**

ii The mean test mark is 5.4. The standard deviation of the test marks is 4.22. Using the dot plot, calculate the percentage of the marks which lie within one standard deviation of the mean. *(2 marks)* **Medium**

iii A student states that for any data set, 68% of the scores should lie within one standard deviation of the mean. With reference to the dot plot, explain why the student's statement is NOT relevant in this context. *(1 mark)* **Medium**

(Q29d, **2017 MS2 HSC**)

21 The speed limit outside a school is 40 km/h. Year 11 students measured the speed of passing vehicles over a period of time. They found the set of data to be normally distributed with a mean speed of 36 km/h and a standard deviation of 2 km/h.

What percentage of the vehicles passed the school at a speed greater than 40 km/h?

A 2.5% **B** 5%
C 47.5% **D** 95% *(1 mark)*

(Q13, **2016 MS2 HSC**) **Easy**

22 A machine produces cylindrical pipes. The mean of the diameters of the pipes is 8 cm and the standard deviation is 0.04 cm.

Assuming a normal distribution, what percentage of cylindrical pipes produced will have a diameter less than 7.96 cm?

A 16% **B** 32%
C 34% **D** 68% *(1 mark)*

(Q20, **2015 MS2 HSC**) **Easy**

23 The weights of 10 000 newborn babies in NSW are normally distributed. These weights have a mean of 3.1 kg and a standard deviation of 0.35 kg.

How many of these newborn babies have a weight between 2.75 kg and 4.15 kg?

A 4985 **B** 6570
C 8370 **D** 8385 *(1 mark)*

(Q24, **2014 MS2 HSC**) **Easy**

24 There are 60000 students sitting a state-wide examination. If the results form a normal distribution, how many students would be expected to score a result between 1 and 2 standard deviations above the mean?

You may assume for normally distributed data that:

- 68% of scores have z-scores between –1 and 1
- 95% of scores have z-scores between –2 and 2
- 99.7% of scores have z-scores between –3 and 3.

A 8100 **B** 16200
C 20400 **D** 28500 *(1 mark)*

(Q20, **2013 MS2 HSC**) Easy

25 Ali's class sits two Geography tests. The results of her class on the first Geography test are shown.

58, 74, 65, 66, 73, 71, 72, 74, 62, 70

The mean was 68.5 for the first test.

i Calculate the standard deviation for the first test. Give your answer correct to one decimal place. *(1 mark)* Easy

ii On the second Geography test, the mean for the class was 74.4 and the standard deviation was 12.4.
Ali scored 62 on the first test. Calculate the mark that she needed to obtain in the second test to ensure that her performance relative to the class was maintained. *(3 marks)* Easy

(Q29b, **2013 MS2 HSC**)

26 Two brands of light bulbs are being compared. For each brand, the life of the light bulbs is normally distributed.

Life of light bulbs (in hours)

	Mean	*Standard deviation*
Brand A	450	25
Brand B	500	50

i One of the Brand B light bulbs has a life of 400 hours.
What is the z-score of the life of this light bulb? *(1 mark)* Easy

ii A light bulb is considered defective if it lasts less than 400 hours. The following claim is made:
'Brand A light bulbs are more likely to be defective than Brand B light bulbs.'
Is this claim correct? Justify your answer, with reference to z-scores or standard deviations or the normal distribution. *(2 marks)* Easy

(Q27c, **2011 MS2 HSC**)

27 The following graph indicates z-scores of 'height-for-age' for girls aged 5–19 years.

i What is the z-score for a six year old girl of height 120 cm? *(1 mark)* Easy

ii Rachel is $10\frac{1}{2}$ years of age.

1 If 2.5% of girls of the same age are taller than Rachel, how tall is she? *(1 mark)* Easy

2 Rachel does not grow any taller.
At age $15\frac{1}{2}$, what percentage of girls of the same age will be taller than Rachel? *(2 marks)* Easy

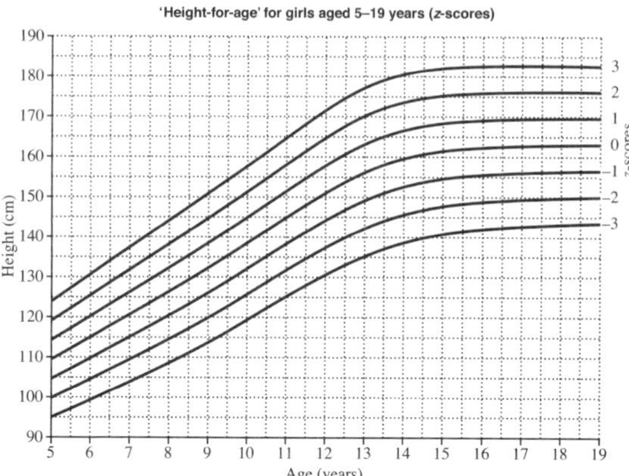

iii What is the average height of an 18 year old girl? *(1 mark)* Easy

iv For adults (18 years and older), the Body Mass Index is given by $B = \frac{m}{h^2}$ where m = mass in kilograms and h = height in metres. The medically accepted healthy range for B is $21 \le B \le 25$.
What is the minimum weight for an 18 year old girl of average height to be considered healthy? *(2 marks)* Easy

v The average height, C, in centimetres, of a girl between the ages of 6 years and 11 years can be represented by a line with equation

$$C = 6A + 79$$

where A is the age in years.

1 For this line, the gradient is 6.
What does this indicate about the heights of girls aged 6 to 11? *(1 mark)* Easy

2 Give ONE reason why this equation is not suitable for predicting heights of girls older than 12. *(1 mark)* Easy

(Q28a, **2008 MS2 HSC**)

28 The results of two class tests are normally distributed. The means and standard deviations of the tests are displayed in the table.

	Test 1	*Test 2*
Mean	60	58
Standard deviation	6.2	6.0

i Stuart scored 63 in Test 1 and 62 in Test 2. He thinks that he has performed better in Test 1. Do you agree? Justify your answer using appropriate calculations.

(2 marks) **Easy**

ii If 150 students sat for Test 2, how many students would you expect to have scored less than 64? *(2 marks)* **Medium**

(Q25d, **2007 MS2 HSC**)

29 The normal distribution shown has a mean of 170 and a standard deviation of 10.

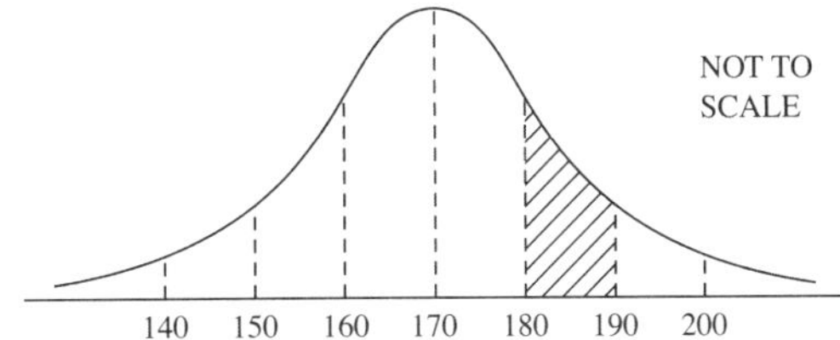

i Roberto has a raw score in the shaded region. What could his z-score be?

(1 mark) **Easy**

ii What percentage of the data lies in the shaded region? *(2 marks)* **Medium**

(Q24c, **2004 MS2 HSC**)

1 $\mu = 840$ and $\sigma = 80$

Using $z = \frac{x-\mu}{\sigma}$:

For $x = 820$: $z = \frac{820-840}{80} = -0.25$

For $x = 920$: $z = \frac{920-840}{80} = 1$ ✓

Given $P(x < 860) = 0.6$, then $P(\mu < x < 860) = 0.6 - 0.5 = 0.1$

Hence by symmetry, $P(840 < x < \mu) = 0.1$

Also as $P(-1 < z < 1) = 0.68$, then $P(0 < z < 1) = 0.34$. ✓

$\therefore P(820 < x < 920) = 0.1 + 0.34 = 0.44$

$\therefore$ 44% of batteries.

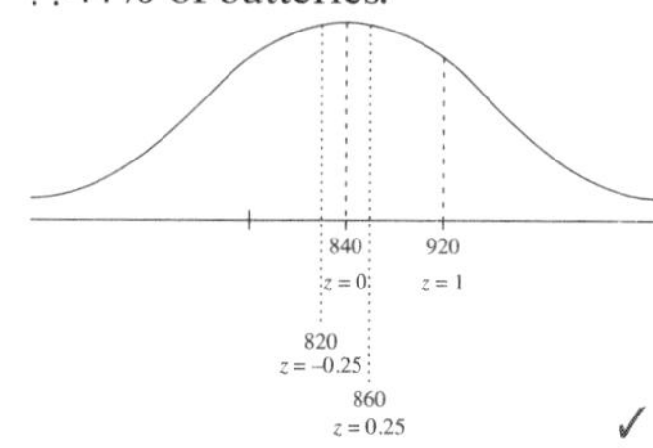

✓ *(3 marks)*

2 **a** $P(0.1 < z < 0.5)$
$= 0.1915 - 0.0398$
$= 0.1517$ ✓ *(1 mark)*

b Using $z = \frac{x-\mu}{\sigma}$:

$z = \frac{3528-3300}{570}$
$= 0.4$

$P(0 < z < 0.4) = 0.1554$ ✓

$\therefore P(z < 0.4) = 0.5 + 0.1554 = 0.6554$

$\therefore P(z > 0.4) = 1 - 0.6554 = 0.3446$ ✓

As $0.3446 \times 1000 = 344.6$, it is expected 344 (or 345) babies have a birth weight greater than 3528 g. ✓ *(3 marks)*

3 As 95% of scores lie within $-2 < z < 2$, then 97.5% of scores when $z < 2$.

As 68% of scores lie within $-1 < z < 1$, then 16% of scores when $z < -1$.

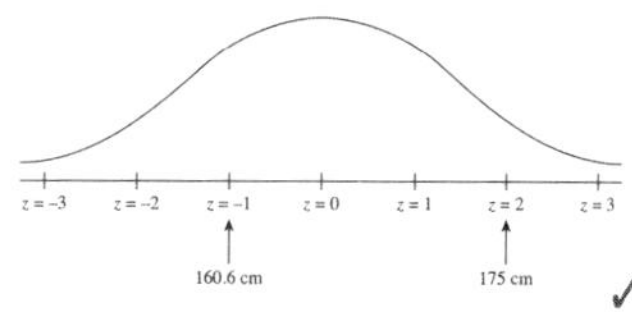

✓

$\therefore$ difference is 3 standard deviations.

As $\frac{175-160.6}{3} = 4.8$, then $\sigma = 4.8$

As $160.6 = \mu - 4.8$, then $\mu = 165.4$ ✓

Now, for males:

$\mu = 1.05 \times 165.4 = 173.67$

$\sigma = 1.1 \times 4.8 = 5.28$ ✓

As 84% of scores lie within $z < 1$, and $173.67 + 5.28 = 178.95$, the selected male is more than 178.95 cm tall. ✓

Alternatively, using z-score $= \frac{x-\mu}{\sigma}$ to form a pair of simultaneous equations:

$2 = \frac{175-\mu}{\sigma}$

$2\sigma = 175 - \mu$ ... ①

$-1 = \frac{160.6-\mu}{\sigma}$

$-\sigma = 160.6 - \mu$ ... ②

① − ②:

$3\sigma = 14.4$

$\sigma = 4.8$

Substitute in ①:

$2(4.8) = 175 - \mu$

$\mu = 175 - 9.6 = 165.4$

$\therefore \mu = 165.4$ and $\sigma = 4.8$

Now, for males:

$\mu = 1.05 \times 165.4 = 173.67$

$\sigma = 1.1 \times 4.8 = 5.28$

As 84% of scores lie within $z < 1$, and $173.67 + 5.28 = 178.95$, the selected male is more than 178.95 cm tall. *(4 marks)*

4 Using $z = \frac{x-\mu}{\sigma}$:

French: $z = \frac{82-70}{8} = 1.5$

Commerce: $z = \frac{80-65}{5} = 3$

Music: $z = \frac{74-50}{12} = 2$

John's strongest subject was Commerce and his weakest was French.

Answer A

5 Below the lower quartile means in the bottom 25%. Consider the percentage of scores less than

$\mu - \sigma = \frac{100\% - 68\%}{2} = 16\%$.

Consider the percentage of scores less than

$\mu - 2\sigma = \frac{100\% - 95\%}{2} = 2.5\%$.

This means the best representation is C.

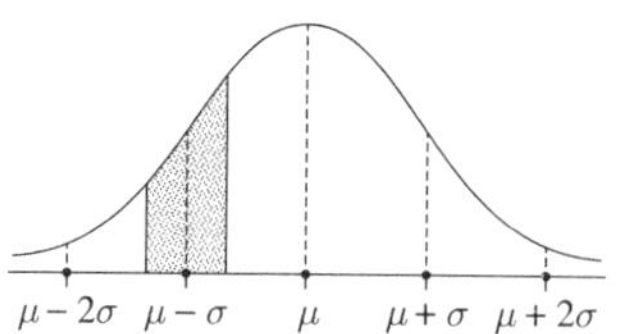

Answer C

6 **a** $\mu = 25$ and $\sigma = 5$ and $z = \frac{x-\mu}{\sigma}$:

When $x = 15$, $z = \frac{15-25}{5} = -2$

When $x = 30$, $z = \frac{30-25}{5} = 1$ ✓

P(between \$15 and \$30)
$= P(-2 < z < 1)$
$= P(z < 1) - P(z < -2)$
$= 0.84 - 0.025$
$= 0.815$

$\therefore P$(does not earn between \$15 and \$30)
$= 1 - 0.815$
$= 0.185$ ✓

P(at least one earns between \$15 and \$30)
$= 1 - P$(none earns between \$15 and \$30)
$= 1 - 0.185^2$
$= 0.965775$
$= 0.966$ (3 dec. pl.) ✓ *(3 marks)*

b $P(\text{adult works}) = \frac{3}{4}$,

$P(\text{earns more than \$25}) = \frac{1}{2}$ ✓

P(adult works and earns more than \$25/h)
$= \frac{3}{4} \times \frac{1}{2}$
$= \frac{3}{8}$ ✓ *(2 marks)*

7 **a** City A: $x = 128$, $\mu = 108$, $\sigma = 10$

$z = \frac{x-\mu}{\sigma}$
$= \frac{128-108}{10}$
$= 2$ ✓

Now, 95% of scores are within two standard deviations of the mean.

So 5% are more than two standard deviations from the mean.

Half of these are above two standard deviations.

So 2.5% of adults will have an IQ higher than Yin's. ✓ *(2 marks)*

b City B: $x = 128$, $\mu = 112$, $\sigma = 16$

$z = \dfrac{x - \mu}{\sigma}$

$= \dfrac{128 - 112}{16}$

$= 1$ ✓

Now, 68% of scores are within one standard deviation of the mean.

Half of these, 34%, are between the mean and one standard deviation above the mean.

50% are below the mean.

So 84% of adults will have an IQ lower than Yin's.

Number of adults = 84% of 1 000 000

840 000 adults would be expected to have an IQ lower than Yin's. ✓

(2 marks)

c Let x be Simon's IQ.

Now $\dfrac{x - 108}{10} = \dfrac{x - 112}{16}$ ✓

$16(x - 108) = 10(x - 112)$

$16x - 1728 = 10x - 1120$

$6x = 608$ ✓

$x = 101.333\,333\ldots$

$= 101.3$ (1 d.p.)

So Simon's IQ would be 101.3 to one decimal place. ✓

(3 marks)

8 i From the table

$P(z \le 1.86) = 0.9686$.

This means

$P(z \ge 1.86)$

$= 1 - 0.9686$

$= 0.0314$,

and by symmetry,

$P(z \ge 1.86) = P(z \le -1.86)$,

hence $P(z \le -1.86) = 0.0314$.

✓ *(1 mark)*

ii $P(-1.86 \le z \le 2.07)$

$- P(z \le 2.07) - P(z \le -1.86)$

$= 0.9808 - 0.0314$

$= 0.9494$ ✓

(1 mark)

iii As top 10% is 90% = 0.9, using the table,

$P(z \le 1.28) \approx 0.9$. ✓

This means the minimum score to be awarded a voucher has a z-score of 1.28. ✓

Using $z = \dfrac{x - \mu}{\sigma}$

$1.28 = \dfrac{x - 78}{6.9}$

$x - 78 = 8.832$

$x = 86.832$

A score of 87 is required. ✓

(3 marks)

9 i $z = \dfrac{x - \mu}{\sigma}$

$= \dfrac{85 - 74.2}{4.5}$

$= 2.4$ ✓ *(1 mark)*

ii Mae's result in the second test needs to have a z-score of 2.4.

$2.4 = \dfrac{x - 78.6}{3.5}$ ✓

$x - 78.6 = 8.4$

$x = 87$

Mae needs a mark of 87. ✓

(2 marks)

10 i Using $z = \dfrac{x - \mu}{\sigma}$:

Freya: $1.4 = \dfrac{81 - \mu}{\sigma}$

$1.4\sigma = 81 - \mu$

$\mu + 1.4\sigma = 81$ …① ✓

Cooper: $-2.2 = \dfrac{63 - \mu}{\sigma}$

$-2.2\sigma = 63 - \mu$

$\mu - 2.2\sigma = 63$ …②

① − ②: $3.6\sigma = 18$

$\sigma = 5$ ✓

Substitute into ①:

$\mu + 1.4(5) = 81$

$\mu + 7 = 81$

$\mu = 74$ ✓

Mean (μ) = 74 and standard deviation (σ) = 5.

(3 marks)

ii As $69 = 74 - 5 = \mu - \sigma$, then 69 has a z-score of −1.

Hence $P(z \ge -1)$

$= \dfrac{68\%}{2} + 50\%$

$= 84\%$ ✓

Number of students

$= 0.84 \times 20\,000$

$= 16\,800$ ✓ *(2 marks)*

11 i $P(-0.93 \le z \le 1.79)$

$= P(z \le 1.79)\ \ P(z \le\ \ 0.93)$

$= 0.9633 - (1 - 0.8238)$

$= 0.7871$ ✓

(1 mark)

ii As 80% = 0.8, using the table $P(z \le 0.84) \approx 0.8$, this means a z-score of 0.84. ✓

Using $z = \dfrac{x - \mu}{\sigma}$

$0.84 = \dfrac{x - 175.6}{7.2}$ ✓

$x - 175.6 = 6.048$

$x = 181.648$

$= 181.6$ (1 dec. pl.)

Jack is 181.6 cm tall. ✓

(3 marks)

12 Let $x = 1031$, $\mu = 940$, $\sigma = 65$:

$z = \dfrac{x - \mu}{\sigma}$

$= \dfrac{1031 - 940}{65}$

$= 1.4$

The table shows

$P(z < 1.4) = 0.9192$

Hence $P(z > 1.4) = 1 - 0.9192$

$= 0.0808$

∴ the probability is 8.08%.

Answer A

13 i As $23.6 = \bar{x} - \sigma$, it has a z-score of −1.

68% of population lie in $-1 \le z \le 1$.

100% − 68% = 32%.

Hence $z \le -1$ is $\dfrac{32}{2}$%, or 16% of population. ✓ *(1 mark)*

ii $z = \dfrac{x - \mu}{\sigma}$

$= \dfrac{22.4 - 26}{2.4}$

$= -1.5$ ✓ *(1 mark)*

iii As 22.4 has z-score of −1.5, and 23.6 has a z-score of −1, then $a = -1.5$ and $b = -1$.

Also, 23 will have a z-score of −1.25. ✓

Temp (°C)	22.4	23	23.6
x	−1.5	−1.25	−1
$\frac{1}{\sqrt{2\pi}}e^{-\frac{x^2}{2}}$	0.1295	0.1826	0.2420
	$f(a)$	$f(x_1)$	$f(b)$

Area $\approx \dfrac{b - a}{2n}\{f(a) + f(b) + 2[f(x_1) + \ldots + f(x_{n-1})]\}$

$= \dfrac{-1 - (-1.5)}{2(2)}\{f(-1.5) + f(-1) + 2f(-1.25)\}$ ✓

$= 0.125(0.1295 + 0.2420 + 2(0.1826))$

$= 0.092\,0875$

$= 9.2\%$ ✓

(nearest tenth of %)

(3 marks)

iv As $P(X \le 23.6) = 16\%$ (from **i**),

and $P(22.4 \le X \le 23.6) \approx 9.2\%$ (from **iii**), then

$P(X \le 22.4)$

$\approx 16\% - 9.2\% = 6.8\%$

The approximate probability is 6.8%. ✓

(1 mark)

14 i $\mu = 70$, $\sigma = \sqrt{100} = 10$

$z = \dfrac{x - \mu}{\sigma}$

$= \frac{66-70}{10}$

$= -0.4$ ✓ *(1 mark)*

ii As $-2 \le z \le 2 = 95\%$ of scores, then $z \ge 2$ is top 2.5% of scores.

$2 = \frac{x-70}{10}$

$x - 70 = 20$

$x = 90$

Students with scores of 90 or above receive a monetary prize. ✓ *(1 mark)*

iii $0 \le z \le 1 = \frac{68}{2} = 34\%$ of scores and $1 \le z \le 2$

$= \frac{95-68}{2} = 13.5\%$ ✓

34% of scores = 816

13.5% of scores

$= 816 \div 34 \times 13.5$

$= 324$

There were 324 distinction certificates awarded. ✓ *(2 marks)*

15 **i** As $169 = 163 + 6 = \bar{x} + \sigma$, it has a z-score of 1.

Hence $z \ge 1$ is $\frac{100-68}{2}\%$

= 16% of the Year 12 girls. ✓ *(1 mark)*

ii $z = \frac{x-\mu}{\sigma}$

$= \frac{160-163}{6}$

$= -0.5$ ✓ *(1 mark)*

iii As 160 cm has a z-score of -0.5, and 169 cm has a z-score of 1, then $a = -0.5$ and $b = 1$. ✓

Temp. (°C)	160 cm	163 cm	166 cm	169 cm
x	-0.5	0	0.5	1
$\frac{1}{\sqrt{2\pi}}e^{-\frac{x^2}{2}}$	0.3521	0.3989	0.3521	0.2420
	$f(a)$	$f(x_1)$	$f(x_2)$	$f(b)$

$$\text{Area} \approx \frac{b-a}{2n}\{f(a)+f(b)+2[f(x_1)+...+f(x_{n-1})]\}$$ ✓

$$= \frac{1-(-0.5)}{2(3)}\{f(-0.5)+f(1)+2[f(0)+f(0.5)]\}$$

$$= \frac{0.5}{2}(0.3521 + 0.2420 + 2(0.3989 + 0.3521))$$

$= 0.524025$

$= 52\%$ (nearest %) ✓ *(3 marks)*

iv Now, P(greater than 169)

$= P(z \ge 1)$

= 16% (from part **i**)

Also, and P(between 160 cm and 169 cm)

$= P(-0.5 \le z \le 1)$

$\approx 52\%$ (from **iii**).

Hence, P(less than 160 cm)

$= 100\% - (16\% + 52\%)$

$= 32\%$ ✓

Number of girls

$= 0.32 \times 0.52 \times 54\,600$

= 9085 (nearest whole)

There were approximately 9085 girls with a height of less than 160 cm. ✓ *(2 marks)*

16 $\bar{x} = 70, \quad s = 6$

So scores between 64 and 76 are within 1 standard deviation of the mean.

Now, 68% of scores are within 1 standard deviation of the mean.

50% lie between the upper and lower quartiles.

So the shaded region can only be that in option A.

Answer A

17 $\bar{x} = 3000$ g

In normally distributed data 95% are within two standard deviations of the mean.

But 95% of these babies lie between 1600 g and 4400 g.

So two standard deviations is 1400 g.

$\therefore s = 700$ g ✓

$z = \frac{x-\bar{x}}{s}$

$= \frac{3497-3000}{700}$

$= 0.71$ ✓ *(2 marks)*

18 $\bar{x} = 48 \quad s = 3$

So 45 is one and 39 is three standard deviations below the mean.

Now, 99.7% of scores lie within three standard deviations of the mean.

68% of scores lie within one standard deviation of the mean.

$\therefore$ 99.7% – 68% or 31.7% lie between one and three standard deviations.

Half of these are below the mean.

Required percentage

$= 0.5 \times 31.7\%$

$= 15.85\%$

Answer A

19 $x = 154, \quad \mu = 165, \quad \sigma = 5.5$

$z = \frac{x-\mu}{\sigma}$

$= \frac{154-165}{5.5}$

$= -2$

Answer A

20 **i** median $= \frac{4+8}{2}$

$= 6$ ✓ *(1 mark)*

ii $\bar{x} = 5.4, \quad s = 4.22$

$\bar{x} + s = 5.4 + 4.22$

$= 9.62$

$\bar{x} - s = 5.4 - 4.22$

$= 1.18$

So scores within one standard deviation of the mean are scores greater than 1.18 but less than 9.62. ✓

From the dot plot there are 13 such scores.

% within 1 standard deviation of the mean

$= \frac{13}{30} \times 100\%$

$= 43\frac{1}{3}\%$ ✓ *(2 marks)*

iii 68% of scores lie within one standard deviation of the mean if the data is normally distributed. Normally distributed data has a bell-shaped graph. From the dot plot it is obvious that the distribution is not bell-shaped and the data is not normally distributed. ✓ *(1 mark)*

21 40 km/h = 36 km/h + 2 × 2 km/h

So it is 2 standard deviations above the mean.

95% lie within 2 standard deviations of the mean so 5% lie outside that. Half of those are above the mean.

So 2.5% of the vehicles will have a speed greater than 40 km/h.

Answer A

22 8 – 0.04 = 7.96

So a diameter of 7.96 cm is one standard deviation below the mean.

Now, 68% of scores are within one standard deviation of the mean.

Percentage more than one standard deviation from the mean = (100 – 68)% or 32%

Half of those will be more than one standard deviation below the mean.

So the percentage with a diameter less than 7.96 cm is (32 ÷ 2)% or 16%.

Answer A

23 $3.1 - 2.75 = 0.35$

So a weight of 2.75 kg is 1 standard deviation below the mean.

Half of 68%, or 34%, of babies will have weights between 2.75 kg and 3.1 kg.

$4.15 - 3.1 = 1.05$

$1.05 \div 0.35 = 3$

So a weight of 4.15 kg is 3 standard deviations above the mean.

Half of 99.7%, or 49.85%, of babies will have weights between 3.1 kg and 4.15 kg.

Now 34% + 49.85% = 83.85%

83.85% of 10 000 = 8385

Answer D

24 68% of scores have z-scores between –1 and 1.

Half of these, 34%, will have z-scores between 0 and 1.

95% have z-scores between –2 and 2.

So 47.5% will have z-scores between 0 and 2.

47.5% – 34% or 13.5% have z-scores between 1 and 2.

13.5% of 60 000 = 8100

So 8100 students would be expected to score a result between 1 and 2 standard deviations above the mean.

Answer A

25 **i** $\sigma_n = 5.220\,153\,254\ldots$

$= 5.2$ [1 d.p.] ✓

(1 mark)

ii Test 1: $x = 62, \mu = 68.5, \sigma = 5.2$

$$z = \frac{x - \mu}{\sigma} = \frac{62 - 68.5}{5.2} = -1.25$$

So Ali's mark in the first test was $1\frac{1}{4}$ standard deviations below the mean. ✓

Test 2: $\mu = 74.4, \sigma = 12.4, z = -1.25$

$$z = \frac{x - \mu}{\sigma}$$

$$-1.25 = \frac{x - 74.4}{12.4}$$ ✓

$$-15.5 = x - 74.4$$

$$x = 58.9$$

$= 59$ [nearest unit]

Ali would need to score 59 in the second test to ensure that her performance relative to the class remained the same.

✓ *(3 marks)*

26 **i** $x = 400, \mu = 500, \sigma = 50$

$$z = \frac{x - \mu}{\sigma} = \frac{400 - 500}{50} = -2$$ ✓

(1 mark)

ii Brand A: $x = 400, \mu = 450, \sigma = 25$

$$z = \frac{x - \mu}{\sigma} = \frac{400 - 450}{25} = -2$$ ✓

The z-scores for light bulbs with a life of 400 hours is the same for both brands.

This means the same percentage of light bulbs are likely to be defective. The claim is not correct. ✓

(2 marks)

27 **i** 1 ✓

(1 mark)

ii **1**

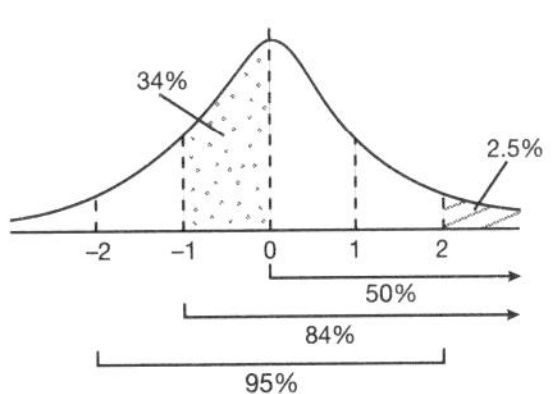

If 2.5% of girls are taller than Rachel, the corresponding z-score is 2.

From the graph, this is a height of 155 cm. Hence Rachel is 155 cm tall. ✓

(1 mark)

2 At age $15\frac{1}{2}$, 155 cm is a z-score of –1. ✓

$\therefore$ From a z-score of 0 to –1, the percentage of girls is 34%.

$\therefore$ (34 + 50)% = 84% of girls are taller than Rachel. ✓

(2 marks)

iii Average height is a z-score of 0.

$\therefore$ Average height is 163 cm. ✓

(1 mark)

iv $B = \dfrac{m}{h^2}$

$B = 21$ and $h = 1.63$

$\therefore 21 = \dfrac{m}{(1.63)^2}$ ✓

$\therefore m = 21 \times (1.63)^2$

$= 55.7949$

$\doteqdot 56$ kg ✓

(2 marks)

v **1** A gradient of 6 means that for girls aged 6 to 11, their growth rate is 6 cm per year. ✓

(1 mark)

2 $C = 6A + 79$

If $A = 12$,

$C = 6 \times 12 + 79$

$= 151$ cm

For any ages > 12 years, the height, C, does not correspond to the graph as girls grow at a slower rate after 12 years of age.

✓ *(1 mark)*

28 **i** $z = \dfrac{x - \mu}{\sigma}$

Test 1: $z = \dfrac{63 - 60}{6.2}$

$= 0.4838\ldots$ ✓

Test 2: $z = \dfrac{62 - 58}{6}$

$= 0.6666\ldots$

Stuart performed better in Test 2 as his z-score is higher. ✓

(2 marks)

ii

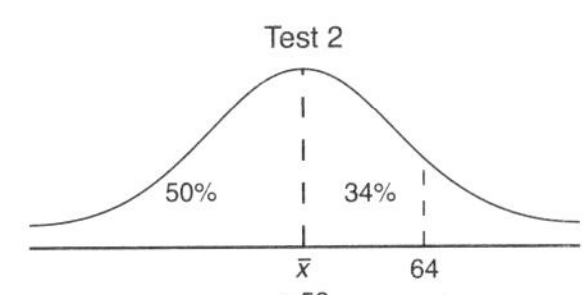

You would expect 84% of students to score less than 64. ✓

i.e. $\dfrac{84}{100} \times 150$

$= 126$ students ✓

(2 marks)

29 **i** Roberto's z-score could be any number between 1 and 2 as this shaded region is the 2nd standard deviation from the mean. ✓

(1 mark)

ii

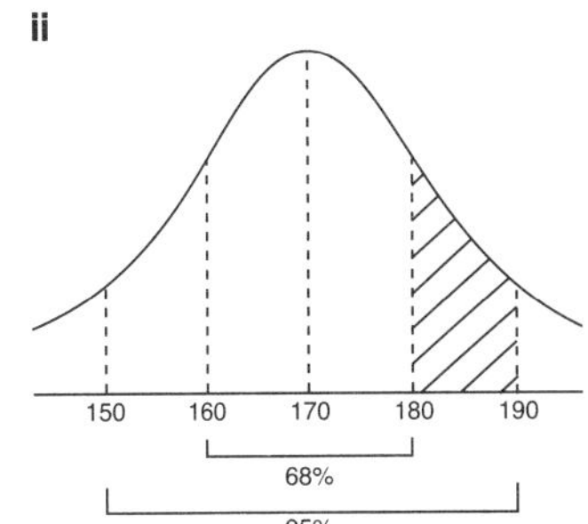

✓

Percentage of data in shaded region

$\doteqdot (95 - 68) \div 2$

$\doteqdot 13.5\%$ ✓

(2 marks)

NSW Education Standards Authority

2022 HIGHER SCHOOL CERTIFICATE EXAMINATION

Mathematics Advanced
Mathematics Extension 1
Mathematics Extension 2

REFERENCE SHEET

Measurement

Length

$$l = \frac{\theta}{360} \times 2\pi r$$

Area

$$A = \frac{\theta}{360} \times \pi r^2$$

$$A = \frac{h}{2}(a + b)$$

Surface area

$$A = 2\pi r^2 + 2\pi rh$$

$$A = 4\pi r^2$$

Volume

$$V = \frac{1}{3}Ah$$

$$V = \frac{4}{3}\pi r^3$$

Functions

$$x = \frac{-b \pm \sqrt{b^2 - 4ac}}{2a}$$

For $ax^3 + bx^2 + cx + d = 0$:

$$\alpha + \beta + \gamma = -\frac{b}{a}$$

$$\alpha\beta + \alpha\gamma + \beta\gamma = \frac{c}{a}$$

and $\alpha\beta\gamma = -\frac{d}{a}$

Relations

$$(x - h)^2 + (y - k)^2 = r^2$$

Financial Mathematics

$$A = P(1 + r)^n$$

Sequences and series

$$T_n = a + (n - 1)d$$

$$S_n = \frac{n}{2}[2a + (n - 1)d] = \frac{n}{2}(a + l)$$

$$T_n = ar^{n-1}$$

$$S_n = \frac{a(1 - r^n)}{1 - r} = \frac{a(r^n - 1)}{r - 1}, \ r \neq 1$$

$$S = \frac{a}{1 - r}, \ |r| < 1$$

Logarithmic and Exponential Functions

$$\log_a a^x = x = a^{\log_a x}$$

$$\log_a x = \frac{\log_b x}{\log_b a}$$

$$a^x = e^{x \ln a}$$

Trigonometric Functions

$$\sin A = \frac{\text{opp}}{\text{hyp}}, \quad \cos A = \frac{\text{adj}}{\text{hyp}}, \quad \tan A = \frac{\text{opp}}{\text{adj}}$$

$$A = \frac{1}{2}ab\sin C$$

$$\frac{a}{\sin A} = \frac{b}{\sin B} = \frac{c}{\sin C}$$

$$c^2 = a^2 + b^2 - 2ab\cos C$$

$$\cos C = \frac{a^2 + b^2 - c^2}{2ab}$$

$$l = r\theta$$

$$A = \frac{1}{2}r^2\theta$$

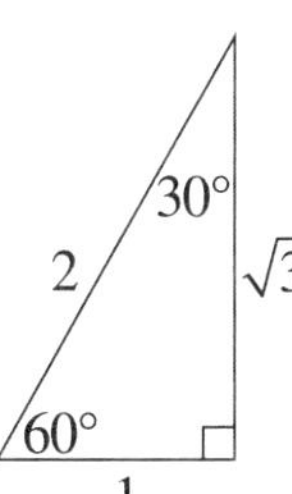

Trigonometric identities

$$\sec A = \frac{1}{\cos A}, \ \cos A \neq 0$$

$$\operatorname{cosec} A = \frac{1}{\sin A}, \ \sin A \neq 0$$

$$\cot A = \frac{\cos A}{\sin A}, \ \sin A \neq 0$$

$$\cos^2 x + \sin^2 x = 1$$

Compound angles

$$\sin(A + B) = \sin A\cos B + \cos A\sin B$$

$$\cos(A + B) = \cos A\cos B - \sin A\sin B$$

$$\tan(A + B) = \frac{\tan A + \tan B}{1 - \tan A\tan B}$$

If $t = \tan\frac{A}{2}$ then $\sin A = \frac{2t}{1+t^2}$

$$\cos A = \frac{1-t^2}{1+t^2}$$

$$\tan A = \frac{2t}{1-t^2}$$

$$\cos A\cos B = \frac{1}{2}\left[\cos(A - B) + \cos(A + B)\right]$$

$$\sin A\sin B = \frac{1}{2}\left[\cos(A - B) - \cos(A + B)\right]$$

$$\sin A\cos B = \frac{1}{2}\left[\sin(A + B) + \sin(A - B)\right]$$

$$\cos A\sin B = \frac{1}{2}\left[\sin(A + B) - \sin(A - B)\right]$$

$$\sin^2 nx = \frac{1}{2}(1 - \cos 2nx)$$

$$\cos^2 nx = \frac{1}{2}(1 + \cos 2nx)$$

Statistical Analysis

$$z = \frac{x - \mu}{\sigma}$$

An outlier is a score
less than $Q_1 - 1.5 \times IQR$
or
more than $Q_3 + 1.5 \times IQR$

Normal distribution

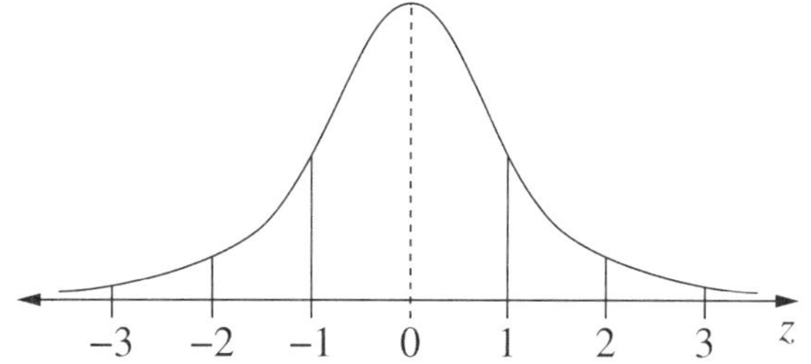

- approximately 68% of scores have z-scores between –1 and 1
- approximately 95% of scores have z-scores between –2 and 2
- approximately 99.7% of scores have z-scores between –3 and 3

$$E(X) = \mu$$

$$\operatorname{Var}(X) = E\left[(X - \mu)^2\right] = E\left(X^2\right) - \mu^2$$

Probability

$$P(A \cap B) = P(A)P(B)$$

$$P(A \cup B) = P(A) + P(B) - P(A \cap B)$$

$$P(A|B) = \frac{P(A \cap B)}{P(B)}, \ P(B) \neq 0$$

Continuous random variables

$$P(X \leq r) = \int_a^r f(x)\,dx$$

$$P(a < X < b) = \int_a^b f(x)\,dx$$

Binomial distribution

$$P(X = r) = {}^nC_r p^r (1 - p)^{n-r}$$

$$X \sim \text{Bin}(n, p)$$

$$\Rightarrow \ P(X = x)$$

$$= \binom{n}{x} p^x (1 - p)^{n-x}, \ x = 0, 1, \ldots, n$$

$$E(X) = np$$

$$\operatorname{Var}(X) = np(1 - p)$$

Differential Calculus

Function	Derivative
$y = f(x)^n$	$\frac{dy}{dx} = nf'(x)[f(x)]^{n-1}$
$y = uv$	$\frac{dy}{dx} = u\frac{dv}{dx} + v\frac{du}{dx}$
$y = g(u)$ where $u = f(x)$	$\frac{dy}{dx} = \frac{dy}{du} \times \frac{du}{dx}$
$y = \frac{u}{v}$	$\frac{dy}{dx} = \frac{v\frac{du}{dx} - u\frac{dv}{dx}}{v^2}$
$y = \sin f(x)$	$\frac{dy}{dx} = f'(x)\cos f(x)$
$y = \cos f(x)$	$\frac{dy}{dx} = -f'(x)\sin f(x)$
$y = \tan f(x)$	$\frac{dy}{dx} = f'(x)\sec^2 f(x)$
$y = e^{f(x)}$	$\frac{dy}{dx} = f'(x)e^{f(x)}$
$y = \ln f(x)$	$\frac{dy}{dx} = \frac{f'(x)}{f(x)}$
$y = a^{f(x)}$	$\frac{dy}{dx} = (\ln a)f'(x)a^{f(x)}$
$y = \log_a f(x)$	$\frac{dy}{dx} = \frac{f'(x)}{(\ln a)f(x)}$
$y = \sin^{-1} f(x)$	$\frac{dy}{dx} = \frac{f'(x)}{\sqrt{1-[f(x)]^2}}$
$y = \cos^{-1} f(x)$	$\frac{dy}{dx} = -\frac{f'(x)}{\sqrt{1-[f(x)]^2}}$
$y = \tan^{-1} f(x)$	$\frac{dy}{dx} = \frac{f'(x)}{1+[f(x)]^2}$

Integral Calculus

$$\int f'(x)[f(x)]^n dx = \frac{1}{n+1}[f(x)]^{n+1} + c$$

where $n \neq -1$

$$\int f'(x)\sin f(x)dx = -\cos f(x) + c$$

$$\int f'(x)\cos f(x)dx = \sin f(x) + c$$

$$\int f'(x)\sec^2 f(x)dx = \tan f(x) + c$$

$$\int f'(x)e^{f(x)}dx = e^{f(x)} + c$$

$$\int \frac{f'(x)}{f(x)}dx = \ln|f(x)| + c$$

$$\int f'(x)a^{f(x)}dx = \frac{a^{f(x)}}{\ln a} + c$$

$$\int \frac{f'(x)}{\sqrt{a^2-[f(x)]^2}}dx = \sin^{-1}\frac{f(x)}{a} + c$$

$$\int \frac{f'(x)}{a^2+[f(x)]^2}dx = \frac{1}{a}\tan^{-1}\frac{f(x)}{a} + c$$

$$\int u\frac{dv}{dx}dx = uv - \int v\frac{du}{dx}dx$$

$$\int_a^b f(x)dx$$

$$\approx \frac{b-a}{2n}\left\{f(a) + f(b) + 2\left[f(x_1) + \cdots + f(x_{n-1})\right]\right\}$$

where $a = x_0$ and $b = x_n$

Combinatorics

$${}^nP_r = \frac{n!}{(n-r)!}$$

$$\binom{n}{r} = {}^nC_r = \frac{n!}{r!(n-r)!}$$

$$(x+a)^n = x^n + \binom{n}{1}x^{n-1}a + \cdots + \binom{n}{r}x^{n-r}a^r + \cdots + a^n$$

Vectors

$$\left|\underset{\sim}{u}\right| = \left|x\underset{\sim}{i} + y\underset{\sim}{j}\right| = \sqrt{x^2 + y^2}$$

$$\underset{\sim}{u} \cdot \underset{\sim}{v} = \left|\underset{\sim}{u}\right|\left|\underset{\sim}{v}\right|\cos\theta = x_1x_2 + y_1y_2,$$

where $\underset{\sim}{u} = x_1\underset{\sim}{i} + y_1\underset{\sim}{j}$

and $\underset{\sim}{v} = x_2\underset{\sim}{i} + y_2\underset{\sim}{j}$

$$\underset{\sim}{r} = \underset{\sim}{a} + \lambda\underset{\sim}{b}$$

Complex Numbers

$$z = a + ib = r(\cos\theta + i\sin\theta) = re^{i\theta}$$

$$\left[r(\cos\theta + i\sin\theta)\right]^n = r^n(\cos n\theta + i\sin n\theta) = r^ne^{in\theta}$$

Mechanics

$$\frac{d^2x}{dt^2} = \frac{dv}{dt} = v\frac{dv}{dx} = \frac{d}{dx}\left(\frac{1}{2}v^2\right)$$

$$x = a\cos(nt + \alpha) + c$$

$$x = a\sin(nt + \alpha) + c$$

$$\ddot{x} = -n^2(x - c)$$

Notes